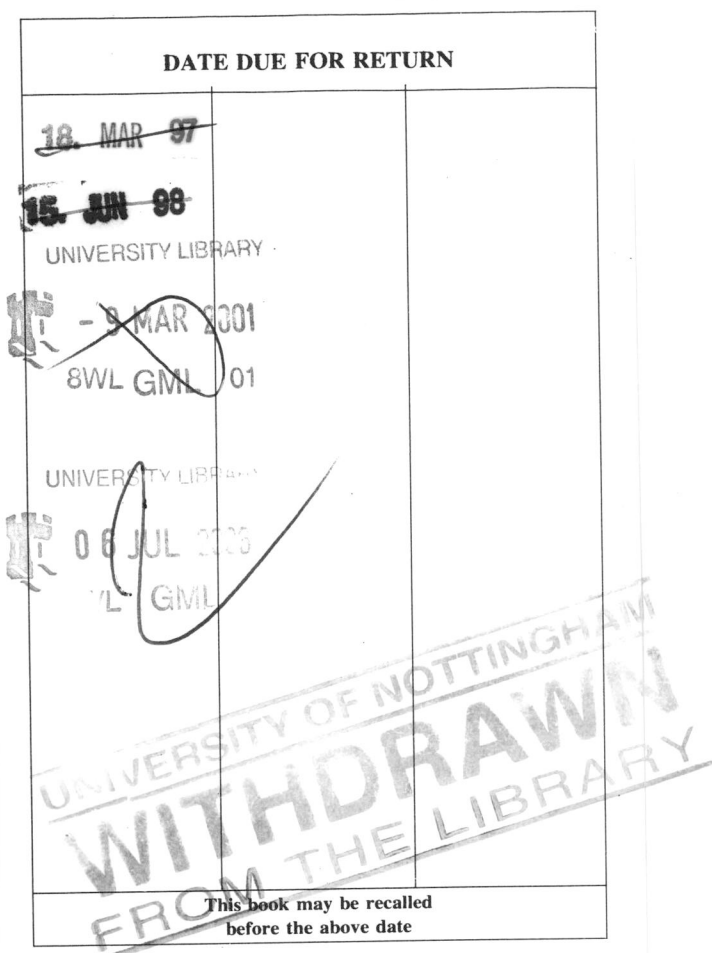

DATE DUE FOR RETURN

18. MAR 97

15. JUN 98

UNIVERSITY LIBRARY

- 9 MAR 2001

8WL GML 01

UNIVERSITY LIBRARY

0 6 JUL 2006

L GML

This book may be recalled
before the above date

90014

PROGRESS IN BRAIN RESEARCH

VOLUME 112

EXTRAGENICULOSTRIATE MECHANISMS UNDERLYING
VISUALLY-GUIDED ORIENTATION BEHAVIOR

PROGRESS IN BRAIN RESEARCH

VOLUME 112

EXTRAGENICULOSTRIATE MECHANISMS UNDERLYING VISUALLY-GUIDED ORIENTATION BEHAVIOR

EDITED BY

MASAO NORITA

*Department of Neurobiology and Anatomy, Niigata University School of Medicine, Asahimachi
Niigata, Japan*

TAKEHIKO BANDO

Department of Physiology, Niigata University School of Medicine, Asahimachi Niigata, Japan

BARRY E. STEIN

*Department of Neurobiology and Anatomy, The Bowman Gray School of Medicine, Wake Forest,
Winston Salem, NC, USA*

ELSEVIER
AMSTERDAM – LAUSANNE – NEW YORK – OXFORD – SHANNON – TOKYO
1996

ISBN 0-444-82347-6 (volume)
ISBN 0-444-80104-9 (series)

Published by:
Elsevier Science B.V.
P.O. Box 211
1000 AE Amsterdam
The Netherlands

1000 858437

Library of Congress Cataloging-in-Publication Data

Extrageniculostriate mechanisms underlying visually-guided orientation
  behavior / edited by Masao Norita, Takehiko Bando, Barry Stein.
      p.   cm. -- (Progress in brain research ; v. 112)
    Includes bibliographical references and index.
    ISBN 0-444-82347-6 (volume : alk. paper). -- ISBN 0-444-80104-9
  (series : alk. paper)
    1. Visual cortex--Congresses. 2. Visual perception--Congresses.
  I. Norita, Masao. II. Bando, Takehiko. III. Stein, Barry E.
  IV. Series.
    [DNLM: 1. Visual Cortex--physiology--congresses. 2. Visual
  Pathways--physiology--congresses. 3. Orientation--physiology-
  -congresses. 4. Superior Colliculus--physiology--congresses.
  5. Behavior, Animal--physiology--congresses. W1 PR667J v.112 1996
  / WL 307 E9515 1996]
  QP376.P7 vol. 112
  [QP383.15]
  612.8'2 s--dc20
  [612.8'4]
  DNLM/DLC
  for Library of Congress   96-41410
                                                             CIP

Printed in The Netherlands on acid-free paper

List of Contributors

T. Bando, Department of Physiology, Niigata University School of Medicine, Asahi-machi, Niigata, Niigata 951, Japan

M. Behan, Department of Comparative Biosciences and Center for Neuroscience, University of Wisconsin, Madison, Wisconsin 53706, USA

Gy. Benedek, Department of Physiology, Albert Szent-Györgyi Medical University, Szeged, Hungary

C.A. Bennett-Clarke, Department of Anatomy and Neurobiology, Medical College of Ohio, Toledo, OH 43699-0008, USA

D. Boire, Département de Psychologie and Centre de Recherche en Sciences Neurologiques, Université de Montréal H3C 3J7, and Department of Neurology and Neurosurgery, Montreal Neurological Institute, McGill University, 3801 University St., Montreal, Canada

S. Carlile, University Laboratory of Physiology, Parks Road, Oxford OX1 3PT, UK

C. Casanova, Departments of Surgery-Ophthalmology and of Physiology and Biophysics, Faculty of Medicine, University of Sherbrooke, Sherbrooke, Quebec, Canada, J1H 5N4

S. Chimoto, Department of Physiology, Institute of Basic Medical Sciences, University of Tsukuba, Tsukuba, Ibaraki 305, Japan

S. Chin, Department of Physiology, Hokkaido University School of Medicine, Sapporo 060, Japan

V.M. Ciaramitaro, Department of Neuroscience, School of Medicine, University of Pennsylvania, Philadelphia, PA 19104-6074, USA

R.L. Djavadian, Department of Anatomy and Histology, Institute for Biomedical Research, The University of Sydney, N.S.W. 2006, Australia

B. Dreher, Department of Anatomy and Histology, Institute for Biomedical Research, The University of Sydney, N.S.W. 2006, Australia

R. Fendrich, Center for Neuroscience, University of California, Davis CA 95616, USA

B.L. Finlay, Department of Psychology, Cornell University, Ithaca, NY 14853, USA

L. Fishcer-Szatmári, Department of Physiology, Albert Szent-Györgyi Medical University, Szeged, Hungary

M.A. Frens, University of Nijmegen, Department of Medical Physics and Biophysics, Geert Grooteplein 21, NL-6525 EZ Nijmegen, The Netherlands and Neurology Department, University Hospital, CH 8091 Zürich, Switzerland

Y. Fukuda, Department of Physiology, Osaka University Medical School, Suita 565, Japan

J. Fukushima, Department of Physiology, Hokkaido University School of Medicine, Sapporo 060, Japan

K. Fukushima, Department of Physiology, Hokkaido University School of Medicine, Sapporo 060, Japan

S. Funaki, Department of Neurobiology and Anatomy, Niigata University School of Medicine, Asahimachi, Niigata 951, Japan

M.S. Gazzaniga, Center for Neuroscience, University of California, Davis CA 95616, USA

N. Hara, Department of Physiology, Niigata University School of Medicine, Asahi-machi, Niigata, Niigata 951, Japan

M. Herbin, Département de Psychologie and centre de recherche en sciences neurologiques, Université de montréal H3C 3J7, and Department of Neurology and Neurosurgery, Montreal Neurological Institute, McGill University, 3801 University St., Montreal, Canada

T.P. Hicks, Tissue Regeneration Group, Institute of Biological Science, National Research Council of Canada, Building M-54, 1200 Montreal Road, Ottawa, Ontario, Canada, K1A OR6

S. Hirano, Department of Neurobiology and Anatomy, Niigata University School of Medicine, Asahimachi, Niigata 951, Japan

K. Hoshino, Department of Neurobiology and Anatomy, Niigata University School of Medicine, Asahimachi, Niigata 951, Japan

X. Huang, Department of Anatomy and Neurobiology, Medical College of Ohio, Toledo, OH 43699-0008, USA

R.-B. Illing, Unit for Morphological Brain Research, Department of Otorhinolaryngology, University of Freiburg, 79106 Freiburg, Germany

Y. Inoue, Neuroscience Section, Electrotechnical Laboratory, Tsukubashi, Ibaraki 305, Japan

S. Ireland, Neuroregeneration Laboratory, Department of Anatomy, The University of Hong Kong, 5 Sassoon Road, Hong Kong

Y. Ito, Department of Neuophysiology, Tokyo Metropolitan Institute for Neuroscience, Musashidai 2-6, Fuchu-shi, Tokyo 183, Japan

Y. Iwamoto, Department of Physiology, Institute of Basic Medical Sciences, University of Tsukuba, Tsukuba, Ibaraki 305, Japan

J.H. Kaas, Department of Psychology, Vanderbilt University, Nashville, TN 37240, USA

S. Kakei, Department of Physiology and Orthopedic Surgery, School of Medicine, Tokyo Medical and Dental University, Tokyo 113, Japan

M. Kase, Department of Neurobiology and Anatomy, Niigata University School of Medicine, Asahimachi, Niigata 951, Japan

Y.Y. Katoh, Department of Anatomy, Fujita-Gakuen Medical University, Toyoake, Japan

K. Kawano, Neuroscience Section, Electrotechnical Laboratory, Tsukubashi, Ibaraki 305, Japan

N.M. Kime, Department of Comparative Biosciences and Center for Neuroscience, University of Wisconsin, Madison, Wisconsin 53706, USA

A.J. King, University Laboratory of Physiology, Parks Road, Oxford OX1 3PT, UK

T. Kitama, Neuroscience Section, Electrotechnical Laboratory, Tsukubashi, Ibaraki 305, Japan

Y. Kobayashi, ART Human Information Processing Research Laboratory, Kyoto 619-02, Japan

G. Kovács, Department of Physiology, Albert Szent-Györgyi Medical University, Szeged, Hungary

S. Kurkin, Department of Physiology, Hokkaido University School of Medicine, Sapporo 060, Japan

J.G. McHaffie, Department of Neurobiology and Anatomy, Bowman Gray School of Medicine, Wake Forest University, Winston-Salem, NC 27157, USA

R. Meguro, Department of Neurobiology and Anatomy, Niigata University School of Medicine, Asahimachi, Niigata 951, Japan

R.R. Mize, Departments of Anatomy and Opthalmology and the Neuroscience Center, Louisiana State University Medical Center, 1901 Perdido Street, New Orleans, LA 70112, USA

N. Mano, Department of Neuophysiology, Tokyo Metropolitan Institute for Neuroscience, Musashidai 2-6, Fuchu-shi, Tokyo 183, Japan

R.D. Mooney, Department of Anatomy and Neurobiology, Medical College of Ohio, Toledo, OH 43699-0008, USA

M. Mustari, University of Texas Medical Branch, Galveston, USA

N. Muto, Department of Physiology and Orthopedic Surgery, School of Medicine, Tokyo Medical and Dental University, Tokyo 113, Japan

K. Naito, Department of Neurophysiology, Tokyo Metropolitan Institute for Neuroscience, Tokyo, Japan

M. Norita, Department of Neurobiology and Anatomy, Niigata University School of Medicine, Asahimachi, Niigata 951, Japan

M. Oka, Department of Neurophysiology, Tokyo Metropolitan Institute for Neuroscience, Tokyo, Japan

S. Onodera, Department of Anatomy, School of Medicine, Iwate Medical University, Morioka 020, Japan

A.J. Van Opstal, University of Nijmegen, Department of Medical Physics and Biophysics, Geert Grooteplein 21, NL-6525 EZ Nijmegen, The Netherlands and Neurology Department, University Hospital, CH 8091 Zürich, Switzerland

G.A. Orban, Laboratorium voor Neuro-en Psychofysiologie, Faculteit der Geneeskunde, KULeuven, Campus Gasthuisberg, B-3000 Leuven, Belgium

C.K. Peck, Schoool of Optometry, University of Missouri-St. Louis, 8001 Natural Bridge Road, St. Louis, MO 63121, USA

J. Perényi, Department of Physiology, Albert Szent-Györgyi Medical University, Szeged, Hungary

A. Ptito, Center for Neuroscience, University of California, Davis CA 95616, USA, and Montreal Neurological Institute and Hospital, Montreal, Quebec H3A 2B4, Canada and Département de Psychologie and centre de recherche en sciences neurologiques, Université de montréal, Montréal H3C 3J7, Canada

M. Ptito, Département de Psychologie and centre de recherche en sciences neurologiques, Université de montréal H3C 3J7, and Department of Neurology and Neurosurgery, Montreal Neurological Institute, McGill University, 3801 University St., Montreal, Canada

J.P. Rauschecker, Georgetown Institute for Cognitive and Computational Sciences 3970, Reservoir Road, NW Washington, DC 20007-2197, and Section on Cognitive neuroscience Laboratory of Neuropsychology, NIH/NIMH, Bldg. 49, Rm. 1B80, Bethesda, MD 20892-4415, USA

R.W. Rhodes, Department of Anatomy and Neurobiology, Medical College of Ohio, Toledo, OH 43699-0008, USA

A. Rosenquist, Department of Neuroscience, School of Medicine, University of Pennsylvania, Philadelphia, PA 19104-6074, USA

S. Sasaki, Department of Neurophysiology, Tokyo Metropolitan Institute for Neuroscience, Tokyo, Japan

T. Savard, Departments of Surgery-Ophthalmology and of Physiology and Biophysics, Faculty of Medicine, University of Sherbrooke, Sherbrooke, Quebec, Canada, J1H 5N4

H. Sawai, Neuroregeneration Laboratory, Department of Anatomy, The University of Hong Kong, 5 Sassoon Road, Hong Kong

J.W.H. Schnupp, University Laboratory of Physiology, Parks Road, Oxford OX1 3PT, UK

M.Y. Shi, Department of Anatomy and Neurobilogy, Medical College of Ohio, Toledo OH 43699-0008, USA

H. Shibutani, Department of Neuophysiology, Tokyo Metropolitan Institute for Neuroscience, Musashidai 2-6, Fuchu-shi, Tokyo 183, Japan

H. Shimazu, Department of Neurophysiology, Tokyo Metropolitan Institute for Neuroscience, 2-6 Musashidai, Fuchu, Tokyo 183, Japan

Y. Shinoda, Department of Physiology, School of Medicine, Tokyo Medical and Dental University, Tokyo 113, Japan

A.L. Smith, University Laboratory of Physiology, Parks Road, Oxford OX1 3PT, UK

P.D. Spear, Department of Psychology and Center for Neuroscience, University of Wisconsin-Madison, 1202 West Johnson St., Madison, WI 53706, USA

J.M. Sprague, Department of Neuroscience, University of Pennsylvania, School of Medicine, Philadelphia, PA 19104-6058, USA

K.-F. So, Neuroregeneration Laboratory, Department of Anatomy, The University of Hong Kong, 5 Sassoon Road, Hong Kong

B.E. Stein, Department of Neurobiology and Anatomy, Bowman Gray School of Medicine, Wake Forest University, Winston-Salem, NC. 27157-1010, USA

M. Takagi, Department of Physiology, Niigata University School of Medicine, Asahi-machi, Niigata, Niigata 951, Japan

A. Takemura, Neuroscience Section, Electrotechnical Laboratory, Tsukubashi, Ibaraki 305, Japan

Y. Tamai, Department of Physiology, Wakayama Medical College, Wakayama 640, Japan

D. Tay, Neuroregeneration Laboratory, Department of Anatomy, The University of Hong Kong, 5 Sassoon Road, Hong Kong

I.D. Thompson, University Laboratory of Physiology, Parks Road, Oxford OX1 3PT, UK

H. Toda, Department of Physiology, Niigata University School of Medicine, Asahi-machi, Niigata, Niigata 951, Japan

K.J. Turlejski, Department of Anatomy and Histology, Institute for Biomedical Research, The University of Sydney, N.S.W. 2006, Australia

J.-G. Villemure, Center for Neuroscience, University of California, Davis CA 95616, USA, and Montreal Neurological Institute and Hospital, Montreal, Quebec H3A 2B4, Canada

R. Vogels, Laboratorium voor Neuro- en Psychofysiologie, Faculteit der Geneeskunde, KU Leuven, Campus Gasthuisberg, B-3000 Leuven, Belgium

M.T. Wallace, Department of Neurobiology and Anatomy, Bowman Gray School of Medicine, Wake Forest University, Winston-Salem, NC. 27157-1010, USA

C. Wang, Department of Anatomy and Histology, Institute for Biomedical Research, The University of Sydney, N.S.W. 2006, Australia

C.M. Wessinger, Center for Neuroscience, University of California, Davis CA 95616, USA, and Montreal Neurological Institute and Hospital, Montreal, Quebec H3A 2B4, Canada

M. Xiong, Department of Psychology, Cornell University, Ithaca, NY 14853, USA

K. Yamamoto, Department of Physiology, Niigata University School of Medicine, Asahi-machi, Niigata, Niigata 951, Japan

K. Yoshida, Department of Physiology, Institute of Basic Medical Sciences, University of Tsukuba, Tsukuba, Ibaraki 305, Japan

Preface

It is, perhaps, no exaggeration to note that many of the most fundamental aspects of our understanding of brain function and maturation have been developed as a result of studies of the visual system. Thus, interest in the visual system derives not only from our interest in visual processes per se, but also from it's utility as a general model of brain organization.

For many years the visual system was thought to be formed by a comparatively small number of distinct pathways and target structures. The visual processes in the midbrain were thought to be involved in reflexive responses to visual stimuli (e.g. eye movements, pupillary changes) which facilitate the immediate responses necessary to shift gaze and accommodate the stimulus and varying light levels. On the other hand, the geniculostriate system and its immediate target structures were viewed as playing the principal roles in all aspects of visual perception. For some time, a straightforward hierarchical model was thought to be the most accurate reflection of the stream of information along thalamocortical and corticocortical pathways. In this model the thalamus directly activates primary sensory cortex (i.e. striate cortex), which in turn activates secondary psychic-related para- and peristriate cortex, which directly activates multi-sensory association, or interpretive, cortex.

Though this sort of uni-directional serial processing model remains an integral part of current concepts of visual cortex organization, the last few years have seen a dramatic increase in the number of areas known to be involved in mammalian vision, a far greater understanding of the importance of reciprocal connections, intrinsic connections, structure-specific modules and modules which span different structures, as well as the introduction of parallel processing models within the thalamocortical and corticocortical streams. At the same time, there have been substantial changes in our views of midbrain functions, as well as the importance of cortical–midbrain (and cortico–striato–nigrotectal) interactions both in terms of sensory processing and sensorimotor transduction.

These changes in our understanding of the visual system have been most extensive in how we view the functional contributions of extrageniculostriate mechanisms to vision and visually-guided behavior. The body of knowledge has become so vast, and is growing so rapidly, that periodic updates are essential even for experts in the field. This was the impetus behind bringing together a group of researchers for a satellite meeting of the Fourth World Congress of Neuroscience which was sponsored by the International Brain Research Organization. The idea was not only to emphasize the most current information regarding midbrain and extrastriate mechanisms underlying vision and visually-guided behavior, but to place these data into the larger context of how interrelated components of the visual system function to produce coherent visual experiences and behavior. The international nature of the conference was underscored by participants from Australia, Canada, China, Eastern and Western Europe, the U.S. and, of course, our host country, Japan.

The various perspectives of those utilizing different scientific methods made the interactions quite lively, and several new collaborative arrangements resulted from these discussions. This volume is, in part, the result of these discussions and interactions. It contains new research findings that are unavailable elsewhere, as well as reviews and broad perspectives in which existing data from multiple sources are brought together in order to help us understand the structure and function of extrageniculostriate visual areas.

The New Hotel Koshiji-So in Muikamachi, Niigata, Japan was a perfect setting for this meeting. The relaxed atmosphere in this lovely mountainous region of Japan was conducive to the active exchange of ideas bearing directly on which of the current functional concepts are likely to endure and which must be discarded. For two-and-a-half days participants attended all-day scientific sessions that were bracketed by wonderful Japanese meals, Japanese baths, and the opportunity to climb any one of several nearby mountains. Two highlights of the many delightful gastronomic experiences were the soba extravaganza prepared by hand by Dr. Norita and his teacher (under the scrutiny and kibitzing of most of us attending the conference), and the extraordinary banquet that marked the official end to the conference. We are deeply indebted to the students from Niigata University who contributed significantly to the organization and coordination of the conference, to Yumiko Norita for organizing day trips, and to the hotel staff for their unflagging patience, cheerful assistance, and gracious attention to every detail.

Barry E. Stein
Niigata 1996

Contents

M. Norita, T. Bando and B. Stein (Eds.)
Progress in Brain Research, Vol 112
© 1996 Elsevier Science BV. All rights reserved.

CHAPTER 1

Neural mechanisms of visual orienting responses

James M. Sprague

Departments of Neuroscience, and Cell and Developmental Biology, School of Medicine, University of Pennsylvania, Philadelphia, PA 19104-6058, USA

Introduction

Visual orienting refers to the coordinated movement of eyes, head and body to explore the environment surrounding the animal or person, and is a response crucial to the development and the survival of the individual and species. Orienting has several components: (1) the reflexive response triggered by the appearance of an object in the visual field; (2) the internally programmed, exploratory motor search which may occur without a specific visual target; and (3) the shift of attention which accompanies both reflex and search. Thus we speak of visuomotor orienting and orienting of attention as parts of the same response, which also in the cat may consist of dilatation of the pupils, widening of the palpebral fissures, pinnal movements, extension of the vibrissa around eyes and mouth, tensing muscles of the body in preparation for movement, general arousal and with it changes in blood pressure and heart rate. Orienting can be considered as the leading edge of the brain's guidance of behavior, and as such it provides a spatial framework upon which the more complex functions of perception, discrimination, recognition, and visuomotor responses operate effectively.

The concept of the orienting reflex originated in Russia in Pavlov's laboratory around the turn of the century (he mentioned it for the first time in a lecture given in St. Petersburg in 1910). He found that unexpected stimuli, frequently audi-

tory, disturbed the salivary conditioned responses he was studying. Initially Pavlov's interest was focused on the motor or directional component of the orienting reflex, but in later years he described a number of unspecific elements: pupillary dilatation, EEG desynchronization, psychogalvanic response, increase in cerebral blood flow. In view of these so-called unspecific factors, Zernicki (1987) proposed the name of 'arousal component' of orienting for them. He went further and distinguished three components of the orienting response: (1) targeting (directional); (2) arousal; and (3) perceptual. The orienting response has greatly interested Russian psychologists and physiologists and has been studied and reviewed by Sokolov (1960, 1963, 1975).

Orienting is a function for exploration of the environment by telereceptors of eyes, ears and nose and by receptors on the body surface; thus it is multimodal in operation-olfaction, vision, acoustic and somesthetic. If the provoking stimuli are repetitive and found to be uninteresting and unimportant, orienting shows rapid habituation, but is it highly sensitive to changes of the stimuli which will re-instate the response?

Orienting can be directed ahead of the animal, above, below or to either side. It is obvious that a response cannot be directed to both sides at the same time, so that the neuronal discharge must be facilitated on one side of the brain (contralateral to the orienting movement) and inhibited on the opposite side. The side to which orienting is directed will be attended to, the opposite area of space or body will be largely unattended or neglected. In the normal animal, orienting responses shift in this reciprocal manner from side to side as the animal responds to various stimuli in the environment and explores the world around it. Loss of orienting may result from lesions in the brain, which if unilateral, cause a neglect of extrapersonal or body space contralateral to the lesion (see Sprague et al., 1973).

Contralateral visual neglect follows lesions in many areas of the brain–prefrontal cortex, frontal eye fields, somesthetic Area I, parietal and occipital lobes, hypothalamus, thalamus, intralaminar nuclei, basal ganglia (striatum, substantia nigra), superior colliculus and tegmentum–although different neurological mechanisms are operative for each area. Therefore, visual orienting responses, including shifts of attention, are generated by widely distributed networks that include all of these regions of the brain (see Mesulam, 1981).

Neuronal processing in the visual cortex is extensively distributed and involves virtually half of the brain of many mammals, including those which have been intensely studied in the laboratory — rat, gerbil, hamster, tree shrew, cat, bush baby, owl monkey and macaque. In the cat some 20 areas of cortex have been identified as involved in vision. Many of these areas are retinotopically organized, and contain spatial maps ranging from fine-grained to coarse in representation of the structure of the environment. It is likely that most of these areas are 'looking' into contralateral visual space and are sensitive to events occurring there. This conclusion is supported by the deficits in orienting which follow lesions in the visual cortical areas of the cat, the results of which are summarized below.

Visuomotor orienting in the cat has been measured in perimetry tests in which the visual fields are divided into segments (15–30°). The food-deprived animal is trained to accept handling and restraint with the eyes aligned along a horizontal line, and to fixate a food stimulus directly ahead at 0° and when released to go rapidly forward to receive a reward. Fixation is strengthened by tapping the forceps holding the food, and once it is predictable, a second food stimulus is introduced into a segment of right or left visual field, which elicits an orientation of head, eyes and body toward the stimulus. The animal is released and rewarded at the site of the 2nd stimulus (see Sprague and Meikle, 1965; Sherman, 1974, 1977; Wallace et al., 1989 for details). In addition to visual perimetry, the following capacities were examined: open field visual following, visual placing, blink-to-threat, ability to localize stationary stimuli, relative neglect of the left or right hemifield to the presentation of two simultaneous stimuli in opposite fields, eye movements, pupillary symmetry and responsiveness, and open field visually guided behavior (jumping to floor, etc.). These auditory capacities were assessed: startle response, accuracy of auditory localization, and pinnae mobility. The following tactile tests were performed: localization of stimuli of the body and limbs and tactile placing of the four limbs. In addition, the animal's spontaneous motor behavior was observed for postural asymmetries, circling, ataxia, etc. Finally, the following standard reflexes were examined: pinnal, vibrissal, buccal, palpebral, righting, hopping, blink-to-bright light, and withdrawal to touch and light pinch of all limbs (see Sprague, 1966b).

Cortex involved in visual orienting

Unilateral lesions were placed in specific cortical areas which have been defined anatomically and physiologically (Otsuka and Hassler, 1962; Sanides and Hoffman, 1969; Tusa et al., 1981), and the animals were rigorously tested over many months. The experimental plan was first to extirpate the so-called 'primary' visual cortex which in the cat consists of areas 17 and 18; this cortex receives an exclusive projection from laminae A and A1 of the dorsolateral geniculate (LGN_d) and a significant projection from lamina C, as well as smaller projections from C1 to C2 and the medial interlaminar nucleus, MIN (see Sprague et al., 1977;

Rosenquist, 1985). These animals make up Group I in the following descriptions.

The next group II had lesions primarily in areas 17–19, which involved also area 21a and part of 7p. These animals had extensive atrophy in all laminae of LGN_d, plus degeneration in the pulvinar (PL) and lateral posterior (LPl) nuclei.

Group III cats had the same lesions as II, plus the medial bank of the suprasylvian sulcus (areas AMLS, PMLS, VLS, which resulted in complete atrophy of LGN_d and extensive loss in PL and LPl.

Group IV animals had the largest lesions which included areas 17, 18, 19, 21a, 21b, 20a, 20b, Ps, 7p, AMLS, PMLS, VLS, ALLS, PLLS, DLS, anterior lateral gyrus (ALG), visual belt of posterior ectosylvian gyrus; anterior ectosylvian visual area (EVA) and the insula (Toga et al., 1979, Tusa et al., 1981; Rosenquist, 1985; Bowman and Olson, 1988). Not included were splenial visual area (SVA), posterior cingulate and frontal eye fields.

Group I

Areas 17 and 18 contain the largest and finest grain retinotopic map for representation of contralateral visual space. Virtually complete unilateral lesion in the right hemisphere resulted initially in failure to orient to stimuli in the left visual fields (LVF) and very slow following into the LVF. No blink to threat or pencil light was present in the left eye. There was also a tactile deficit in the left feet, i.e. absence of tactile placing and poorly localized responses to touch.

After a brief period of retraining (5 days), slow orienting responses returned to the central 60° of the LVF, with improvement of visual following into the LVF. After 3 weeks, no perimetric deficit was present and the only neurological sign which persisted was a slowed following of stimuli moving from the RVF into the LVF.

Group II

When the lesion included 17 and 18 plus adjacent extrastriate areas 19, 21a and 7p, the initial deficits were similar to animals in group I, but recovery was slower and less complete. Five months later, good orienting was present to stimuli in either VF, but when stimuli were presented bilaterally, responses were always ipsilesional with extinction in the contralateral field. Final examination after 7 months showed further recovery, residual deficits consisting of slowness of following contralaterally and slight favoring of orienting to the ipsilesional side with a tendency to circle to that side.

Group III

A large lesion involving the areas in group II plus all of areas 7p, 21b, AMLS, PMLS, VLS, and parts of 20a, 20b, Ps and posterior cingulate gyrus resulted initially in absence in orienting contralaterally, absence of visual placing to that side and no blink to threat in the contralateral eye. There was also a tendency to turn spontaneously to the side of lesion with slow following into the contralateral hemifield and a mild deficit in the contralateral legs of tactile placing and evasion. After 4 months, contralateral following was improved, but still inferior to the ipsilesional side. Contralateral orienting was present only to large movements presented within 30° of fixation, and these responses fatigued rapidly and extinguished to bilateral stimulation.

These three groups of cats all showed initial absence of orienting to stimuli presented in the contralesional visual space. The degree of recovery was related to the size of the lesions, being almost complete in group I and much less in group III.

Group IV

The very large lesions in this group of cats initially included not only known visual areas but also acoustic cortex, and temporal lobe, although later experiments indicated that the same visual deficits followed lesions which spared acoustic cortex as well as insula.

These animals were followed post-operatively

for long periods, some more than a year, and their visual deficits were severe. They circled spontaneously to the lesion side, they followed only to this side; there was total lack of orienting contralaterally; spontaneous eye movements appeared normal; no blink was present to threat in the contralateral field and visual placing contralateral to the lesion was absent. No recovery occurred except for increased ability to turn contralaterally, both spontaneously or in response to acoustic or tactile stimuli.

These animals appeared blind in the contralateral visual field ('cortical blindness') and the deficit was termed hemianopia, which was appropriate in that it reflected the extensive removal of the visual sensory and associational cortices. However, these cats which lacked orienting responses were not completely blind and had something similar to that which has been called 'blindsight' in humans (Weiskrantz et al., 1974). They were capable of simple discriminations based on differences in luminous flux or brightness (Wood, 1975; Loop and Sherman, 1977), an ability most likely mediated by the superior colliculus (Fischman and Meikle, 1965). They were unable to discriminate gratings and shapes. The deficit was therefore, not only an hemianopia due to cortical blindness but included a factor which we called neglect because visual orienting could be restored by subsequent lesion in the midbrain to be described below.

A few animals with very large lesions did show partial and weak recovery of ability to follow slow-moving stimuli into the 'blind' field and to orient slowly toward stimuli introduced into the medial 30° of this field. However, when two stimuli were moved from fixation in opposite directions, or when two stimuli were introduced simultaneously into both hemifields, the animals always followed and oriented to the intact field ipsilateral to the lesion. The parts of cortex spared in these animals varied: (1) peripheral representation 50–90° azimuth$_3$, from −50 to +40° elevation) in area 17, recovery beginning in 1 month; (2) AMLS + ALLS (recovery beginning in 1 month); (3) visual belt in posterior extosylvian

gyrus and area 21b (recovery 3 months); and (4) anterior 7p, recovery 5 months. It appears that many areas of the visual cortex contribute to the orienting response, and that all or most of them must be included in the lesion to result in a total and 'permanent' loss of contralateral orienting.

Superior colliculus in visual orienting

Visual deficits, similar to those already described after cortical lesions, also follow lesion in the superior colliculus; complete unilateral removal, without involvement of the tegmentum, resulted initially in compulsive, ipsilesional circling, hyperactive orienting to the ipsilesional side and absence of orienting contralaterally. There was contralateral hemianopia with a loss of following movements of head and eyes to that side. When acoustic and tactile stimuli were presented contralaterally, the animal was activated but responses were mislocalized to the ipsilesional side. The deficits were diminished over a period of about 1 month and were stable thereafter: circling was much reduced, and the ability to follow slowly moving stimuli in the contralateral VF returned. Tested perimetrically, visual orienting to stimuli in the contralateral, medial 60–90° was irregular and when present was slow and poorly localized. Bilateral stimulation of all modalities invariable led to brisk responses to the ipsilateral side with extinction of those to the contralateral side.

There are many similarities in the initial deficits and their recovery in visuomotor orienting between unilateral collicular lesions and unilateral, subtotal cortical lesions (groups II and III). It was of interest to combine these tectal and cortical lesions placed seriatum on the same side of the brain.

(1) Whether recovery of orienting responses was complete (group I) or incomplete (groups II and III), subsequent ipsilateral lesion of superior colliculus reinstated full, contralateral hemianopia and loss of contralateral orienting which was enduring for the survival of the animals. In addition, marked ipsiversive circling and mislocalization of contralateral tactile and acoustic sti-

muli, characteristic of collicular lesion were also present and enduring. When group II cortical lesions were bilateral, subsequent unilateral collicular lesion abolished the recovery in the contralateral visual field, leaving complete hemianopia and absence of orienting to that field.

An additional cat from group IV showed partial recovery 5 weeks after cortical lesion, a deficit which became total again following ipsilateral collicular lesion, with survival of 7.5 weeks. Histological examination revealed that areas AMLS and ALLS were spared in the cortical lesion.

(2) When the collicular lesion was placed first and time allowed for stable, partial recovery, and an ipsilateral removal of cortex (group II) was subsequently added, full enduring hemianopia with loss of contralateral visual orienting resulted.

Conclusions

(1) Many, probably all, of the cortical areas which process the afferent visual messages from the retina in cats and which therefore 'look' into the contralateral visual space, are involved in generating an orienting response to that visual field.

(2) Surprisingly, areas 17 and 18, which receive a large projection from LGN_d representing the entire visual field, and which contain the finest grain spatial map in the cortex, are the least involved in orienting. This conclusion is based on the rapid and almost complete restoration of orienting after such lesion. Since these areas are responsible for fine resolution of the retinal image (Berkley and Sprague, 1979), it is likely that acuity is not an important factor in orienting.

(3) When additional extrastriate cortex is added to 17–18 lesions, the deficit is potentiated and recovery is incomplete, both depending on the extent of cortical removal. Complete removal of all visual cortices in occipital, parietal and temporal lobes results in complete hemianopia and enduring loss of visual orienting into contralateral visual space. Inadvertent sparing of small cortical islands of visual cortex within this large area sometimes result in partial restoration of orient-

ing after a long recovery period (Sprague, 1966b; Wallace et al., 1989).

(4) Thus, control of orienting is widely distributed in the visual cortex, and the increased severity of the deficits after larger lesions suggests a mechanism of 'mass action'.

(5) Lesion of the superior colliculus also results in marked contralateral neglect and a deficit in orienting which is complete initially, and followed by partial recovery. Interaction between cortex and midbrain in mediating orienting responses is shown by the enduring potentiation of the deficit which follows cortical and tectal lesions placed on the same side of the brain. This effect occurs regardless of the sequence of the lesions (Sprague, 1966b; Sherman, 1977; Wallace et al., 1989, 1990).

After finding interaction between cortical and collicular lesions on the same side of the brain it was a natural step to examine the effects when the lesions were placed sequentially on opposite sides of the brain; the results were surprising and dramatic (Sprague, 1966a). The crucial insight here into the neural mechanism of visual orienting is provided by the large occipito-parietal lesions (group IV) on one side of the brain and following the animals for long periods of time, up to more than 1 year. The initial deficits of total, contralateral hemianopia and absence of visual orienting into the 'blind' field, persist unchanged. Such a deficit is classically ascribed to interruption of the visual radiations from the thalamus to the cortex and is considered permanent (cortical blindness). That such a conclusion is too simplistic and misleading is shown by the following experiment.

Removal of the colliculus opposite the cortical lesion induced marked circling into the previously blind field, so that perimetry testing was initially difficult; this circling was reduced enough in 1–2 weeks for testing to resume. Recovery of orienting began when the cat started showing spontaneous scanning, without stimulation, to the previously blind visual field, closely followed in time (1–3 weeks) by stimulus evoked orienting restricted to the central 5–10°, which expanded to the central 45° within 2–3 weeks, and with improved speed and localization gradually included the entire visual field. Blink to visual threat also returned. Aspects of vision, other than orienting, remained as before the collicular lesion. In short, the aspect of the cortical blindness which was improved by the midbrain lesion was that of neglect; the hemianopia remained unchanged. After time was allowed for stabilization of the orienting responses, a rather remarkable balance was seen in the orienting to the visual field opposite the intact cortex and the visual field opposite the intact colliculus (Sprague, 1966a). Subsequent work (Goodale, 1973; Kirvel et al., 1974; Sherman, 1977; Hardy and Stein, 1988) has confirmed the validity and robustness of this finding, and the mystery of its neurological basis has been considerably clarified by the work of Wallace et al. (1989, 1990) and Durmer et al. (1994).

The initial explanation by Sprague (1966a) for this phenomenon was as follows: in view of the known participation of the SC in orienting behavior, why after cortical lesion, was the ipsilateral colliculus not generating orienting responses since the retinotectal paths which project to it are active (Buchtel et al., 1979) and respond to stimuli introduced into the 'blind' field? However, this colliculus is malfunctioning because of the excitatory deprivation caused by the cortical lesion (McIlwain and Fields, 1971; Berson and McIlwain, 1983; Ogasawara et al., 1984; Deuel, 1987; Hovda and Villablanca, 1990), and because of an inhibitory influx coming from the opposite colliculus (Sprague, 1966a; Hoffmann and Straschill, 1971; Goodale, 1973). The crossed tectal inhibition hypothesis was supported by the comparable recovery of orienting which followed mid-line splitting of the tectal commissure rather than ablation of the opposite colliculus (Sprague, 1966a; Sherman, 1977; Wallace et al., 1989). The commissurotomy has the advantage over the collicular lesion in that no motor asymmetries are introduced which rules out the possibility that the recovery is an artifact due to turning tendencies following collicular lesion (Cooper et al., 1970). Recovery of orienting also occurred when the

commissure was split first, followed later by unilateral cortical lesion (Wallace et al., 1989). The prior commissure split did not however 'protect' against the visual deficits induced by the subsequent cortical lesion and the recovery was somewhat slower then when the cortical lesion was placed first. It should be noted, however, that time of recovery varied widely among animals.

This hypothesis that the crossed-tectal inhibition originated in the opposite colliculus was supported by anatomical studies of the commissural cells described by Edwards (1977), Graham (1977), Magalhaes-Castro (1978) and Behan (1985). These cells lie in the intermediate and deep laminae of the rostral half of the colliculus, and their axons pass in the rostral half to the commisure. This hypothesis of the crossed tectal inhibition had to be modified by the findings of Wallace et al. (1989), that section of the rostral half of the commissure failed to produce recovery of orienting but caudal commissurotomy did. When tectal neurons were destroyed by injection of ibotenic acid into the colliculus, a lesion which spared fibers which originated outside the colliculus and passed into the commissure, no recovery occurred (Wallace et al., 1989); these experiments clearly indicated that the Sprague effect is dependent on neurons lying outside of the colliculus. Caudal section of the tectal commissure caused no neurological sign, whereas rostral section sometimes resulted in a tendency to ventriflex the neck and in decreased excursion of upward eye movements (Matelli et al., 1983; Wallace et al., 1989).

Sherman (1977) has made a very significant contribution to this phenomenon by demonstrating cortical blindness and loss of visual orientation in cats after comparable cortical lesions made bilaterally, followed by recovery of orienting when the tectal commissure was split. This recovery occurred whether the split preceded or followed the cortical lesion. Loop and Sherman (1977) trained a group of cats on three visual discriminations: dark-light (brightness task), horizontal vertical gratings (pattern task) and upright-inverted triangles (form task). (a) Four cats had large, bilateral occipito- parieto-temporal cortical lesions (same as Group IV) described previously); post-operatively they lacked orienting, but could relearn brightness in the same or slightly fewer number of trials required preoperatively. (b) Six other cats, with the same preoperative training had comparable cortical lesions plus section of the tectal commissure. They demonstrated good orienting and relearned the brightness discrimination in about half the number of trials as did the cats in group (a). Neither group (a) nor (b) could relearn gratings or forms. Thus, tectal split restores orienting but does not restore pattern or form discrimination. Interesting studies by Wood (1975) utilized split-chiasm cats to limit the projection from each eye to the ipsilateral hemisphere. The animals were trained pre-operatively in a brightness (dark-light) discrimination monocularly and binocularly, followed by ablation of one superior colliculus and retested. When criterion was again achieved, the opposite visual cortex was removed. Criterion was regained using the eye on the side of decortication in many fewer trials than in other animals in which the tectal lesion followed the cortical lesion. This work agrees with Loop and Sherman (1977) in that the sequence of lesions underlying the Sprague effect on orienting also improves performance in brightness discrimination, a function mediated by the colliculus in decorticated animals.

Hovda and Villablanca (1990) made an important study of cats following unilateral hemispherectomies; they found complete and lasting (14 months) contralateral hemianopia using lick suppression in a perimetric test. The superior colliculus ipsilateral to the cortical lesion was markedly depressed in oxidative metabolism (measured by cytochrome oxidase histochemistry), compared to the contralateral colliculus, a finding which supports the idea of excitatory deprivation in the colliculus after ipsilateral cortical lesions.

Understanding the neural mechanisms of the colliculus in visual orienting was enhanced by the demonstration that this laminated structure can be divided into superficial and deep layers, on the basis of both structure and function in the tree shrew (Casagrande et al., 1972; Harting et al.,

1973) and in cat (Kanaseki and Sprague, 1974; Sprague, 1975). The three superficial layers project forward into the pretectum, LGN_d, LGN_v and the pulvinar-lateral posterior nucleus complex which projects in turn to extrastriate cortex; the three deep layers project to the intralaminar and subthalamic nuclei, inferior colliculus, parabigeminal nucleus, interior olive and brainstem reticular formation (Harting et al., 1973; Graham, 1977; Heurta and Harting, 1984; Yamasaki et al., 1986; Sparks and Hartwick-Young, 1989; Stein and Meredith, 1993). It is chiefly the deeper component in which there is multimodality convergence and integration (Meredith and Stein, 1986; Peck et al., 1993) that is involved in orienting, and the malfunction of the colliculus after cortical lesion is reflected in the altered neural activity in these deep layers, especially the intermediate gray (Wurtz and Goldberg, 1972; Raczowski et al., 1976 ; Graham, 1977; Straschill and Schick, 1977; Wurtz et al., 1980; Grantyn and Grantyn, 1982; Ogasawara et al., 1984; Grantyn and Berthoz, 1985; Karabelas and Moschovakis, 1985; Northmore et al., 1988; Peck et al., 1993). Raczowski et al. (1976) and Sprague et al. (1963), showed that lesions in the efferent tract of these deep laminae (predorsal bundle) resulted in deficits in orienting without sensory or perceptual loss in the tree shrew and cat respectively. Ellard and Goodale (1986) reported that similar lesions in gerbils abolished contraversive head and body orienting movements elicited by collicular stimulation.

There is now considerable evidence that in several species (rat, hamster, cat, macaque) many cells of the deep layers which project to the brainstem are controlled by a neuronal system which originates in the substantia nigra, pars reticulata (SN_R). We owe to Graybiel (1978) in the cat, to Rhoades et al. (1982) in the hamster and to Hikosaka and Wurtz (1983b) in the macaque much credit for the recognition of this system. Lesion in the cat of a small area (called the 'critical zone' by Wallace et al. (1990)) of SN_R contralateral to the cortical lesion, restores the loss of visual orienting caused by large cortical lesions (group IV), similar in all ways to that which follows section of the tectal commissure.

Restoration of orienting also occurs when the large cortical lesion follows the small nigral lesion (Wallace et al., 1990); this nigral lesion alone had no effect on the cat's visual field, but it did produce a syndrome consisting of: (1) compulsive circling/turning and tonic posturing to the contralesional side; (2) compulsive saccades into the contralateral hemifield; and (3) a perioral dyskinesia, more prominent contralaterally. The eye movements disappeared in 1–2 days and the circling in 7–10 days; the dyskinesia sometimes lasted for months. Whether the recovery induced by nigral lesion is mediated monosynaptically by the crossed nigrotectal tract, or polysynaptically via the nigrotegmental path which terminates in the pedunculopontine nucleus (Noda and Oka, 1986; Tokuno et al., 1987) and in turn, projects to the contralateral colliculus (Moon et al., 1983), or both, is not totally clear (Durmer et al., 1994). In any case this small lesion in the 'critical zone' of SN_R destroyed a gabanergic path. Larger SN lesions which also involve the ascending dopaminergic system result in severe and enduring contralateral visual and somesthetic orienting deficits (neglect) and a faciliation of responses to the ipsilesional side (Feeney and Wier, 1979).

The recognition that SN_R is part of a neural orienting system which comprise cortex, striatum, nigra, superior colliculus and thalamus in several species has been building over the past 20 years. The input from the visual cortices to the striatum (both caudate and putamen) originates in a large number of extrastriate areas in the cat (Updike, 1993), in contrast to the minimal projection from, 17 and, 18 (Royce, 1982). The terminals of this extensive cortico-striatal system are excitatory (glutamate) and activate the projection neurons of the striatonigral pathway (Niimi et al., 1970; Williams and Faull, 1985) which inhibit (GABA) the spontaneous firing of cells in SN_R. This part of the nigra projects largely to the intermediate and deep laminae of the superior colliculus where it inhibits (GABA) the firing of the tectal motor units. Unilateral activation of the cortical-

striatonigral path blocks this inhibition allowing the discharge of these tectospinal neurons which results in visuomotor orienting of head and eyes toward contralateral visual space (Hikosaka and Wurtz, 1983a,b monkey; Chevalier et al., 1984, 1985, rat; Joseph and Boussaoud, 1985, cat).

The neurophysiological study of SN_R in cats trained to orient their gaze toward visual and auditory targets by Joseph and Boussaoud (1985) has established that: (1) cells in SN_R have a steady, high rate of spontaneous activity; (2) 60% were responsive to visual stimuli, and 36% were responsive to auditory stimuli, mostly in the contralateral hemifield; (3) 30 cells showed inhibition 50–300 ms prior to the onset of saccades, triggered by visual and auditory stimuli; (4) orienting saccades were accompanied by pinnal movements; and (5) these cells were located in the middle part of SN_R corresponding to the areas labelled by HRP injections in the superior colliculus (Beckstead et al., 1981).

Chevalier et al. (1984) provided direct electrophysiological evidence that tectospinal (TSP) neurons, identified by antidromic activation, are inhibited by stimulation of SN_R in the rat. These neurons, which lie in the lateral part of the intermediate collicular lamina (SGI) had peripherally evoked activity (somatosensory and visual) blocked by the nigrotectal system. This study and a previous one from this laboratory (Deniau et al., 1978) in which tonically active SN_R neurons received a potent inhibitory influence from the striatum, led to the proposal that the striato-nigro-tectal system exerts a gating effect on the excitability of tectospinal neurons via a disinhibitory mechanism. This hypothesis received support from the elegant work reported by Chevalier et al. (1985) in the rat. They supported Joseph and Broussaoud (1985) results on the cat and Hikosaka and Wurtz (1983) on the monkey that nigrotectal cells have a chronically high level of firing which effectively inhibits the tectospinal neurons in SGI. They recorded in SN_R and in tectospinal-diencephalic cells (TSD) while activity in the nigra and striatum was changed by local applications of GABA and glutamate respectively in these two areas. Abolishing the tonic inhibitory discharge of SN_R, either by injection of GABA directly, or by exciting the striatonigral path with glutamate, caused the TSD cells to discharge vigorously and enhanced their responsiveness to somatosensory input. The striatal area which they found to be particularly effective in dis-inhibiting the TSD neurons is that region which when lesioned produces a permanent contralateral sensory neglect and abolishes orienting. Grantyn et al. (1985) showed that TSD cells have membrane properties which generate high frequency group discharge as a result of subtle changes of excitability. Grantyn and Berthoz (1985) studied tecto-reticulo-spinal neurons in alert cats during presentation of moving visual stimuli. Most of them had directionally selective responses in the absence of visuomotor orienting, enhanced firing to stimuli-triggered saccades but no activation for spontaneous saccades. The highest frequency discharges were found in some neurons during active orienting toward novel or 'interesting' objects, i.e. attention neurons.

Chevalier et al. (1985) identified tectospinal-tectodiencephalic neurons because they could be antidromically activated by stimulation of the contralateral spinal cord (C1–C2) and the ipsilateral ventromedial thalamic nucleus. Kemel et al. (1988) found dense projections from SN_R in the cat ventromedial (VM) and ventrolateral (VL) nuclei, and more limited terminal fields in the centrolateral (CL), paracentral (PC) and mediodorsal (DM) nuclei. The nigrotectal system has been the subject of many excellent studies (Graybiel, 1978; Deniau et al., 1978; Bentivoglio et al., 1979; Anderson and Yoshida, 1980; Steindler and Deniau, 1980; Beckstead, 1983; Behan, 1987; Harting et al., 1988; Kemel, 1988 — a partial list), and several of these have demonstrated the presence of branched axons arising in SN_R of cat (Anderson and Yoshida, 1977, 1980; Beckstead, 1983); rat (Deniau et al., 1978; Bentivoglio et al., 1979; Steindler and Deniau, 1980), and monkey (Beckstead, 1983). Moreover, tectospinal neurons themselves have an extensive axonal network in the medullary reticular core

and in the thalamus (Grantyn and Grantyn, 1982); Chevalier and Deniau (1984) have demonstrated that TSP cells send collaterals to at least one of the following thalamic nuclei — DM, CL, PC, VM and ZI (zona incerta). Buse et al. (1986) have studied the interaction which occurs in the ventromedial thalamic nucleus between the inhibitory influence of SN_R and the excitatory driving of the cerebellar nuclei. A GABA induced pause in tonic nigral firing increased the efficacy of cerebellar afferent volleys, but an increase in nigral firing blocked cerebello-thalamo-cortical transmission. Chevalier et al. (1985) have postulated that by activation of this tecto-spinal, tecto-diencephalic system (TSD), the striato-nigro-collicular pathway provides a coordinated influence on orienting at brainstem, thalamic and cortical levels. Thus there appear to be two visual systems which influence the activity of these deep, orienting tectal cells: the direct corticotectal pathways from extrastriate cortex (Palmer et al., 1972; Sprague, 1975; Benson and McIlwain, 1983), and the indirect cortico-striato-nigrotectal path which also originates from extrastriate cortex (see McHaffie et al., 1993).

The intermediate (SGI) and deep (SGP) laminae of the superior colliculus project heavily to the intralaminar nuclei (Graham, 1977; Kaufman and Rosenquist, 1985a; Yamasaki et al., 1986; Krauthamer et al., 1987) which in turn project to the striatum and to all visual cortical areas in the cat except area 17 (Kaufman and Rosenquist, 1985b). Thus, the sensory information (visual, acoustic, somesthetic) reaching the intralaminar nuclei from the superior colliculus can be relayed to the caudate-putamen. 'A potential pathway is therefore available by which the basal ganglia can self-regulate the tectal activity conveyed to the basal ganglia via the intralaminar thalamus', important in the organization and execution of orienting movements (Krauthamer et al., 1987).

The opposite side of the coin of orienting is its absence, a defect called neglect which is different from sensory blindness although they may appear the same. Defects in orienting following unilateral lesions may consist of contralateral sensory neglect (visual, acoustic, somesthetic) ipsiversive circling and hyper-responsiveness, and loss of contralateral following and target tracking, are seen after lesions in neostriatum, substantia nigra, nigrotectal tract, superior colliculus, pulvinar and intralaminar nuclei (Sprague and Meikle, 1965; Orem et al., 1973; Casagrande and Diamond, 1974; Wilburn and Kenne, 1974; Marshall et al., 1974, 1980; Siegfried and Burns, 1975; Goodale and Murison, 1975; Ljungberg and Ungerstedt, 1976; Feeney and Wier, 1979; Zihl and von Cramon, 1979; Schneider, 1984; Peterson et al., 1987; Wallace et al., 1990; Sakashita 1991). We have seen in the cat that after unilateral removal of all retinotopically organized visual cortices (group IV), the animal appears blind in the contralateral visual fields. This severe visual deficit (hemianopia) is the expression of malfunctioning of two interrelated neural mechanisms: one is 'blindness' caused by removal of the visual sensory and association cortex, and the other is the absence of orienting caused by tonic inhibitory influx from the nigrotectal system. The ability to orient can be restored by surgical or pharmacological intervention in nigra and superior colliculus; the cortical, sensory blindness remains.

This aspect of orienting and neglect has been examined further by Hardy and Stein (1988); they found contralateral neglect after small lesions in the posterior region of the lateral suprasylvian cortex of the cat which did not result in hemianopia. This effect, detected in a perimetry test, was present only when the test stimulus was introduced simultaneously with the release of the cat, but was not seen when introduced with the animal standing quietly, i.e. in the absence of an active motor program. The lesion, however, caused a severe decrement in activity of collicular cells in the intermediate and deep laminae, a finding similar to that described by Ogasawara et al. (1984) after cooling the same cortical area. Interestingly the neglect which Harty and Stein found following the cortical lesion was abolished or greatly reduced by ablation of the contralateral superior colliculus.

A larger part of the same extrastriate cortical area of the cat brain (areas 21a, 7p, PMLS, PLLS in the temporo-occipito-parietal junction was cooled and resulted in complete, contralateral loss of orienting in trained animals measured in a perimetry test (Payne et al., 1995). The severity of the deficit matches that induced by cooling the superior colliculus by the same authors. They also make the important point that the deficit is neglect, and not hemianopia, because the cooling does not impair the cat's ability to make pattern discriminations (Lomber et al., 1994). Control animals in which the rostral half of the lateral suprasylvian cortex was cooled failed to show the visual defect. They also demonstrated in one animal that loss of orienting caused by cortical cooling was returned to the neglected field by cooling the contralateral colliculus (unpublished results).

It is beyond the scope of this chapter to discuss orienting and neglect in human patients and the reader is referred to (Posner et al., 1982; Heilman et al., 1987; Kinsbourne, 1987; Peterson et al., 1989; Rizzolatti et al., 1988, 1990 1994). However, a brief comment is important.

Study of these phenomena in human patients has advanced our understanding of neglect following unilateral lesions of dorsolateral (parietal) association cortex. Like cats, the head and eyes deviate toward the side of the lesion and the patient orients briskly toward any stimulus in the ipsilesional visual field. The patient completely neglects the contralateral space and body, but is unaware that half of the world is gone. Yet imaginative and careful testing has revealed that the patient's pre-attentive vision is able to abstract figure from ground, to group objects and to define their primary axis in the neglected field. In short, perceptual processing in the neglected field is intact to an advanced level of semantic classification, and neglect arises quite late at a stage of selection for action. For example, using bilateral stimulation, extinction of the stimulus contralateral to the lesion is more likely to occur when the two objects are the same than when they are different. The stimuli therefore are processed to a level at which their identity and differences are established (Rafal, 1994).

Rafal (1994) has summarized a number of factors which affect neglect in patients with unilateral cortical lesions: (1) hyperreflexive orienting to the visual field on the lesion side; (2) impaired ability to disengage and shift attention from one area to another; (3) not only a change in perception occurs but also a deranged internal representation of space; (4) motor bias toward the ipsilesional side and impaired voluntary orientation contralaterally, causing defective exploratory behavior; (5) defective ability to generate saccades into the neglected visual field; and (6) failure of contralateral stimuli to produce arousal. Thus manifestations of neglect may represent the contribution of each of these mechanisms and the interaction among them (see Kinsbourne, 1993).

The phenomenon of neglect has been given several names, among them is inattention; the previous summary indicates that the deficit is more complex. Two competing theories have been frequently advanced in order to explain neglect — the attentional theory and the representational theory. Rizzolatti and Berti (1990) have recently reviewed the neurophysiological data derived from the monkey and human clinical evidence which can be used to support either of these two theories. They conclude that neglect is not confined to the perceptual-attentional domain, and is primarily a representational deficit which follows lesions in areas of the brain in which space is mapped and coded, and which are responsible for the organization of motor acts. In other words, the neuronal system that controls action is the same that controls spatial attention. This theory as stated by Bisiach et al. (1970) is that representation is necessary for achieving awareness of space and objects both in the presence of sensory stimuli and in their absence. An impairment of the capacity to construct mental images of space by using sensory material or memory engrams is the basic deficit responsible for neglect symptoms. Attentional disturbances are thought by them to be secondary factors which aggravate the deficit.

Acknowledgements

This chapter is an expanded verison of an overview lecture given at the 4th IBRO Satellite Symposium on 'Extrageniculostriate Mechanisms Underlying Visually-Guided Orientation Behavior.' It is not intended as a comprehensive review of this extensive area of research; hence a number of meritorious papers are not cited and various interpretations have not been discussed. For these omissions the author apologizes in advance.

I wish to express my sincere thanks to Wendy Todd, Jeanne Levy and Marvin Jackson for technical assistance, and to acknowledge grant support from the U.S. Public Health Service (EY-02654) and the University of Pennsylvania Research Foundation.

References

Anderson, M.E. and Yoshida, M. (1980) Axonal branching patterns and location of nigrothalamic and nigrocollicular neurons in the cat. *J. Neurophysiol.*, 43: 883–895.

Beckstead, R.M. (1983) Long collateral branches of substantia nigra pars reticulata axons to thalamus, superior colliculus and reticular formation on monkey and cat. Multiple retrograde neuronal labeling with fluorescent dyes. Neuroscience, 10: 767–797.

Beckstead, R.M., Edwards, S.B. and Frankfurter, A. (1981) A comparison of the intranigral distribution of nigrotectal neurons labeled with horseradish peroxidase in the monkey, cat and rat. *J. Neurosci.*, 1: 121–125.

Behan, M. (1985) An EM-Autoradiographic and EM-HRP study of the commissural projection of the superior colliculus in the cat. J. Comp. Neurol., 234: 105–116.

Behan, M., Lin, C.S. and Hall, W.C (1987) The nigrotectal projection in the cat: an electron microscope autoradiographic study. Neuroscience, 21: 529–539.

Bentivoglio, M., Van Der Kooy, H.G. and Kuypers, J.M. (1979) The organization of the efferent projections of the substantia nigra in the rat. A retrograde fluorescent double labelling study. *Brain Res.*, 174: 1–17.

Berkley, M.A. and Sprague, J.M. (1979) Striate cortex and visual acuity functions in the cat. *J. Comp. Neurol.*, 187: 679–702.

Benson, D.M. and McIlwain, J.T. (1983) Visual cortical inputs to deep layers of cat's superior colliculus. *J. Neurophysiol.*, 50: 1143–1155.

Bisiach, E. and Luzzatti, C. (1978) Unilateral neglect of representational space. *Cortex*, 14: 129–133.

12

Bowman, E.M. and Olson, C.K. (1988) Visual and auditory association areas of the cat's posterior ectosylvian gyrus: cortical afferents. *J. Comp. Neurol.*, 272: 30–42.

Buee, J. Deniau, J.M. and Chevalier, G. (1986) Nigral modulation of cerebello-thalamo-cortical transmission in the ventral medial thalamic nucleus. *Exp. Brain Res.*, 65: 241–244.

Buchtel, H.A., Camarda, R., Rizzolatti, G. and Scandolara, C. (1979) The effect of hemidecortication on the inhibitory interactions in the superior colliculus of the cat. *J. Comp. Neurol.*, 184: 795–810.

Casagrande, V.A., Harting, J.K., Hall, W.C. and Diamond, I.T. (1972) Superior colliculus of the tree shrew: a structural and functional subdivision into superficial and deep layers. *Science*, 177: 444–447.

Chevalier, G. and Deniau, J.M. (1984) Spatio-temporal organization of branched tecto-spinal/tecto-diencephalic neuronal system. *Neuroscience*, 12: 427–439.

Chevalier, G., Vacher, S. and Deniau, J.M. (1984) Inhibitory nigral influence on tectospinal neurons, a possible implication of basal ganglia in orienting behavior. *Exp. Brain Res.*, 53: 320–326.

Chevalier, G., Vacher, S., Deniau, J.M. and Desban, M. (1985) Disinhibition as a basic process in the expression of striatal functions I. The striato-nigral influence on tecto-spinal/tecto-diencephlic neurons. *Brain Res.*, 334: 215–226.

Cooper, R.M., Bland, B.H., Gillespie, L.A. and Whitaker, R.H. (1970) Unilateral posterior cortical and unilateral collicular lesions and visually guided behavior in the rat. *J. Comp. Phyisol. Psychol.*, 72: 286–295.

Deniau, J.M., Hammond, C., Riszk, A. and Feger, J. (1978) Electophysiological properties of identified output neurons of the rat substantia nigra (pars compacta and pars reticulata): evidences for the existence of branched neurons. *Exp. Brain Res.*, 32: 409–422.

Deuel, R.K. (1987) Neural dysfunction during hemineglect after cortical damage in two monkey models. In: M. Jeannerod (Ed.), *Neurophysiological and Neuropsychological Aspects of Spatial Neglect*, Elsevier, North Holland.

Durmer, J.S., Ciaramitaro, V., Todd, W.E. and Rosenquist, A.C. (1994) Ibotenic acid lesions of the pedunculopontine nucleus restore visual orienting responses in the previously hemianopic visual field of the cortically blind cat. *Soc. Neurosci. Abstr.*, 20: 1187.

Edwards, S.B. (1977) The commissural projection of the superior colliculus in the cat. *J. Comp. Neurol.*, 173: 23–40.

Ellard, C.G. and Goodale, M.A. (1986) The role of the predorsal bundle in the head and body movements elicited by electrical stimulation of the superior colliculus in the Mongolian gerbil. *Exp. Brain Res.*, 64: 421–433.

Feeney, D.M. and Wier, C.S. (1979) Sensory neglect after lesions of substantia nigra or lateral hypothalamus: differential severity and recovery of function. *Brain Res.*, 178: 329–346.

Fischman, M.W. and Meikle, T.H. Jr. (1965) Visual intensity discrimination in cats after serial tectal and cortical lesions. *J. Comp. Physiol. Psychol.*, 59: 193–201.

Goodale, M.A. (1973) Cortico-tectal and intertectal modulation of visual responses in the rat's superior colliculus. *Exp. Brain Res.*, 17: 75–86.

Goodale, M.A. and Murison, R.C.C. The effects of lesions of the superior colliculus on locomotor orientation and the orienting reflex of the rat. *Brain Res.*, 88: 243–261.

Graham, J. (1977) An autoradiograhic study of the efferent connections of the superior colliculus in the cat. *J. Comp. Neurol.*, 173: 629–654.

Grantyn, A. and Berthoz, A. (1985) Burst activity of identified tecto-reticulo-spinal neurons in the alert cat. *Exp. Brain. Res.*, 57: 417–421.

Grantyn, A. and Grantyn, R. (1982) Axonal patterns and sites of termination of cat superior colliculus neurons projecting in the tecto-bulbo-spinal tract. *Exp. Brain. Res.*, 46: 243–256.

Graybiel, A.M. (1978) Organization of the nigrotectal connection: an experimental tracer study in the cat. *Brain Res.*, 143: 339–348.

Hardy, S.C. and Stein, B.E. (1988) Small lateral suprasylvian cortex lesions produce visual neglect and decreased visual activity in the superior colliculus. *J. Comp. Neurol.*, 273: 527–542.

Harting, J.K., Hall, W.C., Diamond, I.T. and Martin, G.F. (1873) Anterograde degeration study of the superior colliculus in *Tupaia glis*: Evidence for a subdivision between superficial and deep layers. *J. Comp. Neurol.*, 148: 361–386.

Harting, J.K., Huerta, M.F., Hashikawa, T., Weber, J.T. and van Lieshout, D.P. (1988) Neuroanantomical studies of the nigrotectal projection in the cat. *J. Comp. Neurol.*, 278: 615–631.

Heilman, K.M., Bowers, D., Valenstein, E. and Watson, R.T. (1987) Hemispace and hemispatial neglect. In: M. Jeannerod (Ed.), *Neurophysiological and Neuropsychological Aspects of Spatial Neglect*, Elsevier, North Holland.

Hikosaka, O., and Wurtz, R.H. (1983a) Visual and oculomotor functions of monkey substantia nigra pars reticulata. I. Relation of visual and auditory responses to saccades. *J. Neurophysiol.*, 49: 1230–1253.

Hikosaka, O., and Wurtz, R.H. (1983b) Visual and oculomotor functions of monkey substantia nigra pars reticulata. IV. Relation of substantia nigra to superior colliculus. *J. Neurophysiol.*, 49: 1285–1301.

Hoffman, K.P. and Straschill, M. (1971) Influences of cortico-tectal and intertectal connection on visual responses in the cat's superior colliculus. *Exp. Brain Res.*, 12: 120–131.

Hovda, D.A. and Villablanca, J.R. (1990) Sparing of visual field perception in neonatal but not adult cerebral hemispherectomized cats. Relationship with oxidative metabolism of the superior colliculus. *Behav. Brain Res.*, 37: 119–132.

Huerta, M.F. and Harting, J.K. (1984) The mammalian superior colliculus: studies of its morphology and connections.

In: H. Vanegas (Ed.), *The Comparative Neurology of The Optic Tectum*, Plenum, New York.

Joseph, J.P. and Boussaoud, D. (1985) Role of the substantia nigra pars reticulata in eye and head movements I. Neural activity. *Exp. Brain Res.*, 57: 2286–296.

Kanaseki, T. and Sprague, J.M. (1974) Anatomical organization of pretectal nuclei and tectal laminae in the cat. *J. Comp. Neurol.*, 158: 319–338.

Karabelas, A.B. and Moschovakis, A.K. (1985) Nigral inhibitory termination on efferent neurons of the superior colliculus: an intracellular horseradish peroxidase study in the cat. *J. Comp. Neurol.*, 239: 309–329.

Kaufman, E.F.S. and Rosenquist, A.C. (1985a) Afferent connections of the thalamic intralaminar nuclei in the cat. *Brain Res.*, 335: 281–296.

Kaufman, E.F.S. and Rosenquist, A.C. (1985b) Efferent projections of the thalamic intralaminar nuclei in the cat. *Brain Res.*, 335: 257–279.

Kemel, M.L, Desban, M., Gauchy, C., Glowinsky, J. and Besson, M.J. (1988) Topographical organization of efferent projections from the cat substantia nigra pars reticulata. *Brain Res.*, 455: 307–323.

Kinsbourne, M. (1987) Mechanisms of unilateral neglect. In: M. Jeannerod (Ed.), *Neurophysiological and Neuropsychological Aspects of Spatial Neglect*, Elsevier, North Holland, pp. 69–86.

Kirvel, R.D., Greenfield, R.A. and Meyer, D.R. (1974) Multimodal sensory neglect in rats with radical unilateral posterior isocortical and superior colliculus ablations. *J. Comp. Physiol. Psychol.*, 87: 156–162.

Krauthamer, G.M., Yamasaki, D.S. and Rhoades, R.W. (1987) Does the neostriatum self- regulate its sensory input? The role of the superior colliculus. In: J.S. Schneider and T.I. Lidsky (Eds.), *Basal Ganglia and Behavior: Sensory Aspects of Motor Functioning*, Huber, Toronto, pp. 17–26.

Ljungberg, T. and Ungerstedt, U. (1976) Sensory inattention produced by 6-hydroxydopamine degeneration of ascending dopamine neurons in the brain. *Exp. Neurol.*, 53: 585–600.

Lomber, S.G., Cornwell, P. Sun, J.-S., MacNeil, M.A. and Payne, B.R. (1994) Reversible inactivation of visual processing operations in middle suprasylvian cortex of the behaving cat. *Proc. Natl. Acad. Sci.*, 91: 2999–3003.

Loop, M.S. and Sherman, S.M. (1977) Visual discriminations of cats with cortical and tectal-lesions. *J. Comp. Neurol.*, 174: 79–88.

Magalhaes-Castro, H.H., DeLima, A.D., Saraiva, P.E.S. and Magalhaes-Castro, B. (1978) Horseradish peroxydase labeling of cat tectotectal cells. *Brain Res.*, 148: 1–13.

Marshall, J.F. (1978) Comparison of the sensorimotor dysfunctions produced by damage to lateral hypothalamus or superior colliculus in the rat. *Exp. Neurol.*, 58: 203–217.

Marshall, J.F. and Teitlebaum, P. (1974) Further analysis of sensory inattention following lateral hypothalamic damage. *J. Comp. Physiol. Psychol.*, 86: 375–395.

Matelli, M., Oliveri, M.F., Saccani, A. and Rizzolatti, G. (1983) Upper visual space neglect and motor deficits after section of the midbrain commissures in the cat. *Brain Res.*, 10: 263–285.

McHaffie, J.G. Norita, M. Dunning, D.D. and Stein, B.E. (1993) Corticotectal relationships: direct and 'indirect' corticotectal pathways. In: T.P. Hicks, S. Molotchnikoff and T. Ono (Eds.), *Prog. Brain Res.*, 95: 139–150.

McIlwain, J.T. and Fields, H.L. (1971) Interactions of cortical and retinal projections on single neurons of the cat's superior colliculus. *J. Neurophysiol.*, 34: 763–772.

Meredith, M.A. and Stein, B.E. (1986) Visual, auditory and somatosensory convergence on cells in superior colliculus results in multisensory integration. *J. Neurophysiol.*, 56: 640–662.

Mesulam, M.-M. (1981) A cortical network for directed attention and unilateral neglect. *Ann. Neurol.*, 10: 309–325.

Moon Edley, S. and Graybiel, A.M. (1983) The afferent and efferent connections of the feline nucleus tegmenti pedunculopontinus, pars compacta. *J. Comp. Neurol.*, 217: 187–215.

Niimi, K., Ikeda, T., Kawamura, S. and Inoshita, H. (1970) Efferent projections of the head of the caudate nudleus in the cat. *Brain Res.*, 21: 327–343.

Noda, T. and Oka, H. (1986) Distribution and morphology of tegmental neurons receiving nigra inhibitory inputs in the cat: an intracellular HRP study. *J. Comp. Neurol.*, 244: 254–266.

Northmore, D.P.M, Levine, E.S. and Schneider, G.E. (1988) Behavior evoked by electrical stimulation of the hamster superior colliculus. *Exp. Brain Res.*, 73: 595–605.

Ogasawara, L, McHaffie, J.G. and Stein, B.E. (1984) Two visual corticotectal systems in cat. *J. Neurophysiol.*, 52: 1226–1245.

Orem, J., Schlag-Rey, M. and Schlag, J. (1973) Unilateral visual neglect and thalamic intralaminar lesions in the cat. *Exp. Neurol.*, 40: 784–797.

Otsuka, R. aknd Hassler, R. (1962) âber Aufban and Gliederung der corticalen Sehsphère bei der Katze. *Arch. Psychiat. Zeitsch. Neurol.*, 203: 212–234.

Palmer, L.A., Rosenquist, A.C. and Sprague, J.M. (1972) Corticotectal systems in the cat: their structure and function. In: T. Frigyesi, E. Rinvik and M.D. Yahr (Eds.), *Corticothalamic Projections and Sensorimotor activities*, Raven Press, New York, pp. 491–522.

Payne, B.R., Lomber, S.G., Geeraerts, S., Van der Gucht, E. and Vandenbussche, E. (1996) Reversible visual hemieglect. *Proc. Natl. Acad. Sci.*, in press.

Peck, C.K., Baro, J.A. and Warder, S.M. (1993) Sensory integration in the deep layers of superior colliculus. In: T.P. Hicks, S. Molotchnikoff and T. Ono (Eds.) *Prog. Brain. Res.*, 95: 91–102.

14

Peterson, S.E., Robinson, D.L. and Morris, J.D. (1987) Contribution of the pulvinar to visual spatial attention. *Neuropsychologia*, 25: 97–105.

Peterson, S.E., Robinson, D.L. and Currie, J.N. (1989) Infuluences of lesions of parietal cortex on visual spatial attention in humans. *Exp. Brain Res.*, 76: 267–280.

Posner, M.I., Cohen, Y. and Rafal, R.D. (1982) Neural systems control of spatial orienting. *Phil Trans. R. Soc. Lond.* B, 298: 187–198.

Raczkowski, D. Casagrande, V.A. and Diamond, I.T. (1976) Visual neglect in the tree shrew after interruption of the descending projections of the deep superior colliculus. *Exp. Neurol.*, 50: 14–29.

Rafal, R.D. (1994) Neglect. *Curr. Opin. Neurobiol.*, 4: 231–236.

Rhoades, R.W., Kuo, D.C., Polcer, J.D., Fish, S.E. and Voneida, T.J. (1982) Indirect visual cortex input to the deep layers of the hamster's superior colliculus via the basal ganglia. *J. Comp. Neurol.*, 20: 239–254.

Rizzolatti, G. and Gallese, V. (1988) Mechanisms and theories of spatial neglect. In: F. Boller and J. Grafman (Eds.), *Handbook of Neuropsychology*, Vol. 1, Elsevier, Amsterdam, pp. 223–246.

Rizzolatti, G. and Berti, A. (1990) Neglect as a neural representation deficit. *Rev. Neurol. (Paris)*, 146: 626–634.

Rizzolatti, G., Riggio, L. and Sheliga, B.M. (1994) Space and selective attention. In: C. Umilta and M. Moscovitch (Eds.), *Attention and Performance XV*, Erlbaum, Hilsdale, NJ, pp. 231–265.

Rosenquist, A.C. (1985) Connections of visual cortical areas. In: A. Peters and E.G. Jones (Eds.), *Cerebral Cortex*, Plenum, 3: 81–117.

Royce, G.J. (1982) Laminar origin of cortical neurons which project upon the caudate nucleus: a horseradish peroxidase investigation in the cat. *J. Comp. Neurol.*, 205: 8–29.

Sakashita, Y. (1991) Visual attentional disturbance with unilateral lesions in the basal ganglia and deep white matter. *Ann. Neurol.*, 30: 673–677.

Sanides, F. and Hoffmann, J. (1969) Cyto-and myeloarchitecture of the visual cortex of the cat and of the surrounding integration cortices. *J. Hirnforsch.*, 11: 79–104.

Schneider, G.E. (1975) Two visuomotor systems in the Syrian hamster. In: D. Ingle and J.M. Sprague (Eds.), *Sensorimotor Function of the Midbrain Tectum. Neurosci. Res. Prog. Bull.*, 13: 255–257.

Sherman, S.M. (1974) Visual fields of cats with cortical and tectal lesions. Science, 185: 355- 357.

Sherman, S.M. (1977) The effect of superior colliculus lesions upon the visual fields of cats with cortical ablations. *J. Comp. Neurol.*, 172: 211–230.

Sokolov, E.N. (1975) The neuronal mechanisms of the orienting reflex. In: E.N. Sokolov and O.S. Vinogradoova (Eds.), *Neuronal Mechanisms of the Orienting Reflex*, Erlbaum, Hillsdale, New Jersey, pp. 217–235.

Sparks, D.L. and Hartwich-Young, R. (1989) The deep layers of the superior colliculus. In: R.H. Wurtz and M.E. Goldberg (Eds.), *The Neurobiology of Saccadic Eye Movements*, Elsevier, North Holland, pp. 213–255.

Sprague, J.M. (1966a) Interaction of cortex and superior colliculus in mediation of visually guided behavior in the cat. *Science*, 153: 1544–1547.

Sprague, J.M. (1966b) Visual, acoustic and somesthetic deficits in the cat after cortical and midbrain lesions. In: D.P. Purpura and M.D. Yahr (Eds.), *The Thalamus*, Columbia Univ., pp. 391–417.

Sprague, J.M. (1975) Mammalian tectum: intrinsic organization, afferent inputs, and integrative mechanisms. Anatomical substrate. In: D. Ingle and J.M. Sprague (Eds.), *Sensorimotor Function of the Midbrain Tectum. Neurosci. Res. Prog. Bull.*, 13: 204–213.

Sprague, J.M. and Meikle, T.H. Jr. (1965) The role of the superior colliculus in visually guided behavior. *Exp. Neurol.*, 11: 115–146.

Sprague, J.M., Levitt, M., Robson, K., Liu, C.N., Stellar, E. and Chambers, W.W. (1963) A Neuroanatomical and behavioral analysis of the syndromes resulting from midbrain lemniscal and reticular lesions of the cat. *Arch. Ital. Biol.*, 101: 225–295.

Sprague, J.M., Berlucchi, G. and Rizzolatti, G. (1973) The role of the superior colliculus and pretectum in vision and visually guided behavior. In: R. Jung (Ed.), *Handbook Sensory Physiol*, VII/3B: 27–101.

Sprague, J.M., Levy, J., Di Berardino, A. and Berlucchi, G. (1977) Visual cortical areas mediating form discrimination in the cat. *J. Comp. Neural.*, 172: 441–488.

Stein, B.E. and Meredith, M.A. (1993) *The merging of the senses*, Mass Inst. Tech. Press, Cambridge, MA, pp. 1–211.

Steindler, D.A. and Deniau, J.M. (1980) Anatomical evidence for collateral branching of substantia nigra neurons: a combined horseradish peroxidase and {3H} wheat germ agglutinin axonal transport study in the rat. *Brain Res.*, 196: 228–236.

Straschill, N and Schick, F. (1977) Discharges of superior colliculus neurons during head and eye movements of the alert cat. *Exp. Brain Res.*, 27: 131–141.

Toga, A.W., Layton, B.S. and Horenstein, S. (1979) Visually directed behavior is influenced by the posterior ectosylvian gyrus in the cat. *Brain Res.*, 178: 606–608.

Tokuno, H. and Nakamura, Y. (1987) Organization of the nigrotectospinal pathway in the cat: a light and electron microscopic study. *Brain Res.*, 436: 76–84.

Tusa, R.J., Palmer, L.A. and Rosenquist, A.C. (1981) Multiple cortical visual areas. Visual field topography in the cat. In: C.N. Woolsey (Ed.), *Cortical Sensory Organization*, Humana Press, pp. 1–31.

Updike, B.V. (1993) Organization of visual corticostriatal projections in the cat, with observations on visual projections to claustrum and amygdala. *J. Comp. Neurol.*, 327: 159–193.

Wallace, S.F., Rosenquist, A.C. and Sprague, J.M. (1989) Recovery from cortical blindness mediated by destruction of nontectotectal fibers in the commissure of the superior colliculus in the cat. *J. Comp. Neurol.*, 284: 429–450.

Wallace, S.F., Rosenquist, A.C. and Sprague, J.M. (1990) Ibotenic acid lesions of the lateral substantia nigra restore visual orientation behavior in the hemianopic cat. *J. Comp. Neurol.*, 296: 222–252.

Weiskrantz, L. Warringtron, E.K., Sanders, M.D. and Marshall, J. (1974) Visual capacity in the hemianopic field following a restricted occipital ablation. *Brain*, 97: 709–728.

Wertman, K.L., Sinclair, R. aknd Horenstein, S. (1984) Somatosensory Area I is involved in visual neglect in the cat. *Soc. Neurosci. Abstr.*, 10: 731.

Williams, M.N. and Faull, R.L.M. (1985) The striatonigral projection and nigrotectal neurons in the rat. *Neuroscience*, 14: 991–1010.

Wood, B.S. (1975) Monocular relearning of a dark-light discrimination by cats after unilateral cortical and collicular lesions. *Brain Res.*, 83: 156–162.

Wurtz. R.H. and Goldberg, M.E. (1972) The primate superior colliculus and the shift of visual attention. *Invest. Ophthalmol.*, 11: 441–450.

Wurtz, R.H., Goldberg, M.E. and Robinson, D.L. (1980) Behavioral modification of visual responses in the monkey: Stimulus selection for attention and movement. In: J.M. Sprague and A.N. Epstein (Eds.), *Prog. Psychobiol. Physiol. Psychol.*, Vol. 9, Academic Press, New York pp. 44–83.

Yamasaki, D.S.G., Krauthamer, G.M. and Rhoades, R.W. (1985) Superior collicular projection to intralaminar thalamus in rat. *Brain Res.*, 378: 223–247.

Zernicki, B. (1987) Pavolvian orienting reflex. *Acta Neurbiol. Exp.*, 47: 239–247.

Zihl, J. and von Cramon, D. (1979) The contribution of the 'second' visual system to directed visual attention in man. *Brain*, 102: 835–856.

M. Norita, T. Bando and B. Stein (Eds.)
Progress in Brain Research, Vol 112
© 1996 Elsevier Science BV. All rights reserved.

CHAPTER 2

The mosaic architecture of the superior colliculus

R.-B. Illing*

Neurobiological Research Laboratory, Department of Otorhinolaryngology, University of Freiburg, 79106 Freiburg, Germany

The superior colliculus is a midbrain structure serving visual, multisensory and sensorimotor processing. Throughout various collicular layers, visual afferents are linked together with afferents related to other sensory modalities as well as with afferents from sources not easily subsumed under the term 'sensory'. These inputs are orchestrated in a topographic fashion and led to premotor neurons that are important elements in generating saccadic eye movements and orientation movements of other kinds.

Using immunocytochemical techniques to chart the distribution of various substances serving neurotransmission and neuromodulation, it was found that many of them, e.g. acetylcholinesterase (AChE), choline acetyltransferase, the enkephalins, substance P, and parvalbumin, relate to repetitive structural islands, or modules, in the superior colliculus. From studies on the distribution of three further neuroactive substances in rat superior collicular tissue: the calcium binding protein calretinin, the growth and plasticity related protein neuromodulin (GAP-43), and a glutamate receptor of the NMDA-type, we were led to conclude (1) that the intermediate layers of the superior colliculus are composed not of two, but of at least three disjunct types of modules, (2) that not just the intermediate layers but more or less the whole superior colliculus is an assemblage of modules, and (3) that, besides topographic connectivity and laminar structuring, the modules constituting an iterative partitioning represent a third major feature of superior collicular architecture.

The origin of the collicular mosaic is considered under an evolutionary perspective, and a hypothesis is presented stating that the pattern of AChE-rich modules on the level of the multimodal collicular layers can be predicted from retinal ganglion cell topography.

Introduction

Ever since the bony fishes began to populate the seas, the mesencephalon, or midbrain, is a distinguishable part of the vertebrate brain. The ventral portions of the midbrain, also called tegmentum, are considered a derivative of the spinal motor column, whereas its dorsal parts appear to be derived from nervous structures serving sensory function. In agreement to this assignment, the midbrain roof of submammalian vertebrates is called tectum opticum. In mammals, the structure homologue to the optical tectum is referred to as the superior colliculus. Two major structural principles have long been recognized to rule superior collicular architecture. The superior colliculus is built as a stack of layers (Viktorov, 1966; Kanaseki and Sprague, 1974), and its major connections are topographically organized (Robinson, 1972; Lane et al., 1973; Stein et al., 1975; Chalupa and Rhoades, 1977; Shimozawa et al., 1984).

Contemplating possible functional associations of the superior colliculus, several concepts have been brought forward. Many sources describe this midbrain structure as a visual structure, in others it is considered an oculomotor center, and still others characterize it as a region where influences from various sources come together to produce an output that is determined by multisensory integration.

*Corresponding author. Neurobiological Research Laboratory, Univ.-HNO-Klinik, Killianstr. 5, D-79106 Freiburg, Germany; Tel.: +49 761 270 4273; fax: +49 761 270 4104; e-mail: rbi@sun1.ukl.uni-freiburg.de

The first two descriptions of superior collicular function, when taken as excluding definitions, are inadequate. It is true that in submammalian vertebrates the tectum is the almost exclusive recipient of retinal input. But even in these species, nonretinal afferents exist (O'Benar, 1976; Carr et al., 1981; Debski and Constantin-Paton, 1993), indicating that the tectum serves purposes beyond visual perception. It is also true that, upon electrical stimulation of the superior colliculus in rodents, carnivores or primates, saccadic eye movements can be elicited in an intriguingly precise and topographically ordered manner (Robinson, 1972; McHaffie and Stein, 1982). A complementing observation is that many collicular neurons show spike trains that are systematically related to specific parameters of eye movements (Sparks, 1986). But one should not be misled by these intriguing facts. We find an elaborately developed superior colliculus in mammalian species that do not show a rich oculomotor behavior, like the rat (Humphrey, 1968). Even species that hardly use their eyes, like echolocating bats (Wong, 1984) or cave dwelling animals (Cooper et al., 1993), have a colliculus with prominently developed layers. The precise coordination of saccadic eye movements must therefore appear to be a specialization of some more general function that is not restricted to vision.

The proposal that this more general function is orientation (Sprague and Meikle, 1965; Goodale and Murison, 1975; Stein and Meredith, 1991) should be evaluated carefully. The diverse connectional affiliations of the superior colliculus with sources and destinations known to belong to functional systems that are not visual or oculomotor (Huerta and Harting, 1984) are a major argument to support the view that the superior colliculus is associated with functions far beyond monosensory processing. Complex physiological and behavioral phenomena that can be related to tectal function (Dean et al., 1989; Sprague, 1991) point to the same direction.

The functions of the superior colliculus, like those in any other part of the brain, emerge from a coordinated interplay of subcellular, cellular and supracellular machines. The present report focuses on supracellular assemblies of neuronal parts. A heterogeneity of superior collicular tissue perpendicular to its lamination was demonstrated for the first time by Siou in the mouse (1958). Using the technique of acetylcholinesterase (AChE) histochemistry, he stained two zones of high enzyme activity, corresponding to the superficial and intermediate layers, respectively (cf. Fig. 1A of present study). However, although readily visible in his figures, he did not articulate the difference in texture of staining between these two zones. A few years later, Friede (1966) published figures of a very similar pattern in the cat brain, but, like Siou, left it uncommented. Ramon-Moliner (1972) appears to be the first one explicitly noting that, in AChE-stained sections, the deeper collicular layers 'have a characteristic uneven distribution of the stain with regularly spaced dark masses separated by a light stained background'. With this observation, the idea concretized that this pattern might be functionally meaningful (Graybiel, 1978). This idea received strong support when we could show that the periodic patterning of AChE in the intermediate collicular layers corresponded to similarly heterogeneous patterns formed by the terminal fields of several major collicular afferent systems (Illing and Graybiel, 1985). This correspondence proved to be matching in some instances and complementary or non-matching in others (Beninato and Spencer, 1986; Illing and Graybiel, 1986).

Beginning with the observation that the distribution of the enkephalins was similarly heterogeneous in the intermediate collicular layers (Graybiel et al., 1984), and matched in detail zones of high AChE activity (Graybiel and Illing, 1994), a series of discoveries of markers for collicular modules followed (e.g. Wallace, 1986b; Mize, 1989; Miguel-Hidalgo et al., 1989; Illing, 1990; Illing et al., 1990). Eventually, it was also shown that the modules formed by an iteration of terminal fields and of neuroactive substances correspond to clusters of efferent collicular neurons to put afferents and efferents in a specific spatial relationship (Illing, 1992; Jeon and Mize, 1993). A detailed analysis of the heterogeneous architecture revealed that certain parts of the superior colliculus were pervaded by biochemically and connectionally distinguishable, interdigitating domains, and the elements of this compartmentation were occasionally called modules.

With the present report, I intend to further our recognition of collicular structure and function in three ways. I shall first extend neurochemical observations that have been accumulated over the past decade on iteratively distributed substances in various collicular layers. Secondly, I shall speculate on the evolutionary origin of the pattern of collicular modules. Finally, a previously suggested relationship between visual behavior and the geometry of collicular compartmentation (Illing, 1993) will be presented in more detail.

Materials and methods

This study is based on the brains of adult animals of various mammalian species. For AChE-staining, the tissue was fixed at 4°C either by immersion for 48 h (cow and pig) or by perfusion for 1 h (rabbit, monkey, rat and cat). The fixative for immersion consisted of a mixture of paraformaldehyde (4%) and glutaraldehyde (0.5%) in 0.1 M phosphate buffer (pH 7.3). Fixation by perfusion was done with a phosphate buffered solution containing 4% paraformaldehyde, 0.1% glutaraldehyde and 15% saturated picric acid. Following fixation, the colliculus was soaked in 25% sucrose for 12 h and cut on a cryostat in varying planes. If the tissue was not cut frontally, tangential or oblique-sagittal sections were made. Tangential sections were made in a plane tangential to the collicular surface at a point about halfway from the medial to the lateral edge of the colliculus. Depending on the species, this plane was tilted from the horizontal plane along the rostro-caudal axis by 20–40°. Oblique-sagittal sections were cut in a plane oriented rostrocaudally and perpendicular to the tangential plane just described. The histochemical staining procedure for demonstrating AChE-activity was done as previously applied (Illing, 1990).

Immunocytochemistry was done on perfused cat (not illustrated) and rat brains. Antibodies against the calcium-binding proteins parvalbumin (Sigma, 1:15000) and calretinin (SWant, 1:5000), the growth-associated protein GAP-43 (Boehringer-Mannheim, 1:5000), the glycin receptor (Boehringer-Mannheim, not illustrated) and subunit NR1 of the NMDA-receptor (Chemicon, 1:100) were used. Following incubation with one of the primary antibodies for 48 h at 4°C, its localization was detected using the avidin-biotin technique (Vector Laboratories).

Results

We keep finding new markers for collicular modules recurrently. For this study, we began our search by staining sections of the rat superior colliculus for the calcium binding protein calretinin. Fig. 1 shows frontal sections of the rat superior colliculus, one of which was treated for the classical demonstration of AChE-activity, the other for calretinin immunoreactivity. The AChE staining pattern (Fig. 1A) in the intermediate collicular layers shows the well known mosaic of areas high and low in enzyme activity.

Calretinin (Fig. 1B) was localized throughout the superior colliculus in cell bodies that were mostly smaller than those stained with parvalbumin (Illing et al., 1990). Like parvalbumin, calretinin immunocytochemistry stained neuronal processes in considerable detail. The staining pattern of calretinin immunoreactivity was heterogeneous, but it was obviously different from that obtained by AChE histochemistry. The proof for their dissimilarity was furnished by a comparison of adjacent sections (Fig. 1). Calretinin was heterogeneously distributed in the optical layer where AChE-activity is evenly distributed at a low level. Areas heavily stained for calretinin immunoreactivity in the deep collicular layers also failed to show a counterpart in the AChE-staining pattern. Zones most heavily stained for AChE are located in the intermediate layers, but the corresponding

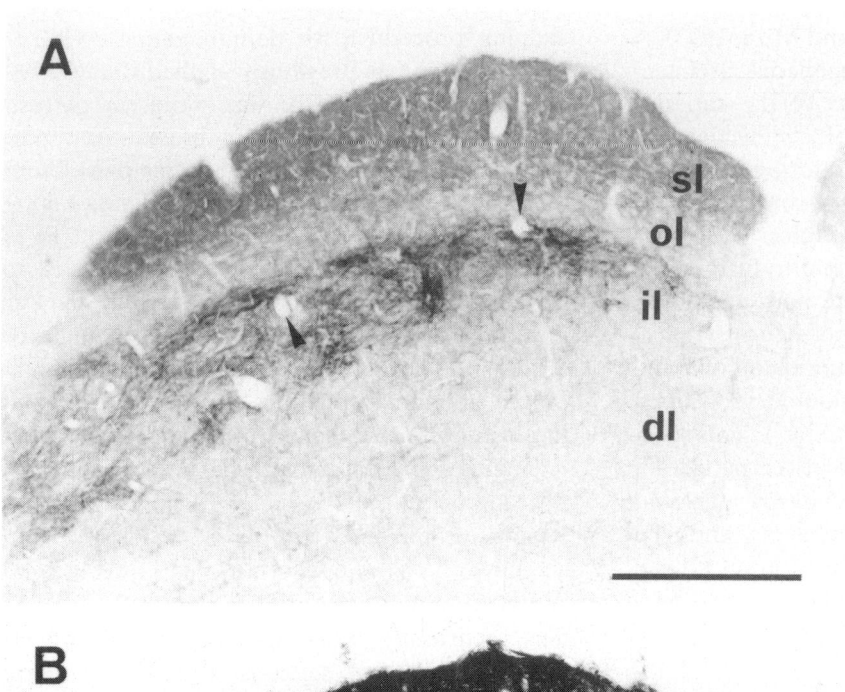

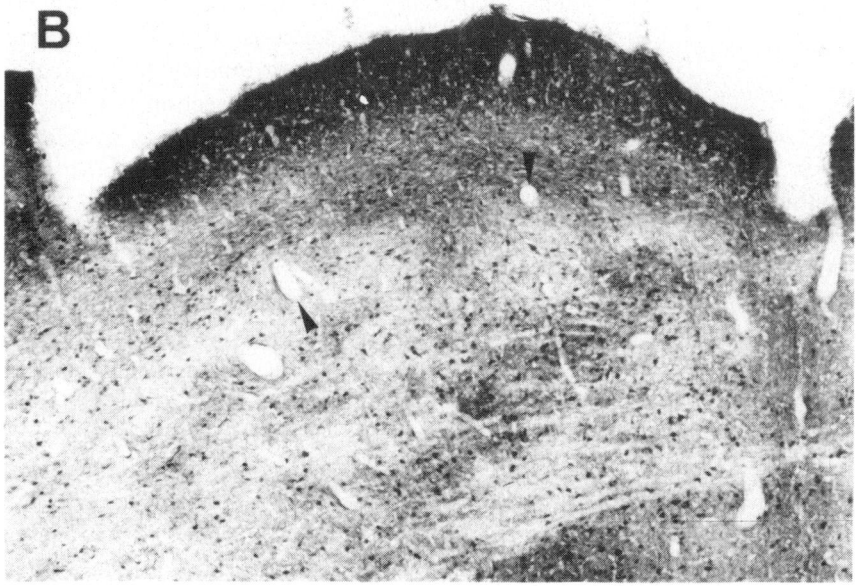

Fig. 1. Frontal sections through the rat superior colliculus. (A) Pattern of AChE-staining. sl, superficial layers; ol, optical layer; il, intermediate layers, dl, deep layers. (B) Adjacent section stained for calretinin immunoreactivity. The arrow head in A and B point to corresponding blood vessel profiles. Below the superficial layers, zones intensely stained for AChE-activity do mostly not match zones of high calretinin-immunoreactivity. Scale bar: 500 μm

zones in the adjacent section were generally low in calretinin immunoreactivity. This comparison led us to conclude that the patterns of AChE histochemistry and calretinin immunoreactivity define mostly different domains, but, judging from adjacent section analysis, some local overlap of the two markers cannot be excluded.

In a previous study (Illing et al., 1990) we reported that the distribution of another calcium binding protein, parvalbumin, is heterogeneously

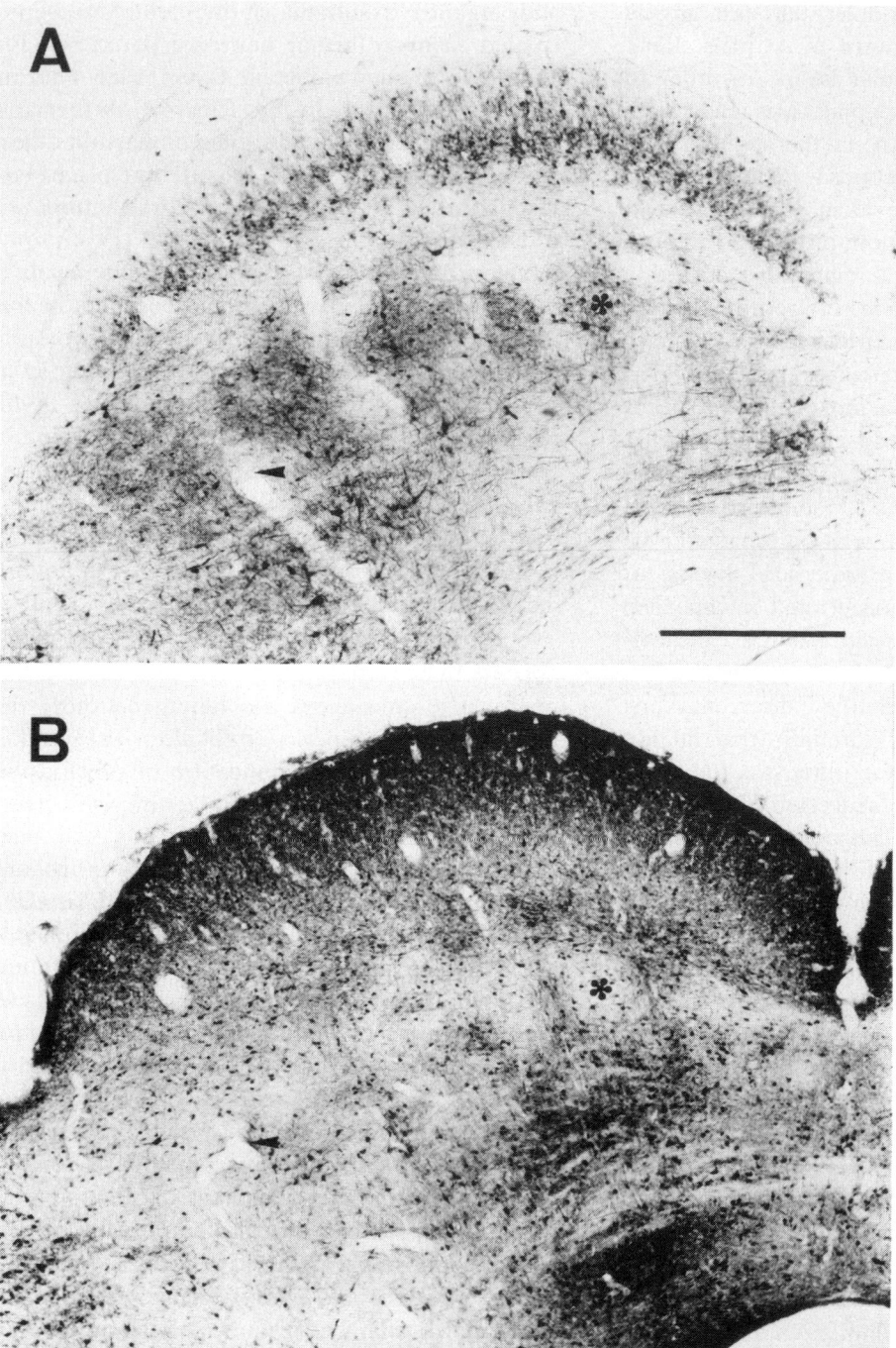

Fig. 2. Frontal sections through the rat superior colliculus: (A) Distribution of parvalbumin immunoreactivity. (B) Staining pattern of calretinin immunoreactivity in an adjacent section. Asterisks indicate corresponding region in these matching sections. Although the distribution of parvalbumin is unlike that of AChE (cf. Fig. 1A and Illing et al., 1990), it is again unlike that of calretinin. Therefore, these three markers define three different modular domains in the superior colliculus. Scale bar: 500 μm.

distributed and forms modules that are largely complementary to the pattern of AChE-staining in the intermediate collicular layers. In order to compare the patterns of parvalbumin and calretinin immunohistochemistry in the superior colliculus, adjacent frontal sections were stained (Fig. 2). It was immediately evident that these two calcium binding proteins were distributed in non-matching patterns. With a remarkable accuracy, parvalbumin-rich zones (Fig. 2A) occupy spaces that are particularly poor in calretinin (Fig. 2B), and vice versa. This distribution indicated that parvalbumin and calretinin mostly occupy spaces that are spared by the respective other calcium binding protein, and this appeared to be valid from stratum opticum down to the periaqueductal gray matter. This spatial relationship could be said to apply also to the superficial layers, as parvalbumin was most concentrated in the deep part of them, while calretinin was most concentrated more superficially.

Since we have now identified three markers: AChE, parvalbumin, and calretinin, that do not match the distribution of the other two, the alternative is not, as previously suggested, AChE compartments vs. non-AChE compartments, but there are at least three mostly disjunct types of compartments in and around the intermediate layers of the superior colliculus.

However, this finding should not lead to the impression that each type of collicular compartment occupies private space. A virtually perfect match between compartments outlined by AChE and choline acetyltransferase immunoreactivity has been reported earlier (Illing, 1990). With the next figures, two other incidences of matching compartments are presented. The first example concerns the growth and plasticity associated protein GAP-43 (Strittmatter et al., 1992). We discovered that this protein is an abundant component of collicular neuropil in adult rats (Fig. 3) and cats. Below the superficial layers GAP-43 was distributed in well-defined collicular compartments. Zones of high GAP-43 content matched in detail zones delineated by elevated calretinin immunohistochemistry (Fig. 3). Secondly, an anti-body against a subunit of the NMDA-receptor labeled many collicular neuronal perikarya (Fig. 4). In the deeper collicular layers, such neurons were seen to form clusters (Fig. 4B). When analyzed in an adjacent section comparison, these clusters were found to be localized at places corresponding to high levels of calretinin immunoreactivity (Fig. 4A).

These findings imply that an iterative matrix of neuronal space, outlined by calretinin, is shared by GAP-43 and the NMDA-receptor, two substances that are known to be closely related to neuronal plasticity (Strittmatter et al., 1992; Malenka and Nicoll, 1993; Hofer et al., 1994).

Discussion

Collicular modules

The major finding of the present report is that it appears to be a rule rather than an exception that neuroactive substances are heterogeneously distributed in the superior colliculus. This finding strongly supports the implication of previous reports that the modular architecture is a major feature of the superior colliculus.

With the newly recognized markers for collicular modules it is possible to relate the modular organization to a dimension of nervous function that was previously not seen in context with collicular architecture. On a supraneuronal level, there are zones in the superior colliculus that are enriched in substances known to be involved in the regulation and expression of plastic changes. With this discovery, we may relate the different classes of collicular modules not only to sensory, motor, or modulatory functions, as suggested before, but, in addition, to plasticity. Since for this study normal adult animals were used, we may go one step further and speculate that the presence of modules characterized by plasticity markers indicates a permanent potential for restructuring synaptic connections in the superior colliculus. To what aspect of collicular function these 'plasticity modules' contribute, and if they are possibly involved in an ongoing tuning between the topo-

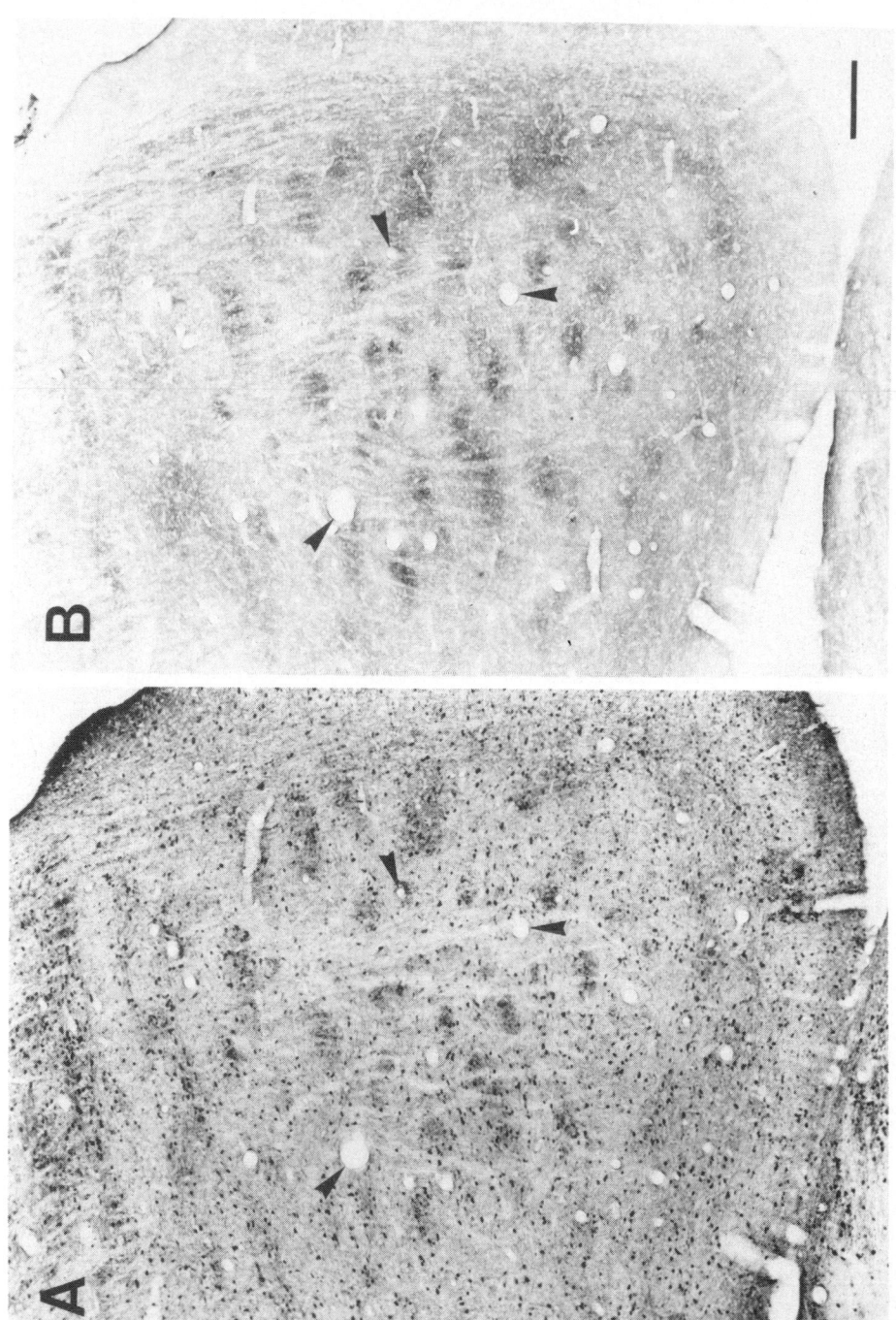

Fig. 3. Tangential sections through the rat superior colliculus, stained for calretinin immunoreactivity (A) and GAP-43 immunoreactivity (B). Rostral is up, medial to the right. Arrowheads in (A) and (B) point to corresponding blood vessel profiles. The positions of the darkly stained patches are almost perfectly matching, indicating that calretinin and GAP-43 are spatially associated in the deeper collicular layers. Scale bar: 200 μm.

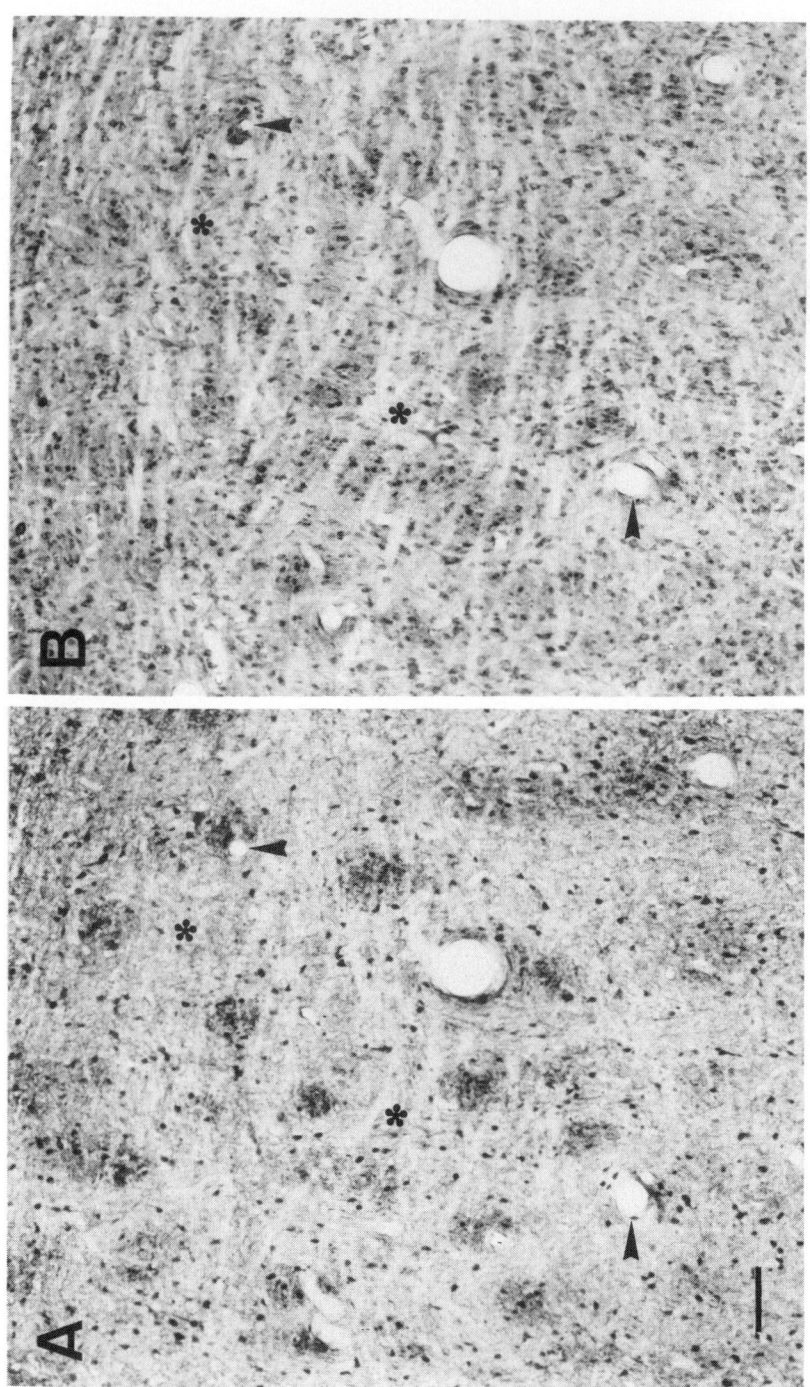

Fig. 4. Tangential sections through the intermediate collicular layers, stained for calretinin (A) and the NMDA-receptor (B). While the arrowheads point to corresponding blood vessel profiles, the asterisks were centered at matching positions into two triangles, the corners of which are marked by elevated staining for calretinin and the NMDA-receptor and match in position and in the relative degree of staining. Scale bar: 100 μm

graphic representations of sensory and motor spaces in the superior colliculus (Knudsen, 1991), remains to be answered.

The collicular mosaic

Not all neuroactive substances demarcate mod-

TABLE 1

Chemoanatomy of the multimodal collicular layers

Substance	Abbreviation	Source
Homogeneously distributed substances		
Pseudocholinesterase	BuChE	Graybiel and Illing, 1994.
Glutamate	Glu	own observations
Gamma-amino-butyric Acid	GABA	Mize, 1988.
Glutamate decarboxylase	GAD	Lu et al., 1985.
Glycine	Gly	own observations
Serotonin	5-HT	Morrison and Foote, 1986; Mize and Horner, 1989
Noradrenalin	NA	Morrison and Foote, 1986.
Calbindin	CB	Mize et al., 1991.
5'-Nucleotidase	-	S.W. Schoen, personal communication
Iteratively distributed substances		
Acetylcholinesterase	AChE	Siou, 1958; Ramon-Moliner, 1972; Graybiel, 1978; Illing and Graybiel, 1986.
Choline acetyltransferase	ChAT	Beninato and Spencer, 1986; Hall et al. 1989; Illing, 1990
Enkephalins	ENK	Graybiel et al., 1984; Miguel-Hidalgo et al., 1989; Graybiel and Illing, 1994.
Nitric oxide synthase (NADPH-diaphorase)	NOS	Wallace, 1986b.
Substance P	SP	Miguel-Hidalgo et al., 1989; Behan et al., 1993.
Cytochrome oxidase	CO	Wallace, 1986a; Illing et al., 1990.
Parvalbumin	PV	Illing et al., 1990.
Calretinin	CR	present study
Somatostatin	SS	Spangler and Morley, 1987.
Glycine receptor	GlyR	own observations
NMDA-receptor	NMDAR1	present study
Arginino succinate synthetase	ASS	Nakamura et al., 1991.
Growth associated protein 43	GAP-43	present study

ules in the superior colliculus. The distribution of several substances gives an indication of vertically piled layers, but does not seem to structure the layers into horizontally adjacent units (Table 1). Listing substances in this table is not meant to exclude the possibility that closer inspection of their distribution will eventually show that they are markers for modules after all. As we have seen, many other neuroactive substances are readily recognizable as module markers (Table 1). Their distribution is patchy and defines an iterative pattern of zones in which the respective substance is more or less concentrated. What has not been fully recognized in previous studies on superior collicular modularity is that modules are not restricted to the intermediate layers of the superior colliculus. It is a major extension of previous descriptions of collicular architecture to state that a modular architecture exists virtually throughout the depth of this structure (Table 2).

This led us to realize that the collicular architecture is characterized not by two, but by at least three organizing principles: topographic connectivity, laminar structuring, and iterative partitioning. These three principles together form the supraneuronal framework for an understanding of the signal flow through superior collicular tissue.

Possible origins of the collicular mosaic

Collicular modules emerge early in ontogeny (Illing and Graybiel, 1994). Their emergence in evolution must have met functional needs. Models of superior collicular functions (Churchland, 1986; Sparks, 1986; Tweed and Vilis, 1990; Knudsen, 1991), were often based on the fact that collicular layers contain matching topographic representations of sensory and motor space. According to these models, a key feature of collicular function is the integrative multimodal communication perpendicular to the layers. Such connections were found to exist (Moschovakis and Karabelas, 1985; Moschovakis et al., 1988; Hilbig and Schierwagen, 1994) and are more elaborate than originally expected (Edwards, 1980). Although the modules appear to reflect a specific arrangement of cell body and dendrite positions (Illing, 1992), we do not yet know their contribution for the shaping of intracollicular communication lines.

By using an evolutionary perspective, we may hope to gain some understanding of the circumstances that could have demanded the superior colliculus to split into compartments, and to introduce this aspect of biological systems into our reasoning about collicular function, I venture a suggestion. Early in mammalian evolution, topographically organized structures were present in cortical and subcortical areas of the brain (Ebbesson, 1984). We may picture such a simply structured area as shown on the left in Fig. 5A. If we accept that the acquisition of additional capacities to analyze sensory stimuli requires the dedication of neuronal mass, then there are in principle three ways in which such a dedication could be achieved (Fig. 5A). The first possibility is that the original area develops layers, one of original functions, whereas the other may perform a derived function. The second possibility is that the original area develops modules. One module may perform the original function and the adjacent module a derived function, both being concerned with the same field within the topographic representation. As compared to the first option, neighborhood relations are different here; the original parts no longer have common borders. The third possibility is multiplication of the original area as a whole, with the original one performing the

TABLE 2

Types of compartments in major collicular layers

Collicular layer	Iteratively distributed substances
Superficial layers (sl)	ChAT
Optical layer (ol)	PV, CR, GAP-43, NMDAR1, SP
Intermediate layers (il)	AChE, ChAT, ENK, NOS, PV
Deep layers (dl)	AChE, ChAT, PV, CR, GAP-43, NMDAR1, SP

A

Structural Differentiation of Topographic Fields

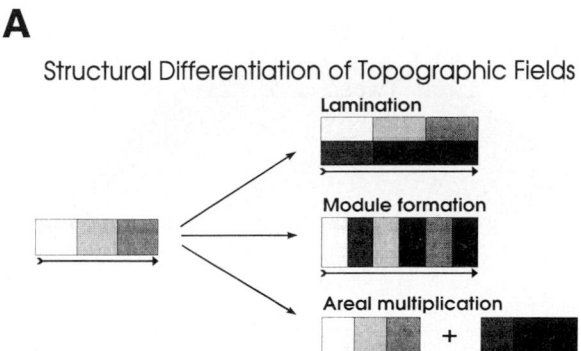

B

Lamination + Modularity → Mosaic-Architecture

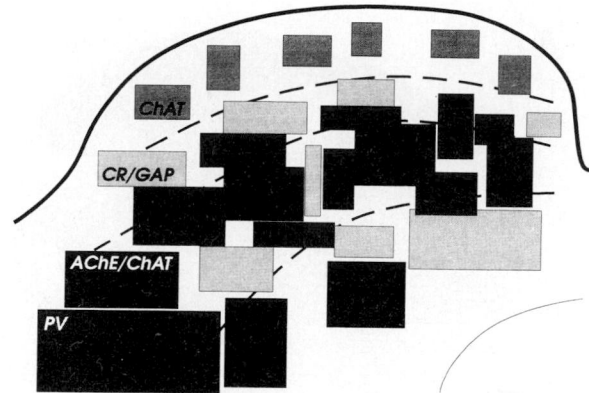

Fig. 6. Schematic view of a frontal section through the superior colliculus, in which some major types of compartments are put at their approximate depth level. This schema illustrates that superior collicular tissue is organized as a mosaic that pervades all layers.

Fig. 5. Models of the evolutionary origin of modules. (A) Part of a primitive vertebrate brain may not show more elaboration than a topographic representation of some sensory space. This topographical organization is indicated on the left by three progressively changing shades of gray and the arrow below them. A structural differentiation of such an area may be achieved by the formation of layers, by the formation of modules, or by multiplication of the original area. (B) If the first two strategies are combined, the formation of layers and the formation of modules together will produce a mosaic architecture.

original functions and the new ones derived functions.

While this third possibility has obviously been extremely successful in the evolution of the neocortex (Cowey, 1981; Barlow, 1986; Kaas, 1989), it did not work in the midbrain. I believe that the reason for this has to do with the topographic matching of sensory and motor maps that lie on top of each other, forming the key aspect of models of collicular function. But no matter if this is a correct conjecture or not, the superior colliculus obviously did without the third option. The two others were, however, available for him

to develop in evolution. If we allow both processes, the formation of layers and the formation of modules, to occur simultaneously, then a mosaic architecture emerges. The merging of lamination and modularity produces, in the example of Fig. 5B, one original and three derivatives for each part of the topographic field. Depending on the number of layers and the spacing of the modules, however, a much greater number of distinct types of modules may emerge. A schematic presentation of some representative types of collicular modules, as provided in Fig. 6, may be taken as suggesting that something similar has happened during evolution of the fish optic tectum into the mammalian superior colliculus. The joined processes of lamination and module formation may have taken place to create an optimized number of locally specialized tissue compartments in a structure that could not give up its principal topographic organization.

Species variations and their possible functional background

Finally, I would like to view collicular modularity from still another perspective and ask the reader to focus again on just one type of collicular modules, the classical AChE-rich patches. It was real-

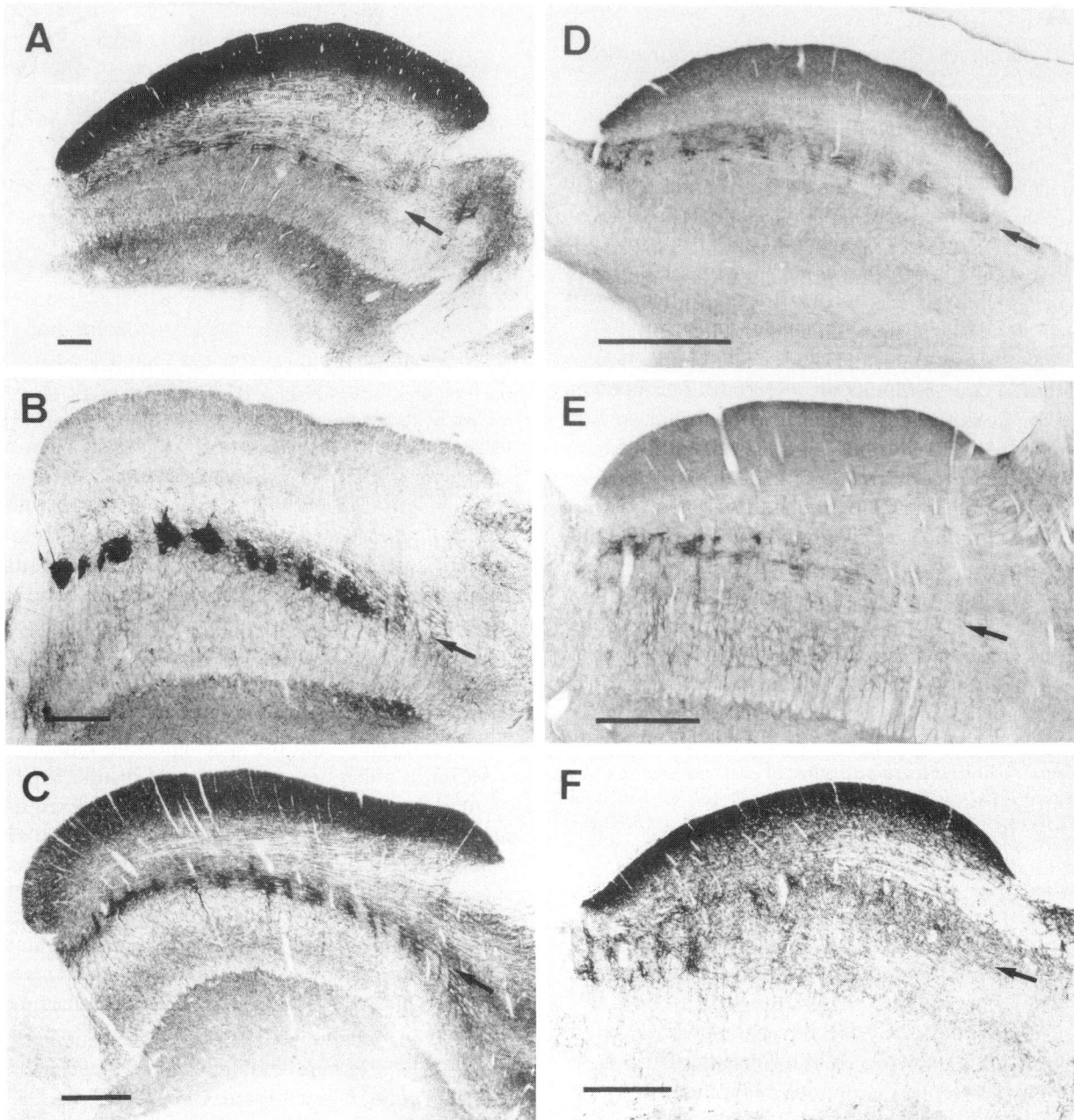

Fig. 7. These colliculi of six species were cut longitudinally through the middle of mediolateral extent of the intermediate layers in a plain perpendicular to the surface, and stained for AChE-activity. Caudal is to the left, rostral to the right. Arrows point to layers of heterogeneous AChE-staining. (A) cow; (B) domestic pig; (C) rabbit; (D) rat; (E) cat; (F) macaque monkey. In the species on the left (A–C) the compartments in the intermediate layers are prominent throughout the rostrocaudal length of the superior colliculus. By contrast, in the species on the right, compartments are prominent caudally but grow weaker more rostrally. While ill-defined AChE-rich compartments are still present at the rostral pole of the rat colliculus (D), they appear to be almost absent at corresponding positions in the cat (E) and monkey (F) colliculus. Scale bar: 1 mm.

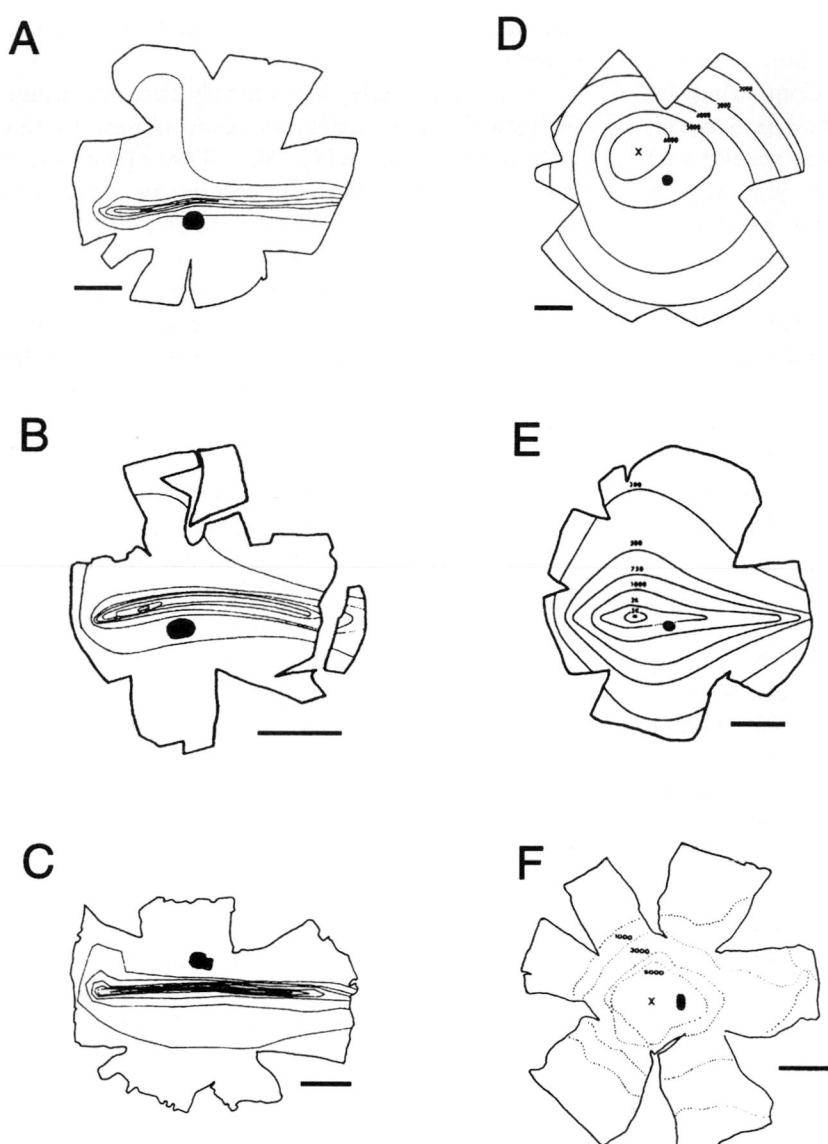

Fig. 8. The topography of ganglion cells in the retinae of six mammalian species illustrated by isodensity contours. For all retinae, superior is up, nasal to the right. Species with sidewards looking eyes and a distribution of retinal ganglion cells that have their highest density along a horizontal streak are illustrated on the left. Species that have a significant binocular field and a retina defining a center of vision are shown on the right side. (A) cow (Hughes, 1977; scale bar: 10 mm); (B) pig (Hughes, 1977; scale bar: 10 mm); (C) rabbit (Hughes, 1971; scale bar: 5 mm); (D) rat (Fukuda, 1977; scale bar: 2 mm); (E) cat (Hughes, 1975; scale bar: 10 mm); (F) macaque monkey (Stone and Johnston, 1981; scale bar: 5 mm).

ized that the pattern of this type of module over the whole superior colliculus differs from species to species (Illing, 1988, 1993). In some species, AChE-rich patches in the intermediate layers are prominent and well delineated through the whole extent of the superior colliculus from caudal to rostral collicular pole. In others, this is not the case. Instead, modules are well defined in the caudal colliculus, but towards the rostral pole the patches indicating tissue specialization grew less intense and less well delineated. Comparing different species, I noticed that the steepness of this structural gradient differs. There are species with patches in the rostral colliculus just slightly blurred as compared to those in the caudal colliculus, but in other species, AChE-rich patches appear to be entirely absent from the rostral colliculus (Fig. 7). Since AChE labels just one type of collicular module, we cannot conclude from the absence of AChE-rich patches from the rostral colliculus in some species that collicular compartmentation is lost there altogether.

Topographically, the rostral colliculus represents forward-directed or central portions of the visual field (Feldon et al., 1970; Hughes, 1971; Lane et al., 1973). In animals with sidewards oriented eyes, the rostral colliculus represents the nasal visual field, and in animals with frontally oriented eyes, this part of the superior colliculus is dedicated to central vision. In the cat (Feldon et al., 1970) and its relatives, but not in primates (Lane et al., 1973), the rostral margin of the superior colliculus represents the ipsilateral visual field. However, the most forward directed part of the visual world is represented in rostral regions of the colliculus in all cases. The retinae of forward-looking species have central specializations with a local maximum in receptor and ganglion cell density and a corresponding maximum in visual acuity. When the structural gradient of AChE-patches in the intermediate collicular layers (Fig. 7) is compared with the retinal ganglion cell topography (Fig. 8), it turned out that there is indeed a remarkable correlation. The nature of this correlation may be stated in two parts. The first is a qualitative argument. It states that species with a streak type retina tend to show a prominently compartmentalized architecture from caudal to rostral colliculus, whereas species with a centralized retina show a structural gradient, with well formed compartments caudally but ill-defined compartments rostrally. The second argument is quantitative and claims that the degree of retinal centralization appears to be reflected in

the steepness of the gradient of collicular AChE-compartments.

Hardly any species has a purely streaked retina. The species that seems to come closest to this type is the rabbit (Fig. 8C). The rabbit has a colliculus that is distinctly compartmentalized in the intermediate layers throughout its rostrocaudal extent (Fig. 7C). Rabbits perform few voluntary saccadic eye movements (Hughes, 1971). The eye position and retinal topography of cows (Fig. 8A) and pigs (Fig. 8B) are not unlike those of the rabbit, but these species have a slightly increased ganglion cell density temporally that provides them with the ability to see objects in front of them with relatively higher resolution. The colliculi of these species (Figs. 7A and B, respectively) are similar to that of the rabbit, but the compartments in the rostral colliculus are not quite as sharply delineated as caudally.

The rat has a clearly concentrically organized retina (Fig. 8D), and its oculomotor system generates saccades at least reminiscent to those observed in cats and monkeys (McHaffie et al., 1982). The ganglion cell density is highest in the center, but the difference to the periphery is moderate. Still, it appears to make sense for the rat to move it's eyes for looking. The centralization is much more pronounced in the cat eye (Fig. 8E) and is extreme in primates (Fig. 8F). Correspondingly, compartment of the rostral colliculi of these species is blurred or absent as judged by the pattern of AChE-staining (Figs. 7E and F, resp.).

Another structural gradient

There is evidence that the gradient seen in the superior colliculus with respect to its compartmental organization is not its only structural gradient. In several mammalian species, a collicular commissure exists that connects the colliculi in both brain hemispheres. This projection originates from collicular neurons that mostly reside in the intermediate layers. Topographic details of the collicular commissure have not been studied in many species. Tectotectal connections have been most thoroughly studied in cat (Magalhaes-

Castro et al., 1978; Behan and Kime, 1996). In this species, neurons that connect both colliculi are abundant in the rostral colliculus, but there are fewer such cells more caudally until, about halfway through the rostrocaudal extent of the colliculi, tectotectal cells were no longer encountered. This pattern appears to mirror the gradient of AChE-rich patches in the intermediate collicular layers. Where AChE compartments are prominent, no collicular commissure exists, but as this compartmentation grew more and more ill-defined towards the rostral pole, the volume of the commissural pathway increases. If this is not incidental, one would predict for species with a purely streak-type retina and, correspondingly, an array of modules of the AChE-type that reaches from caudal to rostral colliculus, that no commissural pathway exists. In fact, attempts to demonstrate a collicular commissure in the rabbit have failed (Nagata et al., 1980). In sharp contrast, an extensive collicular commissure exists in the monkey, sparing only the crisply compartmentalized caudal-most regions (Olivier et al., 1994).

Functional gradients

The hypothesis of a causal connection of retinal topography with the gradient in collicular compartmentation appears to be consistent with observations about the physiology of the superior colliculus. Neurons in the rostral colliculus have been found to be physiologically related to the state of fixation. When a cat (Munoz et al., 1991) or a monkey (Munoz and Wurtz, 1993) fixates an object of interest, these 'fixation cells' discharge tonically. Neurons of this kind were only encountered in the deeper layers. Although details are not yet known to us, we must conclude that the neural apparatus supporting these cells includes a specific connectivity. Indicative of differences in collicular connectivity between caudal and rostral colliculus, is the fact that there is a gradient in their projections to other brainstem nuclei. Huerta et al. (1991) have shown that the primate *Substantia nigra* sends differential projections to caudal and rostral colliculus. Chimoto et al. (1996) report that omnipause neurons in the brainstem

of cats received more excitatory and less inhibitory input from the rostral as compared to the caudal superior colliculus in cat. By contrast, burst neurons were more effectively excited from the caudal superior colliculus.

As explained above, I consider oculomotor functions of the superior colliculus only as a special case in the functional scope of this midbrain structure, but its oculomotor physiology illustrates that specialized functions are exerted by neurons in the rostral colliculus of species with a centralized retina, and it is only a minor step to conclude that this must have consequences for neuronal connectivity and hence tissue architecture surrounding these neurons. The details of the mosaic architecture as considered in this report, including its proven or possible affiliations to connectivity, behavior and evolution, provide evidence that collicular modules, by structuring collicular tissue into relatively separated spaces that could hence assume relative functional autonomy, contribute to extend the computational gamut of the superior colliculus as a whole.

Acknowledgement

This study profitted much from assistance by B. Hobmaier and W. Kuhn, from discussions with Dr. D.M. Vogt Weisenhorn and Ms. C.R. Förster, and from support by Profs. R. Laszig and W.B. Spatz.

References

Barlow, H.B. (1986) Why have multiple cortical areas? *Vision Res.*, 26: 81–90.

Behan, M. and Kime, N.M. (1996) Spatial distribution of tectotectal connections in the cat. In: M. Norita, T. Bando and B. Stein (Eds.), *Extrageniculostriate Mechanisms Underlying Visually-Guided Orientation Behavior. Prog. Brain Res.*, this volume.

Behan, M., Appell, P.P. and Kime, N. (1993) Postnatal development of substance P immunoreactivity in the rat superior colliculus. *Vis. Neurosci.*, 10: 1121–1127.

Beninato, M. and Spencer, R.F. (1986) Cholinergic projections to the rat superior colliculus demonstrated by retrograde transport of horseradish peroxidase and choline acetyltransferase immunohistochemistry. *J. Comp. Neurol.*, 253: 525–538.

Carr, C.E., Maler, L., Heiligenberg, W. and Sas E. (1981) Laminar organization of the afferent and efferent systems of the torus semicircularis of gymnotiform fish: morphological substrates for parallel processing in the electrosensory system. *J. Comp. Neurol.*, 203: 649–670.

Chalupa, L.M. and Rhoades, R.W. (1977) Responses of visual, somatosensory, and auditory neurones in the golden hamster's superior colliculus. *J. Physiol.*, 270: 595–626.

Chimoto, S., Iwamoto, Y., Shimazu, H. and Toda, H. (1996) Roles of the lateral suprasylvian cortex in convergence eye movements in the cats. In: M. Norita, T. Bando and B. Stein (Eds.), *Extrageniculostriate Mechanisms Underlying Visually-Guided Orientation Behavior. Prog. Brain Res.*, this volume.

Churchland, P.M. (1986) Cognitive neurobiology: A computational hypothesis for laminar cortex. *Biol. Philos.*, 1: 25–51.

Cooper, H.M., Herbin, M. and Nevo, E. (1993) Visual system of a naturally microphthalmic mammal: the blind mole rat, Spalax ehrenbergi. *J. Comp. Neurol.*, 328: 313–50.

Cowey, A. (1981) Why are there so many visual areas? In: F.O. Schmidt, F.G. Warden, G. Adelman and S.G. Dennis (Eds.), *The Organization of the Cerebral Cortex*, Cambridge: MIT Press, pp. 395–413.

Dean, P., Redgrave, P. and Westby, G.W.M. (1989) Event or emergency? Two response systems in the mammalian superior colliculus. Trends *Neurosci.*, 12: 137–147.

Debski, E.A. and Constantine-Paton, M. (1993) The development of non-retinal afferent projections to the frog optic tectum and the substance P immunoreactivity of tectal connections. *Dev. Brain Res.*, 72: 21–39.

Ebbesson, S.O.E. (1984) Evolution and ontogeny of neural circuits. *Behav. Brain Sci.*, 7: 321–366.

Edwards, S.B. (1980) The deep layers of the superior colliculus: their reticular characteristics and structural organization. In: J.A. Hobson and M.A.B. Brazier (Eds.), *The Reticular Formation Revisited*, Raven Press, New York.

Feldon, S., Feldon, P. and Kruger, L. (1970) Topography of the retinal projection upon the superior colliculus of the cat. *Vision Res.*, 10: 135–143.

Friede, R.L. (1966) *Topographic Brain Chemistry*, Academic Press, New York.

Fukuda, Y. (1977) A three group classification of rat retinal ganglion cells: histological and physiological studies. *Brain Res.*, 19: 327–344.

Goodale, M.A. and Murison R.C. (1975) The effects of lesions of the superior colliculus on locomotor orientation and the orienting reflex in the rat. *Brain Res.*, 88: 243–261.

Graybiel, A.M. (1978) A stereometric pattern of distribution of acetylcholinesterase in the deep layers of the superior colliculus. *Nature*, 272: 539–541.

Graybiel, A.M. and Illing, R.B. (1994) Enkephalin-positive and acetylcholinesterase-positive patch systems in the superior colliculus have matching distributions but distinct developmental histories. *J. Comp. Neurol.*, 340: 297–310.

Graybiel, A.M., Brecha, N. and Karten, H.J. (1984) Cluster-and-sheet pattern of enkephalin-like immunoreactivity in the superior colliculus of the cat. *Neuroscience*, 12: 191–214.

Hall, W.C., Fitzpatrick, D., Klatt, L.L. and Raczkowski, D. (1989) Cholinergic innervation of the superior colliuclus in the cat. *J. Comp. Neurol.*, 287: 495–514.

Hilbig, H. and Schierwagen, A. (1994) Interlayer neurones in the rat superior colliculus: a tracer study using DiI/Di-ASP. *Neuroreport*, 12: 477–480.

Hofer, M., Prusky, G.T. and Constantine-Paton, M. (1994) Regulation of NMDA receptor mRNA during visual map formation and after receptor blockade. *J. Neurochem.*, 62: 2300–2307.

Huerta, M.F. and Harting, J.K. (1984) Connectional organization of the superior colliculus. *Trends Neurosci.*, 7: 286–289.

Huerta, M.F., van Lieshout, D.P. and Harting, J.K. (1991) Nigrotectal projections in the primate *Galago crassicaudatus*. *Exp. Brain Res.*, 87: 389–401.

Hughes, A. (1971) Topographical relationships between the anatomy and physiology of the rabbit visual system. *Doc. Ophthalmol.*, 30: 33–159.

Hughes, A. (1975) A quantitative analysis of the cat retinal ganglion cell topography. *J. Comp. Neurol.*, 163: 107–128.

Hughes, A. (1977) The topography of vision in mammals of contrasting life style: comparative optics and retinal organization. In: F. Crescitelli (Ed.), *Handbook of Sensory Physiology*, Vol. 7.5, Springer, Berlin.

Humphrey, N.K. (1968) Responses to visual stimuli of units in the superior colliculus of rats and monkeys. *Exp. Neurol.*, 20: 312–340.

Illing, R.B. (1988) Spatial relation of the acetylcholinesterase-rich domain to the visual topography in the feline superior colliculus. *Exp. Brain Res.*, 73: 589–594.

Illing, R.B. (1990) Choline acetyltransferase-like immunoreactivity in the superior colliculus of the cat and its relation to the pattern of acetylcholinesterase staining. *J. Comp. Neurol.*, 296: 32–46.

Illing, R.B. (1992) Association of efferent neurons to the compartmental architecture of the superior colliculus. *Proc. Natl. Acad. Sci.* USA, 89: 10900–10904.

Illing, R.B. (1993) More modules. Trends *Neurosci.*, 16: 179–180.

Illing, R.B. and Graybiel, A.M. (1985) Convergence of afferents from frontal cortex and substantia nigra onto acetylcholinesterase-rich patches in the cat's superior colliculus. *Neuroscience*, 14: 455–482.

Illing, R.B. and Graybiel, A.M. (1986) Complementary and non-matching afferent compartments in the cat's superior colliculus: innervation of the acetylcholinesterase-poor domain of the intermediate layers. *Neuroscience*, 18: 373–394.

Illing, R.B. and Graybiel, A.M. (1994) Pattern formation in the developing superior colliculus: ontogeny of the periodic architecture in the intermediate layers. *J. Comp. Neurol.*, 340: 311–327.

Illing, R.B., Vogt, D.M. and Spatz, W.B. (1990) Parvalbumin in rat superior colliculus. *Neurosci. Lett.*, 120: 197–200.

Jeon, C.J. and Mize, R.R. (1993) Choline acetyltransferase-immunoreactive patches overlap specific efferent cell groups in the cat superior colliculus. *J. Comp. Neurol.*, 337: 127–150.

Kaas, J.H. (1989) Why does the brain have so many visual areas? *J. Cogn. Neurosci.*, 1: 121–135.

Kanaseki, T. and Sprague, J.M. (1974) Anatomical organization of pretectal nuclei and tectal laminae in the cat. *J. Comp. Neurol.*, 158: 319–338.

Knudsen, E.I. (1991) Dynamic space codes in the superior colliculus. *Curr. Opin. Neurobiol.*, 1: 628–632.

Lane, R.H., Allman, J.M., Kaas, J.H. and Miezin, F.M. (1973) The visuotopic organization of the superior colliculus in the owl monkey (*Aotus trivirgatus*) and the bush baby (*Galago senegalensis*). *Brain Res.*, 60: 335–349.

Lu, S.M., Lin, C.S., Behan, M., Cant, N.B. and Hall, W.C. (1985) Glutamate decarboxylase immunoreactivity in the intermediate gray layer of the superior colliculus in the cat. *Neuroscience* 16: 123–131.

Magalhaes-Castro, H.H., de Lima, A.D., Saraiva, P.E.S. and Magalhaes-Castro, B. (1978) Horseradish peroxidase labeling of cat tectotectal cells. *Brain Res.*, 148: 1–13.

Malenka, R.C. and Nicoll, R.A. (1993) NMDA-receptor-dependent synaptic plasticity: multiple forms and mechanisms. *Trends Neurosci.*, 16: 521–527.

McHaffie, J.G. and Stein, B.E. (1982) Eye movements evoked by electrical stimulation in the superior colliculus of rats and hamsters. *Brain. Res.*, 247: 243–253.

Miguel-Hidalgo, J.J., Senba, E., Matsutani, S., Takatsuji, K., Fukuji, H. and Tohyama, M. (1989) Laminar and segregated distribution of immunoreactivities for some neuropeptides and adenosine deaminase in the superior colliculus of the rat. *J. Comp. Neurol.*, 280: 410–423.

Mize, R.R. (1988) Immunocytochemical localization of gamma-aminobutyric acid (GABA) in the cat superior colliculus. *J. Comp. Neurol.*, 276: 169–187.

Mize, R.R. (1989) Enkephalin-like immunoreactivity in the cat superior colliculus: distribution, ultrastructure and colocalization with GABA. *J. Comp. Neurol.*, 285: 133–155.

Mize, R.R. and Horner, L.H. (1989) Origin, distribution, and morphology of serotonergic afferents to the cat superior colliculus: a light and electron microscope immunocytochemistry study. *Exp. Brain Res.*, 75: 83–98.

Mize, R.R., Jeon, C.J., Butler, G.D., Luo, Q. and Emson, P.C. (1991) The calcium binding protein calbindin-D 28K reveals subpopulations of projection and interneurons in the cat superior colliculus. *J. Comp. Neurol.*, 307: 417–436.

Morrison, J.H. and Foote, S.L. (1986) Noradrenergic and serotonergic innervation of cortical, thalamic and tectal visual structures in old and new world monkeys. *J. Comp. Neurol.*, 243: 117–138.

Moschovakis, A.K. and Karabelas, A.B. (1985) Observations on the somatodendritic morphology and axonal trajectory intracellularly HRP-labeled efferent neurons located in the deep layers of the superior colliculus of the cat. *J. Comp. Neurol.*, 239: 276–308.

Moschovakis, A.K., Karabelas, A.B. and Highstein, S.M. (1988) Structure-function relationships in the primate superior colliculus. I. Morphological classification of efferent neurons. *J. Neurophysiol.*, 60: 232–262.

Munoz, D.P. and Wurtz, R.H. (1993) Fixation in monkey superior colliuclus. I. Characteristics of cell discharge. *J. Neurophysiol.*, 70: 559–575.

Munoz, D.P., Guitton, D. and Pélisson, D. (1991) Control of orienting gaze shifts by the tectoreticulospinal system in the head-free cat. III. Spatio-temporal characteristics of phasic motor discharges. *J. Neurophysiol.*, 66: 1642–1666.

Nagata, T., Magalhaes-Castro, H.H., Saraiva, P.E.S. and Magalhaes-Castro, B. (1980) Absence of tectotectal pathway in the rabbit: an anatomical and electrophysiological study. *Neurosci. Lett.*, 17: 125–130.

Nakamura, H., Saheki, T., Ichiki, H., Nakata, K. and Nakagawa, S. (1991) Immunocytochemical localization of argininosuccinate synthetase in the rat brain. *J. Comp. Neurol.*, 312: 652–679.

O'Benar, J.D. (1976) Electrophysiology of neural units in goldfish optic tectum. *Brain Res. Bull.*, 1:529–541.

Olivier, E., Porter, J.D. and May, P.J. (1994) Comparison of the distribution of tectotectal neurons in the cat and macaque. *Soc. Neurosci. Abstr.*, 20: 142.

Ramon-Moliner, E. (1972) Acetylcholinesterase distribution in the brain stem of the cat. *Ergebn. Anat. EntwGesch.*, 46: 1–53.

Robinson, D.A. (1972) Eye movements evoked by collicular stimulation in the alert monkey. *Vision Res.*, 12: 1795–1808.

Shimozawa, T., Sun, X.D. and Jen, P.H.S. (1984) Auditory space representation in the superior colliculus of the big brown bat, Eptesicus fuscus. *Brain Res.*, 311: 289–296.

Siou, G. (1958) Distribution normale et variation experimentale de l'activité cholinesterasique au niveau des tubercules quadrijumeaux anterieurs chez la souris. C.R. *Acad. Sci.* (*Paris*), 246: 315–317.

Spangler, K.M. and Morley, B.J. (1987) Somatostatin-like immunoreactivity in the midbrain of the cat. *J. Comp. Neurol.*, 260: 87–97.

Sparks, D.L. (1986) Translation of sensory signals into commands for control of saccadic eye movements: role of primate superior colliculus. *Physiol. Rev.*, 66: 118–171.

Sprague, J.M. (1991) The role of the superior colliculus in fascilitating visual attention and form perception. *Proc. Natl. Acad. Sci. USA*, 88: 1286–1290.

Sprague, J.M. and Meikle, T.H. (1965) The role of the superior colliculus in visually guided behavior. *Exp. Neurol.*, 11: 115–146.

Stein, B.E. and Meredith, M.A. (1991) Functional organization of the superior colliculus. In: A.G. Leventhal (Ed.), *Vision and Visual Dysfunction*, Vol. 4, Mc Millan Press.

Stein, B.E., Magalhaes-Castro, B. and Kruger, L. (1975) Superior colliculus: visuotopic-somatotopic overlap. *Science*, 189: 224–226.

Stone, J. and Johnston, E. (1981) The topography of primate retina: a study of the human, bushbaby, and new- and old-world monkeys. *J. Comp. Neurol.*, 196: 205–223.

Strittmatter, S.M., Vartanian, T. and Fishman, M.C. (1992) GAP-43 as a plasticity protein in neuronal form and repair. *J. Neurobiol.*, 23: 507–520.

Tweed, D.B. and Vilis, T. (1990) The superior colliculus and spatiotemporal translation in the saccadic system. *Neural Netw.*, 3: 75–86.

Viktorov, I.V. (1966) Neuronal structure of anterior corpora bigemina in insectivora and rodents. *Arkh. Anat. Histol. Embriol.*, 51: 82–89.

Wallace, M.N. (1986a) Lattice of high oxidative metabolism in the intermediate gray layer of the rat and hamster superior colliculus. *Neurosci. Lett.*, 70: 320–325.

Wallace, M.N. (1986b) Spatial relationship of NADPH-diaphorase and acetylcholinesterase lattices in rat and mouse superior colliculus. *Neuroscience*, 19: 381–391.

Wong, D. (1984) Spatial tuning of auditory neurons in the superior colliculus of the echolocating bat, *Myotis lucifugus*. *Hear. Res.*, 16: 261–270.

M. Norita, T. Bando and B. Stein (Eds.)
Progress in Brain Research, Vol 112
© 1996 Elsevier Science BV. All rights reserved.

CHAPTER 3

Neurochemical microcircuitry underlying visual and oculomotor function in the cat superior colliculus

R. Ranney Mize*

Departments of Anatomy and Ophthalmology and the Neuroscience Center, Louisiana State University Medical Center, 1901 Perdido Street, New Orleans, LA 70112, USA

The cat superior colliculus (SC) plays an important role in visual and oculomotor functions, including the initiation of saccadic eye movements. We have studied the organization of neurochemical specific circuits in SC that underly these functions. In this chapter we have reviewed three microcircuits that can be identified by cell type, chemical content, and synaptic input from specific afferents. The first is located within the upper sgl and is related to the W retinal pathway to this region of SC. This circuit includes relay and interneurons that contain the calcium binding protein calbindin (CB), GABA containing presynaptic dendrites, and retinal terminals that have a distribution and size typical of W retinal terminals in the cat SC. This circuit is a typical synaptic triad that mediates feedforward inhibition, possibly to regulate outflow of the W pathway to the lateral geniculate nucleus. CB neurons in SC and other structures may be uniquely related to low threshold calcium currents in these neurons.

The second microcircuit consists of neurons that contain parvalbumin (PV), another calcium binding protein. These neurons are located in a dense tier within the deep sgl and upper ol and they receive input from retinal terminals that are likely from 'Y' retinal ganglion cells. Some of these neurons also project to the lateral posterior nucleus and some colocalize glutamate. We speculate that these neurons also receive cortical 'Y' input although we have yet to prove this experimentally. The role of PV in these cells is unknown, but PV has been shown to be contained in fast spiking, non-accomodating neurons in visual cortex which have very rapid spike discharges that are also characteristic of SC neurons innervated by 'Y' input.

The third microcircuit consists of a group of clustered neurons within the igl of the cat SC that overlaps the patch-like innervation of afferents to this region that come from the pedunculopontine tegmental and lateral dorsal tegmental nuclie, the substantia nigra, and the cortical frontal eye fields. These clustered neurons project through the tectopontobulbar pathway and terminate within the cuneiform region (CFR) of the midbrain tegmentum. They transiently express NOS during development. Ongoing studies in our laboratory suggest that these cells receive synaptic inputs directly from the PPTN and SN and may represent functional modules involved in the initiation of saccadic eye movements.

Introduction

The mammalian superior colliculus (SC) is characterized by very precise patterns of synaptic connections between specific afferents and post-synaptic neurons. This microcircuitry can be defined not only by ultrastructure but also by identification of the specific transmitters, receptors, channels, and other molecules involved in cell function (see Mize, 1992, 1994). We have been studying transmitter specific neural circuits in both the superficial and deep subdivisions of SC in an effort to understand the synaptic connectivity that underlies physiological and behavioral processes in the SC. Recently, we have focussed upon the organization of three chemically defined neural

*Correspondence address. Tel.: +1 504 568-4011; fax: +1 504 568-4392; email: rmize@lsumc.edu

systems that have only recently been recognized. Each system is defined by specific neurotransmitters or neurochemicals, has a selective laminar distribution, overlaps with specific extrinsic afferents, and forms a unique synaptic circuit.

Discovery of these neurochemical specific neural circuits provides a new view of collicular organization. First, they reveal chemically specific sublaminae that bridge the traditional fiber and cell layers of SC. Secondly, they unmask subsets of interneurons and projection neurons that have not previously been identified. Third, they show that calcium binding proteins and nitric oxide, as well as neurotransmitters, are selective markers of specific SC cells. Finally, they provide a basis for studying the receptor and membrane physiology of these cells so that it will be possible to correlate synaptic circuitry with receptor distribution, chemical content, and cell function. Studies describing these circuits are reviewed in this paper.

Materials and methods

The methods used in these studies include antibody immunocytochemistry, electron microscopy, and retrograde and anterograde tracing. Our general techniques for each method are described below.

Tracer procedures

Neurons in SC which project to particular targets were identified using horseradish peroxidase (HRP) as a retrograde tracer. HRP was injected under stereotaxic control into the appropriate target under deep anesthesia (25–40 mg/kg ketamine hydrochloride supplemented with 0.5–1.0 mg/kg xylazine) using aseptic surgical techniques. Following a small craniotomy, a 30-gauge injection needle attached to a 5-μl Hamilton microsyringe was lowered into the brain to inject 0.005–0.1 μl of 10–30% w/v HRP. Following the injection, the wound was sutured, and the animal allowed to recover for 24–72 h prior to perfusion.

Perfusion

Adult cats were deeply anesthetized as above, artificially respirated with 95% O_2 – 5% CO_2, and perfused through the heart with a paraformaldehyde-glutaraldehyde fixative in 0.1 M phosphate buffer (pH 7.4). Glutaraldehyde concentration was varied depending upon the antiserum. Following fixation, the brain was removed, blocked stereotaxically, and the SC cut into 50–100-μm sections on a Vibratome.

Histochemistry procedures

Tracer histochemistry

Cobalt intensification was sometimes used to amplify the HRP label. Sections were incubated in 0.05% 3,3′-diaminobenzidine (DAB) with 0.005% nickel chloride and 0.005% cobalt acetate in 0.1 M phosphate buffer (10 min), followed by the same solution with 0.01% H_2O_2 added (15–30 min). Sections were then stabilized with a 5% ammonium molybdate solution in phosphate buffer for 20 min and in cobalt-intensified DAB with H_2O_2 for 5 min.

NADPH / NOS histochemistry

Animals were perfused using our standard fixative (4% paraformaldehde, 0.1% glutaraldehyde) and the brain removed and cut into 30–50-μm sections on a Vibratome. Sections were then rinsed in Tris buffer (pH 7.1) for 2 × 5 min, incubated in nitro blue tetrazolium and nicotinamide adenine dinucleotide phosphate diaphorase (NADPHd) in Tris buffer for 0.5–2 h, then rinsed again in buffer, mounted on slides, dehydrated, and coverslipped. Our standard immunocytochemistry procedure was used for the bNOS antibodies.

Antibody immunocytochemistry

General processing

The avidin-biotin-peroxidase (ABC) immunohistochemical technique was used for the antibody immunocytochemistry experiments. Sections were: (1) rinsed in several changes of PBS or Tris buffer; (2) incubated in 1% sodium borohydride (NaBH$_4$) for 30 min to reduce cross-linking with glutaraldehyde; (3) incubated in blocking serum for 20–30 min. Blocking serum was also added to the primary and secondary antisera; (4) incubated in buffered primary antiserum. Exposure times and concentrations varied depending upon the antiserum; (5) rinsed in buffer and incubated in secondary antiserum (Vectastain biotinylated immunoglobulin) for 30 min; (6) rinsed again in buffer; (7) incubated in the ABC complex (Vector Laboratories) for 1 h; (8) rinsed again in buffer; (9) reacted with 0.05% 3,3'-diaminobenzidine (DAB) in buffer with 0.003% hydrogen peroxide added for 1–18 min, monitored continuously under a microscope. Sections used for E.M. were then further rinsed and osmicated, dehydrated, and flat mounted on vinyl slides.

Post-embedding immunogold technique

Tissue was fixed and cut as above and then stored in 8% dextrose/phosphate buffer, rinsed three times in 0.1 M sodium cacodylate buffer, osmicated in 1% osmium, 1.5% potassium ferricyanide, and 0.1 M sodium cacodylate for 20 min. The tissue was then embedded in plastic and 600 Å sections were cut on a Reichert Ultracut E and mounted on formvar or parloidin coated grids. We used either 100-mesh gold grids (Fullam) or gold gilded single slot grids (EM Sciences). Sections were processed in an LKB type grid box containing the pertinent solutions, but individually rinsed between steps by immersing in tris buffer or distilled water in 20-ml beakers that were agitated with small stir bars. Grids were then incubated in a humidity chamber at 4°C. The individual steps were: (1) etch grids in 1% aqueous periodic acid for 7 min; (2) etch in 1% aqueous sodium metaperiodate for 7 min; (3) incubate in 0.25% aqueous sodium borohydride for 5 min; (4) incubate in primary antiserum overnight; (5) incubate for 90 min in secondary antiserum conjugated to 15-nm gold particles (Amersham) at dilutions of 1:20–1:30; (6) stain with 4% aqueous

uranyl acetate for 10 min; and (7) stain with Sato's lead citrate for 1.5–2 min.

Double labeling with two antisera

For electron microscopy, we used preembedding immunocytochemistry in which silver-itensified DAB was used to mark one antibody, and post-embedding immunocytochemistry in which a colloidal gold secondary antibody was used to mark the other primary antibody. In these experiments, the two antibodies were raised in different species. For silver intensification, sections were: (1) rinsed in 2% sodium acetate; (2) incubated in 10% thioglycolic acid for 1.5–4 h; (3) rinsed overnight in 2% sodium acetate; (4) incubated in a silver solution of 5% sodium carbonate, 0.2% silver nitrate, 0.2% ammonium nitrate, 1% tungstosilic acid, and 37% formalin for 7–8 min; (5) incubated in 1% acetic acid for 1–2 min to stop the silver reaction; (6) rinsed in sodium acetate; (7) incubated in a 0.05% solution of gold trichloride acid trihydrate yellow dissolved in dH$_2$O for 7–8 min; (8) treated in 3% sodium thiosulfate; (9) rinsed, osmicated, and embedded in plastic prior to treating for post-embedding immunocytochemistry.

Electron microscopy

Most sections were photographed with a JEOL JEM 1210 or a Philips CM-10 electron microscope at a magnification of 3000–20 000 ×. Negatives were printed on Kodak Ektamatic paper.

Microscope computer plots

The positions of cells in single light microscope sections were mapped using an NTS computer-microscope plotting system (Eutectic Electronics). The outline of the section and the position of labeled cells was marked using a joystick. Once digitized, hard copies of the sections and plotted cells were plotted on a laser printer. In some experiments, photographs of labeled cells were taken and their geometries measured using a digitizing tablet interfaced to a MacIntosh IIci running modified MacMeasure software (Wayne Rasband, NIH). The software computed the cross-sectional area, perimeter, maximum, minimum, and average diameters, and shape factor of each cell.

Results

The W cell system: a calbindin and GABA microcircuit

The first cell circuit that we have characterized is closely related to the 'W' retinal pathway that terminates within upper superficial gray layer (sgl) of SC. This circuit includes small cells that contain the calcium binding protein calbindin 28 kDa (CB), a GABA containing interneuron referred to as a P1 presynaptic dendrite, and small retinal terminals that contact both of these cell types. A circuit diagram of the synaptic organization of this system is shown in Fig. 1.

CB cells form three sublaminar tiers of neurons in the SC of the cat (Mize et al., 1991, 1992a), monkey (Mize and Luo, 1992), rabbit (Nunes-Cardozo et al., 1994), and rat (Cork et al., 1995). In cat, each CB tier is located at or near the dorsal margin of one of the cellular layers of SC (Fig. 2A) (Mize et al., 1991, 1992a). The three tiers are even more clearly delineated in the prenatal kitten (Fig. 2B). The most dorsal tier of CB neurons overlaps the dense band of contralateral 'W' retinal input to cat SC (Behan, 1981; Mize, 1983a,b; Berson, 1988). The majority of CB cells in the dorsal tier of cat were small interneurons. Injections of horseradish peroxidase into a number of ascending and descending projection sites labeled fewer than 3% of CB neurons in the dorsal tier (Fig. 2C) (Mize et al., 1991). Those that were double-labeled were found only after injections that involved the C laminae of the dorsal lateral geniculate nucleus (dLGN) or the ventral LGN (Mize et al., 1991). Both of these targets are known to be innervated by 'W' retinal ganglion cells (Sherman and Spear, 1982). Injections into the lateral posterior nucleus as well as injections into descending pathways from SC failed to label any CB immunoreactive neurons (Mize et al., 1991).

CB neurons in the dorsal tier had varying morphologies, including horizontal, pyriform, and stellate shapes (Figs. 2D,E). Two of these varieties of cell — horizontal and pyriform — are also thought to contain gamma aminobutyric acid (GABA) (Mize, 1988, 1992). We used a double label technique in which CB was labeled with one chromagen (diaminobenzidine, DAB) and GABA with another (silver enhanced immunogold). In these studies, approximaately 8% of CB containing neurons were also labeled by antibodies directed against GABA (Fig. 2F) (Mize et al., 1991). In summary, CB cells in the dorsal tier of cat are mostly small interneurons, they overlap the 'W' retinal pathway, and some also contain GABA. The few projection neurons that contain CB all project to 'W'-related regions of the lateral geniculate nucleus.

The synaptic connections onto CB cells have just begun to be studied. Pilot electron microscope data from our laboratory has shown that the dendrites of CB containing cells in the sgl of the rabbit SC do receive synaptic input from small retinal terminals (Figs. 2G,H; Nunes-Cardozo, Butler and Mize, unpublished results). Double labeling studies revealed that some of these CB immunoreactive dendrites also contained GABA, labeled using post-embedding colloidal gold immunocytochemistry (Figs. 2G,I). Small retinal terminals also contacted GABA containing presynaptic dendrites called P1s (Fig. 2G). These presynaptic dendrites had loose clusters of small round or ovoid synaptic vesicles and also received synaptic input from retinal synaptic terminals. These P1 profiles formed synaptic contacts with putative CB containing dendrites (Figs. 2H,I) and thus likely contribute to synaptic triads in this region of SC (Fig. 1; Mize, 1994).

Based upon this data, we hypothesize that the 'W' system circuitry in SC consists of CB containing neurons, some of which project to the lateral geniculate nucleus, P1 GABA containing presynaptic dendrites, and retinal terminals from 'W' retinal ganglion cells (Fig. 1). We suggest that this circuitry forms synaptic triads which mediate feed-forward inhibition and may produce a sec-

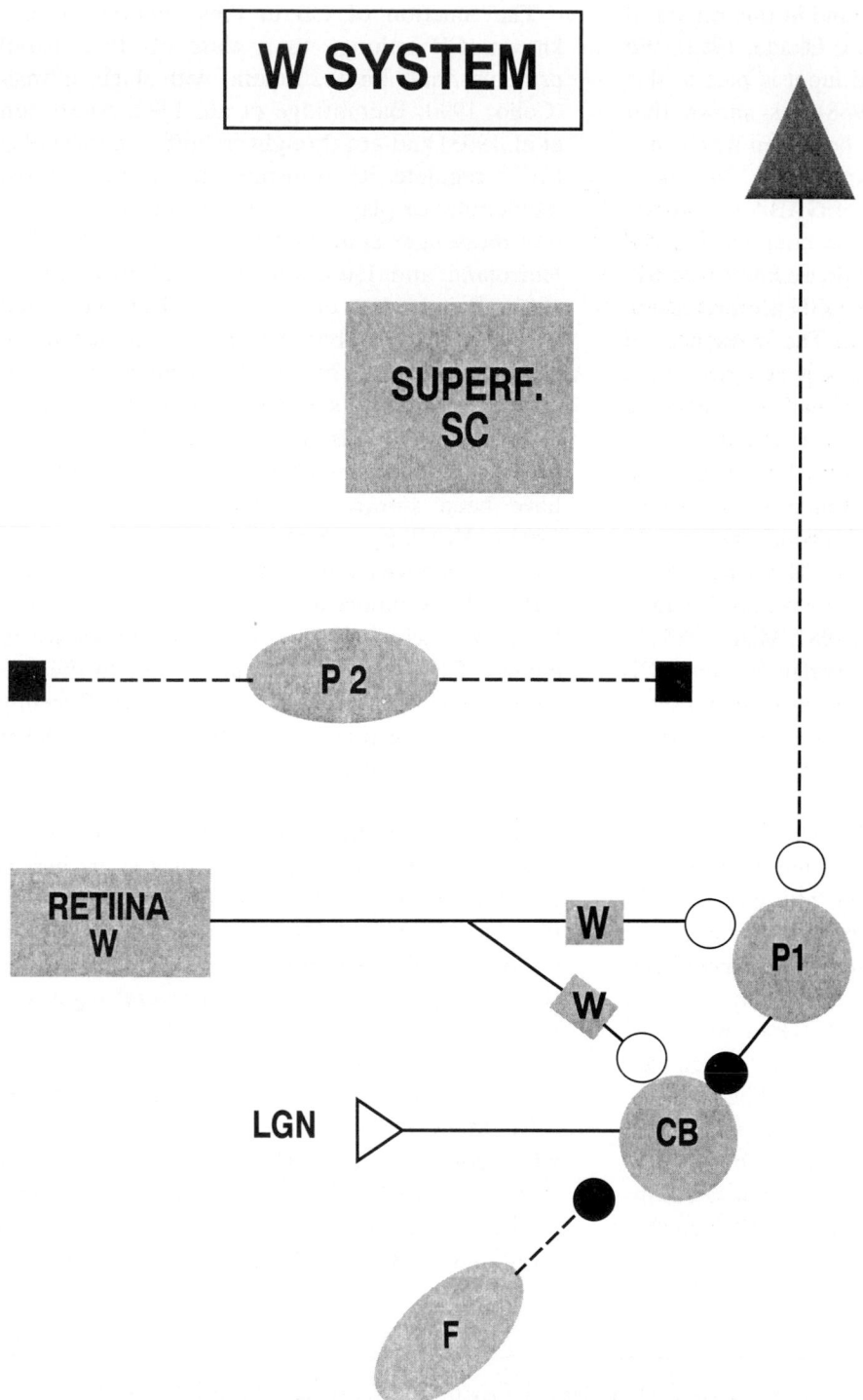

Fig. 1. Diagram of the 'W' system in the upper superficial gray layer of the cat superior colliculus. Synaptic input from 'W' retinal ganglion cells contacts both neurons containing calbindin 28 kDa (CB) and P1 presynaptic dendrites containing gamma aminobutyric acid (GABA). P1 cells also contact CB containing neurons, thus contributing to a synaptic triad. Some CB neurons project to the 'W' innervated portions of the lateral geniculate nucleus (LGN) while others are interneurons. This circuit may also receive

Continued overleaf

ondary response inhibition found in this region of SC (McIlwain and Fields, 1971; Okada, 1992). We do not know whether cortical input is part of this circuit, although Berson (1988) has shown that cells in this region of SC can be driven by cortical stimulation at latencies typical of 'W' cells. Whether or how other types of GABA containing interneurons are related to the circuitry has not been established (Fig. 1). Nor do we know whether CB projection neurons to the LGN are excitatory, inhibitory, or possibly both. The presence of GABA in some of these cells suggests that some must be inhibitory. Additional data is needed to further define the structure of this circuit.

Recent evidence from other laboratories also suggests that the 'W' retinal pathway is synaptically related to CB cells in several species. As noted above, the dorsal tier of CB neurons overlaps the dense band of 'W' contralateral retinal input to the SC of cat (Behan, 1981; Mize, 1983a,b; Berson, 1988). This spatial overlap between CB cells and 'W' retinal input is also seen in the LGN of primates. In galago, CB immunoreactivity is densest in the koniocellular layers 4–5 where 'W' cells predominate (Irvin et al., 1986; Diamond et al., 1993; Casagrande, 1994). In macaque, CB cells are concentrated in the intralaminar zones (Mize et al., 1992b) that also contain cells with 'W'-like response properties (Irvin et al., 1986). CB antibodies also label the koniocellular layers of the tree shrew and the intercalated layers of the owl monkey LGN (Diamond et al., 1993; Casagrande, 1994), both of which are thought to be 'W' related regions. The distribution of CB cellular and neuropil labeling thus overlaps W innervated laminae in both the LGN and SC in several species, a result that is consistent with the hypothesis that CB distinguishes cells that receive W input.

The function of CB in these neurons is unknown. CB belongs to a class of 'E–F hand' proteins that bind calcium with high affinity (Celio, 1990; Baimbridge et al., 1992; Andressen et al.,1993) and are thought to buffer intracellular Ca^{2+}, regulate its transport across the plasma membrane, or play a role in Ca^{2+}-dependent second messenger systems (Baimbridge et al., 1982; Heizmann and Hunziker, 1990; Mattson et al., 1991; Baimbridge et al., 1992; Heizmann and Braun, 1992). CB has also been reported to be contained in cells that display voltage-dependent calcium currents (see Celio, 1990 for review). Non-pyramidal cells in neocortex that have low threshold calcium spikes and rapid adaptation have been shown to contain CB (Kawaguchi, 1993). Kindling, which induces epileptic discharges, reduces both whole cell Ca^{2+} currents and CB immunoreactivity (Kohr et al., 1991). Consistent with this, neurons that are known to contain CB are labeled by antibodies to the calcium P channel (Hillman et al., 1991). Recently we have shown that many cells in the optic layer of rat SC, a layer that contains many CB immunoreactive neurons, have low threshold calcium spikes produced by depolarization, an inward rectifying current (Ih) activated by hyperpolarizing current pulses, and spike accomodation (Lo et al., 1995). Thus CB appears to play an important role in regulating calcium in neurons that display calcium currents in several regions of the CNS.

In summary, CB neurons within the dorsal tier of cat SC receive input from putative retinal 'W' cells. Many CB cells are interneurons, some of which contain GABA. Those that project to other structures send axons to nuclei of the lateral geniculate complex that also receive 'W' retinal input. The CB neurons appear to be a key com-

---

input from corticocollicular neurons with 'W' latencies, but this input has not been demonstrated anatomically. We do not know whether other classes of GABA containing inhibitory input (P2 and F) are involved in this circuit. Open circles depict excitatory synapses; black circles and squares depict inhibitory synapses; solid lines depict known connections; and dotted lines depict unknown connections.

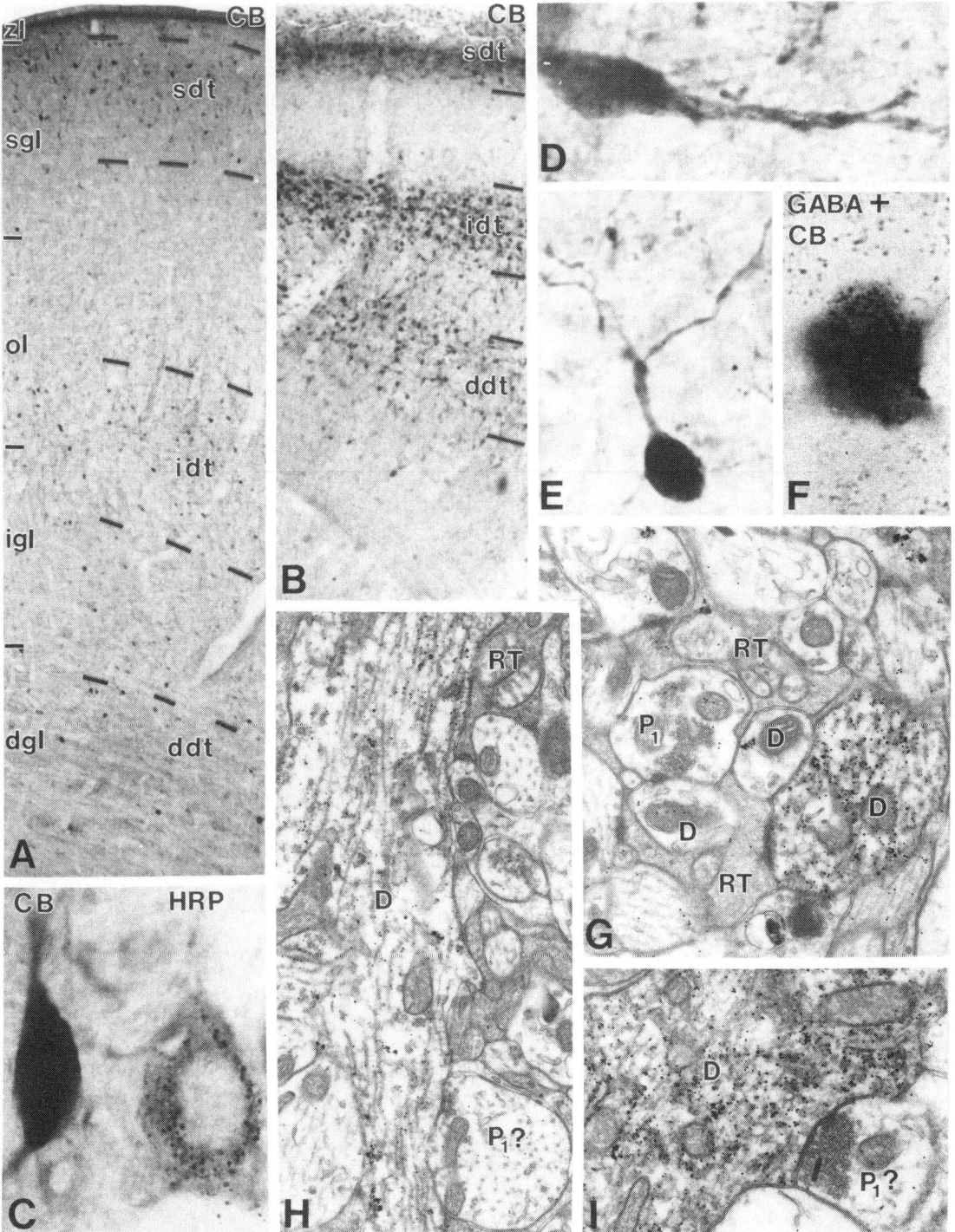

Fig. 2. The calbindin (CB) containing 'W' system in the cat superior colliculus. (A) Distribution of CB containing cells in the adult cat SC. There are three tiers: a superficial dense tier (sdt), an intermediate dense tier (idt), and a deep dense tier (ddt). (B) Distribution of CB-labeled neurons in an E48 prenatal kitten. The three CB tiers are well-demarcated by this age. (C) A bipolar

Continued overleaf

ponent of 'W' retinal synaptic islands that also receive GABAergic input from presynaptic dendrites and form classic synaptic triads which possibly mediate feed-forward inhibition. The CB content of some of these neurons may be related to low threshold calcium currents, but whether this property contributes to the 'W' receptive field properties of these neurons is unknown.

The Y cell system: a parvalbumin and glutamate projection neuron circuit

The second system we call the 'Y' system. Many projection neurons that we propose belong to the 'Y' system have been found to contain parvalbumin (PV), a 12-kDa calcium binding protein. Some of these cells form a dense sublaminar tier of neurons within the deep superficial gray and upper optic layers. This region receives significant input from visual cortex, much of which is thought to be 'Y' input. We also know from double labeling studies that many of the PV neurons in the dense tier project to the lateral posterior nucleus and that some PV neurons also contain the excitatory neurotransmitter glutamate. We hypothesize that 'Y' input from the retina, which is also concentrated in this region of SC, also forms synaptic contacts with PV neurons in the dense tier. The relationship of this circuitry to inhibitory neurons is unknown, but F type synapses do contact some PV neurons. This evidence is consistent with the hypothesis that PV neurons are related to the 'Y' input to SC. This circuitry is illustrated in Fig. 3.

Immunocytochemical studies in our laboratory have shown that PV labeled neurons in the cat SC form a single dense tier of medium sized neurons (Fig. 4A) that lay within the deep SGL and upper optic layers, just between the two dorsal tiers of CB neurons (Mize et al., 1992a). Some PV neurons were also scattered within the deep layers. However, almost no PV-labeled neurons were found within the upper sgl where CB cells are concentrated (Fig. 4A). PV neurons varied substantially in size, but had a mean average diameter of 20.3 μm (Mize et al., 1992a). The morphology of PV neurons also varied. Neurons within the dense band were sometimes vertical fusiform cells with biopolar morphologies (Fig. 4A), while others had multipolar morphologies (Fig. 4D). Many PV cells in the deeper layers were medium to large stellate-shaped neurons (Fig. 4B), typical of neurons that have axons which project through the predorsal bundle.

A large percentage of all PV neurons were projection cells that send their axons to various ascending and descending targets of SC (Mize et al, 1992a). We placed injections of the retrograde tracer HRP into the lateral posterior nucleus (LP), into the two major descending pathways of SC (predorsal bundle and tectopontobulbar pathway), and into brainstem termination sites including the dorsal lateral gray pontine nucleus (PGD) and the opposite colliculus. In three cases analyzed quantitatively, only 16.5% of neurons were labled by HRP that were not also labeled by PV in the same plane of section (Mize et al., 1992a). Thus, most PV neurons in cat SC are projection neurons. PV neurons in the dense tier projected primarily to LP, although some neurons projecting to the PGD were also found within this tier. PV neurons projecting to LP were densely con-

---

vertical fusiform neuron labeled by CB that is adjacent to another neuron labeled by retrograde transport of horseradish peroxidase (HRP). Most CB cells were not double-labeled in retrograde transport studies, suggesting that most CB cells are interneurons. (D) CB-labeled neuron with horizontal morphology in the sdt; (E) CB-labeled neuron with pyriform morphology in the sdt; (F) Neuron-labeled by both CB (black DAB reaction product) and by GABA (silver intensified granular reaction product) antibodies. (G) Electron micrograph of a dendrite (D) labeled by both CB (silver) and by GABA (colloidal gold particles) in the sgl of the rabbit SC. This dendrite receives input from a small retinal terminal (RT) with pale mitochondria. A P1 presynaptic dendrite is labeled by GABA (colloidal gold) but not by CB. (H) Dendrite labeled by both CB (silver) and GABA (colloidal gold). This dendrite receives putative input from both an RT and a P1 presynaptic dendrite. (I) Another double-labeled dendrite with synaptic input from a P1 presynaptic dendrite. C–F modified from Mize et al., 1991.

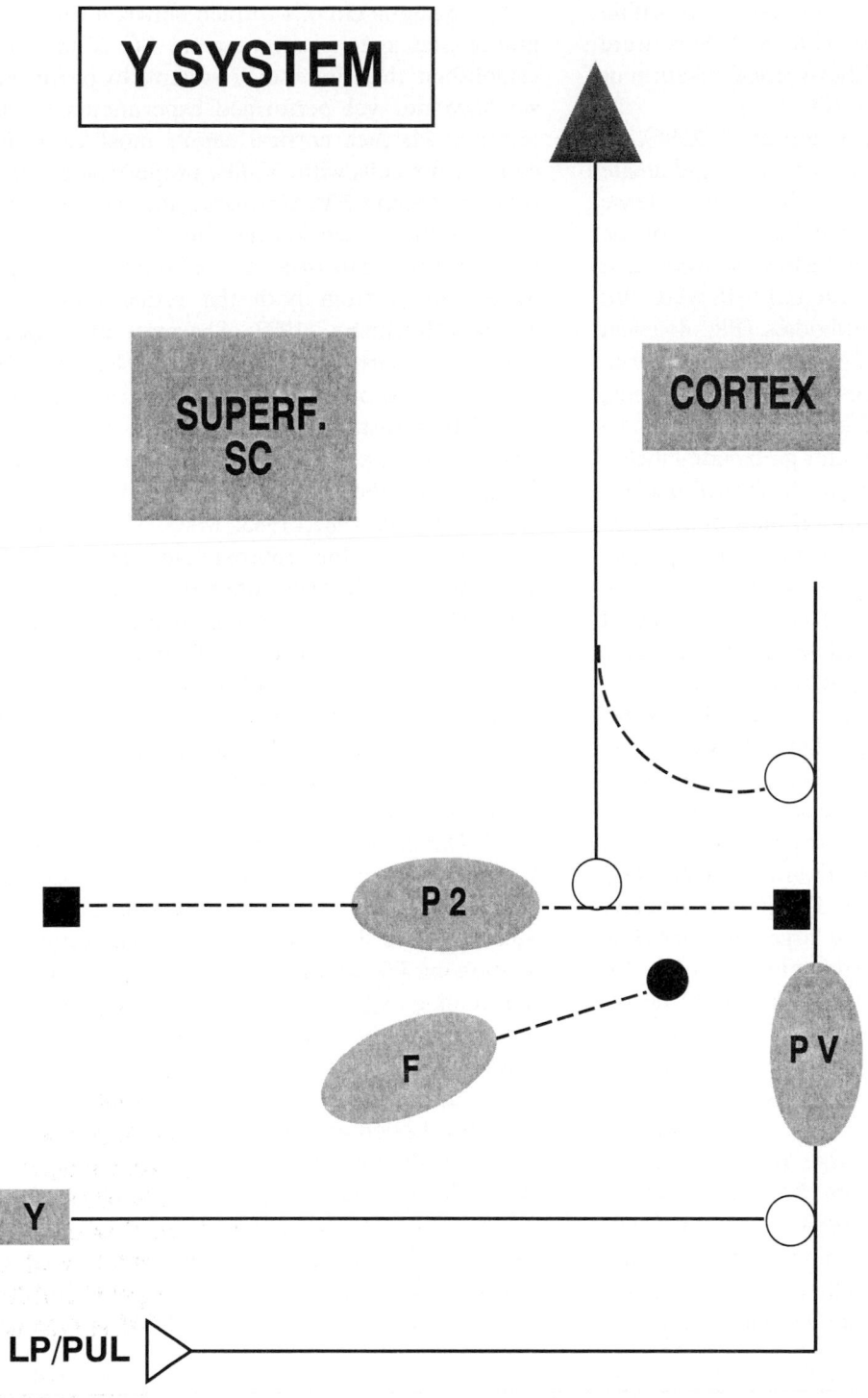

Fig. 3. Diagram of the 'Y' system in the deep superficial gray layer and upper optic layer of the cat superior colliculus. Synaptic input, presumably from 'Y' retinal ganglion cells, contacts neurons containing parvalbumin (PV) within the dense tier in the dsgl, some of which project to the region of the lateral posterior nucleus–pulvinar complex (LP/PUL). This circuit is also thought to receive input from corticocollicular neurons, most of which have 'Y'-like properties. We do not know whether other

Continued overleaf

centrated within the dense tier, especially within the lateral two-thirds of SC (Fig. 4C). Most were medium sized cells with either vertical fusiform or stellate morphologies (Fig. 4D).

From other experiments (Jeon et al., 1996), we know that antibodies that recognize glutamate also label neurons and fibers that form a dense band within the deep superfical gray layer of the cat SC (Fig. 4E). Corticocollicular neurons that project to this region from areas 17–18 were also labeled by these same antibodies (Fig. 4F) and lesions of areas 17–18 of visual cortex reduced the dense tier of glutamate-like antibody labeling (Jeon et al., 1996). In addition, we have shown that about one-half of the anti-glutamate-labeled cells in the dense band project to the LP nucleus (Jeon et al., 1996). Thus, several lines of evidence support the conclusion that at least some projection neurons in the dense tier contain both PV and glutamate and that there is at least a partial overlap within the deep sgl between glutamate containing cells and cells that contain PV.

More recently, pilot studies in the rabbit SC performed in our laboratory have shown directly that some PV neurons colocalize glutamate while none colocalize GABA. We examined colocalization using double labeling techniques in which PV was labeled by silver intensified DAB and glutamate by colloidal gold. The double labeling was particularly prominent in large dendrites (Fig. 4G) and in large myelinated axons (Fig. 4H). We also found medium- to large-sized retinal terminals that were labeled by both PV and glutamate (Fig. 4I) and other retinal terminals in the sgl that contacted dendrites labeled by PV (Fig. 4J). Although this is preliminary data, it is consistent with the hypothesis that large retinal terminals that are thought to arise from 'Y' retinal ganglion cells (Behan, 1981; Mize, 1983) contain PV and that they contact PV containing neurons. In any event, these data show directly that PV and glutamate are colocalized in some neurons in SC.

The synaptic circuits formed between these cell groups and specific afferents to SC is less well established than for the W system. In particular, we have not yet performed experiments to determine whether cortical input, most of which comes from cells with 'Y'-like properties, contacts cells containing PV. Neverless, indirect evidence supports the conclusion that the dense tier of PV cells receive input from the 'Y' system. 'Y' input to SC arises from both the retina and visual cortex (Hoffmann, 1973; Sherman and Spear, 1982). Most corticocollicular neurons have 'Y'-like receptive field properties (Hoffman, 1973; Palmer and Rosenquist, 1974). In addition, the densest distribution of cortical inputs from areas 17–18 is found within the deep sgl (Sterling, 1971; Berson and McIlwain, 1982, 1983; Mize, 1983b; Behan, 1984) which is the approximate region of the dense tiers of PV and glutamate containing neurons. Many cells in this region are also driven synaptically by input with 'Y' type conduction velocities (Berson and McIlwain, 1982, 1983). The dense tier of PV cells also lies in a region where many cells are labeled by the chondroitin sulfate proteoglycan CAT-301 (Mize and Hockfield, 1989) which is a selective marker of 'Y' cells in the LGN (Hendry et al., 1984; Sur et al., 1988). Thus, synaptic connections between PV neurons and 'Y' corticocollicular axons seems likely, although we have yet to show this relationship directly. The relation of PV to the 'Y' system has also been shown in galago where PV immunoreactivity is very dense within the M laminae of the LGN of that species (Casagrande, 1994).

The intracellular physiology of PV cells in SC has not yet been examined. PV is often contained in cells with high frequency, non-adapting firing rates (Celio, 1990; Andressen et al., 1993). Fast-spiking (FS) neurons contain PV in both rat hippocampus (Kawaguchi et al., 1987) and frontal cortex (Kawaguchi, 1993; Kawaguchi and Kubota, 1993). FS cells are activated by NMDA receptor

---

classes of GABA containing inhibitory input (P2 and F) are involved in this circuit. Open circles depict excitatory synapses; black circles and squares depict inhibitory synapses; solid lines depict known connections; and dotted lines depict unknown connections.

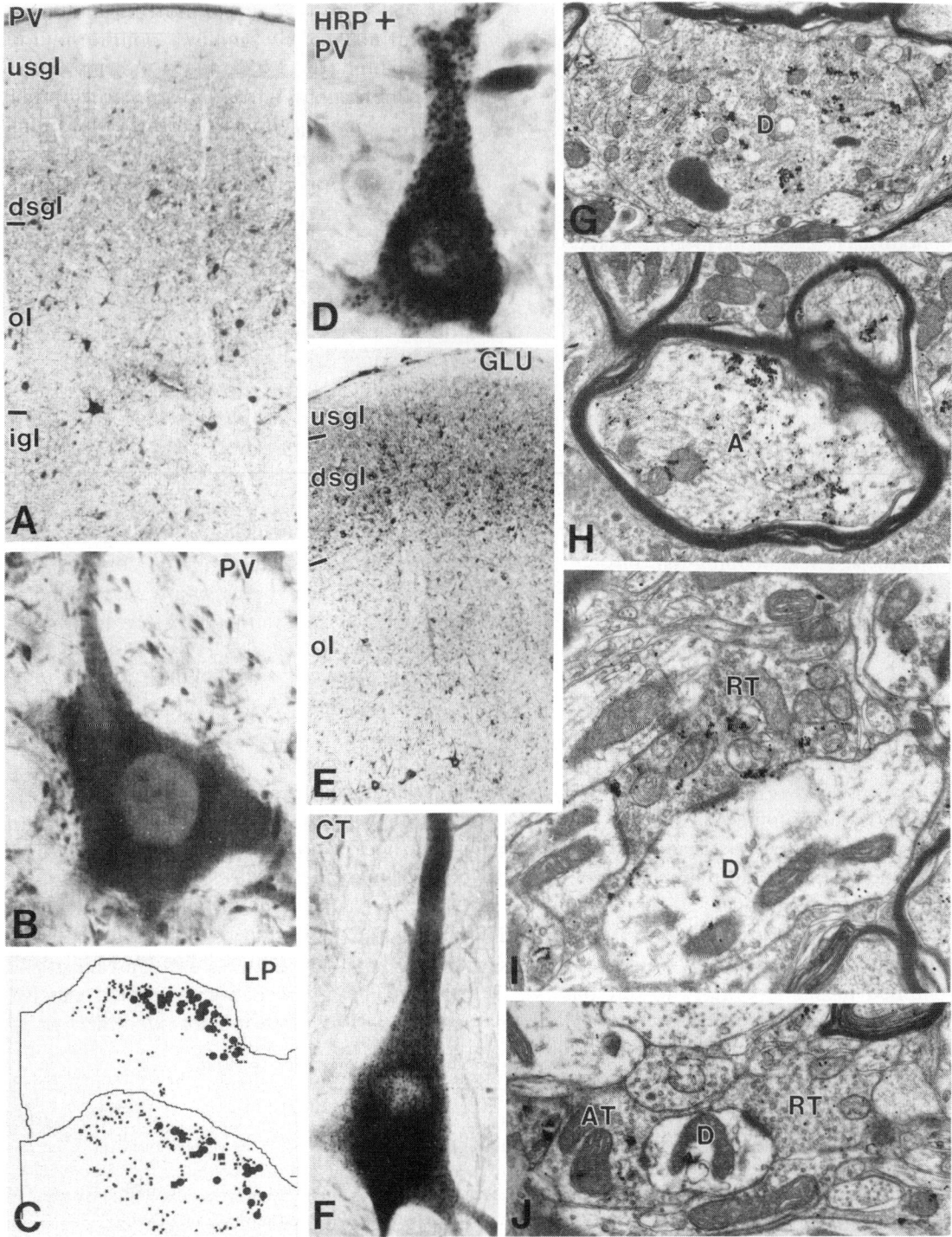

Fig. 4. The parvalbumin (PV) containing 'Y' system in the cat superior colliculus. (A) Distribution of PV immunoreactive neurons within SC. Note the dense tier of neuropil and cellular labeling within the deep superficial gray layer (dsgl) and the upper optic layer (ol). Labeled neurons are also scattered in the deeper layers. (B) Example of a large stellate-shaped neuron within the intermediate gray layer (igl) labeled by PV. (C) Computer generated plots of the distributions of PV-labeled (small squares) and

Continued overleaf

agonists which increase intracellular calcium (Feldman et al., 1990). Repetitive firing in these cells is blocked by NMDA receptor antagonists (Kawaguchi, 1993). Thus, PV in these neurons appears to buffer rapid influxes of Ca^{2+} that occur during NMDA receptor activation and could maintain non-adapting high frequency firing rates by inactivating calcium mediated potassium channels in these neurons (Kawaguchi, 1993). PV could also serve to buffer Ca^{2+} in 'Y' innervated cells in SC. Many 'Y' innervated cells in SC have short latency, high frequency firing rates (Berson and McIlwain, 1982; Berson and McIlwain, 1983) and predorsal bundle neurons in the igl that probably contain PV also discharge at very high frequencies (Grantyn and Berthoz, 1985). Studying the distribution of 'Y' corticocollicular fibers and NMDA receptors onto PV cells will establish whether NMDA and 'Y' inputs are invariably associated with these PV containing neurons.

In summary, PV neurons are mostly medium to large sized projection neurons. A subgroup of these cells forms a dense tier within the deep sgl and upper optic layers. Some of these cells project to the lateral posterior nucleus and their laminar position overlaps the region of densest input from the 'Y' corticocollicular pathway from areas 17–18. Some PV neurons also colocalize glutamate. PV is often contained in neurons that have high frequency discharges that are non-adapting, and neurons in SC with these properties may well be cells that contain PV. There is thus evolving evidence that PV neurons are specifically associated with the 'Y' pathways in SC. Experiments are planned in our laboratory to test this hypothesis more directly.

The patch-cluster system: an Ach and nitric oxide circuit

The third cell system which we have been studying is a group of 'clustered' neurons in the igl that precisely overlap the acetylcholine (ACh) fiber patches in this layer of SC (Jeon and Mize, 1993; Jeon et al., 1993). The cell 'clusters' are groups of neurons that project selectively through the tectopontobulbar pathway to the region of the cuneiform nucleus. Up to seven clusters have been identified in the dorsal igl in single sections, each containing from three to 20 neurons. The clusters appear to overlap ACh containing fibers from the pedunculopontine tegmental (PPTN) and lateral dorsal tegmental nuclei (LDTN) and also fibers from the substantia nigra (SN) and the frontal eye fields (FEF). A circuit diagram of the patch-cluster system is illustrated in Fig. 5.

The fiber patches in the igl of the cat SC were first reported by Graybiel using a histochemical stain for acetylcholinesterase, the degradative enzyme of ACh (Graybiel, 1978, 1979). Subsequently, similar fiber patches within the igl were identified from the PPTN and dorsal lateral tegmental (LDTN) nuclei (Beninato and Spencer, 1986; Hall et al., 1989; Harting and Lieshout, 1991), the frontal eye fields (Leichnetz et al., 1981; Illing and Graybiel, 1985), the nucleus of the posterior commissure (Huerta and Harting, 1982), the spinal trigeminal nucleus (Huerta et al., 1981; McHaffie et al., 1986; Harting and Lieshout, 1991), and portions of the ectosylvian and suprasylvian gyri (Huerta and Harting, 1984). Transmitter or neurochemical specific patches within the igl have also been identified using

---

double-labeled HRP/PV neurons (large circles) after an injection of the retrograde tracer into the lateral posterior nucleus. (D) Neuron labeled by both HRP after retrograde transport from LP and by a PV antibody. The cell has a vertical fusiform shape. (E) Distribution of neurons labeled by a gamma-l-glutamic-l-glutamic acid (gamma glu glu) antibody that recognizes the neurotransmitter glutamate. Note that this antibody also forms a dense band of labeled neurons that partially overlaps the PV dense tier. (F) Corticocollicular neuron labeled by both gamma glu glu and the retrograde tracer HRP injected into SC. (G) Dendrite (D) in the rabbit SC labeled by antibodies to PV (silver intensified DAB) and glutamate (colloidal gold particles). (H) Large myelinated axon (A) also labeled by both PV and glutamate antibodies. (I) Retinal terminal (RT) with round synaptic vesicles and pale mitochondria that is labeled by both PV and glutamate antibodies. (J) An RT and another axon terminal (AT) contacting a dendrite labeled by PV. A–D and E modified from Mize et al. (1992a).

antibodies directed against met and leu enkephalin (Graybiel et al., 1984; Mize, 1989; Berson et al., 1991; Graybiel and Illing, 1994) and histochemical stains and antibodies that recognize nitric oxide synthase (Wallace, 1986, 1988; Arceneaux et al., 1995; Scheiner et al., 1995). The precision of overlap of some of these fiber systems has been studied directly using either adjacent section reconstructions or double-labeling techniques. Based upon these studies, we know that the inputs from the substantia nigra, frontal eye fields, the pedunculopontine tegmental nucleus, and the spinal trigeminal nucleus are clustered in the same regions of the igl (Illing and Graybiel, 1985; Harting and Lieshout, 1991).

Although the origin and neurotransmitter content of some of the patch fibers are now well established, little is known about the neurons which underly these patches. Some cells projecting to the spinal trigeminal nucleus have been reported to cluster in the intermediate and deep layers of SC (Huerta et al., 1981) and injections into the tectopontobulbar pathway also sometimes produce clustered patterns of retrogradely labeled cells (Huerta and Harting, 1982; Redgrave et al., 1986). However, until recently no clustered neurons have been found that directly overlap the ACh containing afferents to the igl patches.

In 1993, our laboratory first reported that a group of clustered neurons do overlap the cholinergic patches in the igl of the cat SC. We placed injections of horseradish peroxidase into both the tectopontobulbar pathway (TPB) and into the cuneiform region (CFR) of the midbrain tegmentum. In three cases, we found that many of the neurons that were retrogradely labeled by HRP were grouped into well-defined clusters within the dorsal igl. In one case, 25 clusters of closely apposed neurons were found in five sections through the caudal one-half of SC, two of which are illustrated in Fig. 6A. The clusters consisted of 3–20 neurons that were separated by characteristic intercluster intervals (Fig. 6A, arrows). The cell clusters were more frequent in the medial half of the caudal SC, but were also distributed across the entire medial-lateral extent of

SC in some sections (Fig. 6A). Up to seven clusters were found in a single 50-μm section (Fig. 6A, lower section). Clustered neurons had relatively uniform morphologies, including round or oval cell bodies with multiple dendrites and a mean cross sectional area of 184.2 μm^2. Thus, the clustered neurons appear to be distinguished by projection site (the cuneiform region), by size, and possibly morphology. A detailed description of these clusters has been published by Jeon and Mize (1993).

To demonstrate directly that these clustered neurons lie within the igl fiber patches, we double-labeled some sections with HRP to mark the clustered neurons and with antibodies to choline acetyltransferase (ChAT), the synthetic enzyme of ACh, to mark the ACh patches. HRP-labeled neurons that formed clusters were found within the ChAT patches in all sections examined. Fig. 6B shows the relationship between ChAT patches and HRP-labeled cells in the most caudal section illustrated in Fig. 6A. In this example, as many as 15 neurons were found within a single patch. Note also that there was a positive correlation between patch size and neuron number. Higher magnification micrographs showed that many of the ChAT-labeled fibers had bouton-like varicosities that were closely apposed to the cell bodies of the HRP-labeled neurons (Figs. 6C,D). Thus, although direct synaptic contacts between the clustered neurons and ACh fibers have not been demonstrated at the electron microscope level, the close apposition between ChAT fiber terminals and HRP-labeled cells makes their synaptic relationship very likely.

Recently, we have shown that the patch cluster system is coincident with fibers and cells labeled for nitric oxide synthase. NOS is known to synthesize nitric oxide (NO), a free radical gas implicated as a retrograde messenger involved in both synaptic development and plasticity in the CNS (Gally et al., 1990; Garthwaite, 1991). In the adult cat SC, NADPH, a histochemical stain for NOS, labeled selected neurons in the periaquaductal gray (PAG) and deep layers of SC. These cells

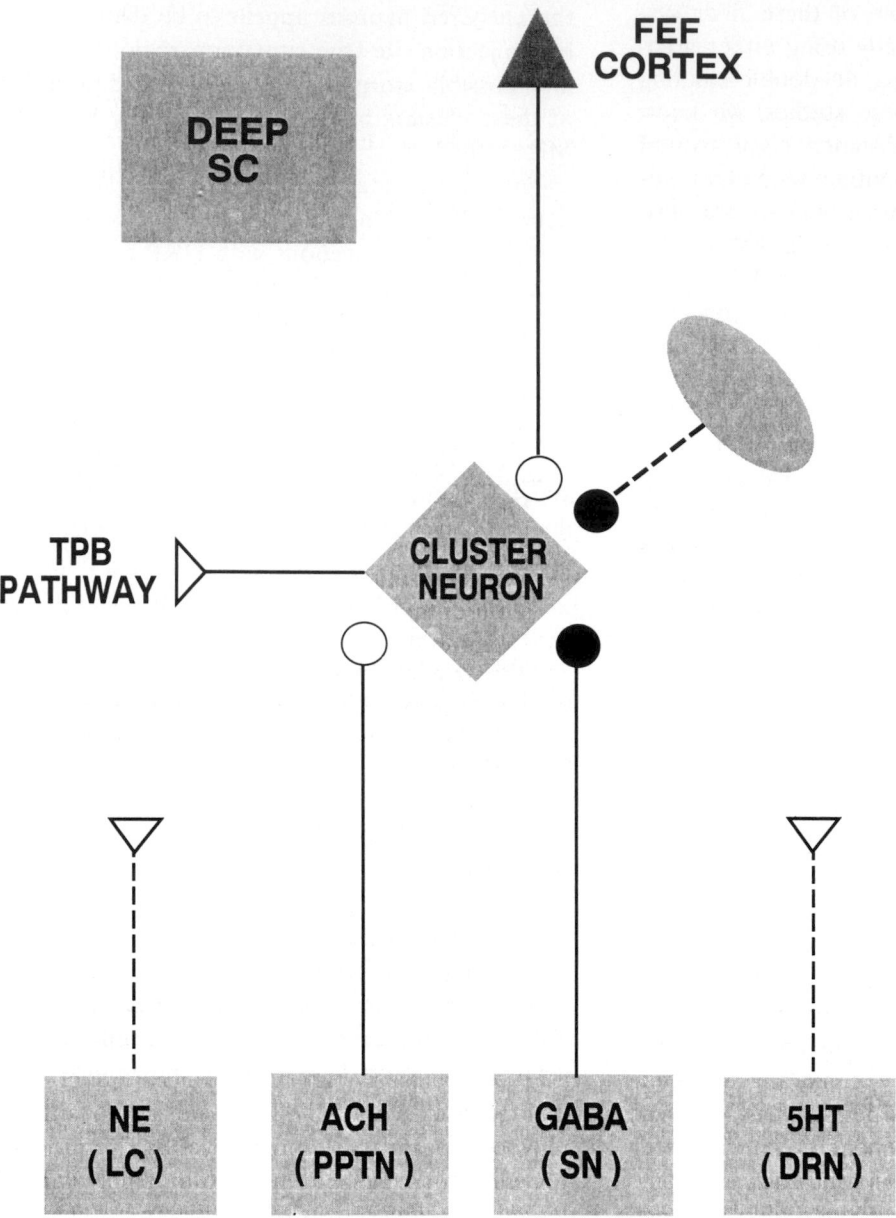

Fig. 5. The patch-cluster system in the intermediate gray layer of the cat superior colliculus. Clustered neurons in the igl project through the tectopontobulbar pathway (TPB) to the region of the cuneiform nucleus. These clusters overlap fiber patches whose axons arise from at least three sources: the pedunculopontine tegmental nucleus (PPTN) which contains acetylcholine (ACh); the substantia nigra (SN) which contains GABA; and cortex of the frontal eye fields (FEF) which probably contains glutamate. Other possible inputs include intrinsic GABA containing neurons as well as input from the locus ceruleus (LC) that contains

were most densely distributed within a wedge-shaped sector in the dorsolateral PAG and also in bundles within the deep gray layer of SC (Fig. 6E). Some scattered neurons labeled by NADPH were also found in the igl as well as in the optic and superficial gray layers (Fig. 6E; Arceneaux et al., 1995). These cells had varying morphologies, including bipolar and stellate shapes (Fig. 6I). Antibodies directed against bNOS showed a similar cellular labeling pattern in the adult SC.

The adult cat SC also contained bundles of fibers that were labeled by NADPH. Within the deepest layers, these fibers formed 'streams' that projected vertically towards the igl (Fig. 6E). Within the igl, NADPH positive fibers formed dense patches that were separated by regions with fewer NADPH-labeled fibers (Fig. 6E,H). Two lines of evidence suggest that these NADPH fibers are the same fibers that contain ACh. First, adjacent sections in which one section had been labeled by ChAT (Fig. 6G) and the other by NADPH (Fig. 6H) showed remarkably similar patterns of fiber labeling. Although the NADPH labeling was sometimes more striking, the same patterns of patches and gaps in labeling could be seen in both sections (asterisks, Fig. 6G,H). Secondly, in sections double-labeled for both NADPH and ChAT, the fiber patches appeared to precisely overlap. In addition, double-labeled sections cut through the brainstem tegmentum revealed that most cells in the PPTN and LDTN contained both labels. Thus, most if not all NOS containing extrinsic afferents terminating in SC also contain ACh and both appear to arise from cells within the brainstem tegmentum, a result also found in the LGN (Bickford et al., 1993).

Neuron labeling by NADPH in the igl was fairly sparse in the SC of the adult cat. However, NADPH was expressed intensely in neurons in the perinatal kitten. Both NADPH histochemistry and bNOS antibody immunocytochemistry pro-duced a pattern of clustered cells in the igl as early E51–E58 (Scheiner et al., 1995). At this age, only a few clusters were seen, but single NOS-labeled cells had characteristic intercluster intervals. Obvious cell clusters containing seven or more labeled neurons could be seen by P3 and clusters were well developed by P14 (Fig. 6F). The clustered pattern and typical intercluster intervals seen at these ages suggest that these are the same cluster cells found within the ACh patches in the adult. NOS is thus a putative marker of the 'clustered' neurons in early postnatal life and of ACh fibers in the adult.

In summary, the patch-cluster system in the igl consists of 'clustered' neurons which are medium sized and project selectively to the cuneiform region of the midbrain tegmentum; they receive transmitter specific afferents from the PPTN and LDTN that contain both ACh and NOS; and they probably also receive GABA containing fibers from the SN and glutamate containing fibers from the FEF.

The function of the clustered neurons is not known. They likely represent a functional unit because they project to a specific site within the brainstem and have unique patterns of synaptic input from specific afferents. These patch-cluster modules probably have an oculomotor function. First, the clustered cells receive inputs that are known to be related to eye movements. The SN input projecting to the igl has long been known to regulate the discharge of saccade-related cells in the monkey SC (Hikosaka and Wurtz, 1983a,d; 1985a,b). SNr cells decrease their response immediately before saccadic eye movements, a decrease that correlates with an increase in the discharge rate of saccade-related cells in the igl (Hikosaka and Wurtz, 1983a,b,c). Inhibition of SNr cells by iontophoretic application of GABA agonists into SN facilitates saccadic eye movements, and application of GABA antagonists in

norepinephrine (NE) and from the dorsal raphe nucleus (DRN) that contains serotonin (5-HT). Open circles depict excitatory synapses; black circles depict inhibitory synapses; solid lines depict known connections; and dotted lines depict unknown connections.

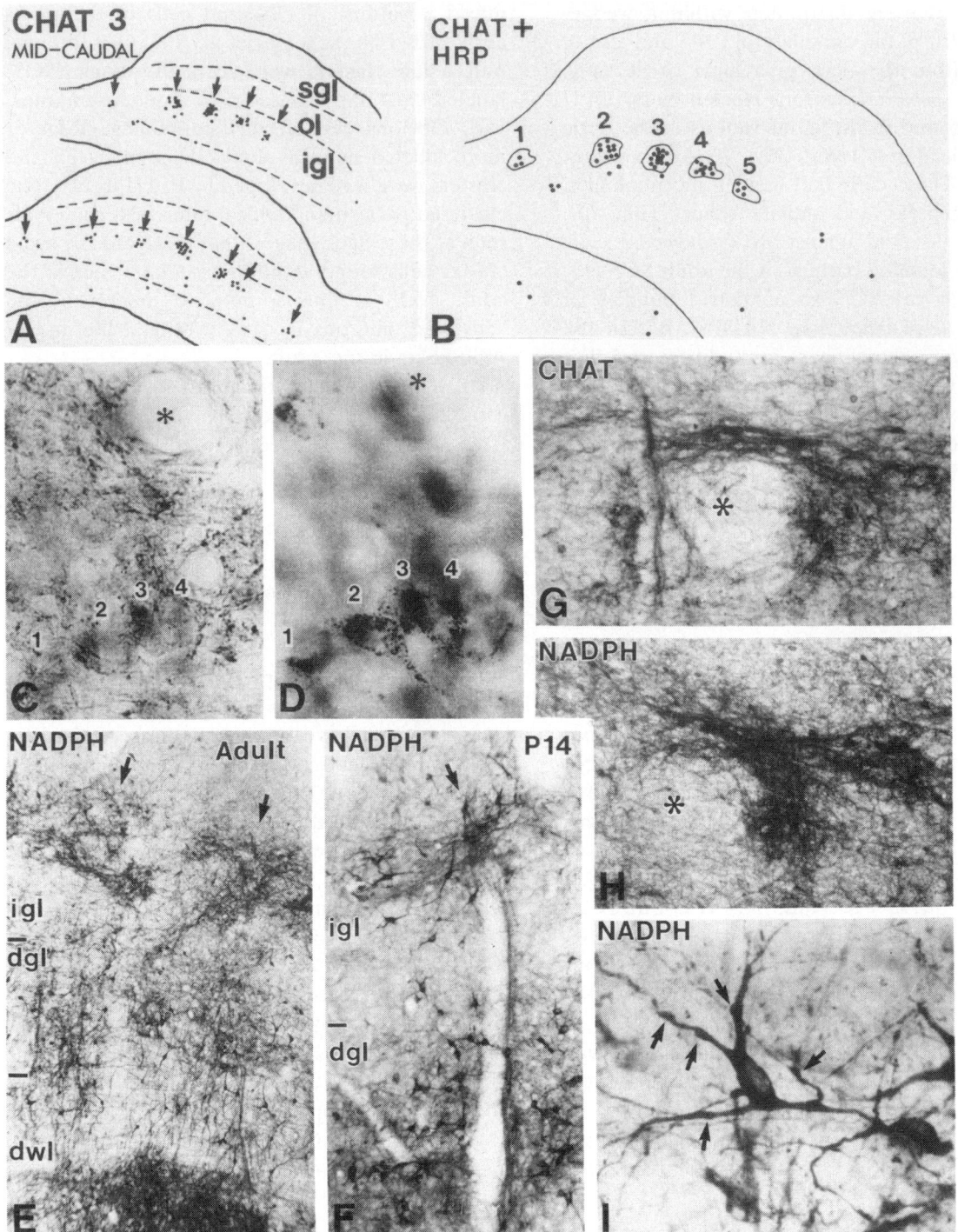

Fig. 6. Patch-cluster system in the cat superior colliculus (SC). (A) Computer generated plots of clustered neurons in the dorsal intermediate gray layer (igl) retrogradely labeled by HRP after injection into the tectopontobulbar pathway. As many as seven clusters (arrows) were identified in a single section. (B) Section through SC double-labeled by retrograde transport of HRP in clustered cells and by an antibody to choline acetyltransferase (ChAT) to label the ACh patches (outline). Note that the vast

the SNr reduces these same saccades (Hikosaka and Wurtz, 1985b). GABA agonists iontophoresed onto saccade-related cells in SC reduce their response and GABA antagonists enhance their response (Hikosaka and Wurtz, 1983d, 1985a). Thus, SNr synapses tonically inhibit saccade-related neurons in the igl of monkey.

The frontal eye field regions of prefrontal cortex are also involved in the generation of saccades in monkeys. Cells in the FEF respond prior to saccades and many of these project to the igl of the monkey SC (Seagraves and Goldberg, 1987). In cat, the FEF has also been shown to project to this region of SC (Leichnitz et al., 1987) and axons from the putative FEF of cat form patches which overlap the ACh patches in the dorsal igl (Illing and Graybiel, 1985). Consistant with this, Guittan and Mandl (1978) have reported that neurons in the cat FEF respond in relation to saccadic eye movements. However, Weyand and Gafka (1996) found few or no corticocollicular neurons in the FEF that were presaccadic. Thus, the presence of saccade-related neurons that project to SC in cat is uncertain, although their presence in monkey is well established. More recently, the PPTN has also been implicated in the control of eye movements in the monkey (Segraves, 1992) and cat (Leichnetz et al., 1987). There is thus solid evidence that the three major inputs that form patches in the cat igl are involved in saccadic eye movements in several species.

Unfortunately, the one physiological study that has investigated whether saccade-related cells form clusters in the monkey SC found no evidence that they were organized into groups or clusters that might match the ACh patches in this species (Ma et al., 1991). On the other hand, Peck (1984) has shown that saccade-responsive neurons in the cat SC do sometimes cluster together in single penetrations. In addition, Cohen and Buttner-Ennever (1984) have shown that igl cells in monkey SC project to a region of the mesencephalic reticular formation (MFR) which is known to generate saccadic eye movements when stimulated electrically. Many of the SC cells in the igl which project to this MRF region are grouped together in a pattern similar to the clustered igl neurons found in cat (see Fig. 3, Cohen and Buttner-Ennever, 1984). Further studies in which clustered neurons have been marked by fluorescent tagged retrograde tracers should determine whether the clustered cells have unique electrophysiological properties that are related to saccadic eye movements.

The role of nitric oxide (NO) in this system is also unknown, but of particular interest. NO is thought to be a diffusable neurotransmitter-like molecule that is released by post-synaptic neurons in order to activate a presynaptic guanylate cyclase-cGMP second messenger system that enhances synaptic transmission during long-term potentiation (Bredt et al., 1990; Dawson et al., 1991; Garthwaite, 1991; Moncada, 1991; Snyder, 1992; Vincent and Hope, 1992). NO has also been implicated in NMDA receptor mediated plasticity that occurs during synaptic development (Gally et al., 1990; Garthwaite, 1991; Haley et al., 1992; Hoydt et al., 1992). In adult rat SC, LTP has been demonstrated in cells within the sgl and NOS may

---

majority of HRP-labeled cells are located within the boundaries of the ACh patches. (C) Example of a ChAT-labeled patch within the dorsal igl. Note that labeled varicose fibers surround clustered cells marked by numbers. (D) Same section as in C in a different plane of focus. HRP-labeled cells are indicated with numbers as in C. Asterisks in both C and D mark the same blood vessel. (E) Nitric oxide synthase (NOS) containing neurons and fibers in the adult cat SC, labeled by NADPH. There is dense labeling of cells and fibers in the dorsolateral periaquaductal gray, streams of neurons and fibers in the deep gray layer (dgl), and fiber patches (arrows) within the dorsal igl. (F) NADPH labeling of cells in the cat SC at postnatal day P14. Note the clustering of NOS cells (arrow) in the dorsal igl at this age. (G) Fiber patch in the igl of the adult cat SC labeled by an antibody to choline acetyltransferase (ChAT). Asterisk marks a region devoid of label. (H) Adjacent section in the igl showing the same fiber patch labeled by NADPH. Note the similarity in labeling in G and H. (I) NADPH-labeled neuron in the adult cat SC. Note the multiple varicose dendrites (arrows) on these cells. A–D modified from Jeon and Mize (1993).

contribute to the regulation of this process (Okada, 1992). The transient expression of NOS in the clustered neurons of perinatal kittens suggests that NO might also serve as a retrograde messenger to stabilize or refine synapses during patch formation in the igl. A variety of experiments are underway in our laboratory to test these hypotheses.

The function of NO in the ACh containing afferents from the midbrain tegmentum has been studied in the LGN. In vivo, NOS inhibitors effectively suppress visual responses in both 'X' and 'Y' cells in the LGN, an effect that is NMDA mediated (Cudeiro et al., 1994). In brain slices, NO generating agents depolarize LGN cells and evoke a voltage-dependent inward current which is mediated by Ih at certain hyperpolarizing membrane currents (Pape and Mager, 1992). Ca^{2+}-mediated oscillatory bursts are also markedly reduced by NO agents. Pape and Mager (1992) therefore suggest that NO acts selectively upon ion channels that promote the onset of desynchronization that occurs in the waking animal. In any case, the intense labeling of fibers by NADPH in both the LGN and SC suggests that NO is an important releasing factor in both of these structures and must be involved in modulating cellular function in the patch cluster system of the cat.

Acknowledgements

Many collaborators contributed to the data analysis contained in this review. C.-J. Jeon, Q. Luo, B. Nabors, G.D. Butler, and B. Nunes-Cardozo performed experiments on the calcium binding proteins. Experiments on GABA and glutamate circuitry were performed by C.-J. Jeon, M. Hartman, M. Gurski, B. Nunes-Cardozo, J. van der Want, R.H. Whitworth, and G. Butler. The patch-cluster system was first described by C.-J. Jeon as his Ph.D. thesis. R.F. Spencer also contributed to these experiments. The recent work on nitric oxide synthase was done by C.A. Scheiner, R. Arceneaux, K. Kratz, W. Guido, and F.T. Banfro. I am especially grateful to B. Nunes-Cardozo and G.D. Butler for their expertise in the double-labeling immunocytochemistry experiments performed while Dr. Nunes-Cardozo was a visiting scholar at Louisiana State University Medical Center, New Orleans. This research has been supported by USPHS Grant EY-02973 from the National Eye Institute, U.S. Army Research and Development Grant DAMD 17-93-V-3013-P40001, and the Neurosicence Center of Excellence at LSU Medical Center.

References

Andressen, C., Blümcke, I. and Celio, M.R. (1993) Calcium-binding proteins: selective markers of nerve cells. Cell Tissue Res., 271: 181–208.

Arceneaux, R.D., Barnes, P.A., Scheiner, C.A., Kratz, K.E., Guido, W. and Mize, R.R. (1995) NADPH-diaphorase histochemistry reveals specific cell types in the cat superior colliculus that contain nitric oxide synthase. Invest. Ophthalmol. Vis. Sci., 36(4): 1345.

Baimbridge, K.G., Miller, J.J. and Parkes, C.O. (1982) Calcium-binding protein distribution in the rat brain. Brain Res., 239: 519–525.

Baimbridge, K.G., Celio, M.R. and Rogers, J.H. (1992) Calcium-binding proteins in the nervous system. TINS, 15: 303–308.

Behan, M. (1981) Identification and distribution of retinocollicular terminals in the cat: An electron-microscopic autoradiographic analysis. J. Comp. Neurol., 199: 1–16.

Behan, M. (1984) An EM-autoradiographic analysis of the projection from cortical areas 17, 18 and 19 to the superior colliculus in the cat. J. Comp. Neurol., 255: 591–604.

Beninato, M. and Spencer, R.F. (1986) A cholinergic projection to the rat superior colliculus demonstrated by retrograde transport of horseradish peroxidase and choline acetyltransferase immunohistochemistry. J. Comp. Neurol., 253: 525–538.

Berson, D.M. (1988) Convergence of retinal W-cell and corticotectal input to cells of the cat superior colliculus. J. Neurophysiol., 60: 1861–1873.

Berson, D.M. and McIlwain, J.T. (1982) Retinal Y-cell activation of deep-layer cells in the superior colliculus of the cat. J. Neurophysiol., 47: 700–714.

Berson, D.M. and McIlwain, J.T. (1983) Visual cortical inputs to deep layers of the cat's superior colliculus. J. Neurophysiol., 50: 1143–1155.

Berson, D.M., Graybiel, A.M., Brown, W.D. and Thompson, L.A. (1991) Evidence for intrinsic expression of enkephalin-like immunoreactivity and opioid binding sites in cat superior colliculus. Neuroscience, 43: 513–529.

Bickford, M.E., Günlük, A.E., Guido, W. and Sherman, S.M. (1993) Evidence that cholinergic axons from the parabrachial region of the brainstem are the exclusive source of nitric oxide in the lateral geniculate nucleus of the cat. *J. Comp. Neurol.*, 334: 410–430.

Bredt, D.S., Hwang, P.M. and Snyder, S.H. (1990) Localization of nitric oxide synthase indicating a neural role for nitric oxide. *Nature*, 347: 768–770.

Casagrande, V.A. (1994) A third parallel visual pathway to primate area V1. TINS, 17(7): 305–310.

Celio, M.R. (1990) Calbindin D-28K and parvalbumin in the rat nervous system. *Neuroscience*, 35: 375–475.

Cohen, B. and Buttner-Ennever, J.A. (1984) Projections from the superior colliculus to a region of the central mesencephalic reticular formation (cMRF) associated with horizontal saccadic eye movements. *Exp. Brain. Res.*, 57: 167–176.

Cork, R.J., Baber, S.Z. and Mize, R.R. (1995) CalbindinD28k and parvalbumin are expressed in complementary patterns in the rat superior colliculus. *Soc. Neurosci. Abst.*, 21: 656.

Cudeiro, J., Grieve, K.L., Rivadulla, C., Rodriguez, R., Martinez-Conde, S. and Acuna, C. (1994) The role of nitric oxide in the transformation of visual information within the dorsal lateral geniculate nucleus of the cat. *Neuropharmacology*, 33: 1413–1418.

Dawson, T.M., Bredt, D.S., Fotuhi, M., Hwang, P.M. and Snyder, S.H. (1991) Nitric oxide synthase and neuronal NADPH diaphorase are identical in brain and peripheral tissues. *Proc. Natl. Acad. Sci. USA*, 88: 7797–7801.

Diamond, I.T., Fitzpatrick, D. and Schmechel, D. (1993) Calcium binding proteins distinguish large and small cells of the ventral posterior and lateral geniculate nuclei of the prosimian galago and the tree shrew (*Tupaia belangeri*). *Proc. Natl. Acad. Sci. USA*, 90: 1425–1429.

Feldman, D., Sherin, J.E., Press, W.A. and Bear, M.F. (1990) *N*-methyl-D-aspartate-evoked calcium uptake by kitten cortex maintained in vitro. *Exp. Brain Res.*, 80: 252–259.

Gally, J.A., Montague, P.R., Reeke, G.N., Jr. and Edelman, G.M. (1990) The NO hypothesis: possible effects of a short-lived, rapidly diffusible signal in the development and function of the nervous system. *Proc. Natl. Acad. Sci. USA*, 87: 3547–3551.

Garthwaite, J. (1991) Glutamate, nitric oxide and cell-cell signalling in the nervous system. *TINS*, 14(2): 60–67.

Grantyn, A. and Berthoz, A. (1985) Burst activity of identified tecto-reticulospinal neurons in the alert cat. *Exp. Brain Res.*, 57: 417–421.

Graybiel, A.M. (1978) A stereometric pattern of distribution of acetylthiocholinesterase in the deep layers of the superior colliculus. *Nature*, 272: 539–541.

Graybiel, A.M. (1979) Periodic-compartmental distribution of acetylcholinesterase in the superior colliculus of the human brain. *Neuroscience*, 4: 643–650.

Graybiel, A.M. and Illing, R.B. (1994) Enkephalin-positive and acetylcholine-esterase-positive patch systems in the superior colliculus have matching distributions but distinct development histories. *J. Comp. Neurol.*, 340: 297–310.

Graybiel, A.M., Brecha, N. and Karten, H.J. (1984) Cluster-and-sheet pattern of enkephalin-like immunoreactivity in the superior colliculus of the cat. *Neuroscience*, 12: 191–214.

Guitton, D. and Mandl, G. (1978) Frontal 'oculomotor' area in alert cat. II. Unit discharges associated with eye movements and neck muscle activity. *Brain Res.*, 149: 313–327.

Haley, J.E., Wilcox, G.L. and Chapman P.F. (1992) The role of nitric oxide in hippocampal long-term potentiation. *Neuron*, 8: 211–216.

Hall, W.C., Fitzpatrick, D., Klatt, L.L. and Raczkowski, D. (1989) Cholinergic innervation of the superior colliculus in the cat. *J. Comp. Neurol.*, 287: 495–514.

Harting, J.K. and Lieshout, D.P.V. (1991) Spatial relationship of axons arising from the substantia nigra, spinal trigeminal nucleus and pedunculopontine tegmental nucleus within the intermediate gray of the cat superior colliculus. *J. Comp. Neurol.*, 305: 543–558.

Heizmann, C.W. and Braun, K. (1992) Changes in Ca^{2+}-binding proteins in human neurodegenerative disorders. *TINS*, 15: 259–264.

Heizmann, C.W. and Hunziker, W. (1990) Intracellular calcium-binding molecules. In: F. Bronner (Ed.), *Intracellular Calcium Regulation*, (1992) Alan R. Liss, New York, pp. 211–248.

Hendry, S.H.C., Hockfield, S., Jones, E.G. and McKay, R. (1984) Monoclonal antibody that identifies subsets of neurones in the central visual system of monkey and cat. *Nature*, 307: 267–269.

Hillman, D., Chen, S., Aung, T.T., Cherksey, B., Sugimori, M. and Llinás, R.R. (1991) Localization of P-type calcium channels in the central nervous system. *Proc. Natl. Acad. Sci. USA*, 88: 7076–7080.

Hikosaka, O. and Wurtz, R.H. (1983a) Visual and oculomotor functions of monkey substanitia nigra pars reticulta. I. Relation of visual and auditory responses to saccades. *J. Neurophysiol.*, 49: 1230–1253.

Hikosaka, O. and Wurtz, R.H. (1983b) Visual and oculomotor functions of monkey substanitia nigra pars reticulta. II. Visual responses related to fixation of gaze. *J. Neurophysiol.*, 49: 1254–1267.

Hikosaka, O. and Wurtz, R.H. (1983c) Visual and oculomotor functions of monkey substanitia nigra pars reticulta. IV. Relation of substantia nigra to superior colliculus. *J. Neurophysiol.*, 49: 1285–1301.

Hikosaka, O. and Wurtz, R.H. (1983d) Effects on eye movements of a GABA agonist and antagonist injected into monkey superior colliculus. *Brain Res.*, 272: 368–372.

Hikosaka, O. and Wurtz, R.H. (1985a) Modification of saccadic eye movements by GABA-related substances. I. Effect of muscimol and bicuculline in monkey superior colliculus. *J. Neurophysiol.*, 53: 266–291.

Hikosaka, O. and Wurtz, R.H. (1985b) Modification of saccadic eye movements by GABA-related substances. II. Effects of muscimol in monkey substantia nigra pars reticulata. *J. Neurophysiol.*, 53: 292–308.

Hoffmann, K.P. (1973) Conduction velocity in pathways from retina to superior colliculus in the cat: a correlation with receptive-field properties. *J. Neurophysiol.*, 36: 409–424.

Hoydt, K.R., Tang, L.-H., Aizenman, E. and Reynolds, I.J. (1992) Nitric oxide modulates NMDA-induced increases in intracellular Ca^{2+} in cultured rat forebrain neurons. *Brain Res.*, 592: 310–316.

Huerta, M.F. and Harting, J.K. (1982) Tectal control of spinal cord activity: neuroanatomical demonstration of pathways connecting the superior colliculus with the cervical spinal cord grey. *Prog. Brain Res.*, 51: 293–328.

Huerta, M.F. and Harting, J.K. (1984) The mammalian superior colliculus: studies of its morphology and connections. In: H. Vanegas (Ed.), *Comparative Neurology of the Optic Tectum*. Plenum Press, New York, pp. 687–773.

Huerta, M.F., Frankfurter, A.J. and Harting, J.K. (1981) The trigeminocolliculuar projection in the cat: patch-like endings within the intermediate gray. *Brain Res.*, 211: 1–13.

Illing, R.B. and Graybiel, A.M. (1985) Convergence of afferents from frontal cortex and substantia nigra onto acetylcholinesterase-rich patches of the cat's superior colliculus. *Neuroscience*, 14: 455–4825.

Irvin, G.E., Norton, T.T., Sesma, M.A. and Casagrande, V.A. (1986) W-like response properties of interlaminar zone cells in the lateral geniculate nucleus of a primate (*Galago crassicaudatus*). *Brain Res.*, 362: 254–270.

Jeon, C.-J. and Mize R.R. (1993) Choline acetyltransferase immunoreactive patches overlap specific efferent cell groups in the cat superior colliculus. *J. Comp. Neurol.*, 337: 127–150.

Jeon, C.-J., Spencer, R.F. and Mize, R.R. (1993) Organization and synaptic connections of cholinergic fibers in the cat superior colliculus. *J. Comp. Neurol.*, 333: 360–374.

Jeon, C.-J., Gurski, M.R., Hartman, M.J. and Mize, R.R. (1996) Evidence that glutamate-like immunoreactivity in the cat superior colliculus is present in bothneurons and in the corticocollicular pathway. *Visual Neurosci.*, in revision.

Kawaguchi, Y. (1993) Groupings of nonpyramidal and pyramidal cells with specific physiological and morphological characteristics in rat frontal cortex. *J. Neurophysiol.*, 69: 416–431.

Kawaguchi, Y. and Kubota, Y. (1993) Correlation of physiological subgroupings of nonpyramidal cells with parvalbumin- and CalbindinD28k-Immunoreactive neurons in layer V of rat frontal cortex. *J. Neurophysiol.*, 70: 387–396.

Kawaguchi, Y., Katsumaru, H., Kosaka, T., Heizmann, C.W. and Hama, K. (1987) Fast spiking cells in rat hippocampus

(CA1 region) contain the calcium binding protein parvalbumin. *Brain Res.*, 416: 369–374.

Köhr, G., Lambert, C.E. and Mody, I. (1991) Calbindin-D28k (CaBP) levels and calcium currents in acutely dissociated epileptic neurons. *Exp. Brain Res.*, 85: 543–551.

Leichnetz, G.R., Spencer, R.F., Hardy, S.G.P. and Astruc, J. (1981) The prefrontal corticotectal projection in the monkey: An anterograde and retrograde horseradish peroxidase study. *Neuroscience*, 65: 1023–1041.

Leichnetz, G.R., Gonzalo-Ruis, A., DeSalles, A.A.F. and Hayes, R.L. (1987) The frontal eye field and prefrontal cortex project to the paramedian pontine reticular formation in the cat. *Brain Res.*, 416: 195–199.

Lo, F.-S., Cork, R.J. and Mize, R.R. (1995) A high-frequency burst firing mode in neurons of the rat superior colliculus. *Neuroscience*, 21: 655.

Ma, T.P., Graybiel, A.M. and Wurtz, R.H. (1991) Location of saccade-related neurons in the macaque superior colliculus. *Exp. Brain Res.*, 85: 21–35.

Mattson, M.P., Rychlik, B., Chu, C. and Christakos, S. (1991) Evidence for Calcium-reducing and excito-protective roles for the calcium-binding protein calbindin-D28k in cultured hippocampal neurons. *Neuron*, 6: 41–51.

McHaffie, J.G., Ogasawara, K. and Stein, B.E. (1986) Trigeminotectal and other trigeminofugal projections in neonatal kittens: An anatomical demonstration with horseradish peroxidase and tritiated leucine. *J. Comp. Neurol.*, 249: 411–427.

McIlwain, J.T. and Fields, H.L. (1971) Interactions of cortical and retinal projections on single neurons of the cat's superior colliculus. *Journal of Neurophysiology*, 34: 763–772.

Mize, R.R. (1983a) Patterns of convergence and divergence of retinal and cortical synaptic terminals in the cat superior colliculus. *Exp. Brain Res.*, 51: 88–96.

Mize, R.R. (1983b) Variations in the retinal synapses of the cat superior colliculus revealed using quantitative electron microscope autoradiography. *Brain Res.*, 269: 211–221.

Mize, R.R. (1988) Immunocytochemical localization of gamma-aminobutyric acid (GABA) in the cat superior colliculus. *J. Comp. Neurol.*, 276: 169–187.

Mize, R.R. (1989) Enkephalin-like immunoreactivity in the cat superior colliculus: distribution, ultrastructure, and colocalization with GABA. *J. Comp. Neurol.*, 285: 133–155.

Mize, R.R. (1992) The organization of GABAergic neurons in the mammalian superior colliculus. *Prog. Brain Res.*, 90: 219–248.

Mize, R. (1994) Conservation of basic synaptic circuits that mediate GABA inhibition in the subcortical visual system. *Prog. Brain Res.*, 100: 123–132.

Mize, R.R. and Hockfield, S. (1989) Cat-301 antibody selectively labels neurons in the Y-innervated laminae of the cat superior colliculus. *Vis. Neurosci.*, 3: 433–443.

Mize, R.R. and Luo, Q. (1992) Visual deprivation fails to reduce calbindin 28kD or GABA immunoreactivity in the rhesus monkey superior colliculus. *Vis. Neurosci.*, 9: 157–168.

Mize, R.R., Jeon, C.-J., Butler, G.D., Luo, Q. and Emson, P.C. (1991) The calcium binding protein calbindin-D28K reveals subpopulations of projection and interneurons in the cat superior colliculus. *J. Comp. Neurol.*, 307: 417–436.

Mize, R.R., Luo, Q., Butler, G., Jeon, C.-J. and Nabors, B. (1992a) The calcium binding proteins parvalbumin and calbindin-D-28K form complementary patterns in the cat superior colliculus. *J. Comp. Neurol.*, 320: 243–256.

Mize, R.R., Luo, Q. and Tigges, M. (1992b) Monocular enucleation reduces immunoreactivity to the calcium-binding protein calbindin 28 kD in the rhesus monkey lateral geniculate nucleus. *Vis. Neurosci.*, 9: 471–482.

Moncada, S., Palmer, R.M.J. and Higgs, E.A. (1991) Nitric oxide: physiology, pathophysiology, and pharmacology. *Pharmacol. Rev.*, 43: 109–1142.

Nunes-Cardozo, B., Mize, R.R. and van der Want, J.J. (1994) GABAergic and non-GABAergic projection neurons from the nucleus of the optic tract project to the superior colliculus: an ultrastructural retrograde tracer and immunocytochemical study in the rabbit. *J. Comp. Neurol.*, 350: 646–656.

Okada, Y. (1992) The distribution and function of gamma-aminobutyric acid (GABA) in the superior colliculus. *Prog. Brain Res.*, 90: 249–262.

Palmer, L.A. and Rosenquist, A.C. (1974) Visual receptive fields of single striate cortical units projecting to the superior colliculus in the cat. *Brain Res.*, 67: 27–42.

Pape, H.C. and Mager, R. (1992) Nitric oxide controls ocscillatory activity in thalamocortical neurons. *Neuron*, 9: 441–448.

Peck, C.K. (1984) Saccade-related neurons in cat superior colliculus: Pandirectional movement cells with postsaccadic responses. *J. Neurophysiol.*, 52: 1154–1168.

Redgrave, P., Odekunle, A. and Dean, P. (1986) Tectal cells of origin of predorsal bundle in the rat: location and segregation from ipsilateral descending pathway. *Exp. Brain Res.*, 63: 279–93.

Scheiner, C.A., Banfro, F.T., Arceneaux, R.D., Kratz, K.E. and Mize, R.R. (1995) Nitric oxide synthase is expressed early in the prenatal development of the cat superior colliculus. *Soc. Neurosci. Abstr.*, 21: 817.

Segraves, M.A. (1992) Activity of monkey frontal eye field neurons projecting to oculomotor regions of the pons. *J. Neurophysiol.*, 68: 1967–1985.

Segraves, M.A. and Goldberg, M.E. (1987) Functional properties of corticotectal neurons in the monkey frontal eye fields. *J. Neurophysiol.*, 58: 1387–1419.

Sherman, S.M. and Spear, P.D. (1982) Organization of visual pathways in normal and visually deprived cats. *Physiol. Rev.*, 63: 738–855.

Snyder, S.H. (1992) Nitric oxide: first in a new class of neurotransmitters? *Science*, 257: 494–496.

Sterling, P. (1971) Receptive fields and synaptic organization of the superficial gray layer of the cat superior colliculus. *Vision Res. (Suppl.)*, 3: 309–328.

Sur, M., Frost, D.O. and Hockfield, S. (1988) Expression of a surface-associated antigen on Y cells in the cat lateral geniculate nucleus is regulated by visual experience. *J. Neurosci.*, 8: 874–882.

Vincent, S.R. and Hope, B.T. (1992) Neurons that say NO. *TINS*, 15: 108–113.

Wallace, M.N. (1986) Spatial relationship of NADPH-diaphorase and acetylcholinesterase lattices in the rat and mouse superior colliculus. *Neuroscience*, 19: 381–391.

Wallace, M.N. (1988) Lattices of high histochemical activity occur in the human, monkey and cat superior colliculus. *Neuroscience*, 25: 569–583.

Weyand, T.G. and Gafka, A.C. (1996) Frontal eye fields of the awake, behaving cat: III. Corticotectal cells. in preparation.

M. Norita, T. Bando and B. Stein (Eds.)
Progress in Brain Research, Vol 112
© 1996 Elsevier Science BV. All rights reserved.

CHAPTER 4

Serotonin modulates retinotectal and corticotectal convergence in the superior colliculus

R.D. Mooney*, X. Huang, M.-Y. Shi, C.A. Bennett-Clarke, and R.W. Rhoades

Department of Anatomy and Neurobiology, Medical College of Ohio, Toledo OH 43699-0008, USA

A dense serotonin (5-HT)-containing projection to the superficial layers of the superior colliculus (SC) has been demonstrated in diverse mammalian species, but how 5-HT may affect visual signals within these laminae is largely unknown. This study undertook to investigate the distribution of 2 types of 5-HT receptors in the SC and to ascertain their physiological effects on transmission of visual signals to the SC from the retinotectal and corticotectual pathways.

Autoradiography of tissue sections exposed to $[^3H]$-8-OH-DPAT (8-hydroxy-dipropylaminotetraline) or to $[^{125}I]$cyanopindolol plus isoproterenol showed that 5-HT_{1A} and 5-HT_{1B} receptors, respectively, were present in the superficial SC layers. In unilaterally enucleated animals, binding of ligand to 5-HT_{1B} receptors was greatly reduced on the deafferented (contralateral) side, which is consistent with the possibility that these receptors are located on preterminal axons. Binding to 5-HT_{1A} receptors was unaltered by enucleation. In recordings of superficial layer neurons from SC slices, application of 5-HT during blockade of 5-HT_{1A} receptors with spiperone reduced the amplitude of EPSPs evoked by stimulation of the optic tract. The 5-HT concentration for a 50% reduction in EPSP amplitude was 6 μM. Under these conditions, there were no significant alterations in either membrane potential or input resistance concurrent with 5-HT mediated reduction in EPSPs. During extracellular in vivo recordings, 5-HT, applied by iontophoresis or micropressure or by endogenous release produced by electrical stimulation of the dorsal raphé nucleus, strongly suppressed visual activity in SC neurons. The effectiveness of 5-HT application was significantly stronger on responses evoked by electrical stimulation of the optic chiasm (an average response decrement of 92.2%) than on these evoked in the same neurons by stimulation of visual cortex (an average response reduction of 32.3%).

These results support the following conclusions. The 5-HT_{1B} receptors are located preferentially on optic axon terminals and exert presynaptic inhibition of retinotectal inputs. Secondly, 5-HT_{1A} receptors probably have a postsynaptic localization and may affect activity of SC neurons irrespective of the source of input. The combined effect of 5-HT at both subtypes would bias SC visual activity toward information received from the corticotectal pathway.

Introduction

The superior colliculus (SC) in hamsters and other rodents is involved in both geniculocortical and extrageniculocortical visual pathways via projections from its two superficial laminae (the *stratum griseum superficial*-SGS and *stratum opticum*-SO) to visual thalamic nuclei (for reviews see Huerta and Harting, 1984; Chalupa, 1984). The superficial layers also convey visual activity to the intermediate and deep SC layers (Mooney et al., 1992), from which arise efferent pathways important in directing orienting and other behavioral responses (for reviews see Sparks, 1986; Grantyn, 1988; Dean et al., 1989). Neurons in the superficial SC layers receive convergent visual signals from the retinotectal and visual corticotectal pathways. Although these neurons are exclusively visual in their sensory processing, they also receive afferents from areas of the brain other than primary visual structures. Little is known about how these

*Corresponding author. Tel.: 419-381-4203; fax: 419-381-4008.

non-visual inputs affect processing of visual information from the retina and cortex.

One such pathway, the serotonergic projection from the brainstem, was first described by Fuxe (1965). Recent immunocytochemical studies in a variety of species have demonstrated a dense innervation of the superficial gray layer of the SC by axons that contain serotonin (5-HT) and originate mainly from the dorsal raphè nucleus (NRD) (e.g. Ueda et al., 1985; Morrison and Foote, 1986; Harvey and MacDonald, 1987; Villar et al., 1988; Mize and Horner, 1989; Rhoades et al., 1990). The serotonergic axons have numerous swellings but make very few conventional synapses in the SC (Arce et al., 1992). Nevertheless, it has been demonstrated in an in vitro preparation that electrical stimulation of NRD causes a Ca^{2+}-dependent 5-HT release within the SC (Wichmann et al., 1989). An extrasynaptic mode of release of 5-HT would emphasize the importance of the distribution and types of 5-HT receptors available, rather than the location of the presynaptic fibers, in determining the effects of this indoleamine in the SC.

The studies reported here were undertaken to determine: upon which structures in the hamster SC were 5-HT receptors located; whether visual activity and responses to electrical stimulation of retinotectal and corticotectal afferents were altered by these receptors following exogenous application of 5-HT; and whether these effects were mimicked by stimulation of the NRD. The results indicated that 5-HT modulated visual activity in SC neurons by at least two actions: presynaptic 5-HT$_{1B}$ receptors selectively control retinotectal transmission, and postsynaptic 5-HT$_{1A}$ receptors alter the overall response strength of many SC neurons.

Methods

Autoradiographic experiments

Four adult hamsters were anesthetized and enucleated as described by Rhoades et al., (1990). These and four normal adult hamsters were killed by rapid decapitation. Brains were removed and blocked in the coronal plane and quickly frozen on dry ice. Frozen tissue was cut into 15-μm sections in the coronal plane. Sections were thaw mounted onto 0.5% gel coated slides and stored at $-20°C$ until used. Autoradiographic assay for 5-HT$_{1B}$ receptors followed the protocol of Manaker and Verdermane (1990). Frozen sections were warmed to room temperature and then immersed in ice-cold buffer (50 mM Tris–HCl, 2.5 mM MgCl$_2$, 10 μM pargyline, 30 μM isoproterenol; pH 7.4) for 10 min. They were then incubated in the same buffer containing 100 pM of 5-HT$_{1B}$ ligand [^{125}I]cyanopindolol at room temperature for 1 h. Preliminary studies verified that 30 μM isoproterenol effectively blocked all [^{125}I]cyanopindolol binding to β-adrenergic receptors. This was demonstrated by adding both 30 μM isoproterenol and 20 μM 5-HT to the incubation solution. The assay for 5-HT$_{1A}$ receptors differed only in that the radioligand employed was [^{3}H]-8-OH-DPAT (8-hydroxy-dipropylaminotetraline) and that isoproterenol was omitted from the buffer solution. As a control, specific 5-HT binding was blocked by the addition of unlabeled 20 μM 5-HT. After incubation, sections were washed in cold buffer, dipped once in cold distilled water, and rapidly dried with a cool air stream. Dried slides were then apposed to LKB Ultrofilm for 48–72 h ([^{125}I]cyanopindolol) or 3 weeks ([^{3}H]-8-OH-DPAT). The exposed film was developed in D-19 for 5 min.

Electrophysiological experiments

The in vitro SC-optic tract preparation

Adult hamsters ($N = 85$) were anesthetized with ether and decapitated. The brain was removed and an oblique slice that included the optic chiasm, diencephalon, and midbrain was dissected and immersed in a modified Krebs solution saturated with 95% O$_2$ and 5% CO$_2$ (5°C). The solution consisted of 127 mM NaCl, 2.0 mM KCl, 1.2 MM KH$_2$PO$_4$, 2.4 mM CaCl$_2$, 1.3 mM MgCl$_2$, 26mM NaHCO$_3$, and 10 mM glucose. For recording, slices were transferred to a small chamber and superfused with the Krebs solution, which was saturated with 95% O_2 and 5% CO_2 and maintained at 21°C.

Intracellular recordings were made with micropipettes filled with 3 M potassium acetate (90–160 MΩ resistance). Electrical activity was recorded with an Axoclamp 2A amplifier in the bridge mode and displayed on a storage oscilloscope and on a Gould chart recorder (bandwidth 0–50 Hz). Neurons were penetrated with brief oscillations via the capacitance compensation circuit of the amplifier. Transynaptic excitation of SC cells was achieved by placing bipolar electrodes in the optic tract just proximal to the optic chiasm and delivering 5–500-μA current pulses.

The neuronal properties tested during 5-HT exposure were: input resistance (obtained from the slope of the regression line over the linear range of current-voltage [I–V] curves), membrane potential and amplitude of synaptic responses to stimulation of the optic tract. In order to isolate the activity of the 5-HT$_{1B}$ receptor, spipirone was added to the perfusate to block 5-HT$_{1A}$ and 5-HT$_2$ receptors (Hamon et al., 1986; Middlemiss and Hutson, 1990; Peroutka et al., 1990).

The in vivo preparation

Adult hamsters ($N = 50$) were anesthetized with sodium pentobarbital (60 mg/kg delivered intraperitoneally along with 0.15 ml of a 1.5% solution of atropine sulphate) and prepared for recording as described by Mooney et al. (1992). All wound edges were infiltrated with a long-lasting local anesthetic (bupivicaine) and anesthesia with urethane (initial intraperitoneal dose = 0.3 g/kg) was initiated. Paralysis was induced with gallamine triethiodide (40 mg/kg, i.p.) and artificial respiration was initiated. Anesthesia and paralysis were maintained with hourly intraperitoneal injections of urethane (0.3 g/kg) and gallamine triethiodide (30 mg/kg), respectively. Heart rate was monitored continuously and rectal temperature was maintained at 37°C with a thermostatically controlling heating pad.

Superior colliculus cells were recorded using multiple barrel micropipettes pulled from thin wall borosilicate capillaries (WP Instruments) to which was glued a recording electrode. Action potentials were recorded with Eutectics 400A electrometer (unity gain, bandpass DC-3 kHz), displayed on a storage oscilloscope and converted to standard pulses for storage by a computer connected to a Cambridge Electronic Designs (CED) 1401 interface.

Serotonin was ejected by iontophoresis from the multibarreled electrodes using a Medical Systems Inc. Neurophore BH-2. One barrel, used for current balancing, was filled with 2 M NaCl. The remaining barrels were filled with 5-HT (5-hyrdoxytryptamine creatine sulfate, Sigma Chemical Co., concentration range 0.5–40 mM, typically 10 or 20 mM, pH = 4.0, ejection currents 20–200 nA). Holding currents from -2 to -5 nA were typically used. In other experiments, 5-HT was applied by micropressure ejections of 1–100 μM 5-HT at pH = 7.6 and in some cases the 5-HT$_{1A}$ $_{\text{and } 2}$ antagonist spiperone (Research Biochemicals Inc) was ejected at concentrations of 1–5 μM from a different barrel.

Electrical stimulation was delivered with concentric bipolar stimulating electrodes positioned in the optic chiasm and in the primary visual cortex. These electrodes were used to present 5–500-μA stimulus pulses (75 μs duration, isolated from ground). Visual stimulation was presented on a screen (10 Cd/m^2) upon which the location of the optic disk was back projected, and receptive field characteristics were initially assessed qualitatively with hand-held stimuli. To produce a uniform series of visual responses, the most effective stimulus was then presented in the receptive field with a Picasso visual stimulator (Innisfree Inc.) driven by a Wavetek signal generator or a CED 1708. A series of responses to a repeated stimulus pattern or to electrical stimulation was then collected during a control period, after 5-HT application, and during recovery. The responses were displayed offline as rasters, summed into poststimulus time histograms (PSTHs) and counted.

Stimulation of the NRD

Twenty-three hamsters were prepared for testing with visual stimulation, as just described, and a concentric bipolar electrode was inserted into the dorsal raphè nucleus using a rostrally angled approach through the cerebellum. Single or trains of electrical pulses (5–500 μA) were presented after control responses had been collected. At the end of an experiment, an electrolytic lesion was made to mark the position of the stimulation electrode. The animals were killed with a lethal dose of sodium pentobarbital and perfused through the heart with normal saline followed by 4% paraformaldehyde in 0.2 M phosphate buffer. A block containing the mid-brain and rostral pons was sectioned at 50 μm. The sections were reacted for the presence of 5-HT as described by Rhoades et al. (1990); plated on slides; and examined with a microscope to confirm the position of the marking lesions.

Results

Effects of stimulation of the dorsal raphè nuclei (NRD)

The function of the projection from the NRD to SC was tested by placement of stimulating electrodes in the NRD and recording of visual responses of SC neurons. Fig. 1A shows rasterized data obtained from a SC neuron: a vertical series of dots represents the response to a single presentation of the visual stimulus, and the horizontal axis shows the time course of the entire experiment. Data collection was suspended during the period of electrical stimulation and resumed just after its cessation. The effects shown here were typical: after a delay of about 1 min, the visual response markedly diminished and showed alternating periods of suppression and partial restoration before final recovery. In some neurons, the alternation of suppression and partial recovery was observed for 3 or 4 cycles. Fig. 1B shows PSTHs, which were summed over the period indicated in the rasters and were used to quantify the

effects. Of 29 cells tested, 21 (73%) showed a visual response decrement $\geq 30\%$ of the spike count in the control PSTH. The average ($\pm$ SD) reduction in response amplitude of this group was 74.2 $\pm$ 37.5%.

The positions of the marking lesions made by the stimulation electrodes (Fig. 1C) indicated that a reduction in visual response amplitude from the SC neurons tested was obtained by placements (filled circles) chiefly within the serotonergic cell groups in the lateral portions of the NRD. The exceptions to this generalization included a placement in the median raphè nucleus and two localized just lateral to the 5-HT containing cells of NRD but in the vicinity of their dorsally projecting axons. Positions of electrodes for which stimulation was ineffective (open circles) were either on the midline in NRD or missed NRD entirely.

Distribution of 5-HT$_{1A}$ and 5-HT$_{1B}$ receptors within superficial SC and effects of enucleation

The 5-HT$_{1A}$ receptors labelled with [^{3}H]-8-OH-DPAT and the 5-HT$_{1B}$ receptors labelled by [^{125}I] cyanopindolol, in the presence of isoproterenol, were densely distributed in the SGS of the SC contralateral to the remaining eye (right side Figs. 2A,C). The density of these receptors was qualitatively much lower in the *stratum opticum* (SO) and in the deep SC laminae. This pattern of label was similar to that in the four intact animals. In the presence of unlabelled 5-HT, little or no binding of the radioligands was present in the superficial SC laminae (Figs. 2B,D).

Enucleation resulted in a marked reduction in the density of the 5-HT$_{1B}$ receptors in the SGS of the contralateral SC and produced no obvious changes in the other layers (Fig. 2C left side). The density of 5-HT$_{1A}$ receptors in the SGS was unaltered by enucleation (Fig. 2A left side). These results were observed in all four animals in which enucleations were carried out. This observation is consistent with the conclusion that 5-HT$_{1B}$ receptors, but not 5-HT$_{1A}$ receptors, are located on the terminals of at least some retinotectal axons.

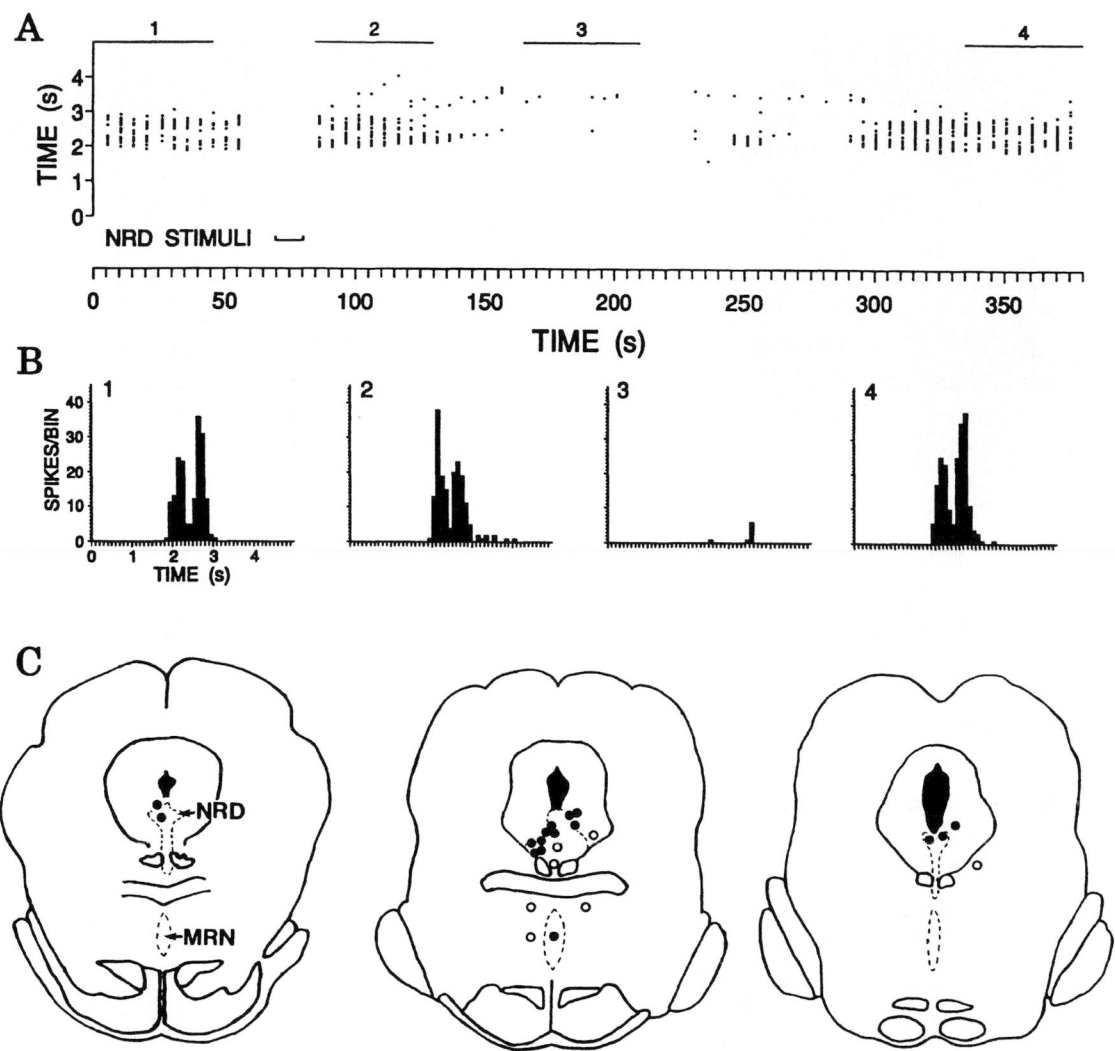

Fig. 1. Suppression of visual activity in the SC by electrical stimulation of NRD. A shows vertical rasters of visual responses; 0 on the ordinate marks movement onset of each successive stimulus presentation on display screen; time on abscissa indicates continuous time throughout the trial. Electrical stimulation of NRD (250 μA; 100-μs pulses at 10 Hz for 20 s) occurred as indicated by bracket; data collection was suspended during this period. B shows PSTHs summed over nine stimulus presentations as denoted by correspondingly numbered lines above the rasters. C shows positions of stimulating electrodes, reconstructed from marking lesions: filled circles are stimulation sites that suppressed visual responses; open circles are ineffective stimulation sites. Abbreviations: NRD, dorsal raphè nucleus; MRN, median raphè nucleus, dashed lines indicate nuclear boundaries of 5-HT immunoreactive perikarya.

Effects of 5-HT on membrane properties of superficial SC neurons and on retinotectal transmission

If a considerable number of serotonin receptors are located presynaptically on optic axons, this fact should be reflected in the function of those receptors. We have examined this issue by recording from SC neurons in vitro in an explant preparation that retained the optic tract from the chi-

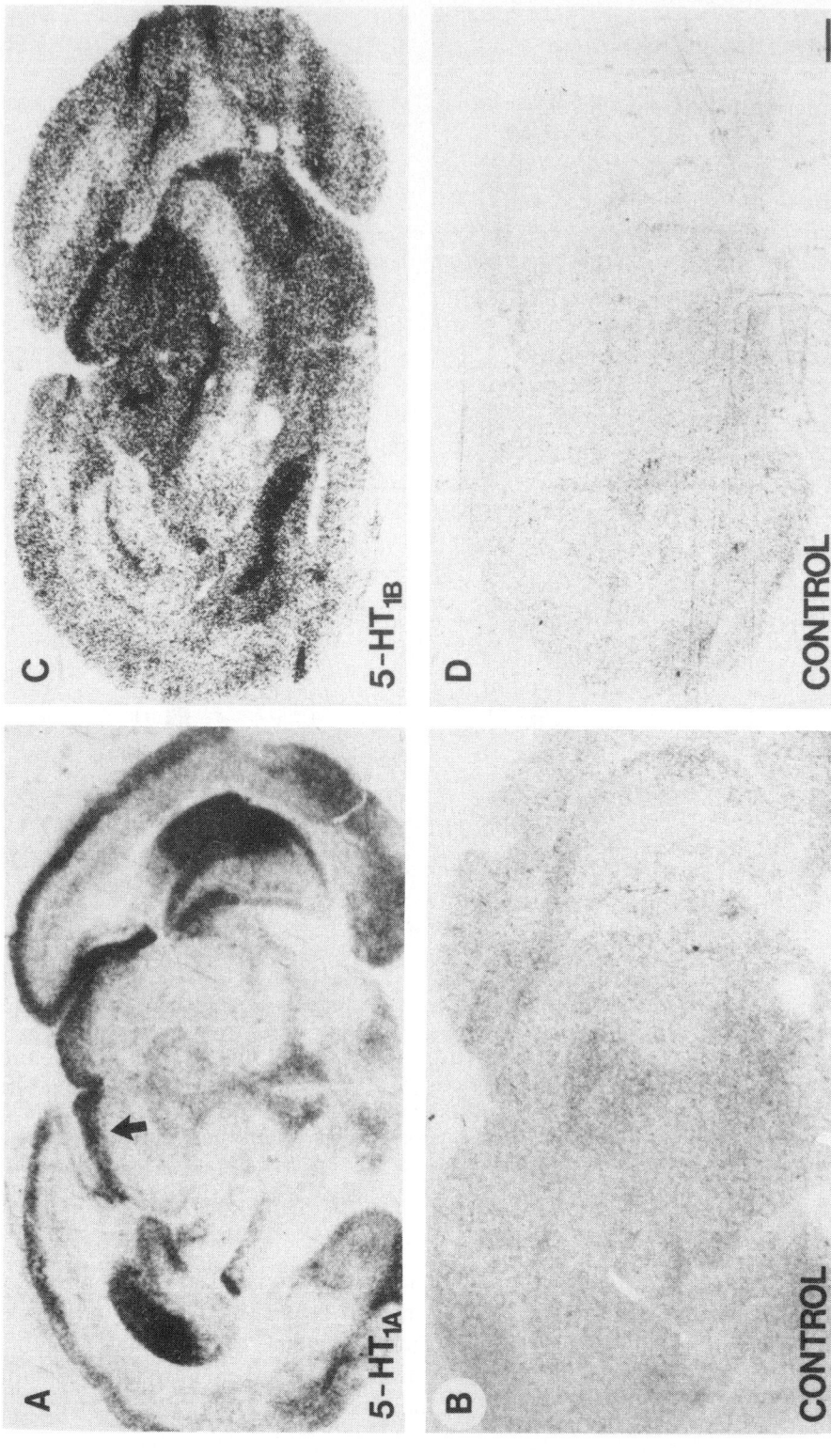

Fig. 2. Autoradiographs of radioligand binding to 5-HT$_{1A}$ and 5-HT$_{1B}$ receptors in the superficial SC layers of monocularly enucleated hamsters. A shows binding of 5-HT$_{1A}$ ligand [^{3}H]-8-OH-DPAT, and B shows low levels of non-specific binding of this agent in the present of unlabelled 5-HT. C shows binding of 5-HT$_{1B}$ ligand [^{125}I] cyanopindolol in the presence of β-adrenergic agonist isoproterenol, and D shows residual binding when unlabelled 5-HT was added to this combination. The left SC (arrows in A and C) was deafferented by enucleation of the contralateral eye. Note that enucleation had no effect on the distribution of 5-HT$_{1A}$ binding in the superficial SC, but it reduced binding of the 5-HT$_{1B}$ ligand to background levels.

asm along with the lateral thalamus, pretectum and SC. Stimulation of the optic tract evoked an EPSP in superficial layer SC neurons.

The effect of 5-HT on SC neurons is typified in the example shown in Figs. 3A–C. The traces shown in A are EPSPs evoked by 4–5 consecutive electrical pulses delivered by a stimulation electrode in the optic tract. Serotonin suppressed the evoked EPSP and produced no observable alteration in the input resistance of the neuron, as

shown in the current-voltage plot in Fig. 3B, nor any change in resting potential of the cell (Fig. 3C). In this instance, spiperone was continuously present in the perfusing solution in order to block any effects of 5-HT_{1A} or 5-HT_2 receptors that might occur in addition to effects mediated by the 5-HT_{1B} receptors.

Similar data for all cells tested, with or without spiperone, are shown in Figs. 3D–F for EPSP amplitude, input resistance, and membrane po-

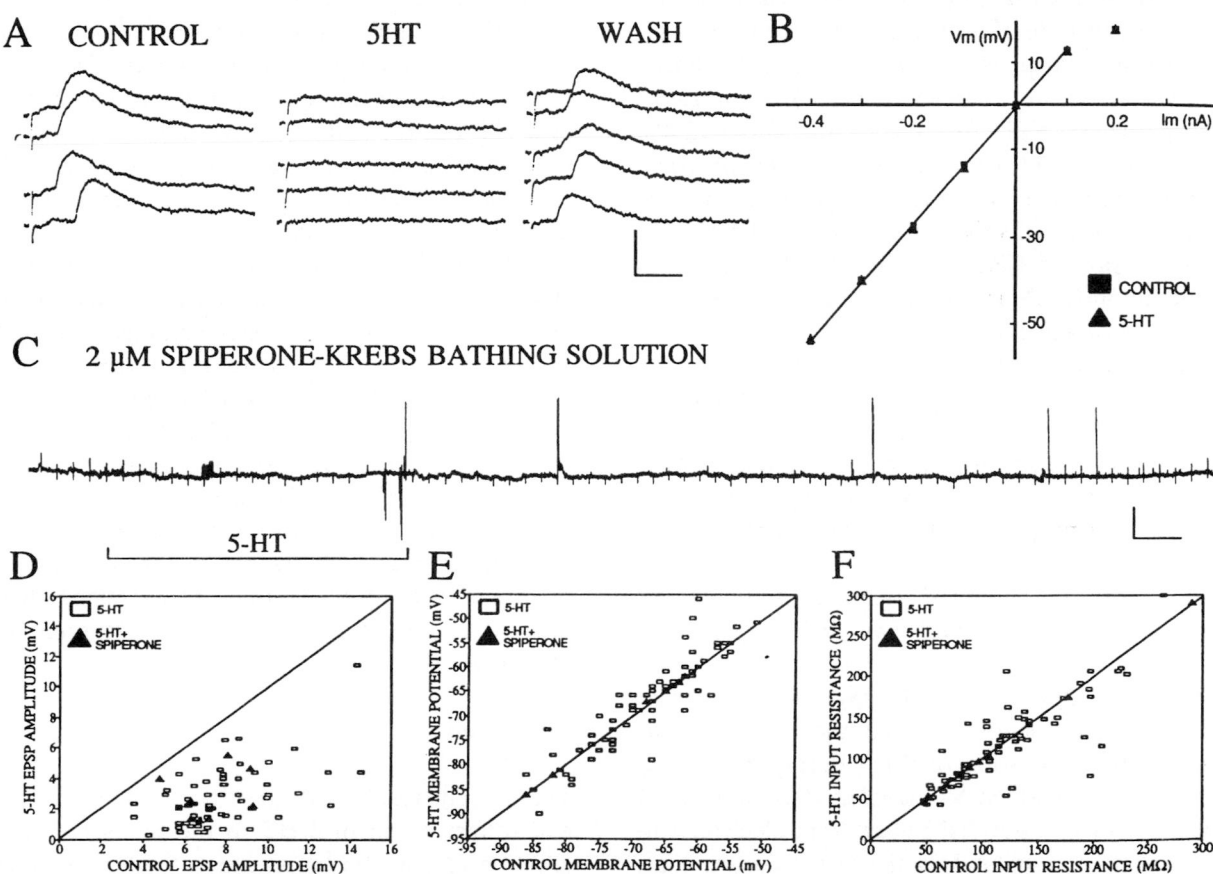

Fig. 3. Effects of 5-HT on superficial layer SC neurons in vitro. A shows vertically offset records of four or five successive EPSPs, evoked by electrical stimulation of the optic tract, and C shows the continuous recording of this neuron. Spiperone (2 μM, an antogonist of 5-HT_{1A} and 5-HT_2 receptors) was present in the perfusion solution throughout this trial, and the concentration of 5-HT was 20 μM. B shows the current-voltage plot for this neuron; its input resistance was 134 MΩ and was unaffected by exposure to 5-HT. Scatterplots D, E and F compare values during control and 5-HT application for EPSP amplitude, membrane potential, and input resistance, respectively. Filled triangles indicate tests in which spiperone was combined with 5-HT. The diagonal line in each of these plots indicates loci of pionts for which 5-HT produced no change in a given parameter. Calibrations are 10 mV by 20 ms in A and 20 mV by 1 min in C.

tential. These graphs are scatterplots with control values on the abscissa and values during 5-HT exposure shown on the ordinate. Points on the diagonal line represent no change in a parameter, and points below this line represent a decrease in its value.

For 5-HT application without spiperone (open squares), values for input resistance (Fig. 3E) and membrane potential (Fig. 3F) cluster along the diagonal with much variability. Cases in which spiperone was also applied (filled triangles) showed no change in these measures. The average differences in values of input resistance and membrane potential during 5-HT exposure compared to control values were small and statistically insignificant. The average reduction in EPSP amplitude by 5-HT for all cells was $50.2 \pm 21.4\%$ ($P < 0.001$) and was unaffected by the presence or absence of the antagonist spiperone. On average, the 50%-effective concentration of 5-HT for suppression of EPSPs was 6 μM. The large reduction in EPSP amplitude in most cases and the absence of any consistent change in passive membrane properties support the conclusion that 5-HT has a strong presynaptic effect on retinotectal transmission. This effect is mediated primarily by the $5HT_{1B}$ receptor subtype, but other, spiperone-sensitive, 5-HT receptor subtypes may alter membrane properties of some SC neurons.

Effects of 5-HT on visual responses and responses evoked by electrical stimulation of afferent pathways

The possibility that 5-HT may modulate visual activity by an action on optic tract terminals was tested by application of 5-HT during in vivo experiments, in which SC neurons were stimulated visually or by electrical pulses from electrodes in the optic chiasm and in the visual cortex. Figs. 4A,B show rasters and PSTHs of visual responses of a superficial layer neuron that responded when the moving stimulus entered and left the receptive field. The responses were suppressed by an ejection of 5-HT, which was delivered by pressure from the multibarrelled pipette. The same result was obtained when 5-HT was concurrently ap-

plied with spiperone, which had no obvious effect by itself.

The effects of 5-HT on all neurons tested during visual stimulation are shown in Fig. 4E. Of 72 neurons tested, the responses of 58 (80.6%) were suppressed by $\geq 30\%$, and the average suppression was $62.6 \pm 25.7\%$ from control levels. All 11 cells tested with 5-HT during spiperone application were suppressed by $\geq 30\%$, and the overall average suppression was $85.5 \pm 15.5\%$.

Figs. 4C and D show the effects of iontophoretically applied 5-HT on responses evoked by stimulation of the optic chiasm or visual cortex (periods of cortical stimulation are denoted by horizontal lines below the rasters). The responses of this cell to stimulation of the optic chiasm were abolished by 5-HT, but the responses to cortical stimulation were unaffected. The effects of 5-HT on electrically evoked responses of all cells sampled are plotted in Figs. 4F and G. Of the 77 neurons tested with 5-HT during stimulation of the optic chiasm, 71 (92.2%) showed a suppressive effect $\geq 30\%$; the average response decrement was $76.4 \pm 24.6\%$. Of the 28 neurons tested during cortical stimulation, 12 (42%) were suppressed by as much as 30%, but the remainder showed little or no effect and the average response suppression was $32.3 \pm 36.4\%$. In the sample of 28 neurons tested under both stimulation conditions, the suppression of responses evoked by stimulation of the optic chiasm was significantly greater than that for responses evoked by cortical stimulation ($P < 0.001$).

Discussion

The present results indicate that the most typical and robust effect of 5-HT in the SC is the suppression of synaptic transmission from optic tract axons to their target neurons. Serotonin produced a substantial reduction in the visual response of most neurons, and in over half of those tested, a relative enhancement of the effectiveness of the corticotectal pathway over the retinotectal projection. The 5-HT_{1B} receptors are likely mediators of this effect since this subtype disappeared after

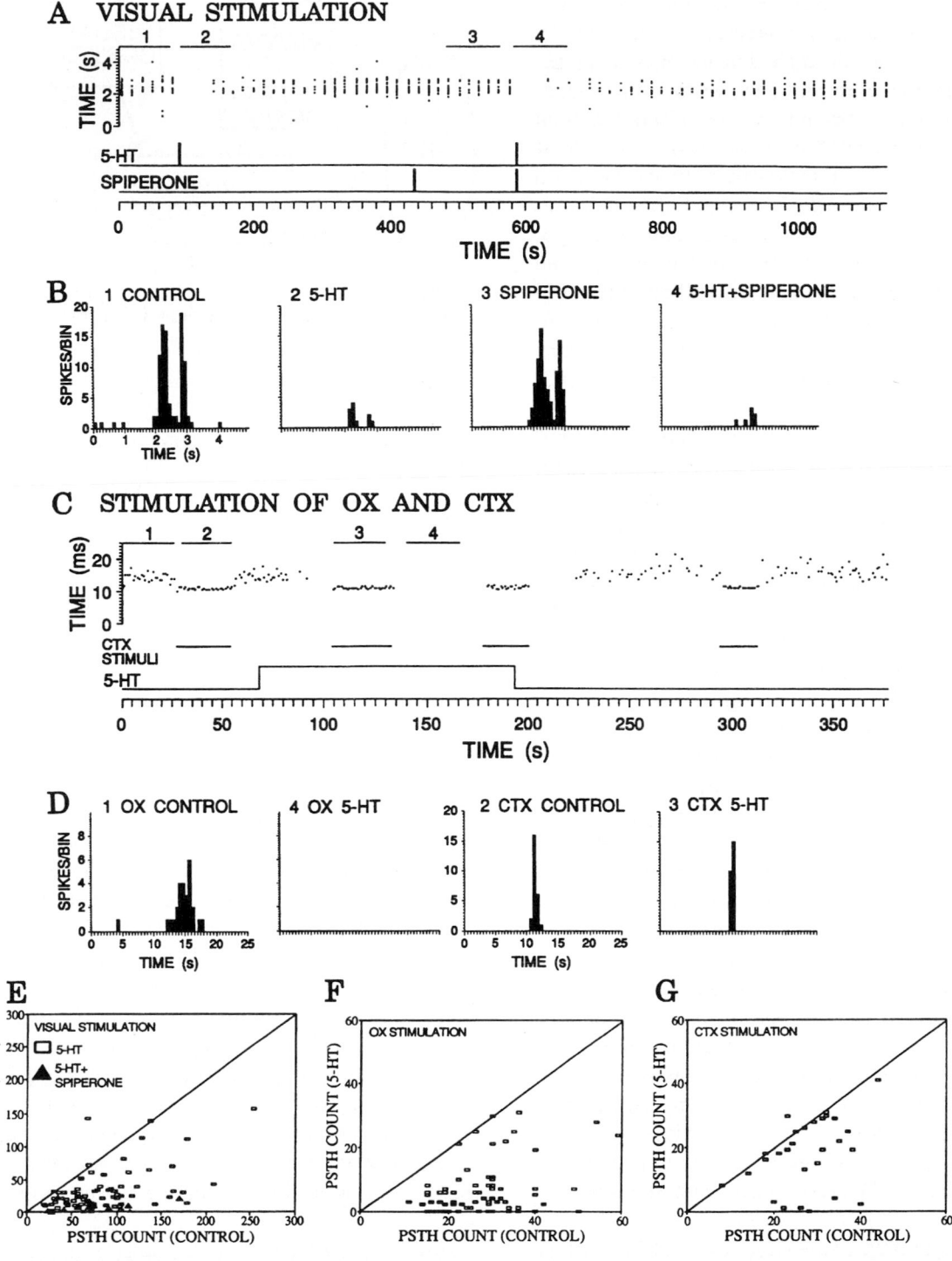

Fig. 4. Effects of 5-HT on responses of SC neurons to visual stimulation and electrical stimulation of afferent visual pathways. A–D follow the conventions in Figs. 1A and B, and E–G follow those of Figs. 3D–F. A and B show the suppressive effect of 5-HT, alone or combined with spiperone, on visual responses (timing of delivery of agents indicated by traces beneath rasters). C and D, taken from a second neuron, show that 5-HT failed to suppress responses that were evoked by electrical stimulation of visual cortex (lines

Continued overleaf

enucleation and since spiperone failed to antagonize suppression of retinotectal transmission by 5-HT. The relatively high density of 5-HT$_{1A}$ receptors in the SGS and the existence of 5-HT$_2$ receptors there (Bennett-Clark, Rhoades and Mooney, unpublished observations) suggest that 5-HT may have actions other than those on retinotectal terminals. These additional influences may account for changes in membrane potential and input resistance observed in some cases during 5-HT application without the addition of spiperone. The reduction in responses to corticotectal stimulation in 42% of the neurons tested presumably involves 5-HT$_{1A}$ or other subtypes of 5-HT receptors.

The localization of 5-HT$_{1B}$ receptors on retinotectal terminals in the hamster is in agreement with previous findings in rats and mice (Segu et al., 1986; Boulenguez et al., 1993; Boschert et al., 1994), and with comparable data for the 5-HT$_{1D}$ receptor in guinea pigs (Waeber and Palacios, 1990). The cellular locations in the SC of the 5-HT$_{1A}$ and 5-HT$_2$ receptors have not yet been identified, however, the relatively even distribution of 5-HT$_{1A}$ receptors throughout the dorsoventral extent of the SGS is incongruent with the localization of corticotectal terminals in the SO and more ventral SGS (Mustari and Lund, 1978; Rhoades and Chalupa, 1978; Rhoades et al., 1991; but see Ramirez et al., 1990 for alternative data). A model of 5-HT circuits in the SC is shown in Fig. 5: 5-HT$_{1B}$ receptors presynaptically inhibit transmission at retinotectal synapses, and the 5-HT$_{1A}$ receptors have been tentatively assigned to a postsynaptic location on membranes of SC neurons. The 5-HT$_{1A}$ receptors are inhibitory (Huang et al., 1993) and their mechanism of action is presumably similar to that observed in the NRD and in the hippocampus (Aghajanian and Lakoski, 1984; Andrade and Nicoll, 1987).

The suppression of visual activity after stimulation of the NRD is consistent with the suppres-

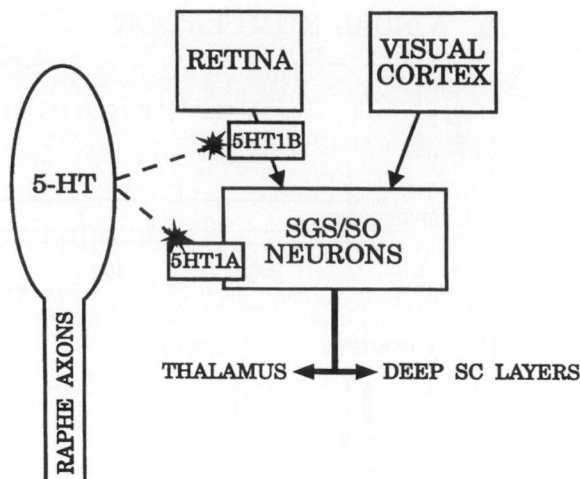

Fig. 5. A schematic diagram of sites of inhibitory actions of 5-HT in the superficial SC layers. The 5-HT$_{1B}$ receptors reduce transmission from retinal axons to SC neurons, and postsynaptic 5-HT$_{1A}$ receptors hyperpolarize some SC neurons and thereby reduce their responses to any excitatory afferent input.

sive effects of 5-HT and the previous findings (see Introduction for references) that the NRD is the main source of serotonergic afferents to the SC. It is certainly possible that substances other than 5-HT are released in the SC by stimulation of NRD, and confirmation of the role of 5-HT must be made in experiments which combine NRD stimulation with blockade of SC 5-HT receptors or depletion of 5-HT within the axon terminals. Analogous findings have been reported in which a reduction of noxious and low threshold somatosensory responses in diencephalic neurons was produced by stimulation of NRD and by application of 5-HT (Ishida and Kitano, 1977; Archer et al., 1986; Prieto-Gomez et al., 1989). The delay in suppression of visual activity after NRD stimulation and the often cyclic time course of recovery of visual responses were not observed when 5-HT was directly applied to SC neurons.

---

below rasters show stimulation periods) but abolished responses to stimulation of the optic chiasm (all times other than during cortical stimulation). E–G show scatterplots of data from all neurons tested with 5-HT during visual or electrical stimulation.

Thus, it may be that mechanisms intrinsic to the NRD or its axon terminals and/or the nature of extrasynaptic release and diffusion of 5-HT within the SC play some role in these phenomena.

Presynaptic effects of 5-HT in other portions of the nervous system

A number of studies have provided data indicating that 5-HT has strong presynaptic inhibitory effects in other structures. North et al. (1980) and Galligan et al. (1988) demonstrated that 5-HT inhibits fast nicotinic epsps in myenteric neurons, and Kilbinger and Pfeuffer-Friederich (1985) showed that 5-HT, acting at 5-HT_1 receptors, inhibited the electrically evoked release of acetylcholine from preparations of the myenteric plexus. In rat spinal cord dorsal horn, membrane potentials and input resistances of most neurons were unaffected by application of 5-HT, but the responses evoked by primary afferent stimulation were depressed by 5-HT in nearly all of these cells (Wu et al., 1991). Similar conclusions have been drawn for primary afferent synapses on frog dorsal horn neurons (Kuraishi et al., 1991; Tan and Miletic, 1992) and sensory pathways onto Rohon-Beard neurons in *Xenopus* (Sillar and Simmers, 1994). Bobker and Williams (1990) have reported that 5-HT presynaptically inhibits a slow synaptic response in guinea pig prepositus hypoglossi neurons. These investigators have also demonstrated that 5-HT can presynaptically inhibit both glutamate- and GABA-mediated synaptic potentials in the locus ceruleus (Bobker and Williams, 1989). The results of the present study are thus consistent with those from a number of previous reports indicating presynaptic inhibitory effects of 5-HT in other portions of the peripheral and central nervous systems.

Role of 5-HT in the functional organization of the superficial SC laminae

The major effect of 5-HT is to inhibit the retinotectal input to SC neurons. Such a selective influence might serve to enhance the relative efficacy of corticotectal or other inputs to SC neurons. Studies in several mammals, including the hamster, have shown that the corticotectal input influences specific response characteristics, and even the visual responsivity, of some superficial layer cells (e.g. Wickelgren and Sterling, 1969; Rosenquist and Palmer, 1971; Stein and Arigbede, 1972; Schiller et al., 1974; Berman and Cynader, 1975; Chalupa and Rhoades, 1977).

It has been previously suggested that brainstem arousal mechanisms, including the NRD and locus ceruleus, enhance sensory responses over background activity (i.e. signal to noise ratios) of visual cortical neurons (Livingstone and Hubel, 1981; de Lima and Singer, 1987; McCormick and Pape, 1990; see review by Steriade and Llinas, 1988). Under these conditions, activity in the NRD may enhance cortical output and simultaneously suppress retinal input to the SC, and thus, bias superficial layer SC neurons toward information received from the output of the visual cortex. In the non-aroused state, retinal input would dominate collicular activity for detection of transient visual signals (Dean and Redgrave, 1984). However, once the animal is aroused, and fully attentive to stimuli in its environment, the cortical input bias may allow the SC to participate more effectively in the elaboration of complex behaviors (Dean et al., 1989).

Acknowledgement

Supported in part by EY 08015, EY 04170, BNS 92-8802 and BNS 93-09597.

References

Aghajanian, G.K. and Lakoski, J.M. (1984) Hyperpolarization of serotonergic neurons by serotonin and LSD: Studies in brain slices showing increased $K^=$ conductance. *Brain Res.*, 305: 181–185.

Andrade, R. and Nicoll, R.A. (1987) Pharmacologically distinct actions of serotonin on single pyramidal neurons of the rat hippocampus recorded in vitro. *J. Physiol. Lond.*, 394: 99–124.

Arce, E.A., Bennett-Clarke, C.A., Mooney, R.D. and Rhoades, R.W. (1992) Synaptic organization of the serotoninergic input to the superficial gray layer of the hamster's superior colliculus. *Synapse*, 11: 67–75.

Archer, T., Johnson, G., Minor, B.G. and Post, C. (1986) Noradrenergic serotonergic interactions and nociception in the rat. *Eur. J. Pharmacol.*, 120: 295–307.

Berman, N. and Cynader, M. (1975) Receptive fields in cat superior colliculus after visual cortex lesions. *J. Physiol. Lond.*, 245: 261–270.

Bobker, D.H., and Williams, J.T. (1989) Serotonin agonists inhibit synaptic potentials in the rat locus ceruleus in vitro via 5-hydroxytryptamine$_{1A}$ and 5-hydroxytryptamine$_{1B}$ receptors. *J. Pharmacol. Exp. Ther.*, 250: 37–43.

Bobker, D.H. and Williams, J.T. (1990) Serotonin-mediated inhibitory postsynaptic potential in guinea-pig prepositus hypoglossi and feedback inhibition by serotonin. *J. Physiol. Lond.*, 422: 447–462.

Boschert, U., Amara, D.A., Segu, L. and Hen, R. (1994) The mouse 5-hydroxytryptamine$_{1B}$ receptor is localized predominantly on axon terminals. *Neuroscience*, 58: 167–182.

Boulenguez, P., Abdelkefi, J., Pinard, R., Christolomme, A. and Segu, L. (1993) Effects of retinal deafferentation on serotonin receptor types in the superficial grey layer of the superior colliculus of the rat. *J. Chem. Neuroanat.*, 6: 167–175.

Chalupa, L.M. (1984) Visual physiology of the mammalian superior colliculus. In: H. Vanegas (Ed.), *Comparative Neurology of the Optic Tectum*, Plenum, New York, pp. 775–815.

Chalupa, L.M. and Rhoades, R.W. (1977) Responses of visual, somatosensory and auditory neurons in the golden hamster's superior colliculus. *J. Physiol. Lond.*, 270: 595–626.

Dean, P. and Redgrave, P. (1984) The superior colliculus and visual neglect in rat and hamster. I. Behavioural evidence. *Brain Res. Rev.*, 8: 129–141.

Dean, P., Redgrave, P. and Westby, G.W.M. (1989) Event or emergency? Two response systems in the mammalian superior colliculus. *T.I.N.S.*, 12: 137–147.

de Lima, A.D. and Singer, W. (1987) The brainstem projection to the lateral geniculate nucleus in the cat: Identification of cholinergic and monoaminergic elements. *J. Comp. Neurol.*, 259: 92–121.

Galligan, J.J., Surprenant, A., Tonini, M. and North, R.A. (1988) Differential localization of 5-HT$_1$ receptors on myenteric and submucosal neurons. *Am. J. Physiol.*, 255: G603–G111.

Fuxe, K. (1965) The distribution of monamine terminals in the central nervous system. *Acta. Physiol. Scand.*, 64 supp., 247: 37–85.

Grantyn, R. (1988) Gaze control through superior colliculus: Structure and function. In: Büttner-Ennever (Ed.), *Neuroanatomy of the Oculomotor System*, Elsevier, Amsterdam, pp. 273–333.

Hamon, M., Cossery, J.-M., Spampinato, U. and Gozlan, H. (1986) Are there selective ligands for 5-HT$_{1A}$ and 5-HT$_{1B}$

receptor binding sites in brain? *Trends Pharmacol. Sci.*, 7: 336–338.

Harvey, A.R. and MacDonald, A.M. (1987) The host serotonin projection to tectal grafts in young rats: An immunohistochemical study. *Exp. Neurol.*, 95: 688–696.

Hoyer, D. and Middlemiss, D.N. (1989) Species differences in the pharmacology of terminal 5-HT autoreceptors in mammalian brain. *Trends Pharmacol. Sci.*, 10: 130–132.

Huang, X., Mooney, R,.D. and Rhoades, R.W. (1993) Effects of serotonin (5-HT) in the hamster's superior colliculus (SC) are mediated by 5-HT$_{1A}$ and 5-HT$_{1B}$ receptors. *Soc. Neurosci., Abstr.*, 19: 742.

Huerta, M.F. and Harting, J.K. (1984) The mammalian superior colliculus: Studies of its morphology and connections. In: H. Vanegas (Ed.), *Comparative Neurology of the Optic Tectum*. Plenum, New York, pp. 687–783.

Ishida, Y. and Kitano, K. (1977) Raphe induced inhibition of intralaminar thalamic unitary activities and its blockade by parachlorophenylanine in cats. *Naunyn Schmiedebergs Arch. Pharmacol.*, 301: 1–4.

Kilbinger, H. and Pfeuffer-Friederich, I. (1985) Two types of receptors for 5-hydroxytryptamine on the cholinergic nerves of the guinea pig myenteric plexus. *Br. J. Pharmac.*, 85: 529–539.

Kuraishi, Y., Minami, M. and Satoh, M. (1991) Serotonin, but neither noradrenaline nor GABA, inhibits capsaicin-evoked release of immunoreactive somatostatin from slices of rat spinal cord. *Neurosci. Res.*, 9: 238–245.

Livingstone, M.S. and Hubel, D.H. (1981) Effects of sleep and arousal on the processing of visual information in the cat. *Nature*, 291: 554–561.

Manaker, S. and Verderame, H.M. (1990) Organization of serotonin 1A and 1B receptors in the nucleus of the solitary tract. *J. Comp. Neurol.*, 301: 535–553.

McCormick, D.A. and Pape, H.-C. (1990) Noradrenergic and serotonergic modulation of a hyperpolarization-activated cation current in thalamic relay neurons. *J. Physiol. Lond.*, 431: 319–342.

Middlemiss, D.N. and Hutson, P.H. (1990) The 5-HT$_{1B}$ receptors. *Ann. N.Y. Acad. Sci.*, 600: 132–148.

Mize, R.R. and Horner, L.H. (1989) Origin, distribution, and morphology of serotonergic afferents to the cat superior colliculus: A light and electron microscope immunocytochemistry study. *Exp. Brain Res.*, 75: 83–98.

Morrison, J.H. and Foote, S.L. (1986) Noradrenergic and serotonergic innervation of cortical, thalamic, and tectal visual structures in Old and New World monkeys. *J. Comp. Neurol.*, 243: 117–138.

Mooney, R.D., Huang, X. and Rhoades, R.W. (1992) Functional influence of interlaminar connections in the hamster's superior colliculus. *J. Neurosci.*, 12: 2417–2432.

Mustari, M.J. and Lund, R.D. (1978) An aberrant crossed visual corticotectal pathway in albino rats. *Brain Res.*, 112: 73–84.

North, R.A., Henderson, G., Katayama, Y. and Johnson, S.M. (1980) Electrophysiological evidence for presynaptic inhibition of acetylcholine release by 5-hydroxytryptamine in the enteric nervous system. *Neuroscience*, 5: 581–586.

Peroutka, S.J., Schmidt, A.W., Sleight, A.J. and Harrington, M.A. (1990) Serotonin receptor 'families' in the central nervous system: An overview. *Ann. N.Y. Acad. Sci.*, 600: 104–113.

Prieto-Gomez, B., Dafny, N. and Reyes-Vázquez, C. (1989) Dorsal raphe stimulation, 5-HT and morphine microiontophoresis effects on noxious and non-noxious identified neurons in the medial thalamus of the rat. *Brain Res. Bull.*, 22: 937–943.

Ramirez, J.J., Jhaveri, S., Hahm, J.-O. and Schneider, G.E. (1990) Maturation of projections from the occipital cortex to the ventrolateral geniculate and superior colliculus in postnatal hamsters. *Dev. Brain Res.*, 55: 1–9.

Rhoades, R.W. and Chalupa, L.M. (1978) Functional and anatomical consequences of neonatal visual cortical damage in the superior colliculus of the golden hamster. *J. Neurophysiol.*, 41: 1466–1494.

Rhoades, R.W., Mooney, R.D., Chiaia, N.L. and Bennett-Clarke, C.A. (1990) Development and plasticity of the serotoninergic projection to the hamster's superior colliculus. *J. Comp. Neurol.*, 299: 151–166.

Rhoades, R.W., Figley, B., Mooney, R.D. and Fish, S.E. (1991) Development of the occipital corticotectal projection in the hamster. *Exp. Brain Res.*, 86: 373–383.

Rosenquist, A.C. and Palmer, L.A. (1971) Visual receptive field properties of cells of the superior colliculus after cortical lesions in the cat. *Exp. Neurol.*, 33: 629–652.

Schiller, P.H., Stryker, M., Cynader, M. and Berman, N. (1974) Response characteristics of single cells in the monkey superior colliculus following ablation or cooling of the visual cortex. *J. Neurophysiol.*, 37: 181–194.

Segu, L., Abdelkefi, J., Dusticier, G. and Lanoir, J. (1986) High-affinity serotonin binding sites: Autoradiographic evidence for their location on retinal afferents in the rat superior colliculus. *Brain Res.*, 384: 205–217.

Sillar, K.T. and Simmers, A.J. (1994) Presynaptic inhibition of primary afferent transmitter release by 5-hydroxytryptamine at a mechanosensory synapse in the vertebrate spinal cord. *J. Neurosci.*, 14: 2636–2647.

Sparks, D. (1986) Translation of sensory signals into commands for control of saccadic eye movements: Role of primate superior colliculus. *Physiol. Rev.*, 66: 118–171.

Stein, B.E. and Argibede, M.O. (1972) Unimodal and multimodal response properties of neurons in the cat's superior colliculus. *Exp. Neurol.*, 36: 179–196.

Steriade, M. and Llinás, R. (1988) The functional states of the thalamus and the associated neuronal interplay. *Physiol. Rev.*, 68: 649–742.

Tan, H. and Miletic, V. (1992) Diverse actions of 5-hydroxytryptamine on frog dorsal horn neurons *in vitro*. *Neuroscience*, 49: 913–923.

Ueda, S., Ihara, N. and Sano, Y. (1985) The organization of serotonin fibers in the mammalian superior colliculus: an immunohistochemical study. *Anat. Embryol.*, 173: 13–21.

Villar, M.J., Vitale, M.L., Hökfelt, T. and Verhofstad, A.A.J. (1988) Dorsal raphe serotoninergic branching neurons projecting both to the lateral geniculate body and superior colliculus: a combined retrograde tracing immunohistochemical study in the rat. *J. Comp. Neurol.*, 277: 126–140.

Waeber, C. and Palacios, J.M. (1990) 5-HT$_1$ receptor binding sites in the guinea pig superior colliculus are predominantly of the 5HT$_{1D}$ class and are presynaptically located on primary retinal afferents. *Brain Res.*, 528: 207–211.

Wichmann, T., Limberger, N. and Starke, K. (1989) Release and modulation of release of serotonin in rabbit superior colliculus. *Neuroscience*, 32: 141–151.

Wickelgren, B.G. and Sterling, P. (1969) Influence of visual cortex on receptive fields in the superior colliculus of the cat. *J. Neurophysiol.*, 32: 16–23.

Wu, S.Y., Wang, M.Y. and Dun, N.J. (1991) Serotonin via presynaptic 5-HT$_1$ receptors attenuates synaptic transmission to immature rat motoneurons *in vitro*. *Brain Res.*, 554: 111–121.

M. Norita, T. Bando and B. Stein (Eds.)
Progress in Brain Research, Vol 112
© 1996 Elsevier Science BV. All rights reserved.

CHAPTER 5

Morphology of single axons of tectospinal and reticulospinal neurons in the upper cervical spinal cord

Yoshikazu Shinoda*, Shinji Kakei and Naoko Muto

Departments of Physiology and Orthopedic Surgery, School of Medicine, Tokyo Medical and Dental University, 1-5-45 Yushima, Bunkyo-ku, Tokyo, 113, Japan

Single axons of tectospinal (TS) and reticulospinal (RS) neurons were stained with intraaxonal injection of HRP after electrophysiological identification, and their axonal trajectory was reconstructed at C1–C3 of the cat. TS neurons were located in the intermediate or deep layers of the caudal two-thirds of the superior colliculus (SC) and had multiple axon collaterals (up to seven collaterals) per stem axon). Collaterals had a simple structure, ramified several times mainly in the transverse plane, and terminated in the lateral parts of laminae V–VIII. More than half also had terminals in lamina IX. Terminals of TS neurons did not appear to make contacts with either the somas or proximal dendrites of retrogradely-labeled motoneurons in lamina IX, but clear contacts were found on counterstained interneurons in the lateral part of laminae V–VIII. Here, we examined three stained spinal interneurons receiving monosynaptic excitation from the SC. These interneurons had multiple axon collaterals mainly in laminae VII–IX, and made extensive contacts with retrogradely-labeled motoneurons of multiple neck muscles. Stem axons of single RS neurons receiving input from the contralateral SC ran in the ventromedial funiculus and gave off multiple axon collaterals to laminae VII–IX over at least several cervical segments. Their terminal boutons appeared to make contact with both the somas and proximal dendrites of retrogradely-labeled neck motoneurons. Single RS neurons made contacts with motoneurons of different neck muscles. These results provide evidence for functional synergies at the level of single RS neurons and spinal interneurons for neck movements. The present finding indicates that the direct TS projection to the spinal cord may influence the activity of multiple neck muscles mainly via spinal interneurons, and plays an important role in control of head movement in parallel with the tecto-reticulospinal system.

Introduction

The neck forms a complicated and redundant mechanical system with many degrees of freedom because of its multi-jointed architecture and the involvement of multiple muscle groups. When a target appears in the visual field, an animal shifts his gaze to that target by coordinated eye and head movements. In this process, the central nervous system transforms the visual information of target location into motor command signals, and generates the spatio-temporal activation pattern of extraocular and neck muscles necessary to acquire the target. The superior colliculus (SC) plays an important role in the control of such visually-triggered orienting movements (Roucoux et al., 1980; Wurtz and Albano, 1980; Grantyn and Grantyn, 1982; Grantyn and Berthoz, 1988; Munoz and Guitton, 1989). Collicular influences onto the spinal cord are conveyed through a direct pathway, the tectospinal (TS) tract, as well as through indirect pathways via reticulospinal

* Corresponding author. Tel: +81 3 5803 5152; fax: +81 3 5803 0118.

(RS) neurons and spinal interneurons. Electrophysiological studies suggested that excitatory postsynaptic potentials (EPSPs) are evoked disynaptically in dorsal neck motoneurons by stimulation of the contralateral SC and are presumed to be mediated by relay neurons in the brain stem (Anderson et al., 1971; Grantyn and Berthoz, 1987; Iwamoto et al., 1990) or spinal cord (Illert et al., 1978; Alstermark et al., 1987). Most morphological studies of the TS tract have agreed that it terminates mainly in the lateral parts of laminae VI–VII, with smaller projections to laminae V and VIII of the cervical spinal cord (Altman and Carpenter, 1961; Nyberg-Hansen, 1964; Petras, 1967; Huerta and Harting, 1982; Holstege, 1988), but recently Rose et al. (1991) reported that terminals were densest in medial lamina VII and dorsal lamina VIII. The projection to lamina IX (neck motor nuclei) is also somewhat controversial (Huerta and Harting, 1982; Redgrave et al., 1987; Rose et al., 1991). Thus, it is clear from a survey of these studies that there is not good agreement on the spinal termination of TS neurons. Moreover, the relationship of axon terminals of TS neurons and their target spinal neurons, most notably neck motoneurons, has never been examined.

This article summarizes our data on the morphology of single neurons in the pathways from the SC to the spinal cord (TS neurons, spinal interneurons and RS neurons receiving TS input) in the cat, and gives evidence of neuronal implementation of functional synergies for neck movements in the branching patterns of single RS axons.

Methods

Experiments were performed in cats anesthetized with pentobarbital sodium (Nembutal; Abott, Switzerland) (initial dose of 35 mg/kg, supplemented as required). A laminectomy was performed between C1 and C2 to permit intraaxonal recording from both TS and RS neurons, as well as from spinal interneurons. A small hole was opened in the parietal bone to place two concentric stimulating electrodes arranged antero-posteriorly (outer diameter, 0.3 mm; inner diameter, 0.1 mm; interelectrode distance, 1.5 mm) in the right SC. Intraaxonal recording and HRP injection were made with a glass microelectrode filled with 7% HRP (Toyobo, Co., Osaka, Japan) in 0.4 M KCl Tris–HCl buffer (pH 8.6). Axons were penetrated in the ventromedial funiculus on the left side between C1 and C2. Axons were identified as TS axons if they could be orthodromically activated directly from the SC (Muto et al., in press) and as RS or spinal interneuron axons if a synapse was deemed to be interposed between the axon and the SC (Kakei et al., 1994). The latencies of synaptically-activated spikes from the SC were less than 1.4 ms and they were regarded as monosynaptic from the SC. These axons were presumed to be either RS axons or spinal interneuron axons. After electrophysiological identification, HRP was injected iontophoretically through the recording electrode. Following intraaxonal injection, HRP was also injected into several peripheral nerves innervating neck muscles at C1 and C2 and the accessory nerves to label motoneurons retrogradely. Animals survived for 24 h, and were deeply anesthetized with pentobarbital sodium and perfused transcardially with a 0.1 M phosphate buffer solution (pH 7.4) followed by a mixture of 1% glutaraldehyde and 2% paraformaldehyde in a 0.1 M phosphate buffer solution containing 4% sucrose. Serial transverse frozen sections of the spinal cord of 100-μm thickness were made on a sliding microtome and treated for HRP by the heavy metal-intensification method of Adams (1981) (see Shinoda et al., 1986 for details). Serial frozen sections of the brain stem of 75-μm thickness were also made and treated for HRP by the TMB method of Mesulam (1978). In three cats, the right SC was exposed and biocytin was injected into multiple sites of the SC. Following survival periods of 2–3 days, the spinal cord was removed after perfusion, and serial frozen sections of 100 μm thickness were treated by the method of King et al. (1989).

Results

Tectospinal axons

To analyze the morphology of single TS axons in the spinal cord, axons traveling in the ventral funiculus between C1 and C2 at depths of 2.5–4.0 mm were identified as TS axons by their direct responses to stimulation of the contralateral SC and were injected intraaxonally with HRP. The latencies of orthodromically-evoked spikes in TS neurons ranged from 0.4 to 0.9 ms. Of the 24 TS axons injected with HRP, 20 were used for the present morphological analysis, and 18 had retrogradely labeled cell bodies in the contralateral SC. All labeled neurons were located in the caudal two-thirds of the SC and were concentrated laterally, and the vast majority ($n = 16$) were located in the stratum griseum intermediale (intermediate) layer (two cells were located in the stratum griseum profundum). The stained length of the stem axons ranged from 11.5 to 27.5 mm (mean $\pm$ SD, 17.8 ± 4.4 mm, $n = 20$) and the diameter of the stem axons ranged from 1.4 to 4.4 μm (mean $\pm$ SD, 2.9 ± 1.1 μm, $n = 20$).

Stem axons of TS neurons gave rise to multiple axon collaterals at C1 and C2 segments (Fig. 1). When viewed in the horizontal plane, primary axon collaterals arose at a more-or-less right angle from the stem axons. When the stem axons were located in the ventral part of the ventral funiculus, collaterals ran dorsolaterally towards the ventral horn. When stem axons were located in the more dorsal part of the ventral funiculus, collaterals ran laterally or ventrolaterally. These collaterals ran in the ventral funiculus for some distance without branching or after bifurcating once, and the primary or secondary collaterals entered the ventral horn at its medial border.

All but one TS axon (19/20) had at least one collateral, and the maximum number of collaterals for a single TS axon was 7 (mean number of collaterals = 2.7 ± 1.6, $n = 19$). The distances between branching points of adjacent primary collaterals from a stem axon ranged from 0.4 to 7.7 mm (2.6 ± 1.7 mm, $n = 36$). The rostrocaudal extension of individual well-stained axon collaterals in the spinal gray matter ranged from 0.2 to 3.1 mm (0.7 ± 0.5 mm, $n = 56$). Since intercollateral intervals were wider than the rostrocaudal spreads of individual axon collaterals, there were usually gaps free from axon terminals between adjacent axon collaterals. After the entrance into the gray matter, some collaterals ran dorsolaterally, passing through lamina IX without any terminal branches, and ramified within the lateral parts of laminae V–VIII. In contrast, other collaterals ran dorsolaterally to the lateral parts of laminae VI–VIII, giving rise to numerous terminal branches in lamina IX on the way.

Most collaterals of TS axons ramified mainly in the transverse plane and had simple structures. Some ramified several times in the gray matter, while giving rise to a number of short side branchlets that bore some swellings. Most short terminal branches were composed of a few en passant structures and a single terminal swelling, and terminated near the parent branch. Collaterals arising from the same axon usually showed a similar trajectory and destination site in the gray matter, although there was individual variability in their branching pattern and exact pattern of terminal distribution. The number of swellings per collateral was small, ranging from 10 to 326 (65.4 ± 75.6, $n = 20$).

Among the 19 TS axons examined, 12 had a considerable number of terminal swellings within lamina IX as well as in laminae V–VIII. Fig. 1 shows an example of a TS axon with terminal branches in lamina IX. Collaterals of this axon had terminal swellings in neck extensor motor nuclei (the ventrolateral part of the ventral horn) and in the spinal accessory nucleus. Conversely, the remaining seven axons had almost no projection to lamina IX.

To investigate whether axon terminals of TS axons actually contact motoneurons in lamina IX, neck motoneurons were retrogradely labeled and

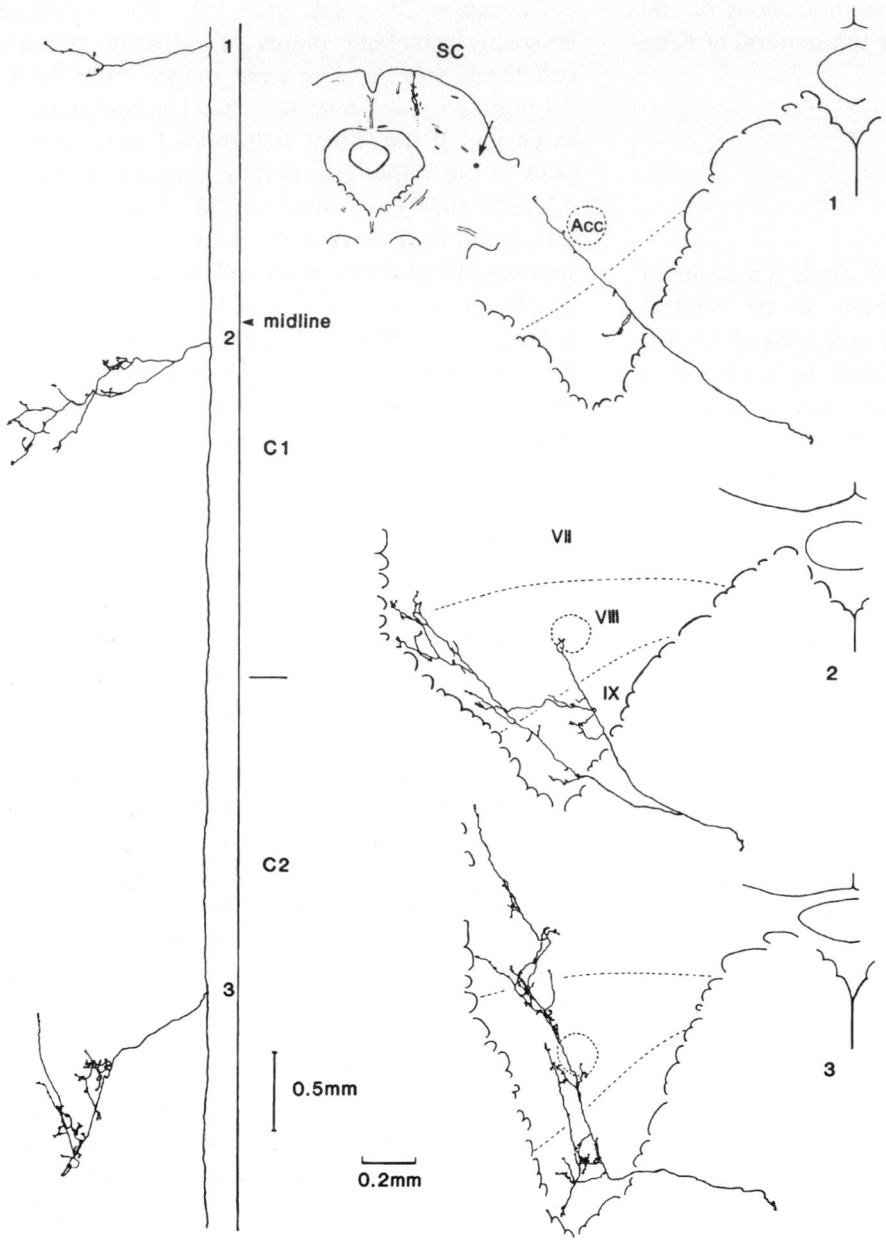

Fig. 1. Reconstructions in the horizontal plane (left) and the transverse plane (right) of a TS axon terminating in the ventral part of the ventral horn (neck extensor motor nuclei) as well as in the lateral regions of laminae V–VIII. The cell body of this axon was located in the caudolateral region of the intermediate layer of the contralateral SC (upper middle inset).

synaptic contacts were searched for in the motor nuclei. With this method, only cell bodies and proximal dendrites of motoneurons were labeled with HRP. Fig. 2 shows an example of a TS axon whose collaterals projected to a wide area of laminae VI, VII, VIII and neck motor nuclei. Despite a number of terminals within the neck motor nuclei, no contacts with retrogradely-

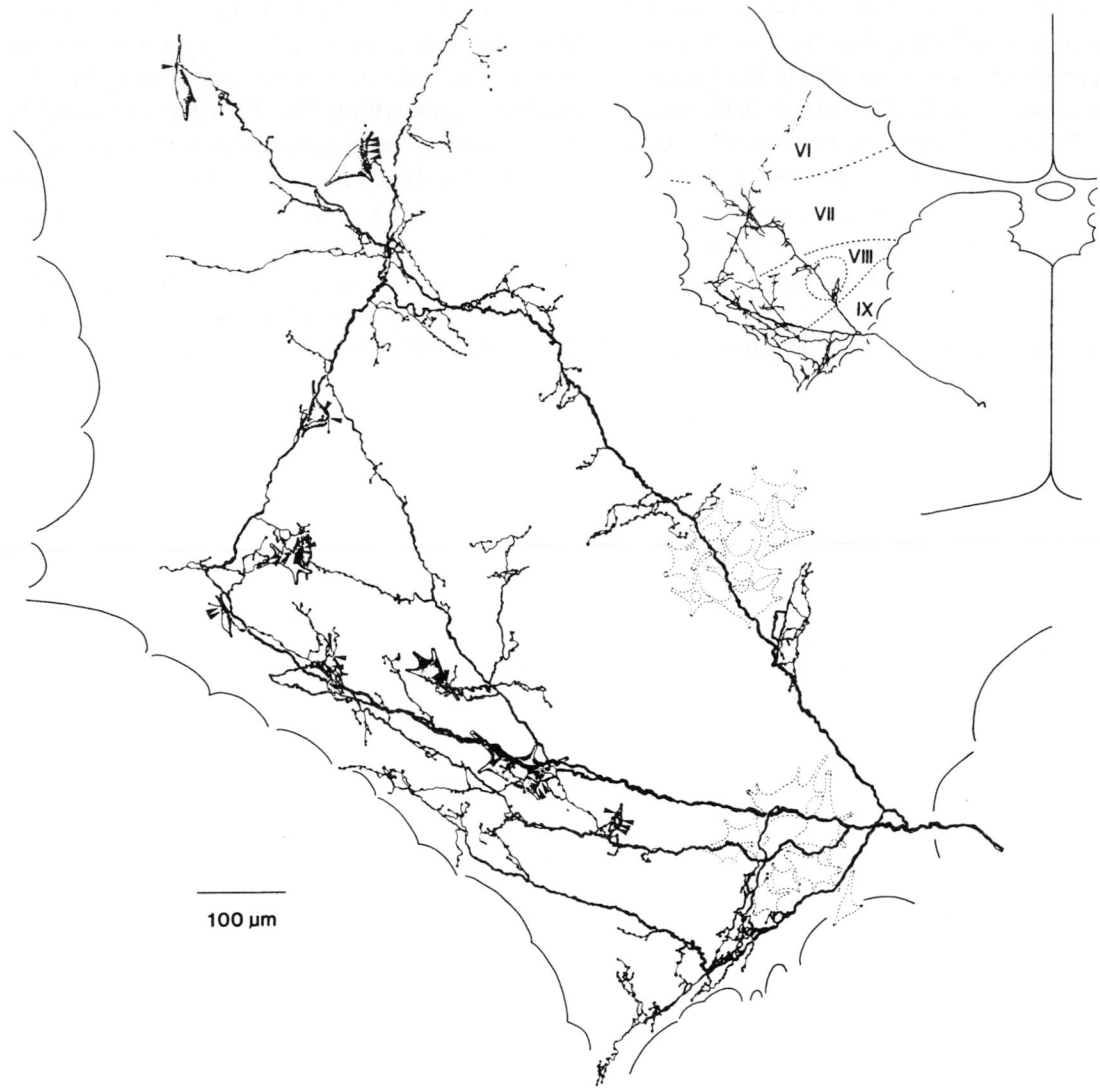

Fig. 2. Camera lucida drawings of synaptic boutons of a TS axon. Neurons indicated by broken lines are neck motoneurons labeled retrogradely with HRP and neurons with dots are counterstained spinal interneurons. Arrowheads indicate apparent synaptic contacts of terminals with counterstained neurons at the light microscopic level. Contacts were detected with ten counterstained neurons in the lateral parts of laminae VI–VIII.

labeled motoneurons were detected. This absence was characteristic of the other 11 TS axons as well.

Although TS axons did not make a contact with retrogradely-labeled neck motoneurons, many terminals were seen in contact with counterstained neurons in the lateral parts of laminae V–VIII (Fig. 2). From one to five TS axon termi-

nals contacted the soma and proximal dendrites of each counterstained neuron. A single axon collateral of a TS axon could contact up to ten interneuronal somata. Such a number likely represents a gross underestimate of the actual number of interneurons contacted, since only cells located near the surface of each spinal cord slice were counterstained due to section thickness.

The distribution of spinal interneurons contacted by TS axons at C1 and C2 is shown on the left of Figure 3. They were mainly located in the lateral parts of laminae V–VIII. The target cells contacted by TS axons were relatively small. The soma area of these neurons varied from 62 to 856 μm^2 (mean $\pm$ SD, 383 $\pm$ 243 μm^2, $n = 31$), and these cells were much smaller than retrogradely-labeled neck motoneurons (689 $\pm$ 309 μm^2, $n = 250$).

To observe the comprehensive projection pat-

tern from the SC to the spinal cord and supplement the single axon study, injection of biocytin into the SC was performed (right panel-Fig. 3). Injection sites within the SC were identified by the presence of well-stained neurons in the intermediate and deep layers, with no stained neurons in the underlying tegmentum. Stem axons were localized in the ventromedial funiculus at C1–C3, contralateral to the injection site and several stained fibers in each experiment were found at C1 on the ipsilateral side. The diameter of stem

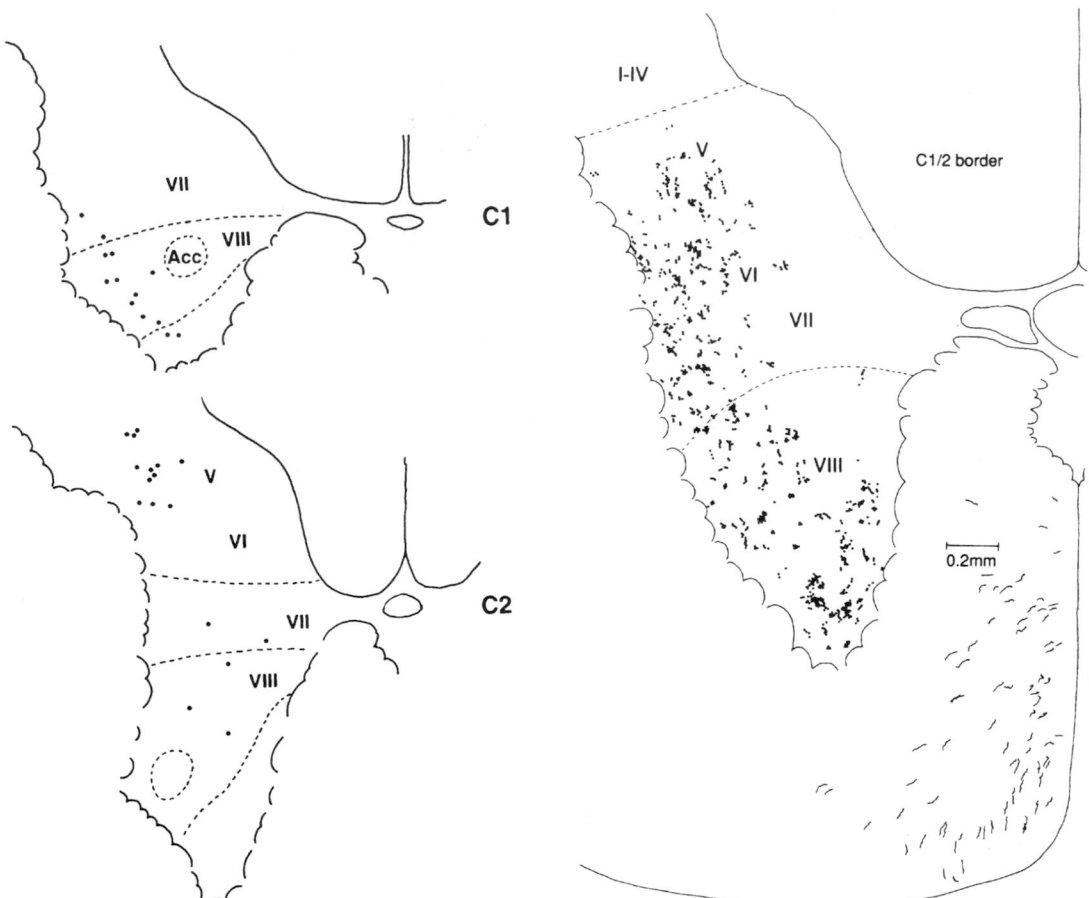

Fig. 3. Left: Distribution of spinal neurons with which TS axon terminals made apparent contacts. Each dot represents a single neuron. Right: Spinal distribution of terminal swellings of TS neurons after biocytin injection in the contralateral SC. Distribution of the swellings is shown in a representative transverse section by superimposing ten serial sections at the C1 level. Note that the swellings are dense in the lateral parts of laminae V–VIII and also in lamina IX. Stem axons in the ventral funiculus were drawn based on a single section.

axons ranged from 0.4 to 5.2 μm. The distribution of terminal swellings was concentrated in the lateral parts of laminae V–VIII and lamina IX. This distribution pattern is consistent with our data on the single TS axons.

Spinal interneurons receiving monosynaptic tectal input

To determine whether interneurons with direct TS input in laminae V–VIII terminate on neck motoneurons, and to provide further insight into the targets of these interneurons, three axons activated by SC stimulation and stained intraaxonally were identified morphologically as spinal interneurons. Reconstruction of the axonal trajectory of such an interneuron is shown in Fig. 4. The cell body of this interneuron had dendrites radiating in the frontal plane and was located in lamina VIII at C1 contralateral to the stimulated SC. The stem axon originating from the cell body ran ventrally through the ventral gray matter to the ventral funiculus. After curving medially, the stem axon ran caudally almost in parallel with the midline at the depth of around 3 mm in the ventral funiculus. Eleven primary collaterals were given off at almost right angles from the stem axon at C1 and C2. Nine of these collaterals ran for some distances laterally to the ventral horn without branching or bifurcating. These collaterals ramified in a delta-like fashion in the ventral horn and terminated in laminae VII–IX. The two most rostral collaterals projected contralaterally through the anterior commissure, and terminated mainly in lamina VIII and only slightly in lamina IX. Compared with the collaterals of TS axons, the collaterals of this interneuron projected extensively to lamina IX and far less to the lateral parts of laminae VII and VIII. In lamina IX, terminals of this interneuron appeared to make contact with cell bodies and proximal dendrites of retrogradely-labeled neck motoneurons. The third axon collateral of this interneuron (Fig. 4) projected to the ventrolateral part of lamina IX (probably the neck extensor motor nucleus) and the spinal accessory nucleus, where terminals

appeared to contact cell bodies and proximal dendrites of retrogradely-labeled neck motoneurons (Fig. 5). Other axon collaterals also terminated in these nuclei. The fourth, seventh and ninth collaterals of the same interneuron had terminals in the ventromedial part of lamina IX, the site of neck flexor motoneurons. The two other labeled interneurons gave rise to 14 and three axon collaterals at C1 and C2, respectively, and terminated most extensively in lamina IX, including the spinal accessory nucleus and its adjacent lamina VIII and in the ventral part of lamina VII. In lamina IX, terminals appeared to contact retrogradely-labeled motoneurons innervating multiple neck muscles.

Reticulospinal axons

Axons were penetrated in the ventromedial funiculus on the left side between C1 and C2. These axons were presumed to be from RS neurons when stimulation of the contralateral SC evoked spikes without any trace of post-synaptic potentials. Usually double- or triple-shock stimuli were needed to evoke spikes. Latencies ranged from 1.0 to 1.5 ms (mean $\pm$ SD = 1.3 $\pm$ 0.2 ms), and were regarded as evoked through a single synapse (Peterson et al., 1974: Grantyn and Berthoz, 1987; Iwamoto et al., 1988). Eleven of 15 RS neurons injected intraaxonally were successfully stained. In nine of these axons, at least two collaterals to the spinal gray matter were reconstructed. In ten, retrogradely-stained cell bodies of origin were found (inset-top right of Fig. 6) in the nucleus reticularis pontis caudalis (NRPC). Axons were stained over distances ranging from 13 to 25 mm rostrocaudally at C1–C3 ($n = 11$). A typical example of the branching pattern of a single RS neuron in the cervical cord is shown in Fig. 6. The stem axon ran in the ventromedial funiculus and gave off multiple axon collaterals at almost right angles over a few cervical segments. Each collateral ramified 3–5 times before its terminal branching which mainly spread in a frontal plane in the ventral horn. Axon terminals were most extensively distributed in lamina IX of

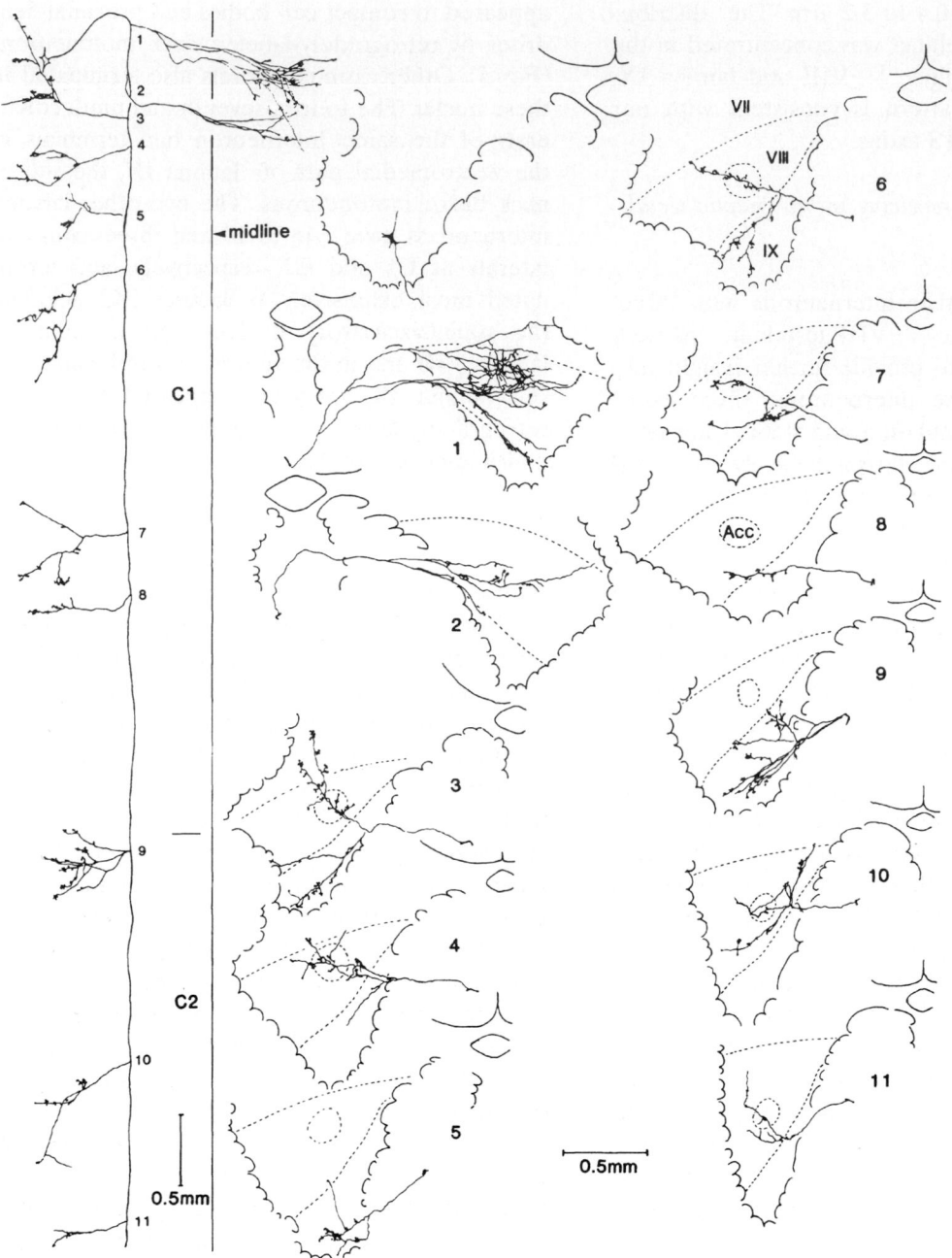

Fig. 4. Reconstruction of a spinal interneuron receiving monosynaptic excitation from the contralateral SC. The cell body is located in lamina VIII at C1 (top middle). The first two collaterals projected to the contralateral side. Note that collaterals of these interneurons project more specifically to neck motor nuclei than TS axon collaterals.

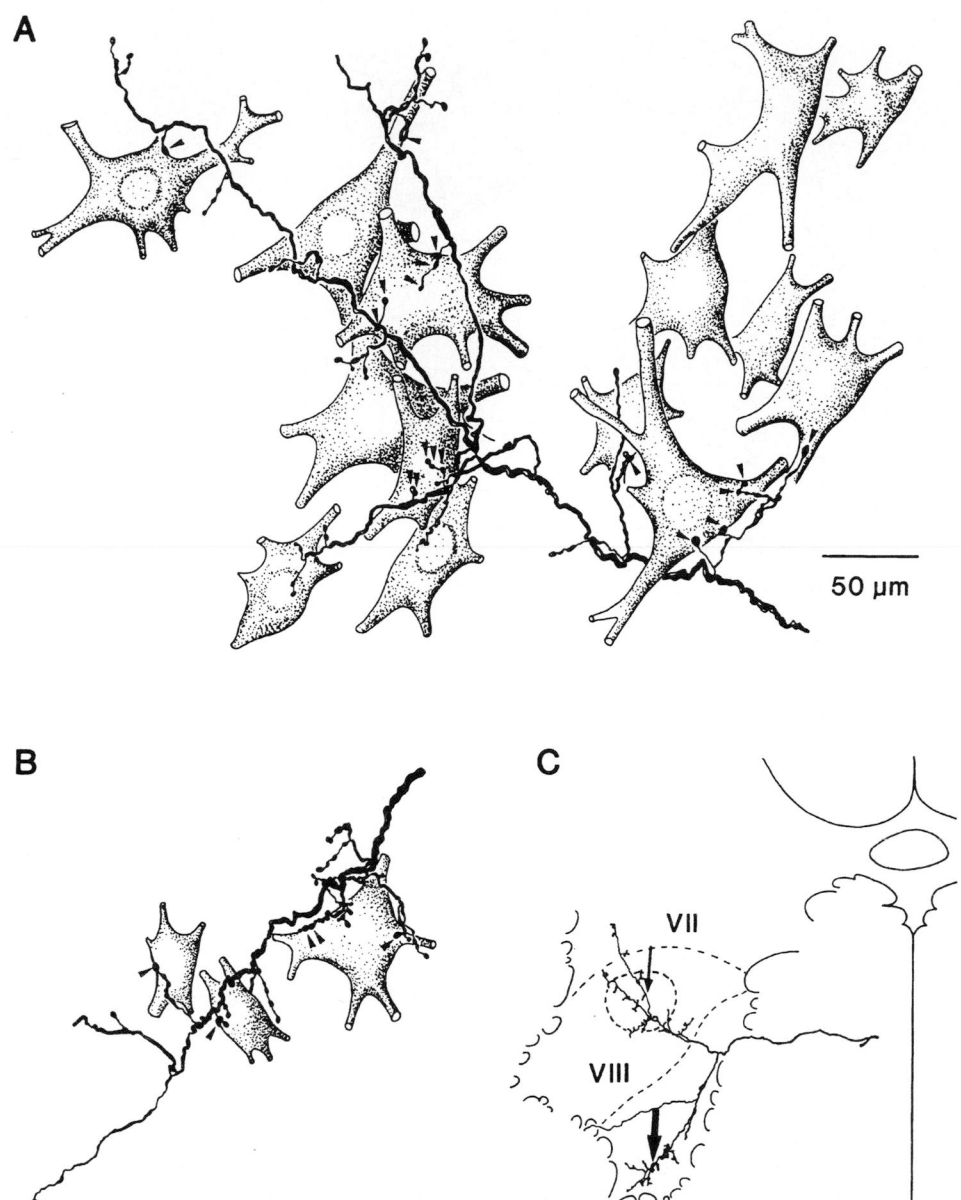

Fig. 5. Camera lucida drawings of synaptic boutons of a spinal interneuron at C1 receiving monosynaptic excitation from the contralateral SC. This axon collateral is the same as the third collateral shown in Fig. 4. All neurons shown in this figure are retrogradely-labeled neck motoneurons. Apparent contacts of the synaptic boutons with retrogradely-labeled neck motoneurons are shown by arrowheads. A: Axon terminals in the spinal accessory nucleus (thin arrow in the inset C). B: Axon terminals in the motor nucleus of the complexus muscle (thick arrow in the inset C).

Rexed (1954). The number of axon collaterals stained per neuron was 2–11 with a mean of 5.4 ($n = 9$). The rostrocaudal extension of individual axon collaterals was relatively narrow (range 0.4–2.5 mm; mean 0.9 $\pm$0.5 mm, $n = 39$) compared to much wider intercollateral intervals

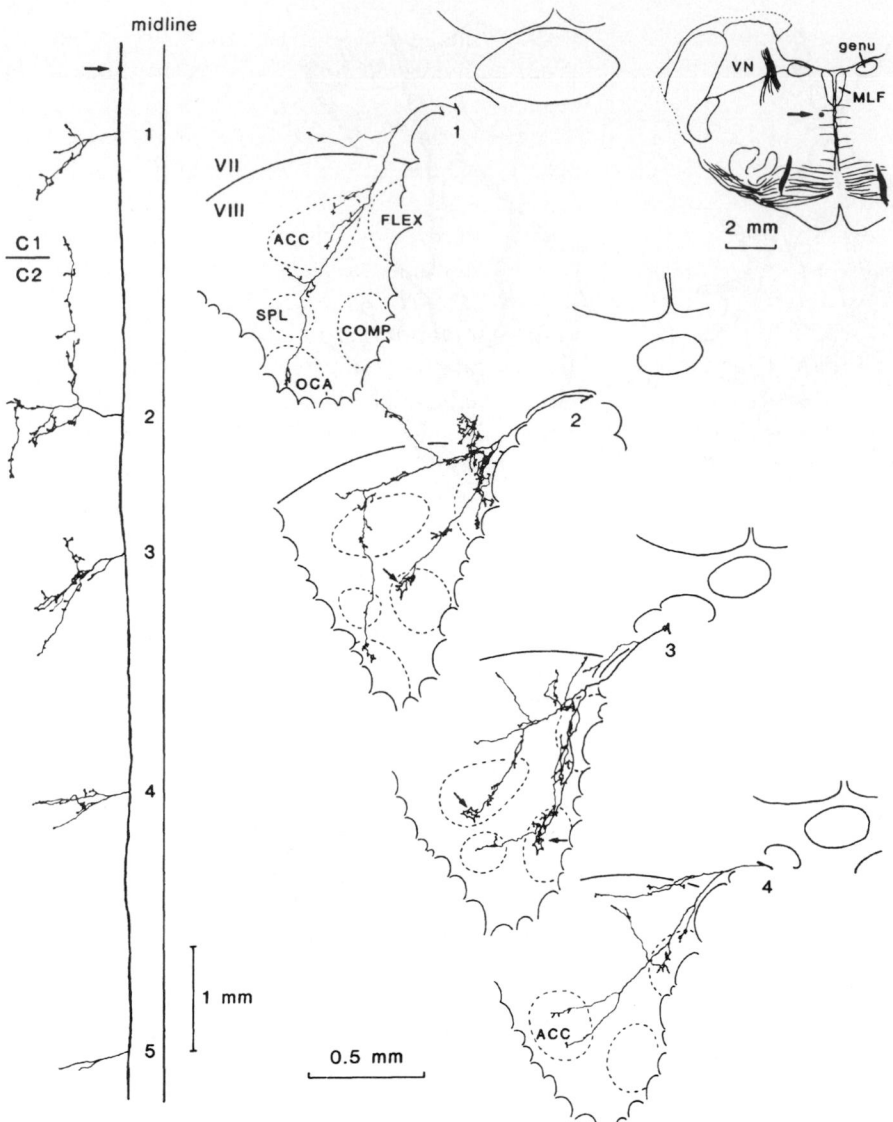

Fig. 6. Reconstruction of a single RS neuron receiving monosynaptic excitatory input from the superior colliculus. The cell body was located in the NRPC (top right: arrow). VN-vestibular nucleus; genu-genu facialis; MLF-medial longitudinal fasciculus. The left inset shows the dorsal view of the reconstructed axonal trajectory of this neuron in the upper cervical spinal cord. An arrow indicates the injection site. Numbers 1–5 represent each collateral. Individual collaterals are shown in representative transverse sections of the spinal cord. Broken lines indicate individual motor nuclei identified by retrograde labeling of motoneurons. ACC represents spinal accessory nucleus. COMP, FLEX, SPL and OCA represent motor nuclei of complexus muscle, flexor muscles, splenius muscle, and the obliquus capitis caudalis muscle, respectively. Only motoneurons contacted by this RS are depicted and indicated by arrows.

(range 0.3–8.0 mm; mean 2.2 ± 1.8 mm, $n = 34$). Thus, there were generally gaps free from axon terminals between adjacent axon collaterals (in-set-left of Fig. 6). The terminals of each primary collateral were localized in the ventral horn and were mainly distributed in laminae IX, VIII and sometimes in lamina VII. Terminal boutons were predominantly distributed in or very close to neck motor nuclei, and some of them appeared to make contact with retrogradely labeled motoneurons. In the neuron shown in Fig. 6, collateral No. 2 appeared to contact a complexus motoneuron, and collateral No. 3 appeared to contact an accessorius motoneuron. Of the nine RS axons with a projection to neck motor nuclei, eight projected to multiple (2–5) neck motor nuclei.

The branching pattern of single RS axons receiving monosynaptic input from the SC could be divided into three groups. One group innervated all neck motor nuclei at C1 and C2, such as the example shown in Fig. 6. The second group innervated motor nuclei of neck extensors in large measure, whereas the third group innervated motor nuclei of neck flexor muscles. In each group, the neck muscles innervated by single RS axons were stereotyped, and comprised a functional group involved in neck movements. Therefore, the present results provide experimental evidence that the branching patterns of single RS neurons determine functional synergies for neck movements. To further this finding, we are in the process of analyzing the detailed branching pattern of numerous RS neurons.

Discussion

The present study has shown that single TS and RS axons have multiple collaterals with narrow rostrocaudal extension at different cervical segments. TS axons terminate not only in the lateral parts of laminae V–VIII, but also in lamina IX. The axon terminals of TS axons appear to make contacts with spinal interneurons, but not with retrogradely-labeled neck motoneurons. In contrast, RS axons receiving tectal input terminate largely in laminae VII–IX, and they appear to terminate on motoneurons which innervate multiple neck muscles.

Our previous studies have revealed the intraspinal morphology of single long descending motor tract axons using an intracellular HRP staining technique. In this work, we have examined corticospinal (Futami et al., 1979; Shinoda et al., 1981), rubrospinal (Shinoda et al., 1982) and vestibulospinal (VS) axons (Shinoda et al., 1986, 1992). Like the TS and RS axons described in this study, virtually all of these long descending tract axons have multiple collaterals at various segments of the spinal cord.

Long descending motor tracts of supraspinal origin are classified into lateral and medial descending groups based on anatomical and behavioral observations after lesions (Kuypers, 1981). The lateral group consists of the corticospinal and rubrospinal tracts, which run in the lateral funiculus of the spinal cord. The medial group consists of the vestibulospinal, the reticulospinal, the tectospinal and the interstitiospinal tracts. The intraspinal branching patterns of corticospinal and rubrospinal axons are similar in that individual collaterals spread widely in a rostrocaudal direction with wide intercollateral intervals, and terminals are mainly localized in lateral laminae V–VII of the spinal gray matter. In contrast, lateral and medial vestibulospinal axons show a narrower collateral distribution in the rostrocaudal direction, have shorter intercollateral intervals and terminate mainly in laminae VII–IX (Shinoda et al., 1994). In the present study, RS axons were found to resemble the latter group, in that they have similar branching and terminal distribution patterns (Kakei et al., 1994). TS axons also resemble the latter group in most respects, although some terminals are distributed in the lateral parts of laminae V-VI. Among the long descending motor tracts, single TS axons have the simplest axonal arborizations and their axon collaterals have the simplest intraspinal structures.

The TS tract has been examined using degeneration methods (Altman and Carpenter, 1961; Nyberg-Hansen, 1964; Petras, 1967), autoradiography (Huerta and Harting, 1982), and PHA-L (Rose et al., 1991). In these studies, there has been a common agreement that TS axons descend in the contralateral ventromedial funiculus of the spinal cord and terminate in the contralateral upper cervical cord. Furthermore, most of these studies have shown that the main termination site in the spinal gray matter is the lateral parts of laminae VI and VII, with weaker projections to lamina V and the dorsal part of lamina VIII. However, the existence of a projection which terminates in lamina IX (neck motor nuclei) is controversial. Huerta and Harting (1982) concluded that the TS tract terminates within lamina IX immediately adjacent to regions of lamina VIII in the cat. Redgrave et al. (1987) also found extensive labeling in lamina VIII and some labeling in medial lamina IX in the hamster. Although it is crucial to distinguish terminal boutons from fine passing axon collaterals in laminae VIII and IX on their way to more dorsal laminae, autoradiographic or WGA-HRP techniques generally lack the resolution needed to make such discriminations. More recently, Rose et al. (1991) used the lectin PHA-L to re-examine the spinal projections of TS neurons in the cat, and reported that boutons are rare in lamina IX, the ventral parts of lamina VIII, and the lateral part of lamina VII, and densest in medial lamina VII and dorsal lamina VIII. In the present study using both intracellular staining and biocytin anterograde labeling techniques, axon terminals of TS neurons were found in the lateral parts of laminae VI, VII and VIII, with less labeling in laminae IX and a sparse projection to lamina V. Terminals were very rare in the medial part of lamina VII. At present, we can not explain this discrepancy between our data and that of Rose and colleagues.

Single TS axons can be divided into two groups on the basis of their projection to lamina IX. Whereas one group has considerable projection to lamina IX, the other group had little or no projection. Although a number of boutons of TS axons were distributed in the vicinity of retrogradely-labeled motoneurons, none were seen to make contact with retrogradely-labeled motoneurons. Anderson et al. (1971) recorded intracellular potentials from neck motoneurons evoked by stimulation of the SC and concluded that 'in few cases' are the latencies of EPSPs compatible with monosynaptic transmission. Gura and Limansky (1986) reported that monosynaptic input from the SC exists in about 14% of the studied

accessory motoneurons. Olivier et al. (1995) also suggested the possibility of a monosynaptic tecto-motoneuronal connection using the method of spike-triggered averaging of rectified neck EMG. Taking both the present morphological data and these electrophysiological data into account, the terminals of some TS axons may terminate on the more distal dendrites of neck motoneurons. Alternatively, TS axons may not terminate directly on motoneurons, but on interneurons there or located in the neighboring area. Spinal commissural neurons of the last order interneurons might be one of the candidates for such interneurons (Sugiuchi et al., 1995). Further analysis is required to determine the existence of this connection.

The terminals of TS axons appeared to make contacts with cell bodies and proximal dendrites of many interneurons in laminae V–VIII. Three intracellularly-stained propriospinal neurons receiving monosynaptic input from the contralateral SC were located in the same area as the counterstained cell bodies with which TS terminals made contacts. Many axon collaterals of these interneurons projected to neck motor nuclei, and appeared to contact retrogradely-labeled motoneurons of multiple neck muscles. Anderson et al. (1971) showed that projections arising in the deep layers of the SC exert a disynaptic excitatory action on contralateral neck motoneurons via the tecto- reticulo-spinal (TRS) pathway, since lesion of the medial longitudinal fasciculus (MLF) had no consistent effect on the amplitude of contralateral EPSPs. However, since a section of the MLF does not eliminate all TS fibers traveling to the cervical cord, the presence of disynaptic EPSPs after sectioning the MLF does not necessarily indicate that the remaining disynaptic responses are relayed through the TRS pathway, but rather that some of the disynaptic EPSPs must be relayed to neck motoneurons via cervical interneurons activated by TS axons. The present finding indicates that the direct TS projection to the spinal cord may influence the activity of multiple neck muscles in large measure via spinal interneurons, and is likely to play an important role in the control of head movement in parallel with the tecto-reticulo-spinal system.

The TRS pathway synapses in the reticular formation and descends to the spinal cord via RS neurons. Previous data suggests that this is a major pathway from the SC to the spinal cord (Anderson et al., 1971; Huerta and Harting, 1982; Gura and Limansky, 1986; Nudo and Masterson, 1989; Rose et al., 1991). Electrophysiological studies have shown that RS neurons receive strong monosynaptic input from the contralateral SC (Grantyn and Berthoz, 1988; Iwamoto et al., 1990). These neurons are mostly located in the NRPC and the nucleus gigantocellularis in the medulla and become active during head movements (Grantyn and Berthoz, 1987, 1988; Isa and Naito, 1995). Iwamoto and Sasaki (1990) reported that connections of single RS neurons with dorsal neck motoneurons were muscle specific. Using the spike triggered averaging method, they found that single fiber EPSPs from a RS neuron were recorded from either biventer cervicis and complexus or splenius motoneurons, but not from both. However, the present study shows that single RS neurons in the NRPC receiving monosynaptic excitation from the SC have multiple axon collaterals in the cervical spinal cord and directly terminate on cell bodies and proximal dendrites of retrogradely-labeled motoneurons of multiple neck muscles (Kakei et al., 1994). Three spatial innervation patterns of single RS neurons on different neck muscles have been identified. These patterns are very similar to the innervation patterns of single vestibulospinal (VS) neurons receiving input from individual semicircular canals on different neck muscles (Shinoda et al., 1992). Thus, the present findings provide strong evidence of neuronal implementation of functional synergies at the level of single RS neurons, and suggest that both the VS and RS systems use the same functional synergies of neck muscles for control of head movements. The presence of the functional synergies determined by supraspinal motor control systems helps the redundant control system of the neck reduce the degrees of freedom involved in head movement.

Acknowledgements

This research was supported by a grant from the Japanese Ministry of Education, Science and Culture for Scientific Research.

References

Adams, J.C. (1981) Heavy metal intensification of DAB-based HRP reaction product. *J. Histochem. Cytochem.*, 29: 775.

Alstermark, B., Lundberg, A., Pinter, M., and Sasaki, S.(1987) Long C3-C5 propriospinal neurons in the cat. *Brain Res.*, 404: 382–388.

Altman, J. and Carpenter, M.B. (1961) Fiber projections of the superior colliculus in the cat. *J. Comp. Neurol.*, 116: 157–177.

Anderson, M.E., Yoshida, M. and Wilson, J. (1971) Influence of superior colliculus on cat neck motoneurons. *J. Neurophysiol.* 34: 898–907.

Futami, T., Shinoda, Y. and Yokota, J. (1979) Spinal axon collaterals of corticospinal neurons identified by intracellular injection of horseradish peroxidase. *Brain Res.*, 164: 279–284.

Grantyn, A. and Grantyn, R. (1982) Axon patterns and sites of termination of cat superior colliculus neurons projecting in the tecto-bulbo-spinal tract. *Exp. Brain Res.*, 46: 243–256.

Grantyn, A. and Berthoz, A. (1987) Reticulo-spinal neurons participating in the control of synergic eye and head movements during orienting in the cat. I.Behavioral properties. *Exp. Brain Res.*, 66:339–354.

Grantyn, A. and Berthoz, A. (1988) The role of the tectoreticulospinal system in the control of head movement. In: B.W. Peterson and F.J. Richmond (Eds.), *Control of Head Movement.* New York: Oxford, pp. 224–244.

Gura, E.V. and Limansky, Y. (1986) Effects of stimulating the superior colliculus on motoneurones of the neck muscles in the cat. *Neurophysiol.*, 18: 145–149.

Holstege, L.B. (1988) Brainstem-spinal cord projections in the cat, related to control of head and axial movements. In: J.A. Buttner-Ennever (Ed.), *Neuroanatomy of the Oculomotor System*, Vol., 2. Amsterdam: Elsevier Reviews in Oculomotor Res., pp. 431–470.

Huerta, M.F. and Harting, J.K. (1982) Tectal control of spinal cord activity: neuroanatomical demonstration of pathways connecting the superior colliculus with the cervical spinal cord gray. *Prog. Brain Res.*, 57: 293–328.

Illert, M., Lundberg, A., Padel, Y. and Tanaka, R. (1978) Integration in descending motor pathways controlling the forelimb in the cat. 5. Properties of and monosynaptic excitatory convergence on C3-C4 propriospinal neurones. *Exp. Brain Res.*, 33: 101–130.

84

Isa, T. and Naito, K. (1995) Activity of neurons in the medial pontomedullary reticular formation during orienting movements in alert head-free cats. *J. Neurophysiol.*, 74: 73–95.

Iwamoto, Y., Sasaki, S. and Suzuki, I. (1988) Descending cortical and tectal control of dorsal neck motoneurons via reticulospinal neurons in the cat. In O. Pompeiano and J.H.J. Allum (Eds.), *Progress in Brain Research*, Vol., 76, Elsevier, Amsterdam, pp. 97–108.

Iwamoto, Y. and Sasaki, S. (1990) Monosynaptic excitatory connexion of reticulospinal neurones in the nucleus reticularis pontis caudalis with dorsal neck motoneurones in the cat. *Exp. Brain Res.*, 80: 277–289.

Iwamoto, Y., Sasaki, S. and Suzuki, I. (1990) Input-output organization of reticulospinal neurones, with special reference to connections with dorsal neck motoneurones in the cat. *Exp. Brain Res.*, 80: 260–276.

Kakei, S., Muto, N. and Shinoda, Y. (1994) Innervation of multiple neck motor nuclei by single reticulospinal tract axons receiving tectal input in the upper cervical spinal cord. *Neurosci. Lett.*, 172: 85–88.

King, M.H., Louis, P.M., Hunter, B.E. and Walker, D.W. (1989) Biocytin: a versatile anterograde neuroanatomical tract-tracing alternative. *Brain Res.*, 497: 361–367.

Kuypers, H.G.J.M. (1981) Anatomy of descending pathways. In: V.B. Brooks (Ed.), *Handbook of Physiology. Section I: The Nervous System*, Vol. II, Motor Control, American Physiological Society, Bethesda, Maryland, pp. 597–666.

Mesulam, M.M. (1978) TMB for HRP neurohistochemistry : a non-carcinogenic blue reaction product with superior sensitivity for visualizing neural afferents and efferents. *J. Histochem. Cytochem.*, 2C: 106–117.

Munoz, D.P. and Guitton, D. (1989) Fixation and orientation control by the tecto-reticulo-spinal system in the cat whose head is unrestrained. *Rev. Neurol.*, 145: 567–579.

Muto, N., Kakei, S. and Shinoda, Y. (1996) Morphology of single axons of tectospinal neurons in the upper cervical spinal cord. *J. Comp. Neurol*, in press.

Nyberg-Hansen, R. (1964) The location and termination of tectospinal fibers in the cat. *Exp. Neurol.*, 9: 212–227.

Olivier, E., Grantyn, A., Kitama, T. and Berthoz, A. (1995) Post-spike facilitation of neck EMG by cat tectoreticulospinal neurones during orienting movements. *J. Physiol.*, 482: 455–466.

Peterson, B.W., Anderson, M.E. and Filion, M., Responses of ponto-medullary reticular neurons to cortical, tectal and cutaneous stimuli. Exp. *Brain Res.*, 21 (1974) 19–44.

Petras, J.M. (1967) Cortical, tectal and tegmental fiber connections in the spinal cord of the cat. *Brain Res.*, 6: 275–324.

Redgrave, P., Mitchell, I.J. and Dean, P. (1987) Descending projections from the superior colliculus in rats: A study using orthograde transport of wheatgerm- agglutinin conjugated horseradish peroxidase. *Exp. Brain Res.*, 68:147–167.

Rexed, B. (1954) A cytoarchitectonic atlas of spinal cord in the cat. *J. Comp. Neurol.*, 100: 297–379.

Rose, P.K., MacDonald, J, and Abrahams, V.C. (1991) Projections of the tectospinal tract to the upper cervical spinal cord of the cat: A study with the anterograde tracer PHA-L. *J. Comp. Neurol.*, 314: 91–105.

Roucoux, A., Guitton, D., and Crommelinck, M. (1980) Stimulation of the superior colliculus in the alert cat. II. Eye and head movements evoked when the head is unrestrained. Exp. *Brain Res.* 39: 75–85.

Shinoda, Y., Kakei, S. and Sugiuchi, Y. (1994) Multisegmental control of axial and limb muscles by single long descending motor tract axons. In: S.P. Swinnen (Eds.), *Interlimb Coordination.* Academic Press, San Diego, pp., 31–47.

Shinoda, Y., Yokota, J. and Futami, T.(1981) Divergent projection of individual corticospinal axons to motoneurons of multiple muscles in the monkey. *Neurosci. Lett.*, 23: 7–12.

Shinoda, Y., Yokota, J. and Futami, T.(1982) Morphology of physiologically identified rubrospinal axons in the spinal cord of the cat. *Brain Res.*, 242: 321–325.

Shinoda, Y., Ohgaki, T. and Futami, T. (1986) The morphology of single lateral vestibulospinal tract axons in the lower cervical spinal cord of the cat. *J. Comp. Neurol.*, 249: 226–241.

Shinoda, Y., Ohgaki, T., Sugiuchi, Y. and Futami, T. (1992) Morphology of single medial vestibulospinal tract axons in the upper cervical spinal cord of the cat. *J. Comp. Neurol.*, 316: 151–172.

Sugiuchi,Y., Izawa, Y.and Shinoda, Y. (1995) Trisynaptic inhibition from the contralateral vertical semicircular canal nerves to neck motoneurons mediated by spinal commissural neurons. *J. Neurophysiol.*, 73: 1973–1987.

Wurtz, R.H. and Albano, J.E.(1980) Visual-motor function of the primate superior colliculus. *Annu. Rev. Neurosci.*, 3: 189–226.

M. Norita, T. Bando and B. Stein (Eds.)
Progress in Brain Research, Vol 112
© 1996 Elsevier Science BV. All rights reserved.

CHAPTER 6

A projection linking motor cortex with the LM-suprageniculate nuclear complex through the periaqueductal gray area which surrounds the nucleus of Darkschewitsch in the cat

Satoru Onodera*[1] and T. Philip Hicks[2]

[1]*Department of Anatomy, School of Medicine, Iwate Medical University, Morioka 020, Japan*
[2]*Tissue Regeneration Group, Institute of Biological Sciences, National Research Council of Canada, Building M-54, 1200 Montreal Road, Ottawa, Ontario K1A OR6, Canada*

Whereas a previous study by one of us (Hicks et al., 1986) suggested that periaqueductal gray (PAG) neurons projecting to the lateralis medialis-suprageniculate (LM-SG) complex might mediate transmission of affective-related nociceptive information, our present work suggests instead, a function in processes related to movement. Cells of the nucleus of Darkschewitsch (ND) are known to have reciprocal projections with the motor cortex (MX), in particular with the hand area of MX, and also to project to the rostral medial accessory olivary (MAO) nucleus (Onodera and Hicks, 1995a). That the ND might be related to saccadic oculomotor function, as well as to the control of hand movements through its connections via the olivo-cerebellar circuit, is indicated by the fact that ND receives a strong projection from the substantia nigra pars reticulata and zona incerta (SNR/ZI) and projects directly and/or indirectly to eye movement nuclei (Onodera and Hicks, 1995b). Thus, ND may function in permitting integration of eye-hand motor coordination. This study focussed on the area of PAG surrounding ND. WGA-HRP was injected into MX and many labelled terminals and large neurones were in ND, with lesser numbers being observed in the area of the PAG surrounding ND. After injections into ND and closely adjacent areas, labelled terminals were observed sparsely distributed within a restricted area of the LM-SG complex. After injections into LM-SG area, small neuronal somata were seen in the area of the PAG surrounding ND, but no labelled somata were detected in ND. Thus if the cells of this PAG area, like those of ND, have similar functions owing to their common reciprocal connections with MX, then the small neurones in PAG projecting to LM-SG may constitute an important link in the circuitry subserving visual processing and/or the regulation of orienting movements of the hand, head and eye.

Introduction

Early studies employing a physiological level of analysis indicated that the nucleus of Darkschewitsch (ND) is probably related functionally to inhibitory regulation of the contraction of extraocular muscles (Szentágotai and Scháb, 1956; Scheibel et al., 1961). It was then considered that the ND was the main centre responsible for the inhibition of eye movements. However, the details of the circuitry through which ND accomplishes this task in concert with the eye movement nuclei remains unknown. Since there was considerable doubt by many investigators regarding the existence of a direct projection from the ND to oculomotor centres (Graybiel and Hartwieg, 1974; Steiger and Büttner-Ennever, 1979; Horn and Büttner-Ennever, 1990), and since the ND was known to receive a prominent projection from the

---

*Corresponding author.

motor cortex (Nakamura et al., 1983a,b; Leichnetz et al., 1984; Leichnetz, 1986; Saint-Cyr, 1987; Rutherford et al., 1989), it was held in many quarters that the ND is related to the somatomotor system, rather than to the oculomotor system. However, our recent work (Onodera and Hicks, 1993, 1995a,b) has shown that the ND and the area that surrounds it, receive projections from the substantia nigra pars reticulata (SNr) and the zona incerta (ZI). These data suggest that the ND and the area of the periaqueductal gray (PAG) which surrounds ND, are related to saccadic eye movement.

Whereas a previous study by one of us (Hicks et al., 1986) suggested that neurones of the caudal PAG projecting to the lateralis medialis-suprageniculate (LM–SG) complex might mediate transmission of affective-related nociceptive information, our present work shows that neurones of the rostral PAG that surround ND, also project to the LM–SG complex and this suggests that these neurones might be related to orienting movements of the hand, head and eye within extrageniculo-striate pathways.

As a first step in the analysis of how the LM–SG complex receives and processes the inputs from the motor cortex and SNr, and how this might be related to visual, auditory and somatosensory inputs, we have chosen fibre-tracing methods using WGA-HRP and immunocytochemical techniques.

Materials and methods

WGA-HRP injection

Nine adult cats (2.8–4 kg) of both sexes were used. After being anaesthetised with sodiumpentobarbital (30 mg/kg), each cat received an injection of 0.007–6.0 μl of 2% WGA-HRP (Toyobo) solution into one of the following: the ND, the LM–SG complex, the motor cortex and the oculomotor nucleus. The injections were done stereotaxically using a glass micropipette connected to a 1-μl Hamilton microsyringe. Animals were allowed to survive for 2–3 days after the injection, and again they were anaesthetised

deeply and perfused transcardially with 500 ml of warm (37°C) physiological saline followed by 2000 ml of fixative containing 1% paraformaldehyde and 2.5% glutaraldehyde in 0.1 M phosphate buffer (PB, pH 7.4), and then by 1000 ml PB. The whole brain was stored overnight at 4°C in 0.1 M PB. Coronal sections (50-μm thick) were cut using a vibratome and collected in 0.1 M PB. Sections were processed for the histochemical demonstration of HRP with the tetramethylbenzidine method (Mesulam, 1978) and counterstained with 1% Neutral Red for light microscopy.

Immunohistochemical method

The distribution of immunoreactivity of GABA in the ND and adjacent area was examined in two cats by the ABC method (rat polyclonal anti-GABA, provided courtesy of Dr. A. Towle, ETI, diluted 1:1000), according to previously described procedures (Onodera and Hicks, 1995a).

Results

WGA-HRP injection studies

LM–SG complex projection from the area of the PAG surrounding ND

Cat ND 11 (0.01-μl injection, killed after 69-h survival post-injection; Figs. 1 and 2). After injection of WGA-HRP into the left ND, almost the entire part of the ND and the adjacent area of the PAG were found to be heavily stained except for the ventromedial edge of the ND, including the partial involvement of the interstitial nucleus of Cajal (Nint) and the surrounding reticular formation (Fig. 1A). Dense terminal labelling was seen ipsilaterally within the inferior olive and this became concentrated in the rostral half of the medial accessory olive (MAO) (Fig. 1B), although the rostrolateral edge of the MAO was labelled only scarcely. The caudal half of the MAO was labelled only weakly, and the ventral lamella of the principal olive (PO) was not labelled at all. Therefore, in this case, it could be determined that WGA-HRP was taken up from the greater

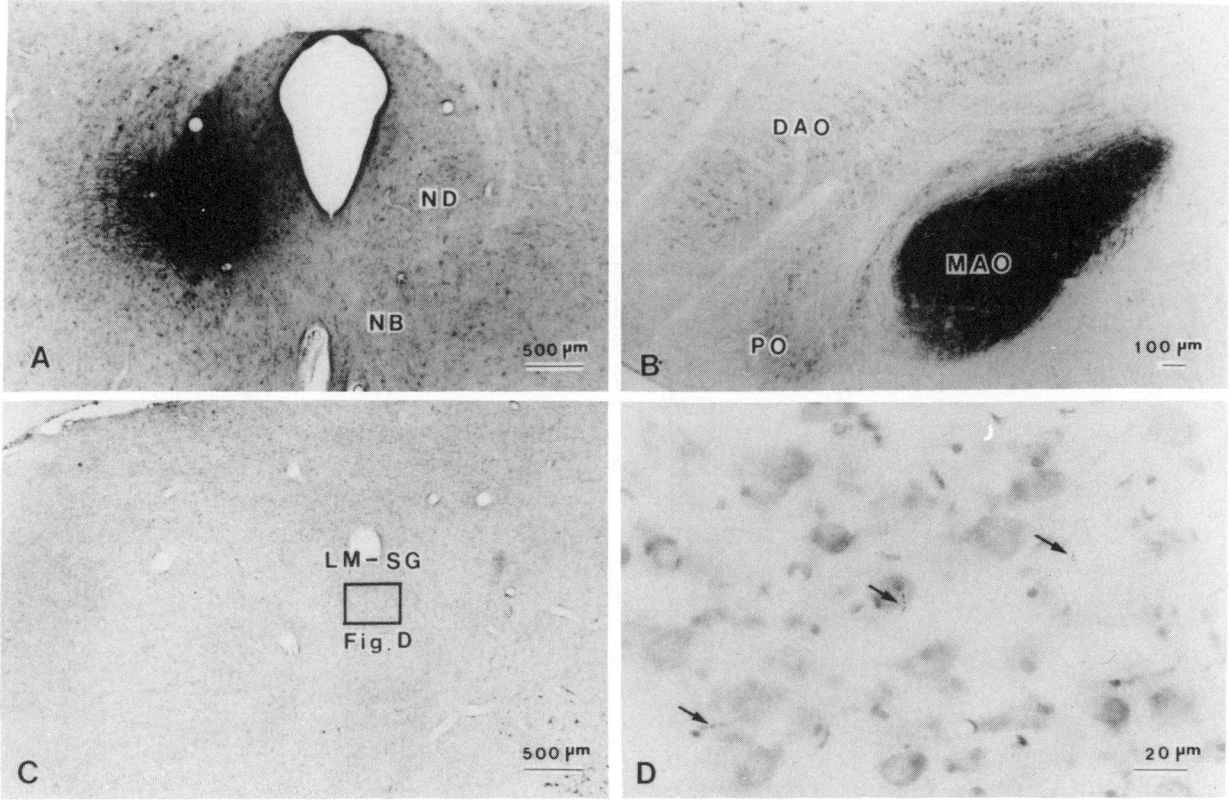

Fig. 1. (A) Photomicrograph showing the maximal extent of injected WGA-HRP in cat ND 11 (see Fig. 2). (B) Photomicrograph from cat ND 11 showing terminal labelling in the rostral part of the MAO. (C,D) Photomicrographs of LM–SG complex in cat ND 11 at low magnification (C); the distribution of WGA-HRP-labelled terminals (arrows) at high magnification (D).

part of the ND, excepting from it's ventromedial edge, and from a portion only of the Nint.

Some labelled terminals were observed ipsilaterally in the motor cortex, the intermediate and deep gray layers of the superior colliculus (SC) and a restricted portion of the LM–SG complex (Figs. 1C,D and 2A). Labelled fibres descended ipsilaterally through the medial tegmental tract (MTT) ventral to the medial longitudinal fasciculus (MLF) and terminated bilaterally in the oculomotor, trochlear and abducens nuclei. Some terminal fibres were observed in the area surrounding the oculomotor nucleus (Fig. 2B). On the ipsilateral side, a number of labelled cells were in layer V of the motor cortex in a very restricted area of the caudal part of the SNr (Fig. 2B). On the contralateral side, many labelled cells

were found in the anterior and posterior interposed cerebellar nuclei (NIA and NIP), as well as in a portion of the lateral cerebellar nucleus (NL), and some labelled cells in the medial cerebellar nucleus (NM). Bilaterally (ipsilaterally dominant), many labelled cells were observed in the medial part of the ZI, the pedunculo-pontine tegmental nuclei, the locus coeruleus, cunieform nucleus, and vestibular nuclei.

Cat ND 28 (0.02-µl injection, killed after 53-h survival post-injection, Figs. 3 and 4). After injection of WGA-HRP into the central region of the LM–SG complex (Figs. 3A and 4), some small labelled neurones were seen ipsilaterally in the rostral PAG surrounding ND, but no labelled somata were detected in the ND (Figs. 3B and 4).

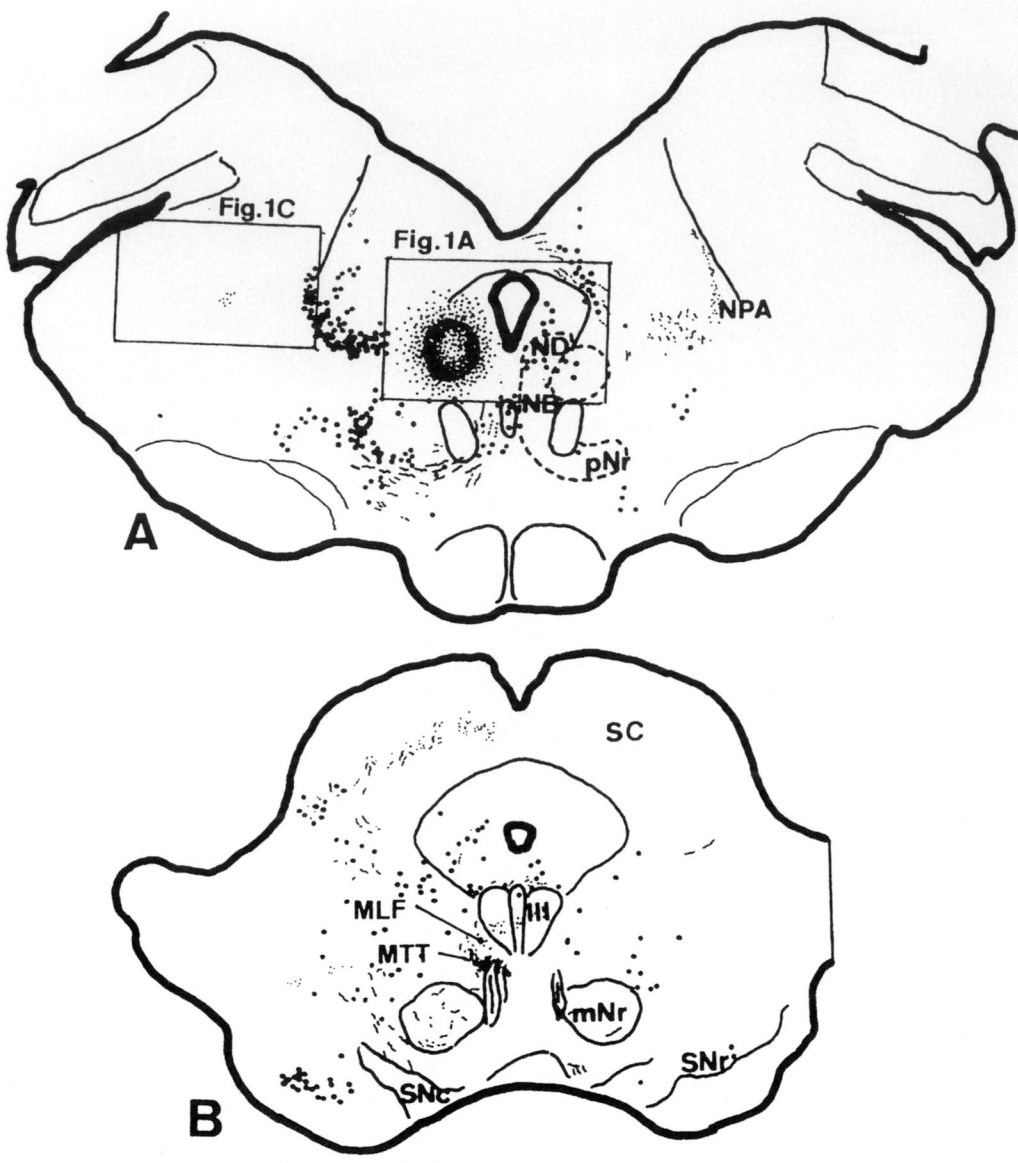

Fig. 2. Scale drawings showing the distribution of WGA-HRP-labelled cells (large dots), labelled fibres (fine lines) and labelled terminals (small dots) in frontal sections of the mesodiencephalic area in cat ND 11. One large dot represents one labelled cell. In the upper drawing (A) the injected WGA-HRP was well restricted to the area comprising the ND plus some small encroachment into a part of Nint (see Fig. 1A). The lower section (B) shows labelled fibres descending through the MTT, labelled terminals in the oculomotor nucleus and the area adjacent to it.

Reciprocal connection between the motor cortex and ND

Cat ND 5 (6.0-μl injection, killed after 40-h survival post-injection, Fig. 5). After the placement of multiple injections of WGA-HRP into motor cortex, a wide-spread area of motor cortex as well as of area 6, was seen to be labelled heavily (Fig. 5A).

Densely labelled terminals were observed in

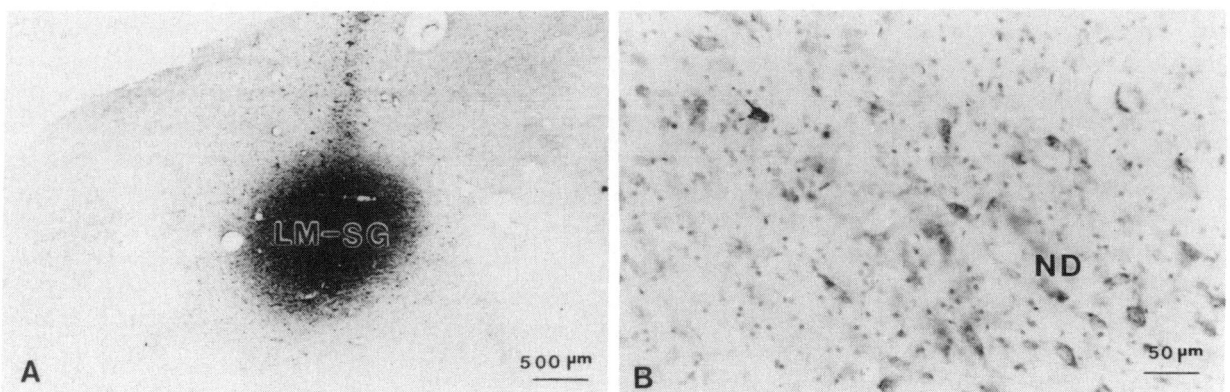

Fig. 3. (A) Photomicrograph showing the maximal extent of injected WGA-HRP in section 26 (top part of Fig. 4) in cat ND 28. (B) Photomicrograph at higher power showing labelled neurones near ND in section 28 of Fig. 4.

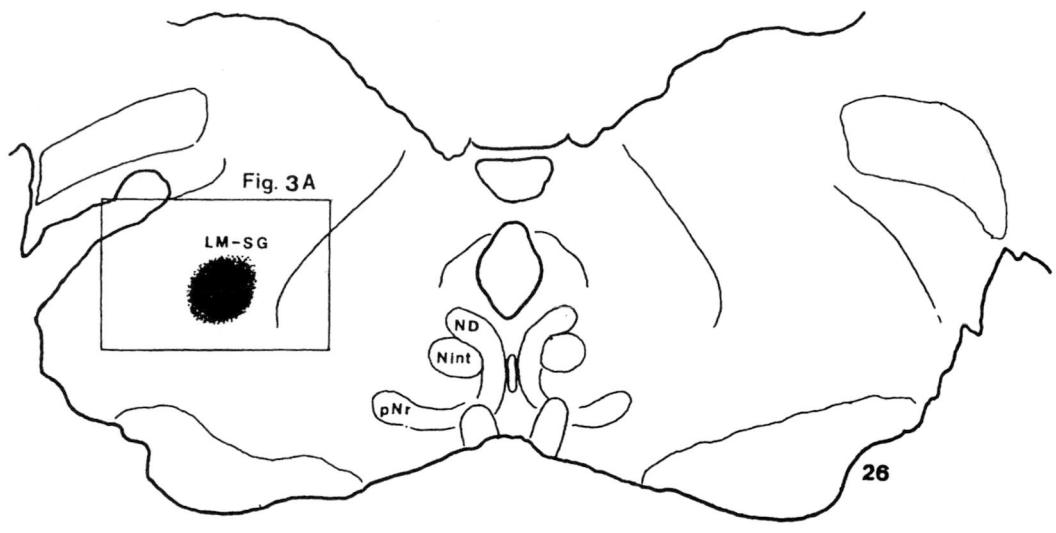

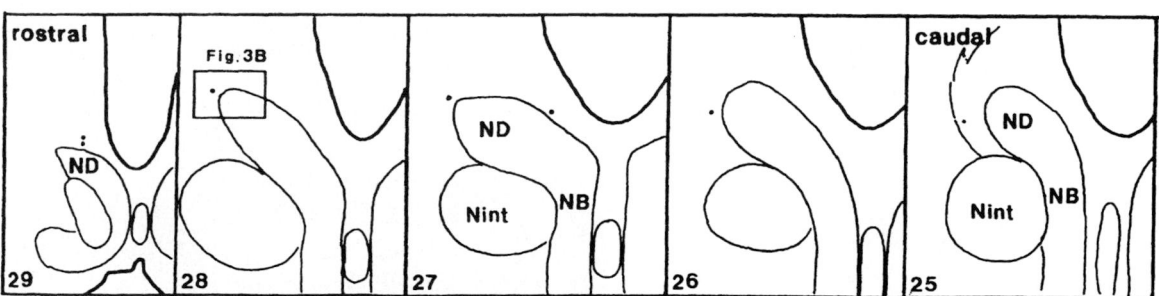

Fig. 4. Upper drawing (section 26) shows the WGA-HRP injection site in the LM−SG complex (transverse section). Lower drawings (transverse sections) through the ND show the distribution of labelled cells (small dots) as seen in a series of five sections.

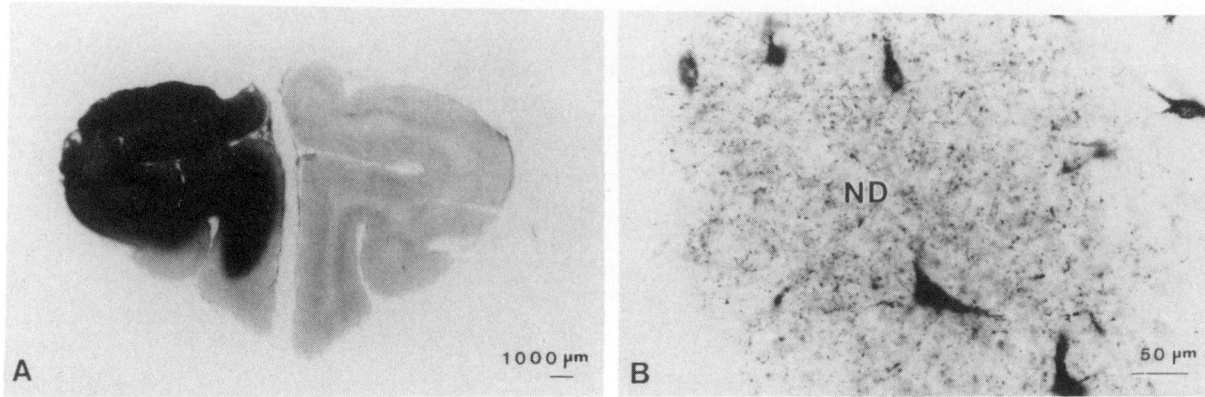

Fig. 5. (A) The maximal extent of the injected WGA-HRP in motor cortex and area 6. (B) WGA-HRP-labelled cells and terminals in the ND and in the area of the PAG surrounding ND.

the ND with lesser numbers being observed in the area of the PAG surrounding ND. Some medium- and large-sized neurones also were noted in the ND and the area surrounding ND (Fig. 5B). Many labelled terminals and some labelled cells were found in the nucleus accessorius medialis of Bechterew (NB) and in the parvicellular red nucleus (pNr).

Oculomotor nucleus and adjacent area projection from ND

Cat ND 25 (0.01-μl injection, killed after 46-h survival post-injection). After injection of WGA-HRP into the left oculomotor nucleus, almost the entire left side of the nucleus appeared stained, except for the midline area; the labelled regions included the supraoculomotor nucleus and a partial involvement of the PAG and reticular formation. The oculomotor nerve fibres were labelled anterogradely. In the ND and the NB, only a few small-sized cells were labelled; most medium- and large-sized neurones were unlabelled (Fig. 6A).

Immunohistochemical studies

GABA

In the ND, NB, pNr and Nint, the labelling with anti-GABA antibody produced an intense pattern of reaction in many terminals, and pro-

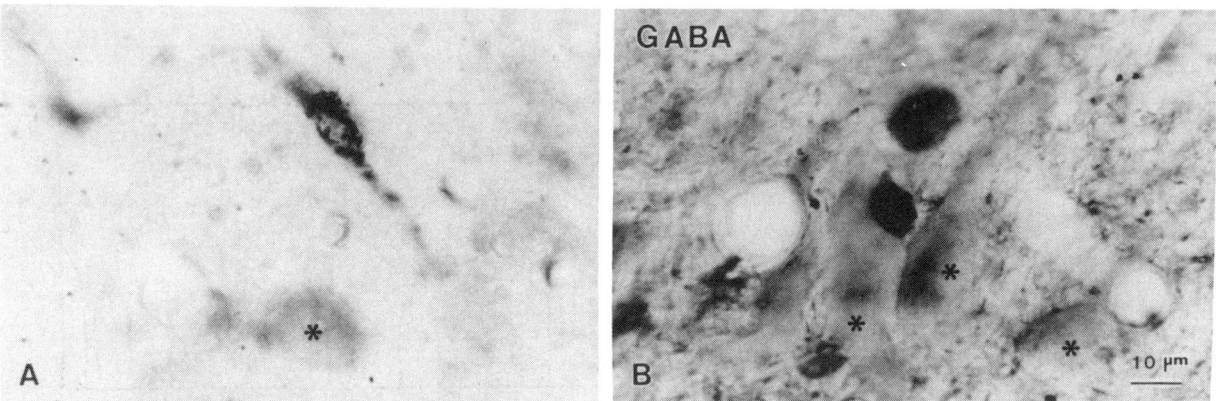

Fig. 6. (A) Photomicrograph showing an unlabelled, medium-sized neurone (*) and small, labelled cells after the injection of WGA-HRP into the oculomotor nucleus and the area adjacent to the nucleus. (B) Photomicrograph showing unlabelled, medium-sized neurones (*) and small, moderately- to strongly-labelled, GABA-positive neurones.

duced moderate to strong patterns of reaction in some somata of the ND and NB (Fig. 6B). This reaction was selective for relatively small-diameter neurones (approx. 10–15 μm) in these areas. Somata of medium diameter that were observable in this material featured only a background level of label. A few similar GABA-positive cells were found in the area of the PAG surrounding ND. In the Nint, no GABA positive neurone was observed and somata of most neurones displayed a background level of label.

Discussion

Many labelled terminals were observed in the intermediate and deep gray layers of the SC in cat ND 11 with ipsilateral labelling being dominant after the injection of WGA-HRP into the ND and the area surrounding ND. Since there is no report known to these authors which describes a direct ND-SC projection (Grofová et al.,1978; Edwards, et al., 1979), we suggested that the neurones of origin of these labelled terminals of the SC are located in a restricted area of the caudal SNr, as shown in Fig. 2B. In support of this suggestion, it is known that SNr neurones located caudo-medially project to the intermediate and deep layers of the SC (Harting et al., 1988) and also project to the ND and the area surrounding it (Onodera and Hicks, 1993). In cat ND 11, labelled-caudal SNr neurones took up WGA-HRP from the injection site, i.e. the ND and the surrounding PAG area, and transported the enzyme to the SC. We would propose that it is collateral branches of this nigro-tectal projection that terminate in the ND and in the region of the PAG which surrounds ND.

Hikosaka and Wurtz (1983) noted that cells of the SNr in monkeys decrease their rates of discharge in relation to saccadic eye movement. Hikosaka and Wurtz (1985a,b) showed that disinhibition mechanism of the nigro-SC pathway produces an excitation of extraocular muscles during saccades. It was known from the work of Matsunami (1972) that neurones of the monkey's ND and the area of the PAG surrounding the ND are activated at high rates of firing, described as

repetitive discharges, prior to and during ocular saccades. The units described by Matsunami (1972) had no ongoing activity whilst the eyes were in a fixed position or were moved slowly, such as in smooth pursuit movements, and their patterns of discharge were independent of the direction of movement of the eye. The burst preceded the onset of saccades by 2–12 ms. He considered that these burst neurones are activated approximately at the same time as the oculomotor neurones, since there was a time lag on the order of 2–12 ms between oculomotor neurone excitation and the start of eye movement. One might propose, therefore, that the ND is involved in producing monosynaptic inhibition of oculomotor neurones during saccades via the nigro-ND-oculomotor pathway. Such an hypothesis corresponds well with the early physiological data (Szentágotai and Scháb, 1956; Scheibel et al., 1961) demonstrating that stimulation of the feline ND plays a unique role in inhibiting extraocular muscles. They showed that the latency of the inhibitory effect on the extraocular muscles is less than 10 ms. This short latency value is similar to the one mentioned by Matsunami (1972) and suggests monosynaptic inhibition between the ND and oculomotor nucleus.

The precise circuitry, whereby this inhibitory process occurs, is unknown. It is interesting to note that labelled descending fibres through the MTT ventral to the MLF terminated in the oculomotor nucleus and the area adjacent to it in cat ND 11 (Fig. 2B). Fibres descending from the ND run through the MTT and those from the Nint run through the MLF (Onodera, 1984). The origin of these labelled terminals, therefore, may not be the Nint, but the ND. The present study has also shown that the Nint has no GABA-positive neurones, but the ND has GABA-positive neurones. Thus, the labelled terminals in the oculomotor nucleus and area surrounding it may convey inhibition directly to the eye movement nuclei from the ND via small, GABA-positive neurones, although it is also possible that these terminals are labelled owing to uptake of WGA-HRP from the Nint and the nucleus of the fields of Forel (other than for ND) (Graybiel and Hartwieg, 1974;

Steiger and Büttner-Ennever, 1979; Horn and Büttner-Ennever, 1990).

Of similar interest in this context is the demonstration of some labelled, small-sized neurones in the ND observed after injection of WGA-HRP into the oculomotor nucleus and the area adjacent to it, in cat ND 25 (Fig. 6A). Similarly small-sized neurones in the ND were GABA positive, as determined by immunohistochemistry (Fig. 6B). Therefore, the possibility exists that the ND may inhibit the oculomotor neurones directly. After injection of cholera toxin-HRP into the extraocular muscles, many labelled aberrant oculomotor neurones were observed in the MLF and PAG between the caudal part of the ND and the oculomotor nucleus (Onodera and Hicks, unpublished observations), and so it is possible that the ND may provide direct inhibitory synaptic contacts with aberrant oculomotor neurones situated outside the oculomotor nucleus. Edwards and Henkel (1978) showed that the dendrites of normal oculomotor neurones extend outside the borders of the oculomotor nucleus. The present study has shown labelled terminals in the area surrounding the oculomotor nucleus (Fig. 2B). Therefore, descending fibres through the MTT ventral to the MLF could made synaptic contacts with normal oculomotor neurones outside the oculomotor nucleus. These data may suggest the reason why many authors have not been able to show evidence for the existence of terminals of the direct ND-oculomotor projection inside the oculomotor nucleus (Szentágotai and Shcab, 1956; Graybiel and Hartwieg, 1974; Steiger and Buttner-Ennever, 1979; Horn and Buttner-Ennever, 1990).

Kase et al. (1986) showed that activity in 'pause' neurones in the monkey's caudal PAG preceded the onset of saccades by nearly 35 ms, and suggested that these caudal PAG cells may regulate presaccadic activity that influences SC cells and/or long-lead burst units in the pontine reticular formation. Hicks et al. (1986) showed in the cat that caudal PAG neurones project to the LM–SG complex. These neurones may be related to saccadic movements rather than to nociceptive

information. There is little information at present about the fibre connections of neurones projecting to the LM–SG complex at the level of the caudal PAG. Thus, the functional significance of these neurones and their projections remains to be studied.

Using the HRP method in the cat, it has been shown quite frequently that the ND, NB and pNr project to the motor cortex (Avendaño, 1976; Itoh and Mizuno, 1976; Rutherford et al., 1989; Onodera and Hicks, 1995a; the present study). Using the same methodologies, Leichnetz (1986) showed quite convincingly the lack of such a motor cortical projection in the monkey. Thus there may be significant species differences in this respect between the cat and the monkey. The ventrolateral nucleus of the cat's thalamus receives inputs from the NL and the interposed cerebellar nuclei (NI), and projects to cortical areas 4 and 6. The terminals of this projection in layer III tend to aggregate into patches, which are arranged in a longitudinal strip oriented in the rostrocaudal direction (Shinoda et al., 1985, 1987; Shinoda and Kakei, 1989). The ND, NB and pNr also receive collateral projections from the NL and NI (Hirai et al., 1982; Kawamura et al., 1982; Sugimto et al., 1982), and these nuclei project to cortical areas 4 and 6 as well (Avendaño, 1976; Itoh and Mizuno, 1976; Rutherford et al., 1989; Onodera and Hicks, 1995a; the present study).

After the injection of WGA-HRP into the SNr, we observed labelled terminals in the NB and pNr (Onodera and Hicks, unpublished data). These anatomical data support the interesting finding (Magariños-Ascone, et al., 1992) that most SNr neurones of the monkey enhance their activity to movements of the limb in the absence of any participation of somatosensory feedback. This report suggests that neurones of the SNr are related not only to eye and head movements, but also to movements of the limbs and body.

Physiological studies (Mano et al., 1989, 1991) showed that there is a somatotopical representation in the monkey's lateral cerebellar hemispheres; lobules III and IV, V, VI, and crus I and IIa, relate to leg, arm, oral, and saccadic eye movements, respectively, since these authors observed limb and saccadic movement-related Purkinje cells. They pointed out that somatotopically restricted areas of the lateral cerebellar hemisphere (zones D1 and D2) play a role in the initiation of preprogramming of each voluntary movement of the limbs, body, head and eyes. Cicirata et al. (1992) showed that a variety of the rat's body segments are activated by stimulation of different zones of the various segments of the NL. This report also suggested the existence of a somatotopical pattern in the lateral cerebellar hemisphere.

There have been described precise topographical olivary projections arising from the cat's ND, NB, pNr and adjacent nuclei (Onodera, 1984; Onodera and Hicks, 1995a), the pretectal nuclei (Itoh, 1977; Itoh et al., 1983; Kawamura and Onodera,1984) and the SC (Weber et al., 1978; Kyuhou and Matsuzaki, 1991). Furthermore, with respect to the next stage of information processing from the olive, the cerebellum is known to be organised into longitudinal strips, these consisting of zones A and B of the medial cerebellum, zones C1, C2 and C3 of the intermediate cerebellum, and zones D1 and D2 of the lateral cerebellum (Growenwegen and Voogd, 1977; Growenwegen et al., 1979; Kawamura and Hashikawa, 1979; Brodal and Kawamura, 1980; Gerrits and Voogd, 1982). These data therefore support the existence of a topographical mesodiencepalo-olivo-cerebellar circuit; the SC projects to zone A (medial cerebellum) via the medial part of the caudal MAO, the ND projects to zone C2 (intermediate cerebellum) via the rostral MAO, the pNr and NB project to zone D1 and D2 (lateral cerebellum) via the dorsal and ventral lamella of the PO, respectively (Onodera, 1984). Recently, Onodera and Hicks (1995a) showed that a certain degree of medio-lateral and dorso-ventral topographic correlation exists within the ND-rostral MAO pathway. This finding strongly suggests a more precisely somatotopical pattern in the motor cortico-ND-rostral MAO-zone C2 projection. The model of the cerebro-cerebellar communication

system proposed by Allen and Tsukahara (1974) suggested that the intermediate cerebellum is related to the actual execution of movements, and of their updating, and the lateral cerebellum — a structure without a source of peripheral inputs — is related to the participation in processes of motor planning.

A diagram of the topographical correlation within the SNr-mesodiencephalo-cerebellar circuit is provided in Fig. 7. The SNr has been recognized as one of major outputs of the basal ganglia, whilst it is known to receive afferents predominantly from the striatum, projecting to the SC as well as to the thalamus. Until recently, no clear connection had been established between the cerebellar and basal ganglionic systems at subcortical levels. Our recent data (Onodera and Hicks, 1993; 1995a,b; the present study) showed that the SNr projects to the ND, NB and pNr as well as to the SC. These data have led us to formulate the novel hypothesis that the caudo-medial SNr may convey information concerning muscle activity, programmed by the basal ganglia in the abscence of peripheral feedback, to the SC, ND, NB and pNr. These latter nuclei then project this information relevant to the motor action to the longitudinal zones of the cerebellum via the inferior olivary complex, as described above.

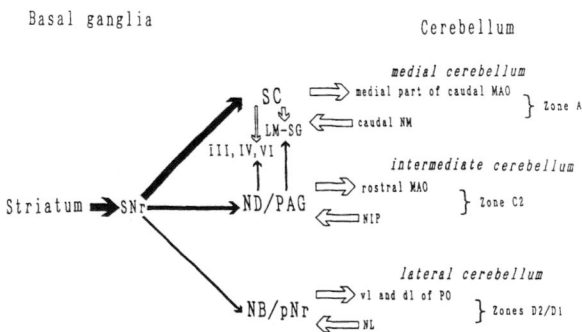

Fig. 7. Schematic presentation summarising the projections from the SNr to the longitudinal cerebellar zones via the mesodiencephalic structures as discussed in the present study. Filled-in arrows show GABAergic inhibitory projections and open arrows indicate excitatory projections.

The medial part of the SNr sends a strong projection to the intermediate and deep layers of the SC (Harting et al., 1988). The LM–SG complex receives inputs from the intermediate and deep layers of the SC and has reciprocal connections with the insular visual area and ectosylvian visual area (Graybiel and Berson, 1980; Mucke et al., 1982; Olson and Graybiel, 1983, 1987; Hicks et al., 1986, 1988a,b; Norita et al., 1986, 1991; Norita and Katoh, 1987; Abramson and Chalupa, 1988; Katoh and Benedek, 1995; Katoh et al., 1995). The substantia nigra pars lateralis projects to the superficial layers of the SC (Harting et al., 1988) and also projects directly to the LM–SG complex (Takada et al., 1984) as well as the part of the inferior colliculus which is related to acoustico-motor processing, such as body orientation to sound (Moriizumi et al., 1991). Katoh and Deura (1993) showed that the caudal part of the NM projects to the LM–SG complex as well as to the SC (Hirai et al., 1982; Kawamura et al., 1982; Sugimoto et al., 1982). This caudal NM receives fibres from the cerebellar visual and auditory areas (Walberg and Jansen, 1964; Courville and Diakiw, 1976; Hirai et al., 1982). The SC projects to the medial part of the caudal MAO (Weber et al.,1978). Therefore, the SC and LM–SG complex are related to the medial cerebellar circuit (zone A). These nuclei have visual, auditory and somatosensory receptive fields and may be related to visual processing and/or to the co-ordination and regulation of orienting movements of the limbs, head and eyes (Hicks et al., 1984; Middlebrooks and Knudsen, 1987; Hutchins and Updyke, 1989; Clemo and Stein, 1991; Meredith et al., 1992). The medial cerebellar circuit may be related to the direction of and size of, the saccadic eye movement, based on peripheral information relevant to the modalities of vision, audition and somatosensation.

The SNr sends a moderate projection to the ND and a more sparse one to the surrounding area in the PAG (Onodera and Hicks, 1993). These areas have reciprocal connections with the motor cortex (especially the hand area: Avendaño,

1976; Itoh and Mizuno, 1976; Nakamura et al., 1983a,b; Saint-Cyr, 1987; Rutherford et al., 1989; the present study). The ND receives inputs mainly from NIP (Hirai et al., 1982; Onodera and Hicks, 1995a) and projects to the rostral MAO (Onodera, 1984; Onodera and Hicks, 1995a). Therefore, the ND and PAG region which surrounds ND, is related to the intermediate cerebellar circuit (zone C2) and may contribute exclusively to the actual execution of movements of the limbs, body, head and saccades (i.e. inhibition of the extraocular muscles during a saccade). PAG neurones surrounding the ND may convey information to the LM–SG complex of the actual execution of movements of the limbs, body, head, and especially of saccadic movements (Fig. 8). As these WGA-HRP-labelled neurones were small-sized, similar

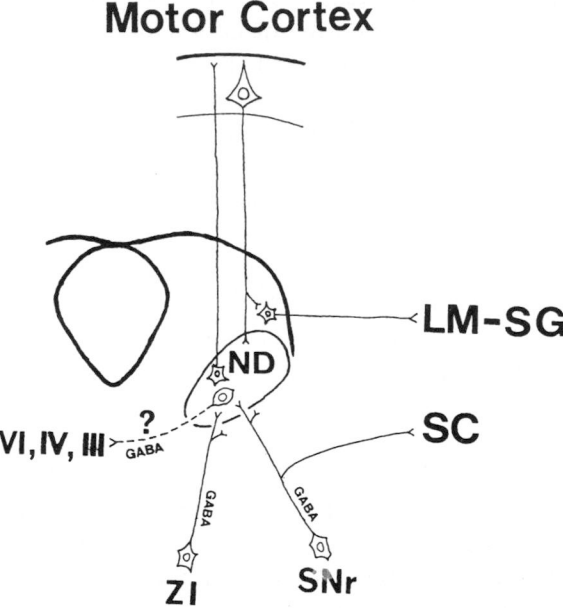

Fig. 8. Diagram depicting the fibre connections of the ND and the area of the PAG surrounding the ND based on both the present findings and on the results from previous studies. The olivary projection from the ND and the ND projections from the cerebellar and pretectal nuclei and the spinal cord are omitted. Question mark refers to a presumed, but not established role for a GABA-containing neuronal projection to oculomotor nuclei III, IV and VI.

to the GABA-positive ones observed in the ND and PAG, this information may be of an inhibitory, rather than of an excitory nature. As many extrinsic GABA-positive terminals were observed in the LM−SG complex (Norita and Kato, 1987; Onodera et al. 1991), rostrally-located PAG neurones may represent a viable candidate for the source of extrinsic GABA-positive terminals in the LM−SG complex. Caudal PAG neurones may convey different information related to presaccades to the LM−SG complex, because these neurones may have a long latency response, as described above. Therefore, both LM−SG complex projections from the rostral and caudal zones of the PAG may constitute important links in the circuitry subserving visual processing and/or the co-ordination and regulation of orienting movements of the limbs, body, head and eyes.

The SNr sends a weak projection to the NB and pNr (Onodera and Hicks, unpublished data). These nuclei receive afferents mainly from the NL (Hirai et al., 1982; Kawamura et al., 1982; Sugimoto et al., 1982), areas 4, 6 and the frontal eye field area (Hartman-von Monakow et al., 1979; Leichnetz, 1982, 1986; Leichnetz et al., 1984; Stanton et al., 1988; Shook et al., 1990) and project to the ventral and dorsal lamellae of the principal olive (PO), respectively (Onodera, 1984). Therefore, the NB and pNr are related to the lateral cerebellar circuit (zone D2,1) and may play a role in the initiation or preprogramming of voluntary movements of the limbs, body, head and of saccades. However, the details of the circuitry through which NB and pNr are related to saccadic eye movements in concert with the eye movement nuclei, remains unknown. After the injection of tritiated amino acids into the rostromedial part of the NL, labelled terminals can be observed in the subnucleus ventrolateralis of the pNr and in the caudal portion of the dorsal lamella of the PO (Onodera and Hicks, unpublished data). Therefore, these data support the existence of a precisely somatotopical relationship in the lateral cerebellar circuit, such as has been described in

recent physiological studies (Mano et al., 1989, 1991).

References

Abramson, B.P. and Chalupa, L.M. (1988) Multiple pathways from the superior colliculus to the extrageniculate visual thalamus of the cat. J. Comp. Neurol., 271: 397−418.
Allen, G.I. and Tsukahara, N. (1974) Cerebrocerebellar communication systems. Physiol. Rev., 54: 957−1006.
Avendaño, C. (1976) Proyecciones desde el nucleo de Darkschewitsch al gyrus sigmoideus en el gato, determinadas mediante el transporte axonal retrogrado de horesradish peroxidase. Ann. Anat., 25: 27−35.
Brodal, A. and Kawamura, K. (1980) Olivocerebellar projection: a review. Adv. Anat. Embryol. Cell. Biol., 64: 1−137.
Cicirata, F., Angaut, P., Serapide, M.F., Panto, M.R. and Nicotra, G. (1992) Multiple representation in the nucleus lateralis of the cerebellum: An electrophysiologic study in the rat. Exp. Brain Res., 89: 352−362.
Clemo, H.R. and Stein, B.E. (1991) Receptive field properties of somatosensory neurons in the cat superior colliculus. J. Comp. Neurol. 314: 534−544.
Courville, J. and Diakiw, N. (1976) Cerebellar corticonuclear projection in the cat. The vermis of the anterior and posterior lobes. Brain Res. 110: 1−20.
Edwards, S.B. and Henkel, C.K. (1978) Superior colliculus connections with the extraocular motor nuclei in the cat. J. Comp. Neurol., 179: 451−467.
Edwards, S.B., Ginsburgh, C.L., Henkel, C.K. and Stein, B.E. (1979) Sources of subcortical projections to the superior colliculus in the cat. J. Comp. Neurol. 184: 309−330.
Gerrits, N.M. and Voogd, J. (1982) The climbing fiber projection to the flocculus and adjacent paraflocculus in the cat. Neuroscience, 7: 2971−2991.
Graybiel, A.M. and Berson, D.M. (1980) Histochemical identification and afferent connections of subdivisions in the lateralis posterior-pulvinar complex and related thalamic nuclei in the cat. Neuroscience, 5: 1176−1238.
Graybiel, A.M. and Hartwieg, E.A. (1974) Some afferent connections of the oculomotor complex in the cat: an experimental study with tracer techniques. Brain Res., 81: 543−551.
Groenewegen, H.J. and Voogd, J. (1977) The parasagittal zonation within the olivocerebellar projection. I. Climbing fiber distribution in the vermis of cat cerebellum. J. Comp. Neurol., 174: 417−488.
Groenewegen, H.J., Voogd, J. and Freedman, S.L. (1979) The parasagittal zonation within the olivocerebellar projection. II. Climbing fiber distribution in the intermediate and hemispheric parts of cat cerebellum. J. Comp. Neurol., 183: 551−602.

Grofová, I., Ottersen, O.P. and Rinvik, E. (1978) Mesencephalic and diencephalic afferents to the superior colliculus and periaqueductal gray substance demonstrated by retrograde axonal transport of horseradish peroxidase in the cat. *Brain Res.*, 146: 205–220.

Harting, J.K., Huerta, M.F., Hashikawa, T., Weber, J.T. and Van Lieshout, D.P. (1988) Neuroanatomical studies of the nigrotectal projection in the cat. *J. Comp. Neurol.*, 278: 615–631.

Hartmann-von Monakow, K., Akert, K. and Künzle, H. (1979) Projections of precentral and premotor cortex to the red nucleus and other midbrain areas in Macaca fascicularis. *Exp. Brain Res.*, 34: 91–105.

Hicks, T.P., Watanabe, S., Miyake, A. and Shoumura, K. (1984) Organization and properties of visually responsive neurones in the suprageniculate nucleus of the cat. *Exp. Brain Res.* 55: 359–367.

Hicks, T.P., Stark, C.A. and Fletcher, W.A. (1986) Origins of afferents to visual suprageniculate nucleus of the cat. *J. Comp. Neurol.*, 246: 544–554.

Hicks, T.P., Benedek, G. and Thurlow, G.A. (1988a) Organization and properties of neurons in a visual area within the insular cortex of the cat. *J. Neurophysiol.*, 60: 397–421.

Hicks, T.P., Benedek, G. and Thurlow, G.A. (1988b) Modality specificity of neuronal responses within the cat's insula. *J. Neurophysiol.*, 60: 422–437.

Hikosaka, O. and Wurtz, R.H. (1983) Visual and oculomotor functions of monkey substantia nigra pars reticulata. I. Relation of visual and auditory responses to saccades. *J. Neurophysiol.*, 49: 1230–1253.

Hikosaka, O. and Wurtz, R.H. (1985a) Modification of saccadic eye movements by GABA-related substances. I. Effect of muscimol and bicuculline in monkey superior colliculus. *J. Neurophysiol.* 53: 266–291.

Hikosaka, O. and Wurtz, R.H. (1985b) Modification of saccadic eye movements by GABA-related substances. II. Effects of muscimol in monkey substantia nigra pars reticulata. *J. Neurophysiol.*, 53: 292–308.

Hirai, T., Onodera, S. and Kawamura, K. (1982) Cerebellotectal projections studied in cats with horseradish peroxidase or tritiated amino acids axonal transport. *Exp. Brain Res.*, 48: 1–12.

Horn, A.K.E. and Büttner-Ennever, J.A. (1990) The time course of retrograde transsynaptic transport of tetanus toxin fragment C in the oculomotor system of the rabbit after injection into extraocular eye muscles. *Exp. Brain Res.*, 81: 353–362.

Hutchins, B. and Updyke, B.V. (1989) Retinotopic organization within the lateral posterior complex of the cat. *J. Comp. Neurol.*, 285: 350–398.

Itoh, K. (1977) Efferent projections of the pretectum in the cat. *Exp. Brain Res.* 30: 89–105.

Itoh, K. and Mizuno, N. (1976) Direct projections from the mesodiencephalic midline areas to the pericruciate cortex in the cat: an experimental study with the hoseradish peroxidase method. *Brain Res.*, 116: 492–497.

Itoh, K., Takada, M., Yasui, Y., Kudo, M. and Mizuno, N. (1983) Direct projections from the anterior pretectal nucleus to the dorsal accessory olive in the cat: an anterograde and retrograde WGA-HRP study. *Brain Res.*, 272: 350–353.

Kase, M., Nagata, R. and Kato, M. (1986) Saccade-related activity of periaqueductal gray matter of the monkey. *Invest. Ophthalmol. Vis. Sci.*, 27: 1165–1169.

Katoh, Y.Y. and Benedek, G. (1995) Organization of the colliculo-suprageniculate pathway in the cat: a wheat germ agglutinin–horseradish peroxidase study. *J. Comp. Neurol.*, 352: 381–397.

Katoh, Y.Y. and Deura, S. (1993) Direct projections from the cerebellar fastigial nucleus to the thalamic suprageniculate nucleus in the cat studied with the anterograde and retrograde axonal transport of wheat germ agglutinin-horseradish peroxidase. *Brain Res.*, 617: 155–158.

Katoh, Y.Y., Benedek, G. and Deura, S. (1995) Bilaterl projectins from the superior colliculus to the suprageniculate nucleus in the cat: a WGA-HRP/ double fluorescent tracing study. *Brain Res.*, 669: 298–302.

Kawamura K. and Hashikawa, T. (1979) Olivocerebellar projections in the cat studied by means of anterograde axonal transport of labeled amino acids as tracers. *Neuroscience*, 4: 1615–1633.

Kawamura, K. and Onodera, S. (1984) Olivary projections from the pretectal region in the cat studied with horseradish peroxidase and tritiated amino acids axonal transport. *Arch. Ital. Biol.*, 122: 155–168.

Kawamura, S., Hattori, S., Higo, S. and Matsuyama, T. (1982) The cerebellar projections to the superior colliculus and pretectum in the cat: an autoradiographic and horseradish peroxidase study. *Neuroscience*, 7: 1673–1689.

Kyuhou, S.-I. and Matsuzaki, R. (1991) Topographical organization of the tecto-olivo-cerebellar projection in the cat. *Neuroscience*, 41: 227–241.

Leichnetz, G.R. (1982) The medial accessory nucleus of Bechterew: a cell group within the anatomical limits of the rostral oculomotor complex receives a direct prefrontal projection in the monkey. *J. Comp. Neurol.*, 210: 147–151.

Leichnetz, G.R. (1986) Afferent and efferent connections of the dorsolateral precentral gyrus (area 4, hand/arm region) in the macaque monkey, with comparisons to area 8. *J. Comp. Neurol.*, 254: 460–492.

Leichnetz, G.R., Spencer, R.F. and Smith, D.J. (1984) Cortical projections to nuclei adjacent to the oculomotor complex in the medial dien-mesencephalic tegmentum in the monkey. *J. Comp. Neurol.*, 228: 359–387.

Magariños-Ascone, C., Buño, W. and García-Austt, E. (1992) Activity in monkey substantia nigra neurons related to a simple learned movement. *Exp. Brain Res.*, 88: 283–291.

Mano, N., Kanazawa, I. and Yamamoto, K. (1986) Complex-spike activity of cerebellar Purkinje cells related to wrist tracking movement in monkey. *J. Neurophysiol.*, 56: 137–158.

Mano, N., Ito, Y. and Shibutani, H. (1991) Saccade-related Purkinje cells in the cerebellar hemispheres of the monkey. *Exp. Brain Res.*, 84: 465–470.

Matsunami, K. (1972) Saccadic eye movement and neurons in the central gray area in awake monkeys. *Brain Res.*, 38: 217–221.

Meredith, M.A., Wallace, M.T. and Stein, B.E. (1992) Visual, auditory and somatosensory convergence in output neurons of the cat superior colliculus: multisensory properties of the tecto-reticulo-spinal projection. *Exp. Brain Res.* 88: 181–186.

Mesulam, M.-M (1978) Tetramethyl benzidine for horseradish peroxidase neurohistochemistry: a non-carcinogenic blue reaction-product with superior sensitivity for visualizing neural afferents and efferents. *J. Histochem. Cytochem.*, 26: 106–117.

Middlebrooks, J.C. and Kundsen, E.I. (1987) Changes in external ear position modify the spatial tuning of auditory units in the cat's superior colliculus. *J. Neurophysiol.*, 57: 672–687.

Moriizumi, T., Leduc-Cross, B. Wu, J.-Y. and Hattori, T. (1991) Separate neuronal populations of the rat substantia nigra pars lateralis with distinct projection sites and transmitter phenotypes. *Neuroscience*, 46: 711–720.

Mucke, L., Norita, M., Benedek, G. and Creutzfeldt, O. (1982) Physiologic and anatomic investigation of a visual cortical area situated in the ventral bank of the anterior ectosylvian sulcus of the cat. *Exp. Brain Res.*, 46: 1–11.

Nakamura, Y., Kitao, Y. and Okoyama, S. (1983a) Projections from the pericruciate cortex to the nucleus of Darkschewitsch and other structures at the mesodiencephalic junction in the cat. *Brain Res. Bull.*, 10: 517–521.

Nakamura, Y., Kitao, Y. and Okoyama, S. (1983b) Cortico-Darkschewitsch-olivary projection in the cat: an electron microscope study with the aid of horseradish peroxidase tracing technique. *Brain Res.*, 274: 140–143.

Norita, M. and Katoh, Y. (1987) The GABAergic neurons and axon terminals in the lateralis medialis-suprageniculate nuclear complex of the cat: GABA-immunocytochemical and WGA-HRP studies by light and electron microscopy. *J. Comp. Neurol.*, 263: 54–67.

Norita, M., Mucke, L., Benedek, G., Albowitz, B., Katoh, Y. and Creutzfeldt, O.D. (1986) Connections of the anterior ectosylvian visual area (AEV). *Exp. Brain Res.* 62: 225–240.

Norita, M., Hicks, T.P., Benedek, G. and Katoh, Y. (1991) Organization of cortical and subcortical projections to the feline visual insular area. *J. Hirnforsch.*, 32: 119–134.

Olson, C.R. and Graybiel, A.M. (1983) An outlying visual area in the cerebral cortex of the cat. In: J.-P. Changeux, J. Glowinski, M. Imbert and F.E. Bloom (Eds.), *Molecular and Cellular Interactions Underlying Higher Brain Functions*,

Prog. Brain Res., Vol. 58, Elsevier, Amsterdam, pp. 239–245.

Olson, C.R. and Graybiel, A.M. (1987) Ectosylvian visual area of the cat: location, retinotopic organization, and connections. *J. Comp. Neurol.*, 261: 277–294.

Onodera, S. (1984) Olivary projections from the mesodiencephalic structures in the cat studied by means of axonal transport of horseradish peroxidase and tritiated amino acids. *J. Comp. Neurol.*, 227: 37–49.

Onodera, S. and Hicks, T.P. (1993) Demonstration of transmitters and connectivity patterns in the nucleus of Darkschewitsch of the cat by immunocytochemistry and WGA-HRP. *Soc. Neurosci. Abstr.*, 19: 749.

Onodera, S. and Hicks, T.P. (1995a) Patterns of transmitter labelling and connectivity of the cat's nucleus of Darkschewitsch: a wheat germ agglutinin-horseradish peroxidase and immunocytochemical study at light and electron microscopical levels. *J. Comp. Neurol.*, 361: 553–573.

Onodera, S. and Hicks, T.P. (1995b) The substantia nigra pars reticulata and zona incerta project to the nucleus of Darkschewitsch in the cat. *Fourth IBRO World Cong. Neurosci. Abstr.*, 323.

Onodera, S., Norita, M., Takeda, K. and Hicks, T.P. (1991) Disposition of amino acid synaptic transmitters, acetylcholine and substance P in the LM-suprageniculate nuclear complex of the cat's thalamus. *Neuroscience Res.*, 11: 134–140.

Rutherford, J.G., Zuk-Harper, A. and Gwyn, D.G. (1989) A comparison of the distribution of the cerebellar and cortical connections of the nucleus of Darkschewitsch (ND) in the cat: a study using anterograde and retrograde HRP tracing techniques. *Anat. Embryol.*, 180: 485–496.

Saint-Cyr, J.A. (1987) Anatomical organization of cortico-mesencephalo-olivary pathways in the cat as demonstrated by axonal transport techniques. *J. Comp. Neurol.*, 257: 39–59.

Scheibel, A., Markham, C. and Koegler, R. (1961) Neural correlates of the vestibulo-ocular reflex. *Neurology*, 11: 1055–1065.

Shinoda Y., Kano, M. and Futami, T. (1985) Synaptic organization of the cerebello-thalamo-cerebral pathway in the cat. I. Projection of individual cerebellar nuclei to single pyramidal tract neurons in areas 4 and 6. *Neurosci. Res.*, 2: 133–156.

Shinoda, Y. and Kakei, S. (1989) Distribution of terminals of thalamocortical fibers originating from the ventrolateral nucleus of the cat thalamus. *Neurosci. Lett.*, 96: 163–167.

Shinoda, Y., Sugiuchi, Y. and Futami, T. (1987) Excitatory inputs to cerebellar dentate nucleus neurons from the cerebral cortex in the cat. *Exp. Brain Res.*, 67: 299–315.

Shook, B.L., Schlag-Rey, M. and Schlag, J. (1990) Primate supplementary eye field: I. Comparative aspects of mesencephalic and pontine connections. *J. Comp. Neurol.*, 301: 618–642.

Stanton, G.B., Goldberg, M.E. and Bruce, C.J. (1988) Frontal eye field efferents in the macaque monkey: II. Topography of terminal fields in midbrain and pons. *J. Comp. Neurol.*, 271: 493–506.

Steiger, H.-J. and Büttner-Ennever, J.A. (1979) Oculomotor nucleus afferents in the monkey demonstrated with horseradish peroxidase. *Brain Res.*, 160: 1–15.

Sugimoto, T., Mizuno, N. and Uchida, K. (1982) Distribution of cerebellar fiber terminals in the midbrain visuomotor areas: an autoradiographic study in the cat. *Brain Res.*, 238: 353–370.

Szentágotai, J. and Scháb, R. (1956) A midbrain inhibitory mechanism of oculomotor activity. *Acta Physiol. Acad. Sci. Hung.*, 9: 89–98.

Takada, M., Itoh, K., Yasui, Y., Sugimoto, T. and Mizuno, N. (1984) Direct projections from the substantia nigra to the posterior thalamic regions in the cat. *Brain Res.*, 309: 143–146.

Weber, J.T., Partlow, G.D. and Harting, J.K. (1978) The projection of the superior colliculus upon the inferior olivary complex of the cat: an autoradiographic and horseradish peroxidase study. *Brain Res.*, 144: 369–377.

Walberg, F. and Jansen, J. (1964) Cerebellar corticonuclear projection studied experimentally with silver impregnation methods. *J. Hirnforsch*, 6: 338–354.

M. Norita, T. Bando and B. Stein (Eds.)
Progress in Brain Research, Vol 112

CHAPTER 7

Firing characteristics of neurones in the superior colliculus and the pontomedullary reticular formation during orienting in unrestrained cats

Shigeto Sasaki*, Kimisato Naito and Mieko Oka

Department of Neurophysiology, Tokyo Metropolitan Institute for Neuroscience, Musashidai 2–6, Fuchu-shi, Tokyo, Japan

Cats were tained to fixate a center LED or light spot for 1–10 s and to orient to a target (LED or light spot), which appeared simultaneous with the disappearance of the fixation light. Firing characteristics of neurones in the superior colliculus (SC) and the pontomedullary reticular formation were examined systematically in cats performing the above orienting task. In the head-free condition, a majority of the superficial neurones in the SC was activated when a target appeared in a particular area in the visual field (receptive field) as in head fixed animals. We identified a new group of neurones in superficial layers of the SC, superficial fixation neurones, which fired tonically when fixated on a target light with little activity at fixating a center fixation light. In the intermediate and deep layers of the SC, many neurones fired when directed to a particular area in the visual field (movement field). In addition to these neurones, we found new groups of neurones fired tonically in a particular phase of orienting. These neurones were divided into three principal types, target activated, target suppressed and center fixation activated neurones. *Target activated* neurones fired tonically during the fixation of a target, but not during fixation of a center spot. *Target suppressed* neurones were characterized by a pause during fixation of targets irrespective of their location. They had clear phasic and tonic components. The phasic component was characterized by a movement field resembling that seen in a typical intermediate layer neurone. *Center fixation activated* neurones discharged during fixation of the center spot and exhibited suppression during fixation of the target. Many neurones in the pontomedullary reticular formation fired during orienting. Four major types of neurones, phasic, phasic sustained, pause and tonic neurones, were differentiated on the basic of their firing patterns. *Phasic units* were characterized by brief phasic firing during orienting and could be divided into four subtypes. *Long lead phasic* neurones began

to fire 50–100 ms prior to the onset of a gaze shift and stopped well before the end of a gaze shift and the total number of firing was best correlated with the total angle of head rotation. *Short lead phasic* neurones started to fire 10–20 ms prior to a gaze shift and were further subdivided into decrement and plateau types. The decrement type exhibited a brief burst just prior to a gaze shift and stopped discharging prior to the end of the gaze shift, while the plateau type fired tonically during the gaze shift and stopped firing after the gaze shift. Total number of discharges of both type was related to maximal angular velocity. The fourth type, *gaze neurone*, began firing coincident with the onset of the gaze shift and stopped firing at the completion of the gaze shift and the total number of discharges was closely related to total angle of head rotation. *Phasic sustained (PS)* neurones fired 50–100 ms before the onset of a gaze shift and their discharges continued after the end of the gaze shift and were divided into three subtypes: augmenting, pause and plateau. The *augmenting type* was characterized by a transient increase of firing during the gaze shift and a maintained slow rate of firing following the gaze shift. Total number of discharges related to head movement amplitude and the sustained activity was correlated with head position. The *pause type* was suppressed during the gaze shift and was tonically active during a head movement. The *plateau type* had a weak phasic component and discharged at a fairly constant rate. The pause and plateau types appeared to be related to head amplitude and head position, respectively. *Tonic units* were classified into several types. *Target fixation* neurones exhibited sustained firing when the cat was fixating a target, irrespective of its location, but did not show sustained firing when the animal fixated the fixation light. *Fixation* neurones fired tonically during the fixation of both the center fixation spot and targets. These activities were suppressed during reward acquisition. *Pause* neurones ceased discharging

---

*Corresponding author.

during the fixation of a target. The suppression of these neurones lasted more than 500 ms, which differentiates them from omnipause neurones. *Omnipause* neurone, characterized by a brief firing pause prior to and during gaze shifts, increased their tonic firing rates during fixation of a target,

indicating that tonic activity of omnipause neurones is modulated during orienting. Location of all these subtypes of neurones were intermingled, though phasic sustained and tonic neurones tended to be located in the more medial part than phasic units in the reticular formation.

Introduction

When an object appears in the visual field, animals direct their eyes and head toward the object to fixate it (orienting). Head movements play a key role in this orienting movement (Fig. 1A). We have been studied the pathways controlling head movements and their functions during orienting. Our results are summarized diagrammatically in Fig. 1B. The superior colliculus (SC) is known to play an important role in orientation, since unilateral lesions of the SC result in a severe impairment in orienting directed to the contralateral side to the lesion (Itouji and Sasaki, 1989; Isa et al., 1992b) as suggested previously (Sprague et al., 1965; Guitton et al., 1980; Roucoux et al., 1980).

We have identified two separate pathways which originate in the SC and ultimately terminate on neck motoneurones. One is a reticulospinal pathway involving reticulospinal neurones (RSNs) in the nucleus reticularis pontis caudalis (NRPC) and nucleus reticularis gigantocellularis (NRG), which project directly to neck motoneurones (Peterson et al., 1978; Iwamoto et al., 1990; Iwamoto and Sasaki, 1990). The other pathway is through Forel's field H (FFH), and projects directly to neck motoneurones as well as to RSNs in the ventral part of the NRPC and NRG (Holstage and Cowie, 1989; Isa and Sasaki, 1992a,b).

In the reticulospinal pathway, we have shown that RSNs can be classified into neck (N-RSNs), cervical (C-RSNs) and lumber (L-RSNs) neurones

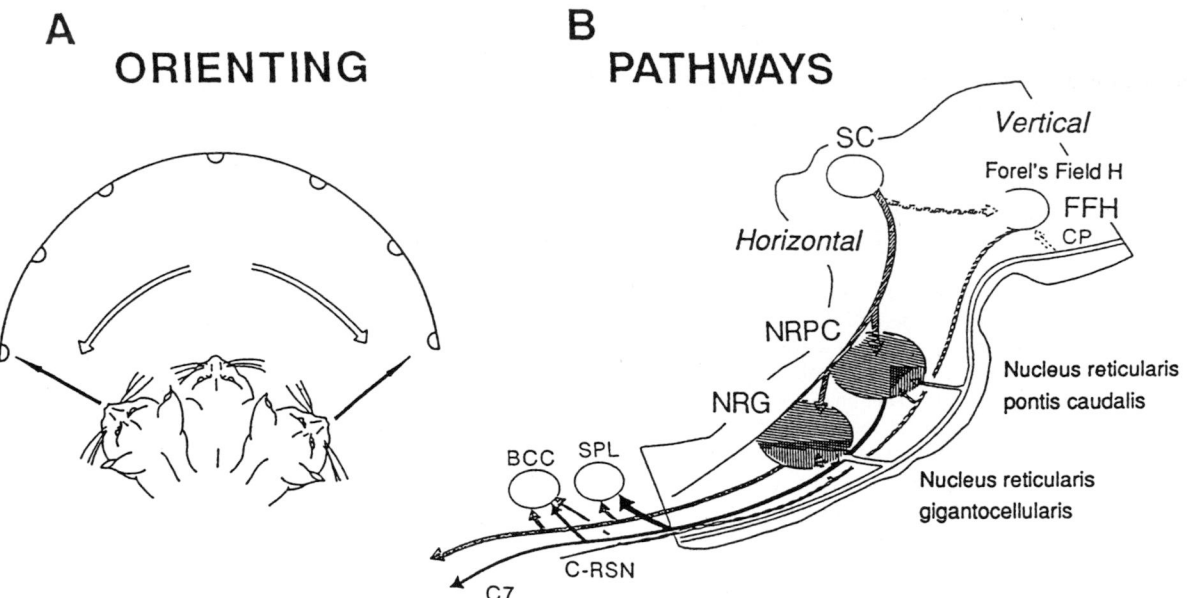

Fig. 1. Schematic drawing of the orientation paradigm. Cats were required to fixate a center LED and then direct the head and eyes to a new LED in the peripheral visual field. B: Schematic drawing of the pathways controlling head orienting. Two pathways from the SC to neck motoneurones are identified. One is through the nucleus reticularis pontis caudalis (NRPC) and nucleus reticularis gigantocellularis (NRG). The other is through Forel's field H (FFH).

according to their level of termination (Iwamoto and Sasaki, 1990). Among these three neuronal types, those projecting to neck motoneurones have been proven to be C-RSNs by a combination of morphological and physiological methods (Iwamoto and Sasaki, 1990). Physiologically, excitatory monosynaptic connections of C-RSNs with dorsal neck motoneurones (splenius, SPL and biventer cervicis and complexus muscles, BCC) were revealed by spike triggered averaging methods as illustrated in Figs. 2C–E. Morphologically, HRP staining has shown that C-RSNs in the NRPC project in the ventral funiculus and give off many collaterals and boutons to neck motor nuclei as well as laminae VI–VIII (Fig. 2A). In comparison, C-RSNs in the NRG project in the ventrolateral funiculus (Iwamoto et al. , 1990) and give off many collaterals to neck motor nuclei and to laminae VIII–VI (Fig. 2B). In addition, C-RSNs in the NRPC and NRG differed in their pattern of axonal trajectories in the brain stem. Whereas many C-RSNs in the NRPC were found to give off collaterals to the abducens nucleus (Fig. 3C) and reticular formation (Fig. 3A and 3C), C-RSNs in the NRG gave off little collaterals in the brain stem (Fig. 3B).

In addition to these brain stem relayed pathways, we have shown that there is a pathway relayed by interneurones in upper cervical spinal cord (Alstermark et al., 1985).

In the FFH pathway, we have shown that large FFH neurones project directly to neck motoneurones, chiefly to BCC motoneurones involved in vertical head movements. In addition, many medium and large FFH neurones project to the ventral part of NRPC and NRG, where the C-RSNs they contact project in turn to neck motoneurones (Isa and Sasaki, 1992a,b).

The functions of these two pathways were assessed by analyzing behavioral deficits following restricted lesions localized within the pontomedullary reticular formation (NRPC and NRG) or FFH in cats trained to perform an orienting task. Following unilateral lesion of NRPC and NRG, orienting to the lesioned side was severely impaired so that cats could either

not rotate the head to the lesioned side at all, or could rotate the head very slowly and not beyond the midsagittal axis of the body. In these animals, orientation toward the intact side and vertical orientation were impaired only slightly (Isa and Sasaki, 1988; Sasaki, 1992). Following lesions of FFH, cats were impaired in their ability to orient vertically, with little change in their ability to orient in the horizontal direction (Isa et al., 1992a). These behavioral results suggest that the reticulospinal (SC to NRPC/NRG to neck motoneurones) pathway is principally involved in the control of horizontal head orienting and the FFH (SC to FFH to neck motoneurones) pathway is principally involved in the control of vertical head orienting.

In this study, we have examined the firing characteristics of neurones in the SC and in the NRPC and NRG during head orienting.

Methods

Behavioral training

Cats were trained to stand in front of a perimetry device 60 cm in diameter, on which three horizontal rows of light-emitting diodes (LEDs) were arranged. Each row contained seven LEDs separated by 7 cm (about 12°) in each row (Fig. 4A). Alternatively, animals were trained to stand in front of a half-transparent flat panel (65 × 30 cm) on which a light spot of various size (2–15 mm in diameter) was back projected. Cats were required to fixate a center LED or light spot for 1–10 s and to orient to a second target LED or light spot, which appeared simultaneous with the disappearance of the fixation light. When cats oriented correctly to the target, they were rewarded with a piece of food given through a small hole (diameter 10 mm) opened below each LED or light spot.

Subjects and surgery

Ten adult cats weighing between 2.5 and 3.5 kg were used for recording from the SC (seven cats)

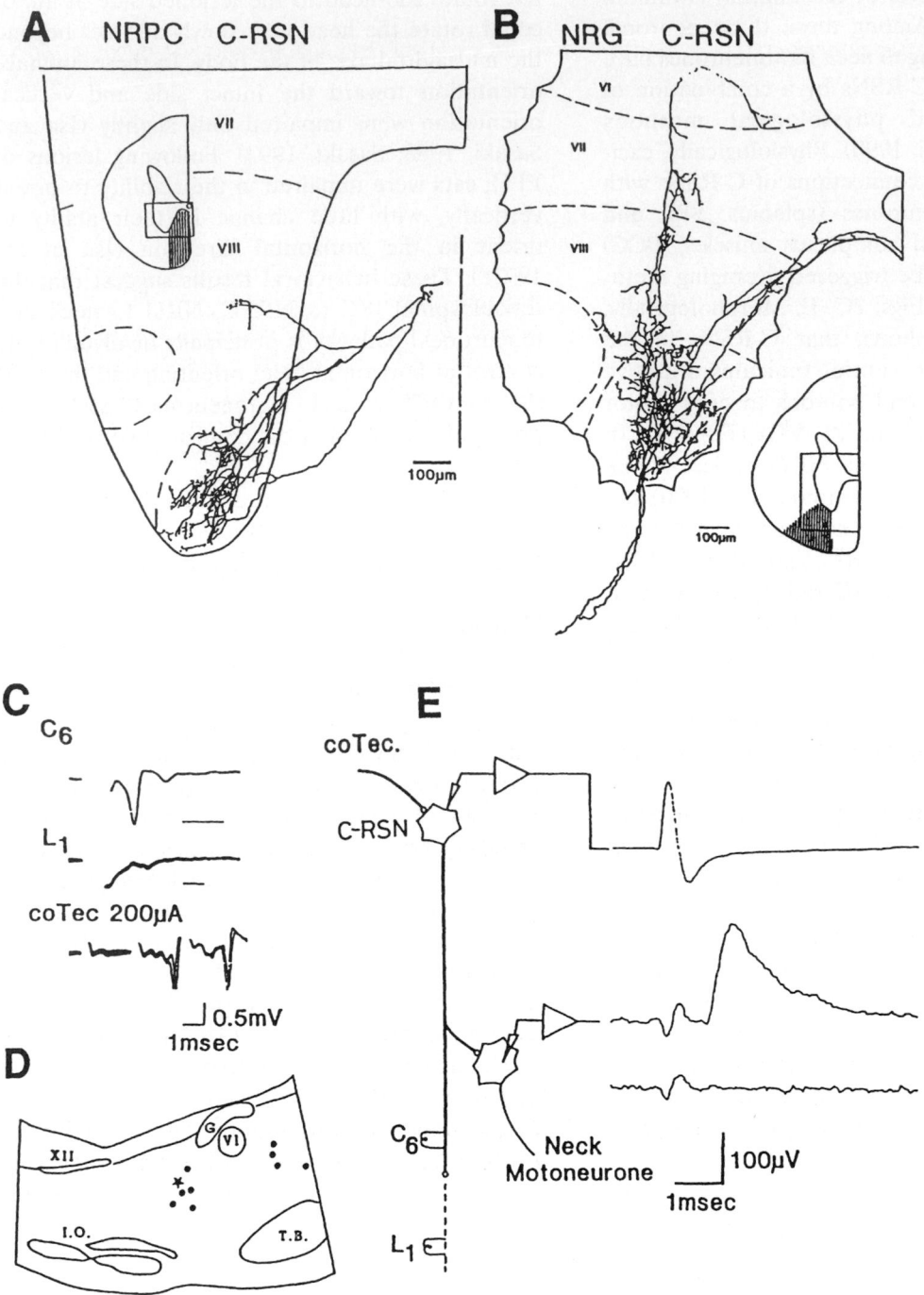

Fig. 2. Axonal trajectories of C-RSNs in the upper cervical segments and single fiber EPSP in neck motoneurones. A–B: an NRPC-C-RSN axon (A) descends in the ventral funiculus and gives off collaterals and boutons to the neck motor nuclei, while an NRG-C-RSN axon (B) descends in the ventrolateral funiculus and gives off many collaterals to neck motor nuclei and laminae

and from the NRPC/NRG (three cats). Head sockets and small manipulators were mounted to the skull under deep anesthesia (sodium pentobarbital, 40 mg/kg). An array of tungsten microelectrodes, mounted on a small manipulator, were placed stereotaxically in the SC. The final positioning of the array was carried out by maximizing orthodromic responses evoked from optic tract stimulation for placing in the superficial SC and antidromic spikes evoked by stimulation of the medial part of the pontine reticular formation close to the predorsal bundle for placing in the deep SC. NRPC and NRG electrodes were placed stereotaxically, and their final position adjusted to position them on the floor of the fourth ventricle by noting an absence of unit activity as well as low stimulus intensities (< 20 μA threshold, 20 pulses at 250 Hz) for evoking neck or trunk movements. Electrode arrays were chronically implanted in the SC or NRPC/NRG.

Recording

The position of the head was recorded by a Sel spot system (Hamamatu Photonics). Two infrared light emitting diodes (LEDs), separate by 8.5 cm and attached to the head socket, were detected by two cameras placed above and to the side of the cats (Figs. 4B and C). Movements were also monitored by video cameras. Fig. 4D–E exemplifies the trajectory of the midsagittal axis of the head, represented by lines connecting the two LEDs (filled circles) in transverse (D) and vertical planes (E). In both the horizontal and vertical planes, the head rotated at a fixed point as shown in Figs. 4D and E. Typically, head movements consisted of both rotational and translational components in the horizontal and vertical planes.

Maximal angular velocities calculated from the trajectories increased with increasing angle of head rotation (Fig. 4F, total angle-maximal angular velocity plot). The slope of the regression line was $8-10°$/s per degree as indicated in Fig. 4F for the horizontal component in three cats. Similar results were obtained with vertical rotation. Head and gaze angles were also measured with the magnetic search coil technique. In these experiments, the chronically implanted electrode array was advanced 100–500 mm in each recording session by the manual rotation of a screw. Typically, it took 1–3 months to complete a record from one 4-mm track. Single or multiple unit activities were recorded for more than 6 months. To minimize the noise which accompanied movements, unit activities were differentially recorded between the electrodes and a reference tungsten electrode placed in the third or fourth ventricle.

Data analysis

Single or multiple units recorded from each electrode were counted by personal computer (PC9801 DA or RA, NEC) and spike density histograms and raster plots were constructed and displayed on-line and stored to hard disk. Head position, head angle and gaze angle data were digitized and displayed on-line and stored to hard disk. All data was stored on a DAT tape recorder (PC216A) for further off-line analyses.

Results

Firing characteristics of neurones in different layers of the SC

It has been well known that the superficial layers of the SC are retinotopically organized (see Stein,

VIII–VI. C: Identification of a C-RSN. The neurone is activated antidromically from a stimulating electrode placed in the ventral funiculus of the C6 segment but not from lumber segment (L1). The neurone is orthodromically activated by stimulation of the contralateral SC. D: location of cells with monosynaptic excitatory connections with neck motoneurones are identified by spike triggered averaging technique. Asterisk is the cell shown in E. E: Single fiber EPSP recorded by spike triggered averaging technique. The postsynaptic effects of a single C-RSN on neck motoneurones are assessed by averaging the latter's membrane potential triggered by C-RSN spikes. Lowest record is extracellular field potential.

1984). The inferior half of the retina represents superior visual space and projects to the medial portion of the SC. Conversely, the superior retina projects to the lateral SC. Central visual space is represented rostrally and temporal space is represented caudally. Many neurones in the intermediate and deep layers of the SC discharge vigorously prior to and during the early phases of an orientation movement (i.e. a saccade) directed to a particular place in the visual field (movement field, which is closely related to visual field of the superficial layer). Although these results give us an important information about the function of the SC, they are based on recordings from head-fixed animals, and are focused mainly on the eye movement role of this structure. In contrast, little has been done to examine the role of the SC in guiding head movements. Given the extensive connections of the SC with the circuitry controlling head movements, we believed it likely

that it's neurones would show activity related to such movements. To examine this, we recorded from SC neurones during orienting head movements in unrestrained cats.

Superficial layer neurones

Fig. 5 exemplifies the response of a typical superficial SC neurone. The neurone (location is indicated in C) discharged vigorously when LED 5H was turned on, but failed to respond to a peripheral LED. Thus, it appears that this neurone had a receptive field centered around LED 5H (about 12° from the central fixation point). We confirmed the existence of a retinotopic map by noting that the receptive fields of superficial SC neurones changed systematically with changing position in the SC.

In addition to this well known class of SC neurones, we identified a second class, superficial gaze neurones. As illustrated in Fig. 6A, this type

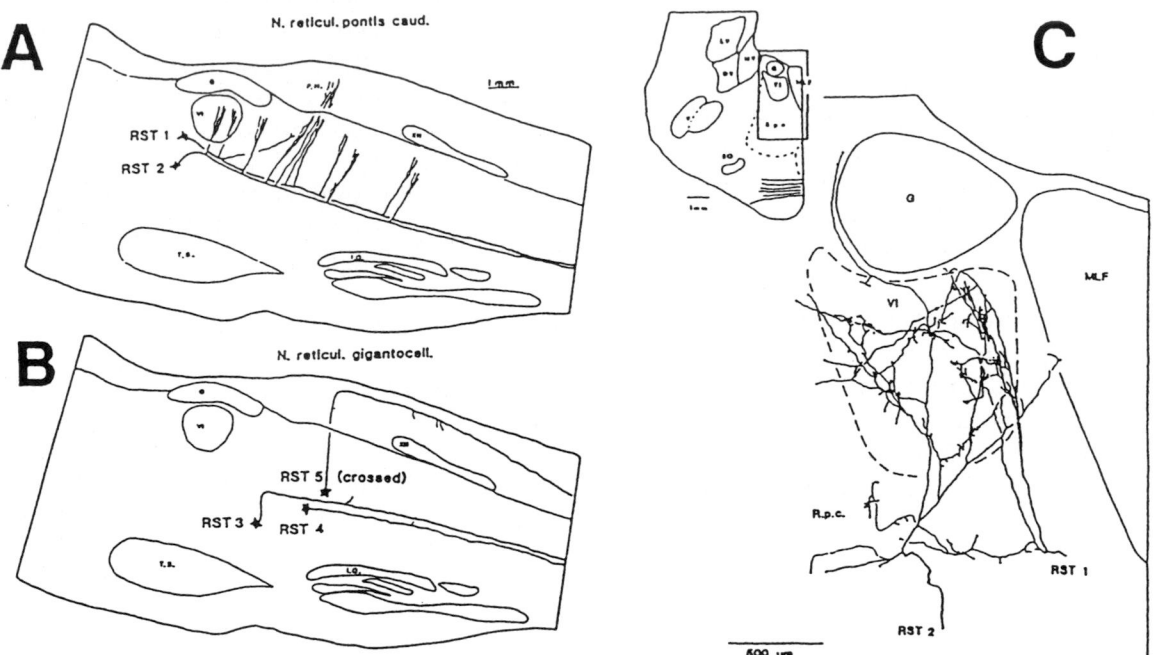

Fig. 3. Axonal trajectories of C-RSNs in the brain stem. A–B: Schematic drawing of two NRPC-C-RSNs (A, RST1–2) and NRG-C-RSNs (B, RST3–5) in the brain stem. Both NRPC-C-RSNs give off many collaterals to the reticular formation in the caudal part of the NRPC and NRG, while NRG-C-RSNs give off few branches to the brain stem. C: Axonal trajectory of the two NRPC-C-RSNs shown in A. From the ascending collaterals, branches and boutons are given off to the abducens nucleus.

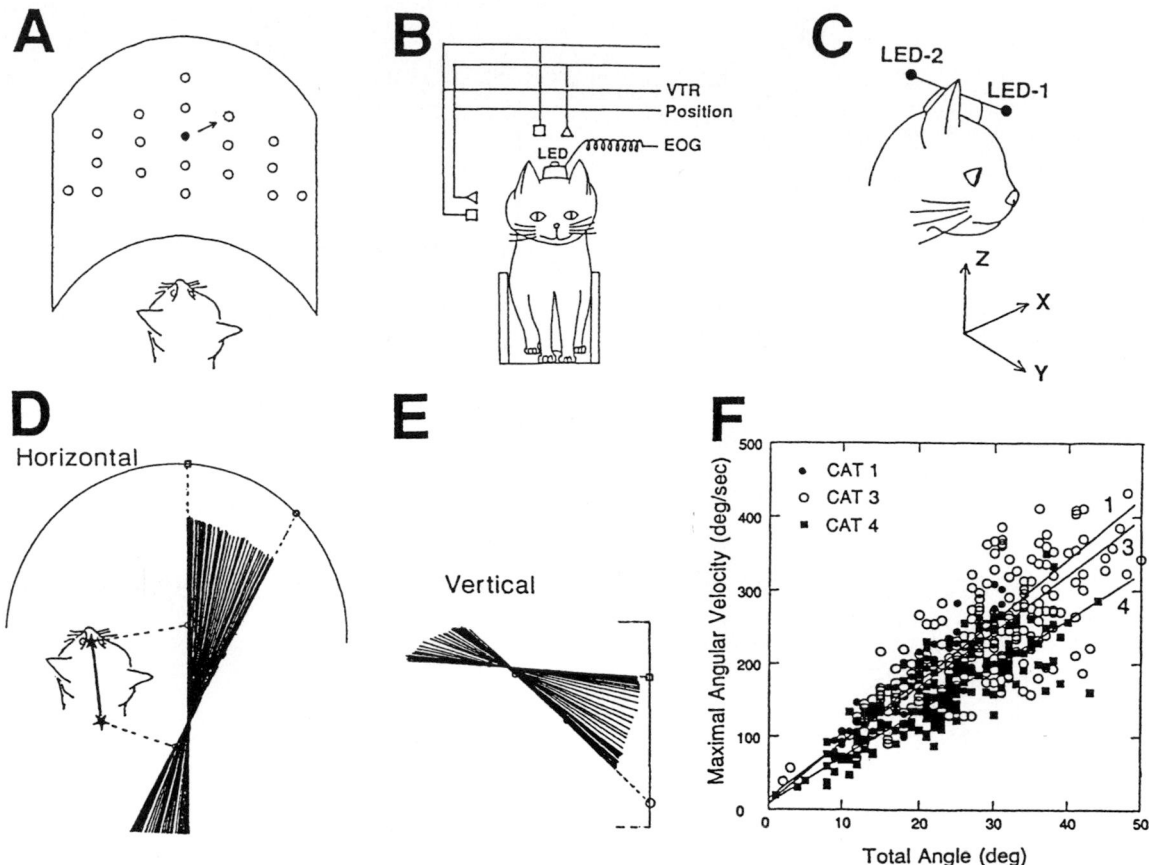

Fig. 4. Orienting task and measurement. A: The cat stands in front of a perimetry device, where 21 LEDs are arranged in three horizontal rows. Cats were required to fixate a center LED and then direct the head and eyes to a new peripheral target. The onset of the target LED was coincident with the offset of the fixation LED. B–C: Head movement was measured by a Sel Spot system, where two infrared-emitting LEDs were placed in parallel to the midsagittal axis of the head and parallel to the horizontal plane of the Horsley Clarke stereotaxic coordinate system. The two-dimensional position of the LEDs are measured with two photosensitive cameras. D–E: Trajectories of the midsagittal axis of the head in the horizontal and vertical planes. Position of two LEDs are indicated by circles (D). F: Angle-angle velocity plot. Maximal angular velocity is plotted against total angle of head rotation in three cats.

of neurone was activated after acquisition of an eccentric target, but not during the initial fixation. The activity of these neurones were dependent on the target acquired (Fig. 6B), and their pattern indicated the existence of a gaze field. A normalized response profile of one of these neurones is shown in Fig. 6D. The gaze field of this neurone is centered around LED 6H. These superficial gaze neurones were distributed widely throughout the superficial SC, and their gaze fields were not related to the visual receptive field of surrounding typical superficial neurones. Although superficial gaze neurones were not recorded in every track, they were always the first visually-responsive neurones encountered. Thus, they appear to be located in the most superficial layer of the SC.

Intermediate and deep layer neurones

Fig. 7 illustrates the response of an intermediate layer SC neurone during orientation to a

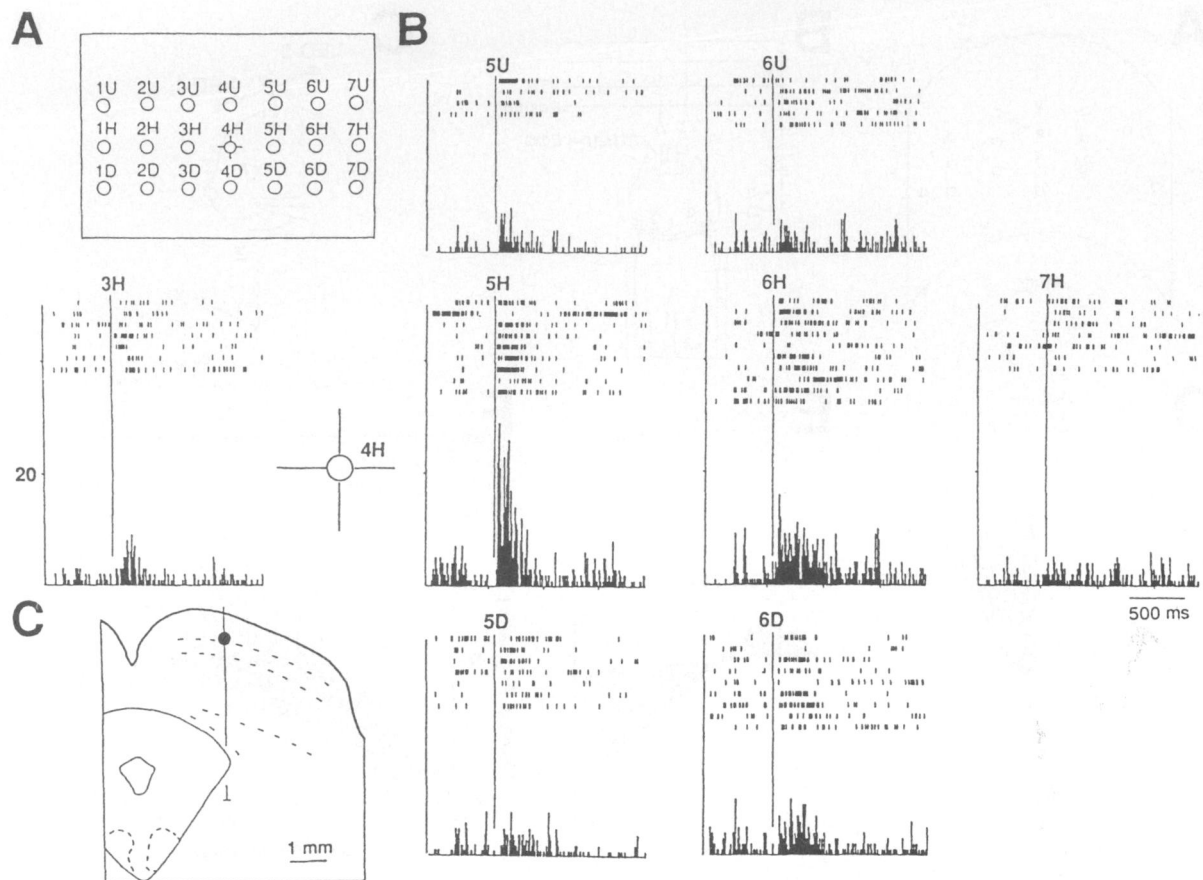

Fig. 5. Superficial SC neuronal responses. A: Arrangement of LEDs. 4H is the center fixation LED. B: Histogram and rasters depict neuronal responses during orientation to a target. C: Location of the cell shown in B.

series of different targets. This neurone fired maximally when the cat directed it's head to LED 3D but very little when it directed it's head to LED 2H. Thus, it appears that the movement field of this neurone was restricted to the area around LED 3D. The firing pattern of this neurone appeared similar to that observed in the head-restrained animal (Peak et al., 1980; Grantyn et al., 1987). In comparison with superficial SC neurones, visual responses were less pronounced in intermediate SC neurones, and the duration of discharge in these neurons was shorter (Figs. 5 and 7). Deep layer neurones resembled intermediate layer neurones in their response profiles.

In addition to these neurones, we found new groups of neurones distinguished by the tonic component of their discharge. These neurones can be divided into three principal types on the basis of their tonic discharge or tonic suppression during particular phases of orienting (Fig. 8). Target activated neurones fired tonically during the fixation of a target, but not during fixation of a center spot (Fig. 8A). Characteristically, the discharge began after fixation of the target (Fig. 8A, lower panel), and could be prolonged by delay of reward. Target suppressed neurones were characterized by a pause during fixation of targets irrespective of their location. They have clear phasic and tonic components (Fig. 8B, bottom panel). The phasic component was characterized by a movement field resembling that seen in a typical intermediate layer neurone. Similar to target acti-

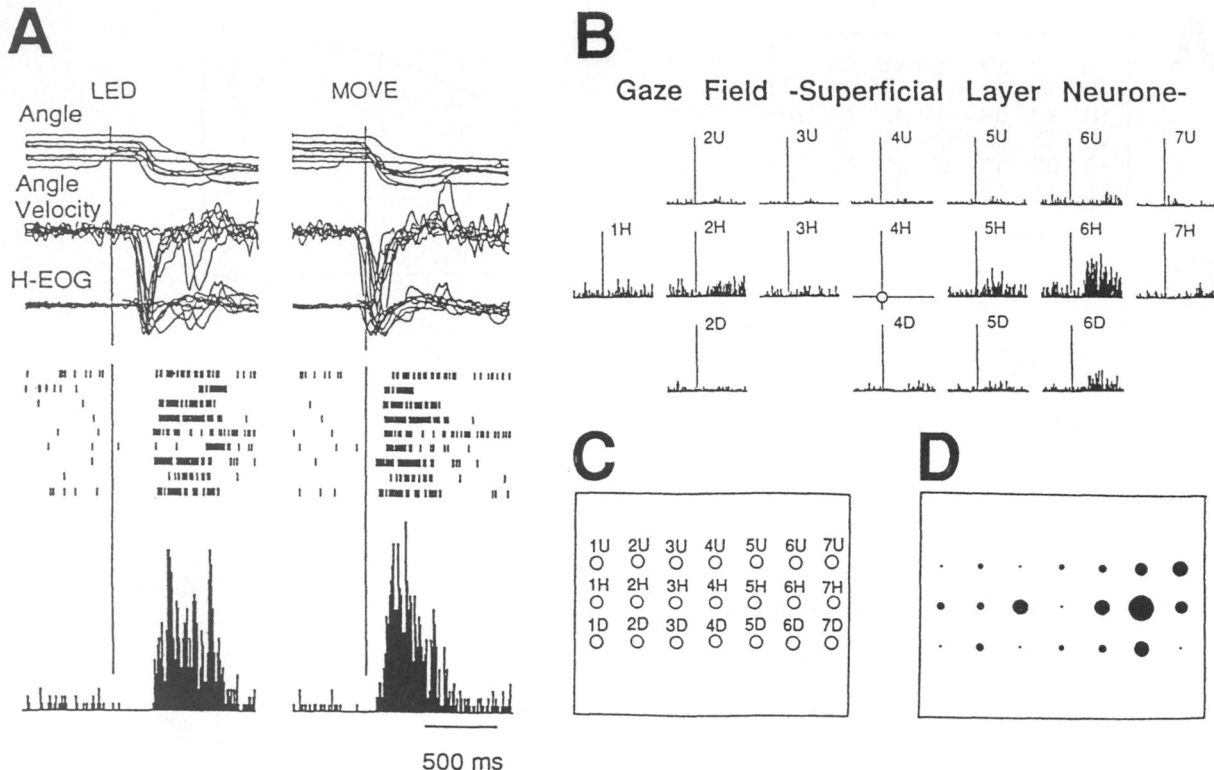

Fig. 6. Response of superficial gaze neurones. A: Histogram and rasters show neuronal activity as it relates to head angle, angle velocity and horizontal EOG. Traces are aligned on the onset of light stimuli (left panel) and on the onset of movement (right). B: Neuronal response during orientation to various targets. C: Arrangement of LEDs. D: Normalized responses. Size of filled circles indicates response magnitude.

vated neurones, the suppression in these neurones could be prolonged with the intentional delay of reward. The third class of neurone, center fixation activated neurones, discharged during fixation of the center spot and exhibited suppression during fixation of the target (Fig. 8C). The duration of tonic firing in these neurones varied among trials.

Firing patterns of neurones in the NRPC and NRG

Types of neurones

The activity of many neurones in NRPC and NRG was closely related to orienting. Four major types of neurones were differentiated on the basis of their firing patterns (Fig. 9). Phasic neurones were characterized by a brief burst of activity (Fig. 9B), while phasic sustained units had both

phasic and sustained components of firing (Fig. 9A). Tonic neurones discharged tonically when the animal oriented in the optimal direction (Fig. 9C). Finally, pause neurones were characterized by a brief pause of activity prior to and at the beginning of the orientation movement (Figs. 12A and C). Each of these major types were further subdivided on the basis of their pattern of discharge (Fig. 9).

Subtypes of neurones and quantitative analysis of their firing pattern

Phasic type. Phasic units could be divided into four subtypes (Figs. 9B and 10). Long lead phasic neurones began to fire 50–100 ms prior to the onset of a gaze shift and stopped well before the end of the gaze shift. Since the head often moved

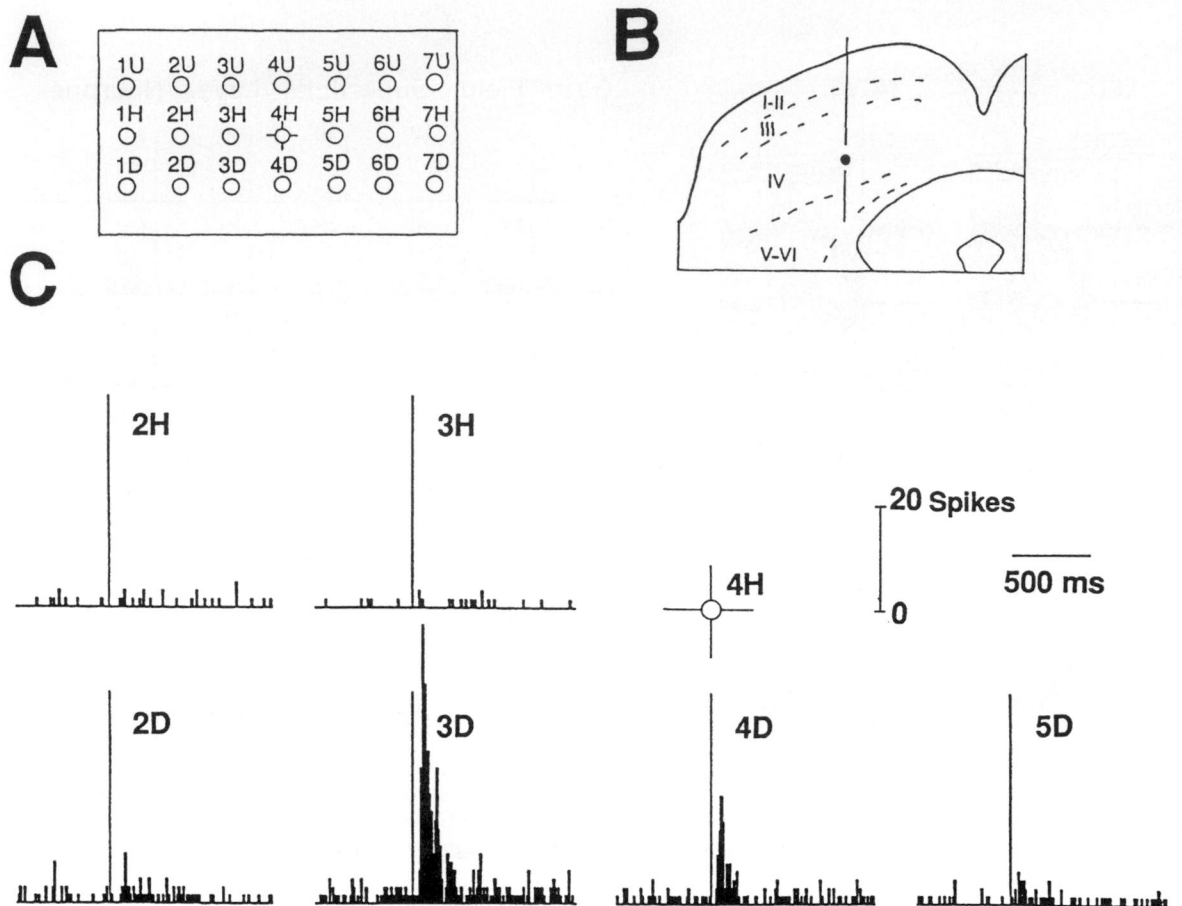

Fig. 7. Response of intermediate layer SC neurones. A: Arrangement of LEDs. B: location of the neurones shown in C. C: Neuronal response during orientation to various targets.

well before the gaze shifted in these trials, the long lead firing of these neurones may be more related to head movement. To test for this possibility, we aligned head angle and gaze shift traces on the onset of firing (Fig. 10Ab). Firing onset correlated fairly well with head movement, suggesting that long lead phasic neuronal activity is more tightly related to head movements rather than gaze shifts. Fig. 10Ac shows the relationship between total number of spikes and total angle of head rotation. The total number of spikes increased with increasing amplitude of head rotation, as well as with maximal angular velocity (not illustrated). Because the activity of these neu-

rones was best correlated with the total angle of head rotation, we suggest that they carry information related to the amplitude of head rotation.

Short lead phasic neurones started to fire 10–20 ms prior to a gaze shift (Figs. 9B and 10B). These neurones could be further subdivided into decrement and plateau types. The decrement type exhibited a brief burst just prior to a gaze shift and stopped discharging prior to the end of the gaze shift (Fig. 9B). In contrast, the plateau type fired tonically during the gaze shift and stopped firing after the gaze shift (Fig. 9B). The clear relationship between maximal angular velocity and total number of discharges (Fig. 10Bc) strongly sug-

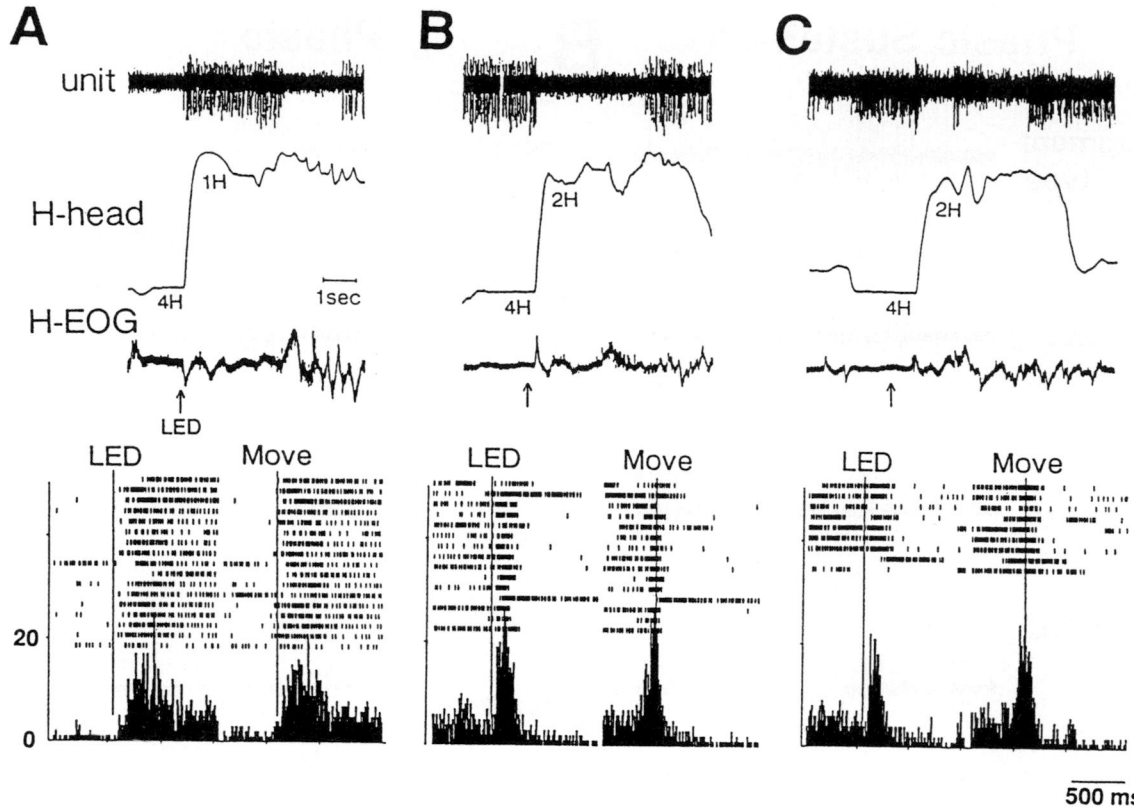

Fig. 8. Response of deep layer SC neurones. A–C: Examples of responses of target activated (A), target suppressed (B) and center fixation activated neurone (C). Lower panels show raster plots and histograms aligned on stimulus and movement onset.

gests that short lead phasic neurones of the decrement type are involved in the generation of rapid head movements. A similar relationship was observed in neurones of the plateau subtype. Thus, short lead burst neurones appear to code information pertaining to maximal angular velocities.

The fourth type of gaze-related neurone began firing coincident with the onset of the gaze shift and stopped firing at the completion of the gaze shift (Figs. 9B and 10Ca). In these neurones, the total number of discharges was closely related to total angle of head rotation (Fig. 10Cc), suggesting that they carry information related to the amplitude of the gaze shift. This was the least frequently encountered phasic subtype.

Phasic sustained (PS) type. PS neurones fired

50–100 ms before the onset of a gaze shift and their discharges continued after the end of the gaze shift (Fig. 9A). PS units were divided into three subtypes: augmenting, pause and plateau (Fig. 9A). The augmenting type was characterized by a transient increase of firing during the gaze shift and a maintained slow rate of firing following the gaze shift. The pause type was suppressed during the gaze shift and was tonically active during a head movement. The plateau type had a weak phasic component and discharged at a fairly constant rate.

Fig. 11A shows the relationship between total number of discharges and head movements in a PS unit of augmenting type. The unit began to fire tonically as the cat made a large orientation movement (Fig. 11B). By plotting the total number of spikes over a period from 100 ms before to

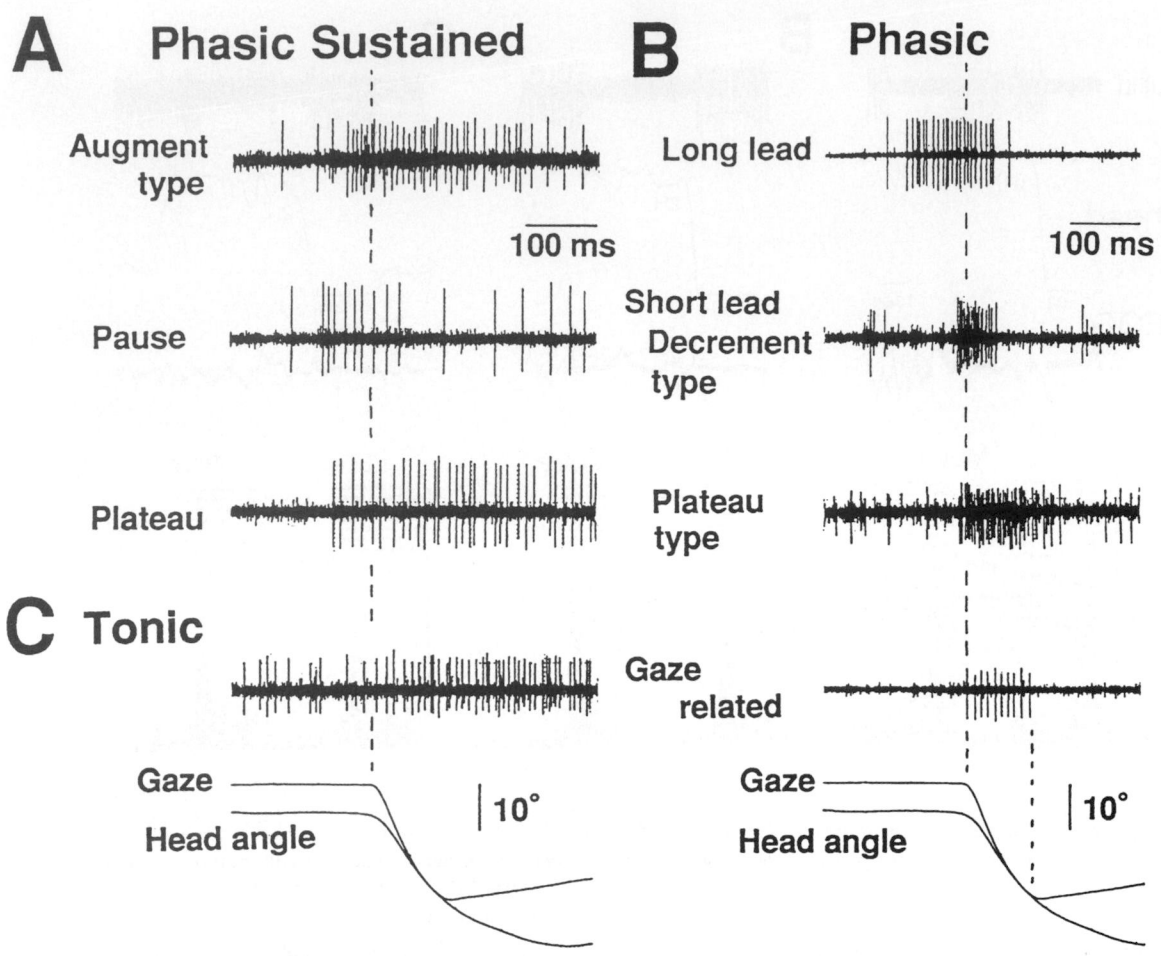

Fig. 9. Four major types of NRPC and NRG neurones and their subtypes. A: Three subtypes of phasic sustained unit. See text for more detailed explanation. B: Four subtypes of phasic units. C: An example of a tonic unit.

800 ms after the onset of movement against target position (after normalization), a clear relation between absolute activity and head movement amplitude was revealed (Fig. 11C, bottom panel). Fig. 11D shows how PS unit activity varies with head movement parameters such as position, angular velocity and trajectory. It can be seen that unit activity outlasted head rotation in these neurones. This sustained activity (Fig. 11B, right upper panel) is well correlated with head position expressed by head angle (Fig. 11E), suggesting that the sustained component of this discharge codes head angle. The second feature of PS neu-

rones is that the phasic component of their discharge is less related to head movement amplitude when compared to phasic neurones (Fig. 11C, 2nd upper panel). The third feature of PS neurones is that they exhibit a short latency (~ 40 ms) visual response (Fig. 11B, left upper panel). Fig. 11C shows the normalized response of the visual response occurring 40–80 ms after the onset of the stimulus (upper pannel), the phasic response (−100–100 ms in relative to the onset of movement, middle panel) and the whole response (−100–800 ms, bottom panel), The differences among them suggested differential pro-

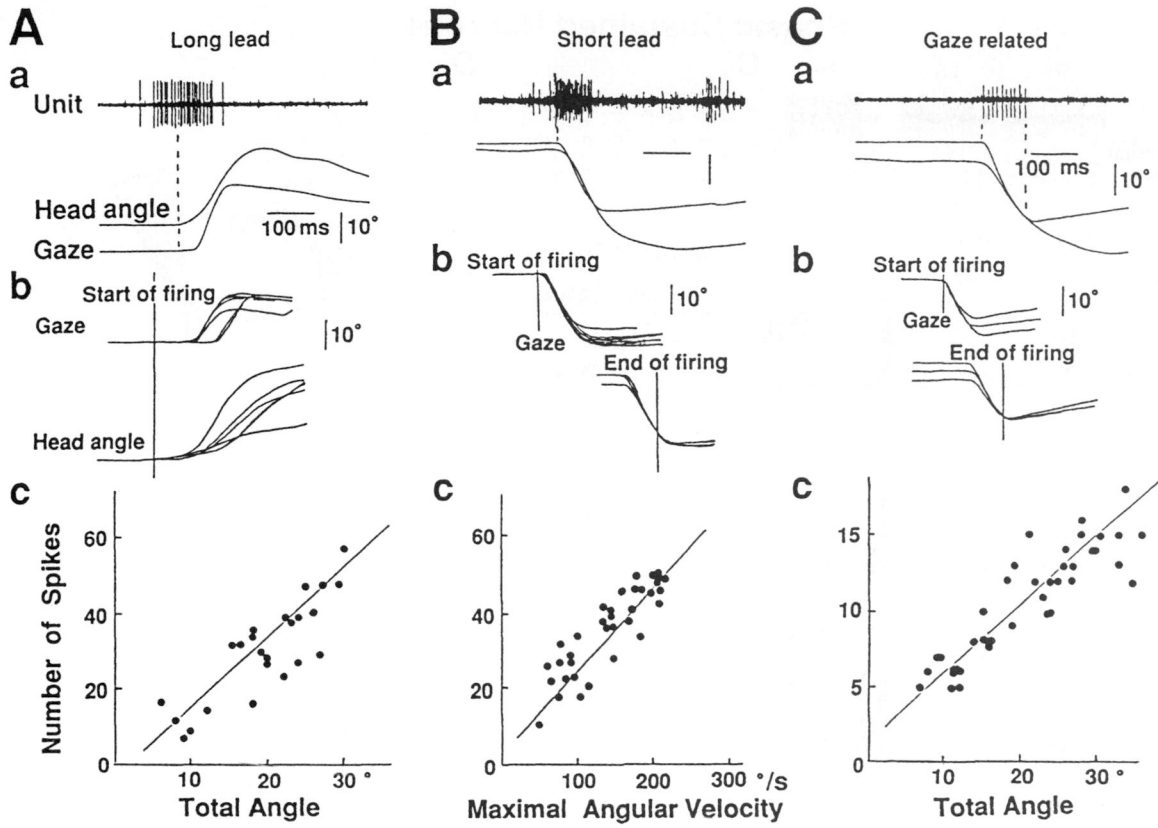

Fig. 10. Relationship between phasic unit subtype activity and the parameters of movement. A: Aa-Relationship between firing of long lead phasic neurone and head angle and gaze shift. Note that firing started about 100 ms before the onset of head movement (dotted line). Ab-Gaze shift and head angle are aligned on the onset of firing in each trial. Ac-Relationship between total angle of head rotation and total number of discharges. B–C: Similar analysis for short lead (B) and gaze (C) units.

jections among SC neurones with different responses properties onto the same PS neurone. These results suggested that PS units of increment type carry information related to both head amplitude and head position. Similarly, pause and plateau types are suggested to be related to head amplitude and head position, respectively.

Tonic type. Tonic units were classified into several types according to their firing pattern. Target fixation neurones exhibited sustained firing when the cat was fixating a target, irrespective of it's location (3H and 5H in Fig. 12B, upper panel), but did not show sustained firing when the animal fixated the fixation light (Fig. 12B, lower panel).

Although the discharge rate of these neurones decreased gradually, their activity often lasted for more than 2 s. This type of neurone also fired during the acquisition of reward. Fixation neurones (not illustrated) fired tonically during the fixation of both the center fixation spot and targets. The activity of these neurones was suppressed during reward acquisition. Pause neurones ceased discharging during the fixation of a target (Fig. 12C). The suppression of these neurones lasted more than 500 ms, a characteristic which differentiates them from omnipause neurones (Fig. 12A). Fig. 12A shows an omnipause neurone which is characterized by a brief firing pause prior to and during gaze shifts. The sup-

112

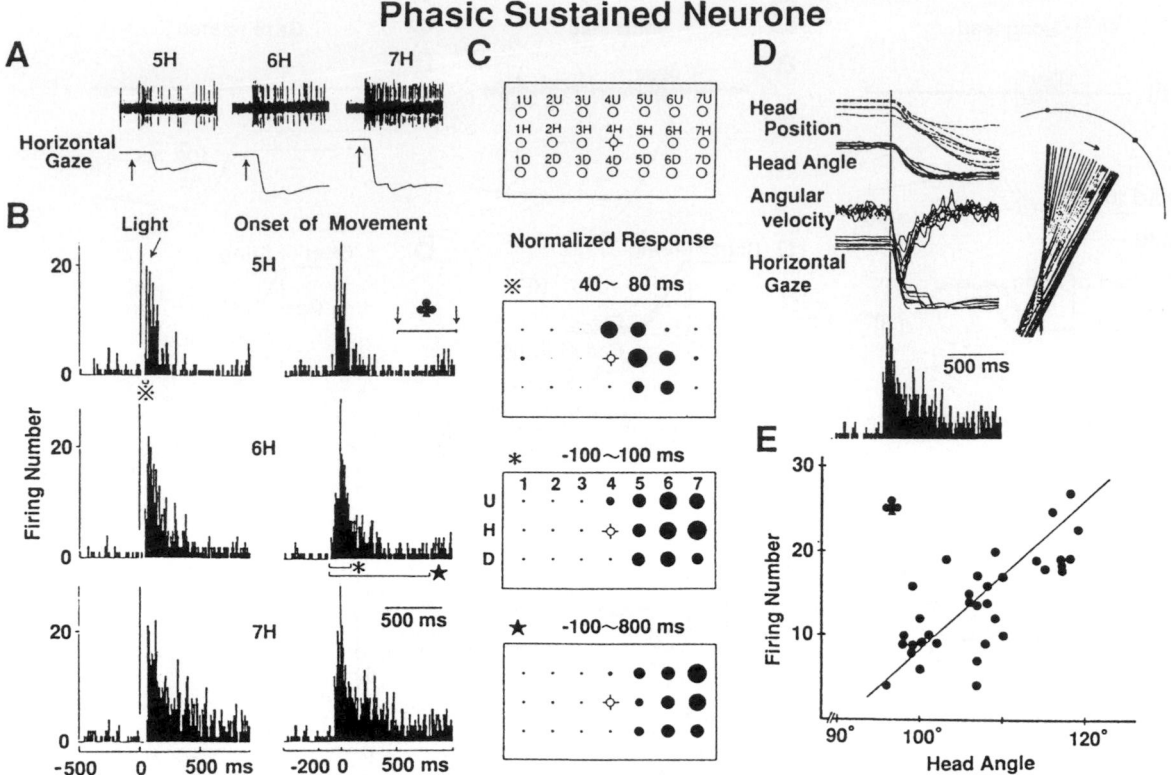

Phasic Sustained Neurone

Fig. 11. Relationship between the firing of a phasic sustained unit (augmenting type) and movement parameters. A: Discharges of phasic sustained unit when the animal oriented to different targets. B: Histograms of firing directed to LEDs 5H, 6H and 7H, aligned on the stimulus (left panel) and onset of the movement (right) [same unit as shown in A]. C: Normalized responses of unit shown in B plotted against target position. D: Relationship between the different movement parameters, trajectory and unit firing. E: Relationship between neuronal activity 500–1000 ms after onset of the movement (indicated in B, upper right panel) and the head angle during this period.

pression occurred for every gaze shift irrespective of direction. We included omnipause neurones in the tonic neurone class, since they increased their firing rate during fixation of a target (Fig. 12A, leftmost panel), indicating that tonic activity of omnipause neurones is modulated during orienting.

Location of neurones

Fig. 13 shows the location of each neuronal subtypes, represented by a different symbol, recorded in one cat with an array of eight electrodes placed bilaterally, four in each side. Note that the penetrations in the left brain stem (Fig.

13B) were located more medially than those in the right (Fig. 13C). All subtypes were intermingled. However, phasic sustained (filled circles) and tonic neurones (open rectangle) have tendency to be located in more medial part than phasic unit in the reticular formation. Similar results were obtained in two other cats.

Discussion

Unit recordings in alert animals have been performed chiefly in the head-fixed condition (Grantyn and Berthoz, 1985, 1987; Peck, 1989), and very rarely in the head-free condition (Isa and Naito,

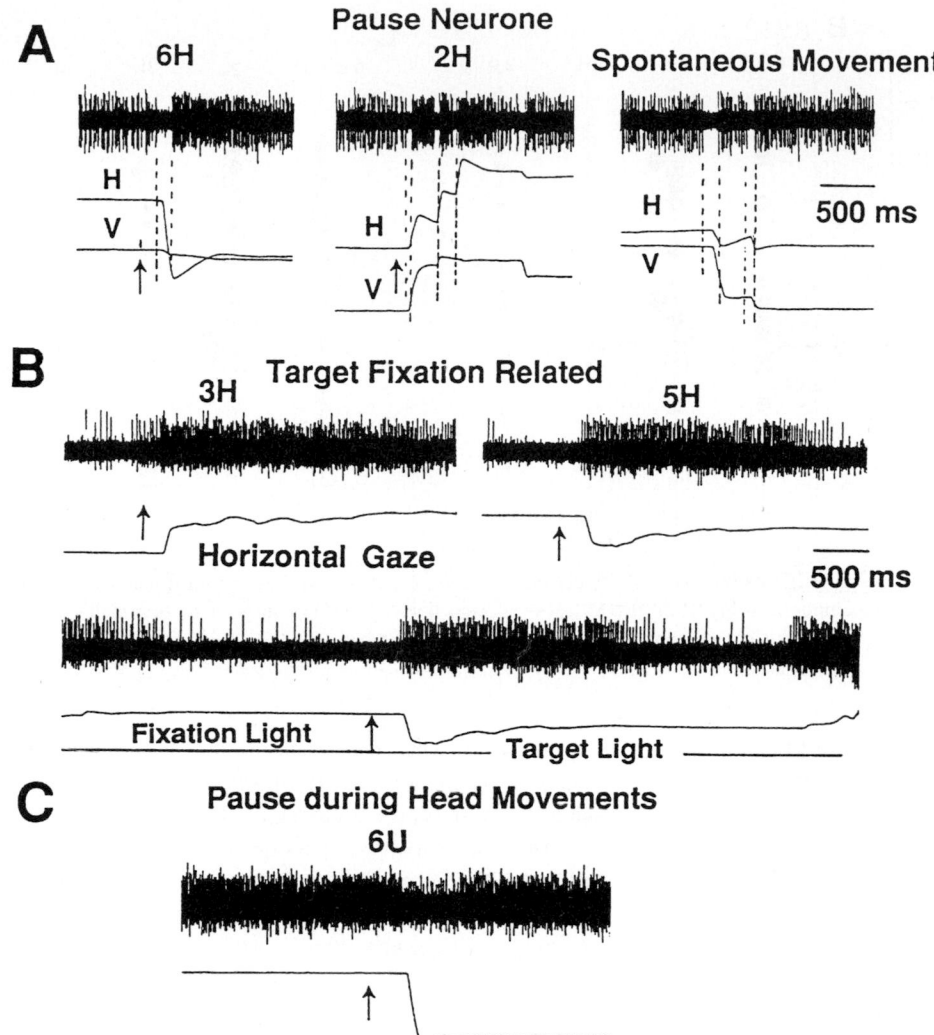

Fig. 12. Examples of firing patterns of omnipause (A), tonic (B) and pause neurones (C).

1955; Guitton and Munoz, 1991; Munoz and Guitton, 1991) Stable recording from unrestrained cats during orientation was an absolute requirement for this study. To meet this objective we developed a micromanipulator which allowed us to record neural activity from multiple foci simultaneously. This system was found to be so stable that we could record from the same unit for more than two days. We advanced electrodes 150–500 μm each day. This recording strategy also minimized the compression effect cause by electrode advancement and allowed us to record superficial gaze neurones which appeared to be small in size and depressed easily by compression.

Neural activity during orienting

Superior colliculus

We identified a new group of superficial neurones, superficial gaze neurones, which differ from previously reported fixation neurones (Munoz and Guitton, 1991; Munoz et al., 1991) in the fol-

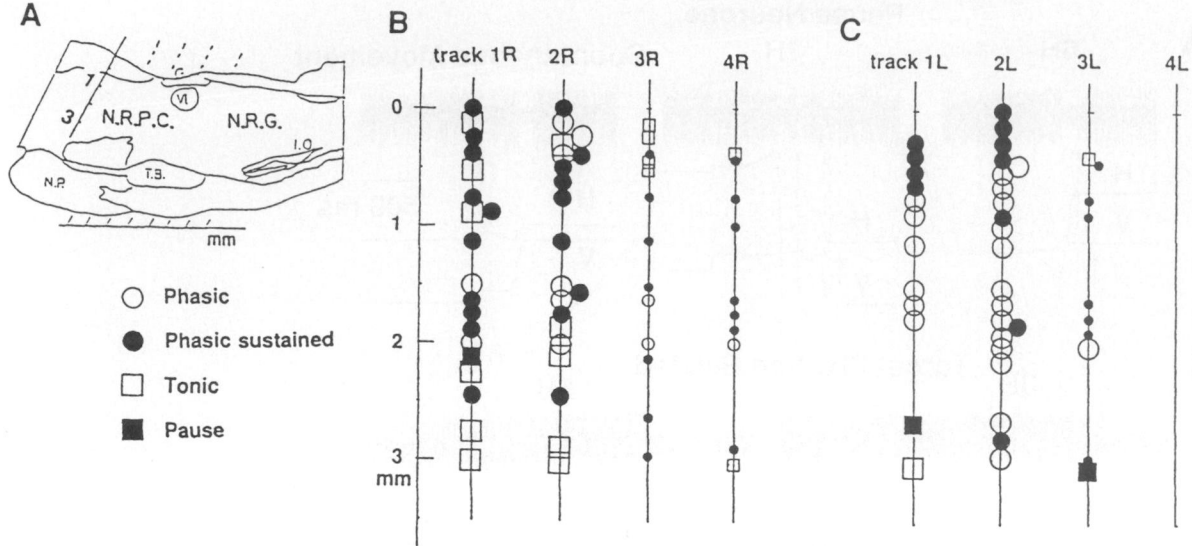

Fig. 13. Location of various types of recorded units (an array of 4 electrodes was implanted in each side). Neuronal subtypes are indicated with different symbols. A: Schematic drawing of electrode tracks (dotted lines). B–C: Location of units. Small symbols indicate locations where multiple units were recorded and single unit analysis was not possible.

lowing ways: (1) superficial gaze neurones were located in the uppermost regions of the superficial layers, and were widely distributed. In contrast, fixation neurones are located in the intermediate and deep layers and are concentrated in the foveal (i.e. rostral) representation; (2) the gaze field (target position dependency) observed in superficial gaze neurones has not been reported in fixation neurones. Superficial gaze neurones are likely to be inhibitory, since GABA neurones are found in the appropriate superficial layers (Mize, 1992), and these neurones suppress activity during target fixation.

We differentiated three major types of tonic neurones in the intermediate and deep layers: target suppressed neurones, target activated neurones and center fixation activated neurones. These neurones exhibited increased or suppressed activity during fixation of a center spot or target. Since the tonic activity level of these neurones was subject to change during different situations, even though the animal continued to fixate the same LED, suggests that the activity of these neurones is under volitional control. Target

suppressed and target activated neurones appeared to be similar to fixation-suppressed and fixation-activated neurones (Peck, 1989). We failed to find fixation cells or orienting tectoreticulospinal neurones (Munoz and Guitton, 1991; Munoz et al., 1991) in the intermediate and deep layers, a difference which may be attributable to the task. Whereas Munoz and colleagues used manually-presented food as a target, we used either a spot of light or an LED.

NRPC and NRG

Unit recordings from the brain stem have been made in both head-free (Siegel and Tomaszewski, 1983; Isa and Naito, 1995) and head-restrained cats (Grantyn and Berthoz, 1987; Grantyn et al., 1987, 1992). In these studies, phasic, phasic sustained and quasi-tonic neurones have been classified (Grantyn et al, 1992). The firing patterns of these neuronal classes were basically similar to the patterns reported in the current study, although we have differentiated these major classes into subtypes. Grantyn and colleagues (1987, 1992) correlated the axonal trajectory of reticulospinal neurons with their firing pattern in alert head-fixed cats. Phasic units, which appear to correspond to short lead phasic neurones in our study, were shown to give off few collaterals in the brain stem (Fig. 3B). Phasic sustained units, which correspond to our augmented subtype of PS unit, have been shown to give off dense collaterals and boutons into the reticular formation (Grantyn et al., 1987, 1992). Their axonal trajectories were similar to those shown in Figs. 3A and C. In the current study, we further differentiated phasic and PS neurones into a number of subtypes, although we did not identify whether they were reticulospinal neurones. The correlation of these physiologically identified neuronal subtypes with their axonal trajectory remains to be done.

We confirmed that NRPC and NRG neurones fire preferentially to head movement with a horizontal component (Isa and Naito, 1995), as suggested in lesion experiments (Isa and Sasaki, 1988). What remains to be determined is how information coding the position of a target in the visual field is directed to intermediate and deep SC neurones and transformed into information coding the speed and amplitude of a directed head movement in the NRPC and NRG.

References

Alstermark, B., Pinter, M. and Sasaki, S. (1985) Pyramidal effects in dorsal neck motoneurones of the cat. *J. Physiol.*, 363: 287–302.

Grantyn, A. and Berthoz, A. (1985) Burst activity of identified rectoreticulospinal neurons in the alert cat. *Exp. Brain Res.*, 57: 417–421.

Grantyn A. and Berthoz, A. (1987) Reticulospinal neurons participating in the control of synergic eye and head movements during orienting in the cat. I. Behavioral properties. *Exp. Brain Res.*, 66: 339–354.

Grantyn, A., Ong-Meang, Jacques, V. and Berthoz, A. (1987) Reticulo-spinal neurons participating in the control of synergic eye and head movements during orienting in the cat. II. Morphological properties as revealed by intra-axonal injections of horseradish peroxidase. *Exp. Brain Res.*, 66: 355–377.

Grantyn, A., Hardy, O., Olivier, E, and Gourdon, A. (1992) Relationships between task-related discharge patterns and axonal morphology of brainstem projection neurons involved in orienting eye and head movements. In: H. Shimazu and Y. Shinoda (Eds.), *Vestibular and Brain Stem Control of Eye and Head and Body Movements*, Japan Scientific Society Press, Tokyo/S, Karger Basel, pp. 255–273.

Guitton, D. and Munoz, D.P. (1991) Control of orienting gaze shifts by the tectoreticulospinal system in the head-free cat. I. Identification, localization, and effects of behavior on sensory responses. *J. Neurophysiol.*, 66: 1605–1623.

Guitton, D., Crommelink, M. and Roucoux, A. (1980) Stimulation of the superior colliculus in the alert cat. I. Eye movements and neck EMG activity evoked when the head is restrained. *Exp. Brain Res.*, 39: 63–73.

Holstage, G. and Cowie, R.J. (1989) Projection from the rostral mesencephalic reticular formation to the spinal cord. An HRP and autoradiographic study in the cat. *Exp. Brain Res.*, 75: 265–279.

Isa, T. and Naito, K. (1995) Activity of neurons in the medial pontomedullary reticular formation during orienting movements in alert head-free cats. *J. Neurophysiol.*, 74: 73–95.

Isa, T. and Sasaki, S. (1992a) Descending projections of Forel's field H neurones to the brain stem and the upper cervical spinal cord in the cat. *Exp. Brain Res.*, 88: 563–579.

Isa, T. and Sasaki, S. (1992b) Mono- and di-synaptic pathways from Forel's field H to dorsal neck motoneurones in the cat. *Exp. Brain Res.*, 88: 580–593.

Isa, T., Itouji, T. and Sasaki, S. (1992a) Control of vertical head movement via Forel's field H. In: A. Berhoz, P.P. Vidal and W. Graf (Eds.), *Head-Neck Sensory Motor System*, Oxford University Press, New York, Oxford, pp. 331–317.

Isa, T., Itouji, T. and Sasaki, S. (1992b) Control of head movement in the cat: two separate pathways from the superior colliculus to neck motoneurones and their roles in orienting movements. In: H. Shimazu and Shinoda, Y. (Eds.), *Vestibular and Brain Stem Control of Eye, Head and Body Movements*, Japan Scientific Societies Press, Tokyo, pp. 275–284.

Itouji, T. and Sasaki, S. (1989) Effects of a lesion of the superior colliculus on orienting movements of the head in the cat. *Neurosci. Res. Suppl.*, 9: S88.

Iwamoto, Y. and Sasaki, S. (1990) Monosynaptic excitatory connexions of reticulo spinal neurones in the nucleus reticularis pontis caudalis with dorsal neck motoneurones in the cat. *Exp. Brain Res.*, 80: 277–289.

Iwamoto, Y., Sasaki, S. and Suzuki, I. (1990) Input-output organization of reticulospinal neurones, with special references to connexions with dorsal neck motoneurones in the cat. *Exp. Brain Res.*, 80: 260–276.

Mize, R.R. (1992) The organization of GABAnergic neurons in the mammalian superior colliculus. *Prog. Brain Res.*, 90: 219–348.

Munoz, D.P. and Guitton, D. (1991) Control of orienting gaze shifts by tectoreticulospinal system in head-free cats. II. Sustained discharge during motor preparation and fixation. *J. Neurophysiol.*, 66: 1624–1641.

Munoz, D.P., Guitton, D. and Pelisson, D. (1991) Control of orienting gaze shifts by tectoreticulospinal system in head-free cats. III. Spatio-temporal characteristics of phasic motor discharge. *J. Neurophysiol.*, 66: 1642–1666.

Peak, C.K. (1989) Visual responses of neurones in cat superior colliculus in relation to fixation of targets. *J. Physiol.*, 414: 301–315.

Peck, C.K., Schlag-Rey, M. and Schlag, J. (1980) Visuo-oculomotor properties of cells in the superior colliculus of the alert cat. *J. Comp. Neurol.*, 194: 97–116.

Peterson, B.W., Pitts, N.G., Fukushima, K. and Mackel, R. (1978) Reticulospinal excitation and inhibition of neck motoneurons. *Exp. Brain Res.*, 32: 471–489.

Roucoux, A., Guitton, D. and Crommelink, M. (1980) Stimulation of the superior colliculus in the alert cat. II. Eye and head movements evoked when the head unrestrained. *Exp. Brain Res.*, 39: 75–85.

Sasaki, S. (1992) Reticulospinal control of head movements in the cat. In: A. Berthoz, P.P. Vidal and W. Grag (Eds.), *Head-Neck Sensory Motor System*, Oxford University Press, New York, Oxford, pp. 331–317.

Siegel, J.M. and Tomazewski, K.S. (1983) Behavioral organization of reticular formation; studies in the unrestrained cats. I. Cell related axial, limb, eye and other movements. *J. Neurophysiol.*, 50: 696–716.

Stein, B.E. (1984) Development of the superior colliculus. *Annu. Rev. Neurosci.*, 7: 95–126.

Wurtz, R.H. and Albano, J.E. (1980) Visuo-motor function of the primate superior colliculus. *Annu. Rev. Neurosci.*, 3: 189–226.

M. Norita, T. Bando and B. Stein (Eds.)
Progress in Brain Research, Vol 112
© 1996 Elsevier Science BV. All rights reserved.

CHAPTER 8

Ibotenic acid lesions of the superior colliculus produce longer lasting deficits in visual orienting behavior than aspiration lesions in the cat

Alan C. Rosenquist*, Vivian M. Ciaramitaro, Jeffrey S. Durmer, Steven F. Wallace and Wendy E. Todd

Department of Neuroscience, School of Medicine, University of Pennsylvania, Philadelphia, PA 19104–6074, USA

We compared the effects of unilateral surgical aspiration and ibotenic acid produced lesions of the superior colliculus (SC) on visual orienting behavior in 20 cats. Four animals with aspiration lesions initially showed an hemianopia in the contralateral hemifield which recovered fully in 4.5 weeks or less. These lesions also destroyed axons in the commissure of the superior colliculus (CSC). In 9 animals we produced complete loss of cells in one SC, with preservation of axons in the CSC, by injections of ibotenic acid. In these animals the contralateral hemianopia persisted for an average of 16.6 weeks, but may have persisted longer had we not intervened by either sacrificing the animal or ablating the visual cortex contralateral to the SC lesion. The cortical lesion produced an immediate hemianopia in the contralateral hemifield and a recovery in the previously hemianopic ('collicular') hemifield. In the remaining 7 animals with attempted ibotenic acid lesions, 5 had incomplete lesions and 2 others sustained major damage to the SC as well as the CSC. These 7 animals recovered visual orienting on an average of 3.0 weeks postoperatively. We conclude that unilateral loss of collicular cell function and the presence of fibers coursing through the commissure of the superior colliculus are both necessary for the prolonged deficit in visual orienting behavior. We suggest that competition between the two hemifields may play a role in the hemianopia caused by collicular manipulations and that the cholinergic pathway from the pedunculopontine nucleus to the contralateral SC via the CSC may be involved.

Introduction

The superior colliculus (SC) plays an important role in mediating visually guided behaviors such as the ability of cats to orient their head and eyes to visual stimuli, i.e. 'the visual orienting response' (Apter, 1946; Hess et al., 1946; Sprague and Meikle, 1965). Following the removal of one SC by surgical aspiration an animal shows a turning bias towards the side of the lesion, and a neglect of stimuli in the visual hemifield con-

tralateral to the lesion (Sprague et al., 1961; Sprague and Meikle, 1965; Sprague, 1966; Wallace et al., 1989; 1990). Within 2–3 weeks after the lesion the motor bias subsides and a residual neglect in the hemifield contralateral the side of the lesion is demonstrable only when two stimuli are presented simultaneously, one in each hemifield. In this test which is termed 'extinction', the animal neglects stimuli in the previously hemianopic hemifield. The considerable recovery that is seen in visual orienting after colliculectomy is likely to be mediated by other intact visual pathways including the retino-geniculo-striate route (Sprague and Meikle, 1965; Sprague, 1966; Mohler and Wurtz, 1977).

*Corresponding author. Tel.: +1 215 8984286; fax: +1 215 5732248 e-mail: rosenqui@mail.med.upenn.edu

In this paper we describe the paradoxical finding that ibotenic acid destruction of one SC produces long lasting losses in orienting to visual stimuli presented in the hemifield contralateral to the SC damage. The excitotoxin, ibotenic acid, destroys cell bodies in the SC but spares fibers of passage through the site of the SC lesion (Schwarcz et al., 1979; Garey and Hornung, 1980; Guldin and Markowitsch, 1981; Kohler and Schwarcz, 1983; Wallace et al., 1989). These remaining axons, which include many of those comprising the commissure of the superior colliculus (CSC), are apparently sufficient to preclude the recovery seen after surgical colliculectomy.

Methods

Twenty adult cats of either sex were used for the present study. Cats were housed on a 12-h light/dark cycle and received food ad libitum except for periods before behavioral testing. They were tested for normal neurological status by means of an extensive neurological examination and for normal binocular visual fields by means of a standard food perimetry test. Cats with abnormal responses in any of these tests were excluded from all studies.

Surgical procedures

Anesthesia was induced with an initial intramuscular (i.m.) injection of ketamine HCl (15–20 mg/kg) and 0.1 mg of atropine sulfate. The femoral vein was then cannulated and an initial dose of 20–50 mg of pentobarbital was administered intravenously (i.v.). Procaine penicillin G (200 000 U) was given i.m. and the trachea was intubated. All subsequent anesthesia was given as a mixture of pentobarbital (17 mg/ml) and thiopental (8 mg/ml) diluted in sterile saline. Supplemental doses of this mixture were given throughout the procedure, if the animal's anesthetic plane appeared too light as evidenced by increased heart or respiratory rate or by withdrawal of the limb to pinch. Body temperature was maintained between 36° and 38°C by means

of a heating pad beneath the animal. The animal's head was then prepared for sterile surgery and positioned in the stereotaxic apparatus.

Collicular aspiration and commissure transection

Bilateral occipital craniotomies were made extending medially to the sagittal sinus and caudally to the boney tentorium. The bone overlying the rostral cerebellum was also removed. The venous sinuses were then freed from the tentorium, which was removed as a whole. To expose the SC, the overlying dura was cut and reflected; and if necessary, the cortex (with dura intact) was gently retracted using protective neurosurgical cottonoid sponges. In most cases, the caudal one-half of both colliculi could be clearly visualized but a small portion of the overlying parasplenial cortex usually had to be aspirated to visualize the entire SC and the CSC. Either the left or right SC was carefully suction aspirated through the blunt end of an 18-g hypodermic needle or the CSC was cut midsagittally using a fine knife. After hemostasis was achieved, Gelfoam (Upjohn) was placed over the midbrain and cerebellum. The animal was then sutured and removed from the stereotaxic apparatus and closely monitored until fully awake.

Ibotenic acid lesion of the superior colliculus

Ibotenic acid (Siris) was dissolved in sterile 1.0 M phosphate buffered saline (pH 7.4) to obtain concentrations of 10–20 μg/μl. Visually guided injections into the SC were made after removal of the tentorium and visualization of the collicular surface as described above. Injections were made by means of a 5-μl Hamilton syringe with a 30-g needle. The syringe was mounted on a stereotaxic carrier inclined at a 45° angle in the sagittal plane and the needle lowered in the space between the anterior lobe of the cerebellum and the posterior parasplenial cerebral cortex. The needle was then visually guided into the SC and 1–3 injections made along each track. Approximately 0.2–0.3 μl of ibotenic acid (10–20 μg/μl) were injected over a 5-min period at each injection site. A total of 10–15 penetrations were made and 30–40 injections were performed. Once injections were complete and hemostasis was achieved, Gelfoam was placed over the midbrain. The animal was then sutured, removed from the stereotaxic apparatus and monitored until fully awake.

Cortical lesions

A midsagittal incision was made and the scalp and temporalis muscle were retracted laterally. A unilateral craniotomy was performed and extended to expose the entire cortical region to be removed. The dura was then cut and reflected and a cortical lesion made by gentle subpial aspiration using landmarks supplied by visual cortical maps (Rosenquist, 1985). The following cortical areas were then removed: 17, 18, 19, 20a, 21a, 21b, DLS, VLS, PS, PMLS, PLLS, AMLS, ALLS, 5, 7, and SVA. Gelfoam was placed over the lesion and the appropriate tissue planes reapproximated and sutured. The animal was then removed from the stereotaxic apparatus and monitored.

Behavioral testing

Animals were food deprived for an 18–48-h period prior to behavioral testing. The length of this period depended upon the motivation of the cat, which was found to vary considerably among animals. Cats were tested at least once per week except in the immediate postoperative period. After cortical lesions cats were typically given a 3-week recovery period during which they were not tested. A 1-week recovery period followed all other surgeries. Over several months of testing cats maintained good health and typically gained weight.

Visual perimetry test

Food deprived cats were tested on a table marked off into 13 lines intersecting at a common origin and radiating at 15° angles from 90° left to 90° right of the 0° fixation line. A detailed description of this test has been presented previously (Wallace et al., 1989).

Prior to each surgery all cats were tested under binocular viewing conditions for a minimum of 242 trials (14 at each angle and 60 blanks) with most cats receiving many more trials. In order to control for possible experimenter bias, cats were occasionally tested by another team of experimenters who were unfamiliar with the previous history of the animal. In no case was a significant discrepancy found between the 'informed' and the 'blind' experimenters owing largely to the robustness of the orienting responses or their clear absence.

Neurological examination

This examination (described in detail in Wallace et al., 1989) was performed as an initial screening procedure following each surgical intervention.

Histology and anatomical reconstruction

After all behavioral data were collected, the cat was given an overdose of pentobarbital and perfused through the heart with physiological saline followed by 10% formalin. The brain was then post-fixed and blocked in the standard stereotaxic frontal plane, dehydrated through an alcohol series, and embedded in celloidin. Forty-eight-micro-meter thick sections were cut with every tenth section Nissl stained with cresyl violet. Adjacent sections were stained for fibers by the Heidenhain-Mahon method. Cytoarchitectural determinations were based primarily on the atlas of Berman (1968), with supplementation by Kanaseki and Sprague (1974) and Updyke (1983). For each experimental paradigm, data were used only from cats that had undergone complete histological examination and reconstruction of lesions.

The extent of each cortical lesion was determined by charting the damage seen in serial coronal sections onto drawings of the hemisphere that were derived either from intraoperative photographs of the hemisphere or from postmortem examination of the opposite intact hemisphere. This reconstructed lesion was then compared to standard physiological and anatomical maps to assess its completeness (Rosenquist, 1985). In addition, the dorsal lateral geniculate nucleus was examined for surviving cells. Any animals demonstrating evidence of cortical or geniculate sparing were excluded from this study.

The location and extent of collicular ibotenic acid lesions were determined by visualization of Nissl stained tissue under high power magnification on a compound microscope. An X–Y plotter was coupled electrically to the stage of the microscope allowing the precise borders of the sections and the location of ibotenic acid lesion to be charted. The lesions were clearly defined as areas devoid of neuronal cells and containing reactive gliosis. The Nissl and adjacent myelin-stained sections were then projected onto the plotter charts.

Results

Surgical aspiration of the SC

In four cats we removed one SC by surgical aspiration. The two largest lesions are illustrated in Fig. 1. In both of these cases (SE 100 and SE 118) and in another case not illustrated (SE 84) the lesion destroyed nearly the entire right SC and damaged the pretectum and posterior thalamus as well. In the final case (SE 40) there was little or no damage to the pretectum or thalamus but removal of the SC was complete. All four of these animals showed immediate post-operative turning biases into the field ipsilateral to the lesion lasting 10–14 days and precluding visual perimetry testing. When the animals were testable they all showed an hemianopia in the contralateral hemifield. This hemianopia completely resolved within 2.5–4.5 weeks postoperatively (see Table 1, Group 1). The only deficit was a relative neglect seen in tests for 'extinction'. Thus, when two stimuli were presented simultaneously, one each in the left and right hemifields, the animal always oriented towards the stimulus in the previ-

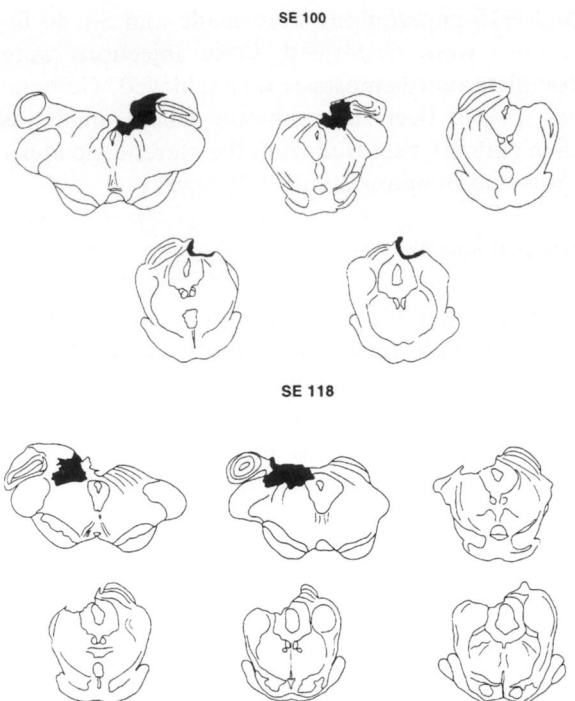

Fig. 1. Reconstructions of the SC surgical aspiration lesions for animals SE 100 and SE 118. Blackened regions on the coronal sections represent areas of severely necrotic tissue. For each animal sections proceed from rostral to caudal.

ously non hemianopic field, i.e. the field ipsilateral to the SC lesion.

Ibotenic acid destruction of the SC

We produced ibotenic acid lesions to one SC in a total of 16 cats. One animal showed no deficits but the lesion was barely visible (SE 110). The 15 remaining animals showed immediate post-operative deficits identical to the cats with SC aspiration lesions. These included turning and an hemianopia in the visual field opposite the SC lesion. In nine of these 15 animals the hemianopia persisted for over 8 weeks (see Table 1, Group II) after which time animals were either sacrificed or given a contralateral visual cortical lesion. Fig. 2 illustrates one case (SE 99) in which an homonymous hemianopia persisted for 26 weeks before the visual field recovered spontaneously. This animal remained hemianopic longer than any other

TABLE 1

Surgical history and visual field status

Cat	Procedure performed	Time (weeks)	Visual field status

Group I: Surgical aspiration of the SC

SE 40	Right SC lesion		Left HH (homonymous hemianopia)
	Postoperative period	4.5	Left field recovery
SE 84	Right SC lesion		Left HH
	Post-operative period	4.5	Left field recovery
SE 100	Right SC lesion		Left HH
	Post-operative period	2.5	Left field recovery
SE 118	Right SC lesion		Left HH
	Post-operative period	4.3	Left field recovery

Mean time to recovery for Group I Cases: 4.0 weeks

Group II: Ibotenic acid lesion of the SC and no recovery before 8 weeks

SE 21	Left SC ibotenic lesion		Right HH
	Post-operative period	8.1	Right field recovery
SE 38	Left SC ibotenic lesion		Right HH
	Post-operative period	10.8	Right HH persists
	Right visual cortical lesion		Right field recovery PD (postoperative day) 1
			Left HH PD 1
	Post-operative period	9.2	Left HH persists
SE 47	Right SC ibotenic lesion		Left HH
	Post-operative period	12.7	Left HH persists
	Left visual cortical lesion		Left field recovery PD 8
			Right HH
	Post-operative period	12.5	Right HH persists
	CSC transection		Right field recovery PD 14
	Post-operative period	5.3	Full fields
SE 49	Left SC ibotenic lesion		Right HH
	Post-operative period	11.7	Right HH persists
	Right visual cortical lesion		Right field recovery PD 5
			Left HH
	Post-operative period	14.7	Left HH persists
	CSC transection		Left field recovery PD 13
	Post-operative period	5.5	Full fields
SE 59	Right SC ibotenic lesion		Left HH
	Post-operative period	20.0	Left HH persists
	Left visual cortical lesion		Left field recovery PD 8
			Right HH
	Post-operative period	15.3	Right HH persists
	CSC transection		Right field recovery PD 6
	Post-operative period	3.7	Full fields

TABLE 1 (*continued*)

SE 83	Right SC ibotenic lesion		Left HH
	Post-operative period	18.0	Left HH persists
	Left visual cortical lesion		Left field recovery PD 2
			Right HH
	Post-operative period	13.5	Right HH persists
SE 99	Left SC ibotenic lesion		Right HH
	Post-operative period	26.0	Right HH persists until recovery PD 182
SE 108	Left SC ibotenic lesion		Right HH
	Post-operative period	23.0	Right HH persists
SE 111	Left SC ibotenic lesion		Right HH
	Post-operative period	19.0	Right HH persists

Mean time to recovery for Group II Cases: 16.6 weeks

Group III: Ibotenic acid lesion of the SC and recovery within 6 weeks

SE 39	Left SC ibotenic lesion		Right HH
	Post-operative period	9.0	Right field recovery PD 11
	Right visual cortical lesion		Left HH
	Post-operative period	8.0	Left HH persists
SE 43	Right SC ibotenic lesion		Left HH
	Post-operative period	5.0	Left field recovery PD 24
	Left visual cortical lesion		Right HH
	Post-operative period	9.5	Right HH persists
SE 72	Left SC ibotenic lesion		Right HH
	Post-operative period	6.3	Right field recovery PD 39
	Right visual cortical lesion		Left HH
	Post-operative period	9.7	Left field recovery PD 37
SE 91	Left SC ibotenic lesion		Right HH
	Post-operative period	7.0	Right field recovery PD 33
SE 92	Right SC ibotenic lesion		Left HH
	Post-operative period	10.3	Left field recovery PD 6
SE 101	Left SC ibotenic lesion		Right HH
	Post-operative period	4.7	Right field recovery PD 19
SE 110	Left SC ibotenic lesion		No deficit at first test PD 17
	Post-operative period	3.3	Full fields

Mean time to recovery for Group III cases: 3.0 weeks

case. As the histological reconstruction shows there was virtually complete destruction of all cells in all laminae of the SC with the only sparing limited to the far caudal pole.

Either of two factors distinguish the nine cases with long term hemianopia from the seven cases in which there was either no deficit or in which the hemianopia recovered in less than 6 weeks (see Table 1, Group III). First is the failure of the attempted ibotenic acid lesion to produce extensive SC cellular loss (SE 39, SE 43, SE 92, SE 101, SE 110). Fig. 3 illustrates one case in which the lesion was very incomplete (SE 101). This animal recovered visual orienting in the affected field within 2.5 weeks of the ibotenic injection. The second factor leading to a rapid recovery is the presence, along with SC cellular loss, of extensive fiber damage in the CSC (SE 72, SE 91). This loss of fibers appears to be the result of ischemia or trauma caused by multiple needle penetrations into the SC. Fig. 4 illustrates cases in which extensive cellular loss is present in the SC along with major fiber damage to the CSC (SE 72 and SE 91), and one case in which the CSC was relatively spared (SE 99).

Further evidence that axons in the CSC were destroyed in case SE 72 comes from the observation that the hemianopia produced by a subsequent visual cortical lesion recovered after 37

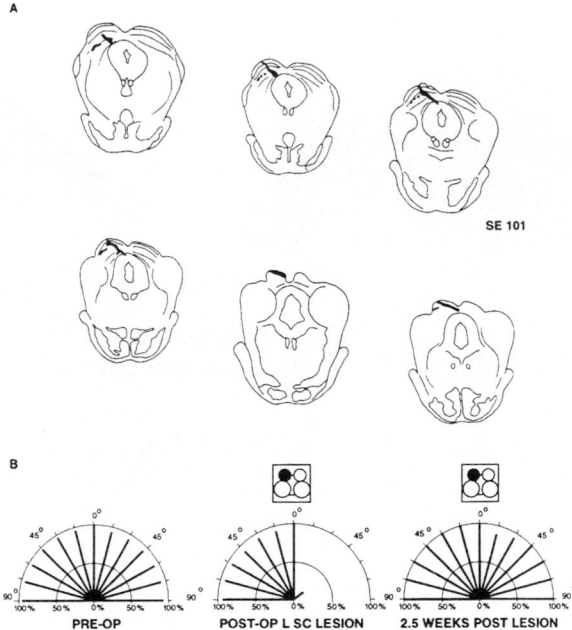

Fig. 3. Animal SE 101. A: Reconstruction of an incomplete SC ibotenic acid lesion. Blackened regions indicate zones of complete neuronal cell loss. B: Summary of the cat's visual field status at each stage of the experiment. Note the full recovery of visual orienting within 2.5 weeks post-injection.

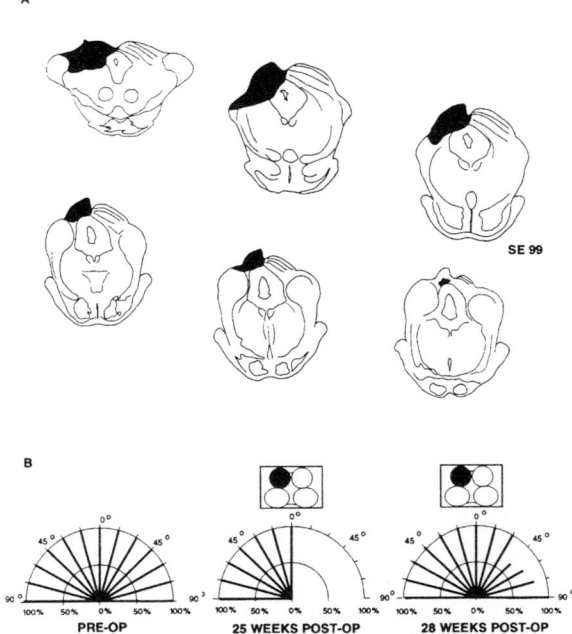

Fig. 2. Animal SE 99. A: Reconstruction of an SC ibotenic acid lesion. Blackened regions indicate zones of complete neuronal cell loss. B: Summary of the cat's visual field status at each stage of the experiment. The normalized response levels for each of the 13 perimetry radii are presented as bold lines. The two concentric semicircles represent the 50% and 100% response levels. Insets and labels indicate the surgical status of the animal. Note that this animals remained hemianopic for over 25 weeks post-injection.

days. This 'Sprague Effect' recovery is seen in animals after CSC transections but not after SC ibotenic acid produced lesions that spare CSC fibers (Wallace, et al., 1989). By comparison, animals in which the CSC was spared did not show a recovery from hemianopia after a subsequent visual cortical lesion (SE 39 and SE 43). Thus, paradoxically, ibotenic acid lesions which destroy nearly all of the cell bodies in one SC but spare fibers of passage result in an hemianopia of much greater duration than surgical aspiration lesions of the entire SC.

All cats in Groups II and III whose hemianopia recovered spontaneously after damage to the SC continued to show marked asymmetries on a test for extinction except for cases SE 92, SE 101 and SE110 in which there was relatively little SC damage.

Five cats in Group II of Table 1 (SE 38, SE 47, SE 49, SE 59, and SE 83) underwent a visual

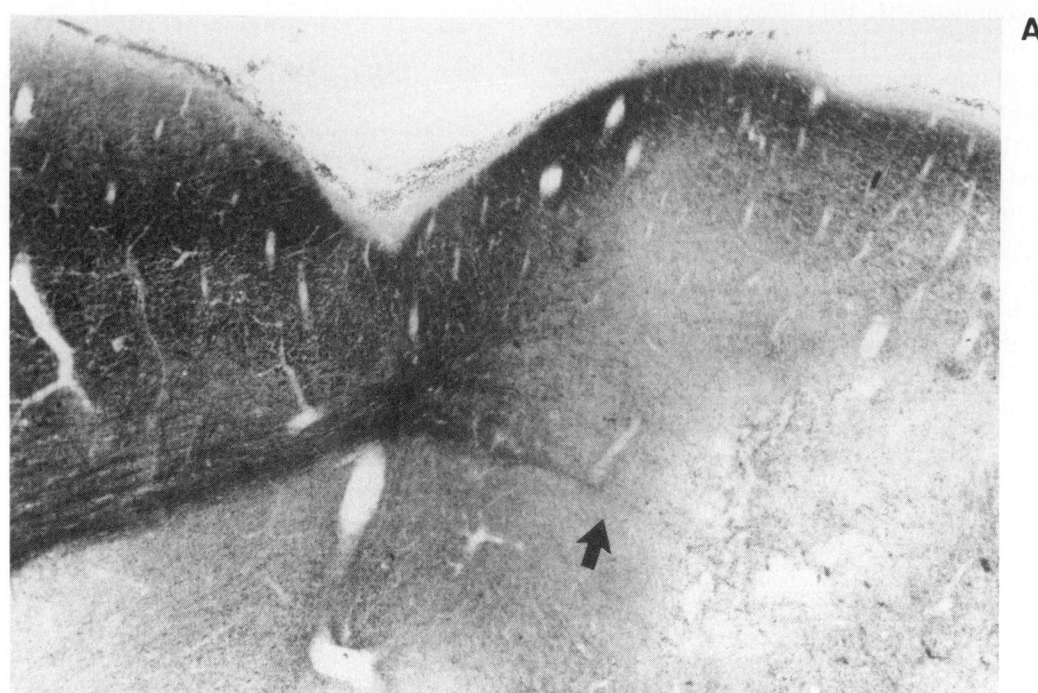

A

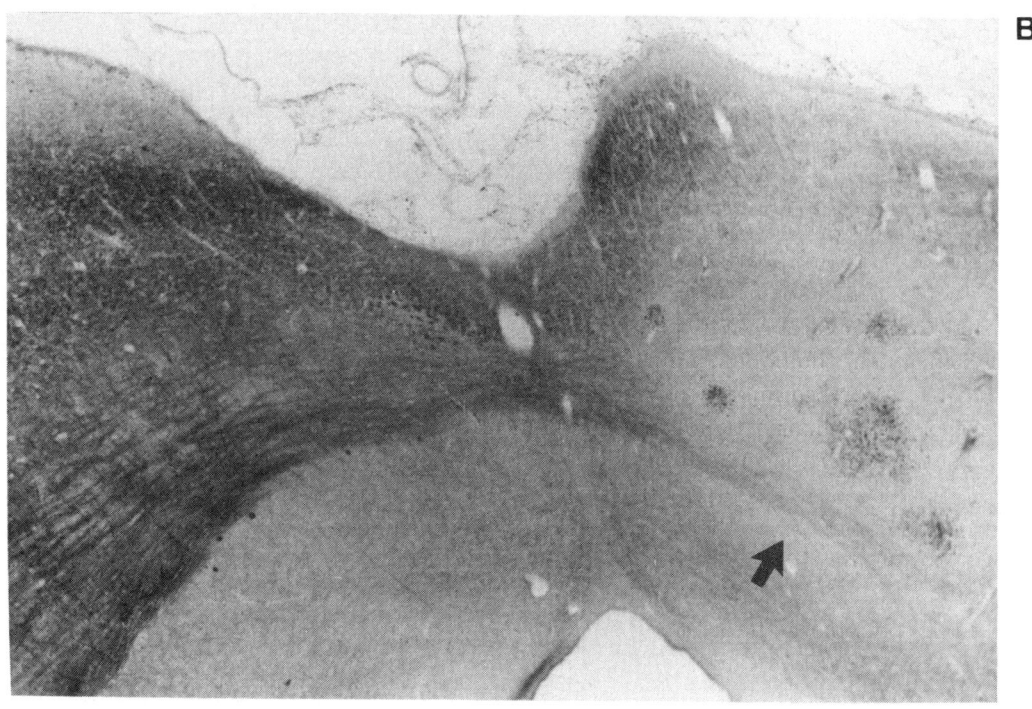

B

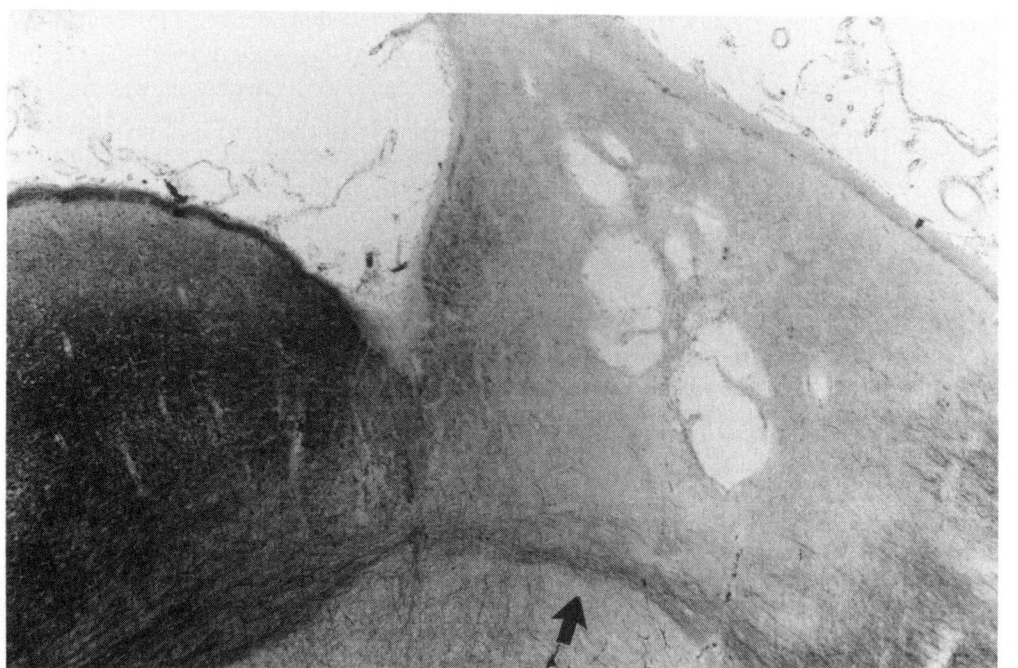

C

Fig. 4. Photomicrographs of myelin stained sections of the SC in three animals. A: Animal SE 91 shows severe loss of CSC axons (arrow). B: Animal SE 72 shows a similar loss of CSC axons (arrow). C: Animal SE 99 shows a relatively intact CSC (arrow).

cortical ablation on the side opposite the SC ibotenic acid lesion. At the time of the cortical ablation each animal had been hemianopic in the field opposite the SC lesion for a period of 10–20 weeks. The cortical lesion produced an immediate recovery of visual orienting in the 'collicular' hemifield and an immediate hemianopia in the 'cortical' hemifield. These effects were seen on the first postoperative testing day for each animal. These five animals displayed reduced postoperative turning behavior compared to cats sustaining only a visual cortical lesion. Thus we could conduct visual perimetry testing earlier: from postoperative day 1 to day 8. Finally, in three of the five animals (SE 47, SE 49, and SE 59) with SC ibotenic acid lesions and a contralateral visual cortical lesion we placed a third and final lesion: to the CSC. Each of these 3 animals recovered visual orienting in the cortical hemifield, i.e. they showed a 'Sprague Effect'. Fig. 5 depicts these results from animal SE 59.

Discussion

In our four animals that sustained unilateral surgical aspiration lesions of the SC we confirmed earlier reports of ipsiversive circling and head posturing which lasted for as long as two weeks post-operatively (Sprague and Meikle, 1965). This motor behavior prevented reliable visual perimetry testing. When animals could be tested for visual orienting, all cases showed an homonymous hemianopia in the visual field contralateral to the aspirated SC. The hemianopia resolved within 4.5 weeks and all animals again displayed full visual fields. The only residual deficit was a relative neglect to stimuli in the previously hemianopic field, as demonstrated on an extinction task, a test of relative spatial 'neglect' of stimuli (Mesulam, 1985).

Similar to aspiration lesions, ibotenic acid lesions of one SC produced the same ipsiversive posturing and circling behaviors. These motor

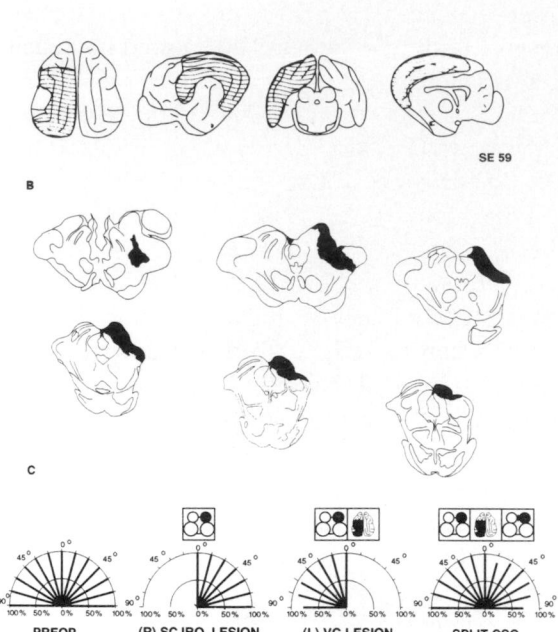

Fig. 5. Animal SE 59. A: Reconstruction of the left visual cortical lesion. B: Six drawings of coronal midbrain sections demonstrating the areas of complete neuronal cell loss resulting from the ibotenic acid lesion (black). Note also the CSC cut. C: Summary of the cat's visual field status at each stage in the experiment. Note that the hemifield of the hemianopia following the SC ibotenic lesion was reversed by the visual cortical lesion and that the hemianopia resulting from the cortical lesion was abolished by the CSC transection.

behaviors subsided in 2–4 weeks and perimetry testing was then possible. Nine animals with extensive cellular loss in the SC and little or no damage to the CSC showed an homonymous hemianopia in the visual field contralateral to the SC lesion for 8 weeks up to 26 weeks. This hemianopia may have persisted longer if we had not intervened with a second surgery or if the animals had been maintained in the protocol longer.

Seven other animals either showed no deficit ($N = 1$) after attempted SC ibotenic destruction or recovered spontaneously ($N = 6$) from 6 days to 6 weeks post-operatively. These animals are distinguished from the nine animals with a long

term hemianopia by two factors: (1) they either had very incomplete lesions ($N = 5$) or; (2) the extensive SC cellular destruction was accompanied by massive damage to fibers coursing through the CSC ($N = 2$). This is identical to the four cases of surgical aspiration lesions to one SC, the CSC was always destroyed along with cells in the SC, and visual orienting promptly recovered.

Thus, after complete ibotenic acid destruction of cells in the SC, the hemianopia persists for a much longer duration than after seemingly more severe surgical aspirations of the SC. The crucial difference, as demonstrated by ibotenic acid lesions resulting in fiber damage, is that surgical aspiration destroys axons of passage, including those of the CSC. We conclude that it is the presence of CSC fibers that prevents rapid recovery of hemianopia after SC cellular damage. This conclusion is supported by three additional observations. First, we previously reported that ibotenic acid lesions to the SC contralateral to a visual cortical lesion failed to produce a Sprague Effect, i.e. a recovery of visual orienting in the 'cortical' hemifield. Recovery was only seen if the CSC was also cut or damaged (Wallace et al., 1989). Second, in the present study we observed extensive fiber damage in two animals with complete SC cellular loss subsequent to an ibotenic acid lesion. These animals (SE 72 and SE 91) recovered visual orienting in the 'collicular' hemifield. In one of these animals (SE 72) we made a cortical lesion contralateral to the damaged SC. The cortical lesion immediately restored visual orienting to the previously blind 'collicular' hemifield and we also observed recovery in the 'cortical' hemifield, i.e. a Sprague Effect (Sprague, 1966). Since the Sprague Effect recovery is the result of CSC damage, the results from this animal confirm that the CSC was inadvertantly damaged by the earlier SC lesion. Third, in three animals (SE 47, 49, and 59) that were still hemianopic after an SC ibotenic lesion we placed a contralateral cortical lesion and observed no Sprague Effect. Animals remained hemianopic in the 'cortical' hemifield.

This is interpreted as evidence for the integrity of the CSC and is consistent with the conclusion that the presence of the CSC is responsible for both the longstanding hemianopia and the failure to produce a Sprague Effect. In each of these three animals once the CSC was transected visual orienting was quickly restored to the 'cortical' hemifield. This indicates that the CSC had been intact during the period of prolonged hemianopia after SC ibotenic acid cellular damage.

Finally, as with ibotenic acid lesions, we have shown (Ciaramitaro et al., 1994) that reversible pharmacological inactivation of SC neurons, which also fails to affect fibers of passage, produces a transient yet complete loss of visual orienting in the contralateral hemifield for several hours post-injection. Control injections of comparable volumes and pH of saline were ineffective (see Fig. 6). Again we see the loss of visual orienting subsequent to an intervention which decreases the activity of cells in the SC. It remains to be determined whether such a loss in visual orienting would occur if the CSC had been previously destroyed.

The assessment of visual orienting in the cat is highly dependent upon the testing method employed. Hardy and Stein (1988) showed no deficits in visual orienting contralateral to a small lesion in the posterior suprasylvian cortex when they tested their animals with a method similar to that used in the present study, i.e. animals remained stationary during the presentation of the target stimulus. Yet, they were able to demonstrate a 'complete contralateral visual neglect' when their paradigm required the animal to be in forward motion when the target stimulus was presented. These results imply that the test of visual orienting used in the present study is less sensitive to unilateral injury. This makes the demonstration of a longstanding hemianopia after SC ibotenic acid damage using the stationary method of testing even more compelling.

Our results do not bear directly on the issue of which lamina or laminae of cells in the SC are responsible for mediating visual orienting behavior in the cat. The preponderance of evidence points to cells in the intermediate and deep layers as the most likely candidates. In the cat and

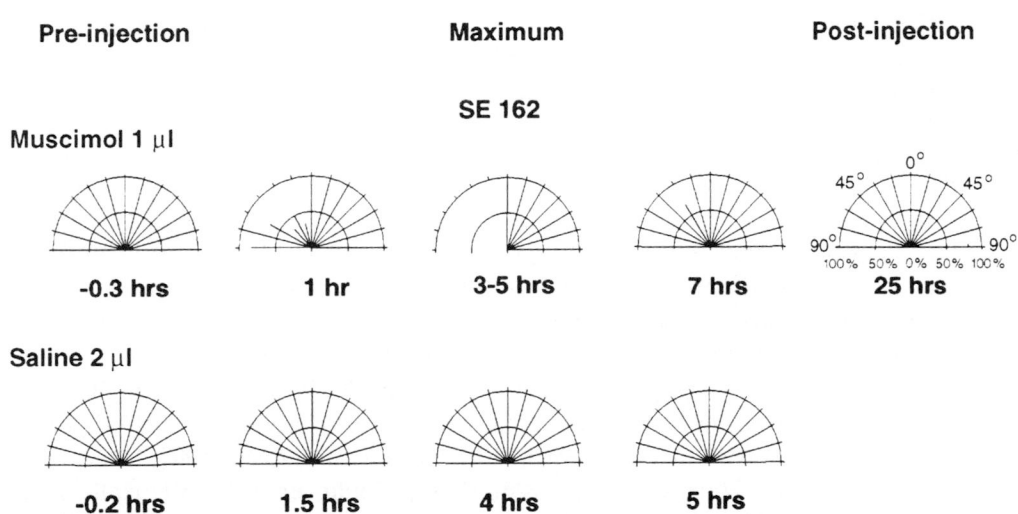

Fig. 6. Visual perimetry results for animal SE 162 before and following either a 1-μl muscimol injection (1 μg/μl) or a 2-μl saline control injection into the right SC. Injections were made through a chronic indwelling cannula on two separate days. Note that the animal had full visual fields but lost visual orienting to stimuli in the left hemifield for several hours after the muscimol but not the saline injection. Injections were made at 0.0 h.

monkey neurons in these layers project axons to regions of the brainstem that control head and eye movements (Edwards, 1978; Wurtz and Albano, 1980; Huerta and Harting, 1984). In the cat it is also known that small posterior suprasylvian lesions that result in deficits in visual orienting in certain testing paradigms (Hardy and Stein, 1988) reduce the visually driven activity of intermediate and deep layer SC cells, but not cells in the superfical layers (Ogasawara et al., 1984). Electrical stimulation of the SC in the cat and monkey initiates contralateral head and eye movements (Hess et al., 1946; Schiller and Stryker, 1972; Schiller, 1984; Stein, 1984; Peck, 1990). Many of the intermediate layer cells fire prior to eye movements and have spatially overlapping visuosensory and visuomotor fields (Schiller and Koerner, 1971; Schiller and Stryker, 1972). The result of this overlap is that a visual stimulus excites cells whose firing produces a coordinated head and eye movement that brings the stimulus onto the center of gaze. This is the visual orienting response.

Why does a relatively intact CSC appear to prolong the recovery time after unilateral SC cellular damage? One hypothesis is that projections coursing through the CSC serve to activate cells in the undamaged SC, making the 'collicular' hemifield hyperattentive. If a competition between the two hemifields is heightened by this SC activation then recovery, presumably mediated by cortical regions, may be delayed. What nuclei projecting through the CSC might serve this function? One possibility is the extensive projection from the pedunculopontine nucleus (PPN). Several studies have shown that acetylcholinergic cells in the PPN send axons which terminate in the ipsilateral and, via the CSC, the contralateral colliculus as patches in the intermediate gray layer (Edwards et al., 1979; Edley and Graybiel, 1983; Jones and Beaudet, 1987; Jones and Webster, 1988; Hall et al., 1989; Harting and Lieshout, 1991; McHaffie et al., 1991; Jeon et al., 1993).

The function of the acetylcholinergic PPN projections is not understood but preliminary evidence from our laboratory indicates that an acetylcholine agonist injected into the intermediate layers of one SC, in an intact cat, results in effects similar to electrical stimulation, i.e. the immediate and prolonged deviation of the eyes, head and body into the contralateral hemifield (Durmer et al., unpublished observations). In this animal we noted a prolonged hyperattentiveness to visual stimuli in the hemifield contralateral to the injection for several hours. Thus, in a cat with an SC ablation, damage to the CSC presumably destroys PPN projections to the undamaged SC and thereby renders the undamaged SC less active and its hemifield less attentive. The reduced competition between the intact and hemianopic fields could lead to a faster recovery.

The PPN hypothesis, however, ignores other pathways that contribute axons to the CSC and that may be involved in recovery (Edwards, 1977; Jayaraman et al., 1977; Grofova et al., 1978; Edwards et al., 1979; Appell and Behan, 1990). A more complete explanation of our results will undoubtedly involve understanding how several pathways interact to affect SC neurons. For example, a GABAergic and putatively inhibitory pathway from the contralateral substantia nigra pars reticulata (SNpr) is also a component of the CSC (Edwards et al., 1979; Beckstead, 1981; 1983; Gerfen et al., 1982; May and Hall, 1986; Harting et al., 1988; Appell and Behan, 1990; Wallace et al., 1990; Redgrave et al., 1992; Savaki, et al., 1992). Damage to the SNpr projection through the CSC would be expected to increase activity in the intact SC and lead to effects opposite to those observed and predicted by focusing on the PPN pathway exclusively. Another complexity is evident when we consider that the SNpr also projects to the ipsilateral PPN and therefore potentially influences the contralateral SC via a direct and an indirect route (Gerfen, 1982; Beckstead, 1983; Noda and Oka, 1986; Jones and Webster, 1988; Gould, et al., 1989; Hall et al., 1989; Kang and Kitai, 1990; Garcia-Rill, 1991). The modulatory nature of these influences on SC function is likely to contribute to the competition between visual fields and the consequent inattentiveness to visual stimuli after cortical and sub-cortical interventions.

References

Appell, P.R. and Behan, M. (1990) Sources of subcortical GABAergic projections to the superior colliculus in the cat. *J. Comp. Neurol.*, 302: 143–158.

Apter, J.T. (1946) Eye movements following strychninization of the superior colliculus of cats. *J. Neurophysiol.*, 9: 73–86.

Beckstead, R. (1983) Long collateral branches of substantia nigra pars reticulata axons to thalamus, superior colliculus and reticular formation in monkey and cat. Multiple retrograde neuronal labeling with fluorescent dyes. Neurosci., 10: 767–779.

Beckstead, R.M., Edwards, S.B. and Frankfurter, A. (1981) A comparison of the intranigral distribution of nigrotectal neurons labeled with horseradish peroxidase in the monkey, cat, and rat. *J. Neurosci.*, 1: 121–125.

Behan, M. (1985) An EM-autoradiographic and EM-HRP study of the commissural projection of the superior colliculus in the cat. *J. Comp. Neurol.*, 234: 105–116.

Berman, A.L. (1968) *The Brainstem of the Cat: A Cytoarchitectonic Atlas with Stereotaxic Coordinates.* University of Wisconsin Press, Madison, WI.

Ciaramitaro, V., Durmer, J., Todd, W. and Rosenquist, A. (1994) $GABA_A$ mediated disinhibition of the superior colliculus restores visual orienting responses in the previously hemianopic visual field of the cortically blind cat. *Soc. Neuro. Abstr.*, 20: 1186.

Edley, S. and Graybiel, A. (1983) The afferent and efferent connections of the feline nucleus tegmenti pedunculopontis, pars compacta. *J. Comp. Neurol.*, 217: 187–215.

Edwards, S.B. (1977) The commissural projection of the superior colliculus in the cat. *J. Comp. Neurol.*, 173: 23–40.

Edwards, S.B. and Henkel, C.K. (1978) Superior colliculus connections with the extraocular motor nuclei in the cat. *J. Comp. Neurol.*, 179: 451–468.

Edwards, S.B., Ginsburg, C.L., Henkel, C.K. and Stein, B.E. (1979) Sources of subcortical projections to the superior colliculus in the cat. *J. Comp. Neurol.*, 184: 309–330.

Garcia-Rill, E. (1991) The pedunculopontine nucleus. *Prog. Neurobiology*, 36: 263–389.

Garey, L.J. and Hornung, J.P. (1980) The use of ibotenic acid lesions for light and electron microscopy of anterograde degeneration of the visual pathway of the cat. *Neurosci. Lett.*, 19: 117–123.

Gerfen, C., Staines, W., Arbuthnott, G. and Fibiger, H.C. (1982) Crossed connections of the substantia nigra in the rat. *J. Comp. Neurol.*, 207: 283–303.

Grofova, I., Ottersen, O.P. and Rinvik, E. (1978) Mesencephalic and diencephalic afferents to the superior colliculus and periaqueductal gray substance demonstrated by retrograde axonal transport of horseradish peroxidase in the cat. *Brain Res.*, 146: 205–220.

Gould, E., Woolf, N. and Butcher, L. (1989) Cholinergic projections to the substantia nigra from pedunculopontine and laterodorsal tegmental nuclei. *Neuroscience*, 28: 611–623.

Guldin, W.O. and Markowitsch, H. (1981) No detectable remote lesions following massive intrastriatal injections of ibotenic acid. *Brain Res.*, 225: 446–451.

Hall, W., Fitzpatrick, D., Klatt, L. and Raczkowski, D. (1989) Cholinergic innervation of the superior colliculus in the cat. *J. Comp. Neurol.*, 287: 495–514.

Harting, J. and Lieshout, D. (1991) Spatial relationships of axons arising from the substantia nigra, spinal trigeminal nucleus and pedunculopontine tegmental nucleus within the intermediate gray of the cat superior colliculus. *J. Comp. Neurol.*, 305: 543–558.

Harting, J.K., Huerta, M.F., Hashikawa, T., Weber, J.T. and Van, L.D.P., (1988) Neuroanatomical studies of the nigrotectal pathway in the cat. *J. Comp. Neurol.*, 278: 615–631.

Hardy, S.C. and Stein, B. (1988) Small lateral suprasylvian cortex lesions produce visual neglect and decreased visual activity in the superior colliculus. *J. Comp. Neurol.*, 273: 527–542.

Hess, W.R., Burgi, S. and Bucher, V. (1946) Motorische Funktion des Tektal- und Tegmentalgebietes. *Monatschr. Psychiatr. Neurol.*, 112: 1–52.

Huerta, M.F. and Harting, J. (1984) The mammalian superior colliculus: Studies of its morphology and connections. In: H. Vanegas (Ed.), *The Comparative Neurology of the Optic Tectum.* Plenum Press, New York, 687–773.

Jayaraman, A. and Carpenter, M. (1977) Nigrotectal projections in the monkey: An autoradiographic study. *Brain Res.*, 135: 147–152.

Jeon, C., Spencer, R. and Mize, R. (1993) Organization and synaptic connections of cholinergic fibers in the cat superior colliculus. *J. Comp. Neurol.*, 333: 360–374.

Jones, B. and Beaudet, A. (1987) Distribution of acetylcholine and catecholamine neurons in the cat brainstem: a choline acetyltransferase and tyrosine hydroxylase immunohistochemical study. *J. Comp. Neurol.*, 261: 15–32.

Jones, B. and Webster, H. (1988) Neurotoxic lesions of the dorsolateral pontomesencephalic tegmentum cholinergic cell area in the cat. 1. Effects upon cholinergic innervation of the brain. *Brain Res.*, 451: 13–32.

Kanaseki, T. and Sprague, J. (1974) Anatomical organization of pretectal nuclei and tectal laminae in the cat. *J. Comp. Neurol.*, 158: 319–338.

Kang, Y. and Kitai, S. (1990) Electrophysiological properties of pedunculopontine neurons and their postsynaptic responses following stimulation of substantia nigra reticulata. *Brain Res.*, 535: 79–95.

Kohler, C. and Schwarcz, R. (1983) Comparisons of ibotenic and kainic neurotoxicity in rat brain: A histological study. *Neuroscience*, 8: 819–835.

May, P.J. and Hall, W.C. (1986) The sources of the nigrotectal pathway. *Neuroscience*, 19: 159–180.

McHaffie, J., Beninato, M., Stein, B. and Spencer, R. (1991) Postnatal development of acetylcholinesterase in and cholinergic projections to, the cat superior colliculus. *J. Comp. Neurol.*, 313: 113–131.

Mesulam, M.M. (1985) Attention, Confusional States and Neglect. In: M.M. Mesulam (Ed.), *Principles of Behavioral Neurology*, Vol. 26. F.A. Davis, Philadelphia, pp. 125–169.

Noda, T. and Oka, H. (1986) Distribution and morphology of tegmental neurons receiving nigral inhibitory inputs in the cat: an intracellular HRP study. *J. Comp. Neurol.*, 244: 254–266.

Ogasawara, K., McHaffie, J.G. and Stein, B.E. (1984) Two visual corticotectal systems in the cat. *J. Neurophysiol.*, 52: 1226–1245.

Peck, C.K. (1990) Neuronal activity related to head and eye movements in cat superior colliculus. *J. Physiol.*, 421: 79–104.

Redgrave, P., Marrow, L. and Dean, P. (1992) Topographical organization of the nigrotectal projection in the rat: evidence for segregated channels. *Neuroscience*, 50: 571–595.

Rosenquist, A.C. (1985) Connections of visual cortical areas in the cat. In: A. Peters and E.G. Jones (Eds.), *Cerebral Cortex*, Vol. 3. Plenum Press, New York, pp. 81–117.

Savaki, H., Raos, V. and Dermon, C. (1992) Bilateral cerebral metabolic effects of pharmacological manipulation of the substantia nigra in the rat: unilateral intranigral application of the inhibitory GABA$_A$ receptor agonist muscimol. *Neuroscience*, 4: 781–794.

Schiller, P.H. (1984) The superior colliculus and visual function. In: P. Brookhart and V. B. Mountcastle (Eds.), *Handbook of Physiology*. Waverly Press, Baltimore, pp. 457–505.

Schiller, P. and Koerner, F. (1971) Discharge characteristics of single units in the superior colliculus of the alert Rhesus monkey. *J. Neurophysiol.*, 34: 920–936.

Schiller, P.H. and Stryker, M. (1972) Single unit recording and stimulation in superior colliculus of the alert rhesus monkey. *J. Neurophysiol.*, 35: 915–924.

Schwarcz, R., Hokfelt, T., Fuxe, K., Jonsson, G., Goldstein, M. and Terenius, L. (1979) Ibotenic acid- induced neuronal degeneration: A morphological and neurochemical study. *Exp. Brain Res.*, 37: 199–216.

Sprague, J.M. (1966) Interaction of cortex and superior colliculus in mediation of visually guided behavior in the cat. *Science*, 153: 1544–1547.

Sprague, J.M. and Meikle, T. (1965) The role of the superior colliculus in visually guided behavior. *Exp. Neurol.*, 11: 115–146.

Sprague, J.M., Chambers, W.W. and Stellar, E. (1961) Attentive, affective and adaptive behavior in the cat. *Science*, 133: 165–173.

Stein, B.E. (1984) Multimodal representation in the superior colliculus and optic tectum. In: H. Vanegas (Ed.), *Comparative Neurology of the Optic Tectum*. Plenum, New York, pp. 819–841.

Updyke, B.V. (1983) A reevaluation of the functional organization and cytoarchitecture of the feline lateral posterior complex, with observations on adjoining cell groups. *J. Comp. Neurol.*, 219: 143–181.

Wallace, S.F., Rosenquist, A.C. and Sprague, J. (1989) Recovery from cortical blindness mediated by destruction of nontectotectal fibers in the commissure of the superior colliculus in the cat. *J. Comp. Neurol.*, 284: 429–450.

Wallace, S.F., A.C. Rosenquist and J.M. Sprague (1990) Ibotenic acid lesions of the lateral substantia nigra restore visual orientation behavior in the hemianopic cat. *J. Comp. Neurol.*, 296: 222–252.

Wurtz, R.H. and Albano, J. (1980) Visual-motor function of the primate superior colliculus. *Annu. Rev. Neurosci.*, 3: 189–226.

M. Norita, T. Bando and B. Stein (Eds.)
Progress in Brain Research, Vol 112
© 1996 Elsevier Science BV. All rights reserved.

CHAPTER 9

Spatial distribution of tectotectal connections in the cat

M. Behan* and N. M. Kime

Department of Comparative Biosciences and Center for Neuroscience, University of Wisconsin, Madison, WI 53706, USA

In mammals, the paired superior colliculi of the midbrain play a significant role in the generation and guidance of eye movements that enable an animal to orient to novel visual stimuli. In several species including monkey, cat and hamster, the paired colliculi are connected by a commissure. In the cat, many commissural axons arise from tectotectal neurons located in the deep layers in the rostral two-thirds of the colliculus. The role of these tectotectal neurons is unclear, but it is likely that they play some role in eye movement control.

In this study, the neuroanatomical tracer Biocytin was used to make small, localized injections into the deep layers of the cat superior colliculus at a variety of different locations in nine animals. The distribution of tectotectal synaptic terminals in the opposite colliculus was then plotted. Regardless of which layers were included the injection site, labelled boutons were most dense in the deep layers in the contralateral colliculus. There was a striking point-to-point organization in the tectotectal projection such that terminals were concentrated at an almost mirror-symmetrical region to the injection site in the rostrocaudal plane. In the majority of cases, however, the focus of terminal boutons was shifted medially by 1–2 mm.

These results suggest that tectotectal connections may influence select populations of neurons in the contralateral colliculus. By coupling specific groups of neurons in the two colliculi, their effectiveness in sensory motor processing could be enhanced. At this time it is not clear whether specific commissural terminals contain excitatory or inhibitory neurotransmitters, and our ongoing studies are addressing this question.

Introduction

In most vertebrate species thus far examined, the paired superior colliculi are connected by a commissure that includes, amongst others, the axons of tectotectal neurons. In many species the commissure does not extend fully rostrocaudally; in cats for example, it is confined to the rostral two-thirds of the colliculus. Most tectotectal axons synapse in the contralateral colliculus, although a small number extend further in the contralateral midbrain and pons. Tectotectal neurons are generally found in the deep layers of the superior colliculus, and there is considerable diversity in both their anatomy and physiology (Martin, 1969; Edwards, 1977; Magalhaes-Castro et al., 1978; Raczkowski and Diamond, 1978; Rhoades et al., 1981; Fish et al., 1982; Grantyn and Grantyn, 1982; Moschovakis and Karabelas, 1982; Yamasaki et al., 1984; Moschovakis and Karabelas, 1985; Rhoades et al, 1986; Moschovakis et al., 1988a). Many tectotectal neurons project to other target areas via ascending or descending pathways. The best characterized of these are the T-type neurons of both cat and rat superior colliculus that, in addition to having a commissural branch, also contribute one or more branches to the medial and lateral ipsilateral de-

---

*Corresponding author. Department of Comparative Biosciences, School of Veterinary Medicine, 2015 Linden Drive West, Madison, WI 53706, USA. Tel.: +1 608 2639833; fax: +1 608 2633926; e-mail: behanm@svm.vetmed.wisc.edu

scending pathways and the predorsal bundle (Moschovakis and Karabelas, 1985; Moschovakis et al., 1988a).

The function of the tectotectal projection is unclear. Based on electrophysiological studies in cat, Antonini et al. (1978) suggested that the rostral part of each colliculus functions as an integrated whole to 'subserve straight-ahead attention and orientation'. They further suggested that the posterior portions of each superior colliculus act independently of one another to subserve visually-guided contraversive turning of the head and body. In cat, tectotectal neurons have been shown to receive visual input, possibly by an indirect Y-cell pathway through the visual cortex (McIlwain, 1991). It is likely that they also receive input from other sensory modalities, as has been shown in hamster (Rhoades et al., 1986). In monkey, some tectotectal cells have been shown to discharge prior to saccades, implying a role in eye movement control (Moschovakis et al., 1988a,b). However, since each of the colliculi operates to orient the animal to the opposite hemifield, mutual interactions between the colliculi would of necessity have to be balanced such that signals descending to brainstem oculomotor nuclei were also in accord.

There are well-defined topographic maps in the superior colliculus, and early anatomical experiments suggested that a mirror-symmetric arrangement existed between tectotectal neurons in the rostral half of one colliculus and their terminal field in the opposite colliculus (Edwards, 1977; Magalhaes-Castro et al., 1978; Fish et al., 1982). In contrast, studies of intracellularly labelled tectotectal neurons in some cases showed a more widespread arrangement of labelled axons and terminals in the contralateral colliculus (Moschovakis and Karabelas, 1985; Rhoades et al., 1986; Moschovakis et al., 1988a; Grantyn, 1988), implying that tectotectal interactions are more broadly distributed. In light of these conflicting findings, we decided to investigate the tectotectal connections of a small group of neurons in the cat colliculus. As it is the activity of a group of neurons in the colliculus that determines

its motor output, we thought that tracing the axonal terminations from these small injections would provide a more realistic picture of the activated target area in the contralateral colliculus (Lee et al., 1988; Sparks and Mays, 1990). In this study we asked the following question: are the connections between one colliculus and the other organized topographically, such that similar areas are connected, or is there a random topography such that a localized group of neurons in one colliculus can potentially influence neurons throughout the opposite colliculus? We made very small injections of Biocytin into the deep layers of one colliculus and mapped the distribution of tectotectal terminals in the contralateral colliculus. Our results demonstrate a striking precision in the tectotectal projection, and suggest that interactions occur between similar populations of neurons in the two colliculi.

Materials and methods

All experimental protocols in the present study were approved by the Institutional Animal Care and Use Committee at the University of Wisconsin School of Veterinary Medicine, and complied with the guidelines of the Public Health Service policy on the Humane Care and Use of Laboratory Animals.

Surgery and general preparation

Nine adult cats were anesthetized with ketamine (25 mg/kg i.m.), followed by an intravenous injection of thiopental sodium (8 mg/kg). Following intubation, cats were maintained with 1–2% halothane in oxygen (2 l/min). Glass micropipettes with an internal tip diameter of 10 μm were backfilled with a solution of 5% Biocytin in 0.1 M Tris buffer (pH 7.4). The surface of the colliculus was identified by recording visually evoked potentials through the micropipette. Thereafter, the pipette was advanced a further 1–2 mm into the deep collicular layers. A single injection of Biocytin was made using positive current pulses (2.5 μA DC, 7 s on, 7 s off, for 7 min).

The electrode was held in place for 20 min and then withdrawn under a holding current. After a 48-h survival, animals were overdosed with sodium pentobarbital (100 mg/kg i.p.), and perfused transcardially with 400 ml of heparinized saline followed by 4 l of 2% paraformaldehyde and 0.1% glutaraldehyde in 0.1 M sodium phosphate buffer (pH 7.2). Coronal 70-μm sections through the superior colliculi were cut into cold buffer. Every section was processed immunocytochemically for the presence of Biocytin as follows: sections were washed in buffer with added Triton (0.3%), incubated in Avidin D-HRP (1:500, Vectastain), then washed and reacted with 0.05% DAB and 0.01% hydrogen peroxide in buffer.

Light microscopy

Biocytin-labelled cells and their terminals in the contralateral colliculus were plotted with a microscope-driven digitizer linked to a microcomputer (MDPlot, Minnesota Datametrics, Inc.). Every second section through the superior colliculus was plotted, and counts of labelled neurons and terminals were obtained from these two-dimensional plots. Where the colliculi were not cut precisely in the coronal plane, plots were corrected rostrocaudally. The size and location of each injection site was determined as follows. The mediolateral extent of the superior colliculus, and the distance of the most medially and laterally located Biocytin-labelled neuron from the midline were incorporated into a schematic representation of the superior colliculus. The boundaries of the superior colliculus and of the injection site were then depicted as viewed from above. Laminar boundaries, as defined by Kanaseki and Sprague, (1974) were identified by comparing sections to a reference series of myelin-stained coronal sections through the cat colliculus. By convention, the superficial layers included stratum zonale (SZ), stratum griseum superficiale (SGS), and stratum opticum (SO). The deep layers included stratum griseum intermedium (SGI), stratum album intermedium (SAI), stratum griseum profundum (SGP), and stratum album

profundum (SAP). The SGI and SAI are often referred to as the intermediate layers. The mean size of labelled neurons was obtained from digitized measurements of every Biocytin-filled neuron in a section through the center of the injection site in each case; a total of 180 neurons were measured in nine cases and the data pooled.

In order to verify that Biocytin labelled swellings were synaptic boutons, a few sections from one case (DL43) were embedded and examined at the electron microscopic level. Tissue to be processed for electron microscopy was perfused with 2% paraformaldehyde and 0.5% glutaraldehyde, reacted for the presence of Biocytin, post-fixed in 1% osmium tetroxide and 1.5% potassium ferrocyanide in 0.1 M phosphate buffer, dehydrated through a graded series of alcohols and propylene oxide, and flat embedded in resin between sheets of Aclar plastic. Ultrathin sections were cut from regions of colliculus containing labelled boutons, counterstained, and viewed in a Philips 410 electron microscope.

Results

Nine cases that included injection sites in several different areas of the rostral two-thirds of the colliculus were analyzed in this study (Fig. 1). Five injection sites were located close to the representation of the area centralis. Cases DL27 and DL28 were centered at the representation of the area centralis and differed primarily in the depth of the injection site. The injection sites in DL33, DL31 and DL41 were centered slightly caudal and medial to the area centralis representation. In four other cases, the injection sites were at a distance from the area centralis representation. DL47 was centered laterally on the vertical meridian. Injection sites in cases DL43 and DL39 were in the rostral and rostrolateral pole of the colliculus respectively, in the representation of the ipsilateral hemifield. DL44 contained the most caudally located injection site, in the representation of the peripheral visual field. The anteroposterior location, laminar depth, and number of labelled neurons at each injection site is shown in

Table 1. A variety of different neuronal morphologies were labelled at each injection site; Biocytin did not appear to label any neuronal type selectively. The size distribution of labelled neurons at the injection site was not significantly different in any of the cases, as determined by *t*-test. An example of a typical injection site and axonal arborization are shown in Fig. 2.

From the group of five cases with injections sites located close to the representation of the area centralis, the most significant findings were as follows: (1) there was a strong point-to- point organization in the tectotectal projection. Few tectotectal boutons were located caudal to the level of the injection site. In cases where the injection site was centered in the deepest collicular layers, few terminals were found rostral to the level of the injection site. In those cases where the injection site included neurons in the stratum griseum intermedium (SGI), many boutons were present rostral to the level of the injection site; (2) in the majority of cases, the focus of the terminal boutons was shifted medially; (3) regardless of which layers were included the injection site, labelled boutons were most dense in the deep collicular layers.

In case DL27, there was a striking mirror-symmetry in the distribution of labelled neurons in the left superior colliculus and the distribution of boutons in the right colliculus (Figs. 3 and 4). This correspondence was most obvious in the rostrocaudal plane (Figs. 4 and 5a). However, the point-to-point organization of the commissural projection was not entirely precise, as terminals were present across the mediolateral extent of the colliculus at the level of the injection site (Fig. 3). Furthermore, the correspondence between neurons and boutons in the mediolateral plane was shifted medially by approximately 1 mm (Figs. 4 and 5b). Almost no labelled boutons were found in the caudal half of the colliculus (Fig. 3, sections 13, 25), and few were found rostral to the level of the injection site (Fig. 3, sections 61, 73). While most labelled tectotectal terminals were found in the deep layers of the colliculus, (SGI, SAI, SGP, SAP), an occasional terminal was found in the superficial layers.

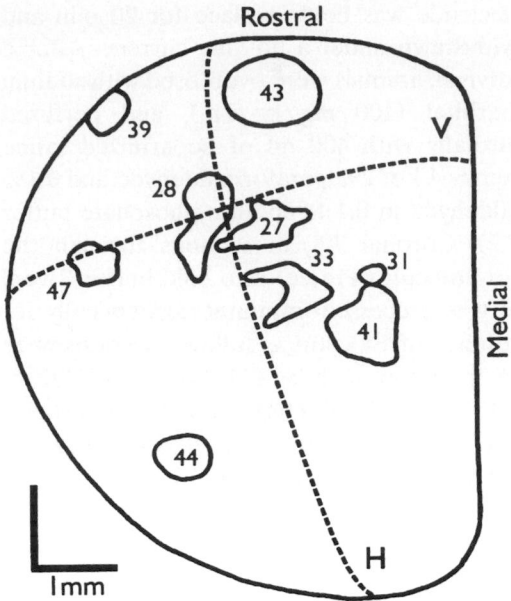

Fig. 1. Dorsal view of the cat superior colliculus showing the location and dimensions of injection sites. The vertical (V) and horizontal (H) meridians are represented as dashed lines.

As axons could be traced from the commissure into the contralateral colliculus, their arborization patterns could be examined in detail. Some axons extended horizontally into the collicular layers while others branched at right angles from the parent commissural fiber in the SAP, and ascended into the collicular layers. In general, axons were of small diameter and branched infrequently (Fig. 2b). Many axons had en passant boutons of various sizes and, in addition, small clusters of terminal boutons of various sizes were frequently found. Infrequently, labelled boutons were clustered around cell bodies. Electron microscopic analysis of several labelled boutons in DL43 confirmed that they contained vesicles and formed synaptic specializations.

In contrast to DL27 where the injection site was centered in the deepest of the collicular layers (SGP and SAP), the injection site in DL28 was located in the intermediate layers, SO/SGI/SAI (Table 1). As with DL27, there was a clear mirror-symmetry in the distribution of boutons. However, in DL28 the distribution of

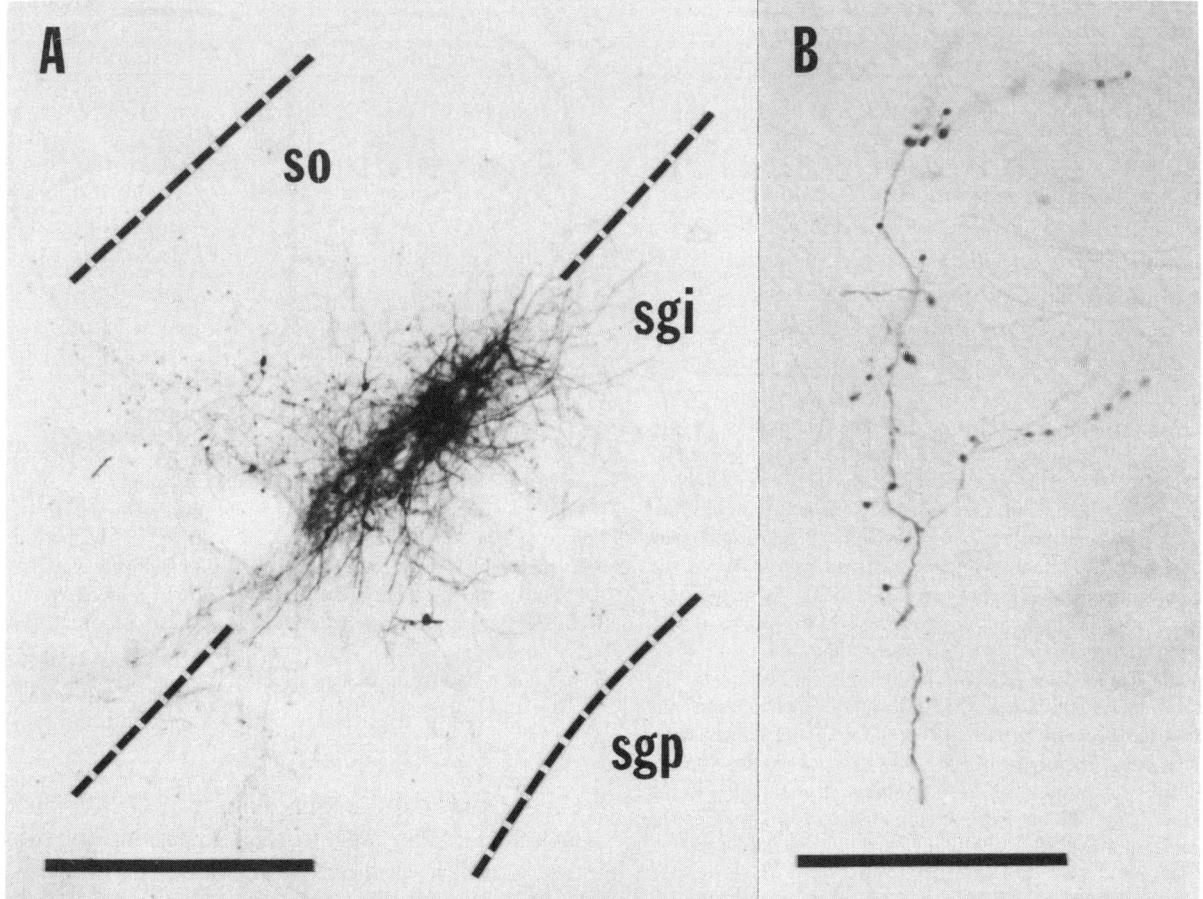

Fig. 2. Photomicrographs of injection site and Biocytin-labelled terminals. A: Injection site in case DL 39. B: Tectotectal axonal arborization in the deep layers of the colliculus. Arborizations are relatively simple with both en passant and terminal boutons. Scale bar for A = 500 μm. Scale bar for B = 25 μm. so, stratum opticum; sgi, stratum griseum intermedium; sgp, stratum griseum profundum.

terminal label extended into the rostral pole of the colliculus as well as the pretectum (Fig. 6). As with DL27, most terminals were concentrated in the medial half of the colliculus. The distribution of tectotectal terminals in cases DL41 and DL33 was similar to DL27 and DL28. The fifth case in which the injection site was located at the area centralis representation consisted of only three neurons (DL31; Table 1). Because of this, it was relatively easy to trace the axonal arborization pattern in the contralateral colliculus. These three neurons gave rise to a total of 180 labelled tecto-tectal terminals in the contralateral colliculus.

There was a medially-displaced cluster of termi-nals at precisely the same rostrocaudal level as the injection site, and there was also a slightly more caudally located cluster.

In four additional cases, the injection site was located rostral, rostrolateral, or caudolateral to the area centralis representation. Nonetheless, these cases shared many of the same features as those previously described. The point-to-point or-ganization characteristic of injection sites at the area centralis representation was still present. Also present was a shift or spread in the distribu-tion of labelled terminals medially. Furthermore,

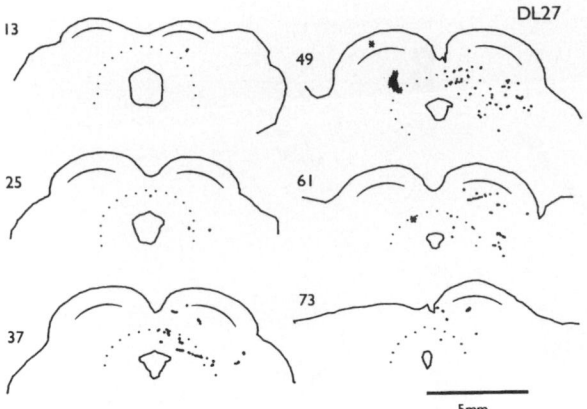

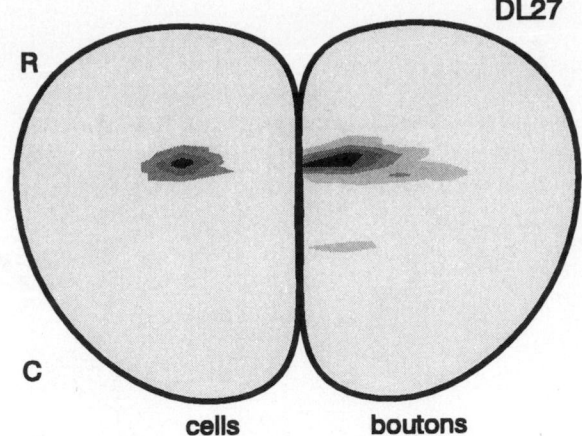

Fig. 3. Computer generated plots of the distribution of Biocytin-labelled neurons (shown as asterisks in the left colliculus) and boutons (shown as dots in the right colliculus) in selected sections from case DL 27. Section numbers are indicated on each of the six plots with 13 being the most caudal and 73 the most rostral. Left and right sides of the colliculus are not in perfect register, and this was taken into consideration in the quantitative anaylsis of these data in Figs. 4 and 5. The curved line represents the border between superficial and deep layers (stratum opticum above, stratum griseum intermedium below). The lower dotted line represents the border between the superior colliculus and the periaqueductal grey. The injection site was located in the stratum griseum profundum and stratum album profundum, and the majority of terminals were found in the deep layers of the contralateral colliculus in the region of the injection site.

Fig. 4. Quantitative analysis of the distribution of labelled cells and boutons in case DL 27. The paired superior colliculi are represented from a dorsal perspective, with the rostral pole (R) at the top and the caudal pole (C) at the bottom. Every second section through the paired colliculi was analyzed. In the left panel, the density of Biocytin-labelled neurons is shown on a scale from pale grey (least dense) to black (most dense). In the right panel, the density of labelled boutons is shown on a scale from pale grey (least dense) to black (most dense).

the majority of labelled terminals were found in the deeper layers of the colliculus. However, there were some differences. For example, a unique feature of DL47, where the injection site was centered at the lateral margin of the colliculus in the deep layers, was that labelled tectotectal terminals spanned the entire mediolateral extent of the contralateral colliculus at the level of the injection site (Fig. 6). In DL 39, with a rostrolaterally located injection site on the SO/SGI border, many terminals were found in superficial as well as deep layers in the contralateral colliculus. But at the level of the injection site, boutons were found only in the same layers as the injection site. In DL43 many labelled terminals were found caudal to the level of the injection site, which was at the rostral pole of the colliculus.

Finally, the most striking feature of DL44, which had a caudolaterally located injection site, was that very few terminals were found rostral to the level of the injection site (Fig. 6).

Discussion

In this study, we made a detailed analysis of the spatial distribution of tectotectal connections in the cat. We found a striking precision in the tectotectal projection at all injection site locations, suggesting that tectotectal connections may influence select populations of neurons in the contralateral colliculus.

Methodological considerations

Biocytin can be taken up by fibers of passage, but it's uptake appears to be dependent on the method of delivery, the size of the injection, and the area of the brain being injected. In previous studies, tracer delivered by iontophoresis through small diameter micropipettes rarely labelled any fibers of passage, although pressure injection through large needles did result in some labelling (King et al., 1989; Izzo, 1991; Smith, 1992). Chevalier et al. (1992) reported uptake of Biocytin by axons following iontophoretic injections of Biocytin in the corpus callosum of rats. Comparing our injection

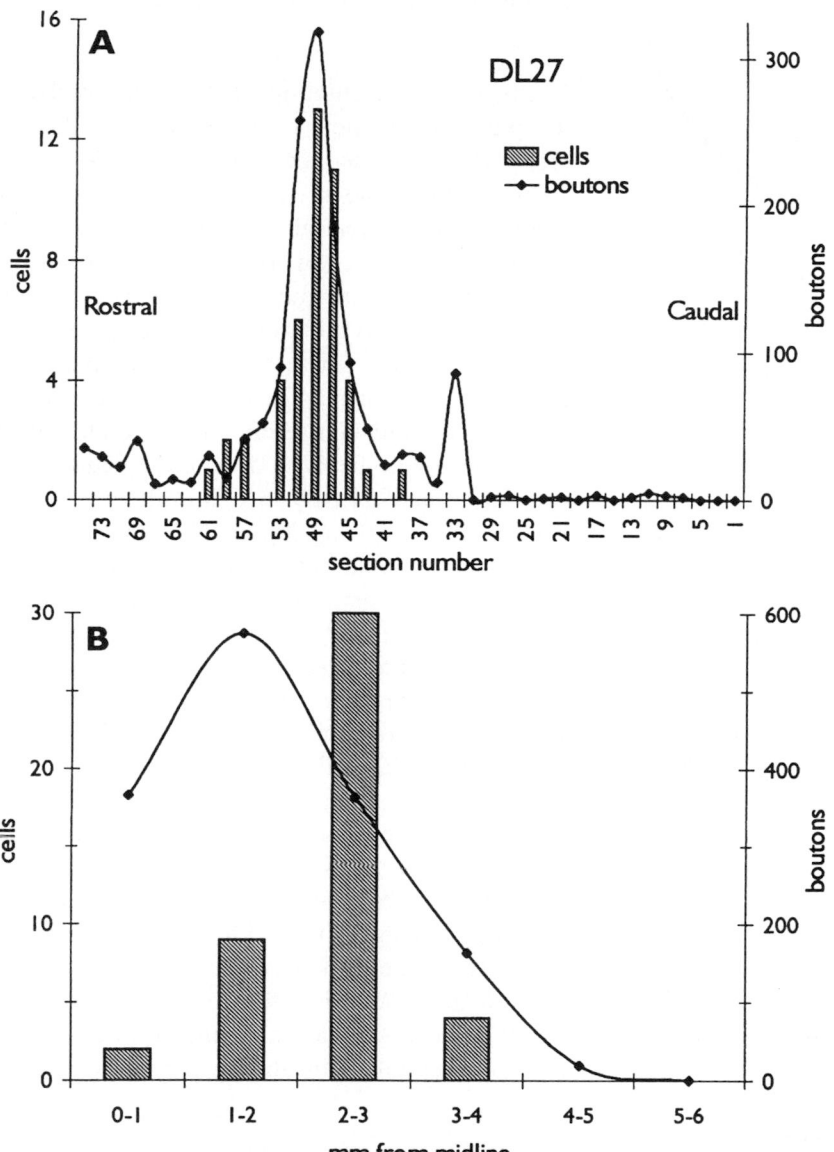

Fig. 5. A: Graph of the rostrocaudal distribution of Biocytin-labelled cells and boutons in case DL 27. This injection site was at the representation of the area centralis, and there is a very precise correspondence between the site of labelled neurons in one colliculus and the associated boutons in the contralateral colliculus. B: Graph of the mediolateral distribution of Biocytin-labelled cells and boutons in case DL 27. There is a 1-mm shift medially in the correspondence between the maximum density of labelled neurons in one colliculus and the maximum density of associated boutons in the contralateral colliculus.

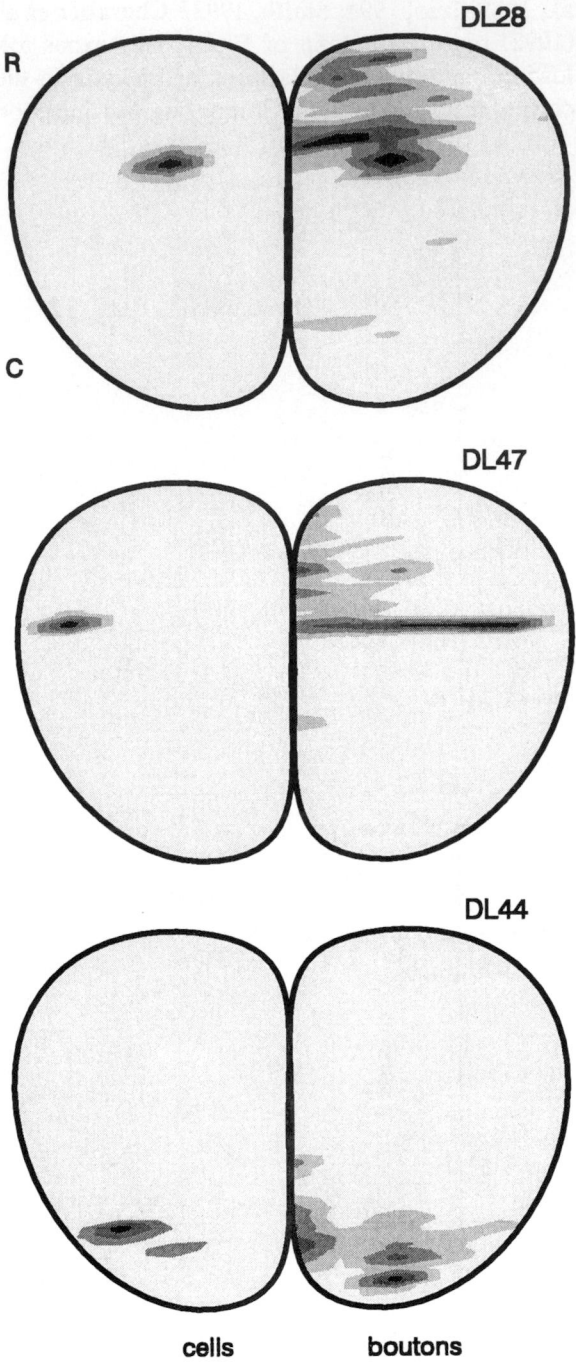

R

C

DL28

DL47

DL44

cells   boutons

Fig. 6. Quantitative analysis of the distribution of labelled cells and boutons in cases DL 28, DL47 and DL 44. The paired superior colliculi are represented from a dorsal perspective, with the rostral pole (R) at the top and the caudal pole (C) at the bottom. Every second section through the paired colliculi was analyzed. In each of the three cases

parameters with those of Chevalier et al. (1992), we used consistently smaller micropipette tips, 50% less current, and over 50% shorter injection delivery times.

Tectotectal connections at the fixation zone

In cat, the representation of the area centralis is located approximately 1 mm posterior to the rostral pole of the superior colliculus (Feldon et al., 1970). While none of the injection sites in this study was physiologically defined, five were centered very close to this location in the deep layers. There are numerous published maps for the physiological representation of visual space in the superficial layers, but there is only one published report of a visual map in the deep layers of the cat superior colliculus (Stein and Meredith, 1993). Visual receptive fields of deep layer neurons are very large and not easy to relate directly to neurons in the overlying layers. Nonetheless, there are distinct similarities to the superficial layers: ipsilateral and central visual space are represented in the rostral and rostrolateral regions of the deep layers, and temporal visual space is represented caudally; upper visual fields are represented medially and lower visual fields laterally (Stein and Meredith, 1993). Furthermore, the closest alignment between superficial and deep layers is at the representation of central visual space. There is evidence from studies in cat that deep layer neurons in this region of the colliculus discharge during periods of active fixation and do not increase their discharge rate prior to saccades (Munoz and Guitton, 1989, 1991). If neurons at the fixation zone have axon collateral branches in the opposite colliculus, during fixation a group of

represented, on the left the density of Biocytin-labelled neurons is shown on a scale from pale grey (least dense) to black (most dense). On the right, the density of labelled boutons is shown on a scale from pale grey (least dense) to black (most dense).

TABLE 1

Injection site parameters

Case Number	Stereotaxic coordinates for center of injection site	Laminar location of injection site	Estimated total number of Biocytin labeled neurons at injection site
DL44	A1.0, L3.0	SGP/SAP	55
DL41	A2.3, L1.2	SGI/SAI/SGP	272
DL33	A2.5, L2.0	SO/SGI	55
DL47	A2.5, L5.5	SAI/SGP/SAP	34
DL31	A2.8, L1.4	SGI/SAI	3
DL27	A3.2, L2.3	SGP/SAP	123
DL28	A3.3, L2.8	SO/SGI/SAI	148
DL39	A4.2, L4.2	SO/SGI	91
DL43	A4.3, L1.75	SGP/SAP	*

*Number of labeled neurons not available; most sections were embedded for electron microscopy

contralateral neurons would also be activated. According to our results, neurons at the fixation zone have a projection to precisely the mirror symmetrical region in the opposite colliculus (Figs. 3–5). The precision of this projection is all the more surprising considering the large receptive fields of neurons in the deep layers. The presence of such precise connections between the representation of central visual space in each colliculus suggests that there is mutual facilitation of distinct populations of efferent neurons during fixation.

Tectotectal connections in non-fixation areas

The direction and amplitude of electrically-evoked eye movements elicited in the deep layers of the colliculus are in register with the overlying visual map (Robinson, 1972; McIlwain, 1990). Thus, small and large saccades are represented rostrally and caudally while upward and downward eye movements are represented in medial and lateral halves of the colliculus respectively. Since the superior colliculi serve to orient the animal to the opposite hemifield, during an eye movement mutually excitatory tectotectal connections could create an imbalance in the signals descending to

brainstem nuclei. For example, during a small downward saccade to a target on the vertical meridian, a population of neurons in the lateral region of one colliculus would be active. In the contralateral colliculus, activity at the fixation zone, the ipsilateral hemifield representation, and the mirror-symmetrical region, would all be counter productive to the eye movement, and would need to be suppressed. This raises the possibility that some tectotectal neurons are inhibitory or exert an inhibitory effect in the contralateral colliculus. Two pieces of physiological evidence from studies in cat lend support to this hypothesis. Infante and Leiva (1986) have shown an increase in the frequency of discharge in one colliculus associated with a decrease of the frequency of discharge in the opposite colliculus. Furthermore, there is evidence that burst neurons in the deep layers of the cat colliculus with high spontaneous activity pause during saccades in the null direction (Peck, 1990).

In light of the above, it is interesting to examine the tectotectal connections of neurons in DL47. This injection site was located on the vertical meridian about two millimeters lateral to the representation of the area centralis. There is a clear projection to the mirror-symmetrical region of the opposite colliculus at the same level as the injection site, with a medial shift or smearing in the focus of terminals. More striking, however, is the dense projection to the rostral one-third of the contralateral colliculus. It is conceivable that a population of tectotectal neurons at the injection site could inhibit neurons in the ipsilateral hemifield representation in the contralateral colliculus by way of this rostral tectotectal projection. A similar rostral spread of terminal label was also seen in DL28.

Are tectotectal neurons inhibitory?

There is both physiological and anatomical evidence that some tectotectal cells are inhibitory. Monosynaptic inhibition has been recorded in collicular neurons in cat and squirrel monkey following stimulation of the contralateral colliculus (Maeda et al., 1979; Moschovakis and Karabelas, 1982). Electron microscopic analysis of tectotectal terminals in cat has shown that they can terminate directly on large efferent neurons whose axons travel in the predorsal bundle (Behan, unpublished observations), and many terminals have a morphology consistent with inhibition, i.e. flattened vesicles and symmetrical synapses (Behan, 1985). Furthermore, in a study of subcortical GABAergic inputs to the cat superior colliculus, tectotectal neurons stained immunocytochemically with an antibody to GABA (Appell and Behan, 1990). However, not all tectotectal neurons were immunoreactive for GABA, raising the possibility that they are neurochemically diverse. Recent studies in our lab indicate that glutamate is present in some tectotectal terminals (Behan, unpublished observations). Considering that many tecto-reticulo-spinal neurons have commissural collaterals, it is not surprising that glutamate, an excitatory neurotransmitter, would be present in some of these efferent neurons (Grantyn and Grantyn, 1982; Moschovakis and Karabelas, 1985; Moschovakis et al., 1988a). Further support for the presence of excitatory as well as inhibitory connections between the two colliculi comes from an electron microscopic study in which two morphologically distinct types of tectotectal terminals were observed (Behan, 1985). Excitatory and inhibitory tectotectal cells may be differentially distributed in superior colliculus, in the same or different layers.

Tectotectal projections and multisensory convergence

We have suggested that the tectotectal projection may be involved in coordinating the output of tectal neurons in each colliculus during fixation, and possibly during saccadic eye movements. But the superior colliculus is also involved in other orienting behaviors, including movements of the head, neck and ears. Most descending efferent neurons in the cat colliculus receive convergent sensory inputs, conveying visual, auditory, and somatosensory information (Meredith and Stein, 1985). Many of these efferent neurons have commissural axons, and, as we have shown, these axons are remarkably precise in their termination pattern in the opposite colliculus. By coupling the two colliculi, stimulus effectiveness could potentially be enhanced. This type of enhancement has already been shown for visual and auditory inputs onto individual deep layer neurons in the cat colliculus (Stein and Meredith, 1993). Not surprisingly, these authors also found that spatially disparate stimuli exert a depressive effect on the output of collicular neurons. In the same way, tectotectal projections could enhance or depress the output of neurons in the opposite colliculus, depending on the spatial relationships of the stimuli, and their significance to the animal. It is worth pointing out that the tectotectal projection is found primarily in the rostral half of the colliculus in cat, where there is a disproportionate representation of central sensory space. Thus, central visual and auditory fields, the head and forelimbs could preferentially participate in this modulation of stimulus efficacy, as opposed to peripheral visual and auditory fields, the body, hindlimbs and tail.

It is not clear why there would be a medial shift in the distribution of labelled tectotectal boutons in the colliculus, which was seen in all but the most rostrolaterally located injection site (DL39). This shift could not be explained by a disproportionate estimation of the number of terminals in the more compressed lamina at the medial border of the colliculus, for it was seen in lateral (DL47), caudal (DL44), and centrally placed (DL27, DL28) injection sites. It was also seen in case DL31, in which only three neurons were labelled. Nor can the shift be explained by the presence of terminals in the periaqueductal grey, as these were excluded from the quantitative analysis. However, in light of the previous discussion, this selective projection to the medial colliculus suggests that the tectotectal projection has a particular role in orienting to stimuli located directly above the animal, a region of considerable importance to cats.

The projection from the deep layers of the superior colliculus to the contralateral periaqueductal grey (PAG) was remarkably robust, particularly in cases where the injection site was located in the deepest layers. As with the tectotectal projection, tecto-PAG boutons seemed to be most concentrated at the level of the injection site. This projection has not been described previously in any detail, and nothing is known of its function (Huerta and Harting, 1984). However, it may play a role in aversive behaviors (Schmitt et al., 1984), as nociceptive neurons have been identified in the deep layers of the rat colliculus and their distribution is limited to the rostral region (McHaffie et al., 1989).

Conclusion

The organization of the tectotectal projection may have been influenced by a changing role for the superior colliculus in evolution. In the pigeon, there does not appear to be any homotopic relationship between the two sides of the optic tectum (Voneida and Mello, 1975). In rodents, the superior colliculus probably has several functions including orienting behavior, attention, and defensive and autonomic responses (Dean et al., 1988). While there is some degree of mirror-symmetry, the topographic organization of the tecto-

tectal projection in rats and hamsters does not appear to be very precise (Yamasaki et al., 1984; Rhoades et al., 1986). In this study in cat, we have shown a distinct homotopic relationship between the two colliculi, particularly at the representation of central visual space, that suggests a role in the coordination of orienting movements, in particular eye movements.

Acknowledgements

We are grateful to Tony Patenaude for excellent technical assistance. Thanks are also due to Drs. Alex Meredith and Paul May for reading an earlier draft of the manuscript. This work was supported by NIH EY04478 to MB.

References

Antonini, A., Berlucchi, G. and Sprague, J.M. (1978) Indirect, across-the-midline retinotectal projections and representation of ipsilateral visual field in superior colliculus of the cat. *J. Neurophysiol.*, 41: 285–304.

Appell, P.P. and Behan, M. (1990) Sources of subcortical GABAergic projections to the superior colliculus in the cat. *J. Comp. Neurol.*, 302: 143–158.

Behan, M. (1985) An EM-autoradiographic and EM-HRP study of the commissural projection of the superior colliculus in the cat. *J. Comp. Neurol.*, 234: 105–116.

Chevalier, G., Deniau, J.M. and Menetrey, A. (1992) Evidence that biocytin is taken up by axons. *Neurosci. Lett.*, 140: 197–199.

Dean, P., Mitchell, I.J. and Redgrave, P. (1988) Contralateral head movements produced by micro-injection of glutamate into superior colliculus of rats: evidence for mediation by multiple output pathways. *Neuroscience*, 24: 491.

Edwards, S.B. (1977) The commissural projection of the superior colliculus in the cat. *J. Comp. Neurol.*, 173: 23–40.

Feldon, S., Feldon, P. and Kruger, L. (1970) Topography of the retinal projection upon the superior colliculus of the cat. *Vis. Res.*, 10: 135–143.

Fish, S., Goodman, D., Kuo, D., Polcer, J. and Rhoades, R. (1982) The intercollicular pathway in the golden hamster: an anatomical study. *J. Comp. Neurol.*, 204: 6–20.

Grantyn, R. (1988) Gaze control through superior colliculus: structure and function. In: Buttner-Ennever (Ed.), *Neuroanatomy of the Oculomotor System*, Elsevier, Amsterdam, pp. 273–333.

Grantyn, A. and Grantyn, R. (1982) Axonal patterns and sites of termination of cat superior colliculus neurons projecting in the tecto-bulbo-spinal tract. *Exp. Brain Res.*, 46: 243–256.

Huerta, M.F. and Harting, J.K (1984) The mammalian superior colliculus: Studies of its morphology and connections. In: H. Vanegas (Ed.), *Comparative Neurology of the Optic Tectum*, Plenum, New York, pp. 687–773.

Infante, C. and Leiva, J. (1986) Simultaneous unitary activity in both superior colliculi and its relation to eye movements in the cat. *Brain Res.*, 381: 390–392.

Izzo, P.N. (1991) A note on the use of biocytin in anterograde tracing studies in the central nervous system: Application at both light and electron microscopic level. *J. Neurosci. Methods* 36: 155–166.

Kanaseki, T. and Sprague, J.M. (1974) Anatomical organization of pretectal nuclei and tectal laminae in the cat. *J. Comp. Neurol.*, 158: 319–338.

King, M.A., Louis, P.M., Hunter, B.E. and Walker, D.W. (1989) Biocytin: a versatile anterograde neuroanatomical tract-tracing alternative. *Brain Res.*, 497: 361–367.

Lee, C., Rohrer, W.H. and Sparks, D.L. (1988) Population coding of saccadic eye movements by neurons in the superior colliculus. *Nature*, 332: 357–360.

Maeda, M., Shibazaki, T. and Yoshida K. (1979) Monosynaptic inhibition evoked in S.C. neurons following contralateral collicular stimulation. In: M. Ito (Ed.), *Integrative Control Functions of the Brain*, Vol. II, Elsevier, Amsterdam, pp. 68–71.

Magalhaes-Castro, H.H., Dorbela Da Lima, A., Saraiva, P.E.S. and Magalhaes-Castro, B. (1978) Horseradish peroxidase labeling of cat tecto-tectal cells. *Brain Res.*, 148: 1–13.

Martin, G.F. (1969) Efferent tectal pathways of the opossum (Didelphis virginia). *J. Comp. Neurol.*, 135: 209–224.

McHaffie, J.G., Gao, C.-Q. and Stein, B.E. (1989) Nociceptive neurons in rat superior colliculus: Response properties, topography, and functional implications. *J. Neurophysiol.*, 62: 510–525.

Meredith, M.A. and Stein, B.E. (1985) Descending efferents from the superior colliculus relay integrated multisensory information. *Science*, 227: 657–659.

McIlwain, J.T. (1990) Topography of eye position sensitivity of saccades evoked electrically from the cat's superior colliculus. *Vis. Neurosci.*, 4: 289–298.

McIlwain, J.T. (1991) Visual input to commissural neurons of the cat's superior colliculus. *Vis. Neurosci.*, 7: 389–393.

Moschovakis, A.K. and Karabelas, A.B. (1982) Tectotectal interactions in the cat. *Soc. Neurosci. Abstr.*, 8: 293.

Moschovakis, A.K. and Karabelas, A.B. (1985) Observations on the somatodendritic morphology and axonal trajectory of intracellularly HRP-labelled efferent neurons located in the deeper layers of the superior colliculus of the cat. *J. Comp. Neurol.*, 239: 276–308.

Moschovakis, A.K., Karabelas, A.B. and Highstein, S.M. (1988a) Structure-function relationships in the primate superior colliculus. I. Morphological classification of efferent neurons. *J. Neurophysiol.*, 60: 232–262.

Moschovakis, A.K., Karabelas, A.B. and Highstein, S.M. (1988b) Structure-function relationships in the primate superior colliculus. II. Morphological identity of presaccadic neurons. *J. Neurophysiol.*, 60: 263–302.

Munoz, D.P. and Guitton, D. (1989) Fixation and orientation control by the tecto-reticulo-spinal system in the cat whose head is unrestrained. *Rev. Neurol. (Paris)*, 145: 567–579.

Munoz, D.P. and Guitton, D. (1991) Control of orienting gaze shifts by the tectoreticulospinal system in the head free cat: II. Sustained discharges during motor preparation and fixation. *J. Neurophysiol.*, 66: 1624–1641.

Peck, C.K. (1990) Neuronal activity related to head and eye movements in cat superior colliculus. *J. Physiol.*, 421: 79–104.

Raczkowski, D. and Diamond, I.T. (1978) Cells of origin of several efferent pathways from the superior colliculus in Galago senegalensis. *Brain Res.*, 146: 351–357.

Rhoades, R.W., Fish, S.E. and Voneida, T.J. (1981) Anatomical and electrophysiological demonstration of tectotectal pathway in the golden hamster. *Neurosci. Lett.*, 21: 255–260.

Rhoades, R.W., Mooney, R.D., Szczepanik, A.M. and Klein, B.G.(1986) Structural and functional characteristics of commissural neurons in the superior colliculus of hamster. *J. Comp. Neurol.*, 253: 197–215.

Robinson, D.A. (1972) Eye movements evoked by collicular stimulation in the alert monkey. *Vis. Res.*, 12: 1795–7808.

Schmitt, P., DiScala, G., Janck, F. and Sandner, G. (1984). Periventricular structures, elaboration of aversive effects and processing of sensory information. In: R. Bandler (Ed.), *Modulation of Sensorimotor Activity During Alterations in Behavioral States*, Alan R. Liss, New York, pp. 393–414.

Smith, Y. (1992) Anterograde tracing with PHA-L and biocytin at the electron microscopic level. In: J.P. Bolam (Ed), *Experimental Neuroanatomy: A Practical Approach*, Oxford Press, New York, pp. 661–679.

Sparks, D.L. and Mays, L.E., (1990) Signal transformation required for the generation of saccadic eye movements. *Annu. Rev. Neurosci.*, 13: 309–336.

Stein, B.E. and Meredith, M.A. (1993) *The Merging of the Senses*, MIT Press, Cambridge. pp. 87–156.

Voneida, T.J. and Mello, N.K. (1975) Interhemispheric projections of the optic tectum in pigeon. *Brain Behav. Evol.*, 11: 91–108.

Yamasaki, D.S., Krauthamer, G. and Rhoades, R.W. (1984) Organization of the intercollicular pathway in rat. *Brain Res.*, 300: 368–371.

M. Norita, T. Bando and B. Stein (Eds.)
Progress in Brain Research, Vol 112
© 1996 Elsevier Science BV. All rights reserved.

CHAPTER 10

Roles of the lateral suprasylvian cortex in convergence eye movement in cats

Takehiko Bando*, Naoto Hara, Mineo Takagi, Kenji Yamamoto
and Haruo Toda

Department of Physiology, Niigata University School of Medicine, Asahi-machi, Niigata 951, Japan

Ocular convergence and lens accomodation were evoked by microstimulation in the lateral suprasylvian area (LS cortex) in the parieto-occipital cortex in the cat. Electrolytic lesions in LS cortex reduced the amplitude and velocity of ocular convergence. Neurons in LS cortex discharged in relation to ocular convergence and/or lens accommodation. These results support the hypothesis that the LS cortex plays an important role in controlling ocular convergence

The LS cortex receives visual inputs from cortical visual areas 17, 18 and 19, and in addition from the superior colliculus through the LP nucleus of the thalamus. Electrophysiological recordings have revealed that these visual inputs, which include cues about 3-dimensional target motion, are integrated in the LS cortex. The integrated output from LS cortex may provide the brainstem motor centers with the neural signals that facilitate eye movements, especially when the target is moving at high speeds.

Outputs from the LS cortex travel directly to brainstem structures including the superior colliculus and pretectum. Evidence from monkey suggests that information may also travel to the mesencephalic reticular formation, where neurons have been recorded that are related to ocular convergence, lens accomodation or both. Although comparable data is lacking in the cat, it is suggested that the efferent circuit from the LS cortex to the motor nuclei in the brainstem includes both the superior colliculus and the mesencephalic reticular formation. It is also suggested that this pathway is rather short, given that the mean latency of the early component of evoked disjunctive eye movements was approximately 60 ms.

Introduction

Convergent eye movements are triggered by changes in the visual cues which are related to target motion in three-dimensional space, especially changes in binocular disparity and target size (Westheimer and Mitchell, 1969; Erkelens and Regan, 1986). Target information is obtained by early cortical processing of visual information in primary and secondary visual areas, as well as in extrastriate visual areas (Poggio and Fisher, 1977; Cynader and Regan, 1978; Regan et al., 1979; Ferster, 1981; Regan and Cynader, 1982; Toyama et al., 1986). Electrical stimulation of extrastriate cortex affects convergence eye movement and lens accommodation, a finding first reported by Jampel (1960) in the monkey. Recently, and in agreement with this observation, cortical neurons in the banks of the superior temporal sulcus and intraparietal sulcus have been found to discharge in relation to ocular convergence and lens accommodation (Gnadt and Mays, 1989; Takemura et al., 1995).

Cats also have the ability to perform convergent eye movements (Hughes, 1972; Stryker and Blakemore, 1972) and lens accommodation (Elur

* Corresponding author. Tel.: +81 25 223 6161, ext. 2230; fax: +81 325 229 3077; e-mail: bando@med.niigata-u.ac.jp

and Marchiafava, 1964). The lateral suprasylvian area (the LS cortex) is one of the best studied areas of extrastriate cortex in the cat. Neurons in the LS cortex, especially those in its posteromedial part (the PMLS area, Palmer et al., 1978) have visual properties resembling those of neurons in area MT of the monkey. The PMLS area has been shown to play an important role in analyzing three-dimensional stimulus motion relative to background (Spear and Baumann, 1975; Camarda and Rizzolatti, 1976; Toyama et al., 1986; Rauschecker et al., 1987).

The PMLS area is also related to lens accommodation and pupillary constriction. Lens accommodation and/or pupillary constriction have been evoked by microstimulation in PMLS (Bando et al., 1984b, 1989; Bando, 1985, 1987; Yoshizawa et al.; 1991 Sawa et al., 1992). Neurons which discharged prior to lens accommodation and pupillary constriction were also found in this area (Bando et al., 1984b; Bando et al., 1988).

Because ocular convergence and lens accommodation are strongly linked with the near response of the eye, it seems likely that PMLS may also play an important role in controlling convergent eye movements. In the present study, we performed three types of physiological experiments to confirm this expectation: intracortical microstimulation, electrolytic lesions and extracellular recordings from neurons in the LS cortex. We found that microstimulation in circumscribed regions of PMLS resulted in ocular convergence (Toda et al., 1991). In the same regions, neurons discharged in relation to ocular convergence (Takagi et al., 1992, 1993). Finally, the amplitude and peak velocities of ocular convergence were decreased by lesion of PMLS (Hara et al., 1992).

In this paper, we review some of our published results, adding some new unpublished data, and attempt to integrate this information into a clear picture of the functional role of PMLS in controlling ocular convergence and related eye movements. It is suggested that the LS cortex plays an important role in facilitating ocular convergence in association with related eye movements (Bando and Toda, 1991; Bando et al., 1992).

General methods

The general methods used in each of the experimental series are described here, and the specific methods used for each experiment are described in the results.

Preparation and training

Adult cats weighing 2–3 kg were used. All surgical procedures were done under pentobarbital anesthesia (Nembutal, Abott, 35–40 mg/kg). A head holding device was attached to the skull, and eye coils were sutured to the sclera of both eyes (see Takagi et al., 1993). Routine medication was given including local and general application of antibiotics.

Cats were trained for 3–10 days to lie quietly in a cloth bag with their head restrained (Stryker and Blakemore, 1972), and to track a moving visual target under computer control. Cats were given food rewards for successful ocular convergence. Upon attaining a successful convergence rate of 70–80% to moving targets, a craniotomy in the parieto-occipital bone overlying the lateral suprasylvian (LS) cortex was performed under Nembutal anesthesia. A plastic chamber was implanted to cover and seal the opening (10-mm length and 5-mm width) (Takagi et al., 1993). Cats were allowed to recover for a minimum of 3 days following surgery. All experiments were performed in the alert condition, given that ocular convergence can not be evoked by intracortical microstimulation under anesthesia.

Measurement of eye movement and lens accommodation

Eye movements were continuously monitored by the magnetic search-coil method (Robinson, 1963). The noise associated with this method was less than 0.1°. Lens accommodation was monitored by use of an infrared optometer (Campbell and Robson, 1959; Bando et al., 1984a). Lens accommodation could be measured accurately only when the eye positions were kept within a limited range. This limit was estimated at less than 2° in control experiments in which lens accommodation was paralyzed by using cyclopentate hydrochloride (Cyplegin 1% eye drops, Santen, Osaka) (Takagi et al., 1993).

Visual stimulation

A target was moved under computer control along a guide rail placed in front of the animal from the far position (a distance of 870 mm from the animal's nose) to the near position (at 160 mm) at a constant speed of 550 mm/s. The horizontal and vertical components of eye movement were monitored, and the movement of the target was initiated when the gaze of the cat was directed to it and stabilized for 1–2 s. The target was a white circle (diameter, 30 mm; luminance, 20 Cd/m^2) positioned at the center of a black square board (70 mm × 70 mm; luminance, 0.5 Cd/m^2), and was presented against a white background (luminance, 20 Cd/m^2) in the earlier experiments. In later experiments, the target was an LED (diameter, 10 mm) which was illuminated for 1 s and then turned off for 0.5 s before the target began it's approach.

Data analyses

Records of neuronal activity and eye movements were stored on digital tapes, and off-line analyses were performed (sampling frequency, 1 kHz). The amplitude of ocular convergence was measured as the sum of the amplitudes of the horizontal components of left and right eye movements. The onset of ocular convergence was determined by fitting a line (for 1 s), followed by a parabola (for 300 ms) to the trace of eye movement using the least squares method (Bando et al., 1984a). The velocity of convergence eye movement was calculated as the slope of the third-order polynomial curve fitted to the trace of convergence eye movement (equivalent to a digital filter with cut-off frequency (−3 dB) of 5 Hz).

Histological confirmation

At the end of the final experiment, cats were deeply anesthetized with sodium pentobarbiturate (50 mg/kg body wt.) and were perfused with saline and 10% formalin transcardially. The brain was cut coronally to a thickness of 50–100 μm, by using a freezing microtome, and stained with neutral red.

Results

In this section, results of three experimental series are reviewed. These include both published and unpublished data. The Discussion section presents an integrated interpretation of these results.

Intracortical microstimulation in the PMLS area

A tungsten-in-glass microelectrode was inserted into the medial bank of the middle suprasylvian sulcus (MSs) in an area corresponding to PMLS. The electrode was inserted at an angle of 20–30° from the vertical axis in the frontal plane by a motor-driven micromanipulator (ME-71, Narishige, Tokyo). The insertion point of the electrode was determined by a stereotaxic reference mark on the chamber (AP coordinate of A5.0). Extracellular recordings were done during insertion to confirm that the electrode was in MSs. Following proper positioning of the electrode, microstimulation was accomplished using a train of 50-μA current pulses (200 bipolar pulses; a bipolar pulse consisted of a 0.2-ms negative-going pulse followed by a 0.2-ms positive-going pulse; intratrain frequency, 300 Hz).

Microstimulation in PMLS evoked disjunctive eye movements similar to those seen in ocular convergence (Fig. 1) (Toda et al., 1991). Although the amplitude of these microstimulation evoked disjunctive eye movements, they were much smaller than the amplitude of eye movements associated with ocular convergence movements to approaching targets, the relationship between amplitude and peak velocity was very similar as shown in C.

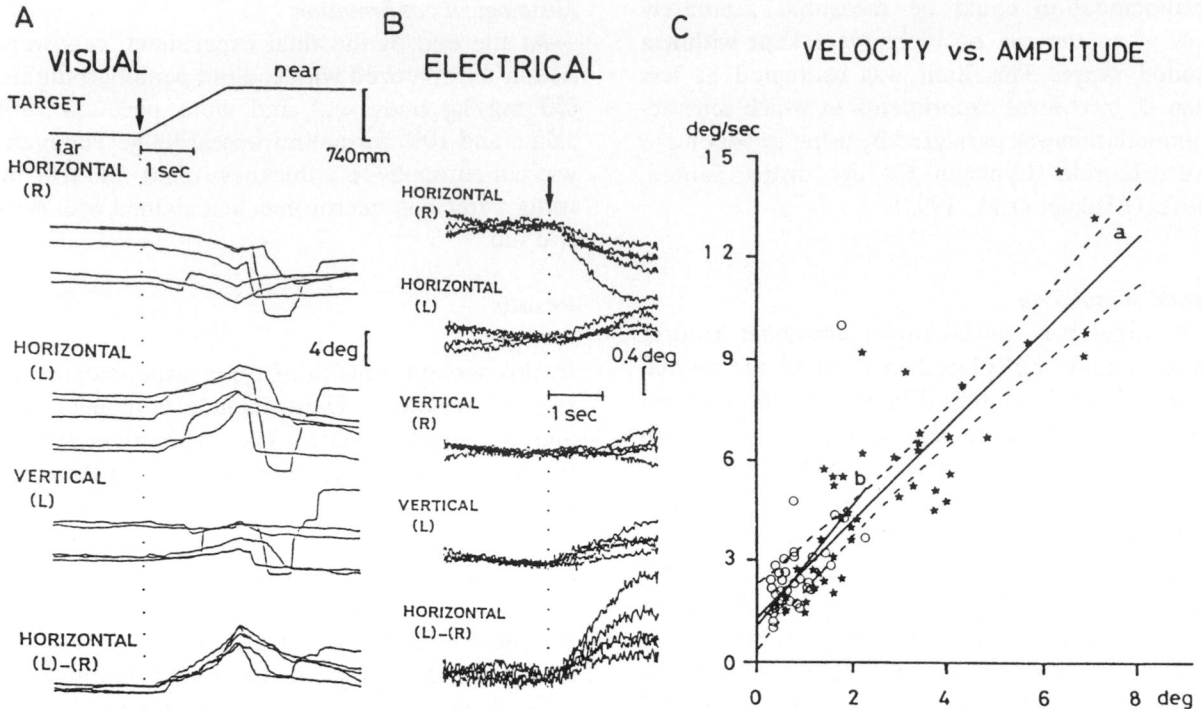

Fig. 1. (A) Convergence eye movements evoked by movement of a target in depth. Shown from top to bottom are: target movement, horizontal components of right and left eye movements, vertical component of left eye movement, and the amplitudes of ocular convergence. The upward deflections indicate rightward movements in the horizontal components and upward movements in the vertical components. (B) Disjunctive eye movements evoked by microstimulation in the postero-medial lateral suprasylvian (PMLS) cortex. Shown from top to the bottom are: horizontal components of right and left eye movements, vertical components of the right and left eye movements, and the amplitudes of the disjunctive eye movements. (C) Relationship between amplitude and peak velocity of ocular convergence evoked by visual stimulation (solid stars) and disjunctive eye movements evoked by intracortical microstimulation (open circles). The rigid lines indicate the regression lines for solid stars (a, $y = 1.42x + 1.37$) and open circles (b, $y = 1.74x + 1.12$), and the dotted lines indicate the confidence intervals (95%) for the population regression line of solid stars. From H. Toda et al. (1991) *Neurosci. Res.*, 12: 300–306.

Areas which were effective in evoking disjunctive eye movements were found both in the rostral and caudal parts of the PMLS area and the rostral PLLS area (Toda et al., 1991). No disjunctive eye movements were evoked from the middle part of the PMLS area. The mean latency of disjunctive eye movements evoked from each of the effective areas were: caudal PMLS, 150 ms ± 125 ms (mean and SD, $n = 176$); rostral PMLS, 149 ± 94 ms ($n = 116$); and rostral PLLS, 353 ± 125 ms ($n = 56$). In the caudal PMLS, the distribution of latencies for evoked disjunctive eye movements was clearly bimodal (ref. Fig. 2A, abs-

cissa), no latencies of 130–170 ms were seen. We distinguish the disjunctive eye movements, corresponding to these two peaks, as the early and late components. The mean latency of the early component was 63 ms (Table 1), and that of the late component was 282 ms. Although the latencies of the disjunctive eye movements evoked from the rostral PMLS area had a wider distribution (Fig. 2B, abscissa), they tended to also have early and late components. The mean latency of these components was 119 and 338 ms, respectively (Table 1), each of which was approximately 60 ms longer than those evoked from the caudal PMLS.

147

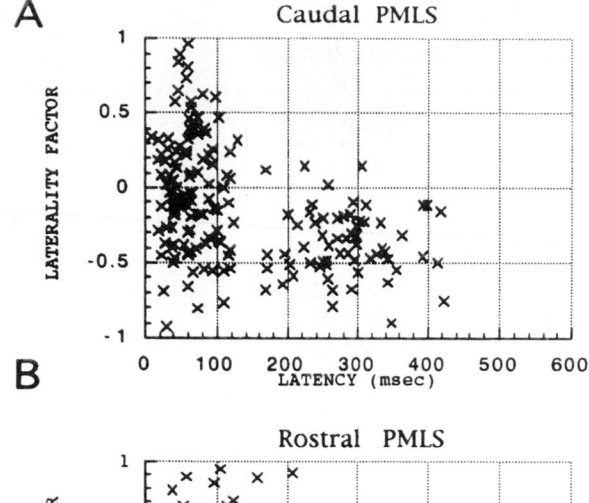

A   Caudal PMLS

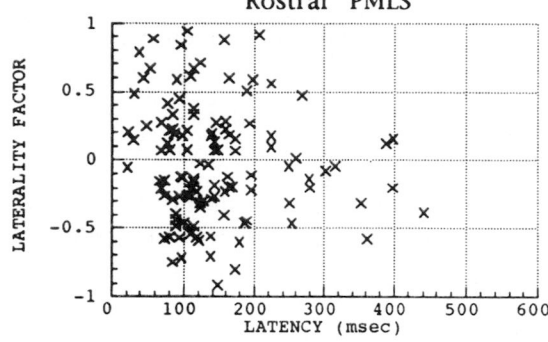

B   Rostral PMLS

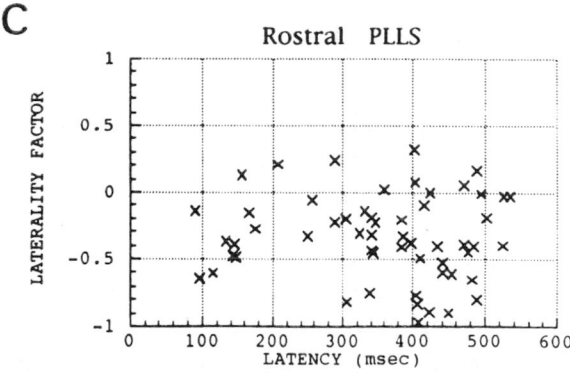

C   Rostral PLLS

Fig. 2. The relationship of the latencies of disjunctive eye movements evoked by microstimulation and the laterality factors defined in the text. Disjunctive eye movement were evoked in the caudal PMLS (A), the rostral PMLS (B) and the rostral PLLS (C) areas.

Some of the evoked disjunctive eye movements had symmetrical right and left eye movements, while others had larger contralateral eye movements. In order to quantitatively describe the

symmetry of the disjunctive eye movements, we have defined a laterality factor (Lf) as follows:

$$Lf = (Ai - Ac)/(Ai + Ac)$$

where Ai and Ac are the amplitudes of the ipsilateral and contralateral components of the eye movement, respectively. The amplitude of the disjunctive eye movement is given as Ai + Ac.

In Figs. 2A–C, we see that the laterality factors of evoked disjunctive eye movements are related to latency. Whereas the early components of disjunctive eye movements evoked from caudal PMLS had a symmetrical distribution (laterality factor not significantly different from zero; the sign test, $P > 0.05$) (Table 1), the late components showed a contralateral predominance (significantly different from zero; $P < 0.01$). Although less prominent, disjunctive eye movements evoked from rostral PMLS also showed a similar tendency: those with shorter latencies had a symmetrical distribution ($P > 0.05$) and those with longer latencies had an asymmetrical distribution with contralateral predominance ($P < 0.05$) (Table 1). Disjunctive eye movements evoked from rostral PLLS were asymmetrical (Fig. 2C).

It is suggested from this that the late components of disjunctive eye movement evoked from caudal PMLS may be more closely related to visual processing, while the early components may be more closely related to the motor-related signals transferred from the PMLS area to the brainstem. Judging from these latency differences, it is also likely that the rostral PMLS is upstream of caudal PMLS.

The lesion study

A tungsten microelectrode coated with polyurethane lacquer within 1 mm of the tip was used both for extracellular recording and for electrocoagulation. The microelectrode was inserted into the medial bank of the MSs at an angle 20–30° from the vertical axis in the frontal plane. After confirming that the electrode was in the gray matter, by recording neuronal activity, electrolytic lesions were made at six sites by passing a

TABLE 1

Properties of disjunctive eye movements evoked from the caudal PMLS area and the rostral PMLS area

	Caudal PMLS		Rostral PMLS	
	Early	Late	Early	Late
Number	113	63	100	16
Latency (ms)	63 ± 28	282 ± 63	119 ± 47	338 ± 94
Laterality factor	0.0 ± 0.4	−0.4 ± 0.2	0.0 ± 0.4	−0.1 ± 0.3
Amplitude (degrees)	0.8 ± 0.5	0.8 ± 0.4	0.9 ± 0.5	0.7 ± 0.4
Peak velocity (degrees/s)	2.8 ± 0.9	2.1 ± 0.6	3.7 ± 1.3	2.8 ± 1.3

0.5-mA current for 5 min. Lesions were made at two depths (4 and 5 mm from the surface of the sulcus) in each track and at three antero-posterior stereotaxic coordinates (A1, A3 and A5). In two cats, electrolytic lesions were made bilaterally (total of 12 sites).

Fig. 3 shows the relationship between the amplitudes and peak velocities of convergence eye movements before (A) and after (B) PMLS lesions. Both amplitude and peak velocity were significantly decreased following lesions (see below). Nonetheless, the correlation between these parameters was not significantly altered (Hara et al., 1992).

The time course of changes in ocular convergence amplitude and velocity are shown in Fig. 4. It can be seen that these parameters are significantly altered in cases of both bilateral (A) and unilateral (B) lesions.

Another effect of PMLS lesion is a change in the symmetry of eye movements in ocular convergence (Fig. 5). After a unilateral lesion, the frequency distribution of the laterality factors shifted to the side ipsilateral to the lesion, i.e. ipsilateral eye movements had greater amplitude (B). On the other hand, the laterality was not changed after the bilateral lesions (A). This result is in agreement with the hypothesis based on the results obtained by intracortical microstimulation. In this hypothesis, PMLS receives visual inputs in large measure from the contralateral visual field, and then integrates these inputs into symmetric motor-related signals, most likely via interhemispheric connections.

Neuronal activity

It is known that a small group of neurons in PMLS respond to stimuli moving in depth. These approach/recess cells (Toyama et al., 1986), which code motion disparity, are the most plausible source of the visual information needed to control ocular convergence. Therefore, we attempted to isolate approach/recess cells by manually moving a visual target in depth. Once an approach/recess cell was isolated, the target was moved under computer control. We recorded neural activity, eye movements and lens accommodation in these experiments (see general methods). The target was moved at two speeds: 550 and 420 mm/s.

Eighteen of the 659 cells recorded (3%) exhibited activity related to ocular convergence (Fig. 6) (Takagi et al., 1993). These PMLS neurons responded to an approaching target and discharged during ocular convergence. They did not discharge prior to the onset of ocular convergence, but discharged before the time at which ocular convergence attained the peak velocity.

Because target movement in depth results in ocular convergence, correlating neuronal activity with ocular convergence does not necessarily reveal a causal relationship. In order to determine such a relationship, we did repetitive experiments

VELOCTIY-AMPLITUDE RELATIONSHIP

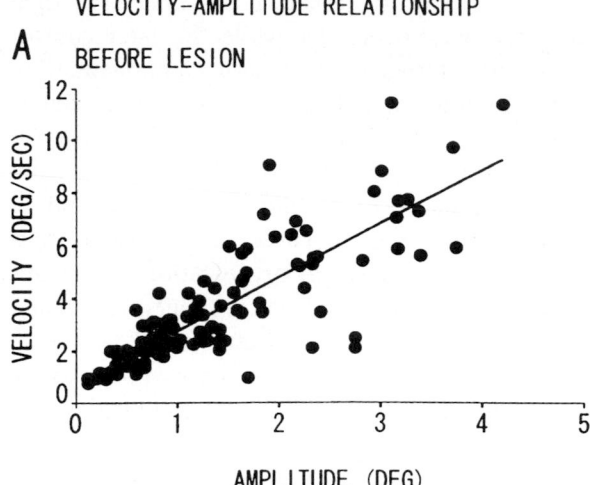

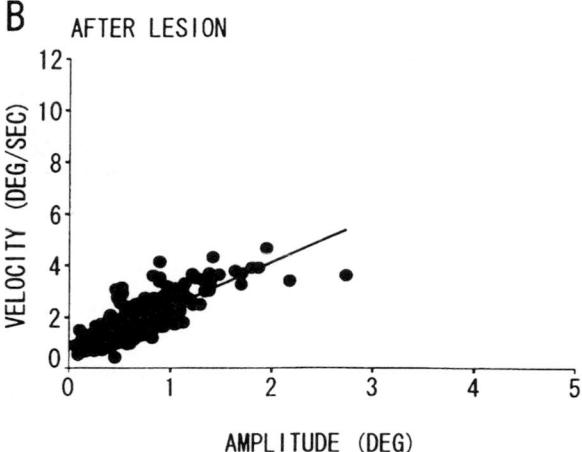

Fig. 3. The amplitude vs. peak velocity relation of ocular convergence before and after electrolytic lesions in the PMLS area. Abscissae: eye movement amplitude in degrees. Ordinate: peak velocity (deg/s). The solid lines are the regression lines before and after lesion, $y = 0.72 + 2.04x$ and $y = 0.62 + 1.74x$, respectively.

using identical visual stimulation. In these experiments, neuronal responses and eye movements varied from trial to trial. If neuronal activity has a causative relation to the eye movement, the two should co-vary from trial to trial (Takagi et al., 1993). Also, neural activity in each trial should correlate with the peak velocity and amplitude of ocular convergence (Fig. 7).

Twelve out of the 18 PMLS neurons which

discharged in a manner correlated with the peak velocity of ocular convergence were tested at both higher and lower target speeds (Fig. 8). Seven of these 12 neurons are classified as type 1 neurons. Their activity showed a significant correlation with peak velocity of ocular convergence at both target speeds. Five other neurons, classified as type 2

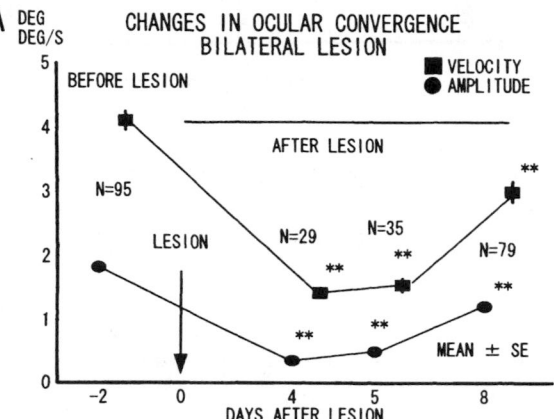

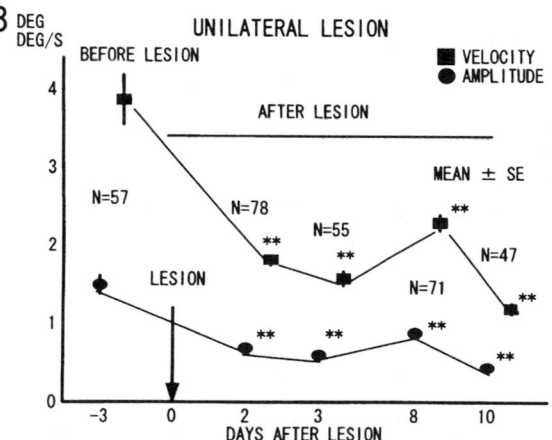

Fig. 4. Changes in the amplitude and peak velocity of ocular convergence before and after bilateral (A) and unilateral (B) PMLS lesion, respectively. Filled circles indicate amplitudes, and filled squares indicate peak velocities. Abscissae: days after lesion. The lesion was made at day 0 (downward arrows, minus signs show the days before the lesion). The horizontal lines show the range after lesion. Vertical lines indicate standard errors. Ordinates: amplitudes (deg) and peak velocities (deg/s). ** Significantly changes from the values before lesion (the Mann-Whitney U-test, $P < 0.001$). N is the number of data sampled at each day.

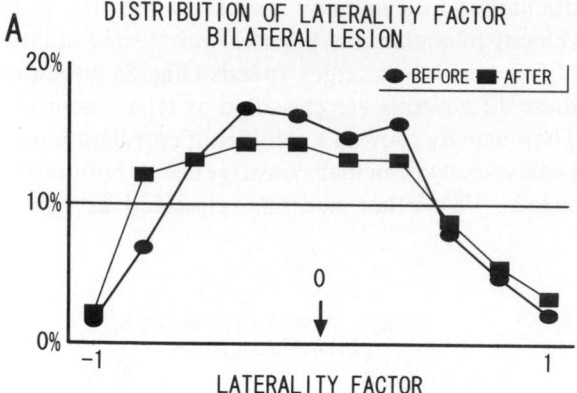

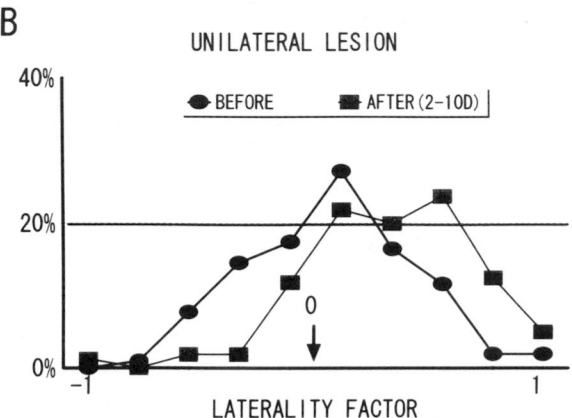

Fig. 5. The frequency distribution of laterality factors before and after bilateral (A) and unilateral (B) PMLS lesions. Data obtained after lesions are pooled. Abscissae: laterality factor (downward arrows indicate symmetric movements). Ordinates: probability of occurrence of each laterality factors in percentage.

neurons, showed a significant correlation only at the higher target speed. They differ from type 1 neurons in that they discharge in correlation with lens accommodation at both target speeds.

Discussion

Hypothesis on the role played by the LS cortex in controlling ocular convergence

Two components of disjunctive eye movements were evoked by microstimulation in the PMLS area. The early component had symmetrical right and left eye movements, while the later component had a larger contralateral component. This latter component is likely related to the sensory organization of PMLS, where contralateral visual space is represented. Conversely, the early component is likely to be more closely related to ocular convergence, because of it's rather short latency and symmetrical organization. From this evidence, it is suggested that a functional organization which serves to convert visual signals to motor-related output exists in the PMLS area. The mean latencies of both early and late components evoked from the caudal PMLS area were shorter than those evoked from the rostral PMLS area by about 60 ms, suggesting that rostral PMLS lies 'upstream' of caudal PMLS.

By lesioning PMLS, the peak velocity and amplitude of evoked eye movements were significantly reduced, suggesting that the PMLS area may play it's principal role in ocular convergence to targets moving in depth at rather high speeds. This suggestion is in agreement with the visual properties of the most PMLS neurons, which responded well to moving stimuli and poorly to a stationary flashing target.

Recording studies found two populations of neurons in PMLS: type 1 and type 2 neurons. A diagram of the presumptive circuit by which these cells work is shown in Fig. 9. It is presumed that type 1 neurons are closely related to ocular convergence, and type 2 neurons are closely related to lens accommodation. The PMLS area exerts its facilitatory influence on the motor centers of ocular convergence and/or lens accommodation in the brainstem. When a target approaches the animal at high speed, it is postulated that a gate is opened, bringing type 2 neurons into the control of ocular convergence. This cooperation between neurons participating in ocular convergence and those participating in lens accommodation may facilitate the effective link in the near response.

The LS cortex receives visual inputs from cortical visual areas 17–19, as well as from the superior colliculus through the LP nucleus of the

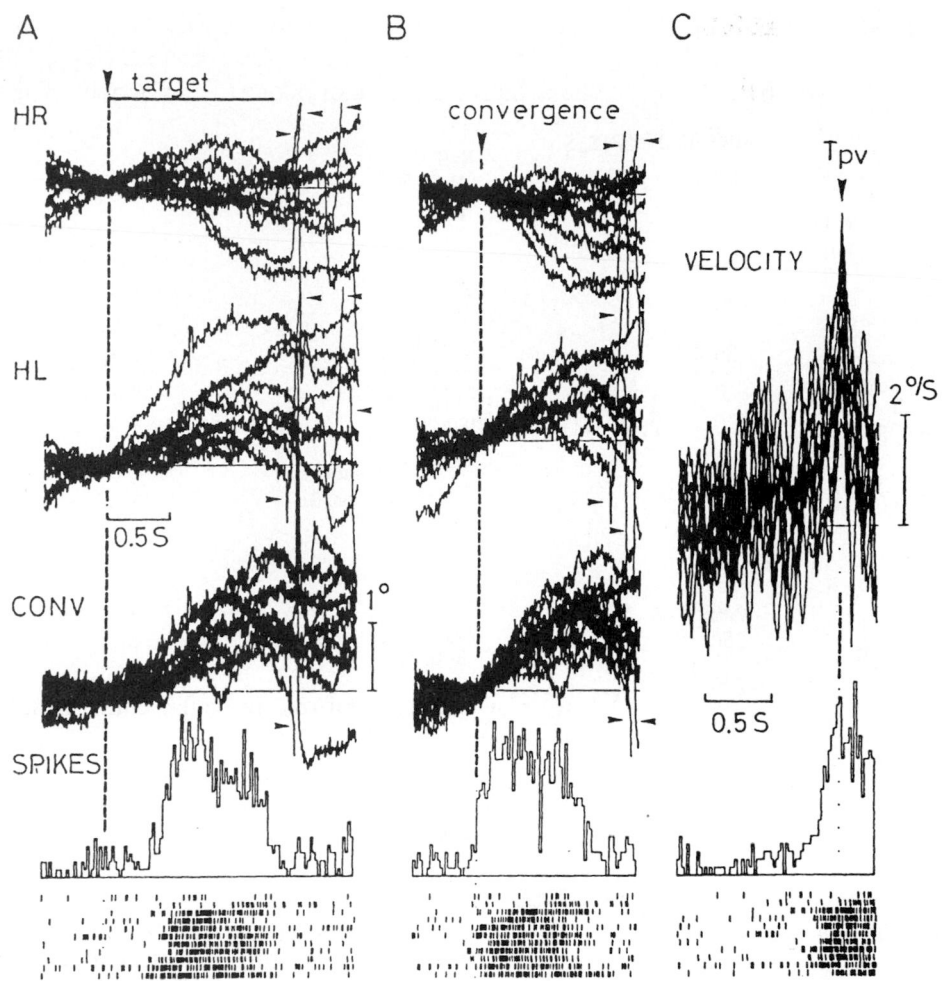

Fig. 6. Neuronal activity correlated with convergence eye movements evoked by target movement in depth. (A) From top to bottom: horizontal components of right (HR) and left (HL) eye movements, ocular convergence (CONV), peri-event time histogram for a PMLS neuron (SPIKES), and a raster display of times of occurrence of spikes for this neuron. The arrow and vertical dotted lines indicate the onset of target movement, and the horizontal line at the top shows the time period during which the target approached. The horizontal arrows indicate the saccadic eye movement. (B) Data in A are rearranged in reference to the onset time of ocular convergence (arrow and dotted line). Time and amplitude calibration is given in A. (C) The vergence traces of the same data synchronized with the time at which the peak velocity (Tpv) of each trace was attained, and the histogram and the raster display of the activity of the same neuron as in A and B. From Takagi et al., (1993) *Neurosci. Res.*, 17: 141–158.

thalamus. These visual inputs are integrated in the LS cortex, where information concerning the three-dimensional movement of visual stimuli is extracted. It seems likely that LS uses this information to provide brainstem motor centers with the neural signals to facilitate eye movements, particularly to targets moving at high speeds. As a result, the visual target can be attained by convergence eye movements which would normally be too slow to capture the target.

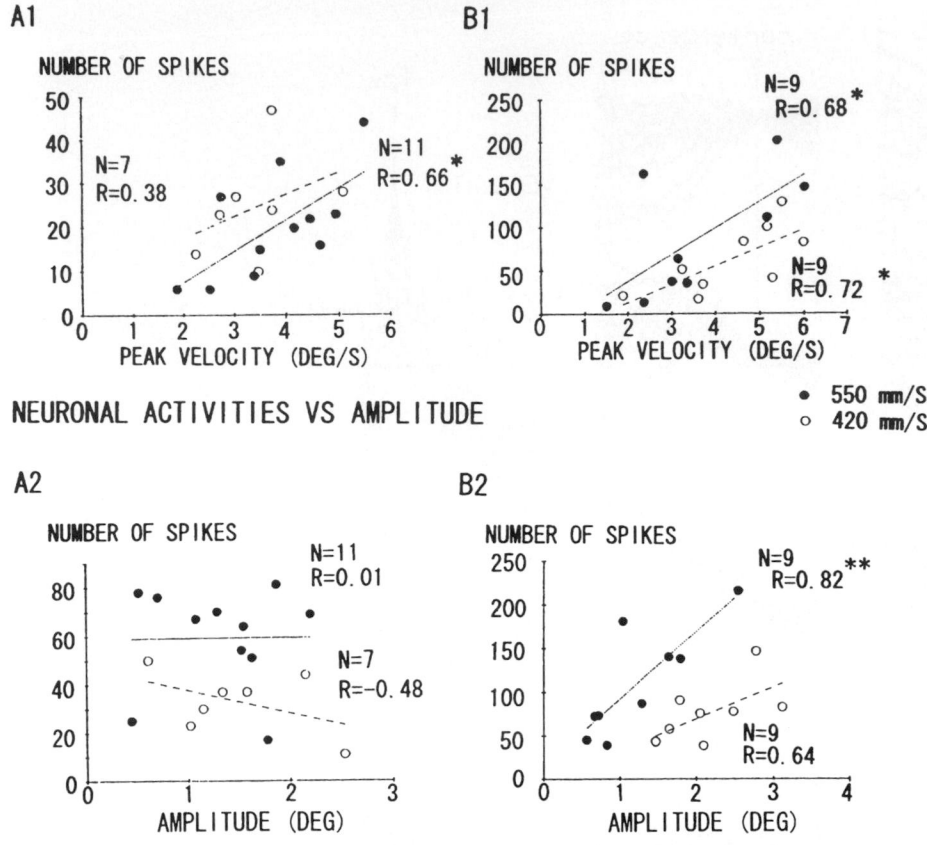

NEURONAL ACTIVITIES VS PEAK VELOCITY

NEURONAL ACTIVITIES VS AMPLITUDE

Fig. 7. Correlation of neuronal activity with peak velocity and amplitude of ocular convergence obtained for each trial of visual stimulation. A1 and A2 show data obtained from a PMLS neuron, and B1 and B2 show data from a different PMLS neuron. *N* is the number of trials; *R* is the correlation coefficient (**P < 0.05; *P < 0.01). Solid circles represent data obtained in response to target motion at a speed of 550 mm/s, and open circles data obtained for targets moving at 420 mm/s. Solid and dotted lines are the regression lines for 550 mm/s and 420 mm/s, respectively. Ordinates: number of spikes in 1 s. Abscissae: peak velocity of convergence eye movement attained in each trial (A1 and B1) and the amplitude of convergence eye movements in each trial (A2 and B2).

The link in the near response

Since Jampel observed ocular convergence, lens accommodation and pupillary constriction, the parieto-occipital border region has been considered one of the central areas involved in the mediation of these three components of the near response. In a series of studies, we have found LS cortex in cat to be involved in this triad of responses, and that different regions of LS play a different role in these responses. From the distribution of response latencies and laterality factors, it is suggested that a hierarchical structure exists in the LS cortex itself as well as in it's connections with other extrastriate cortices and the thalamus.

Another candidate structure which appears to play a role in the near response is the mesencephalic reticular formation in the monkey (Mays et al., 1984; Mays et al., 1986; Judge et al., 1986;

TARGET SPEED	CORRELATION WITH CONVERGENCE				CORRELATION WITH LENS ACCOMMODATION			
	AMPL		VEL		AMPL		VEL	
	L	H	L	H	L	H	L	H
TYPE I NEURON	◎	◎	◎	○	—	—	—	—
	◎	○	○	○	X	X	X	X
	○	X	◎	◎	X	X	X	X
	X	X	○	○	X	X	X	X
	X	X	○	○	○	X	X	X
	X	X	○	◎	—	—	—	—
	X	X	○	◎	○	◎	X	X
TYPE II NEURON	X	○	X	◎	◎	○	X	X
	○	◎	X	○	○	◎	X	X
	X	◎	X	○	○	X	X	X
	X	○	X	○	X	X	◎	○
	○	X	X	◎	—	—	—	—

TARGET SPEED: L. 420 mm/S; H. 550 mm/S

Fig. 8. The properties of PMLS neurons whose activity was well correlated with ocular convergence and/or lens accommodation. Circles: significant correlations (double circles: *P* < 0.01; single circle: *P* < 0.05); crosses: no correlation; short horizontal bar: case not tested. For details, see text.

Zhang et al., 1991, 1992; Morley et al., 1992). Near response neurons here discharge in a highly correlated fashion with ocular convergence and/or lens accommodation. Although the relationship of these neurons with lens accommodation and ocular convergence is rather complicated, these results suggest that neural information related to lens accommodation and ocular convergence are not fully integrated in the cerebral cortex. The cerebral cortex is also essential in the processing of the visual cues for the elementary movements in the near response. Thus, it is suggested that the link in the near response may take place at several sites along the pathway from the cerebrum to the oculomotor neurons.

Relationship with psychophysical studies

Convergence eye movements are classically known as slow eye movements, which have been postu-

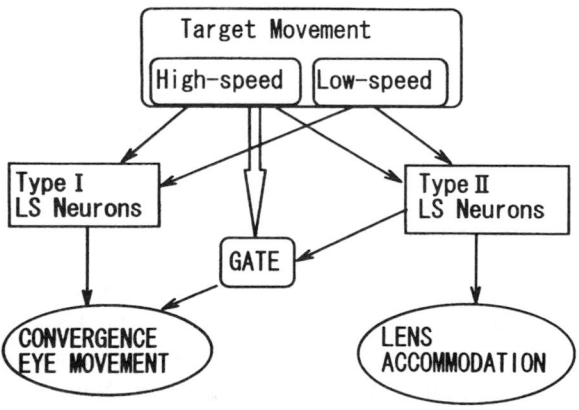

Fig. 9. A presumed role for type 1 and type 2 neurons. See text for details.

Judge, S.J. and Cumming, B.G. (1986) Neurons in the monkey midbrain with activity related to vergence eye movement and accommodation. *J. Neurophysiol.*, 55: 915–930.

Krishnan, V.V. and Stark, L. (1977) A heuristic model for the human vergence eye movement system. *IEEE Trans. Biomed. Engng.*, 24: 44–49.

Mays, L.E. (1984) Neuronal control of vergence eye movements: convergence and divergence neurons in midbrain. *J. Neurophysiol.*, 51: 1091–1108.

Mays. L.E., Porter, J.D., Gamlin, P.D.R. and Tello, C.A. (1986) Neural control of vergence eye movements: neurons encoding vergence velocity. *J. Neurophysiol.*, 56: 1007–1021.

Morley, J.W., Judge, S.J. and Lindsey, J.W. (1992) Role of monkey midbrain near-response neurons in phoria adaptation. *J. Neurophysiol.*, 67: 1475–1492.

Naito, K. and Sasaki, S. (1991) Roles of area 17–18 and lateral suprasylvian area in visual discrimination in the cat. *Neurosci. Res.*, S14: S72.

Palmer, L.A., Rosenquist, A.C. and Tusa, R.J. (1978) The retinotopic organization of lateral suprasylvian visual areas in the cat. *J. Comp. Neurol.*, 177: 237–256.

Poggio, G.F. and Fischer, B. (1977) Binocular interaction and depth sensitivity in the striate and prestriate cortex of the behaving monkey. *J. Neurophysiol.*, 40: 1392–1405.

Rashbass, C. and Westheimer, G. (1961) Disjunctive eye movements. *J. Physiol.*, 159: 339–360.

Rauschecker, J.P., Von Grünau, M.W. and Poulin, C. (1987) Centrifugal organization of direction preferences in the cat's lateral suprasylvian visual cortex and its relation to flow field processing. *J. Neurosci.*, 7: 943–958.

Regan, D. and Cynader, M. (1982) Neurons in cat visual cortex tuned to the direction of motion in depth: effect of stimulus speed. *Invest. Ophthalmol. Vis. Sci.*, 22: 535–550.

Regan, D., Beverley, K.L. and Cynader, M. (1979) Stereoscopic subsystems for position in depth and for motion in depth. *Proc. R. Soc. Lond., B.*, 204: 465–501.

Robinson, D.A. (1963) A method of measuring eye movement using a scleral search coil in magnetic field. *IEEE Trans. Biomed. Engng.*, BME-10: 137–145.

Sawa, M. and Ohtsuka, K. (1994) Lens accommodation evoked by microstimulation of the superior colliculus in the cat, *Vision Res.*, 34:975–981.

Sawa, M., Maekawa, H. and Ohtsuka, K. (1992) Cortical area related to lens accommodation in cat. *Jpn. J. Ophthalmol.*, 36: 371–379.

Semmlow, J.L., Hung, G.K. and Ciuffreda, K.J. (1986) Quantitative assessment of disparity vergence components. *Invest. Ophthalmol. Vis. Sci.*, 27: 558–564.

Spear, P.D. (1991) Functions of extrastriate visual cortex in non-primate species. In: A.V. Leventhal (Ed.), *Vision and Visual Dysfunction*, Vol. 4, *The Neural Basis of Visual Function*, Macmillan, Basingstroke, pp. 339–370.

Spear, P.D. and Baumann, T.P. (1975) Receptive-field characteristics of single neurons in lateral suprasylvian visual area of the cat. *J. Neurophysiol.*, 38: 1403–1420.

Stryker, M. and Blakemore, C. (1972) Saccadic and disjunctive eye movements in cats. *Vision Res.*, 12: 2005–2013.

Takagi, M., Toda, H., Yoshizawa, T., Hara, N., Ando, T., Abe, H. and Bando, T. (1992) Ocular convergence-related neuronal responses in the lateral suprasylvian area of alert cats. *Neurosci. Res.*, 15: 229–234.

Takagi, M., Toda, H. and Bando, T. (1993) Extrastriate cortical neurons correlated with ocular convergence in the cat. *Neurosci. Res.*, 17: 141–158.

Takemura, A., Inoue, Y., Kawano, K. and Miles, F.A. (1995) Short-latency discharges in medial superior temporal area of alert monkey to horizontal disparity steps. *Abstracts of 4th IBRO World Congress of Neuroscience*. p. 335.

Toda, H., Takagi, M., Yoshizawa, T. and Bando, T. (1991) Disjunctive eye movement evoked by microstimulation in an extrastriate cortical area of the cat. *Neurosci. Res.*, 12: 300–306.

Toyama, K., Komatsu, Y. and Shibuki, K. (1984) Integration of retinal and motor signals of eye movements on striate cortex cells of the alert cat. *J. Neurophysiol.*, 51: 649–665.

Toyama, K., Komatsu, Y. and Kozasa, T. (1986) The responsiveness of Clare-Bishop neurons to motion cues for motion stereopsis. *Neurosci. Res.*, 4: 83–109.

Tusa, R.J., Demer, J.L. and Herdman, S.J. (1989) Cortical areas involved in OKN and VOR in cats: Cortical lesions. *J. Neurosci.*, 9: 1163–1178.

Westheimer, G. and Mitchell, G. (1969) The sensory stimulus for disjunctive eye movements. *Vision Res.*, 9: 749–755.

Yoshizawa, T., Takagi, M., Toda, H., Shimizu, H., Norita, M., Hirano, T., Abe, H., Iwata, K. and Bando, T. (1991) A projection of the lens accommodation-related area to the pupillo-constrictor area in the posteromedial lateral suprasylvian area in cats. *Jpn. J. Ophthal.*, 35: 107–118.

Yin, T.C.T. and Greenwood, M. (1992) Visuomotor interactions in responses of neurons in the middle and lateral suprasylvian cortices of the behaving cat. *Exp. Brain Res.*, 88: 15–32.

Zhang, Y., Gamlin, P.D.R. and Mays, L.E. (1991) Antidromic identification of midbrain near response cells projecting to the oculomotor nucleus. *J. Neurophysiol.*, 67: 944–960.

Zhang, Y., Mays, L.E. and Gamlin, P.D.R. (1992) Characteristics of near response cells projecting to the oculomotor nucleus. *J. Neurophysiol.*, 67: 944–960.

Zuber, B.L. and Stark, L. (1968) Dynamic characteristics of the fusional vergence eye movement system. *IEEE Trans. Sys. Sci. Cybern.*, SSC-4: 72–79.

M. Norita, T. Bando and B. Stein (Eds.)
Progress in Brain Research, Vol 112
© 1996 Elsevier Science BV. All rights reserved.

CHAPTER 11

Functional connectivity of the superior colliculus with saccade-related brain stem neurons in the cat

S. Chimoto[1], Y. Iwamoto[1], H. Shimazu[2] and K. Yoshida[1]

[1]*Department of Physiology, Institute of Basic Medical Sciences, University of Tsukuba, Tsukuba, Ibaraki 305, Japan*
[2]*Department of Neurophysiology, Tokyo Metropolitan Institute for Neuroscience, 2–6 Musashidai, Fuchu, Tokyo 183, Japan*

Effects of stimulation of the superior colliculus on saccade-related brain stem neurons were studied in the alert cat. Extracellular recordings were made from medium-lead burst neurons (MLBNs), omnipause neurons (OPNs) and burster-driving neurons (BDNs) in the paramedian pontomedullary region rostral and caudal to the abducens nucleus. MLBNs were activated from the contralateral superior colliculus with monosynaptic latencies when single-pulse stimulation was given during saccades or ipsilateral head rotation, although this activation was not observed during fixation periods. The caudal SC was more effective than the rostral SC in monosynaptic activation of MLBNs. Most OPNs were also activated monosynaptically from the SC. In contrast to MLBNs, the activation of OPNs was more frequently induced from the rostral SC than from the caudal SC. Stimulation of the caudal SC often induced suppression of spikes in OPNs. BDNs received excitation from the ipsilateral SC through a di- or trisynaptic pathway. Like MLBNs, BDNs tended to receive stronger input from the caudal SC than the rostral SC. Results indicate the existence of tectofugal excitatory pathways to MLBNs and BDNs and an inhibitory pathway to OPNs. It seems likely that these pathways originate from saccade-related burst cells in the SC. Since excitation of BDNs and inhibition of OPNs increase the excitability of MLBNs, all of these pathways may contribute to burst activity in MLBNs and thereby saccade generation. Results also support the current idea that cells in the rostral SC may participate in fixation by activating OPNs.

Introduction

The superior colliculus (SC) is an important element of the neural circuitry that initiates and controls saccadic eye movements (Wurtz and Albano, 1980; Grantyn, 1988; Sparks and Mays, 1990; Grantyn et al., 1993, for review). Neurons in the intermediate and deep layers of the SC project to brain stem preoculomotor structures. There are several distinct groups of neurons in the pons and medulla that are involved in saccade generation: burst neurons (BNs), omnipause neurons (OPNs), and burster-driving neurons (BDNs). BNs

are further subdivided into medium-lead burst neurons (MLBNs) and long-lead burst neurons (LLBNs). The firing characteristics and efferent projection patterns of each group of these neurons have been extensively studied.

MLBNs are silent during fixation and exhibit a high frequency burst during saccades (Luschei and Fuchs, 1972; Cohen and Henn, 1972; Keller, 1974). The burst contains appropriate information about amplitude, velocity and duration of either the horizontal or vertical component of saccades (King and Fuchs, 1979; Kaneko et al., 1981; Yoshida et al., 1982; Scudder et al., 1988). These neurons are immediate premotor neurons and consist of excitatory and inhibitory burst neurons (EBNs and IBNs). In the horizontal system, it is known that EBNs and IBNs are located in

---

*Corresponding author: Tel.: +81 0298 533303; fax: +81 0298 533039; e-mail: kyoshida@md.tsukuba.ac.jp

the pontine and medullary reticular formation, respectively. EBNs project to ipsilateral abducens motoneurons (Igusa et al., 1980; Strassman et al., 1986a), and IBNs to contralateral abducens motoneurons (Hikosaka and Kawakami, 1977; Hikosaka et al., 1978; Yoshida et al., 1982; Strassman et al., 1986b). The EBNs and IBNs on one side of the brain burst simultaneously, thereby inducing ipsilateral saccades. LLBNs are distinguished from MLBNs by the low-frequency prelude of spike activity preceding vigorous bursts (Luschei and Fuchs, 1972; Keller, 1974).

OPNs are located near the midline of the pons. They discharge tonically during intersaccadic intervals and exhibit a pause in activity during saccades in all directions (Luschei and Fuchs, 1972; Cohen and Henn, 1972; Keller, 1974; Evinger et al., 1982). Morphophysiological evidence indicates that OPNs project to the MLBN area (Nakao et al., 1980; Langer and Kaneko, 1983; Curthoys et al., 1984; Ohgaki et al., 1987; Strassman et al., 1987). It has been suggested that OPNs tonically inhibit MLBNs during fixation and slow eye movements, and that the pause in their activity causes disinhibition of MLBNs, thereby controlling burst duration (Keller, 1974; Robinson, 1975).

BDNs are located within and immediately below the prepositus hypoglossi nucleus, discharge tonically during fixation and slow eye movements, and exhibit a burst of spikes in association with contralateral saccades (Kitama et al., 1988, 1995). Electrophysiological evidence shows that BDNs terminate in the contralateral EBN and IBN areas and make monosynaptic excitatory connections with BNs (Ohki et al., 1988). Because the pause in OPN activity is omnidirectional, MLBNs are released from inhibition for saccades in all directions. To generate a burst of spikes for saccades in a particular direction, MLBNs must receive a direction-specific, saccade-related excitatory input. Quantitative analysis of firing characteristics suggests that BDNs are a likely source of this direction-specific input that determines the amplitude and velocity of saccades (Kitama et al., 1995).

There have been several studies that revealed collicular input to various elements of the saccadic burst generator. Raybourn and Keller (1977) have found monosynaptic excitation of monkey LLBNs following electrical stimulation of the SC. OPNs have also been shown to receive monosynaptic excitation from the colliculus (Raybourn and Keller, 1977; Kaneko and Fuchs, 1982; Paré and Guitton, 1994). MLBNs, on the other hand, were not excited from the SC until multiple shocks reached the threshold for saccades in the monkey (Raybourn and Keller, 1977).

The SC has recently been suggested to be involved not only in the generation of saccades but also in the control of fixation. The rostral part of the SC contains neurons that discharge tonically during fixation and pause for saccades (Munoz and Guitton, 1991; Munoz and Wurtz, 1993), whereas the rest of the SC contains pre-saccadic burst neurons which discharge for saccades of a particular range of amplitude and direction (Schiller and Koerner, 1971; Wurtz and Goldberg, 1971, 1972; Sparks, 1975, 1978, Sparks and Mays, 1980). It seems important, therefore, to examine whether the rostral and caudal parts of the SC make different connections with elements of the burst generator. To elucidate the functional connectivity of the SC with the premotor circuitry for saccade generation, we examined the effects of stimulation of rostral and caudal areas of the SC upon saccade-related brain stem neurons in the alert cat.

Methods

Experiments were performed on ten adult cats. Each animal underwent surgical procedures with aseptic conditions under pentobarbital sodium anesthesia (initial dose, 40 mg/kg i.p., supplemented by 2–5 mg/kg per h i.v.). Surgical procedures and recording and stimulation conditions were similar to those described previously (Kitama et al., 1995). In brief, a coil of Teflon-coated stainless steel wire was implanted in one eye to measure eye movements using the search-coil method. The tympanic bulla on each side was opened, and a silver ball electrode was placed on the round window to stimulate the vestibular nerve. Microelectrodes made of Elgiloy wire insulated with solder glass were used to stimulate the rostral and caudal parts of the SC. The optimum site of each electrode tip was determined with the cat alert by recording neuronal responses to visual stimulation and observing saccadic eye movements evoked by stimulation with a short-pulse train at currents less than 30 μA. Histological reconstruction located stimulation sites in the intermediate or deep layer in the SC. During recording sessions the animal was placed in a stereotaxic apparatus mounted on a turntable that could be rotated about the earth's vertical axis. The head was fixed in a 26.5° nose-down position so that rotations would activate the horizontal canals maximally. Single neuron activity was recorded extracellularly with glass micropipettes filled with 2 M NaCl solution or glass-coated tungsten microelectrodes. Electrical stimulation of the SC (single or double pulses, 0.1–0.2-ms duration, intensity < 200 μA) was used to examine responses of brain stem neurons.

Results

Effects of SC stimulation on MLBNs

Horizontal MLBNs were identified by their high frequency burst occurring immediately before and during ipsilateral saccades (Fig. 1A). The onset of the burst was abrupt, and the duration of the burst was approximately equal to the duration of the horizontal component of the saccade. Stimulation of the SC was delivered under two sets of different conditions: (1) with the head fixed in space vs. during head rotation and (2) during intersaccadic intervals vs. during saccades. Effects of single- and double-pulse stimulation were compared.

Fig. 1B shows the effects of caudal SC stimulation on a single MLBN located in the IBN area. When single-pulse stimulation was applied to the contralateral SC during the intersaccadic interval with the head stationary, no response was evoked

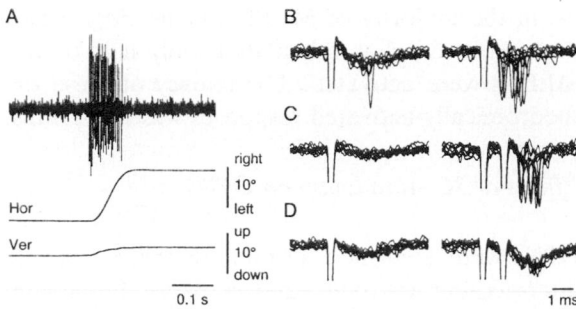

Fig. 1. Effects of contralateral SC stimulation on a medium lead burst neuron (MLBN). A: a burst of spikes associated with an ipsilateral saccade. From top to bottom, filtered spike activity of the MLBN recorded on the right side of the brain stem and horizontal and vertical components of eye position. B: effects of single-pulse stimulation (200 μA) of the caudal SC with the head stationary (left) and during horizontal head rotation directed to the recording side (right). C: effects of caudal SC stimulation (100 μA) with single (left) and double pulses (right). Head stationary. D: single- and double-pulse stimulation (200 μA) of the rostral SC.

in most trials (Fig. 1B, left). In contrast, when SC stimulation was delivered during horizontal head rotation directed to the recording side, this MLBN was activated with short latencies ranging from 0.8 to 2.0 ms (Fig. 1B, right). The shortest latency, 0.8 ms, suggests monosynaptic excitation. No collicular activation was obtained with head rotation in the opposite direction. The short-latency excitation was also induced when the SC was stimulated at the beginning of saccades even with the head stationary (Chimoto et al., 1996). Temporal facilitation by double-pulse stimulation of the caudal SC was effective in evoking spikes in this MLBN (Fig. 1C, right), the shortest latency after the effective second shock being 0.9 ms. In contrast to the clear activation from the caudal SC, rostral SC stimulation did not evoke spikes in this MLBN, even at stronger currents (Fig. 1D). Similar results were obtained in other MLBNs. All MLBNs examined were activated with short latency after stimulation of the caudal SC, when the stimulus was given during ipsilateral head rotation or at the beginning of saccades, or when double-pulse stimulation was applied. The latency following SC stimulation ranged from 0.8 to 1.4

ms in the majority of MLBNs. Conversely, when the rostral SC was stimulated, only half of the MLBNs were activated. The latency of these orthodromically-activated responses was 1.0–2.0 ms.

Effects of SC stimulation on OPNs

OPNs were identified by tonic discharges during intersaccadic intervals and a pause in activity during saccades in all directions (Fig. 2A). Stimulation of the rostral and caudal SC tended to exert opposite effects on OPNs. In Fig. 2B, stimulation of the rostral SC evoked spikes in an OPN. The shortest latency of evoked spikes was 0.9 ms, suggesting monosynaptic activation. In contrast, stimulation of the caudal SC suppressed spikes in this OPN (Fig. 2C). Fig. 2D shows this suppression more clearly, plotting the responses of another OPN to caudal SC stimulation in a number of superimposed traces. It is noteworthy that this suppression is not preceded by excitatory responses, suggesting that the observed suppression is indeed synaptic inhibition and not an event related to refractoriness. Latencies of spike suppression in individual neurons ranged from 1.3 to 2.2 ms after stimulation, suggesting that the shortest pathway is disynaptic. The short-latency activation of OPNs was more frequently induced

from the rostral SC (24/39) than from the caudal SC (7/36). In contrast, the suppression of OPN spikes was more frequently observed with caudal SC than rostral SC stimulation.

Effects of SC stimulation on BDNs

BDNs exhibited irregular tonic discharges during intersaccadic intervals and generated a high-frequency burst of spikes before and during saccades directed to the contralateral side (Fig. 3A). These neurons displayed vestibular type II responses during horizontal head rotation: the firing rate increased with contralateral head rotation and decreased with ipsilateral rotation and was approximately proportional to and in phase with contralateral head velocity (Fig. 3B). The number of spikes in the burst was linearly related to the amplitude of the contralateral component of the saccade (Figs. 4A and B). The mean burst

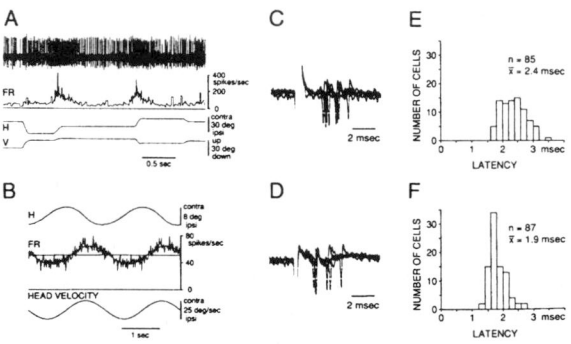

Fig. 3. Discharge pattern of BDNs and their response to stimulation of the SC and the vestibular nerve. A: from top to bottom, spike activity of a BDN associated with saccades, it's instantaneous firing rate (FR), and horizontal (H) and vertical (V) eye position. B: type II response of a BDN to sinusoidal horizontal head rotation. Horizontal eye position (H) and firing rate (FR) were averaged over 8 stimulus cycles. Sine wave superimposed on the firing rate trace indicates best-fitting response fundamental calculated by a least-squares method. C: response of a BDN to stimulation of the ipsilateral SC. D: response of the same BDN shown in C to stimulation of the contralateral vestibular nerve. E: latency histogram of excitation from the ipsilateral SC. F: latency histogram of excitation induced from the contralateral vestibular nerve (From Kitama et al., 1995).

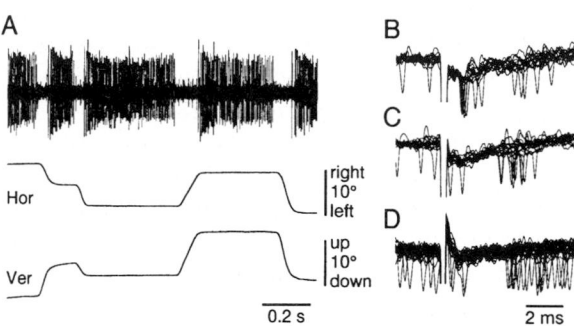

Fig. 2. Effects of SC stimulation on OPNs. A: discharge pattern of an OPN associated with saccades. From top to bottom, filtered spike activity and horizontal and vertical components of eye position. B: excitation of an OPN induced by rostral SC stimulation. C: suppression of the same OPN after caudal SC stimulation. D: suppression of another OPN after caudal SC stimulation.

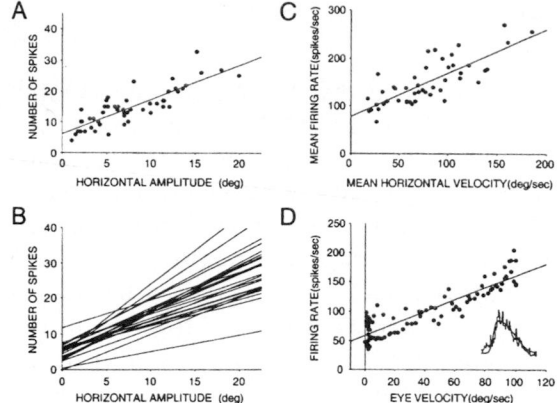

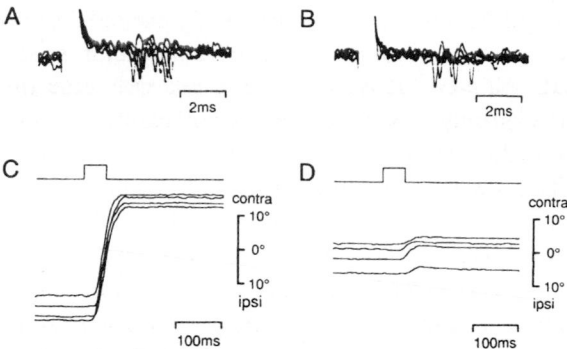

Fig. 4. Characteristics of BDN activity in relation to amplitude and velocity of saccades. A: relationship between the number of spikes in the burst and the amplitude of the contralateral component of the saccade for a BDN. B: regression lines obtained from 25 BDNs examined for saccades. C: relationship between the mean firing rate in the burst and the mean velocity of contralateral horizontal component of saccades for a BDN. D: instantaneous firing rate plotted as a function of instantaneous component velocity. Same cell as in C. Inset: instantaneous firing rate trace is obtained by averaging 19 traces, corrected for the lead time of 25 ms, and superimposed on the velocity trace (From Kitama et al., 1995).

Fig. 5. Effects of caudal and rostral SC stimulation on a BDN. A: effects of single-pulse stimulation of the caudal SC on a BDN. B: same as in A, but the stimulus was applied to the rostral SC. C: horizontal component of saccades induced by short pulse train (400 Hz, 20 pulses) applied to the caudal SC. D: same as in C, but the stimulus was applied to the rostral SC.

firing rate was proportional to the mean velocity of the contralateral component of the saccade (Fig. 4C). A highly significant correlation was also found between intraburst firing rate and instantaneous eye velocity (Fig. 4D).

All BDNs investigated were transsynaptically excited by single-pulse stimulation of the ipsilateral SC as shown in Fig. 3C. The threshold for the activation was less than 50 μA, and double or triple spikes were often evoked as stimulus intensity was increased. The latency of induced spikes ranged from 1.7 to 3.5 ms (Fig. 3E), suggesting that SC activation of BDNs was di- and/or trisynaptic. All BDNs were excited by stimulation of the contralateral vestibular nerve with latencies of 1.4–2.7 ms (Figs. 3D and F). It was regularly observed that collicular excitation of BDNs was facilitated by contralateral horizontal head rotation. Effects of stimulating the rostral and caudal SC were compared for 21 BDNs. An example is shown in Fig. 5, in which more vigorous responses

were evoked from the caudal SC (Fig. 5A) as compared to the rostral SC (Fig. 5B). Correspondingly, a short pulse train applied to the caudal SC induced saccades with a larger horizontal component (Fig. 5C) than saccades induced from the rostral SC (Fig. 5D). Stimulation of the caudal SC activated all of the 21 BDNs, and stimulation of the rostral SC activated 18 of the 21.

Discussion

In the present study, we have shown that MLBNs and OPNs receive monosynaptic excitation and disynaptic inhibition, respectively, from the SC. The present study has also revealed a difference between caudal and rostral SC stimulation in terms of it's effects on saccade-related brain stem neurons. We will first discuss our findings in relation to the results of previous studies, and then the implications of the present study for the circuitry underlying saccade generation.

Burst neurons

Raybourn and Keller (1977) studied the connectivity of the SC with burst neurons in alert monkeys. After single-pulse stimulation of the

SC, LLBNs were monosynaptically excited, but no response was found in MLBNs. In their study, activation of MLBNs was observed with multiple shocks only at current levels sufficient to evoke saccadic eye movements. In the present study, when single-pulse stimulation was applied to the caudal SC during head rotation directed toward the recording side or at the beginning of saccades, MLBNs were activated with short latencies. The responses with latencies less than 1.4 ms are suggested to be monosynaptically induced, as estimated for monosynaptic activation of LLBNs and OPNs (Raybourn and Keller, 1977; Paré and Guitton, 1994). These results were in striking contrast to the lack of responses seen upon SC stimulation during intersaccadic intervals with the head stationary. The question arises as to why SC stimulation was effective during head rotation or saccades. In addition to collicular input, MLBNs have been shown to receive excitatory input from BDNs (Ohki et al., 1988) and inhibitory input from OPNs (Keller, 1974; Nakao et al., 1980). Therefore, the excitability of MLBNs may vary with the activity of these afferent neurons. The excitatory input from BDNs should increase during horizontal head rotation because of their type II activation. This may result in increased excitability of MLBNs, and SC stimulation will become effective in evoking spikes. During saccades, inhibitory input from OPNs is removed due to a pause in their activity. This disinhibition will also increase excitability of MLBNs and make collicular stimulation effective. If the cells in the SC that project to MLBNs discharge during saccades, they would directly contribute to the generation of burst activity in MLBNs. The caudal SC was more effective than the rostral SC in monosynaptic activation of MLBNs, suggesting denser projections from this region. This may be related to the well-known finding that saccades evoked from the caudal SC have a larger horizontal component (Robinson, 1972). It seems reasonable to suggest that the responses to SC stimulation were probably the result of activation of collicular burst cells rather than fixation cells.

OPNs

It has been found that OPNs are monosynaptically excited from the SC (Raybourn and Keller, 1977). The excitation originates mainly from the fixation area in the rostral SC and is suggested to play a role in active fixation by suppressing saccade generation (Paré and Guitton, 1994). The present finding that monosynaptic activation of OPNs is more frequently induced from the rostral SC is consistent with these previous studies. It has also been reported that the monosynaptic excitation is followed by suppression of spikes in OPNs (Raybourn and Keller, 1977; Kaneko and Fuchs, 1982). There has been no study that shows SC-induced suppression of OPN activity that is not preceded by excitation. In the present study, suppression of spike activity was observed following stimulation of the caudal SC in about one-third of OPNs even without earlier excitation. The shortest latency of suppression was 1.3 ms, 0.4 ms longer than the shortest latency of excitation. This suggests that the observed suppression of OPNs is due to disynaptic inhibition.

BDNs

Discharges of BDNs associated with saccades are similar to those of MLBNs in many respects. The burst activity of BDNs was most vigorous for contralaterally directed saccades. Similar direction specificity has been found in MLBNs (Luschei and Fuchs, 1972; Keller, 1974; Kaneko et al., 1981; Yoshida et al., 1982; Scudder et al., 1988). The on-direction of MLBNs is ipsilateral, which is consistent with crossed projection of BDNs to MLBNs. The number of spikes in the burst of BDNs was proportional to the amplitude of the contralateral horizontal component. Similar relationships have been found for horizontal MLBNs (Kaneko et al., 1981; Yoshida et al., 1982; Strassman et al., 1986a,b; Scudder et al., 1988). The firing rate in the burst of BDNs was proportional to the velocity of the contralateral component of the saccade. A significant correlation between burst firing rate and eye velocity has also been found for horizontal MLBNs (Kaneko et al., 1981; Yoshida et al., 1982; Scudder et al., 1988). These similarities suggest that BDNs are not only anatomically connected to MLBNs, but also provide them with appropriate saccade signals.

BDNs received excitatory input from the ipsilateral SC via a di- or tri-synaptic pathway. Neurons that transmit collicular effects to BDNs have not been identified, but may be located in the pontomedullary reticular formation or the prepositus hypoglossi nucleus where collateral axons of tecto-reticulo-spinal neurons terminate (Grantyn and Grantyn, 1982; Olivier et al., 1993). The collicular activation of BDNs was probably not relayed by MLBNs, which did not respond to SC stimulation during fixation. LLBNs that are known to exhibit monosynaptic responses in the monkey could be a possible relay cell. Whatever they may be, the relay neurons linking the SC and BDNs could be an important element of the burst generator. The present study has revealed that BDNs receive stronger input from the caudal SC than the rostral SC, which is also the case with MLBNs. This suggests that cells in the caudal SC exert stronger actions on both MLBNs and BDNs than more rostrally-located cells. It is possible that this may be part of the mechanism by which saccade signals in the SC are transformed to final command signals that directly drive motoneurons.

When the animal makes a rapid gaze shift to acquire a visual target, it usually makes simultaneous eye and head movements in the same direction. The tecto-reticulo-spinal pathway is known to be involved in the control of synergistic eye and head movements (Grantyn et al., 1993; Olivier et al., 1993). In the present study, single-pulse stimulation of the SC usually did not evoke responses in MLBNs when the head was stationary. During head rotation to the side of MLBNs, however, the same stimulus was effective in evoking spikes. This suggests that excitation of the ipsilateral horizontal canal increased the excitability of MLBNs probably through BDNs, which receive disynaptic vestibular input. Similarly, if BDNs show vestibular responses during

eye-head gaze shifts as well, increased activity of BDNs will facilitate MLBNs. For a given collicular input, MLBNs may therefore exhibit a higher discharge frequency and a larger number of spikes in the burst when the head moves than when the head is fixed. In agreement with this hypothesis, the amplitude of the horizontal component of SC-induced saccades at various phases of sinusoidal head rotation is modulated approximately in phase with head velocity in the direction of saccades (Kitama et al., 1992). These effects of head rotation on SC-induced saccades are likely to be mediated by the pathway through BDNs.

Acknowledgements

This study was supported by Intramural Grant from the University of Tsukuba Project Research.

References

Chimoto, S., Iwamoto, Y., Shimazu, H. and Yoshida, K. (1996) Monosynaptic activation of medium-lead burst neurons from the superior colliculus in the alert cat. *J. Neurophysiol.*, 75: 2658–2661.

Cohen, B. and Henn, V. (1972) Unit activity in the pontine reticular formation associated with eye movements. *Brain Res.*, 46: 403–410.

Curthoys, I.S., Markham, C.H. and Furuya, N. (1984) Direct projection of pause neurons to nystagmus-related excitatory burst neurons in the cat pontine reticular formation. *Exp. Neurol.*, 83: 414–422.

Evinger, C., Kaneko, C.R.S. and Fuchs, A.F. (1982) The activity of omnipause neurons in alert cats during saccadic eye movements and visual stimuli. *J. Neurophysiol.*, 47: 827–844.

Grantyn, R. (1988) Gaze control through superior colliculus: structure and function. In: J.A. Büttner-Ennever (Ed.), *Neuroanatomy of the Oculomotor System, Reviews of Oculomotor Research*, Vol. 2, Elsevier, Amsterdam, pp. 273–333.

Grantyn, A. and Grantyn, R. (1982) Axonal patterns and sites of termination of cat superior colliculus neurons projecting in the tecto-bulbo-spinal tract. *Exp. Brain Res.*, 46: 243–256.

Grantyn, A., Olivier, E. and Kitama, T. (1993) Tracing premotor brain stem networks of orienting movements. *Curr. Opin. Neurobiol.*, 3: 973–981.

Hikosaka, O. and Kawakami, T. (1977) Inhibitory reticular neurons related to the quick phase of vestibular nystagmus — their location and projection. *Exp. Brain Res.*, 27: 377–396.

Hikosaka, O., Igusa, Y., Nakao, S. and Shimazu, H. (1978) Direct inhibitory synaptic linkage of pontomedullary reticular burst neurons with abducens motoneurons in the cat. *Exp. Brain Res.*, 33: 337–352.

Igusa, Y., Sasaki, S. and Shimazu, H. (1980) Excitatory premotor burst neurons in the cat pontine reticular formation related to the quick phase of vestibular nystagmus. *Brain Res.*, 182: 451–456.

Kaneko, C. and Fuchs, A.F. (1982) Connections of cat omnipause neurons. *Brain Res.*, 241: 166–170.

Kaneko, C.R.S., Evinger, C. and Fuchs, A.F. (1981) Role of cat pontine burst neurons in generation of saccadic eye movements. *J. Neurophysiol.*, 46: 387–408.

Keller, E.L. (1974) Participation of medial pontine reticular formation in eye movement generation in monkey. *J. Neurophysiol.*, 37: 316–332.

King, W.M. and Fuchs, A. (1979) Reticular control of vertical saccadic eye movements by mesencephalic burst neurons. *J. Neurophysiol.*, 42: 861–876.

Kitama, T., Ohki, Y., Shimazu, H. and Yoshida, K. (1988) Physiological and morphological properties of brain stem neurons related to vestibular and saccadic eye movements in the cat. In: J.C. Hwang, N.G. Daunton, and V.J. Wilson (Eds.), *Basic and Applied Aspects of Vestibular Function*, Hong Kong University Press, Hong Kong, pp. 45–53.

Kitama, T., Shimazu, H., Tanaka, M. and Yoshida, K. (1992) Vestibular and visual interaction in generation of rapid eye movements. *Ann. NY Acad. Sci.*, 656: 396–407.

Kitama, T., Ohki, Y., Shimazu, H., Tanaka, M. and Yoshida, K. (1995) Site of interaction between saccade signals and vestibular signals induced by head rotation in the alert cat: functional properties and afferent organization of burster-driving neurons. *J. Neurophysiol.*, 74: 273–287.

Langer, T.P. and Kaneko, C.R.S. (1983) Efferent projections of the cat oculomotor omnipause neuron region: an autoradiographic study. *J. Comp. Neurol.*, 217: 288–306.

Luschei, E.S. and Fuchs, A.F. (1972) Activity of brain stem neurons during eye movements of alert monkeys. *J. Neurophysiol.*, 35: 445–461.

Munoz, D.P. and Guitton, D. (1991) Control of orienting gaze shifts by the tectoreticulospinal system in the head-free cat. II. Sustained discharges during motor preparation and fixation. *J. Neurophysiol.*, 66: 1624–1641.

Munoz, D.P. and Wurtz, R.H. (1993) Fixation cells in monkey superior colliculus. I. Characteristics of cell discharge. *J. Neurophysiol.*, 70: 559–575.

Nakao, S., Curthoys, I.S. and Markham, C.H. (1980) Direct inhibitory projection of pause neurons to nystagmus-related pontomedullary reticular burst neurons in the cat. *Exp. Brain Res.*, 40: 283–293.

Ohgaki, T., Curthoys, I.S. and Markham, C.H. (1987) Anatomy of physiologically identified eye- movement-related pause neurons in the cat: pontomedullary region. *J. Comp. Neurol.*, 266: 56–72.

Ohki, Y., Shimazu, H. and Suzuki, I. (1988) Excitatory input to burst neurons from the labyrinth and it's mediating pathway in the cat: location and functional characteristics of burster-driving neurons. *Exp. Brain Res.*, 72: 457–472.

Olivier, E., Grantyn, A., Chat, M. and Berthoz, A. (1993) The control of slow orienting eye movements by tectoreticulospinal neurons in the cat: behavior, discharge patterns and underlying connections. *Exp. Brain Res.*, 93: 435–449.

Paré, M. and Guitton, D. (1994) The fixation area of the cat superior colliculus: effects of electrical stimulation and direct connection with brainstem omnipause neurons. *Exp. Brain Res.*, 101: 109–122.

Raybourn, M.S. and Keller, E.L. (1977) Colliculoreticular organization in primate oculomotor system. *J. Neurophysiol.*, 40: 861–878.

Robinson, D.A. (1972) Eye movements evoked by collicular stimulation in the alert monkey. *Vision Res.*, 12: 1795–1808.

Robinson, D.A. (1975) Oculomotor control signal. In G. Lennerstrand and P. Bach-y-Rita (Eds.), *Basic Mechanisms of Ocular Motility and their Clinical Implications*, Pergamon, New York, pp. 337–378.

Schiller, P.H. and Koerner, F. (1971) Discharge characteristics of single units in superior colliculus of the alert rhesus monkey. *J. Neurophysiol.*, 34: 920–936.

Scudder, C.A., Fuchs, A.F. and Langer, T.P. (1988) Characteristics and functional identification of saccadic inhibitory burst neurons in the alert monkey. *J. Neurophysiol.*, 59: 1430–1454.

Sparks, D.L. (1975) Response properties of eye movement-related neurons in the monkey superior colliculus. *Brain Res.*, 90: 147–152.

Sparks, D.L. (1978) Functional properties of neurons in the monkey superior colliculus: coupling of neuronal activity and saccade onset. *Brain Res.*, 156: 1–16.

Sparks, D.L. and Mays, L.E. (1980) Movement fields of saccade-related burst neurons in the monkey superior colliculus. *Brain Res.*, 190: 39–50.

Sparks, D.L. and Mays, L.E. (1990) Signal transformations required for the generation of saccadic eye movements. *Annu. Rev. Neurosci.*, 13: 309–336.

Strassman, A., Highstein, S.M. and McCrea, R.A. (1986a) Anatomy and physiology of saccadic burst neurons in the alert squirrel monkey. I. Excitatory burst neurons. *J. Comp. Neurol.*, 249: 337–357.

Strassman, A., Highstein, S.M. and McCrea, R.A. (1986b) Anatomy and physiology of saccadic burst neurons in the alert squirrel monkey. II. Inhibitory burst neurons. *J. Comp. Neurol.*, 249: 358–380.

Strassman, A., Evinger, C., McCrea, R.A., Baker, R.G. and Highstein, S.M. (1987) Anatomy and physiology of intracellularly labelled omnipause neurons in the cat and squirrel monkey. *Exp. Brain Res.*, 67: 436–440.

Wurtz, R.H. and Albano, J.E. (1980) Visual-motor function of the primate superior colliculus. *Annu. Rev. Neurosci.*, 3: 189–226.

Wurtz, R.H. and Goldberg, M.E. (1971) Superior colliculus cell responses related to eye movements in awake monkeys. *Science*, 171: 82–84.

Wurtz, R.H. and Goldberg, M.E. (1972) Activity of superior colliculus in behaving monkey. III. Cells discharging before eye movements. *J. Neurophysiol.*, 35: 575–586.

Yoshida, K., McCrea, R., Berthoz, A. and Vidal, P.P. (1982) Morphological and physiological characteristics of inhibitory burst neurons controlling horizontal rapid eye movements in the alert cat. *J. Neurophysiol.*, 48: 761–784.

M. Norita, T. Bando and B. Stein (Eds.)
Progress in Brain Research, Vol 112
© 1996 Elsevier Science BV. All rights reserved.

CHAPTER 12

Visual-auditory integration in cat superior colliculus: implications for neuronal control of the orienting response

Carol K. Peck*

School of Optometry, University of Missouri St. Louis, 8001 Natural Bridge Road, St. Louis, MO 63121, USA

Previous physiological studies have demonstrated that inputs from different sensory modalities converge on individual neurons in the superior colliculus. Moreover, in anesthetized, paralyzed animals, those tectal neurons which are most directly connected to brain stem circuits mediating orienting eye and head movements are highly likely to exhibit significant integration of sensory inputs from multiple modalities. The purpose of the present study was to examine the responses of tectal neurons in the alert cat when visual and auditory stimuli were presented as targets for ocular fixation and orienting responses. For comparison to previous work in anesthetized, paralyzed animals, we also examined the responses of tectal neurons to the presentation of these stimuli during periods when the cats voluntarily maintained their eyes near primary position in the absence of a fixation target. Under these conditions, there were significant differences between the strength of the response to the simultaneous presentation of visual and auditory targets and the strength of response to the most effective unimodal stimulus in about 40% of the cells tested. Many tectal neurons also responded tonically during fixation of visual, auditory and bimodal targets, and some of these also exhibited significant bimodal interactions. However, among individual neurons which responded phasically to stimulus onset or offset and tonically during fixation, there was only a weak correlation between the extent of bimodal interaction under the two conditions. Finally, among saccade-related neurons, the magnitude of saccade-related activity was only slightly affected when a biomodal target was used to elicit a saccade, and the extent of bimodal interactions was generally less than was found for the onset and offset of sensory targets. Such multisensory interactions can be significant for behavior. Indeed, simply using a multisensory target has been shown to influence the probability and latency of overt orienting responses, although the extent of such effects will probably vary across both tasks and stimulus conditions. Strong multisensory interactions are most likely to occur when low intensity stimuli are used. Our use of moderately intense sensory stimuli probably accounts for our finding of a relatively small percentage of cells in which bimodal responses were greater than the sum of their unimodal responses.

Introduction

A basic question in the neuronal control of orienting movements is how different sensory inputs, based on different coordinate systems, are reconciled so that accurate movements can be made regardless of the modality of the target. This is not an easy task. For example, while the positions visual stimuli are initially coded in terms of their retinal locations, and topographic representations of the retinocentric coordinates of visual stimuli are maintained centrally by point-to-point mapping, the locations of sounds in space are encoded with respect to the external ears (pinnae).

In the midbrain, the superior colliculus (SC) or optic tectum is well known to be involved in integrating visual, auditory, and somatosensory inputs and in re-orienting gaze and attention. In

---

*Corresponding author. Tel.: +1 314 5165812; e-mail: sckpeck@umslvma.umsl.edu

many species, the tectum has a neural map of auditory space as well as a neural map of the retina (Knudsen and Brainard, 1995). The auditory map cannot be simply conveyed to the SC from the sensory epithelium but must be constructed from differences in the time and intensity of sounds arriving at the two ears (Knudsen et al., 1987). Integration of visual and auditory space seems to require transformations between eye- and head/pinna-centered coordinates. Otherwise, whenever the eyes are deviated in the head, the orderly topographic maps of sensory space in the SC, will be misaligned (Poppel, 1973), and misalignment of the maps of sensory space would be likely to produce inappropriate responses or confusion on the part of the organism. We have recently shown that changes in eye position modulate the responses of the majority of sound-sensitive neurons in the cat's SC (Peck et al., 1995). The receptive fields of many neurons appear to shift with the direction of the visual axis so that alignment between visual and auditory maps is maintained. Sparks and colleagues have interpreted the effects of eye position on auditory responses of SC neurons to reflect a transformation from head/pinna coordinates to retinal error coordinates (e.g. Jay and Sparks, 1987; Sparks and Mays, 1990). However, many auditory receptive fields of some neurons shift only partially. Other neurons modulate their response strength, but not their preferred spatial location, with changes in eye position, and some neurons are unaffected by eye position. This variety of effects of eye position on the auditory responses of SC neurons is surprising. One possibility is that this transformation occurs within the SC and that neurons which exhibit incomplete compensation of changes in eye position are at an intermediate stage in the transformation of coordinates (Jay and Sparks, 1987; Knudsen and Brainard, 1995). This would imply that tectal output neurons should exhibit the most complete compensation. Ideally, this proposition should be tested in alert animals which are trained to redirect their eyes in a variety of contexts. Although antidromic identi-

fication of tectal output neurons in alert animals is possible, it is quite difficult to obtain a large sample of antidromically-identified neurons and to study those neurons in a variety of behavioral contexts. In the present work, we have studied the responses of tectal neurons which exhibit significant integration of visual and auditory information. The vast majority of tectal output neurons have been shown, in anesthetized animals (Stein and Meredith, 1993), to exhibit multisensory integration. On the assumption that a similar correspondence holds in alert animals, neurons exhibiting multisensory integration are highly likely to be tectal output neurons.

Methods

Physiological and behavioral methods were used to study the SC in awake, behaving cats. Most of our methods are similar to those described recently (Peck et al., 1995) and will therefore be described briefly. Our methods for animal care and use conformed to the regulations of the United States Public Health Service, and all of the procedures used in this study were approved by the University of Missouri-St. Louis Institutional Animal Care and Use Committee.

The cats were trained on several tasks which required either fixation of sensory targets or saccadic eye movements. Eye position was recorded by the scleral search coil technique. Single-unit activity was recorded extracellularly with parylene-insulated tungsten microelectrodes. Responses to sensory stimuli were monitored by presenting visual and auditory stimuli, singly and in combination, at locations within the oculomotor range of the cat (approx. $\pm 25°$). Each cell's receptive fields for visual, auditory, and bimodal stimuli were mapped, and the responses of the cell were evaluated with controlled, reproducible stimuli. Visual stimuli were spots of light (luminance 0.34 Cd/m^2 on a background of < 0.005 Cd/m^2), produced by amber light-emitting diodes. Auditory stimuli were broad-band noise bursts (200–20 000 Hz, 52 dB SPL re: 35 dB background), produced by a custom-built noise generator and delivered by one of an array of speakers located on a horizontally-oriented semicircular track at the level of the interaural axis. For bimodal stimulation, visual and auditory stimuli were presented simultaneously at the same spatial location. Bimodal and unimodal stimuli were presented in interleaved blocks of trials in order to control for long-term changes in excitability.

Visual receptive fields were mapped by subtracting the position of the eye in the orbit from the head-centered position of the visual stimulus. Auditory receptive fields were initially mapped by varying the position of the target when the eye was within $\pm 5°$ of primary position. If the position of the eye in the orbit had no effect on the cell's response, data from all eye positions were pooled to plot the receptive field for auditory stimuli. Otherwise, only trials during which the eye remained near primary position were used. Some bimodal neurons did not have a clear receptive field for the less effective unimodal stimulus and in these cases, the optimal location for the other modality was used to estimate the center of both receptive fields.

The time of occurrence of each cell spike was recorded to the nearest 0.1 ms. Other events (horizontal and vertical coordinates of eye position, the times of onset and offset of sensory stimuli, and the location of each stimulus) were digitized at 200 Hz. Spike counts, instantaneous frequencies, latencies, and other measures of cell activity were obtained for each trial by setting an appropriate time window. Most measures of sensory responses were taken over 200-ms windows synchronized on the stimulus event (onset or offset). For cells with very brief responses to sensory stimuli, shorter windows were sometimes used.

Neurons were first classified according to their responses to stimulus onset and offset during periods when the eyes were maintained near primary position in the absence of a fixation target. Cells were categorized as 'bimodal' if their response in the bimodal condition was significantly different from their response to the most effec-

tive unimodal stimulus with a probability of < 0.05. Bimodal interactions were assessed by calculating the percent of response enhancement or depression by the formula

$$100 \times (CM - SM_{max})/SM_{max}$$

where CM is the response to the combined modality stimulus at the most effective location and SM_{max} is the response to the more effective single modality (i.e. visual alone or auditory alone) at the same location. This measure has been used in previous work, and thus permits direct comparison to earlier studies. It does not produce symmetrical percentage changes for equal increases or decreases in response: there is no upper bound on the percent of response enhancement, but response depression cannot be less than -100% if CM and SM_{max} are positive numbers (e.g. when the measure of response is the number of spikes in a given interval).

Whenever possible, cells were further studied while the cat fixated and made saccades to unimodal and bimodal targets. For neurons exhibiting sustained patterns of activation during fixation, fixation-related discharge was measured over an interval of 200 ms, beginning with the onset of fixation. Because saccade-related activity is phasic, it was measured over shorter intervals which were synchronized on the peak of saccade velocity. In this paper, we will report results measured over an interval of 20 ms. The optimal direction and amplitude of saccades was determined with visual targets, and then auditory and bimodal targets were used to obtain a sample of saccades which were matched, in amplitude and direction, to those evoked by visual targets.

Cats were euthanized with an overdose of sodium pentobarbital and perfused with saline, followed by buffered formalin. Frozen coronal sections were stained with thionin for reconstruction of recording sites marked with electrolytic lesions. The stereotaxic and microdrive coordinates of identified penetrations allowed reconstruction of the locations of unmarked sites.

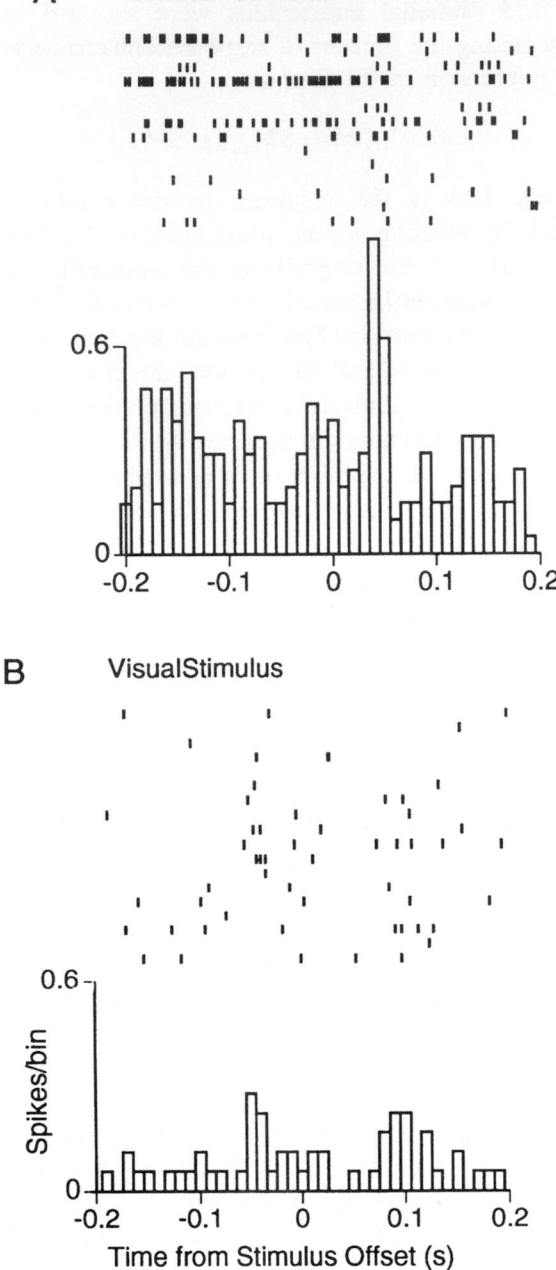

A Bimodal Stimulus

0.6

0

-0.2   -0.1   0   0.1   0.2

B VisualStimulus

0.6

Spikes/bin

0

-0.2   -0.1   0   0.1   0.2

Time from Stimulus Offset (s)

Fig. 1. Bimodal enhancement of neuronal responses. A: Responses to the offset of a bimodal target. Rasters and histograms are aligned on stimulus offset at 0 ms. Each fine vertical bar in the rasters represents one action potential, and each raster represents the response of this cell on one trial. The histograms sum the responses in the complete data set. The vertical scale is mean spikes per 10-ms bin. B: Responses to the offset of the best unimodal stimulus (a visual target with retinal and spatial locations matched to those for the

Results

Responses to the onset and offset of bimodal targets

The data presented in this section were obtained during periods when the cat voluntarily maintained it's eyes near primary position in the absence of a fixation target. In agreement with previous work, bimodal neurons were found only in the deeper layers of the SC and were of two types: some cells exhibited significantly stronger responses to bimodal targets (bimodal enhancement), while others showed significantly weaker responses to the combined modality target (bimodal depression). The responses of a neuron which showed bimodal enhancement are illustrated in Fig. 1 and those of a neuron which exhibited bimodal depression are illustrated in Fig. 2. The neuron whose responses are illustrated in Fig. 1 was best activated by bimodal stimuli (Fig. 1A). Visual and auditory stimuli, presented alone, produced weak responses, as illustrated by the responses depicted in Fig. 1B for visual stimuli (which were more effective than auditory stimuli). Similarly, Fig. 2 illustrates bimodal depression in the responses of another unit. This cell responded less to bimodal stimuli (Fig. 2A) than to either of the unimodal stimuli. Its responses to the stronger of the unimodal stimuli (in this case, auditory) are illustrated in Fig. 2B.

Fig. 3 compares the responses of 4 typical SC neurons to the onset and offset of visual, auditory and bimodal stimuli. Most SC neurons exhibited similar bimodal interactions to stimulus onset and offset. Responses to the onset and offset of sensory stimuli have been analyzed in 73 cells, and intersensory interactions ranged from +1575% to −100%. Fig. 4 shows the response of each neuron to the onset of bimodal stimuli in comparison

data in A). In both A and B, the eye was stationary and centered in the orbit (within ±5° of primary position).

A Bimodal Stimulus

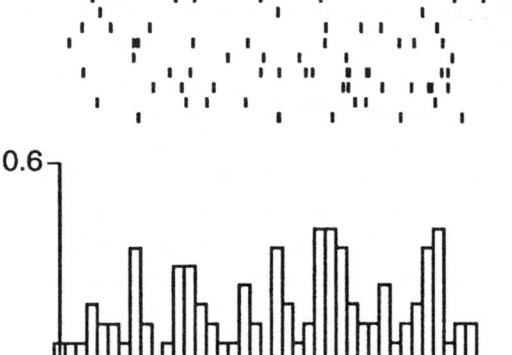

0.6

0

-0.2   -0.1   0   0.1   0.2

B Auditory Stimulus

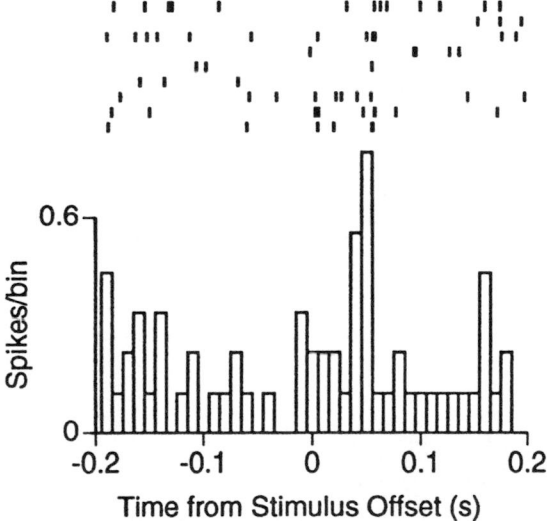

Spikes/bin

0.6

0

-0.2   -0.1   0   0.1   0.2

Time from Stimulus Offset (s)

Fig. 2. Bimodal suppression of neuronal responses. A: Responses to the offset of a bimodal target. B: Responses to the offset of the best unimodal stimulus (an auditory target). Conventions as in Fig. 1.

to their responses to visual and auditory stimuli alone. The diagonal line has a slope of 1, and if there were no difference in the responses to unimodal and bimodal stimuli, the symbols would

be expected to be clustered about the line. The cells which exhibit bimodal enhancement lie above the line, while those which exhibit bimodal depression lie below it. More cells lie above the line than below it, indicating that we encountered somewhat more cells that exhibited response enhancement. Statistically significant differences between the strength of bimodal responses and the strength of the response to the most effective unimodal stimulus were found in 29 of the 73 cells tested (39.7%).

In 40 neurons, bimodal interactions were assessed as a function of eye position. Thirty-nine of these neurons (97.5%) showed the same sign of bimodal interaction (enhancement or depression) at each position of the eye in the orbit. Changes in eye position had a statistically significant effect on the strength of response to bimodal stimuli in only 4/40 neurons (10%). Of 12 neurons which exhibited significant effects shifts in the optimal location of auditory stimuli with changes in eye position and for which bimodal tests were complete, 9 (75%) also exhibited significant bimodal interactions.

Responses during fixation of unimodal and bimodal targets

In addition to their relative transient responses to the onset and offset of sensory stimuli, as illustrated in Figs. 1 and 2, some SC neurons exhibit more sustained responses during fixation of visual targets (Peck, 1989; Munoz and Wurtz, 1993). We identified putative 'fixation neurons' in the superior colliculus by their tonic discharge patterns during sustained fixation of visual targets. In 57 neurons, we obtained measurements of responses during fixation of visual, auditory and bimodal stimuli as well as to the onset and offset of these stimuli. As shown in Fig. 5A, the extent of bimodal interaction during fixation was not well correlated with the extent of bimodal interaction to stimulus onset. The diagonal line has a slope of 1, and if there were no difference in the extent of bimodal interaction in the two tasks, the symbols would be expected to be tightly clustered about the line. Instead, there is considerable scatter and the overall correlation between the two measures ($r = 0.11$) is not significant.

In order to determine whether the activity seen during fixation was dependent on the presence of the stimulus, we assessed the discharge of some fixation neurons during brief interruptions (blinks) of the fixation stimulus. Cell responses during the blink varied: some cells virtually ceased discharging, while others continued to discharge at appreciable rates. To compare the discharge of these neurons with the fixation stimulus present and absent, we calculated their firing rates during maintained fixation of a visual stimulus and during the last 100 ms of the blink, thereby avoiding changes in activity related to the offset of the fixation stimulus. Approximately half (14/30) of

the neurons tested in the blink paradigm continued to discharge at a rate of at least 10 spikes/s during the period of the blink and are therefore regarded as more strongly related to active fixation than to the presence of the fixation stimulus (Munoz and Wurtz, 1993).

We analyzed bimodal interactions in 12 neurons which maintained a discharge rate above 10 spikes/s during interruptions of the fixation target. In these neurons, we expected that the modality of the fixation stimulus would have little effect, and that the extent of bimodal interactions would be reduced. Because the rate at which these neurons discharged during fixation was relatively independent of the presence of the fixation stimulus, we expected to find an extremely weak relationship between the extent of bimodal inter-

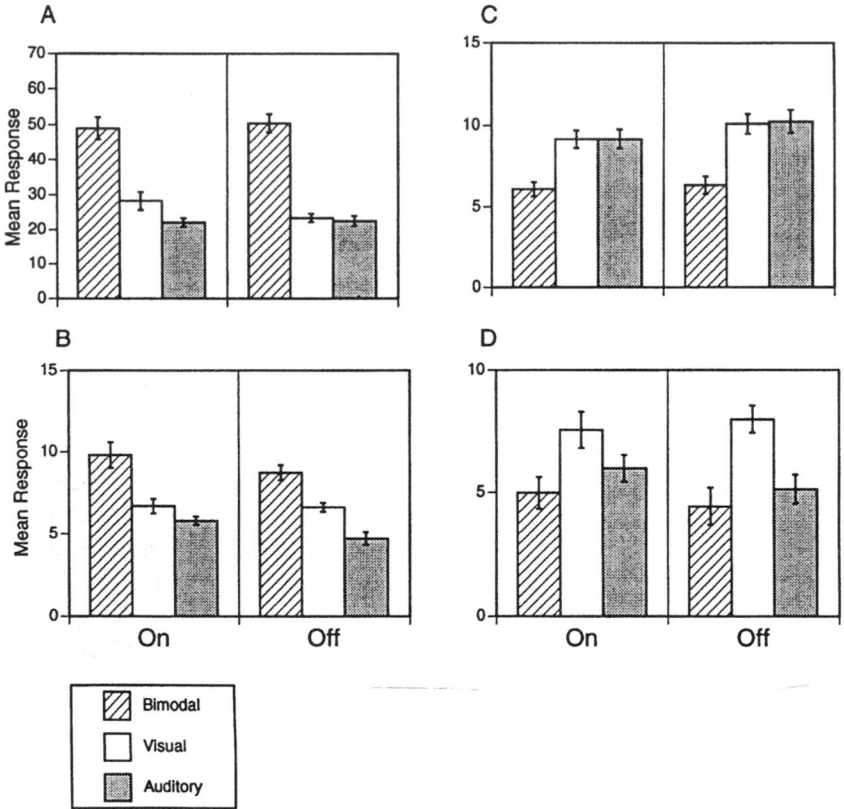

Fig. 3. Mean number of spikes evoked by stimulus onset and offset, as a function of target modality (visual only, auditory only or bimodal) for four SC neurons. A and B: bimodal enhancement of neuronal discharge for stimulus onset and offset. C and D : bimodal depression of neuronal discharge for stimulus onset and offset. Vertical axis is the average number of spikes in the 200-ms interval following stimulus onset or offset, respectively. Error bars indicate ±1 standard error of the mean.

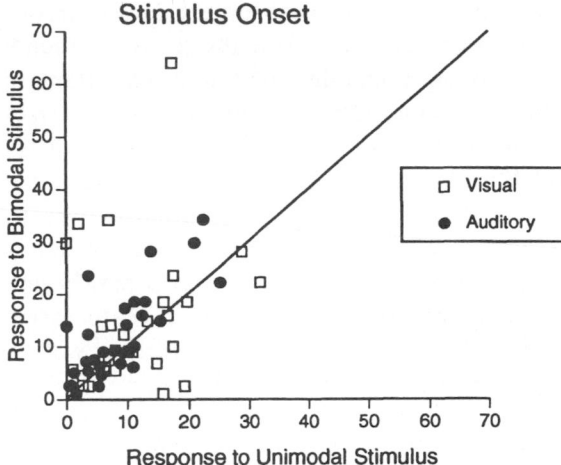

Fig. 4. Responses to unimodal stimuli plotted against response to a bimodal stimulus at the same location in the receptive field. Each data point represents a single neuron. The response is the mean number of spikes in the 200-ms period following the onset of the stimulus. The diagonal line represents equal activation in the two conditions. Every point above the line represents a cell with a greater response to the bimodal stimulus than to the unimodal stimulus, and every point below the line represents a cell with a weaker response to the unimodal stimulus than to the bimodal stimulus. Open squares represent cells for which the best unimodal stimulus was visual. Filled circles represent cells for which the best unimodal stimulus was auditory.

action to stimulus onset and during active fixation. Indeed, as shown in Fig. 5B, there was little correlation between the two measures of bimodal interaction ($r = 0.09$).

Saccade-related responses

We plotted bimodal interactions for saccade-related neurons by examining their discharge in the 20-ms interval preceding the peak saccade velocity when visual, auditory and bimodal targets were used to elicit saccades. Because saccades to auditory targets may have lower peak velocity than saccades to visual targets, and because the level of tectal activity may be related to saccade velocity, for these analyses we eliminated data from cells which exhibited a significant difference in the slopes of the amplitude-velocity functions for auditory and visual saccades. Fig. 6 plots the

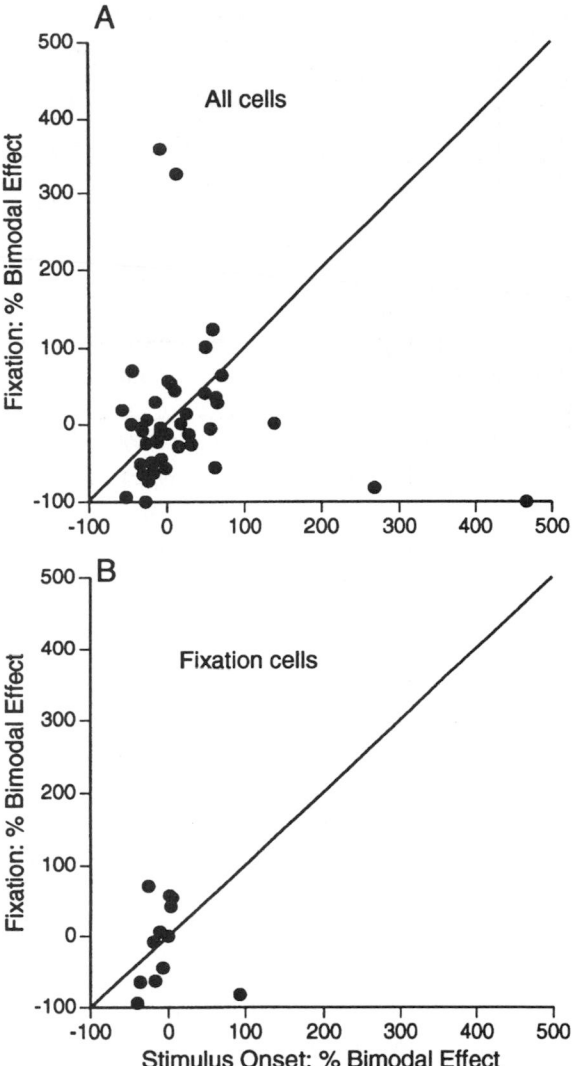

Fig. 5. Bimodal index measured during fixation plotted against bimodal index measured from the response to stimulus onset. Each data point represents a single neuron. The bimodal index is $100 \times (CM - SM_{max})/SM_{max}$, where CM is the response to the combined modality stimulus and SM_{max} is the response to the more effective single modality stimulus (visual or auditory). A: Responses of all cells studied during fixation. B: Responses of 'fixation' neurons, that is, of neurons which maintained a sustained discharge rate of at least 10 sp/s when the fixation target was briefly extinguished. The diagonal line indicates equal effects of bimodal stimulation in the two conditions.

saccade-related responses of 18 SC saccade-related neurons when saccades were made to bi-

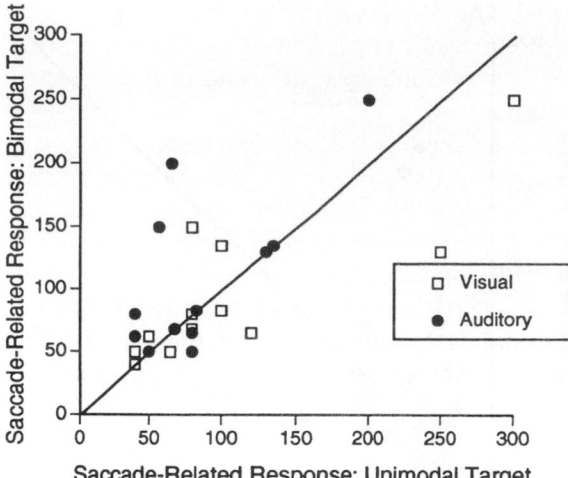

Fig. 6. Saccade-related responses when the saccade target was unimodal plotted against saccade-related responses when the saccade target was bimodal. The saccade-related response is the mean number of spikes in the 20-ms period preceding the peak velocity of the saccade. Other conventions as in Fig. 4.

modal targets in comparison to their responses when saccades were made to visual and auditory targets. The diagonal line has a slope of 1, and if there were no difference in the saccade-related responses as a function of target modality, the symbols would be expected to be clustered about the line. Cells which exhibit bimodal enhancement lie above the line, while those which exhibit bimodal depression lie below it. For most neurons, the magnitude of saccade-related responses was only slightly affected when a bimodal target was used to elicit a saccade. Moreover, among neurons which exhibited significant multisensory interactions, the extent of interaction was less than was found for the onset and offset of sensory stimuli: the largest enhancement was 87.5%, and the greatest depression was − 48.0%. Finally, the extent of bimodal interaction in saccade-related activity was poorly correlated with the extent of bimodal interaction in the response to stimulus onset ($r = -0.29$, not illustrated).

Discussion

Most of our knowledge of neuronal responses relevant to the control of orienting behavior has

been obtained using visual targets. However, orienting movements are normally made to stimuli from various modalities, and it is important to know if the same rules are obtained across target modalities. In addition, the visual and auditory modalities often provide complementary information about the presence and location of targets, and when visual information is reduced or defective, both perception and movements may be improved by use of other sensory cues to target location. Thus, there may be practical as well as theoretical benefits to increasing our increased knowledge of the neural networks involved in integrating auditory and visual information.

The onset and offset of spatially-coincident bimodal stimuli are known to produce enhancement or depression in different neurons. Depression is more likely when one of the stimuli is not aligned with the excitatory center of it's receptive field. The present study addressed only the situation in which visual and auditory stimuli are presented simultaneously at the same spatial location. For that condition, the basic patterns of our results confirm previous findings in both paralyzed/anesthetized and alert animals (e.g. Meredith and Stein, 1983; Peck, 1987; Meredith et al., 1993; Stein and Meredith, 1993; Frens and van Opstal, 1995; Peck et al., 1995; van Opstal and Frens, this volume). In many cells, the simultaneous onset and offset of spatially-coincident bimodal stimuli produced relatively modest enhancement or depression of neuronal discharge relative to that produced by the more effective single unimodal stimulus. About one-fourth of our cells showed interactions which were too large to be the result of simple additive effects of their unimodal inputs. Previous work has shown that multisensory interactions are most likely to be 'supra-additive' when low intensity stimuli are used and that the degree of bimodal interaction decreases as stimulus intensity increases (Stein and Meredith, 1993). In order to maintain a relatively consistent state of performance on the behavioral tasks, we used moderately intense sensory stimuli, and this is probably one of the reasons why a relatively small percentage of cells showed bimodal responses which were greater than the sum of their unimodal responses.

Almost all SC neurons whose axons project through the tecto-reticulo-spinal tract are multi-sensory, and it has been suggested that such neurons may form an integrated multisensory network which provides access to the neuronal circuits mediating orienting behavior (Meredith et al., 1992). When we controlled for the effects of combined visual/auditory stimulation on saccade velocity, we found that bimodal targets could enhance or depress premotor discharge in some SC neurons, as shown in Fig. 6, although the extent of the interactions was modest in the sample of cells studied to date. The fact that multisensory stimuli also influence the probability of overt orienting responses (Stein and Meredith, 1993; Hughes et al., 1994; Baro et al., 1995) indicates that these multisensory interactions can be significant for behavior. However, under some conditions, multisensory stimuli are likely to be processed independently and, thus, it should not be surprising to find that both behavioral and neurophysiological measures of multisensory interaction vary across tasks (Baro et al., 1995) as well as across stimulus conditions (Stein and Meredith, 1993).

How can visual and auditory stimuli remain integrated within SC when the eyes move in the orbit? Three alternative interpretations of the effects of eye position on responses to bimodal stimuli can be advanced. Within the population of light- and sound-sensitive neurons, changes in eye position might modulate response strength, without changing the size or location of the population, as illustrated in Fig. 7B1. Alternatively, the size of the responsive population might change, as illustrated in Fig. 7B2, or the location of activity might shift across the collicular map, as illustrated in Fig. 7B3 (Jay and Sparks, 1987). Because some sound-sensitive neurons in the SC do not show such shifts (Harris et al., 1980; Nelson et al., 1986; Jay and Sparks, 1987; Peck et al., 1995), and because these shifts are not necessarily equal to the change in eye position, the extent to which visual and auditory receptive fields remain in register appears to vary considerably across the

population. For some cells, then, changes in eye position could allow spatially-aligned targets to fall into the inhibitory surround of one receptive field. For such cells, changes in eye position would reduce bimodal enhancement or change it to bimodal depression. Our results indicate that such changes are very rare. However, if the amplitude of the response or the size of the population of responsive neurons varied systematically with changes in eye position, as postulated in Fig. 7B1 and 7B2, multisensory integration would be possible over a large extent of the collicular map

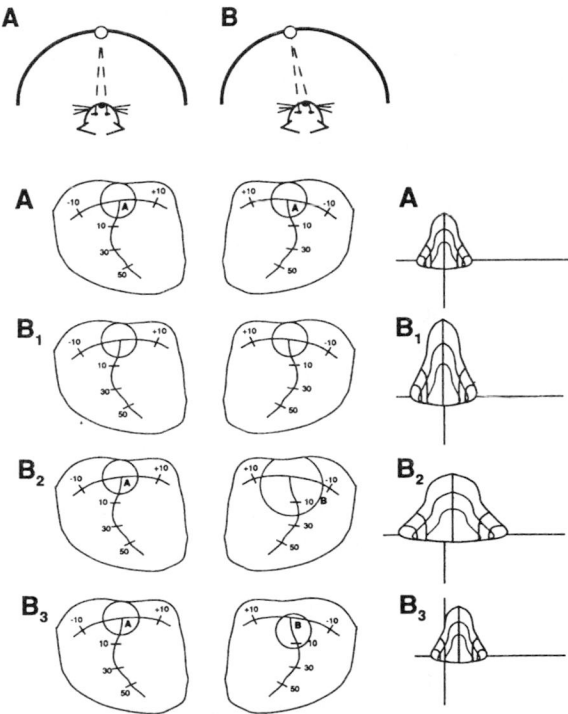

Fig. 7. Proposed changes in the activity patterns of SC neurons as a function of eye position. Neuronal activity is represented schematically (on the far right) as a hill protruding out of the map of the right SC. Schematic maps of the visual field on left and right SC are shown to the left. A: when a cat fixates a target at 0, a moderate level of activity is produced in a population of neurons in both SC. B: when a cat shifts fixation to a target at 10 to the left, the size of the population could remain fixed but the level of activity could increase, as shown in B1. Alternatively, the level of activity could remain fixed but the size of the population could increase, as shown in B2, or the population could shift across the representation of the visual field within SC, as shown in B3.

and relatively independent of changes in eye position. Determining which of these alternatives is true will require simultaneous recordings from several bimodal neurons.

In a number of areas of the brain, visual responses of individual neurons appear to be modulated by the position of the eye in the orbit and may form a distributed representation of the spatial location of objects. Within certain subdivisions of the posterior parietal cortex (particularly areas 7A and LIP), the majority of visually-responsive neurons have been shown to have 'gain fields' in which the strength of the visual response is modulated by eye position (Andersen, 1995). There are also recent reports that the visual responses of neurons in the primary visual pathways may be gated by the position of the eyes in the orbit (e.g. Weyand and Malpeli, 1993), suggesting that many levels of the brain may participate in integrating visual and eye position signals. In addition, there are reports of space-fixed receptive fields in premotor cortical areas. In some neurons these area's visual receptive fields shift with movements of the body in order to code the same location in space, regardless of the retinal location of the stimulus. Thus, there appear to be different representations of space in different parts of the brain. In the cat's SC, there does not seem to be a population of visually-responsive cells with receptive fields in head, body or spatial coordinates. However, in a substantial minority of visually-responsive tectal neurons, eye position does modulate amplitude of the visual response (Peck et al., 1980; Peck et al., 1995). Such modulation has been found for neurons in the superficial layers as well as for neurons in the intermediate and deep layers, suggesting that there is not a simple anatomical segregation of eye position signals in the layers where movement-related activity is encountered.

Despite considerable discussion about whether orienting eye movements are organized in retinocentric, head, body or spatial coordinates (Andersen, 1995; Colby et al., 1995), no clear consensus has emerged. Although it is generally assumed that tectal output neurons always code in retinocentric coordinates, it is also known that the amplitude and direction of saccades evoked by electrical stimulation of the intermediate and deep layers of the SC vary with eye position (in cat: Roucoux and Crommelinck, 1976; McIlwain, 1986; in monkey: Segraves and Goldberg, 1992). Moreover, the size and direction of the effects of initial eye position depend on the location of the site of stimulation in SC. The horizontal components of saccades are least sensitive to initial eye position when eye movements are evoked from rostral SC, near the representation of the vertical meridian, and the vertical components of saccades are sensitive to initial eye position when evoked from sites near the representation of the horizontal meridian (McIlwain, 1988). Modulation of visual responses by eye position is a necessary step in transforming retinocentric coordinates to head-centered, body-centered, or spatial coordinates. Although spatial location may be explicitly represented at the single cell level in cortical premotor areas, examples of distributed coding are found in other cortical and subcortical visual areas. It is possible that the tectum is another level at which a distributed code exists for the location of targets in head- or body-centered coordinates.

Acknowledgements

This work was supported by the U.S. Public Health Service (NINDS grant NS-21238) and by the University of Missouri Research Board. I would like to thank Diann Prescott for animal care and assistance with surgery, John Baro for developing the computer programs used to analyze these data, and Stephanie Warder for assistance with all aspects of this work.

References

Andersen, R.A. (1995) Encoding of intention and spatial location in the posterior parietal cortex. *Cerebral Cortex*, 5: 457–469.
Baro, J.A., Hughes, H.C. and Peck, C.K. (1995) Express saccades in cat: effects of task and target modality. *Exp. Brain Res.*, 103: 209–217.

Colby, C.L., Duhamel, J.-R. and Goldberg, M.E. (1995) Oculo-centric spatial representation in parietal cortex. *Cerebral Cortex*, 5: 470–481.

Frens, M.A. and van Opstal, A.J. (1995) Audio-visual integration in the superior colliculus of the awake monkey. *Soc. Neurosci. Abstr.*, 21: 1194.

Harris, L.R., Blakemore, C. and Donaghy, M. (1980) Integration of visual and auditory space in the mammalian superior colliculus. *Nature*, 288: 56–59.

Hughes, H.C., Reuter-Lorenz, P.A., Nozawa, G. and Fendrich, R. (1994) Visual-auditory interactions in sensorimotor processing: saccades versus manual responses. *J. Exp. Psychol.: Human Percep. Perf.*, 20: 131–153.

Jay, M.F. and Sparks, D.L. (1987) Sensorimotor integration in the primate superior colliculus. II. Coordinates of auditory signals. *J. Neurophysiol.*, 57: 35–55.

Knudsen, E.I. and Brainard, M.S. (1995) Creating a unified representation of visual and auditory space in the brain. *Annu. Rev. Neurosci.*, 18: 19–43.

McIlwain, J.T. (1986) Effects of eye position on saccades evoked electrically from superior colliculus of alert cats. *J. Neurophysiol.*, 55: 97–112.

McIlwain, J.T. (1988) Saccadic eye movements evoked by electrical stimulation of the cat's visual cortex. *Vis. Neurosci.*, 1: 135–143.

Meredith, M.A. and Stein, B.E. (1983) Interactions among converging sensory inputs in the superior colliculus. *Science*, 221: 389–391.

Meredith, M.A., Wallace, M.T. and Stein, B.E. (1992) Visual, auditory and somatosensory convergence in output neurons of the cat superior colliculus: multisensory properties of the tecto-reticulo-spinal projection. *Exp. Brain Res.*, 88: 181–186.

Meredith, M.A., Wallace, M.T. and Stein, B.E. (1993) Integration of multisensory information in superior colliculus neurons in alert cats. *Soc. Neurosci. Abstr.*, 19: 768.

Munoz, D.P. and Wurtz, R.H. (1993) Fixation cells in monkey superior colliculus. I. Characteristics of cell discharge. *J. Neurophysiol.*, 70: 559–575.

Nelson, J.S., Meredith, M.A. and Stein, B.E. (1986) The influence of passive eye rotation on cat superior colliculus neurons. *Soc. Neurosci. Abstr.*, 12: 1538.

Peck, C.K. (1987) Visual-auditory interactions in cat superior colliculus: their role in the control of gaze. *Brain Res.*, 420: 162–166.

Peck, C.K. (1989) Visual responses of neurones in cat superior colliculus in relation to fixation of targets. *J. Physiol. (Lond.)*, 414: 301–315.

Peck, C.K., Schlag-Rey, M. and Schlag, J. (1980) Visuo-oculomotor properties of cells in the superior colliculus of the alert cat. *J. Comp. Neurol.*, 194: 97–116.

Peck, C.K., Baro, J.A. and Warder, S.M. (1995) Effects of eye position on saccadic eye movements and on the neuronal responses to auditory and visual stimuli in cat superior colliculus. *Exp. Brain Res.*, 103: 227–242.

Poppel, E. (1973) Comment on 'visual system's view of acoustic space'. *Nature*, 243: 231.

Roucoux, A. and Crommelinck, M. (1976) Eye movements evoked by superior colliculus stimulation in the alert cat. *Brain Res.*, 106: 349–363.

Segraves, M.A. and Goldberg, M.E. (1992) Properties of eye and head movements evoked by electrical stimulation of the monkey superior colliculus. *The Head-Neck Sensory-Motor System*. New York, Oxford University Press. 292–295.

Sparks, D.L. and Mays, L.E. (1990) Signal transformations required for the generation of saccadic eye movements. *Annu. Rev. Neurosci.*, 13: 309–336.

Stein, B.E. and Meredith, M.A. (1993) *The Merging of the Senses*. Cambridge, MA, MIT Press.

van Opstal, A.J. and Frens, M.A. (1996) Superior colliculus mechanisms in the control of saccades. (this volume)

Weyand, T.G. and Malpeli, J.G. (1993) Responses of neurons in primary visual cortex are modulated by eye position. *J. Neurophysiol.*, 69: 2258–2260.

M. Norita, T. Bando and B. Stein (Eds.)
Progress in Brain Research, Vol 112
© 1996 Elsevier Science BV. All rights reserved.

CHAPTER 13

Task-dependence of saccade-related activity in monkey superior solliculus: implications for models of the saccadic system

A.J. Van Opstal[1,*] and M.A. Frens[1,2]

[1]*University of Nijmegen, Dept. of Medical Physics and Biophysics, Geert Grooteplein 21, NL-6525 EZ Nijmegen, The Netherlands,*
[2]*Neurology Department, University Hospital, CH 8091 Zürich, Switzerland*

Current models assign a crucial role to the deep layers of the Superior Colliculus (SC) in the dynamic feedback control of saccadic eye movements. However, if the SC is to be part of the local feedback loop for saccades, it is expected that the movement-related firing patterns of deep layer SC cell maintain a fixed relation with the instantaneous saccade trajectory, regardless of the conditions that evoked the saccade.

In this paper we provide three different lines of evidence, suggesting that the *movement* activity of SC burst cells may change as a function of the *sensory* conditions evoking the saccade. First, it is shown that bimodal (visual-auditory) stimulation may markedly enhance (up to about 350%) or suppress (on average down to 70%) SC motor bursts when compared to the activity for unimodal visual stimulation. Second, the movement activity associated with auditory-evoked saccades ap-

peared to be reduced by almost 60% relative to visually-evoked saccades of the same metrics (tested in one monkey). However, for both paradigms, these relatively large changes in movement activity went without a concomitant change in the saccade properties. Third, a short-term saccadic adaptation paradigm produced saccades with a smaller amplitude (gain about 0.7) upon presentation of the adapting visual stimulus. However, we found that the movement-related activity of SC burst cells did not change in this paradigm.

These findings suggest that the SC cells do neither encode saccade kinematics, nor the precise components of the saccade vector. Rather, we propose that the motor SC issues a crude desired eye displacement signal to the brainstem that is transformed into the appropriate movement signals by downstream or parallel mechanisms.

Introduction

The midbrain Superior Colliculus (SC) is an important sensorimotor interface for the control of saccadic eye movements. Cells in its intermediate and deep layers produce an intense burst of spikes that is tightly linked to the onset of a saccade. Different types of saccade-related burst neurons (SRBNs) are encountered in these layers: purely

movement-related cells which lack a sensory-evoked response, and sensorimotor cells, in which the perisaccadic activity is preceded by a sensory-evoked burst (see e.g. Sparks, 1986, for an extensive review). Moreover, the SC is involved in multisensory aspects of movement control, since the visual modality converges with auditory and somatosensory signals at the level of single cells in the SC (Meredith and Stein, 1986a; Stein and Meredith, 1993, for review).

The saccade-related movement activity of intermediate and deeper layer SC cells is character-

---

* Corresponding author.

ized by their movement field, which is the restricted range of saccade amplitudes (R) and directions (ϕ) for which a given cell is recruited (Sparks and Mays, 1980). Small saccadic eye movements are represented by cells in the rostral and central SC, whereas the caudal part is involved in the generation of large saccades. Since anatomically nearby cells have overlapping movement fields, the SC cells are arranged such as to form a topographic motor map (Robinson, 1972; see also Fig. 2). In this way, it is thought that the generation of a saccadic eye movement involves the recruitment of a large population of SC cells, having a bell-shaped activity profile the center of which neatly corresponds to the representation of the saccade vector in the motor map (McIlwain, 1982; Ottes et al., 1986; Lee et al., 1988).

Until recently it was generally held that the motor map encodes a fixed saccadic eye displacement vector by the location of the active neural population, rather than by the temporal firing patterns of the neurons. However, there is a controversy in the current literature concerning the collicular mechanisms subserving saccade generation. For example, recent experimental evidence has suggested that the collicular output may also determine the kinematics and trajectory of the saccade (Berthoz et al., 1986; Lee et al., 1988; Van Opstal et al., 1990; Munoz et al., 1991; Waitzman et al., 1991; Wurtz and Optican, 1994). Quantitative models have been proposed that assume different mechanisms to incorporate a dynamic collicular role. The majority of these models now place the SC inside the so-called local feedback loop that is thought to control the saccade trajectory by a continuous comparison of the desired eye displacement signal ('motor error') with the actual eye displacement during the saccade (e.g. Droulez and Berthoz, 1991; Lefèvre and Galiana, 1992; Van Opstal and Kappen, 1993; Arai et al., 1994; Optican, 1995).

Note, that one of the main assumptions underlying these models is the existence of a fixed relation between the spatial-temporal firing patterns in the SC motor map and the properties of the ensuing eye movement. Thus, changes in the firing patterns of these cells (e.g. lowering of the firing rates) would lead to a change in movement parameters (e.g. a smaller saccade, or a lower eye velocity). However, there are some indications that such an assumption may be an oversimplification. For example, saccades towards remembered visual targets are slower, more variable, and endowed with a systematic vertical error, when compared to visually-triggered saccades (Gnadt et al., 1991). Recordings from single neurons in the SC have indicated that the movement field of SRBNs during the execution of remembered saccades is shifted with respect to the movement field associated with visual saccades, by an amount that corresponds to the systematic vertical error. Thus, the activity of collicular SRBNs may, under certain conditions, be dissociated from the actual saccade vector (Stanford and Sparks, 1994).

In an interesting series of experiments by Stein and coworkers it has also been shown that the presentation of multisensory stimuli has a strong and systematic influence on the sensory-evoked responses of cells in the deep layers of the anesthetized cat (e.g. Meredith and Stein, 1986a,b) and guinea pig SC (King and Palmer, 1985; see Stein and Meredith, 1993, for an extensive review). So far, the influence of multisensory stimuli on saccade-related activity has not been investigated in the behaving monkey.

Despite these examples, not much is known on the relation between SC firing patterns, the resulting saccadic eye movement, and the task-related conditions underlying these patterns. Therefore, in this paper we report on the involvement of SRBNs of the monkey SC during different sensorimotor paradigms. We recorded SC activity while eliciting saccades toward visual, auditory, as well as to combined visual-auditory stimuli.

Auditory-evoked saccades have been reported to be different from visually driven saccades in the sense that the former tend to be less accurate and slower (monkey: Jay and Sparks 1990; humans: Frens and Van Opstal 1995a). In the light of the hypotheses described above it is therefore of interest to know whether such differences may be reflected in the saccade-related activity of DLSC neurons. In order to further assess the influence of different sensory conditions on the saccade-related activity, we also recorded from these cells while presenting combined visual-auditory stimuli to the monkey.

Finally, in an attempt to systematically dissociate the coordinates of the visual target from the oculomotor response, we applied the so-called saccadic adaptation paradigm (McLaughlin, 1967; see Methods). By recording from SRBNs during this paradigm, it is possible to investigate the collicular role in both the adaptation process and in the visuomotor transformation underlying saccades.

Our data show that the movement-related activity of collicular SRBNs can be substantially different for these different sensorimotor tasks. Often, however, these differences are not reflected in the oculomotor output. We therefore believe that these findings have implications for models on the role of the SC in saccadic control.

Methods

General methods

We report on the activity of collicular SRBNs obtained from two adult male rhesus monkeys (*Macaca mulatta*; Sa and Pj). In Fig. 2, we have also included the movement field data from four other monkeys (Ce, Ca, Cr and Yu) of the Zürich laboratory, recorded during a spontaneous eye movement paradigm in the light (see also Van Opstal et al., 1995).

Monkey Sa was specifically trained to make saccades towards auditory targets in the dark throughout the two-dimensional oculomotor range. Both monkeys were trained in the three other sensorimotor paradigms described below.

All surgical procedures for implanting the head-stabilization device, the stainless-steel recording chamber and a gold-plated copper eye ring, respectively, were performed under standard asceptic conditions and full anesthesia. The specifics are provided in recent papers from this laboratory (Ottes et al., 1986; Van Opstal et al., 1990).

Eye movements were recorded with the double-magnetic induction technique described by Bour et al. (1984). This technique allows the registration of the horizontal and vertical components of eye position at a precision of $< 1°$ and a temporal resolution of 2 ms. The principal analysis of the eye movement data (i.e. calibration of eye position signals and subsequent saccade detection) has been described elsewhere (Van Opstal et al., 1990, 1995).

Apparatus

Experiments were performed in a sound-isolated room, in which the very dim background illumination was about 10^{-3} cd/m^2. The walls, ceiling and floor of the room were covered with black, sound-absorbing foam that eliminated acoustic echoes above 500 Hz. During the experiment, the monkey sat in a primate chair with its head fixed in the center of the room.

Visual stimuli were delivered by red LEDs (intensity: 0.15 cd/m^2; visual diameter: 0.3 deg), mounted on an acoustically transparent wireframe that constituted the frontal part of a sphere at a distance of 85 cm from the monkey's eyes. A thin black cloth, attatched to this frame, completely blocked vision of the auditory stimulus apparatus.

Auditory stimuli consisted of broad-band noise (150 Hz–20 kHz) at a typical level of 60 dB SPL. The speaker was mounted on a robot arm that possessed two rotatory joints, each equipped with a stepping motor (Frens and Van Opstal, 1995a). Between trials, the acoustic stimulus was rapidly positioned to a randomly selected location on the surface of a virtual sphere just distal from the visual stimuli. The sounds emanating from the stepping motors could not provide the monkey with any cues as to the actual location of the speaker.

Experimental paradigms

During the recording of an isolated SRBN, different experimental paradigms were employed:

1. The visual-movement (VM) scan: The visual and saccade-related response characteristics of the cell were measured, when the monkey made saccades towards unimodal visual stimuli. First, a series of targets was presented in the contralateral hemifield, to estimate the extent of the cell's response field. Then, targets in and near the visual-movement field were presented at a higher spatial resolution (down to 1°). A first estimate of the center of the cell's movement field was then made, on the basis of which the target positions for the other paradigms were planned.

2. The audio-motor (AM) paradigm: In this task, monkey Sa was rewarded for making accurate saccades from the central visual fixation spot towards a peripheral auditory target presented at randomly chosen positions in the contralateral oculomotor field. In this way, the auditory-movement field of the cell was measured in a similar way as in the VM paradigm.

3. The visual-auditory interaction (VA) paradigm: In this task, a visual target was presented at the estimated center of the cell's movement field (see above), in combination with an auditory distractor (white noise). The monkey was rewarded for making an accurate saccade towards the visual target while ignoring the acoustic stimulus. The sound was positioned anywhere on the radial line through the central fixation spot and the visual target. In this way, spatial stimulus disparities from 0 up to 60° were realized.

4. The short-term adapatation (AD) experiment: In the adaptation experiment we induced short-term changes of saccade amplitude. In a typical trial, a visual fixation spot appeared at the straight ahead position. After extinguishing this spot, a visual target appeared at the center of the cell's movement field, e.g. at $\vec{T}_1 = (20,0)$ (in deg). When the monkey made a goal-directed eye movement, the computer changed the position of the target,

e.g. to $\vec{T}_2 = (14,0)$ (in deg), as soon as eye velocity exceeded a preset criterion. Initially, the monkey makes a saccade towards $\vec{T}_1$, followed by a back-step saccade towards T_2. Repeated presentation of a sequence of such trials ($N \sim 400$) yields a gradual decrease of saccade amplitude, in which the initial saccade is more and more directed at T_2.

Data analysis

The movement-related activity of an SRBN for a given saccade was quantified by the average firing rate from 50 ms before saccade onset to saccade offset. We first directly compared the average firing rates for the different paradigms during saccades confined to a narrow amplitude-direction bin near the estimated center of the movement field. Significance of a difference was then assessed by a one-sided t-test.

In our quantitative analysis of cell responses we have also used a model description of collicular movement fields that takes into account the logarithmic nature of the collicular motor map (see Fig. 2, and Ottes et al., 1986, for extensive details). Assuming a gaussian spatial sensitivity in this map for saccades, one obtains an accurate parametrization of a cell's movement field (Ottes et al., 1986; Van Opstal et al., 1990, 1995).

The movement field is described by four parameters: F_o, in [spikes/s], which is a measure of the cell's peak firing rate associated with the optimal saccade vector; $[u_o, v_o]$ (in millimeters), are the coordinates of the center of the cell's Gaussian activation profile in the collicular motor map that correspond to the polar coordinates $[R_o, \Phi_o]$ (in degrees), of the optimal saccade vector, $\vec{S}$ (see Fig. 2A). Finally, σ_o (in millimeters), is the width of the Gaussian activation profile, which relates to the tuning width of the movement field.

With this movement field model it is possible to predict the cell's activity for any saccadic eye displacement vector and to deal with the inherent variability of the cell's response (which is averaged out) as well as the saccadic eye movement vectors (which is taken into account). We applied this description to the saccades of the VM paradigm (see above) in order to enable a quantitative comparison with the activity evoked by the other paradigms. We have also applied the same model to obtain the movement field parameters for auditory-evoked saccades. In this way, one may compare the movement field properties of the cell for saccades evoked by different sensory modalities.

Results

Properties of VM movement fields

In this section, we report on the properties of 50 visual-movement cells from monkeys Sa and Pj, as well as from an additional 55 SRBNs from the other four monkeys (see Methods). These neurons could all be well described by the movement field model (correlations between fit and data from 0.70 to 0.96).

A typical example of the results of the move-

ment field fit is given in Fig. 1 (cell sa6303). Panel 1A shows the predicted vs. the actually measured activity for saccades in and near the cell's movement field. As is illustrated in Fig. 1B, this SRBN is optimally recruited for leftward saccades near $(R_o, \Phi_o) = (21.6, 174)$ (in degrees). The peak activity is estimated at $F_o = 301$ spikes/s, and the width of the movement field is 0.49 mm (see Methods). The latter means, for example, that the activity of the cell exceeds 85% of its peak firing rate when the amplitude is between 17 and 27°, and saccade direction falls between 164 and 184° (inner contour in Fig. 1B). Note the asymmetric shape of the movement field for saccade amplitude variations, which is due to the logarithmic nature of the motor map (see also Fig. 2A).

In Fig. 2A, the movement field centers, $[R_o, \Phi_o]$, of all recorded SRBNs ($N = 105$) have been mapped onto the representation of the motor map in collicular coordinates $[u_o, v_o]$. The other two movement field parameters of the cells, F_o and σ_o, have been plotted against each other in panel 2B. Note that these parameters scatter over

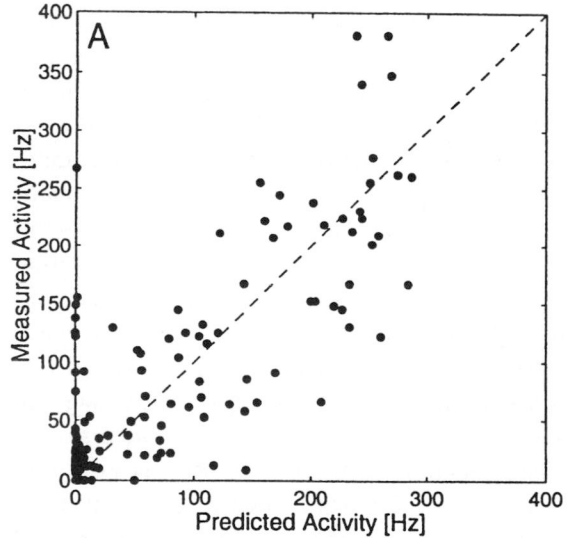
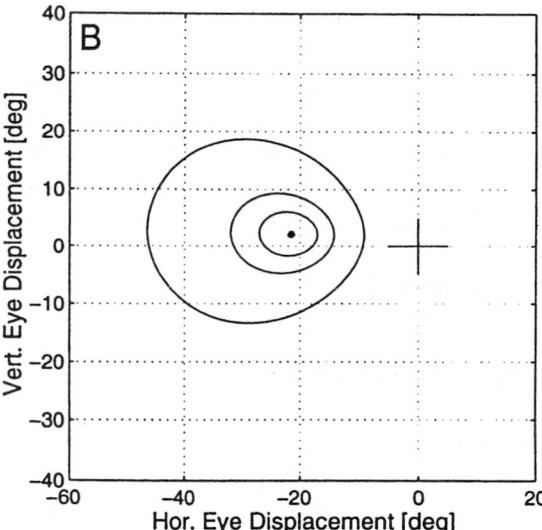

Fig. 1. Movement field fit of a typical SRBN (sa6303). (A) Predicted vs. measured activity. The prediction is based on the movement field model (see Methods). Note good correspondence between fit and data ($r = 0.89$; $N = 150$). (B) The spatial extent of the cell's movement field. The cross corresponds to fixation, contours indicate the range of saccade vectors at three levels of activity: 14% (2 σ_o), 61% (σ_o), and 85% of the fitted peak activity of 310 spikes/s. The dot reflects the cell's optimal saccade: $(R_o, \phi_o) = (21.6, 174)$ (degrees). Note the asymmetric shape of the movement field along the radial dimension.

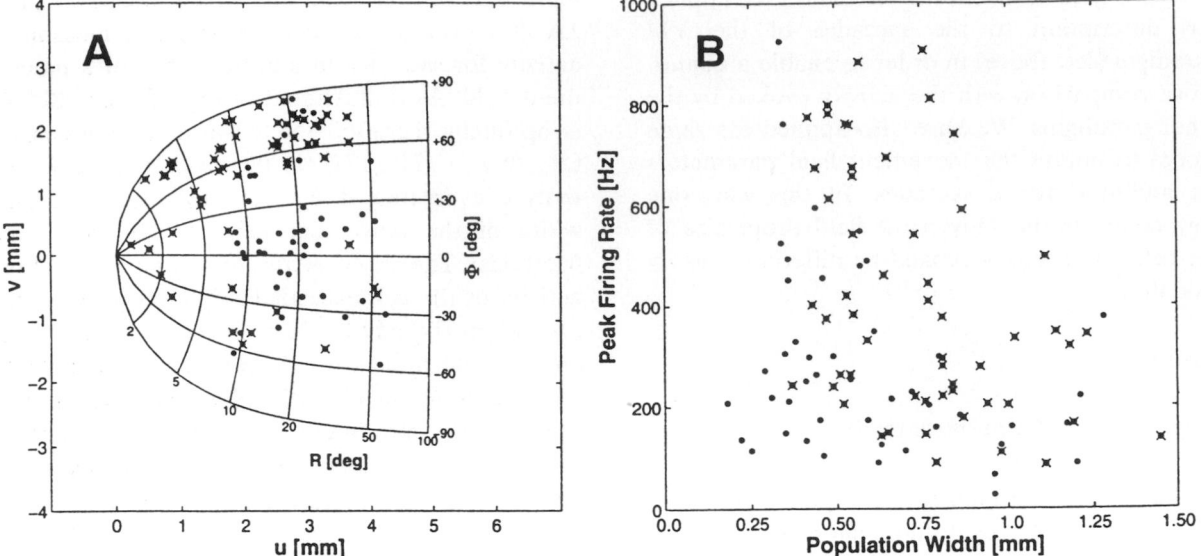

Fig. 2. Results of the visuomotor movement field fits of 105 SRBNs. Crosses: neurons recorded in Zürich during voluntary saccades (see Methods). Dots: neurons of Sa and Pj recorded in the VM paradigm. (A) Optimal saccade vectors (R_o, ϕ_o) are mapped onto the collicular motor map, which is spanned by cartesian coordinates (u,v) (millimeters). Superimposed on the motor map are saccade iso-direction (ϕ) lines and iso-amplitude (R) lines. Note the logarithmic representation of saccade amplitude. (B) Fitted peak firing rate, F_o, against the estimated tuning width, σ_o (see Methods). Note the large parameter ranges. No systematic relation was found between these, and any of the other movement field parameters.

a very wide range, and that they do not crosscorrelate (see Discussion). None of these parameters is related to the location of the cell within the motor map, although we obtained a weak, but significant, positive correlation of the movement field width, σ_o, with recording depth. On average, the width of the movement field increased by about 0.2 mm per mm change in penetration depth (not shown).

Influence of sensory conditions

AM paradigm

Monkey Sa generated goal-directed saccades towards auditory targets in the dark throughout the oculomotor range. Fig. 3 compares the performance of the monkey during a typical recording session (sa4702) in the unimodal visuomotor (filled symbols) and audiomotor (open symbols) tasks. The upper panels (Fig. 3A,B) allow for a comparison of saccade accuracy, the lower panels (Fig. 3C,D) of saccade kinematics. Note, that although the correlation between auditory target azimuth and first-saccade azimuth (Fig. 3A; $r = 0.76$) was about the same as for the sound elevation component (Fig. 3B; $r = 0.74$), the monkey systematically undershot the target in the elevation direction (gain for azimuth close to 1.0, for elevation about 0.5). No such horizontal/vertical dichotomy exists for the visual saccades, for which undershoots typically occurred in the radial (i.e. amplitude) direction, thus equally affecting the horizontal and vertical saccade components.

The kinematic properties of the auditory saccades (Fig. 3C,D) were slightly different from visually driven movements. The peak velocity of auditory saccades saturated at a lower level (853°/s) than the visual saccades (1074°/s), and auditory saccade durations were often slightly longer. Nevertheless, the fastest auditory evoked

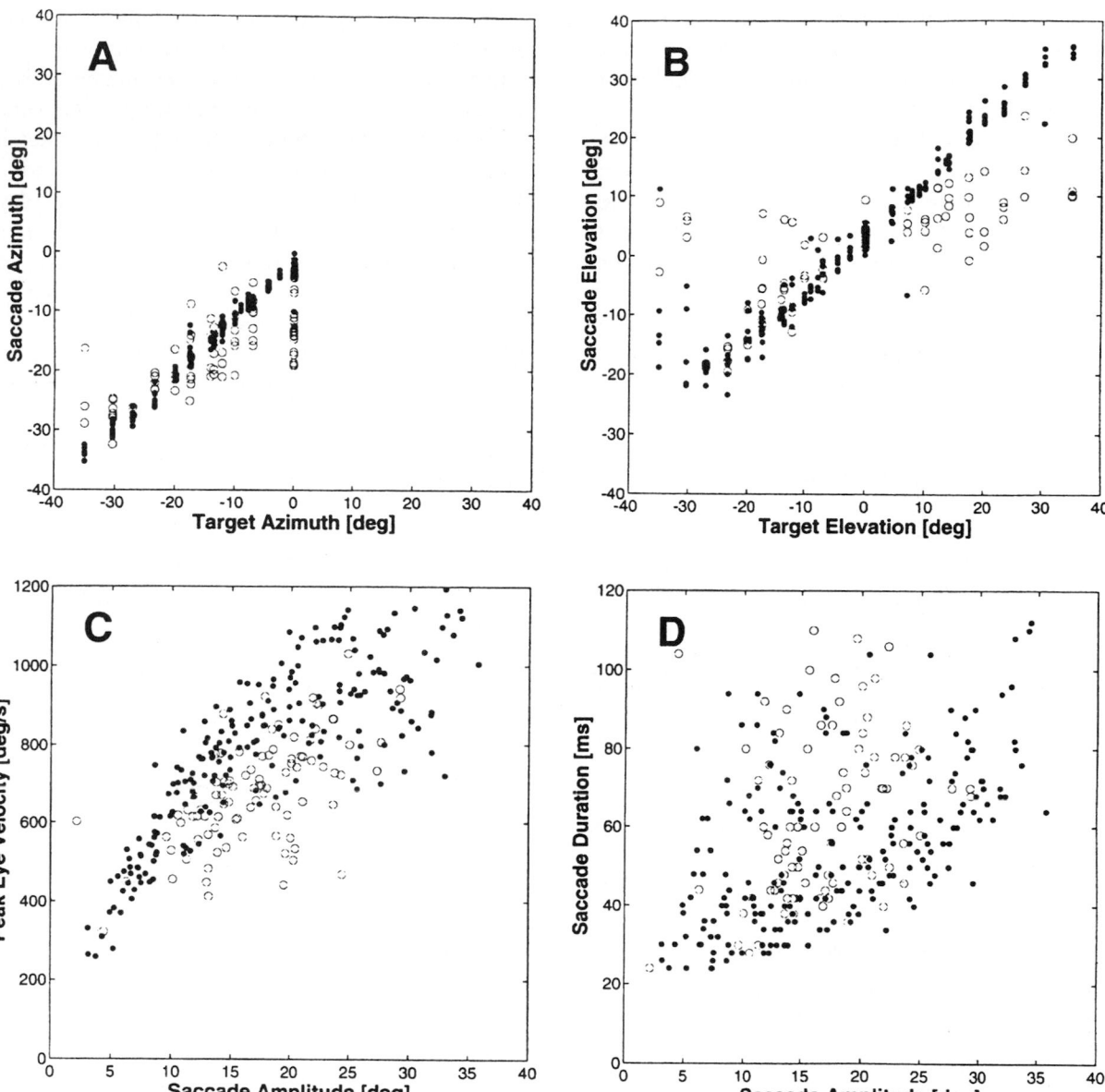

Fig. 3. Properties of saccades evoked by visual (closed symbols) and auditory (open symbols) stimuli. (A) Accuracy of the horizontal saccade components (only the left ocumolotor field was tested in this recording). Although visual accuracy is obviously higher, the gain of the auditory saccades is close to 1.0 ($r = 0.76$). Largest scatter occurs near zero azimuth. (B) Despite some obvious errors, there is also a highly significant correlation for saccade elevation as a function of auditory stimulus elevation ($r = 0.74$), although the gain is substantially lower (about 0.5) than for visually-evoked saccades. (C) Peak velocity as a function of saccade amplitude for visual and auditory saccades. Auditory saccades in darkness are somewhat slower and more variable (saturation levels 853 vs. 1074 °/s), although the fastest auditory saccades are almost indistinguishable from the visually-evoked movements. (D) Saccade duration as a function of amplitude is roughly similar for the two types of saccades.

saccades were almost indistinguishable from their visually-evoked counterparts. Also the latencies of auditory and visual saccades had similar distributions (not shown).

We recorded the auditory-evoked sacade-related activity in 22 SRBNs of monkey Sa. All cells that were endowed with a clear saccade-related burst in the VM task, were also recruited prior to auditory saccades. Two of these cells also had very short-latency (approx. 10 ms) auditory-evoked sensory activity. As a general finding, which occurred in 20/22 cells, the saccade-related activity in the AM task was markedly suppressed when compared to the responses for equal saccades in the VM paradigm. This feature is illustrated in Fig. 4 (left and center columns), which displays the saccade-related activity of cell sa6303 (same neuron as in Fig. 1) during these tasks for se-

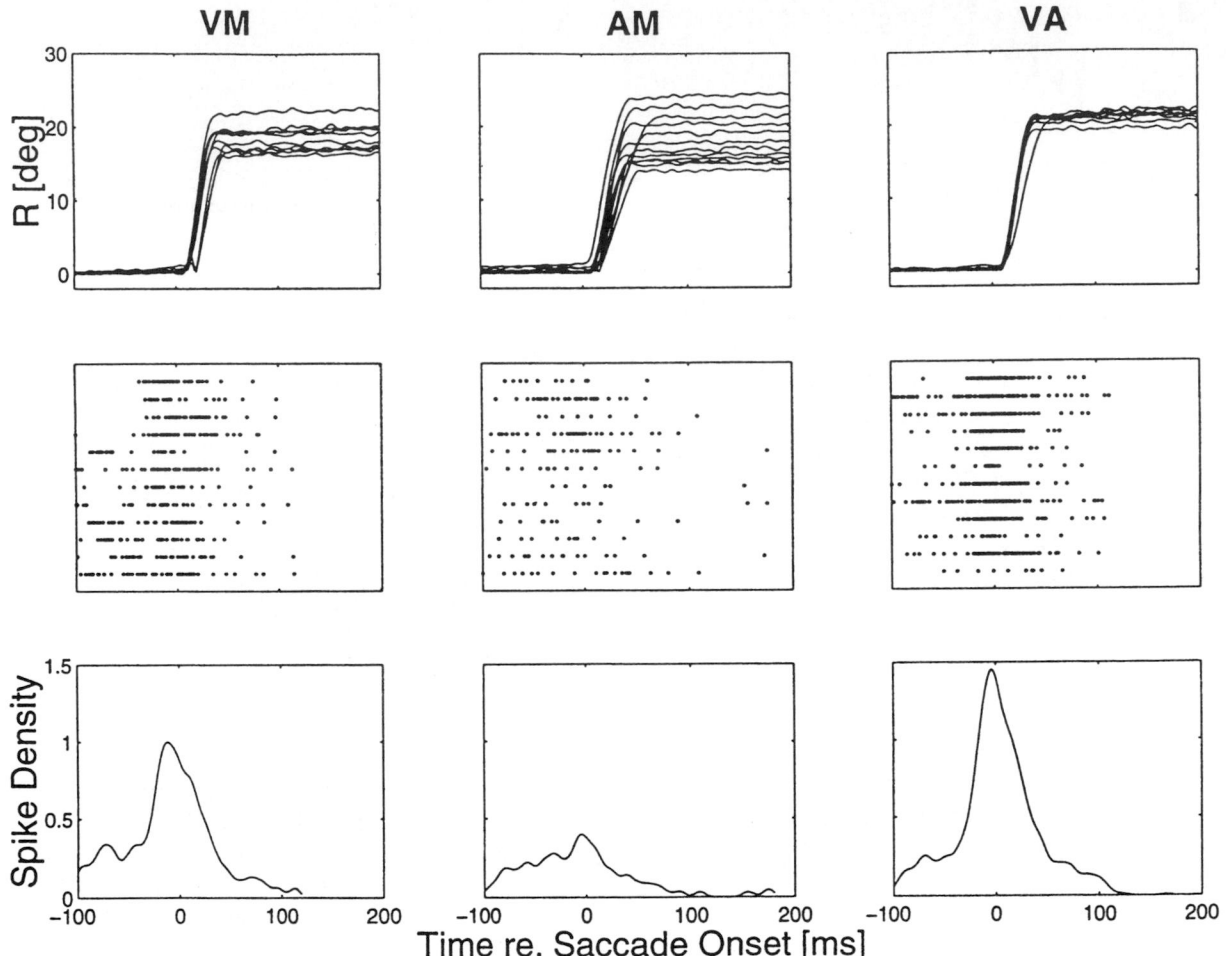

Fig. 4. Activity of the same collicular SRBN as in Fig. 1 (sa6303) during three different sensorimotor paradigms for saccades near the center of its movement field. Left: activity associated with saccades in the VM paradigm. Center: activity during the AM paradigm. Right: activity during the VA paradigm. Top row: saccade traces ($N = 12$; amplitude as a function of time relative to saccade onset). Center row: dot displays of the cell's activity. Each dot represents one action potential. Bottom: Spike density functions. The peaks are normalized with respect to the VM result. Note, that activity is strongly reduced during the AM paradigm (peak of the normalized spike density function about 0.4), although the saccades are all close to the movement field's center. For the VA paradigm, however, the cell's activity is enhanced by almost 50% (peak close to 1.5).

lected saccades with amplitudes between 17 and 23°, and directions between 174 and 184°. Note that these selected saccades all belong to the center of the cell's movement field (compare with Fig. 1B). Notwithstanding, the response during the auditory movements is strongly reduced. The bottom panels, which show the normalized spike density functions of the spike rasters, indicate a reduction in auditory-evoked activity by almost 60%. Note, that this decrease is not due to variations in saccade metrics (all saccades belong to the center of the movement field), nor to a strong reduction in saccade velocity.

In principle, the differences in firing rate could be attributable to a systematic shift of the cell's movement field (see Introduction). In that case, the center panel of Fig. 4 would not show the cell's responses for optimal auditory saccades. In order to investigate this point in detail, we have also applied the movement field model to the auditory-evoked movement activity. A quantitative comparison between the four parameters (see Methods) for 18/22 collicular neurons is made in Fig. 5. From this figure, it can be deduced that the 'visuomovement' field and the 'audiomovement' field have the same tuning width (no systematic discrepancies between the respective σ_o values; Fig. 5B) and that the two movement field centers are at the same position (no systematic shift of neither the horizontal (Fig. 5C), nor the vertical (Fig. 5D) components of the optimal saccade displacement vectors). In contrast, however, the peak of the mean firing rate, F_o, appeared to be substantially lower for the audiomovement fields than for the visuomovement fields (Fig. 5A). The mean ratio between the auditory-related value for F_o and the visual value is 0.4 ± 0.2, indicating, on average, a 60% decrease.

VA paradigm

Interestingly, in the visual-auditory interaction experiments (see Methods), we obtained a substantial influence of the auditory distractor on the movement-related burst in 90% (45/50) of the cells tested (Frens, 1995; Frens and Van Opstal, 1995b). When the visual and auditory stimulus were spatially and temporally aligned, two types of interactions were obtained: response enhancement (26/50) and response suppression (19/50). In enhancement cells, the mean firing rate was on average almost twice the level of the unimodal visual condition, whereas for suppression cells the auditory co-stimulus led to a decrease of the cell's activity by about 30%. It was also observed (not shown here), that the audio-visual interactions were systematically dependent on the spatial configuration of the targets, such that a cell's activity was always highest in the spatial alignment condition (Frens, 1995; Frens and Van Opstal, 1995b).

An example of the influence of the acoustic distractor on the saccade-related burst is given in Fig. 4C, where the same neuron (sa6303) was also tested in the VA paradigm. As can readily be observed from the spike-density plots (bottom panels), activity has increased by about 50% with respect to the default visual-only condition. A quantitative analysis of the cell's movement field shows that this change in activity cannot be attributed to slight differences in the saccade vectors (see above and legend Fig. 1).

Relation with saccade kinematics

The finding that the sensory conditions have a substantial influence on the cell's presaccadic activity (illustrated in Fig. 4), although the saccadic responses were quite similar under these conditions, implies that the instantaneous firing rate of an SRBN may not be uniquely related to the movement trajectory. This is more explicitly illustrated in Fig. 6, where the normalized spike density has been plotted against instantaneous eye motor error (see e.g. Waitzman et al., 1991). Clearly, instead of one unique curve, expected when the neuron's activity would relate to the saccade kinematics, the three sensory conditions lead to markedly different relations. This result applies to all neurons for which a significant (multi-)sensory influence was observed.

Dissociation of SRBN activity from saccade vector

As was explained in the Introduction and Methods, the short-term saccadic adaptation paradigm is a straightforward method for dissociating the coordinates of the visual stimulus from the coor-

dinates of the targeting eye movement in a well-controlled way. We have applied the AD paradigm while recording from 30 additional SRBNs in monkeys Sa and Pj. Fig. 7 shows the effect of this procedure on the saccade-related burst of one of these neurons (pj6001). The left-hand column

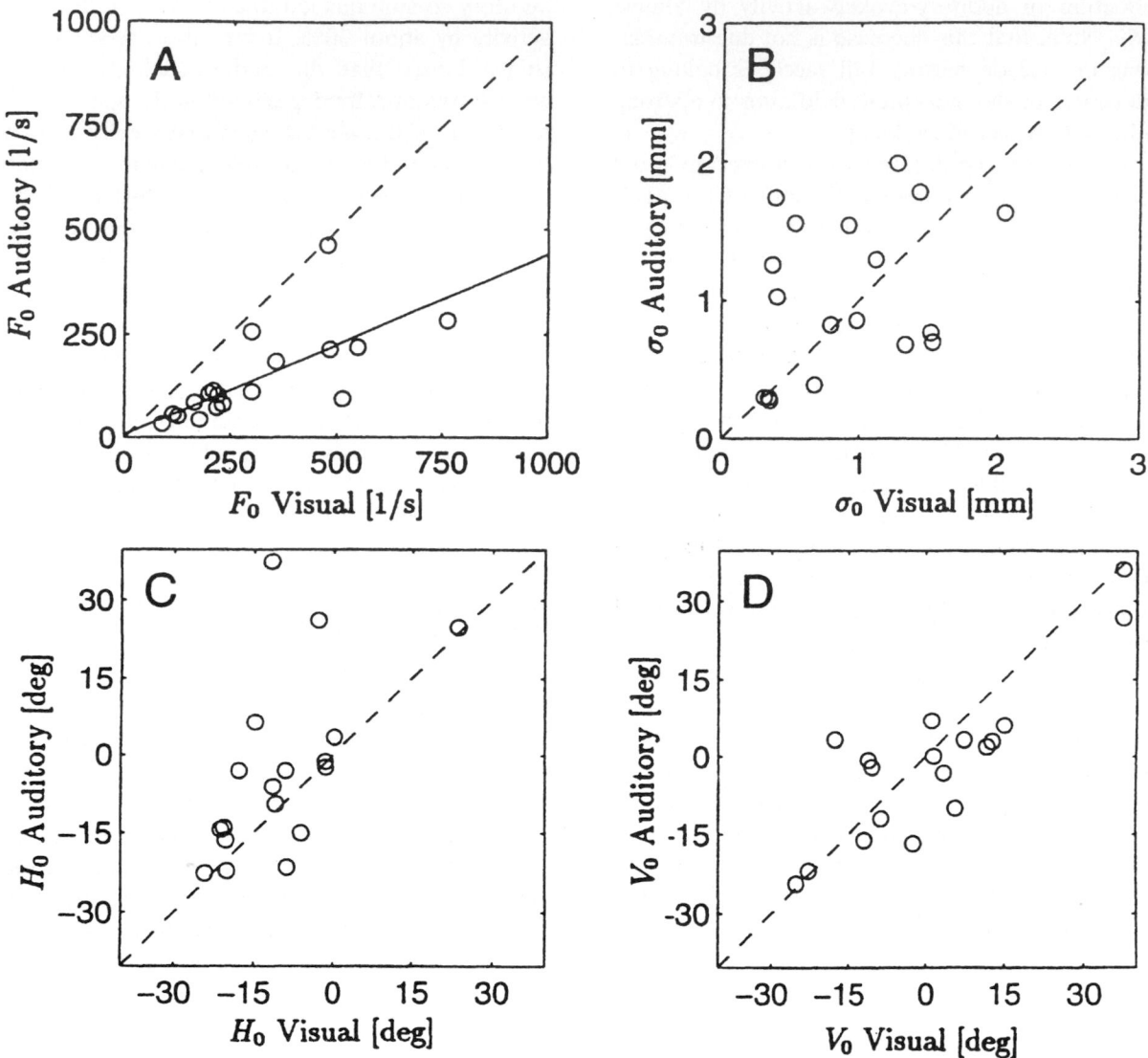

Fig. 5. Result of the movement field fits for 18/22 cells recorded from during visual (ordinate) and auditory (abscissa) evoked saccades. The fit parameters of the remaining four cells were unreliable and have not been included in this figure. However, their results qualitatively agree with the main finding presented here. (A) Fitted peak activity of the movement fields indicates that, with the exception of two cells, the auditory peak is reduced by almost 60% with respect to the visually-evoked peak activity. (B) The width of the auditory and visual movement fields is similar since no systematic discrepancy did arise. (C,D) The centers of the movement fields corresponded quite well. No systematic shift of neither the horizontal nor the vertical component was obtained.

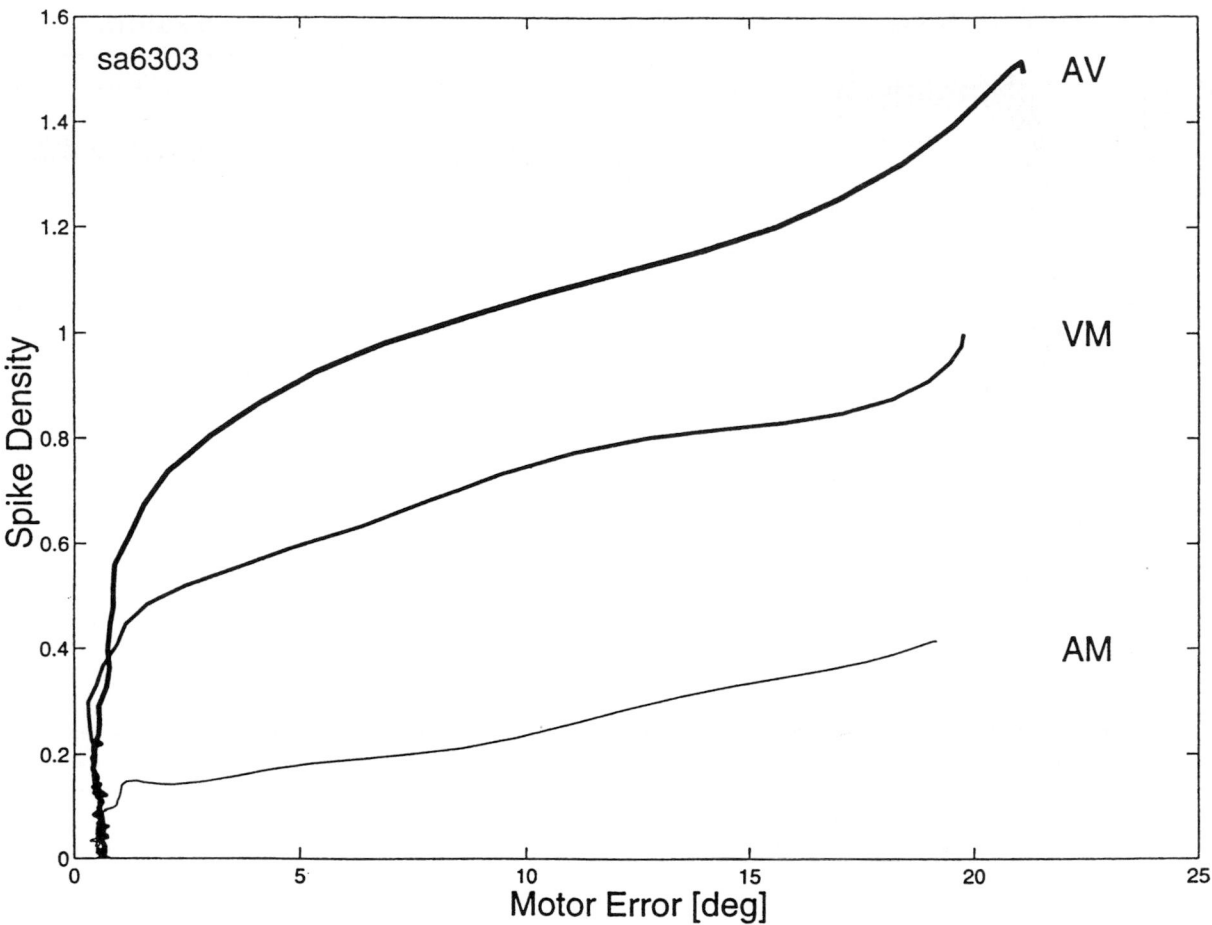

Fig. 6. Instantaneous normalized spike density (re. peak of the VM condition) as a function of the mean dynamic radial motor error (averaged over 12 similar saccades, and shifted in time with respect to the spike density by 6 ms; see Waitzman et al., 1991) for cell sa6303 during the three different sensorimotor paradigms. This cell could be characterized as a so-called 'clipped cell', since a monotonic relation emerged for the VM paradigm. However, the traces for the other two paradigms are very dissimilar, indicating that the neuron's activity does not uniquely signal the instantaneous motor error of the eyes (see also Fig. 4).

shows the strong recruitment of the cell for saccades towards a target near the center of its movement field ($\vec{T}_1$ = [27,60] deg) during the pre-adaptation phase. In the center column, the cell's activity for saccades towards a target near the edge of the movement field, at $\vec{T}_2$ = [20,60] deg, is shown. Then, about 400 trials of adaptation ensured that, upon presentation of the target at $\vec{T}_1$, the monkey responded with a single saccade towards $\vec{T}_2$. Interestingly, as is shown in the right-hand column, the activity associated with these smaller saccades is now almost indistinguishable from the activity for the saccades pre-

sented in the left-hand column. Therefore, the movement field of the cell has substantially changed as a result of the adaptation process: its new center now lies approximately at T_2. However, the cell's sensitivity for the initial motor error, T_1, has not changed. Comparable results were obtained for the other SRBNs.

Discussion

Summary and conclusions

Our main findings are the following: First, in line

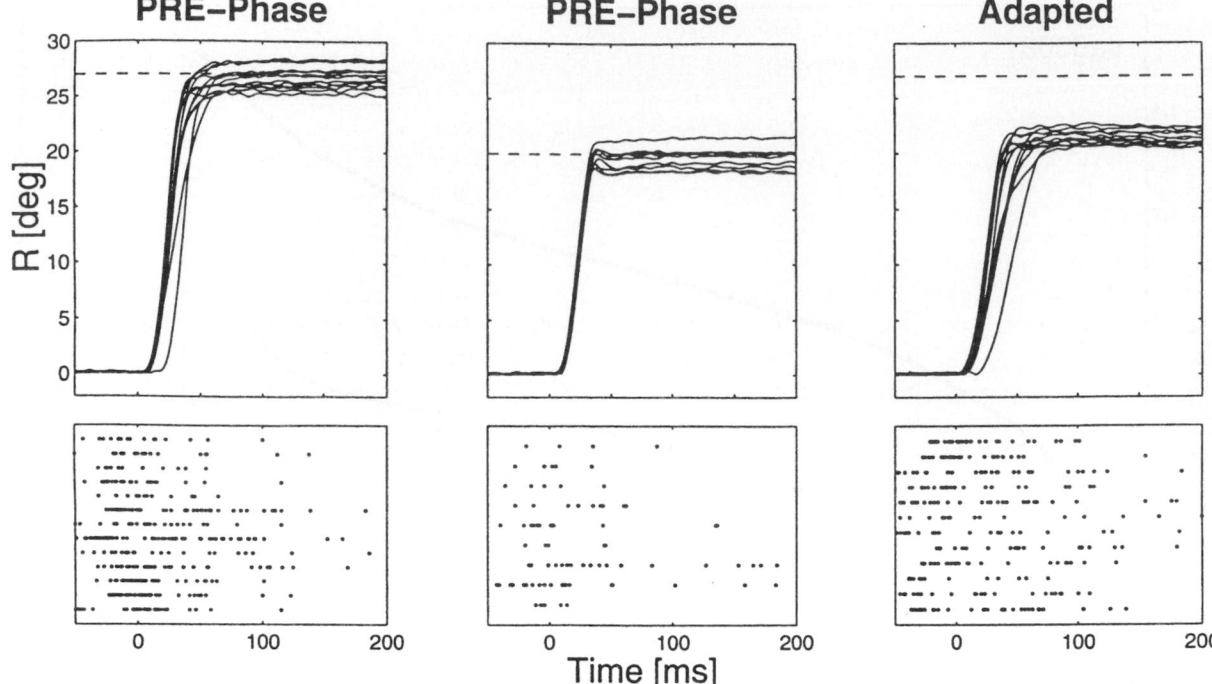

Fig. 7. Activity of neuron pj6001 before (left-hand and center columns) and after (right-hand column) inducing short term adaptation which reduced the saccade amplitude from 27° (the pre-adaptation center of the movement field) to 20° upon presentation of a target at 27° (see Methods). Target position is indicated by a dashed line. The center column shows that, before the adaptation, the cell is only marginally involved in saccade responses of 20°, evoked by a target at 20°. However, when the same saccades of 20° result from a target presented at 27°, the cell is now vigorously activated. Thus, after the adaptation, the center of the cell's movement field has shifted from 27 to 20° eccentricity.

with earlier studies, the movement-related activity of deep-layer SC neurons is well-described by the Ottes et al. (1986) model. We observed, however, that the cell-to-cell variation of the movement field parameters (F_o, σ_o) covers a full order of magnitude.

Second, we found (tested in one monkey only) that the movement fields of collicular SRBNs for visual and auditory-evoked saccades are identical, both in location and in tuning width. However, the peak firing rate associated with auditory saccades was strongly reduced, whithout a concomitant large change in saccade properties (neither latency, nor kinematics).

Third, we obtained a strong inluence of an auditory distractor on the movement-related activity of SC neurons during visually-evoked sac-

cades. Again, no change in saccade properties was observed. However, saccade latency could be reduced by almost 50 ms, provided that both stimuli were spatially aligned (see Frens, 1995; Frens and Van Opstal, 1995b).

Finally, in the short-term adaptation paradigm the visuomovement field systematically shifted with respect to the pre-adaptation condition, such as to remain aligned with the retinocentric motor error vector during the adaptation process in all cells tested.

In what follows, we will further comment on the various issues raised by these findings.

Movement field properties

The movement field data presented in Fig. 2 indicate that the population of SRBNs is not well described by a smooth, bell-shaped, mountain of activity. Since the tuning width of a particular cell may be very large (> 1 mm, see Fig. 2B), the region of recruited cells during any saccade is much larger than is generally assumed in quantitative models of the SC (e.g. Van Gisbergen et al., 1987; Van Opstal and Kappen 1993; Arai et al., 1994). Because the movement field parameters are uncorrelated, the size and shape of the collicular population cannot be readily estimated on the basis of a single-unit recorded during many saccades. If the population of SRBNs transmits information on the coordinates of the motor error vector (see above), the irregularly-peaked activity profile has to be weighted appropriately. Thus, a simple weighting scheme that relates the position of a cell in the motor map to its efferent connection strength, as proposed in most models (e.g. Van Gisbergen et al., 1987; Van Opstal and Kappen, 1993; Arai et al., 1994), may not provide the neural code needed by the brainstem (see also Van Opstal and Hepp, 1995).

Auditory saccades

The accuracy of auditory evoked saccades was found to be inferior to visually-driven eye movements. Specifically, the elevation of the auditory target was localized less precisely than the azimuth component (Fig. 3A,B). This feature is consistent with earlier reports about monkeys (Jay and Sparks, 1990) and humans (Frens and Van Opstal, 1995a), and illustrates nicely the different monaural (i.e. spectral) and binaural (intensity and timing differences) stages underlying primate auditory localization (reviewed in Blauert, 1983).

All SRBNs in this study that were recruited for saccades towards visual targets were also active in the audiomotor task. Although the movement-related activity for each neuron was considerably reduced for an acoustically driven saccade (by almost 60%; Figs. 4,5A), neither the center of the movement fields nor their tuning width to saccade vectors were systematically different for visual or auditory conditions (Figs. 5B–D). Based on these findings we conclude that the population of recruited cells was identical for both sensorimotor

conditions. This conclusion does not seem to support an earlier suggestion that fewer collicular SRBNs are involved in the generation of auditory than of visual saccades (Jay and Sparks, 1987). This proposal, however, was mainly based on their finding that specifically visuo-movement cells with small optimal saccade vectors failed to be activated for auditory saccades. In our data base, the range of optimal saccade amplitudes for the 22 cells on which auditory tests were performed, extended from 9.5° onwards.

It is quite remarkable, that the observed reduction of SRBN activity during auditory saccades had a relatively small effect on the kinematic properties. Large auditory evoked eye movements had peak velocities typically in the order of 85% of matched visual saccades (Fig.4C), whereas no systematic difference was obtained for smaller movements ($R < 15°$).

Multisensory integration

For the far majority (90%) of SRBNs tested in the VA paradigm, the visuomotor activity was significantly modulated by the auditory distractor (e.g. Fig. 4C). Some cells underwent a response enhancement (up to 350%) of the unimodal visuomovement activity; others were suppressed by the acoustic costimulus. We observed that the strength of the bimodal interaction depended systematically on the spatial configuration of the stimuli (not shown here, but see Frens, 1995; Frens and Van Opstal, 1995b).

So far, most experiments studying multimodal integration in the DLSC have been carried out with anesthetized preparations (for an excellent review, see Stein and Meredith, 1993). In these studies, similar multimodal interactions (enhancement/suppression; dependence on stimulus configuration) were observed in the sensory-evoked activity of many DLSC neurons.

Our finding, that some cells increase their activity (ME-cells) and others decrease their firing rate (MS-cells), may result from a change in balance of different sensory inputs to a particular neuron. The results indicate that these sensory influences may have an opposite sign (in the case of suppression), and that they are spatially weighted.

Saccadic adaptation

The result that collicular movement fields systematically change during saccadic adaptation provides strong evidence that the signal responsible for the change in saccade amplitude exerts its influence at a site downstream from (or parallel to) the collicular SRBNs. This conclusion is further supported by the finding (not shown here, but qualitatively evident from Fig. 7) that also the velocity of the adapted saccades is significantly reduced when compared to matched saccades from the pre-adaptation condition. It is important to note that despite these vectorial and kinematic changes, the activity of the SRBNs remained unaffected when expressed in motor error coordinates (Fig. 7). These data therefore suggest that the output of the SC may be a *desired* motor command, rather than an *actual* movement command.

Implications for SC models

As was mentioned in the Introduction, most models assigning a role to the SC in saccade generation assume some invariant relation between the collicular activity patterns and the resulting eye movement. However, the data presented in this paper indicate that saccades with similar properties may be associated with very different collicular activity patterns (e.g. Figs. 4 and 6). In addition, the observed sensory influences on the saccade-related activity are quite reproducible from neuron to neuron and are independent of the way in which the activity is quantified (instantaneous firing rate, mean firing rate, or the number of spikes in the presaccadic burst). Therefore, these data do not seem to support the hypothesis that the activity of invidual SRBNs reflects the kinematics of the ongoing saccade (see Introduction).

One way to circumvent this problem would be to assume that the efferent contribution of all recruited neurons is somehow normalized, which would necessite a non-linear weighting scheme. The actual activity levels may thus be averaged out accross the population ensuring a fixed representation of the collicular output (e.g. Lee et al., 1988; Waitzman et al., 1991). Such a hypothesis thus regards the systematic modulations of SRBN activity by sensory cues as an epiphenomenon which is weeded out by efferent mechanisms. However, it cannot explain why movement fields of SRBNs may be subject to change (e.g. Fig. 7).

Alternatively, the systematic sensory modulations could be regarded as an efficient way of representing different sensorimotor signals by a population code. In a recent model study it was shown that small, but systematic, modulations of SRBN activity by changes in initial eye position (Van Opstal et al., 1995), could in principle be used by the premotor circuitry to extract accurate information on both the position of the target with respect to the head, as well as on the desired oculocentric movement coordinates of the eye (Van Opstal and Hepp, 1995). One may speculate, that both signals are needed, e.g. in the generation of combined eye-head movements, since the eye and head motor systems require different inputs expressed in different motor frames. Similar encoding principles could underlie the systematic changes in activity patterns accross the population resulting from combined visual-auditory stimulation. In this way, the collicular population could transmit accurate spatial information on the multimodal target configuration.

In this respect, the recent findings of Werner (1993) may be of importance. It was shown that a large population of pre-saccadic SC cells is also involved in the guidance of arm movements towards visual targets. Even more remarkably, the temporal activity patterns of the SRBNs often correlated with the EMG patterns of proximal arm muscles, rather than with parameters of the hand trajectory. Thus, the output of similar collicular cell populations appears to be used by both the spinal cord and brainstem for very different movement strategies and motor programs. Although quite speculative at present, a population scheme as mentioned above could embody such different motor events.

Taken together, the strong influence of the sensory input on the movement-related activity of SRBNs suggests that these cells do not determine the detailed properties of the actual motor response (neither the exact coordinates of the eye displacement vector, nor the saccade kinematics). Rather, the invariance of the tight temporal link between the onset of the pre-saccadic activity and saccade latency indicates a strong involvement of the SC in the initiation of saccadic eye movements. The site of the active population appears to reflect the coordinates of the desired movement ('motor error'), rather than of the actual eye displacement vector. In this sense, 'motor-error field' would be a more appropriate term to denote SRBN sensitivity than 'movement field'.

Acknowledgements

Supported by the University of Nijmegen (AJVO), the Swiss National Science Foundation (MAF) and the European Esprit program (Mucom 6615; MAF and AJVO). We acknowledge the assistance of G. Windau, C. Van der Lee, H. Kleijnen, and T. Van Dreumel. Furthermore, we express our gratitude to T. Arts and F. Philipsen from the Central Animal Lab. in Nijmegen and to dr. R. Dirksen from the Anesthesiology Dept. of the Nijmegen University Hospital, for their expertise in taking care of our monkeys.

References

Arai, K., Keller, E.L. and Edelman, J.A. (1994) Two-dimensional neural network model of the primate saccadic system. *Neural Networks*, 7: 115–1135.

Berthoz, A., Grantyn, A. and Droulez, J. (1986) Some collicular efferent neurons code saccadic eye velocity. *Neurosci. Lett.*, 72: 289–294.

Blauert, J. (1983) *Spatial hearing. The psychophysics of human sound localization.* MIT Press, Cambridge, MA.

Bour, L.J., Van Gisbergen, J.A.M., Bruijns, J. and Ottes, F.P. (1984) The double-magnetic induction method for measuring eye movement: results in monkey and man. *IEEE Trans. Biomed. Eng.* 31: 419–427.

Droulez, J. and Berthoz, A. (1991) A neural network model of sensoritopic maps with predictive short-term memory properties. *Proc. Natl. Acad. Sci. USA*, 88: 9653–9657.

Frens, M.A. (1995) *Multisensory Control of Orienting Movements.* Ph.D. Thesis, University of Nijmegen, NL.

Frens, M.A. and Van Opstal, A.J. (1995a) A quantitative study of auditory-evoked saccadic eye movements in two dimensions. *Exp. Brain Res.*, 107: 103–117.

Frens, M.A., and Van Opstal, A.J. (1995b) Audio-visual integration in the superior colliculus of the awake monkey. *Soc. Neurosci. Abstr.*, 21: 1194.

Gnadt, J.W., Bracewell, R.M. and Andersen, R.A. (1991) Sensorimotor transformation during eye movements to remembered visual targets. *Vis. Res.*, 31: 693–715.

Jay, M.F. and Sparks, D.L. (1987) Sensorimotor integration in the primate superior colliculus. II. Coordinates of auditory signals. *J. Neurophysiol.*, 57: 35–55.

Jay, M.F. and Sparks, D.L. (1990) Localization of auditory and visual targets for the initiation of saccadic eye movements. In: M.A. Berkley and W.C. Stobbins (Eds.), *Comparitive Perception. Vol. I. Basic Mechanisms.* Wiley and Sons.

King, A.J. and Palmer, A.R. (1985) Integration of visual and auditory information in bimodal neurones in the guinea-pig superior colliculus. *Exp. Brain. Res.*, 60: 492–500.

Lee, C., Rohrer, W.H. and Sparks, D.L. (1988) Population coding of saccadic eye movements by neurons in the superior colliculus. *Nature*, 332: 357–360.

Lefèvre, P. and Galiana, H.L. (1992) Dynamic feedback to the superior colliculus in a neural network model of the gaze control system. *Neural Networks*, 5: 871–890.

McIlwain, J.T. (1982) Lateral spread of neural excitation during microstimulation in the intermediate gray layer of cat's superior colliculus. *J. Neurophysiol.*, 47: 167–178.

McLaughlin S.C. (1967) Parametric adjustment in saccadic eye movements. *Percept. Psychophys.*, 2: 359–362.

Meredith, M.A. and Stein, B.E. (1986a) Visual, auditory and somatosensory convergence on cells in superior colliculus results in multisensory integration. *J. Neurophysiol.*, 56: 640–662.

Meredith, M.A. and Stein, B.E. (1986b) Spatial factors determine the activity of multisensory neurons in cat superior colliculus. *Brain Res.*, 365: 350–354.

Munoz, D.P., Pèlisson, D. and Guitton, D. (1991) Movement of neural activity on the superior colliculus map during gaze shifts. *Science* 251: 1358–1360.

Optican, L.M. (1995) A field theory of saccade generation: temporal-to-spatial transformation in the superior colliculus. *Vis. Res.*, 35: 3313–3320.

Ottes, F.P., Van Gisbergen, J.A.M. and Eggermont, J.J. (1986) Visuomotor fields of the superior colliculus: a quantitative model. *Vis. Res.*, 26: 857–873.

Robinson, D.A. (1972) Eye movements evoked by collicular stimulation in the alert monkey. *Vis. Res.*, 12: 1795–1808.

Sparks D.L. (1986) Translation of sensory signals into commands for control of saccadic eye movements: role of primate superior colliculus. *Physiol. Rev.*, 66: 118–171.

Sparks, D.L. and Mays, L.E. (1980) Movement fields of saccade-related burst neurons in the monkey superior colliculus. *Brain Res.*, 190: 39–50.

Stanford, T.J. and Sparks, D.L. (1994) Systematic errors for saccades to remembered targets: Evidence for a dissociaton between saccade metrics and activity in the superior colliculus. *Vis. Res.*, 34: 93–106.

Stein, B.E. and Meredith, M.A. (1993) *The merging of the senses*. MIT Press, Cambridge, MA.

Van Gisbergen, J.A.M., Van Opstal, A.J. and Tax, A.A.M. (1987) Collicular ensemble coding for saccades based on vector summation. *Neurosci.*, 21: 541–555.

Van Opstal, A.J. and Hepp, K. (1995) A novel interpretation of the collicular role in saccade generation. *Biol. Cybern.*, 73: 431–445.

Van Opstal, A.J. and Kappen H. (1993) A two-dimensional ensemble coding model for spatial-temporal transformation of saccades in monkey superior colliculus. *Network*, 4: 19–38.

Van Opstal, A.J., Smit, A.C., and Van Gisbergen, J.A.M. (1990) Comparison of saccades evoked by visual stimulation and collicular electrical stimulation in the alert monkey. *Exp. Brain Res.*, 79: 299–312.

Van Opstal, A.J., Hepp, K., Suzuki, Y. and Henn, V. (1995) Influence of eye position on activity in monkey superior colliculus. *J. Neurophysiol.*, 74: 1593–1610.

Waitzman, D.M., Ma, T.P., Optican, L.M. and Wurtz, R.H. (1991) Superior colliculus neurons mediate the dynamic characteristics of saccades. *J. Neurophysiol.*, 66: 1716–1737.

Werner, W. (1993) Neurons in the primate superior colliculus are active before and during arm movements to visual targets. *Eur. J. Neurosci.*, 5: 335–340.

Wurtz, R.H. and Optican, L.M. (1994) Superior colliculus cell types and models of saccade generation. *Curr. Opin. Neurobiol.*, 4: 857–861.

M. Norita, T. Bando and B. Stein (Eds.)
Progress in Brain Research, Vol 112
© 1996 Elsevier Science BV. All rights reserved.

CHAPTER 14

Coding of stimulus invariances by inferior temporal neurons

Rufin Vogels* and Guy A. Orban

Laboratorium voor Neuro- en Psychofysiologie, Faculteit der Geneeskunde, KULeuven, Campus Gasthuisberg, B-3000 Leuven, Belgium

Primates are able to recognize a particular object despite major differences in the retinal images of the object. Inferior temporal cortex is suggested to be involved in this invariant object recognition. Neurons of this cortex show shape selectivity and it has been shown that this shape selectivity is relatively invariant for changes in the position and size of the shape. We show that macaque inferior temporal cortical neurons may respond to shapes that are defined by relative motion or by texture differences. The degree of shape selectivity can vary for the different visual cues, but overall shape preference is invariant for shapes defined either by luminance, by relative motion or by texture. Also, we found that shape selective inferior temporal neurons respond to partial occluded shapes and that their shape selectivity is similar with and without the partial occlusion. These results suggest that inferior temporal neurons, as a population, can code for an abstract, stimulus invariant shape or object part.

Introduction

We have no difficulty recognizing objects despite profound changes in the retinal image of these objects. For instance, we can identify a particular object at different distances (yielding different sized retinal images) and at different positions in the visual field. Thus, object identification is invariant for changes in the retinal size (size invariance) and position (position invariance) of the object. The same invariances hold for more simplified stimuli such as outlines of objects or shapes. The question thus arises whether our brain also builds a representation of a shape which is invariant for changes in size and position in order to identify the object (i.e. for memory retrieval), or whether, instead, one has a different representation for each retinal size and position of the object. The latter possibility seems to be ineffi-

cient since it requires a large database of object representations for different object positions and retinal sizes.

The first single unit studies in cat visual cortex (Hubel and Wiesel, 1962) showed that some position invariance is already present at the single cell level in area 17. Indeed, Hubel and Wiesel defined their complex cell as having the same type of response (On or Off) within their receptive field, thus allowing some, albeit modest, stimulus position invariance. Later studies in the macaque showed that receptive field size progressively increases as one moves from V1 to extrastriate areas V2, V4 and IT. In fact, the pioneering studies of Gross and colleagues (Gross et al., 1972) showed that IT neurons have large receptive fields (mean size of 409 deg^2) and can even include the ipsilateral field. It was suggested that these large, bilateral receptive fields underly the behavioral position invariance and the interocular transfer of visual discriminations as seen in chiasm-sectioned monkeys (Rocha-Miranda et al.,

---

* Corresponding author.

1975). Later studies (Schwartz et al., 1983; Komatsu et al., 1992; Sary et al., 1993; Lueshow et al., 1994; Tovee et al., 1994; Ito et al., 1995; Logothetis et al., 1995) replicated and extended Gross' findings, showing that stimulus selectivity is largely invariant for changes in stimulus position within the receptive field. Several studies also reported that in general the shape (Sato et al., 1980; Schwartz et al., 1983; Sary et al., 1993; Lueshow et al., 1994; Ito et al., 1995; Logothetis et al., 1995) and color selectivity (Komatsu et al., 1992) of IT cells is invariant for changes in the size of the stimulus. This size invariance of neuronal response selectivity may underly the behavioral size invariance.

Objects need to be segregated from their background in order to be recognized. Many perceptual studies have shown that discontinuities in the distribution of several parameters lead to a salient segregation of figure and background, e.g. discontinuities in luminance, depth, color, texture and relative motion. Thus, the same object can be segregated from its background using several visual cues and be identified as that object irrespective of its defining cue. This cue invariance of object recognition has survival value since the same object (e.g. a predator) must be recognized under different environmental conditions (e.g. hidden in bushes) in some of which it will stand out from its background by virtue of only a single visual cue (e.g. relative motion or texture). Thus, the question arises whether the brain represents these objects in a cue-independent way, and then, whether the shape selectivity of IT neurons is invariant for changes in the defining visual cue. We examined this issue by recording from single IT neurons in alert fixating monkeys and by measuring the response selectivity for shapes defined by any one of three visual cues: luminance, relative motion and texture.

In our environment, many objects are partly occluded by other objects. Despite their partial occlusion, we are able to recognize these objects, which, again, has obvious survival value. Thus, object recognition is invariant with respect to changes in the retinal image due to partial occlusion of the object. This type of invariance for changes in the retinal input can also be demonstrated in macaque monkeys (Kovacs et al., 1995; Schiller, 1995) using simple geometrical shapes as stimuli. In a second series of experiments, we determined whether the shape selectivity of single IT neurons is invariant for partial occlusion of the shapes, a property that may underly the perceptual immunity for partial shape occlusion.

Methods

We will only give a brief summary here since a detailed account of the methods have been published earlier (Sary et al., 1993; Kovacs et al., 1995). In each experiment, two male rhesus monkeys (Macaca mulatta) served as subjects. The recording sites were histologically confirmed and included the posterior and central parts of area TE and may have extended into anterior parts of TEO, as defined by Iwai and Mishkin (1968). Neurons were recorded either in the lower bank the Superior Temporal Sulcus, close to the lateral convexity, or in the lateral part of the IT itself. The single cell recordings were made using conventional techniques for recordings in behaving macaques. Eye position was measured using the scleral search coil technique.

Stimuli were generated on a BARCO CD233 monitor (frame rate 50 Hz; P22 Phosphor) using dedicated hardware developed in the Biophysics Institute (J.J. Koenderink) of Utrecht, Nederland. A set of eight different shapes (see Figs. 10 and 11 for illustration of the shapes) were used as stimuli. The shapes were presented on a random-dot texture pattern (density of 50% and dot size of 3 or 5 arc/min) which was replaced on each trial. The different shapes covered the same area, approximately 10 deg^2. In the visual cue invariance experiment (Fig. 1), the eight shapes were either defined by a difference in luminance of figure and background dots, or by the relative motion of dots inside and outside the shape, or by a difference in dot size (texture cue). In the case of shapes defined by relative motion, the dots inside the shape moved in a direction opposite to those in the background (speed 3 degrees/s), yielding a vivid percept of the shape. In both

LUM. KIN. TEX.

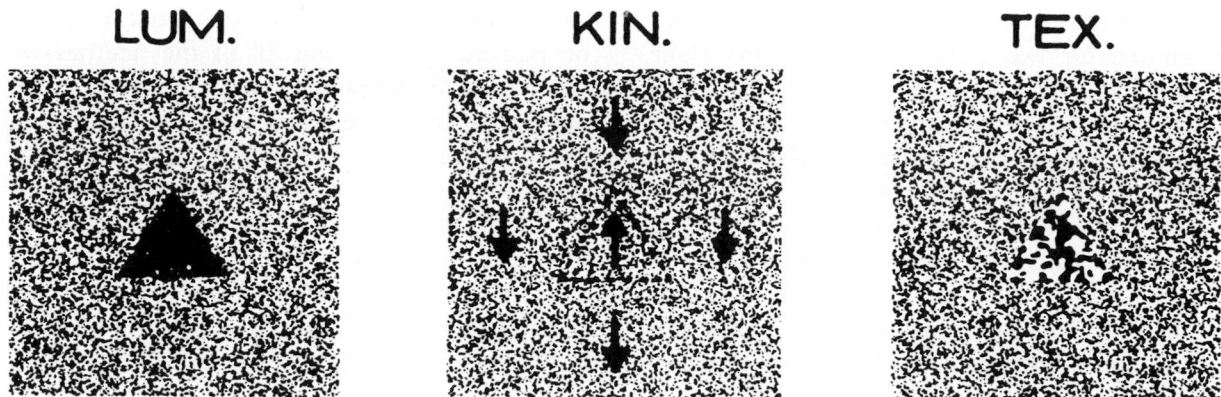

Fig. 1. Illustration of the stimuli used in the cue-invariance experiment. A triangle is defined by a difference in luminance between figure and background (LUM), by relative motion of the dots (up versus down) of the figure and background (KIN), or by a difference in dot size (texture) between figure and background (TEX). For the middle figure (KIN), the arrows show the direction of motion of the dots and the hatching indicates the virtual borders, which are rendered visible solely by virtue of the motion. The direction of motion (up or down) was randomized over trials.

relative motion and texture defined stimuli, the luminance of the dots inside the figure was the same as that of the dots outside the figure, so that there was no measurable average luminance difference between shape and background.

In the occlusion experiments, the same shapes, but only luminance-defined, were used except that on half of the trials the shapes were covered by a static or moving (3 degrees/s) pattern that partially occluded the shapes. The occluding pattern consisted of randomly positioned squares (size 0.6 degrees) with a density of 50%. These squares were opaque having the same luminance as the shape outlines. The occluding pattern filled the screen completely and its position with respect to the shape was randomized between trials. Under these partial occlusion conditions both human and monkey subjects could discriminate the shape, despite a 50% occlusion (on average).

The monkeys were trained to fixate a spot. Immediately after fixation the background pattern was shown followed by the shape 700 or 1000 ms later. This shape was presented for 300 or 500 ms, depending on the experiment. The background pattern was always moving when shapes defined by motion were to be presented, and in case of occlusion trials, the occluding pattern was presented immediately at the onset of fixation by

the monkey. Only trials in which the monkey did not break fixation during stimulus presentation were analysed and each stimulus condition was presented, in an interleaved fashion, for at least ten completed trials. Offline, the net responses were computed by trialwise subtraction of the activity preceding stimulus presentation from the activity during stimulus presentation. The spike counts for this subtraction were counted in bins with the same width as the stimulus duration. The significance of shape responses and selectivity were tested by Analysis of Variances (ANOVA) for each cell.

Results

We recorded from 372 IT units that were responsive to at least one of the shape stimuli and found, as have several preceding studies, that IT cells can be highly selective for the simple geometrical shapes used in our experiment. In fact, many units responded by excitation to a few shapes and were at the same time inhibited by other shapes. Also, in agreement with previous studies, the shape selectivity was usually not affected by a 16-fold variation in stimulus size. An example of this size invariance of a single IT neuron is shown in Fig. 2. The net response level is plotted for two different sizes of the eight shapes. Note that the size affects the response level: responses to the larger stimuli are higher than responses to the smaller stimuli. Thus, response magnitude is not size invariant in this cell. However, the shape selectivity is size invariant: the relationship between response and shape is very similar for the two sizes. This is a typical finding: the response magnitude of IT units can depend, in some cases even strongly, on size, but their shape preference is largely invariant to changes in stimulus size.

We will now show that the same principle holds for cue invariance of IT neurons. First, it should be pointed out that not all of the neurons we recorded from responded to shapes defined by any one of the three visual cues. For instance in Fig. 3 we show two cells which respond well to shapes defined by one cue but only marginally or not at all when the same shapes were defined by other cues. The neuron of the upper part of Fig. 3 shows stimulus selectivity for the luminance defined shapes, but does not respond when the same shapes were defined by either motion or texture cues. The neuron in the lower part of Fig. 3 responds to shapes defined by motion, but little to shapes defined by luminance or texture. How-

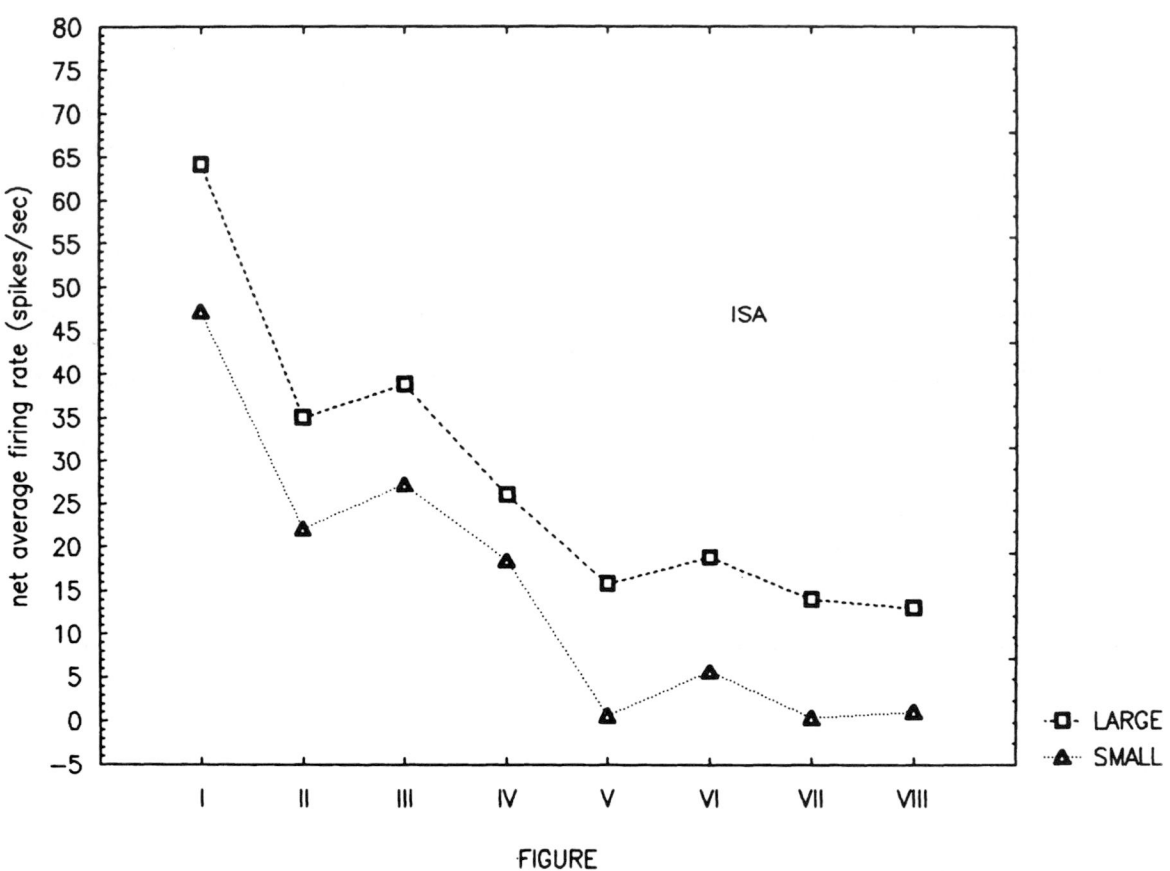

Fig. 2. Response of an IT neuron to eight different shapes of two different sizes. The figures were eight luminance-defined shapes and the area of the large figures was 16 times larger than the area of the small figures. The net average firing rate was higher for the large than for the small figures but the shape selectivity was invariant to the change in size.

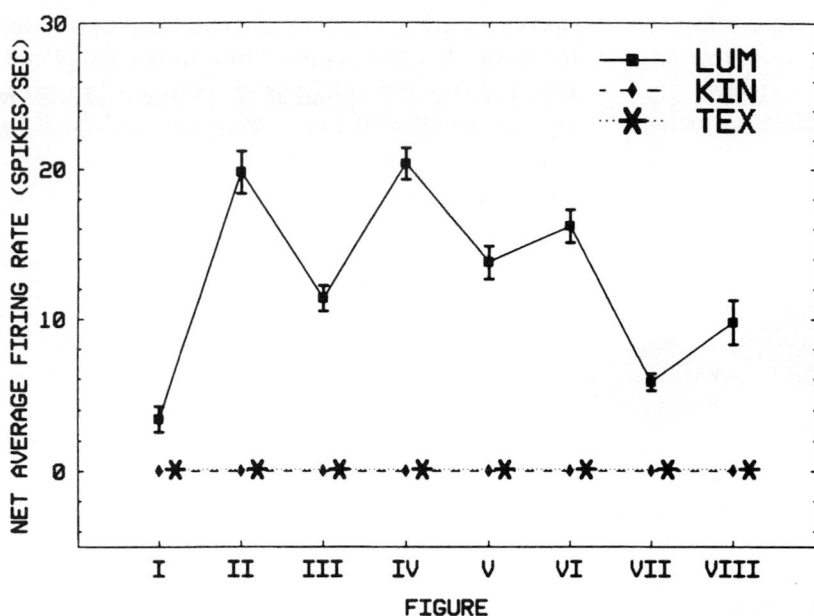

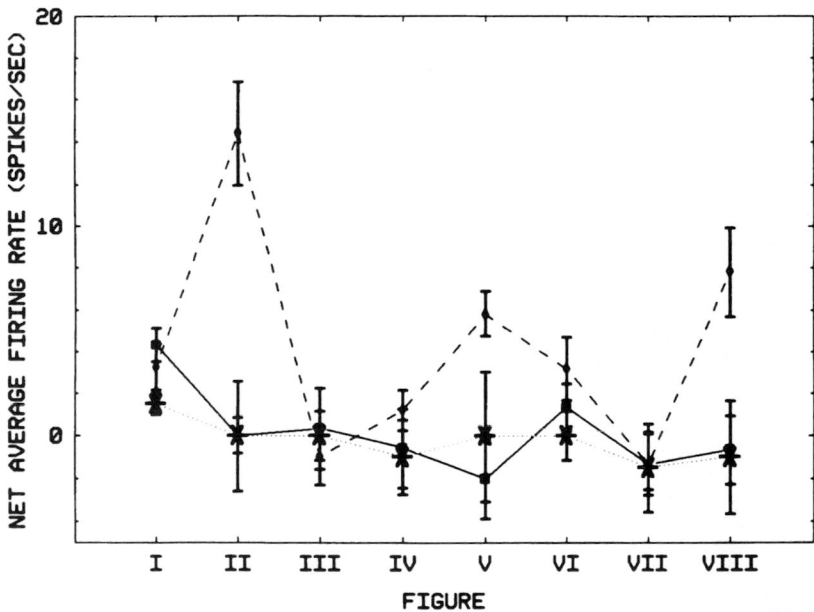

Fig. 3. Responses of two IT neurons to eight shapes defined by luminance (LUM), relative motion (KIN) and dot size (TEX). Standard errors of the mean are indicated by the vertical bars.

ever, most IT cells respond to more than one visual cue: 65% of the responsive neurons ($n = 261$) responded to each of the three types of cues, showing 'convergence' of multiple cues at the single cell level.

Second, 28% of those neurons responding to

200

each of the three cues were also shape selective for each of these cues. This proportion also implies that many neurons can be selective for shapes defined by one visual cue, while not being shape selective, but nonetheless responsive, for the same shapes as defined by another visual cue. Fig. 4 shows the relationship between the presence or absence of shape selectivity (as tested by

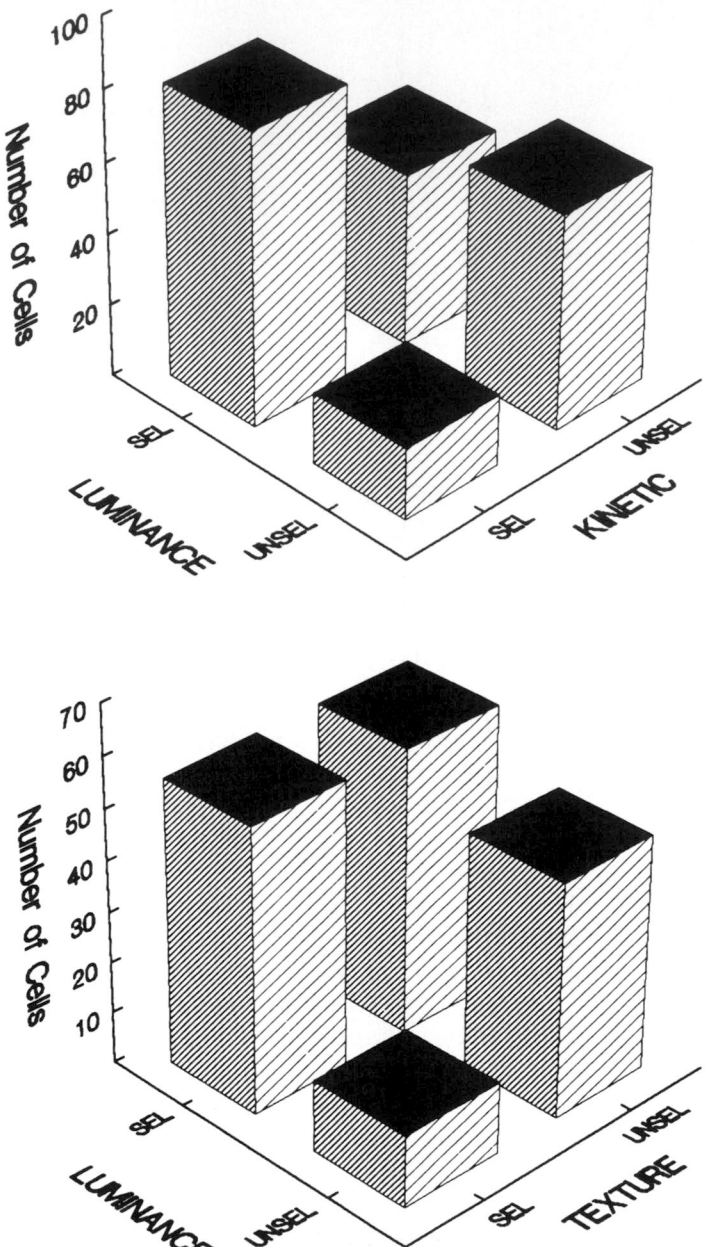

Fig. 4. Shape selectivity compared for IT neurons responding both to luminance and motion cues (Top) and for neurons responding to luminance and texture cues (Bottom). The presence of shape selectivity (SEL) was determined for each responding neuron and visual cue by Analysis of Variance.

ANOVA) for one cue and the shape selectivity for another cue, and this for all cells responsive to both visual cues. Only data for the combination of the luminance and motion cue and for the luminance and texture cue are plotted. In both cases, as well as in the case of the combination of motion and texture cues, the presence or absence of selectivity correlated with the presence or absence of selectivity for the other cue (each of the three Chi Square tests $P < 001$). However, it is clear from Fig. 4 that this correlation is far from absolute, e.g. in case of the luminance and motion cues, 32% of the cells were selective for one cue but not for the other. Also, the degree of correlation depended on the particular cue used to predict the shape selectivity for the other cue, e.g. 63% of the units selective for the luminance-defined shapes were also selective for the motion cue, whereas 80% of the units selective for the motion defined shapes were selective for the luminance shapes. The same trend was present in the selectivities for luminance and texture defined shapes (Fig. 4, bottom). Thus, these data suggest that neurons selective for shapes defined by iso-luminant cues are also likely to be selective for luminance-defined shapes, while many neurons selective for luminance defined shapes are not selective to shapes defined by other cues.

Finally, for those IT neurons showing shape selectivity for more than one visual cue, shape preference is largely independent of the defining visual cue. An example of such a neuron is shown in Fig. 5. This neuron was strongly shape selective, responding only to the star-shaped figure. The preference was the same regardless the visual cue. We also positioned the stimuli at 4 degrees eccentricity, but this displacement had no effect on the selectivity of the unit for the luminance and motion cue (the texture cue was not tested at that position), showing position invariance of the shape selectivity for the motion-defined shapes.

It should be noted that in most neurons selective for shapes defined by each of the three visual cues, the response level depended on the visual cue. The response was usually larger for the luminance defined shape than for the same motion defined shape (median ratio, 1.4) or texture defined shape (median ration, 1.5). Despite this dependency of the response strength on the visual cue, the shape preference was largely independent of the visual cue. In order to demonstrate this visual cue invariance of the preferred shape we performed the following analysis. For each neuron selective for at least the luminance and one other visual cue, we ranked the eight shapes according to their net response for the luminance cue, where rank 1 corresponds to the preferred luminance shape and rank 8 to the non-preferred luminance shape. Then the responses for each shape were normalized with respect to the maximum response and for each cue. The normalized responses of the different neurons were averaged for each rank and visual cue. The results for the comparisons between luminance- and motion- or texture-defined shapes are shown in Figs. 6 and 7, respectively. If the shape preference for the motion or texture cue was independent of the shape selectivity for the luminance cue, than the curve describing the relationship between normalized response for the motion (Fig. 6) and texture cue (Fig. 7) and shape rank would be flat. However, as the figures show, the latter was not the case: both curves are similar to those for the luminance defined shapes indicating that overal shape preference was similar for the two visual cues. The same analysis using those neurons selective for all three cues produced the same results (Fig. 4 of Sary et al., 1993). Thus, shape preference of IT neurons is cue invariant.

We found that 43% of the neurons we tested were selective for shapes defined by motion. However, our motion display contained another cue besides the relative motion cue. Flicker is present at those borders not parallel to the direction of motion due to the appearance and disappearance of the dots at these borders. This dynamic occlusion cue (see Sary et al., 1994 for more discussion of this confounding cue) could contribute to the responses to the motion defined shapes. This possibility was examined in control displays in which the direction of motion inside and outside the virtual shape was identical (Fig. 8). In this control display in which the relative motion cue is absent, only flickering dots at borders not parallel to the motion direction are perceived. Thus, we determined the contribution of the relative motion cue by comparing the response to the shapes defined by motion (the standard motion display) and shapes defined without the relative motion (the control display). For each neuron tested we subtracted the responses to the control display from the responses to the standard motion display and divided this difference by the sum of both responses. The results shown in Fig. 9 indicate that the average net response was larger in the displays with than those without

relative motion (paired *t*-test; $P < 001$). In fact, for all shape selective neurons tested ($n = 15$; indicated in black in Fig. 9) the response in the condition with relative motion was always larger than the response without relative motion (average ratio, 2.6). Hence, the relative motion cue by itself contributes significantly to the responses of these IT neurons.

In another experiment, we determined the effect of partial occlusion of the shape by another moving pattern. In order to maximize the effect of partial occlusion, the shape stimuli consisted of shape outlines only. The occluding pattern was

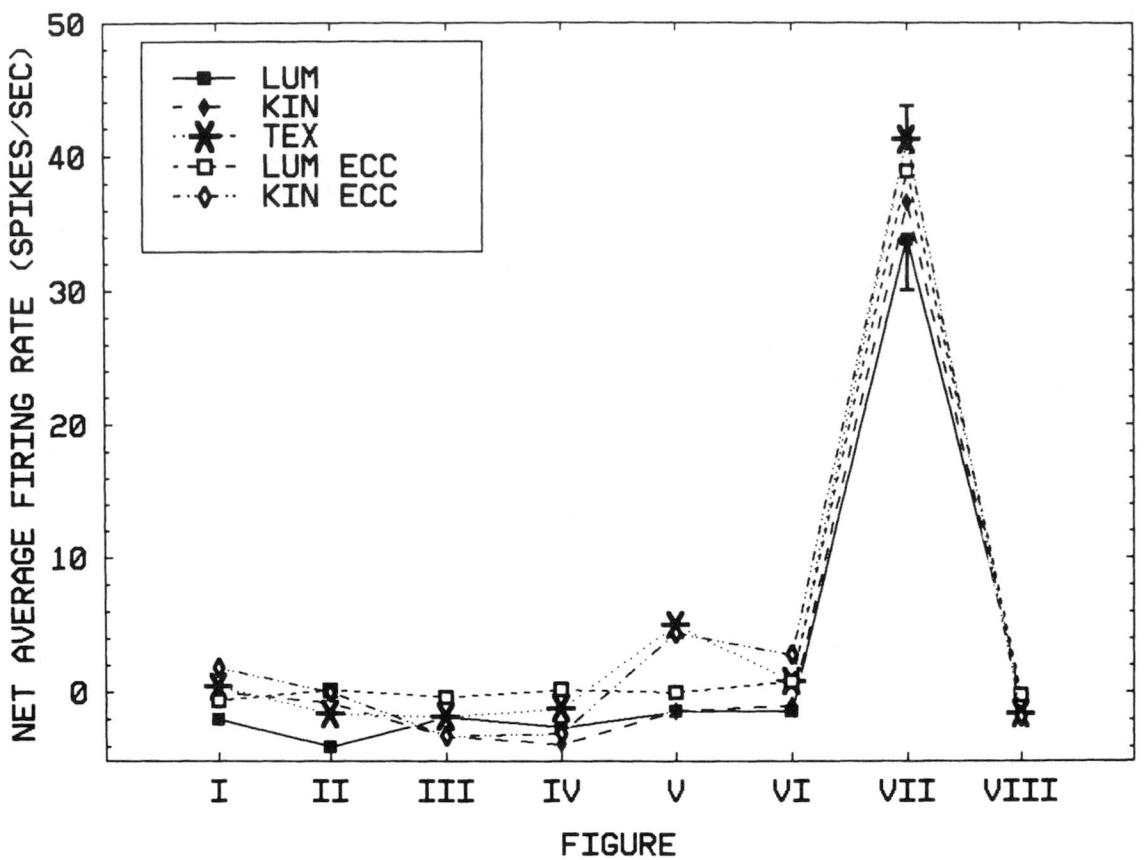

Fig. 5. Shape selectivity of an IT neuron selective for each of the three visual cues. Fig. 7 was a star-like shape. LUM: foveal presentation of luminance-defined shape; KIN: foveal presentation of motion-defined shape; TEX: foveal presentation of texture-defined shape; LUM ECC: presentation of the luminance-defined shape at 4 degrees eccentricity on the horizontal meridian in the contralateral visual field; LUM KIN: peripheral presentation of the motion-defined shape (same position as LUM ECC). All standard errors had a magnitude similar to the three shown.

presented when the monkey started to fixate, that is 700 ms before the shape was presented. Of 80 neurons tested, 21% no longer responded to the shapes when they were partially occluded. One of these neurons that stopped responding under the partial occlusion condition is shown in Fig. 10. However, most neurons continued to respond when the shapes were partially occluded. Some of these units (31%) also responded to the onset of the occluding pattern, but the majority of the neurons were not responsive to the occluding pattern itself.

An example of a neuron responding to occluding pattern onset and which also responded to the shapes under conditions of partial occlusion is shown in Fig. 11. Three interesting observations, that also proved to be true for our entire population of neurons, can be made regarding the responses of this neuron during partial shape occlusion. First, the responses are smaller with than without the partial occlusion. In our sample of neurons, the responses in the no-occlusion condition were on average 1.7 times larger than those in the occlusion condition. Second, the response latency is longer in the occlusion compared to the no-occlusion condition. This latency difference (on average 50 ms) was also present for the other neurons tested under identical conditions. Third, the response selectivity was very similar with and without partial occlusion: the shape eliciting the

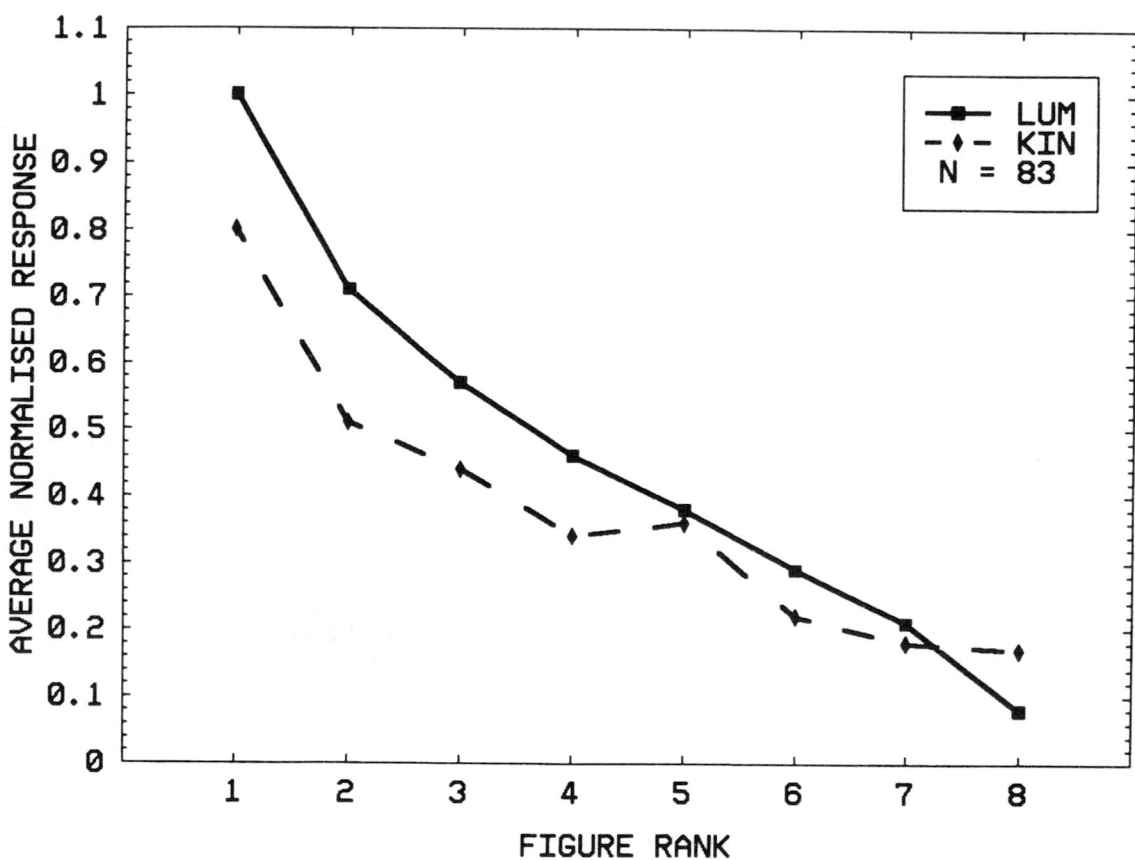

Fig. 6. The average normalised response plotted as a function of shape rank for those neurons showing shape selectivity for the luminance and motion cue. The figures were ranked according to their response to the luminance-defined shapes (LUM). The same ranking was applied for the motion-defined (KIN) figures.

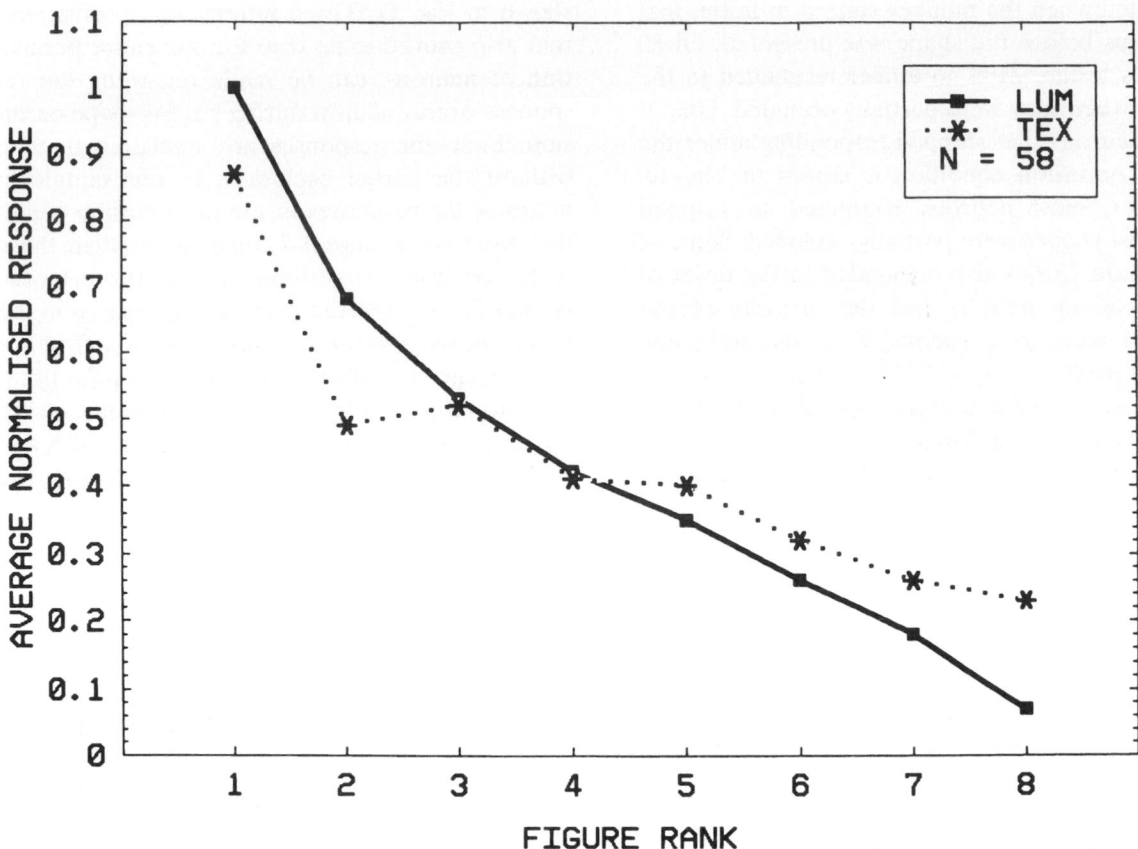

Fig. 7. The average normalised response plotted as function of shape rank for those neurons showing shape selectivity for the luminance and texture cue. The figures were ranked according to their response to the luminance defined shapes (LUM). The same ranking was applied for the texture-defined (TEX) figures.

largest response without occlusion also produced the largest response in the partial occlusion condition. This was further examined for our sample of cells by using a shape ranking procedure similar to the one described above for the demonstration of visual cue invariance. In short, for each shape selective neuron, the shapes were ranked according to their responses in the no-occlusion condition and then the average net responses were averaged over neurons for the occlusion and no-occlusion conditions. The results, shown in Fig. 12, demonstrate that the shape preference, at least on average, is identical with and without partial occlusion. Fig. 12 also shows the difference in response magnitude between the occlusion and no-occlusion conditions. Thus, despite a

reduction in response, IT neurons remain selective for shapes partially occluded by another pattern. Further analysis (see Kovacs et al., 1995) showed that the degree of shape selectivity itself is weakly affected by occlusion. Also, similar results were obtained when the occluding pattern was stationary instead of moving. Results of these and other experiments on the effect of partial occlusion on IT shape selectivity are reported in more detail in Kovacs et al. (1995).

Discussion

These results show that single IT units can show stimulus invariances similar to those present behaviorally. The selectivity of single IT neurons is largely invariant for changes in the retinal input caused by differences in position, size, defining visual cue, and partial occlusion of the shape. A recent report (Ito et al., 1995) has stressed that shape preference may depend on position or size of the shape. However, these changes in shape preference were, as described by these authors, rather 'subtle'. We have also seen effects on shape preference from the visual cue or partial occlusion, but these are minor compared to the overall similarity of shape preference in these variant stimulus conditions, as evident from the similarity of the curves relating normalized response and shape rank for the different cues. Of course, it is presently impossible to determine whether these exceptions to invariance have some

functional purpose or merely reflect the inevitable imperfections of a biological system such as the brain.

One feature common to the different kinds of stimulus invariances of these IT neurons is that the response magnitude of most neurons is not usually stimulus invariant. In the case of cue invariance, even the degree of selectivity can vary largely for different cues. The invariance is chiefly manifested in the shape preference: in general shape preference is invariant for changes in position, size and visual cue. Thus, the response level can depend on the position, size or (saliency of) the visual cue without affecting the selectivity, akin to the effect of contrast on orientation selectivity in the cat (Sclar and Freeman, 1982). Fur-

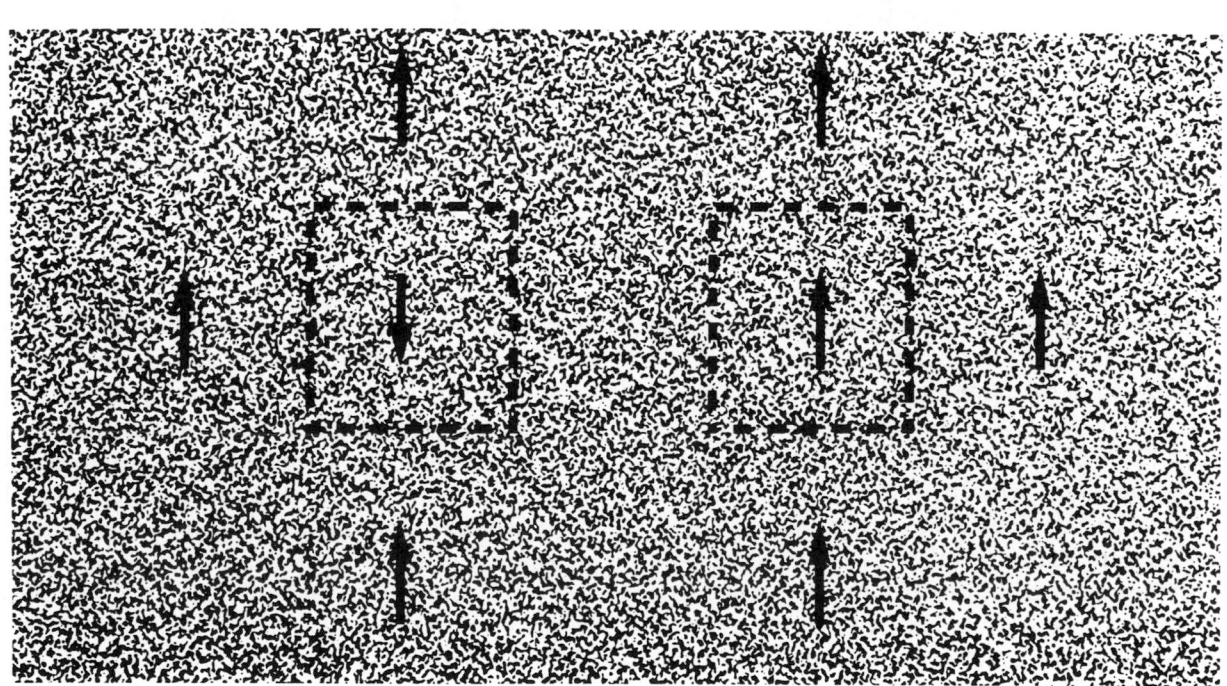

Fig. 8. Illustration of the relative motion display (left) and control display (right) used to determine the contribution of the relative motion cue to the neuronal responses. The control display is identical to the relative motion display except that in the former display the motion of the dots in figure and background have the same direction (no relative motion). In the case of the illustrated square, its horizontal borders will be visible as flickering dots due to the presence of the dynamic occlusion cue (appearance and disappearance of the dots at that virtual border). Other conventions as for Fig. 1.

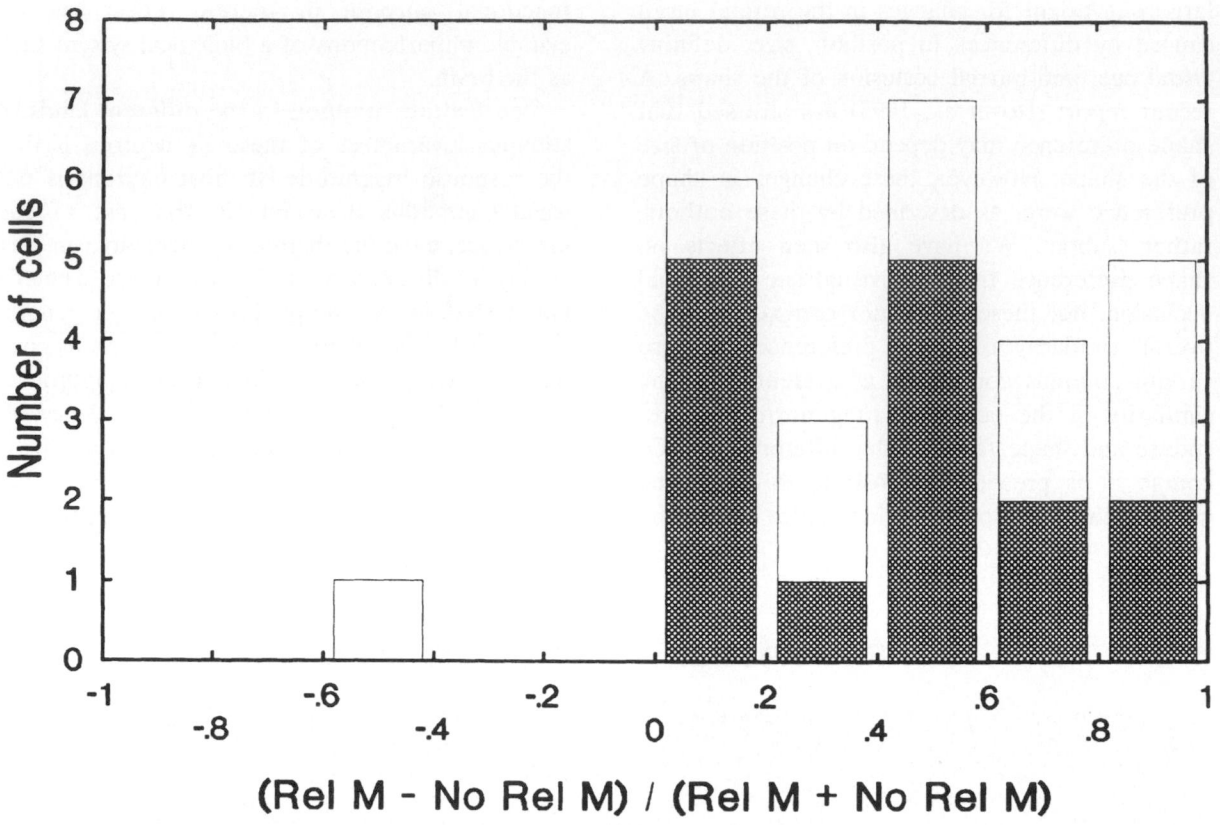

Fig. 9. Distribution of the difference in response to the relative motion and control display. The difference was expressed as an index defined by subtracting the response to the control display (No Rel M) from the response to the relative motion display (Rel M) and dividing the result by the sum of both responses. Hatched bars indicate shape selective neurons.

thermore, the degree of selectivity can depend on size or visual cue, analogous to the effect of line length on the orientation selectivity of neurons in cat (Henry et al., 1973) and monkey (Schiller et al., 1976) visual cortex: the degree of orientation selectivity of a neuron, but not its orientation preference, can depend strongly on line length.

The dependence of the response strength on variations in stimulus parameters other than shape combined with the invariance of the shape preference has repercussions for how the brain may code invariant representations: it suggests that it is the relative and not the absolute activity of a set of neurons, each prefering a different shape, that signals the presence of a shape (or object part), independently of its position, size, and vi-

sual cue. As long as the preferred shape of the neurons does not vary with changes in other parameters such as size, position, visual cue and partial occlusion, those neurons whose preferred shape matches the stimulus shape, will be, on average, the most active and will be able to signal that particular shape under all conditions. The dependency of the response magnitude on position, size, visual cue, and other parameters unrelated to shape (such as color (Komatsu et al., 1992)), on the other hand, can be useful in signaling other stimulus aspects such as relationships between different object parts or object surface properties. Indeed, one should not forget that besides being able to identify, e.g. a square independently of its defining visual cue, we also

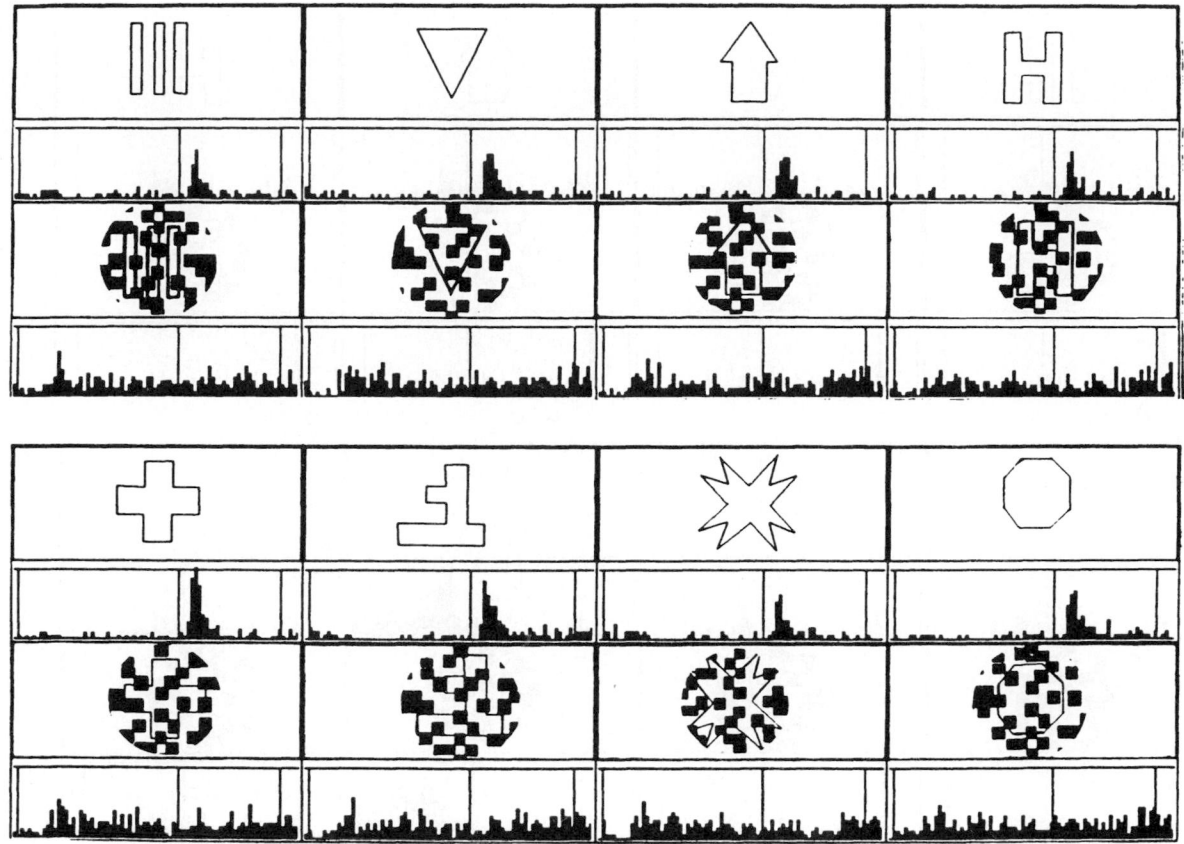

Fig. 10. Responses of an IT neuron with and without partial occlusion of the shapes. The peristimulus time histograms of the spikes for each of the eight shape outlines shown above the histograms in the no-occlusion (1st and 3th row) and occlusion condition (2nd and 4th row). Presentation of the shape is indicated by the second vertical bar in each histogram. This neuron failed to respond when the shapes were partially occluded.

perceive that the luminance defined square stimulus differs from the motion defined one in other aspects. Thus, it is not unexpected that the response level of the neurons, and even their degree of selectivity, can depend strongly on the defining visual cue. These considerations also suggest that behavioral stimulus invariances are not coded at the level of a single cell — since its response strength is not invariant, but is distributed over a population of cells with different degrees of response generalization for position (receptive field), size (selectivity) and visual cue (convergence), but with an invariant shape preference over the range of the stimulus conditions

(sizes, visual cues, etc.) in which they respond.

It has been reported (Miyashita and Chang, 1988) that neurons in the anterior medial temporal cortex show stimulus invariance not only in their pattern selectivity but also in their response magnitude, unlike cells in more posterior parts of IT. However, another study in the anterior medial temporal cortex (Lueshow et al., 1994) reported that some cells were indeed invariant in their absolute response magnitude, while the response level of others depended on the position (69% of the neurons tested) and size (two-fold size variation: 43%) of the stimulus. Of course it is very likely that the more anterior neurons in temporal

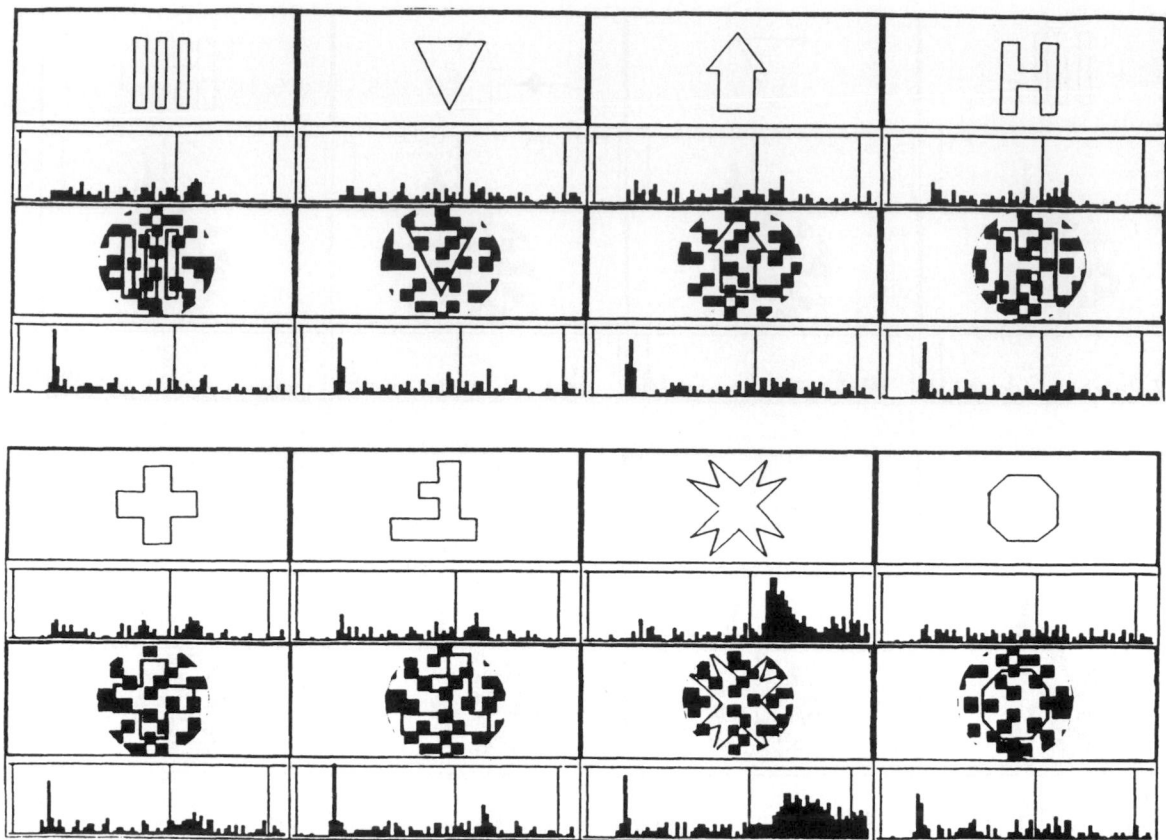

Fig. 11. Example of an IT neuron showing similar shape selectivity with and without partial occlusion. Note that this neuron responded transiently to the onset of the occluder, which was presented when the monkey started to fixate (indicated by the first vertical line in each histogram). Conventions are the same as for Fig. 10.

cortex have a broader range of sizes, positions and other stimulus parameters to which they give equivalent responses. Also, some of these neurons will show little sensitivity for some stimulus dimensions (such as size) and thus possess some degree of response magnitude invariance for those stimulus dimensions. However, the crucial question here is whether this response magnitude invariance holds for the entire range over which the perception or identification of the stimulus is invariant for changes in those dimensions. Indeed, to our knowledge, no neuron has been reported that showed complete invariance in its response magnitude for the entire range of sizes, positions, lightness conditions, etc., for which the

perception of that stimulus shows invariance. On the contrary, current data suggests that, as discussed above, temporal cortical neurons vary widely in their range of sensitivity to these stimulus manipulations (e.g. size, position, etc.), and that only their stimulus preference is kept largely independent of changes in these stimulus dimensions.

Single IT neurons show considerable stimulus invariance in their shape selectivity, as shown by both our results and those of others. As already noted in the introduction, single units of 'lower' cortical areas also show some degree of stimulus invariance, suggesting that the invariances we observed in IT emerge gradually within the visual

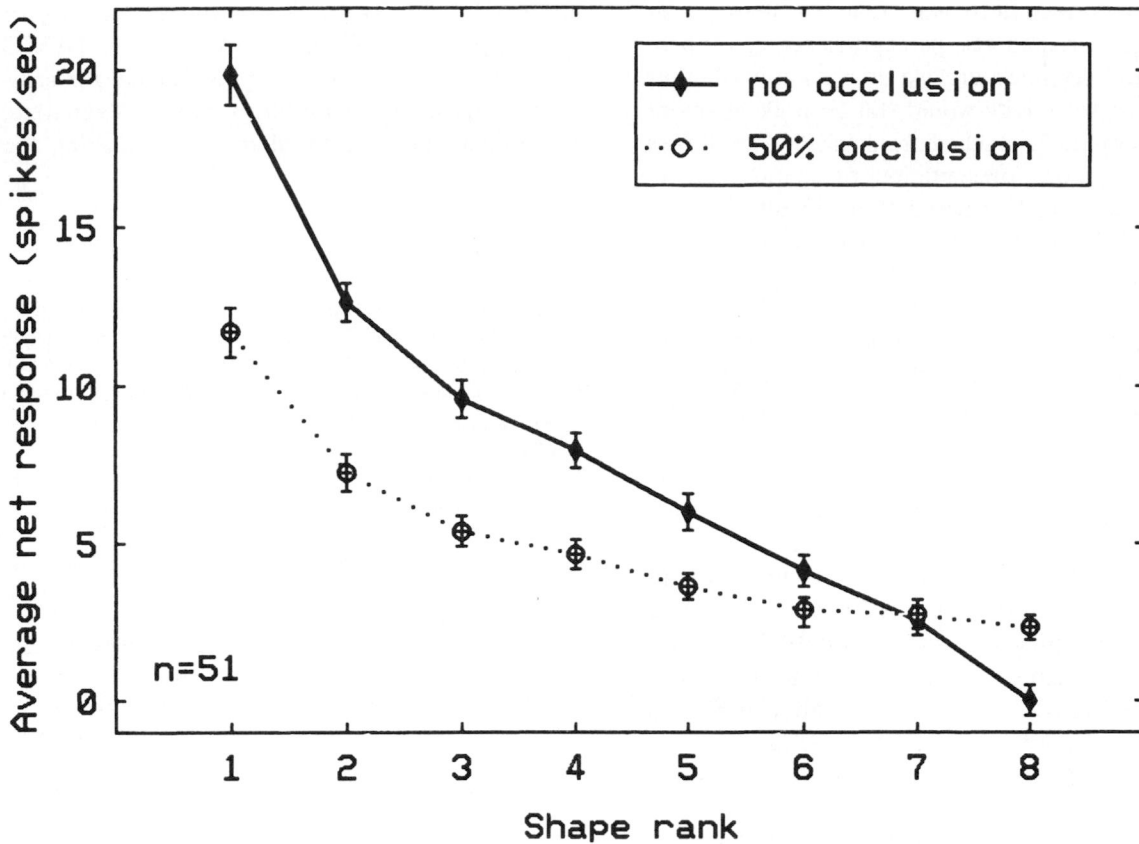

Fig. 12. Average net response and standard errors as a function of shape rank with and without partial shape occlusion. The shape rank was determined using the shape selective responses in the no-occlusion condition only and the same ranking was applied to average the responses in the occlusion condition.

system. For instance, receptive field size increases gradually from V1 to IT, and in V1 (complex cells), V2 (complex cells), V4 (Gallant et al., 1993) and IT, stimulus selectivities show position invariance within the receptive field. Visual cue invariances similar to the one we observed in IT (Sary et al., 1993, 1995) have also been reported in area V2 (illusory contours (von der Heydt et al., 1984); stereo boundaries (von der Heydt et al., 1995); kinetic boundaries (Marcar et al., 1994)) and V4 (kinetic boundaries (Logothetis and Charles, 1990)). These data suggest that relatively early in the visual system (V2), several visual cues already 'converge' on single neurons. These neurons could signal the presence of a boundary irrespective of

its defining cue. Convergence of information from multiple cues helps to distinguish real boundaries from accidental, noise induced signals, since discontinuities in many cues (e.g. luminance, depth, color) usually coincide spatially for real but not 'noise' boundaries. This could explain why cue convergence is found at early stages of the visual system. The IT neurons showing cue invariance likely receive input from the invariant 'boundary' neurons of lower visual areas.

Our results also show that IT units are selective for partially occluded shapes and that this selectivity is similar to that for un-occluded shapes. We should point out that in our experiments there was a time lag between the presentation of the occluding pattern and occluded shape. This time lag increases the perceptual segregation of shape and occluder, and we do not know whether the shape selectivity would still be present under the perceptually much more stringent conditions of simultaneous presentation of shape and occluding pattern. Nonetheless, our results (Kovacs et al., 1995) clearly show that under at least some conditions in which monkeys are able to discriminate partially occluded shapes, IT cells remain shape selective. Under partial occlusion conditions, humans and probably also monkeys, perceive the partially occluded shapes as single, complete forms (amodal completion). Kovacs et al. (1995) were unable to demonstrate a correlation of this amodal completion at the single cell level in IT: cells responded on average as well to shape parts that were perceived as isolated, disconnected units, as they did to shape parts that were seen as belonging to a single Gestalt. It is possible that neurons more anteriorely than our recording sites respond better under occlusion conditions yielding amodal completion than to occlusion conditions that produce the perception of disconnected parts.

These results indicate that the selectivity of IT neurons shows stimulus invariances and that the activity of these IT units, as a population, can code for an abstract, stimulus invariant shape or object part. This kind of abstract representation will reduce the space required for storing objects and make object recognition and identification more efficient since these processes will act upon an already abstract representation of the stimulus. In this context, it should be noted that size and position invariance is not only a property of the selectivity for physically present stimuli but also of the neuronal activity that reflects preceding, but physically absent, memorized, stimuli (Lueshow et al., 1994).

Acknowledgements

The research described in this chapter was done in collaboration with Gy. Sary and Gy. Kovacs and supported by ESPRIT BRA Insight, IUAP 22 and FGWO 9.0039.90. The writing of the chapter was supported by GSKE. R. Vogels is a NFWO research associate. We thank S. Raguel for critical reading of the manuscript and G. Vanparrijs, A. Coeman and G. Meulemans for making the figures.

References

Gallant, J.L., Braun, J. and Van Essen, D.C. (1993) Selectivity for polar, hyperbolic, and cartesian gratings in macaque visual cortex. *Science*, 259: 100–103.

Gross, C.G., Rocha-Miranda, C.E. and Bender, D.B. (1972) Visual properties of neurons in the inferotemporal cortex of the macaque. *J. Neurophysiol.*, 35: 96–111.

Henry, G.H., Bishop, P.O., Tupper, R.M. and Dreher, B. (1973) Orientation specificity and response variability of cells in the striate cortex. *Vis. Res.*, 13: 1771–1779.

Hubel, D. and Wiesel, T. (1962) Receptive fields, binocular interaction and functional architecture in the cat's visual cortex. *J. Physiol. (London)*, 160: 106–154.

Ito, M., Tamura, H., Fujita, I. and Tanaka, K. (1995) Size and position invariance of neuronal responses in monkey inferior temporal cortex. *J. Neurophysiol.*, 73: 218–226.

Iwai, E. and Mishkin, M. (1968) Two visual foci in the temporal lobe of monkeys. In: N. Yoshii and N.A. Buchwald (Eds.), *Neurophysiological basis of learning and behavior*. Osaka Univ. Press, Osaka, pp. 1–8.

Komatsu, H., Ideura, Y., Kaji, S. and Yamane, S. (1992) Color selectivity of neurons in the inferior temporal cortex of the awake macaque monkey. *J. Neurosci.*, 12: 408–424.

Kovacs, Gy., Vogels, R. and Orban, G.A. (1995) Selectivity of macaque inferior temporal neurons for partially occluded shapes. *J. Neurosci.*, 15: 1984–1997.

Logothetis, N.K. and Charles, E.R. (1990) V4 responses to gratings defined by random textured motion. *Invest. Opthalmol. Visual Sci.*, 31, 444.

Logothetis, N.K., Pauls, J. and Poggio, T. (1995) Shape representation in the inferior temporal cortex of monkeys. *Curr. Biol.*, 5: 552–563.

Lueshow, A., Miller, E.K. and Desimone, R. (1994) Inferior temporal mechanisms for invariant object recognition. *Cerebral Cortex*, 5: 523–531.

Marcar, V.L., Xiao, D.K., Raiguel, S.E. and Orban, G.A. (1994) Selectivity of area V2 of the macaque to kinetic and other types of boundaries. *Soc. Neurosci. Abstr.*, 20: 1740.

Miyashita, Y. and Chang, H.S. (1988) Neuronal correlates of pictorial short- term memory in the primate temporal cortex. *Nature*, 331: 68–70.

Rocha-Miranda, C.E., Bender, D.B., Gross, C.G. and Mishkin, M. (1975) Visual activation of neurons in inferotemporal cortex depends on striate cortex and forebrain commisures. *J. Neurophysiol.*, 38: 475–491.

Sary, Gy., Vogels, R. and Orban, G.A. (1993) Cue-invariant shape selectivity of macaque inferior temporal neurons. *Science*, 260: 995–997.

Sary, Gy., Vogels, R. and Orban, G.A. (1994) Orientation discrimination of motion-defined gratings. *Vis. Res.*, 34: 1331–1334.

Sary, Gy., Vogels, R.,. Kovacs, Gy. and Orban, G.A. (1995) Responses of monkey inferior temporal neurons to luminance-, motion-, and texture-defined gratings. *J. Neurophysiol.*, 73: 1341–1354.

Sato, T., Kawamura, T. and Iwai, E. (1980) Responsiveness of inferotemporal single units to visual pattern stimuli in monkeys performing discrimination. *Exp. Brain Res.*, 38: 313–319.

Schiller, P.H. (1995) Effect of lesions in visual cortical area V4 on the recognition of transformed objects. *Nature*, 376: 342–344.

Schiller, P.H., Finlay, B.L. and Volkman, S.F.(1976) Quantitative studies of single-cell properties in monkey striate cortex. II. Orientation specificity and ocular dominance. *J. Neurophysiol.*, 39: 1320–1333.

Schwartz, E.L., Desimone, R., Allbright, T.D. and Gross, C.G. (1983) Shape recognition and inferior temporal neurons. *Proc. Natl. Acad. Sci. USA*, 80: 5776–5778.

Sclar, G. and Freeman, R.D. (1982) Orientation selectivity in the cats striate cortex is invariant with stimulus contrast. *Exp. Brain Res.*, 46: 457–462.

Tovee, M.J., Rolls, E.T. and Azzopardi, P. (1994) Translation invariance in the responses of faces in the temporal visual cortical areas of the alert macaque. *J. Neurophysiol.*, 72: 1049–1060.

Von der Heydt, R., Peterhans, E. and Baumgartner, G. (1984) Illusory contours and cortical neuron responses. *Science*, 224: 1260–1262.

Von der Heydt, R., Zhou, H., Friedman, H. and Poggio, G.F. (1995) Neurons of area V2 of visual cortex detect edges in random-dot stereograms. *Soc. Neurosci. Abstr.*, 21: 18.

M. Norita, T. Bando and B. Stein (Eds.)
Progress in Brain Research, Vol 112
© 1996 Elsevier Science BV. All rights reserved.

CHAPTER 15

Theories of visual cortex organization in primates: areas of the third level

Jon H. Kaas*

Department of Psychology Vanderbilt University Nashville, TN, USA

Introduction

The importance of good maps was obvious to early explorers, and valid maps are equally important for explorers of the brain. Most investigators accept the prevailing view, well-established from the time of Brodmann (1909), that the neocortex of mammals is divided into some number of functionally distinct areas or 'organs of the brain', each with more or less sharp and definable boundaries. We are all familiar with the summary maps of proposed subdivisions of Brodmann (1909) for humans and many other species of mammals, and with the later summary of von Economo (1929) for the human brain, which also achieved considerable use. We also recognize some dis-enchantment with both of these proposals. Results of modern experimental and histological approaches clearly indicate that neither proposal correctly defines many subdivisions of neocortex, and other proposals for how extrastriate cortex is subdivided into visual areas in primates have evolved. While modern proposals clearly have more validity than early proposals, if only because they are based on more types of observations, there are still a great number of uncertainties. Differences and similarities in the current proposals need to be carefully evaluated in order to ascertain parts that are well supported and parts that are more problematic, and require further study, additional data, and considerations of alternatives.

In the most elaborate portrayal of visual cortex organization in primates to date, Felleman and Van Essen (1991) argued for the validity of 32 visual areas in the neocortex of macaque monkeys, while recognizing that the evidence for specific areas was variable. To address the issue of uncertainty, they assigned 'confidence values' of 1–3 to each area, with areas that are well-defined rating 1 and areas with significant uncertainty rating 3. In their scheme, only five areas received a rating of 1, while 17 areas scored a 3. Since only three of these areas (V1, V2 and MT) are components of all current proposals (see Kaas, 1995a), one could even take a more conservative position that only three of the proposed areas rate a confidence value of 1. Given this recognition of a high degree of uncertainty, how do we get more reliable maps, and what parts of current maps are most reliable?

A related issue is the variability in visual cortex organization across primates, but this issue cannot be adequately addressed in these few pages. However, there is little justification for considering proposals based on Old World monkeys (macaques) as applicable, with little modification,

---
*Corresponding author. 301 Wilson Hall Department of Psychology Vanderbilt University Nashville, TN 37240, USA Tel.: +1 615 3438449; fax: +1 615 3434342.

to human cortex, while at the same time considering differences in proposals for Old and New World monkeys as reflecting valid species differences (see Preuss, 1995). Clearly, we need to carefully consider what components of extrastriate cortex are common to all primates, and what components are variable and how. Comparative studies have further value in that proposals for brain organization for any given species should be consistent with observations from related species within the framework of evolution (see Kaas, 1993). For example, since all or most mammals appear to have the second visual area V2, proposals of extrastriate cortex organization in rats that do not include a V2 require special scrutiny (see Kaas and Krubitzer, 1991).

The focus in this review is on extrastriate visual areas that are early in the cortical processing hierarchy where we have the most extensive experimental data. While some or most of these areas are likely to be basic features of primate brains, from prosimians to humans, this review is of proposals and experimental studies on monkeys, since this is where most of the experimental data have been collected. While the review does not directly address the issue of extrastriate mechanisms of visually guided behavior, extrastriate cortex, even in primates, is not completely dependent on the geniculostriate relay (e.g. Rodman et al., 1989; Girard et al., 1991) and considerable vision, including visual tracking, appears to be mediated by extrastriate cortex even in the absence of striate cortex (see Cowey and Stoerig, 1991). In addition, even cortical areas that are dependent on V1 for above threshold activation, such as V2 (Girard and Bullier, 1989; Schiller and Malpeli, 1977), may be significantly modulated by a superior colliculus to pulvinar to cortex relay (see Kaas and Huerta, 1988).

Early theories based on cortical architecture

Current concepts of visual cortex organization in primates have been greatly influenced by the proposals of Brodmann (1909). Rather than replicate his familiar maps of cortical organization in humans and monkeys, where much cortex is hidden in fissures and not shown in the surface-view summaries, it is more productive here to show McCulloch's (1944) much later summary view of how Brodmann's areas relate to each other in cortex that has been schematically flattened into an unfolded sheet (for a more realistic unfolding, see Jouandet et al., 1989). Such unfolded maps are very useful in that they present proposals of cortical organization in a way that allows easier evaluation and testing. As is clear from the summary map (Fig. 1), primary visual cortex, striate cortex or area 17 of Brodmann, was seen as surrounded by two ring-like areas, 18 (parastriate cortex) and 19 (preoccipital cortex). The major output target of area 17 was thought to be area 18, and the major target of area 18 was (in addition to feedback to area 17) area 19. When it

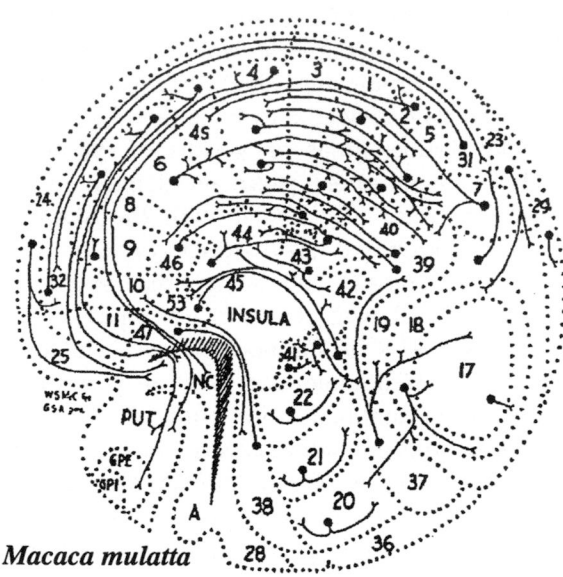

Macaca mulatta

Fig. 1. A summary diagram from McCulloch (1944) showing his view of how Brodmann's areas are arranged in the flattened cortex of macaque monkeys. The basal ganglia are also included in this diagram. The illustrated connections were based on experimental results of recording after local strychninization ('physiological neuronography'). Note that area 17 is surrounded by ring- like areas 18 and 19 and that these three fields form three successive stages of processing. This view, stemming from Brodmann (1909), has greatly influenced subsequent theories of extrastriate processing in primates.

became clear that areas 17, 18, and 19, as defined in cats, form successive, mirror-reversal representations of the contralateral visual hemifield, termed visual areas V1, V2, and V3 (Hubel and Wiesel, 1965), it was natural to assume a similar arrangement pertains in monkeys, with each of the proposed architectonic fields corresponding to a retinotopic map. In most current proposals, areas 18 and 19 of Brodmann's scheme have been modified in shape and extent, but they persist in most current proposals as V2 and V3.

Although Brodmann's proposal has had great impact, there were early as well as subsequent reasons to question the specifics. Most notably, Brodmann's contemporaries, using the same architectonic criteria, divided extrastriate cortex differently (e.g. Campbell, 1905; Elliot Smith, 1906; von Economo, 1924). Of course all these investigators agreed on the location and extent of striate cortex, the most distinctive subdivision of cortex, but none illustrated areas comparable to Brodmann's areas 18 and 19. Campbell bordered striate cortex with a large single 'visuopsychic' band of cortex; Elliot Smith divided extrastriate cortex quite differently into parastriate and peristriate areas, while von Economo's areas OA and OB were much narrower than Brodmann's areas 19 and 18. These differences in opinion, which continued with later investigators (e.g. Bailey and von Bonin, 1951), led to serious questions about the usefulness of architectonic features for revealing functionally significant subdivisions of visual cortex (see Lashley and Clarke, 1946).

A second reason to question the scheme of Brodmann is that we can now identify a second visual area, V2, in primates with assurance, using the banding pattern that appears in brain sections processed for cytochrome oxidase (e.g. Tootell et al., 1983; Wong-Riley and Carroll, 1984; Livingstone and Hubel, 1984; Krubitzer and Kaas, 1990). Because this V2 corresponds to an orderly second representation of the contralateral visual hemifield, a retinotopic pattern of input from V1, and other distinctive features, V2 is now well-defined and there is widespread agreement over its existence and exact location. This V2, however, does not correspond to area 18 of Brodmann as depicted in either Old World monkeys (the common model for humans) or humans, although his area 18 in marmosets closely conforms to present descriptions of V2 in marmosets. Since Brodmann's area 18 is much wider than V2 in some but not all primates, it does not consistently reflect a functional subdivision of visual cortex. Instead, his area 18 in macaque monkeys and humans contains more than one visual area. In a similar manner, the wide expanse of cortex that Brodmann included in area 19 clearly contains parts or all of a number of visual areas, although proposals on how to divide this region vary. In summary, differences in proposals based on cortical architectonics led to questions about the validity of any of these proposals, and modern studies reveal that Brodmann's areas 18 and 19 do not reflect valid subdivisions of extrastriate cortex in higher primates.

Early theories based on modern methods

The 1960s brought anatomical procedures for demonstrating patterns of connections and microelectrode mapping methods into widespread use as tools for revealing the organization of visual cortex. The experimental methods yielded new results that investigators tended to interpret, with some modification, within the framework of V1, V2, and V3 based on areas 17, 18, and 19 of Brodmann.

Myers (1965) provided the forerunner of new interpretations based on connection patterns in macaque monkeys (Fig. 2). Callosal connections were found to be most dense just outside the border of V1 (also see Myers, 1962; Cusick and Kaas, 1986), and the projection pattern of V1 revealed a second, mirror-image representation of the visual hemifield outside of V1 that is split so that the zero horizontal meridian forms the outer boundary. As a result of this type of organization (also see Cowey, 1964), locations along the horizontal meridian in V1 project to two opposite locations along the outer border of the second representation. Another observation was the narrowing of the second representation across from the representation of the fovea in V1, a feature now well-established for V2 in macaques and several other primates. Thus, Myers provided accurate information about the organization of the

part of extrastriate visual cortex that is the second visual area, V2.

What is of additional interest in his scheme, however, is his effort to include the areas of Brodmann, although with considerable modifica-

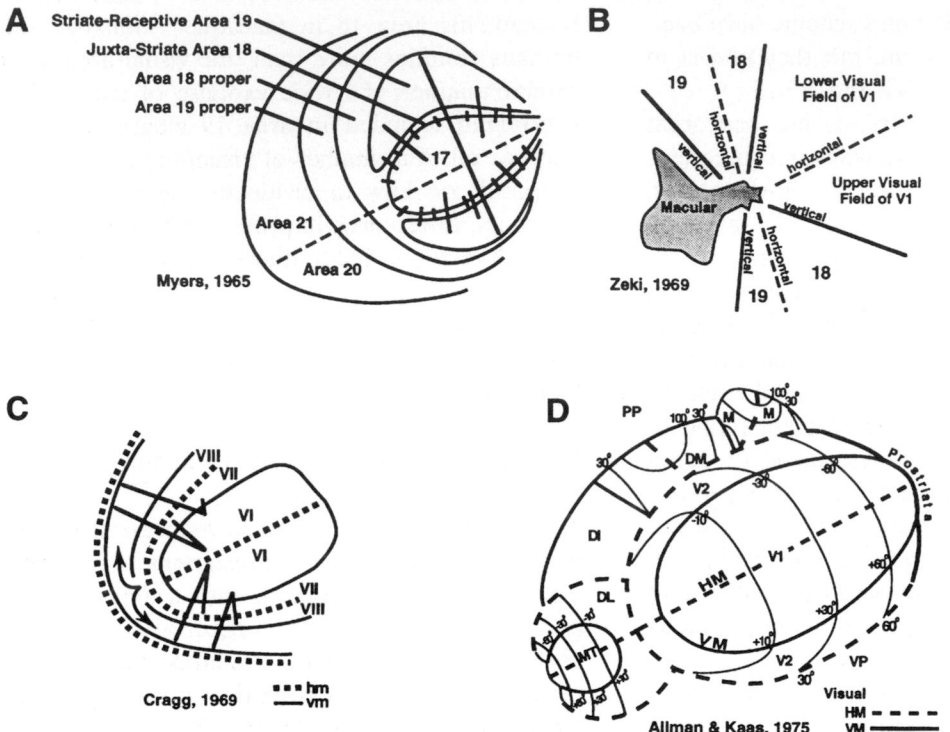

Fig. 2. Early modern portrayals of the postulated organization of extrastriate cortex in monkeys. Each scheme shows cortical fields from a surface-view of flattened cortex, with caudal to the right and medial above. Schemes A–C were based on connection patterns in macaque monkeys, while scheme D was based on microelectrode recordings in owl monkeys. A. The proposal of Myers (1965). Lines mark the projections of area 17 or V1 to what we now term V2. The juxta-striate area 18 not only received inputs from the border region of area 17, thus providing evidence for the representation of the zero vertical meridian, but also it received callosal connections related to the vertical meridian (also see Myers, 1962). The dashed line marks the zero horizontal meridian. Since the middle of area 17, corresponding to the horizontal meridian, projects to the outer border of striate-receptive area 19, this cortex was also seen as representing the horizontal meridian. B. The scheme of Zeki (1969) was based on the projection pattern of V1 (area 17). V1 was found to project not only to V2 (area 18), as found by Myers (see A), but also to more rostral cortex termed area 19 or V3. The proposed projections to ventral 'area 19' have not been found and the projections to dorsal area 19 have been described as to DM (see Krubitzer and Kaas, 1993). C. The scheme of Cragg (1969) was also based on V1 projections (lines). Projections are again summarized as forming two ring-like patterns, V2 (VII) and V3 (VIII), although a complete pattern of projections from V1 to the proposed V3 has not been found. Cragg also illustrated a third ring-like projection zone of converging inputs from parts of V1 representing the horizontal and vertical meridians. Subsequent investigations indicate that projections are to the middle temporal visual area, MT, and not other parts of this proposed ring. When examined in greater detail (e.g. Weller and Kaas, 1983), the projections from different parts of V2 to MT do not converge as proposed. D. A scheme for visual cortex organization in owl monkeys from Allman and Kaas (1975). Besides V1 and V2, proposed areas include the dorsolateral (DL), the dorsointermediate (DI), and the dorsomedial (DM) visual areas as well as the medial (M), Middle Temporal (MT) ventroposterior (VP) and posterior parietal (PP) areas. Some of the representations of visual field coordinates in degrees are indicated.

tion. V2, as currently defined, includes both the 'juxta-striate 18' and the 'striate-receptive 19' of Myers. His '18 and 19 proper' remained as further ring-like fields while 'areas 21 and 20' gained new dimensions as parts of another large ring. Thus, Brodmann's areas were retained by Myers, but greatly modified (cf. Figs. 1 and 2A).

Related schemes of visual cortex organization based on connection patterns soon followed (Fig. 2B,C). Both Zeki (1969) and Cragg (1969) described a similar V2 with the outer border formed by the representation of the zero horizontal meridian, with Zeki terming the field 'area 18', while recognizing it corresponded to only part of Brodmann's area 18, and Cragg (1969) calling the field VII (V2). Both investigators realized that V1 also projects to at least some of the cortex just outside V2, and both investigators interpreted these additional projections within conceptual framework of Brodmann's rings. Thus, Cragg postulated a VIII (V3) as mirroring the retinotopic organization of V2 along the complete outer border of V2. Zeki described a similar field, but referred to it as a redefined 'area 19'. The terms V2 and V3 subsequently came into preferred use. In summary, the strong influence of Brodmann's descriptions of areas 18 and 19 seems clear in all three early interpretations of experimental results.

A quite different proposal for extrastriate areas resulted from early microelectrode mapping studies of extrastriate visual cortex in New World owl monkeys, which also provided the first complete maps of V1 and V2 in primates (Allman and Kaas, 1971b; 1975). Initially, the influence of Brodmann's scheme was strong here also, for the terms areas 18 and 19 were used for regions of cortex between V1 and a newly discovered visual representation, MT (Allman and Kaas, 1971a; Kuypers et al., 1965 had already described V1 projections to the region of MT), with area 18 corresponding to V2 and area 19 corresponding to the largely unexplored region of cortex between V2 and MT. Since Brodmann's proposed extent of area 18 in another New World monkey (marmoset) did approximate that of V2 in owl

monkeys, the use of the term 'area 18' for V2, which was distinguished architectonically, was perhaps appropriate. However, further results from microelectrode mapping in cortex just rostral to V2 soon seemed incompatible with the concept of a ring-like area 19 or V3, and this region of cortex was considered instead to be a 'third tier' containing at least six visual areas (Fig. 2D), rather than a single V3. Areas were named by location, since it seemed unlikely that they could be assigned to successive levels of visual processing, as implied by the sequence V1, V2, V3, V4, etc. Each area was presumed to contain a complete representation of the visual hemifield, although extensive evidence was obtained for only DL (Allman and Kaas, 1974b) DM (Allman and Kaas, 1975) and M (Allman and Kaas, 1976). Subsequently, more complete results have led to a division of the DL region into three fields (see below and Fig. 3B).

Contemporary theories

Widespread agreement on the nature of the organization of extrastriate visual cortex in primates has not yet been achieved. Differences in proposals based on recordings in New World monkeys and connection patterns in Old World monkeys (Fig. 2) typically were dismissed by the premise that these monkeys have basically different plans of organization. However, if such major differences in the arrangement of early stages of cortical processing stations exist in New and Old World monkeys, it would seem that great caution should prevail in attempts to use any of these proposals for modeling the organization of extrastriate cortex in humans. Alternately, as argued here, organization of at least the early stages of cortical processing is highly similar across primate taxa and the differences in proposals reflect the sparseness and ambiguousness of collected data (Kaas, 1993, 1995a, 1995b; Kaas and Preuss, 1993). In either case, we could profit from further study and discussion.

Major differences between current proposals involve the existence or not of an area V3, and

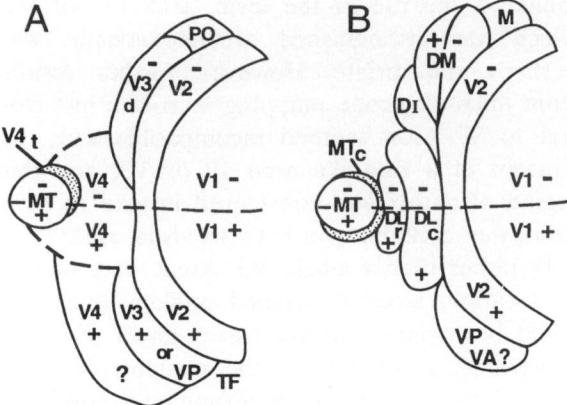

Fig. 3. Current views of visual cortex organization in primates. A. Most current proposals include a V3 and a V4, although in different ways. Earlier portrayals of V4 showed a smaller field with nearly equal representations of the lower (V4 −) and upper (V4 +) visual quadrants (e.g. Van Essen, 1985), but later summaries showed V4 as extending ventrally far into the temporal lobe, thus having a disproportionate representation of the upper quadrant (see Desimone and Ungerleider, 1989; Felleman and Van Essen, 1991) V3 was split into dorsal and ventral parts representing the lower (V3 −) and upper (V3 +) quadrants by Desimone and Ungerleider (1989), while Felleman and Van Essen (1991; also see Van Essen, 1985) concluded that only dorsal V3, representing only the lower quadrant, was V3. This V3 was also separated into two parts along the V2 border by another proposed visual area, PIP (not shown) Ventral V3 was considered to be another visual area, ventral posterior or VP. Both groups of investigators also included a 'transitional' area, V4t, along the part of the middle temporal area, MT, that represents the lower quadrant. PO is a visual area that appears to correspond to M (see Colby et al., 1988), B. An alternative view of how visual cortex is organized in macaques and other higher primates. According to this view, there is no area V3. Instead, the region of V3 is occupied by a series of visual areas, M, DM, DI, DL and VP (plus a VA?) as in owl monkeys (Fig 2D) DM, DI, and M all represent both the upper and lower quadrants. The crescent around the middle temporal area, MTc, is longer than V4t and it represents both visual quadrants. According to this view, both New and Old World monkeys have similar areas, and it is likely that the scheme applies to humans as well (see Kaas, 1995b) The summary is based on Krubitzer and Kaas (1993), Stepniewska and Kaas (1996), and earlier reports.

the extent and subdivisions of the 'V4' or DL region (Fig. 3). The V3 issue is perhaps the most interesting. Early proposals for macaque monkeys (Fig. 2), as noted previously, included a wide and

nearly or completely continuous V3 along the outer border of V2. In contrast, early proposals for New World monkeys included a series of visual areas along the outer boundary of V2 (Fig. 2D). At first it seemed that both proposals could be valid, but for different taxonomic groups. However, a wide and continuous V3 along the outer border of V2 has now been featured in a scheme for cortical organization in New World monkeys (Sousa et al., 1991), and a series of bordering areas, as in owl monkeys, has now been directly postulated for Old World monkeys (Krubitzer and Kaas, 1993; Beck and Kaas, 1995; Stepniewska and Kaas, 1996). Thus, the differences in proposals appear to stem from different data sets and different interpretations of data, rather than from fundamentally different types of cortical organization in primate taxa.

The current proposals that retain a V3 in macaque monkeys have either been unclear about the extent of the area, or have portrayed the field as smaller and more discontinuous than in early proposals. Desimone and Ungerleider (1989), for example, illustrate V3 as composed of narrow ventral (V3v) and dorsal (V3d) parts that are separated by a wide expanse of V4. In addition, they conclude that parts of V2 are bordered dorsally by a parietal-occipital area, PO (possibly the homologue of area M), and ventrally by an area 'TF' (Fig. 3A). Thus, five or more visual areas form the outer border of V2, making this proposal for macaque monkeys somewhat similar to that for owl monkeys. The well-known scheme of Felleman and Van Essen (1991) departs even more from the early portrayals for macaque monkeys in that ventral and dorsal halves of V3 are considered to be different areas, and dorsal V3 is shown as discontinuous. On the basis of evidence that ventral V3 has different connections and neurons with different properties than dorsal V3, ventral V3 was considered to be another field, VP (Fig. 3), as in owl monkeys. The remaining 'V3,' thereby became a very strange visual area, one that is separated into two regions by a proposed posterior interparietal visual area,

PIP, and one that represents only the lower visual quadrant. According to this scheme, the outer border of V2 is formed by six visual areas.

In our alternative proposal, dorsal V3 is part of a larger dorsomedial area, DM, bordered laterally by DI and medially by M (Fig. 3B). Parts of the proposal are well supported and parts are not. Connection patterns with other visual areas support the contention that a complete representation of the visual hemifield, DM, exists along the border of V2 in both New and Old World monkeys (Krubitzer and Kaas, 1993; Beck and Kaas, 1995; Stepniewska and Kaas, 1996). M and PO may be different terms for the same area (see Colby et al., 1988), and thus the field may not be a point of disagreement. The evidence for DI is limited in both New and Old World monkeys, and more study is needed (however, see Rosa and Schmid, 1995). Likewise, the VP region has not been adequately studied (see Newsome et al., 1986), although the VP region clearly has different connections than dorsal V3 (Burkhalter et al., 1985; also see Weller and Kaas, 1983). However, the important distinction of this scheme is that the concept of a V3, rather than being substantially modified from a ring-like area 19, is fully eliminated. V3 is replaced by a sequence of visual areas, as proposed for owl monkeys.

Other differences in the proposals in Fig. 3 are also significant because they involve cortex formerly included in Brodmann's areas 18 and 19. V4 and DL are terms for what seemed to be the same field in Old and New World monkeys. The smaller V4 of Fig. 3A reflects earlier concepts of the location, and extent and retinotopic organization of area V4 (see Van Essen, 1985) that were very similar to early views of DL (Fig. 2A). Mapping studies of Gattas et al., (1988) led to modifications that extended V4 into the temporal lobe (Fig. 3A). In addition, part of V4 was renamed V4t as a separate visual area representing only the lower visual quadrant (Desimone and Ungerlieder, 1986). Our studies of connections (Cusick and Kaas, 1988; Kaas and Morel, 1993; Stepniewska and Kaas, 1996) suggest a different organization. First, V4t seems to be more exten-

sive so that it represents both upper and lower quadrants. We refer to this field as MTc (crescent) after its earlier architectonic identification as a crescent around MT (Tootell et al., 1985). The rest of DL appears to consist of at least two visual areas of different sizes, DLc and DLr (also see Steele et al., 1991; Allman et al., 1994). We conclude that divisions of the V4-DL region are similar in Old and New World monkeys, and we use the same terms.

Summary

This brief review has a few main points.

(1) Early proposals on how extrastriate cortex is subdivided were inconsistent with each other, and differences in interpretation were not resolved.

(2) Brodmann's proposal of two ring-like areas, 18 and 19, surrounding primary visual cortex gained great acceptance despite the lack of agreement among different investigators considering the same evidence.

(3) The concepts of areas 18 and 19, transposed to signify V2 and V3, have had great impact on recent and even current theories of extrastriate visual cortex organization in primates.

(4) Nevertheless, Brodmann's areas 18 and 19, as defined in humans and Old World monkeys, correspond to none of the fields currently proposed for these primates.

(5) All or most mammals appear to have a V2, and there is now widespread complete agreement over the extent and organization of this area in all studied primates. V2 is commonly referred to as area 18 because of its correspondence to area 18 as defined by Brodmann in some mammals. Yet, we should recognize that V2 is about half the size of Brodmann's area 18 in Old World monkeys and humans.

(6) Current concepts of V3 differ greatly from the ring-like area 19 of Brodmann. We question the validity and usefulness of retaining

the concept of V3 in primates. Our proposal for DM and other visual areas along the outer border of V2 seems more consistent, not only with the evidence from New World monkeys, but with evidence from Old World and prosimian primates, and even mammals most closely related to primates (see Kaas and Preuss, 1993). In all of these primates and close relatives of primates, the evidence indicates that more than one field forms the outer border of V2.

References

Allman, J.M. and Kaas, J.H. (1971a) A representation of the visual field in the caudal third of the middle temporal gyrus of the owl monkey (*Aotus trivirgatus*). *Brain Res.* 31: 85–105.

Allman, J.M. and Kaas, J.H. (1971b) Representation of the visual field in striate and adjoining cortex of the owl monkey (*Aotus trivirgatus*). *Brain Res.* 35: 89–106.

Allman, J.M. and Kaas, J.H. (1974a) The organization of the second visual area (V-II) in the owl monkey: a second-order transformation of the visual hemifield. *Brain Res.* 76: 247–265.

Allman, J.M. and Kaas, J.H. (1974b) A crescent-shaped cortical visual area surrounding the middle temporal area (MT) in the owl monkey (*Aotus trivirgatus*). *Brain Res.* 81: 199–213.

Allman, J.M. and Kaas, J.H. (1975) The dorsomedial cortical visual area: a third tier area in the occipital lobe of the owl monkeys (Aotus trivirgatus). *Brain Res.* 100: 473–487.

Allman, J.M. and Kaas J.H. (1976) Representation of the visual field in the medial wall of the occipital-parietal cortex in the owl monkey. *Science* 191: 572–576.

Allman, J., Jeo, R., and Sereno, M. (1994) The functional organization of visual cortex in owl monkeys. In: J.F. Baer, R.E. Weller, and I. Kakoma, (Eds.), *Aotus: The Owl Monkey*. Academic Press, Orlando, FL, 1994, pp. 287–320.

Bailey, P. and von Bonin, G. (1951) The isocortex of Man., Urbana, University of Illinois Press.

Beck, P.D. and Kaas, J.H. (1995) Evidence for the presence of the dorsomedial visual area (DM) in five primate species. *Soc. Neurosci. Abstr.*. 21: 1275.

Brodmann, K. (1909). *Vergleichende Lokalisationslehre der Grosshirnrinde*. Leipzig: Verlag Barth.

Burkhalter, A., Felleman, D.J., Newsome, W.T. and Van Essen, D.C. (1985) Anatomical and physiological asymmetries related to visual areas V3 and VP in macaque extrastriate cortex. *Vis. Res.* 26: 63–80.

Campbell, A., W. (1905). *Histological studies on the localization of cerebral function*. Cambridge, England: Cambridge University Press.

Colby, C.L., Gattass, R., Olson, C.R., and Gross, C.G. (1988) Topographic organization of cortical afferents to extrastriate visual area PO in the macaque: a dual tracer study. *J. Comp. Neurol.* 269: 392–413.

Cowey, A. (1964) Projection of the retina on to striate and prestriate cortex in the squirrel monkey (*Saimiri Sciureus*). *J. Neurophysiol.* 27: 366–396.

Cowey, A. and Stoerig, P. (1991) The neurobiology of blindsight. *TINS* 14: 140–145.

Cragg, B.G. (1969) The topography of the afferent projections in the circumstriate visual cortex of the monkey studied by the Nauta method. *Vis. Res.* 9: 733–747.

Cusick, C.G. and Kaas, J.H. (1986) Interhemispheric connections of cortical, sensory and motor maps in primates. In: F. Lepore, M. Ptito, and H., H. Jasper (Eds.), *Two Hemispheres—One Brain*, New York: Alan, R. Liss, pp. 83–102.

Cusick, G.G. and Kaas, J.H. (1988) Cortical connections of area 18 and dorsolateral visual cortex in squirrel monkeys. *Vis. Neurosci.* 1: 211–233.

Desimone, R. and Ungerleider, L.G. (1986) Multiple visual areas in the caudal superior temporal sulcus of the macaque. *J. Comp. Neurol.* 248: 164–189.

Economo, von, C. (1929). *The cytoarchitectonics of the human cortex*. Oxford, England: Oxford University Press.

Felleman, D.J. and Van Essen, D.C. (1991) Distributed hierarchical processing in primate cerebral cortex. *Cerebral Cortex* 1: 1–47.

Gattass, R., Sousa, A.P.B., and Gross, C.G. (1988) Visuotopic organization and extent of V3 and V4 of the macaque. *J. Neurosci.* 8: 1831–1845.

Girard, P. and Bullier, J. (1989) Visual activity in area V2 during reversible inactivation of area 17 in the macaque monkey. *J. Neurophysiol.* 62: 1287–1302.

Girard, P. Salin, P.A., and Bullier, J. (1991) Visual activity in areas V3a and V3 during reversible inactivation of area V1 in the macaque monkey. *J. Neurophysiol.* 66: 1493–1503.

Hubel, D.H. and Wiesel, T.N. (1965) Receptive fields and functional architecture in two nonstriate visual areas (18 and 19) of the cat. *J. Neurophysiol.* 30: 1561–1573.

Jouandet, M.I., Tramo, M.J., Herron, D.M., Hermann, A., Loftus, W.C., Barzell, J., and Gazzaniga, M.S. (1989) Brain Prints: Computer-generated two-dimensional maps of the human cerebral cortex in vivo. *J. Cog. Neurosci.* 1: 88–117.

Kaas, J.H. (1993) The organization of visual cortex in primates: problems, conclusions, and the use of comparative studies in understanding the human brain. In: B. Gulyβa, D. Ottoson, and P.E. Roland (Eds.) The functional organization of the human visual cortex. Pergamon Press, Oxford, 1993, pp. 1–11.

Kaas, J.H. (1995a) The evolution of isocortex. *Brain Behav. Evol.* 46: 187–196.

Kaas, J.H. (1995b) Human visual cortex; Progress and Puzzles. *Current Biology* 5: 1126–1128.

Kaas, J.H. and Huerta, M.G. (1988) Subcortical visual system of primates. In: H.P. Steklis (Ed.), *Comparative Primate*

Biology, Vol. 4: Neurosciences. Alan R. Liss, New York, 1988, pp. 327–391.

Kaas, J., H. and Krubitzer, L.A. (1991) The organization of extrastriate visual cortex. In: B. Dreher and S., R. Robinson (Eds.), *Neuroanatomy of Visual Pathways and their Retinotopic Organization*, Vol. III of *Vision and Visual Dysfunction*, J. Cronly-Dillon (Gen. Ed.), The MacMillan Press, London, 1991, pp. 302–359.

Kaas, J.H. and Morel, A. (1993) Connections of visual areas of the upper temporal lobe of owl monkeys: the MT crescent and dorsal and ventral subdivisions of FST. *J. Neurosci.* 13: 534–546.

Kaas, J.H., and Preuss, T.M. (1993) Archontan affinities as reflected in the visual system. In: F. Szalay, M. Novacek and M. McKenna. *Mammal Phylogeny*, Springer-Verlag. New York, pp. 115–128.

Krubitzer, L.A. and Kaas, J.H. (1990) Cortical connections of MT in four species of primates: areal, modular, and retinotopic patterns. *Vis. Neurosci.* 5: 165–204.

Krubitzer, L.A., and Kaas, J.H. (1993) The dorsomedial visual area of owl monkeys: connections, myeloarchitecture, and homologies in other primates. *J. Comp. Neurol.* 334: 497–528.

Kuypers, H.G.J.M., Szwarcbart, M.K., Mishkin, M., and Rosvold, H.E. (1965) Occipitotemporal corticocortical connections in the rhesus monkey. *Exp. Neurol.* 11: 245–262.

Lashley, K.S. and Clark, G. (1946) The cytoarchitecture of the cerebral cortex of ateles: a critical examination of architectonic studies. *J. Comp. Neurol.* 85: 223–305.

Livingstone, M.S. and Hubel, D.H. (1984) Anatomy and physiology of a color system in the primate visual cortex. *J. Neurosci.* 4: 309–356.

McCulloch, W.S. (1944) Functional organization of cerebral cortex. *Physiol. Rev.* 24: 390–407.

Myers, R.E. (1962) Commissural connections between occipital lobes of the monkey. *J. Comp. Neurol.* 118: 1–16.

Myers, R.E. (1965) Organization of visual pathways. In: E.G. Ettlinger (Ed.), *Functions of the corpus callosum.* Churchill, London., p. 133.

Newsome, W.T., Maunsell, J.H.R., and Van Essen, D.C. (1986) Ventral posterior visual area of the macaque: visual topography and areal boundaries. *J. Comp. Neurol.* 252: 139–153.

Preuss, T.M. (1995) The argument from animals to humans in cognitive neuroscience. In: *The Cognitive Neurosciences*, M.S. Gazzuniga, (Ed.), MIT press, Boston, pp. 1227–1241.

Rodman, H.R., Gross, C.G. and Albright, T.D. (1989) Afferent basis of visual response properties in area MT of the macaque: I. Effects of striate cortex removal. *J. Neurosci.* 9: 2033–2050.

Rosa, M.G.P. and Schmid, L.M. (1995) Visual areas in the dorsal and medial extrastriate cortices of the marmoset. *J. Comp. Neurol.* 359: 272–299.

Schiller, P.H. and Malpeli, J.G. (1977) The effect of striate cortex cooling on area 18 cells in the monkey. *Brain Res.* 126: 366–369.

Smith, G.E. (1906) A new topographic survey of human cerebral cortex, being an account of the distribution of the anatomically distinct cortical areas and their relationship to the cerebral sulci. *J. Anat. Physiol.* 42: 237–254.

Sousa, A.P.B., Carmen M., Pinon, G.P., Gattas, R. and Rosa, M.G.P. (1991) Topographic organization of cortical input to striate cortex in the cebus monkey: a fluorescent tracer study. *J. Comp. Neurol.* 308: 665–682.

Steele, G.E., Weller, R.E. and Cusick, C.G. (1991) Cortical connections of the caudal subdivision of the dorsolateral area (V4) in monkeys. *J. Comp. Neurol.* 306: 495–520.

Stepniewska, I. and Kaas, J.H. (1996) Topographic patterns of V2 cortical connections in macaque monkeys. *J. Comp. Neurol.* in press.

Tootell, R.B.H., Silverman, M.S., Devalois, R.L. and Jacobs, G.H. (1983) Functional organization of the second cortical visual area in primates. *Science* 220: 737–739.

Tootell, R.B.H., Hamilton, S.L. and Silverman, M.S. (1985) Topography of cytochrome oxidase activity in owl monkey cortex. *J. Neurosci.* 5: 2786–2800.

Van Essen, D.C. (1985) Functional organization of primate visual cortex. In: Jones, E.G., and A. Peters (Eds.): *The Cerebral Cortex*, Vol. 3., New York: Plenum Press, pp. 259–329.

Weller, R.E. and Kaas, J.H. (1983) Retinotopic patterns of conections of area 17 with visual areas V-II and MT in macaque monkeys. *J. Comp. Neurol.* 220: 253–279.

Wong-Riley, M.T.T. and Carrol, E.W. (1984) Quantitative light and electron microscopic analysis of cytochrome oxidase-rich zones in V-II prestriate cortex of the squirrel monkey. *J. Comp. Neurol.* 222: 18–37.

Zeki, S.M. (1969) Representation of central visual fields in prestriate cortex of monkey. *Brain Res.* 14: 271–291.

M. Norita, T. Bando and B. Stein (Eds.)
Progress in Brain Research, Vol 112

CHAPTER 16

Afferent and developmentally inherent mechanisms of form and motion processing in cat extrastriate cortex

Peter D. Spear*

Department of Psychology and Center for Neuroscience, University of Wisconsin-Madison, 1202 West Johnson St., Madison, WI 53706, USA

Introduction

Every mammalian species that has been studied has been shown to have multiple extrastriate cortical areas devoted to vision (see Spear, 1991, for a review). Each of these extrastriate areas receives multiple sources of afferent information, including inputs from other visual cortical areas and from several thalamic nuclei. The recipient neurons somehow process and integrate these multiple sources of input to produce their own output visual receptive-field properties. How is this done? That is, how do the recipient neurons use the multiple sources of afferent information? Do they combine specific input receptive-field properties from each afferent source to produce the output receptive-field? Or do some inputs have only modulatory functions that are not reflected in specific output receptive-field properties? During development, do the extrastriate neurons have their receptive-field properties imposed on them by their inputs, or do the neurons themselves have inherent properties that determine their output receptive-field characteristics? For example, if the inputs to extrastriate neurons are changed during development, will the output receptive fields also change, or will the neurons use the different inputs to produce the same output receptive-field properties that they normally would have?

My laboratory has addressed many of these questions in the course of our studies of the mechanisms of compensation for visual cortex damage in adult cats and newborn kittens. As part of these studies, we have investigated the effects of both acute and long-term removal of inputs from areas 17, 18, and 19 (referred to collectively as visual cortex, or VC) on the receptive-field properties of neurons in the posteromedial lateral suprasylvian (PMLS) extrastriate visual area of cortex. This has allowed us to determine the role of various inputs in the elaboration of PMLS receptive fields in normal adult animals. In addition, it has provided evidence that the development of certain receptive-field properties depends upon intrinsic characteristics of the PMLS neurons, not simply upon the particular inputs to the neurons.

Normal motion and form processing in PMLS cortex

It has long been known that PMLS cortex is involved in processing information about stimulus

---

*Corresponding author. Tel.: +1 303 492 7294; fax: +303 492 4944; email: peter.spear@colorado.edu

motion. Thus, the vast majority of the neurons are motion sensitive and direction selective. They respond better to moving than to stationary flashing stimuli, and their responses depend upon the direction of stimulus movement (e.g., Hubel and Wiesel, 1969; Wright, 1969; Spear and Baumann, 1975; Turlejski, 1975; Camarda and Rizzolatti, 1976). This direction-selective motion processing can be quite complex. For example, many cells have opponent direction-selectivity between the receptive-field center and surround regions (von Grunau and Frost, 1983). In addition, the direction selectivity extends beyond simple sensitivity to the direction of motion in the fronto-parallel plane, and many cells respond differentially to stimuli moving toward or away from the animal (Toyama et al., 1985).

PMLS neurons also are sensitive to various aspects of stimulus form. For example, many of the receptive fields have inhibitory surrounds (e.g., Spear and Baumann, 1975; Camarda and Rizzolatti, 1976), which makes the cells sensitive to the size of a stimulus. Tests with sine-wave grating stimuli also demonstrate that PMLS cells are sensitive to the spatial-frequency content of visual stimuli (e.g., Morrone et al., 1986; Zumbroich and Blakemore, 1987; Gizzi et al., 1990; Guido et al., 1990b). However, both the optimal spatial frequencies and the spatial resolutions of PMLS neurons tend to be quite low. Thus, the spatial-frequency sensitivity is very crude compared to that in striate cortex.

Recent studies also have shown that PMLS neurons are sensitive to stimulus orientation (Blakemore and Zumbroich, 1987; Hamada, 1987; Gizzi et al., 1990; Danilov et al., 1995a), another aspect of form processing. Like spatial-frequency sensitivity, however, the orientation sensitivity is very crude compared to that of neurons in striate cortex. For example, PMLS cells respond to a much wider range of stimulus orientations than do striate cortex neurons.

Thus, normal PMLS neurons are sensitive to stimulus form as well as to stimulus motion and direction. However, the form processing is fairly crude, and the neurons do not appear to be providing information for detail vision. Correspondingly, the receptive fields tend to be quite large, and the topographic organization of PMLS cortex is quite course (e.g., Spear and Baumann, 1975; Palmer et al., 1978; Zumbroich et al., 1986). These and other properties of PMLS neurons have led to suggestions that PMLS cortex is involved in the processing of image motion, to specialized aspects of attention and orientation, or to the near response (lens accommodation, pupillary constriction, and convergent eye movements) that occurs when an object moves closer to the eyes (see Spear, 1991, for a review).

Influence of afferents to PMLS cortex in adults

The PMLS cortex receives convergent inputs from several pathways (for reviews see Rosenquist, 1985; Spear, 1985; Bullier, 1986; Dreher, 1986). The main visual inputs are summarized in Fig. 1. There is a direct retinogeniculate pathway to PMLS cortex via the geniculate wing (GW), medial interlaminar nucleus (MIN), and C layers of the dorsal lateral geniculate nucleus (LGN). This pathway includes inputs from LGN Y-cells and W-cells, but not X-cells. A second visual pathway comes from the tectothalamic system. The upper layers of the superior colliculus (SC) project to portions of the LGN, the lateral posterior nucleus (LP), and the posterior nucleus (PN) of the thalamus that project to PMLS cortex. A third pathway consists of a projection from the pretectal nucleus of the optic tract (PRE-TECT) to portions of the LGN and pulvinar (PUL) that project to PMLS cortex. PMLS cortex also receives direct cortico-cortical inputs from areas 17, 18, and 19 of both hemispheres and from a variety of other extrastriate cortical areas. Unlike areas 17 and 18 (Fig. 1, right), PMLS cortex does not receive projections from layers A and A1 of the LGN.

If one surgically removes the inputs from ipsilateral visual cortex (areas 17, 18, and 19), the receptive-field properties of PMLS cells change dramatically. As shown in Fig. 2 (cf. A and B), the percentage of direction-selective cells decreases

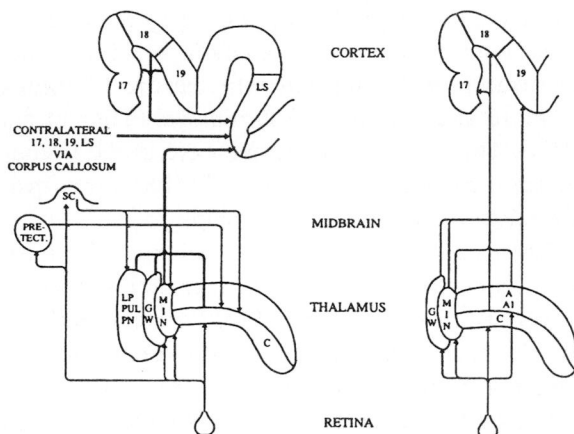

Fig. 1. Diagrams showing the 'primary' retino-geniculo-cortical pathways to areas 17, 18, and 19 (right) and the projections to the posteromedial lateral suprasylvian (LS) extrastriate visual area of cortex (left). A and A1 refer to the corresponding layers of the dorsal lateral geniculate nucleus (LGN). C refers to the C complex of the LGN, which includes four hidden layers (C, C1, C2, and C3). MIN, medial interlaminar nucleus. GW, geniculate wing. LP, lateral posterior nucleus. PUL, pulvinar nucleus. PN, posterior nucleus. SC, superior colliculus. PRE-TECT, pretectal nuclei.

from about 80% of the responsive cells in normal cats to about 20% in cats with ipsilateral VC inputs removed. This decrease is accompanied by an increase in the percentage of cells that respond best to moving stimuli but lack direction selectivity (cells in the movement-sensitive class). However, the increase in movement-sensitive cells corresponds to only about half of the decrease in direction-selective cells. There also is an increase in the percentage of cells that respond as well to stationary flashed stimuli as to stimulus movement (stationary class). Thus, the VC removal appears to produce a reduction in both direction selectivity and movement sensitivity. These effects occur within hours of removing inputs from ipsilateral VC, and no further changes occur during a period of more than a year after the lesion. Thus, the effects are due to removal of inputs from VC and not to secondary consequences of the lesion, such as retrograde degeneration in the thalamus or in PMLS cortex itself (which projects back to areas 17, 18, and 19).

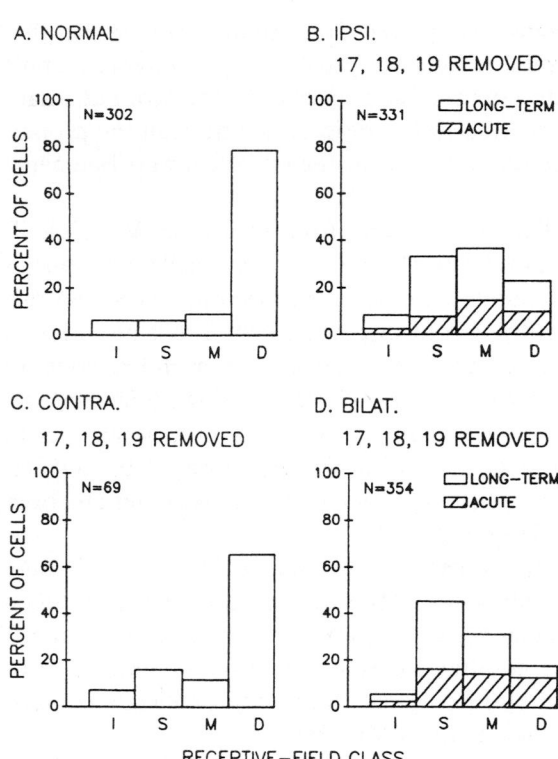

Fig. 2. Summary of acute and long-term effects of adult VC removal on the receptive-field properties of PMLS neurons. Panels show the percentages of PMLS cells with different receptive-field properties in normal cats (A) and cats with ipsilateral (B), contralateral (C), or bilateral (D) removal of areas 17, 18, and 19. Cross-hatched portions of each bar are results from cats studied within a day of the cortical lesion (acute lesion). Open portions of each bar are results from cats studied 2 weeks to more than a year after the lesion (long-term lesion). N is the number of cells in each condition. Receptive-field classes are: D, direction selective; M, movement sensitive; S, stationary; I, indefinite. Data in A are combined from Spear and Baumann (1975), Smith and Spear (1979), Spear et al. (1985), and McCall et al. (1988). Data in B are combined from Spear and Baumann (1979a), Spear et al. (1980, 1988), and Tong et al. (1984). Data in C are from Spear and Baumann (1979a). Data in D are combined from Spear and Baumann (1979a,b) and Spear et al. (1988). Figure from Spear (1988).

Fig. 2C shows that unilateral VC removal has little or no effect on the receptive-field properties of the PMLS neurons in the contralateral hemisphere. In addition, bilateral removal of VC produces effects that are very similar to ipsilateral removal (Fig. 2D). Other studies have shown that similar effects are produced by removal of inputs from areas 17 and 18, and that removal of inputs from area 19 has little or no effect on the properties that have been studied (Spear and Baumann, 1979a; Spear, 1988).

Removing inputs from ipsilateral VC also affects the orientation sensitivity of PMLS neurons (Danilov et al., 1995a,b). In normal cats, approximately 70% of responsive PMLS cells are orientation sensitive to a grating stimulus independent of the spatial phase of the grating. Following removal of ipsilateral VC inputs, only about 15% of the cells are orientation sensitive. Thus, orientation sensitivity also depends upon inputs from ipsilateral VC.

In contrast, the spatial-frequency tuning properties of PMLS neurons do not depend upon cortico-cortical inputs. Removal of ipsilateral VC has no effect on spatial resolution (see Fig. 4) or the optimal spatial frequency of PMLS neurons (Guido et al., 1990b, 1992).

Taken together, the results indicate that in normally reared adult cats, PMLS neurons use inputs from ipsilateral areas 17 and 18 for the elaboration of motion sensitivity, direction selectivity, and orientation selectivity. Many other properties that are not affected by VC removal must therefore be elaborated on the basis of thalamic or other cortical inputs, independent of areas 17, 18, or 19 (see Spear, 1988). These properties include responses to flashed stimuli, receptive-field size, spatial summation, surround inhibition, responses to different stimulus velocities, and spatial-frequency tuning. Some inputs, such as those from the contralateral areas 17, 18, and 19 and ipsilateral area 19, do not appear to be used for receptive-field formation. Presumably, these inputs have modulatory functions that are yet to be studied.

Changing the inputs during development

When inputs from ipsilateral areas 17, 18, and 19 are removed on the day of birth, the subsequent development of projections from the thalamus is altered. This originally was shown by a transneuronal anterograde tracing study in which radioactive tracers were injected into the eye 6 months or more after neonatal VC removal (Tong et al., 1984). This study found an increase in the retino-thalamo-cortical projection to PMLS cortex ipsilateral to the neonatal VC lesion.

Enhanced projections from thalamus also have been shown by retrograde tracing studies in which tracers were injected into PMLS cortex (Kalil et al., 1991; Lomber et al., 1995). These studies found a 5–10-fold increase in the numbers of C-layer LGN cells that project to PMLS cortex ipsilateral to a neonatal VC lesion. In addition, projections from the LGN A-layers, which are present at birth and normally withdraw and disappear during development (Kato et al., 1986; Bruce and Stein, 1988; Tong et al., 1991), stabilize and remain into adulthood following neonatal removal of VC inputs. No such enhanced projections to PMLS cortex are present following removal of VC in adults, even after long survival times (Tong et al., 1984; Kalil et al., 1991; Lomber et al., 1995).

Thus, PMLS cortex cells receive very different inputs in normal adult animals and adult animals that had VC removed at birth. Normal PMLS cells receive inputs from ipsilateral areas 17, 18, and 19, whereas PMLS cells in animals with neonatal VC removal do not. Normal PMLS cells receive weak inputs from the LGN C-layers and none from the A-layers, whereas PMLS cells in animals with neonatal VC removal receive strong inputs from the C-layers and a significant anomalous projection from the A-layers. The question that arises is, How do the receptive-field properties compare in these two groups of PMLS cells with their very different sets of inputs?

Fig. 3 shows the results for motion sensitivity and direction selectivity (Spear et al., 1980; see also Tong et al., 1984, 1987; Guido et al., 1990a). Animals that grow up following VC removal at birth have a normal complement of direction-selective cells. Moreover, the directional tuning of the direction-selective cells is very similar to that of normal-adult PMLS cells. This is quite different than what happens after VC removal in adults, where there is a marked reduction in motion and direction sensitivity (Fig. 2). Thus, physiological compensation is seen in PMLS cortex following neonatal VC removal, and the cells develop direction-selective properties like those seen in normal adults.

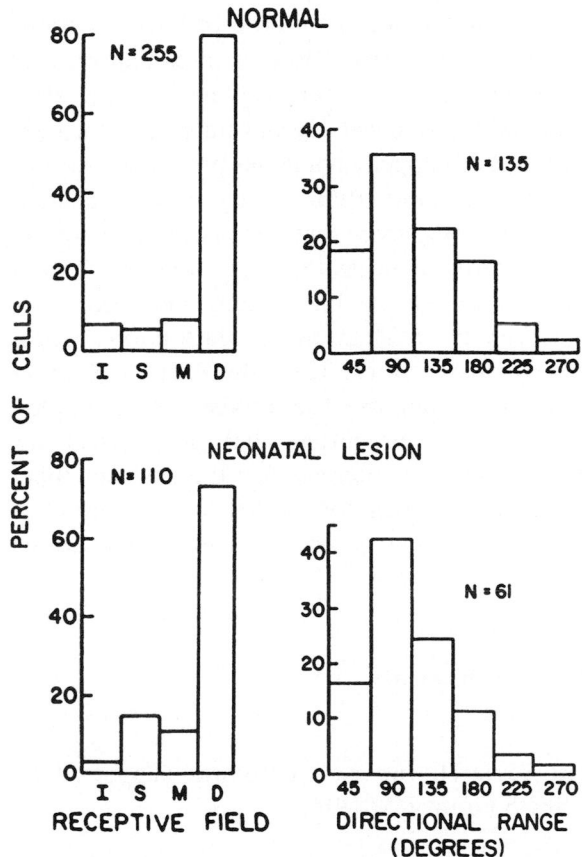

Fig. 3. Summary of the long-term effects of neonatal VC removal on receptive-field properties of PMLS neurons. Panels on the left show the percentages of PMLS cells with different receptive-field properties. Receptive-field classes are: D, direction selective; M, movement sensitive; S, stationary; I, indefinite. Panels on the right show the directional tuning for direction-selective cells in each condition. The total range of directions of stimulus movement to which each cell responded is shown on the abscissa. N is the number of cells studied. Top shows the results in normal adult cats; bottom shows results from adult cats that had received a VC lesion on the day of birth. Data from Smith and Spear (1979), Spear and Baumann (1975) and Spear et al. (1980).

Similar results are observed for orientation sensitivity (Danilov, Moore, King and Spear, unpublished). Following VC removal at birth, the animals grow up to have about 42% orientation-sensitive cells in PMLS cortex. Although this is not as high a percentage as in normal cats (70%), it is higher than in cats with VC inputs removed as adults (15%). Thus, there is a partial compensation of orientation sensitivity in PMLS cortex following neonatal removal of inputs from VC. The orientation tuning is quite broad among the orientation-sensitive cells in cats with a neonatal lesion, just like that in normal adult PMLS cortex. Thus, there is no evidence that the PMLS cells develop sharply tuned orientation sensitivity like that of striate cortex.

Fig. 4 shows results for spatial resolution. As already noted, neither acute nor long-term removal of VC in adult cats has any effect on the spatial resolution PMLS cells. Very similar results are seen in cats with neonatal removal of VC (Guido et al., 1992). The distribution of spatial resolutions is similar to that in normal PMLS cortex and following adult VC removal. The same results were seen for optimal spatial frequency (Guido et al., 1992).

Conclusions

Fig. 5 presents a simplified summary of the results and conclusions. In normal adult cats, PMLS neurons respond selectively to stimulus motion (MS), direction (DS), orientation (OS), and spatial-frequency content (SF). The orientation tuning is broad, and spatial resolution and optimal spatial frequency are low. Experiments in which VC is removed in adult cats indicate that the motion, direction, and orientation sensitivities are elaborated on the basis of inputs from ipsilateral areas 17 and 18. The spatial-frequency tuning is elaborated on the basis of remaining inputs, perhaps those from the C-layers of the LGN. Removal of ipsilateral area 19 and contralateral areas 17, 18, and 19 have no obvious effect on the

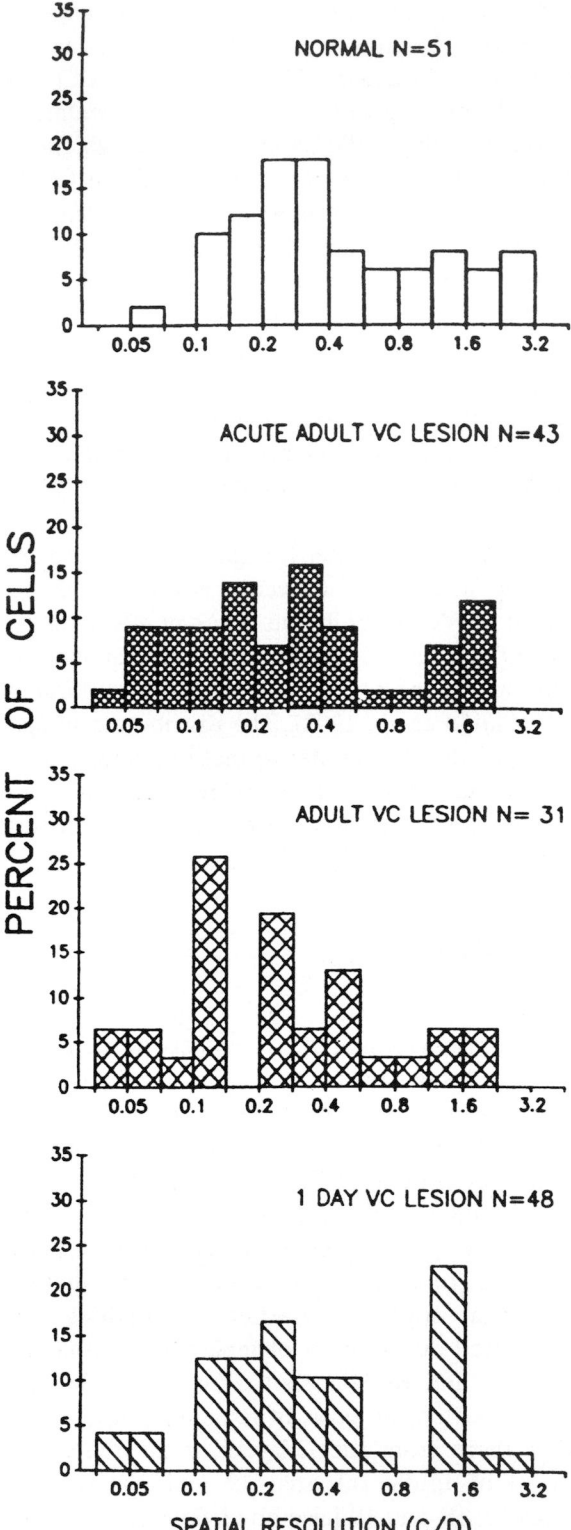

receptive-fields of PMLS neurons, and their roles remain a mystery.

Following neonatal VC damage, PMLS cells have markedly different inputs than in normal adults. They are missing their ipsilateral inputs from areas 17, 18, and 19, they have heavier than normal inputs from the LGN C-layers, and they have anomalous inputs from the LGN A-layers. Yet the receptive-field properties of the neurons are virtually identical to those of PMLS neurons in normal adult cats. The properties that are lost following adult VC removal are present after neonatal VC removal. Equally important, the cells have not developed anomalous properties, such as the narrow orientation tuning or high spatial-frequency sensitivity that are characteristic of the striate cortex neurons that were removed. This is so even though the PMLS cells now receive inputs from the LGN A-layers, which normally project to striate cortex. Thus, the PMLS cells have used a vastly abnormal set of afferents to develop receptive-field properties that are normal for PMLS cells. This suggests that there is something inherent in PMLS cortex that leads neurons to develop particular properties based on whatever inputs are available.

Acknowledgements

The research reported here was supported by USPHS Grant EY01916.

Fig. 4. Spatial resolution (visual acuity) of PMLS neurons in normal adult cats (NORMAL), cats that were studied within 1–3 days of a VC lesion that was received as an adult (ADULT ACUTE VC LESION), cats that were studied ≥ 6 months after a VC lesion received as an adult (ADULT VC LESION), and cats that were studied ≥ 6 months after a VC lesion received on the day of birth (1 DAY VC LESION). *N*, number of cells studied in each group of cats. Spatial resolution was defined as the highest spatial frequency (cycles/deg, C/D) to which the cell gave a statistically significant response. Data from Guido et al. (1992).

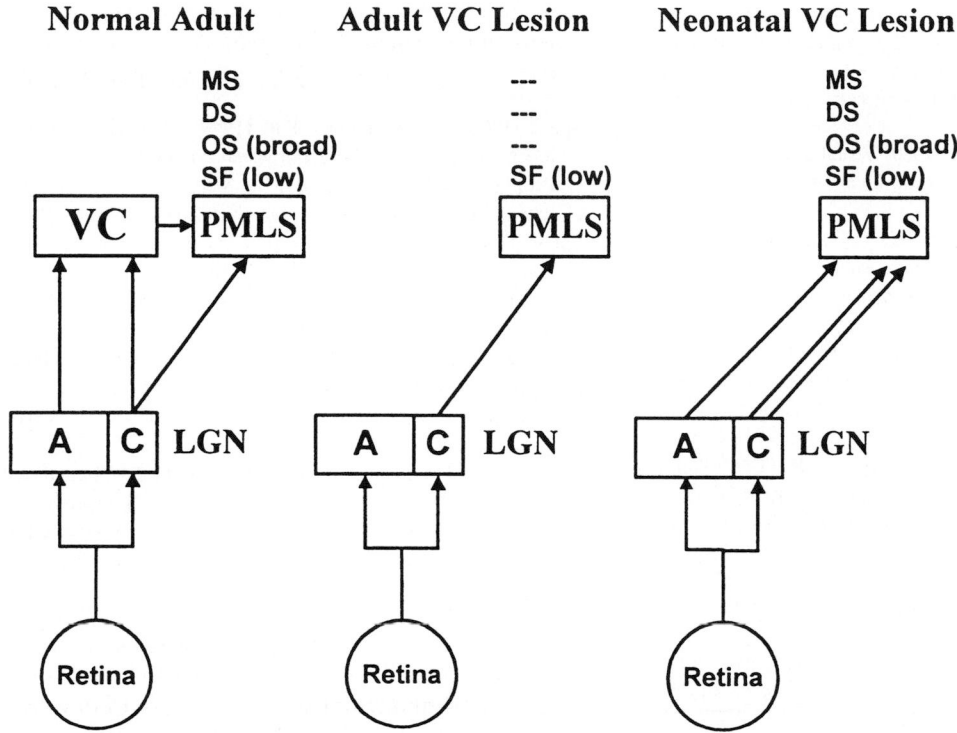

Fig. 5. Simplified diagrams of inputs to posteromedial lateral suprasylvian (PMLS) cortex in normal adult cats, cats with VC removed as adults (Adult VC Lesion), and adult cats that had VC removed on the day of birth (Neonatal VC Lesion). LGN, lateral geniculate nucleus. A, C, the corresponding groups of LGN layers. VC, visual cortex (areas 17, 18, and 19). MS, motion sensitive. DS, direction selective. OS, orientation selective. SF, spatial-frequency sensitive.

References

Blakemore, C. and Zumbroich, T.J. (1987) Stimulus selectivity and functional organization in the lateral suprasylvian visual cortex of the cat. *J. Physiol., Lond.*, 389: 569–603.

Bruce, L.L. and Stein, B.E. (1988) Transient projections from the lateral geniculate to the posteromedial lateral suprasylvian visual cortex in kittens. *J. Comp. Neurol.*, 278: 287–302.

Bullier, J. (1986) Axonal bifurcation in the afferents to cortical areas of the visual system. In: J.D. Pettigrew, K.J. Sanderson and W.R. Levick (Eds.), *Visual Neuroscience*, Cambridge University Press, Cambridge, pp. 239–259.

Camarda, R. and Rizzolatti, G. (1976) Visual receptive fields in the lateral suprasylvian area (Clare-Bishop area) of the cat. *Brain Res.*, 101: 427–443.

Danilov, Y., Moore, R.J., King, V.R. and Spear, P.D. (1995a) Are neurons in cat posteromedial lateral suprasylvian visual cortex orientation sensitive? Tests with bars and gratings. *Visual Neurosci.*, 12: 141–151.

Danilov, Y., Moore, R.J., King, V.R. and Spear, P.D. (1995b) Function of feedback from extrastriate to striate cortex in the cat: comparison of cooling and lesion effects. *Soc. Neurosci. Abs.*, 359.9.

Dreher, B. (1986) Thalamocortical and corticocortical interconnections in the cat visual system: relation to the mechanisms of information processing. In: J.D. Pettigrew, K.J. Sanderson and W.R. Levick (Eds.), *Visual Neuroscience*, Cambridge University Press, Cambridge, pp. 290–314.

Gizzi, M.S., Katz, E., Schumer, R.A. and Movshon, J.A. (1990) Selectivity for orientation and direction of motion of single neurons in cat striate and extrastriate visual cortex. *J. Neurophysiol.*, 63: 1529–1543.

Guido, W., Spear, P.D. and Tong, L. (1990a) Functional compensation in the lateral suprasylvian visual area following bilateral visual cortex damage in kittens. *Exp. Brain. Res.*, 83: 219–224.

Guido, W., Tong, L. and Spear, P.D. (1990b) Afferent bases of spatial-frequency and temporal-frequency processing by neurons in the cat's posteromedial lateral suprasylvian cortex — effects of removing areas 17, 18, and 19. *J. Neurophysiol.*, 64: 1636–1651.

Guido, W., Spear, P.D. and Tong, L. (1992) How complete is

physiological compensation in extrastriate cortex after visual cortex damage in kittens. *Exp. Brain Res.*, 91: 455–466.

Hamada, T. (1987) Neural response to the motion of textures in the lateral suprasylvian area of cats. *Behav. Brain Res.*, 25: 175–186.

Hubel, D.H. and Wiesel, T.N. (1969) Visual area of the lateral suprasylvian gyrus (Clare-Bishop area) of the cat. *J. Physiol., Lond.*, 202: 251–260.

Kalil, R.E., Tong, L. and Spear, P.D. (1991) Thalamic projections to the lateral suprasylvian visual area in cats with neonatal or adult visual cortex damage. *J. Comp. Neurol.*, 314: 512–525.

Kato, N., Kawaguchi, S. and Miyata, H. (1986) Postnatal development of afferent projections to the lateral suprasylvian visual area in the cat: An HRP study. *J. Comp. Neurol.*, 252: 543–554.

Lomber, S.G., MacNeil, M.A. and Payne, B.R. (1995) Amplification of thalamic projections to middle suprasylvian cortex following ablation of immature primary visual cortex in the cat. *Cerebral Cortex*, 2: 166–191.

McCall, M.A., Tong, L. and Spear, P.D. (1988) Development of neuronal responses in cat posteromedial lateral suprasylvian visual cortex. *Brain Res.*, 447: 67–78.

Morrone, M.C., Di Stefano, M. and Burr, D.C. (1986) Spatial and temporal properties of neurons of the lateral suprasylvian cortex of the cat. *J. Neurophysiol.*, 56: 969–986.

Palmer, L.A., Rosenquist, A.C. and Tusa, R.J. (1978) The retinotopic organization of lateral suprasylvian visual areas in the cat. *J. Comp. Neurol.*, 177: 237–256.

Rosenquist, A.C. (1985) Connections of visual cortical areas in the cat. In: A. Peters and E.G. Jones (Eds.), *Cerebral Cortex*, Plenum Publishing Corporation, New York, pp. 81–117.

Smith, D.C. and Spear, P.D. (1979) Effects of superior colliculus removal on receptive field properties of neurons in lateral suprasylvian visual area of the cat. *J. Neurophysiol.*, 42: 57–75.

Spear, P.D. (1985) Neural mechanisms of compensation following neonatal cortex damage. In: C.W. Cotman (Ed.), *Synaptic Plasticity and Remodeling*, Guilford Press, New York, pp. 111–167.

Spear, P.D. (1988) Influence of ares 17, 18, and 19 on receptive-field properties of neurons in the cat's posteromedial lateral suprasylvian visual cortex. In: T.P. Hicks and G. Benedek (Eds.), *Progress in Brain Research: Vision Within Extrageniculo-Striate Systems*, Elsevier, Amsterdam, pp. 197–210.

Spear, P.D. (1991) Functions of Extrastriate Visual Cortex. In A. Leventhal (Ed.), *The Neural Basis of Visual Function*, Macmillan Press, England, pp. 339–370.

Spear, P.D. and Baumann, T.P. (1975) Receptive-field characteristics of single neurons in lateral suprasylvian visual area of the cat. *J. Neurophysiol.*, 38: 1403–1420.

Spear, P.D. and Baumann, T.P. (1979) Effects of visual cortex removal on receptive-field properties of neurons in lateral suprasylvian visual area of the cat. *J. Neurophysiol.*, 42: 31–56.

Spear, P.D. and Baumann, T.P. (1979) Neurophysiological mechanisms of recovery from visual cortex damage in cats: Properties of lateral suprasylvian visual area neurons following behavioral recovery. *Exp. Brain Res.*, 35: 161–176.

Spear, P.D., Kalil, R.E. and Tong, L. (1980) Functional compensation in lateral suprasylvian visual area following neonatal visual cortex removal in cats. *J. Neurophysiol.*, 43: 851–869.

Spear, P.D., Tong, L., McCall, M.A. and Pasternak, T. (1985) Developmentally induced loss of direction-selective neurons in the cat's lateral suprasylvian visual cortex. *Devel. Brain Res.*, 20: 281–285.

Spear, P.D., Tong, L. and McCall, M.A. (1988) Functional influence of areas 17, 18, and 19 on lateral suprasylvian cortex in kittens and adult cats: implications for compensation following early visual cortex damage. *Brain Res.*, 447: 79–91.

Tong, L., Kalil, R.E. and Spear, P.D. (1984) Critical periods for functional and anatomical compensation in lateral suprasylvian visual area following removal of visual cortex in cats. *J. Neurophysiol.*, 52: 941–960.

Tong, L., Spear, P.D. and Kalil, R.E. (1987) Effects of corpus callosum section on functional compensation in the posteromedial lateral suprasylvian visual area after early visual cortex damage in cats. *J. Comp. Neurol.*, 256: 128–136.

Tong, L., Kalil, R.E. and Spear, P.D. (1991) Development of thalamic projections to the cat's lateral suprasylvian visual area of cortex. *J. Comp. Neurol.*, 314: 526–533.

Toyama, K., Komatsu, Y., Kasai, H., Fujii, K. and Umetani, K. (1985) Responsiveness of Clare-Bishop neurons to visual cues associated with motion of a visual stimulus in three-dimensional space. *Vision Res.*, 25: 407–414.

Turlejski, K. (1975) Visual responses of neurons in the Clare-Bishop area of the cat. *Act. Neurobiol. Exp.*, 35: 189–208.

von Grunau, M. and Frost, B.J. (1983) Double-opponent-process mechanism underlying RF-structure of directionally specific cells of cat lateral suprasylvian visual area. *Exp. Brain Res.*, 49: 84–92.

Wright, M.J. (1969) Visual receptive fields of cells in a cortical area remote from the striate cortex in the cat. *Nature*, 223: 973–975.

Zumbroich, T.J. and Blakemore, C. (1987) Spatial and temporal selectivity in the suprasylvian visual cortex of the cat. *J. Neurosci.*, 7: 482–500.

Zumbroich, T.J., von Grunau, M., Poulin, C. and Blakemore, C. (1986) Differences of visual field representation in the medial and lateral banks of the suprasylvian cortex (PMLS/PLLS) of the cat. *Exp. Brain Res.*, 64: 77–93.

M. Norita, T. Bando and B. Stein (Eds.)
Progress in Brain Research, Vol 112
© 1996 Elsevier Science BV. All rights reserved.

CHAPTER 17

Extrinsic and intrinsic connections of the cat's lateral suprasylvian visual area

M. Norita[1,*], M. Kase[1], K. Hoshino[1], R. Meguro[1], S. Funaki[1], S. Hirano[1] and J.G. McHaffie[2]

[1]*Department of Neurobiology and Anatomy, Niigata University, School of Medicine, Asahimachi, Niigata 951, Japan*
[2]*Department of Neurobiology and Anatomy, Bowman Gray School of Medicine, Wake Forest University, Winston-Salem, NC 27157, USA*

The lateral suprasylvian visual area (LS) is known to have numerous interconnections with visual cortical areas as well as with subcortical structures implicated in visually-guided behaviors. In contrast, little data is available regarding connections within the LS itself. In order to obtain information about intra-areal connections and to re-investigate LS connectivity with various cortical and subcortical areas, the tracer (biocytin or WGA-HRP) was injected into various loci along the medial and lateral banks of the LS. The anterograde tracer, biocytin injections into both medial and lateral bank produced label contained within the respective bank that extended rostrally and caudally from the injection site. In addition, following medial bank injections, considerable label was distributed throughout the fundus and, to a lesser extent, in the lateral bank. In contrast, no label could be detected in the medial bank after lateral bank injections, and, although label was observed in the fundus, it was restricted to the most lateral aspects. Moderate labeling could be observed in the medial bank following the tracer injection into the most rostral aspect of the lateral bank. It is likely that input derived from various visual cortical areas which project to the medial bank of the LS has access to this intra-areal circuitry. This may provide a route by which visual cortical information can be relayed to other cortical and subcortical structures involved in visually-guided behaviors such as the anterior ectosylvian visual cortex, striatum, and the deep layers of the superior colliculus, despite the fact that these structures themselves do not receive substantial direct projections from the visual cortical areas that are associated with the medial bank.

Examination of the laminar location of the cells-of-origin of striate and extrastriate projections to LS using retrograde tracer, WGA-HRP, revealed that the supragranular laminae of areas 17, 18 and 19 were the source of LS afferents whereas afferents from the other cortical areas (e.g., 20a, 20b, 21a, 21b, 7 and anterior ectosylvian visual area) were from both supra- and infragranular laminae. In addition, all LS subregions received intra-areal afferent projections from all LS cortical laminae. Thus, although rather clear hierarchical relationship between LS and visual cortical areas appears to exist, the interconnections among LS subregions provide no clear evidence of simple hierarchial relationships between regions LS or may have feed-forward and feed-back pathways.

Introduction

Regions along the cat's lateral suprasylvian (LS) sulcus have long been recognized as being responsive to visual stimuli (Marshall et al., 1943; Clare and Bishop, 1954); more contemporary physiological and anatomical studies have revealed the complex heterogeneous nature of this region of extrastriate visual cortex. Electrophysiological mapping studies (Palmer et al., 1978) have delineated six [anterolateral lateral suprasylvian cortex (ALLS), posterolateral lateral suprasylvian cortex (PLLS), anteromedial lateral suprasylvian cortex (AMLS), posteromedial lateral suprasyl-

*Corresponding author. Tel.: +81-25-223-6161, ext. 2210; fax: +81-25-222-5642; e-mail: mnorita@med.niigata-u.ac.jp

vian cortex (PMLS), dorsal lateral suprasylvian cortex (DLS), and ventral lateral suprasylvian cortex (VLS)], variable representations of the visual world, each with somewhat different patterns of afferent and efferent connectivity (see Grant and Shipp, 1991 and references therein). While the specific role of these individual representations remain to be determined, it is generally assumed that LS is implicated in visually-guided orientation behaviors and may be particularly important for detecting novel visual stimuli as the animal moves through it's environment. Consistent with this notion, lesions of posterior aspects of the LS cortex have no little effect on orientation of the head and eyes when whole body movements are not required for orientation (Spear et al., 1983; Hardy and Stein, 1988). However, when a LS-lesioned animal is moving toward a initial fixation stimulus and is required to redirect ongoing orientation toward a second stimulus that is introduced into the visual field, a marked visual neglect is produced (Hardy and Stein, 1988).

LS receives visual inputs via two major afferent pathways, one arising directly from the thalamus and the other indirectly from striate and extrastriate visual cortex. The transthalamic pathways eminent from both the dorsal lateral geniculate (LGNd) complex, [C-lamina, medial interlaminar nucleus (MIN), geniculate wing (GW) of Guillery et al., 1980] and the lateral posterior (LP)–pulvinar complex (Raczkowski and Rosenquist, 1983; Dreher, 1986). Transcortical afferents are received from primary visual cortex by some, but not all, of the subdivisions of LS (Symonds and Rosenquist, 1984a; Spear, 1991). In addition, LS is known to have numerous connections with the extrastriate visual cortical areas (Symonds and Rosenquist, 1984a; Rosenquist, 1985; Norita et al., 1986) as well as with subcortical structures implicated in visually-guided behaviors (Kawamura et al., 1974; Marcotte and Updyke, 1982; Norita et al., 1991; Harting et al., 1992; Updyke, 1993). Despite the wealth of information on cortical and thalamic afferents to LS, however, there is little data available regarding connections within LS itself (Symonds and Rosenquist,

1984a,b; Dreher, 1986). In the present report, we have re-investigated LS connectivity with various cortical and subcortical areas, as well as providing new data on the intra-areal connections within LS itself in order to clarify how the visual information is processed within LS.

Materials and methods

Injections of anterograde and retrograde tracers (see below) were made in 14 adult cats of either sex weighing between 2.3 and 4.0 kg (Table 1). Injections were made either into electrophysiologically defined regions of the LS cortex (cases 1–6) or stereotaxically (cases 7–14) without prior mapping. For electrophysiological recording, animals were anesthetized, paralyzed, artificially respired, and placed in a stereotaxic head-holder. The refractive state of the eyes was measured by direct ophthalmoscopy, and contact lenses of the appropriate powers (0 to +3 diopters) fitted to focus the eyes on a translucent plexiglass hemisphere placed 60 cm in front of the animal's eyes. The position of the optic discs was plotted onto the hemisphere by means of reverse ophthalmoscopy, and the location of the area centralis was estimated according to Bishop et al. (1962). Extracellular recordings were made with tungsten microelectrodes ($Z = 1$–4 MΩ). As the electrode was advanced parallel to the sulcal wall, receptive fields of single- or multi-unit activity, evoked by visual stimulation through the contralateral eye, were delineated by moving spots or bars of light projected from a hand-held pantascope across the hemispheric screen at various directions and velocities.

Following electrophysiological mapping, the electrode was removed and replaced with a 1-μl Hamilton syringe fitted with a glass micropipette (tip diameter 60–80 μm). The tip of the micropipette was re-positioned at a selected site, and pressure injections of 0.03–0.05 μl 2% WGA-HRP (Sigma) were made. The remaining animals (cases 7–14) received 1.0–1.5 μl of 5% biocytin (Sigma) injected into either the medial or lateral bank at AP levels from +13 to −2. After

TABLE 1

Summary of experimental cases

Animal	Tracer	Injection site (AP level)	Amount of tracer (μl)
Case 1	WGA-HRP	ALLS ($+10$)	0.04
Case 2	WGA-HRP	PLLS ($+4$)	0.05
Case 3	WGA-HRP	PLLS (-2)	0.03
Case 4	WGA-HRP	PMLS ($+2$)	0.05
Case 5	WGA-HRP	AMLS ($+10$)	0.03
Case 6	WGA-HRP	PLMS (-2)	0.03
Case 7	biocytin	AMLS ($+13$)	1.5
Case 8	biocytin	PMLS (-2)	1.5
Case 9	biocytin	ALLS ($+12$)	1.5
Case 10	biocytin	PLLS (-2)	1.5
Case 11	biocytin	AMLS/PMLS ($+9$)	1.0
Case 12	biocytin	Fundus ($+3$)	1.0
Case 13	biocytin	PLLS ($+7$)	1.5
Case 14	biocytin	Fundus ($+7.5$)	1.0

2 days survival, animals were sacrificed with an overdose of sodium pentobarbital and perfused transcardially with phosphate-buffered 0.9% saline followed by a fixative containing 2% paraformaldehyde and 2% glutaraldehyde in 0.1 M phosphate buffer at pH 7.4. The brains were removed immediately, blocked, and stored for 1 day in 30% sucrose. Blocks were sectioned in the coronal plane at 40 μm on a freezing microtome. Sections were processed for the visualization of biocytin according to the procedures of King et al. (1989) and for WGA-HRP according to Mesulam et al. (1980) using tetramethylbenzidine as the chromogen and counterstained with neutral red. Delineation of LS and other visual cortical areas was done according to Rosenquist (1985) and thalamic nomenclature was done according to Graybiel and Berson (1980).

Results

Extrinsic connections of LS

WGA-HRP was used to determine the distribution of cortical and thalamic neurons projecting to electrophysiologically-defined loci within LS, as well as to document LS efferents to the striatum, pretectal areas, accessory optic nuclei, and superior colliculus (SC).

Cortical interconnections

WGA-HRP injections resulted in retrograde label in the following visual cortical areas: area 17, 18, 19, 20a, 20b, 21a, 21b, 7, anterior ectosylvian visual area (AEV, Mucke et al., 1982; Olson and Graybiel, 1983), posterior suprasylvian sulcus area (PS, Heath and Jones, 1971), and within LS itself. In all instances, corticocortical connections from a given site were reciprocal. Specific regional variations within LS are described below (see Table 2).

Connections of ALLS and PLLS. In case 1, the deposit of WGA-HRP was centered at approximately AP $+10$ (ALLS), which was electrophysiologically defined as the lower visual field and 20–40° of azimuth. In cases 2 and 3 (PLLS), the injection sites were centered at about AP $+4$ (horizontal meridian, 20–80° of azimuth) or at about AP -2 (central visual fields), respectively.

TABLE II

Pattern of interconnections between LS (ALLS, PLLS, AMLS, PMLS) and 10 cortical areas

	17	18	19	20a	20b	21a	21b	7	AEV	PS
ALLS	?	+	+	+	+	+	+	+ +	+ + +	+
PLLS	+ +	+ +	+ +	+	+	+ +	+ +	+	+ +	+ +
AMLS	+	+ +	+	+ +	+ +	?	+ +	+	+	+
PMLS	+ +	+ +	+ +	+ +	?	+	+ +	?	+	+

The relative density of interconnections is indicated by +. A question mark (?) indicates a very sparse or non-existent interconnection.

Both ALLS and PLLS maintained reciprocal connections with areas 18, 19, 20a, 20b, 21a, 21b, 7, PS, and AEV. ALLS/PLLS also received substantial afferent projections from AMLS/PMLS. Although the caudal-most portion of PLLS, corresponding to more centrally located visual fields, was densely interconnected with area 17, rostral PLLS was only sparsely connected with area 17. No projections could be observed between ALLS and area 17 (Figs. 1a, 2 and 3). Sparse to moderate interconnections were seen between area 7 and ALLS/PLLS with ALLS connections more apparent than those with PLLS; the most caudal portion of PLLS had very sparse, if any, connections with area 7 (Figs. 2 and 3). While ALLS and the more rostral aspects of PLLS had dense interconnections with AEV, the caudal-most portions of PLLS were only sparsely interconnected (cf. cases 1 and 2 to case 3, Figs. 2 and 3).

Connections of AMLS and PMLS. In case 5 (AMLS), the tracer deposit was centered at about AP +10 (vertical meridian, 0–20° of azimuth), while in case 4 (PMLS), the tracer was centered at about AP +2 (central visual field). Both AMLS and PMLS interconnected with visual cortical areas 17, 18, 19, 20a, 20b, 21a, 21b, 7, PS, AEV and within LS itself. AMLS/PMLS had very sparse interconnections with area 7 and AEV (Fig. 4). Although AMLS shared most afferent sources in common with PMLS (Fig. 4), there appeared to be notable distinctions. For example, while primary visual cortex had denser connections with

PMLS than with AMLS, PS had more dense connections with AMLS than with PMLS. Both AMLS and PMLS maintain dense intra-regional and inter-regional connections with PMLS and AMLS, respectively. In addition, while both AMLS and PMLS sent fibers across the fundus that terminated within the opposite bank, these afferents were relatively weak compared to other regions (see below).

Examination of the laminar location of the cells-of-origin of striate and extrastriate projections to LS revealed that the supragranular layers of areas 17, 18, and 19 were the source of LS afferents (Figs. 1b,c) whereas afferents from the remaining cortical areas (20a, 20b, 21a, 21b, 7, PS and AEV) were from both supra- and infragranular layers (Fig. 1e). In addition, ALLS, PLLS, AMLS, and PMLS all received intra-areal afferent projections from all LS cortical layers (Fig. 1d).

Thalamic interconnections

Although WGA-HRP deposits inevitably traverse rather side and often disparate parts of the visual field representations in LS, the interconnections between LS subregions and LP-pulvinar complex showed a crude retinotopic organization (cf. Hutchins and Updyke, 1989). Tracer deposits into ALLS or PLLS resulted in both retrogradely labeled neurons and anterogradely labeled terminals in the LP-pulvinar complex, as well as with the lateral medial-suprageniculate nuclear complex (LM-Sg), posterior nuclear group, the in-

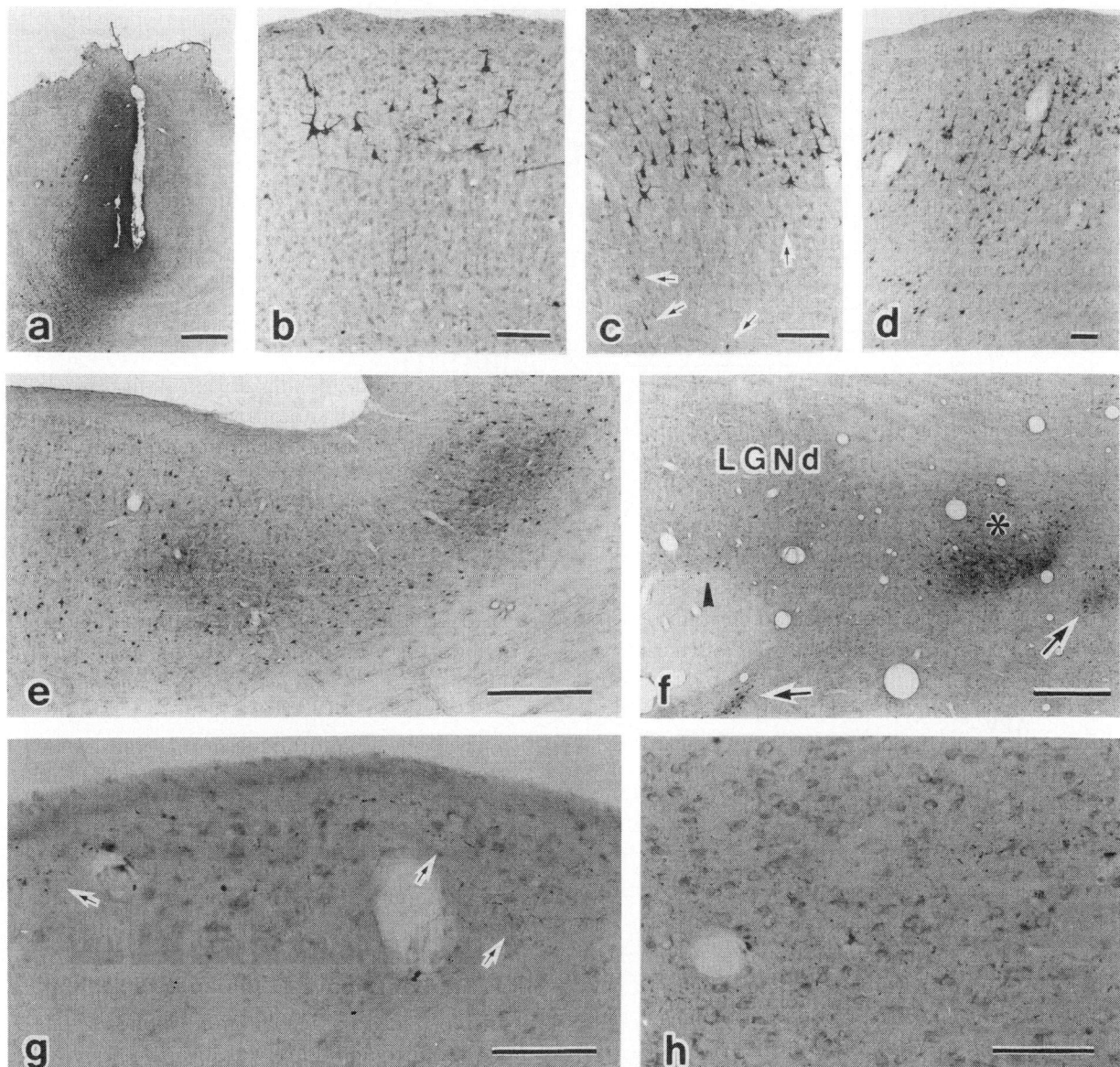

Fig. 1. Photomicrographs of a representative WGA-HRP injection site and patterns of retrograde and anterograde labeling. (a) Tracer deposit in PLLS (case 1). (b) Retrograde neuronal labeling in area 17 after an injection into AMLS (case 5). Labeled neurons are found only in the supragranular layers. (c) Retrograde neuronal labeling in area 18 after an injection into AMLS (case 5). Most labeled neurons are located in the supragranular layers although some were also found in the infragranular layers (arrows). (d) Retrograde neuronal labeling in PMLS after an injection into AMLS (case 5). Labeled neurons are present in both the supragranular and infragranular layers. (e) Retrograde neuronal labeling in AEV after an injection into PLLS (case 2). Labeled neurons are observed in both the supragranular and infragranular layers. (f) Retrograde neuronal labeling in LGNd (C-lamina, arrowhead), LPl (asterisk) and LPm (arrows) after an injection into PMLS (case 4). (g) Anterograde axonal labeling (arrows) in LTN after an injection into PMLS (case 4). (h) Anterograde axonal labeling in the putamen after an injection into ALLS (case 1). Scale bars: a = 1 mm; b, c, d, g and h = 100 μm; e and f = 500 μm.

236

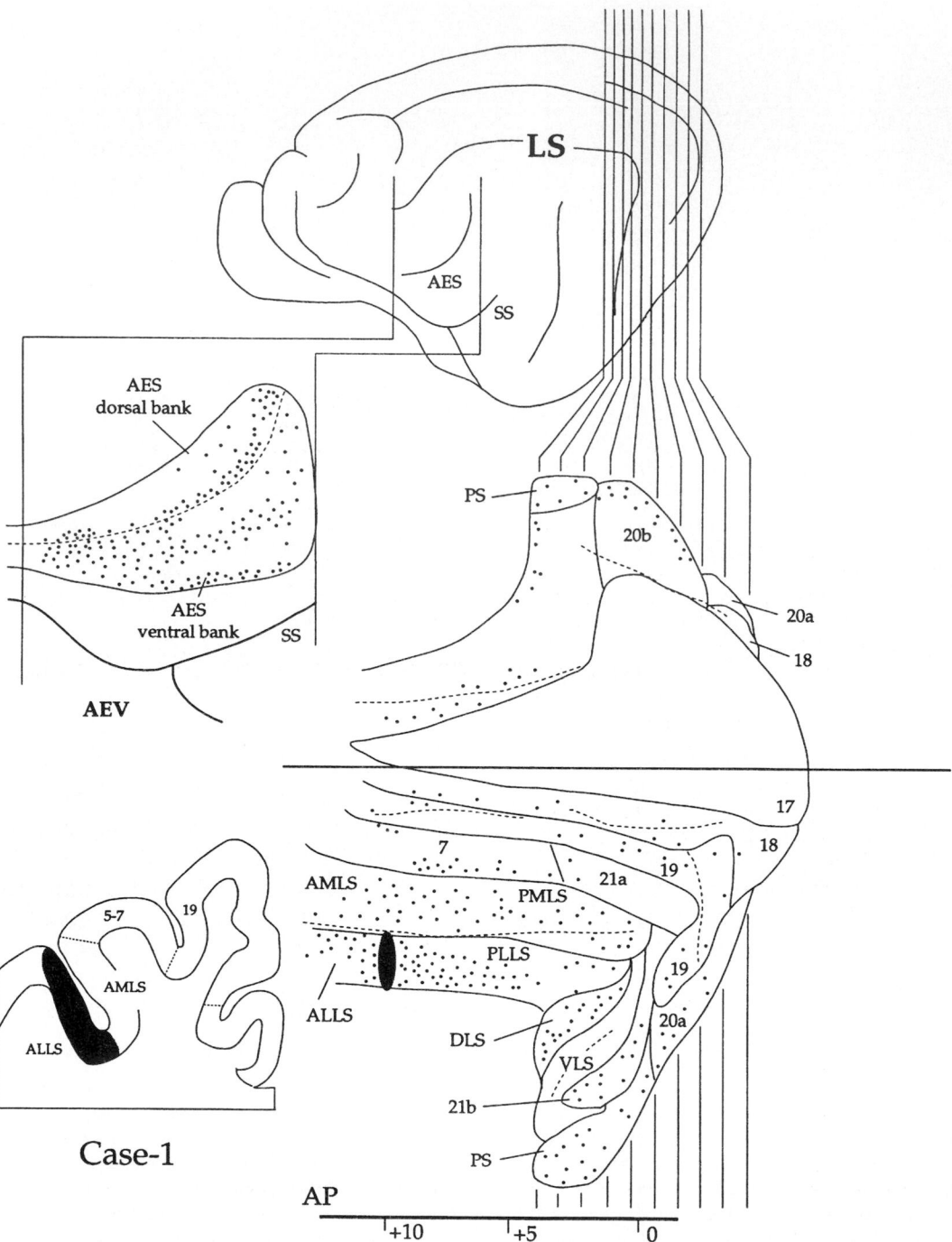

Fig. 2. The distribution of retrogradely-labeled neurons resulting from a WGA-HRP injection into ALLS (case 1). In this, and the following two figures, the location of the injection site (blackened/stipple) and positions of labeled neurons (dots/crosses) is displayed on an unfolded, two-dimensional cortical map (modification of Dr. S. Kawamura's original map) by reference to standard Horsy-Clarke coordinates along the anteroposterior axis of the cerebrum (see Palmer et al., 1978). The solid lines indicate the histological border of the cortical areas while the dotted lines indicate the fundus.

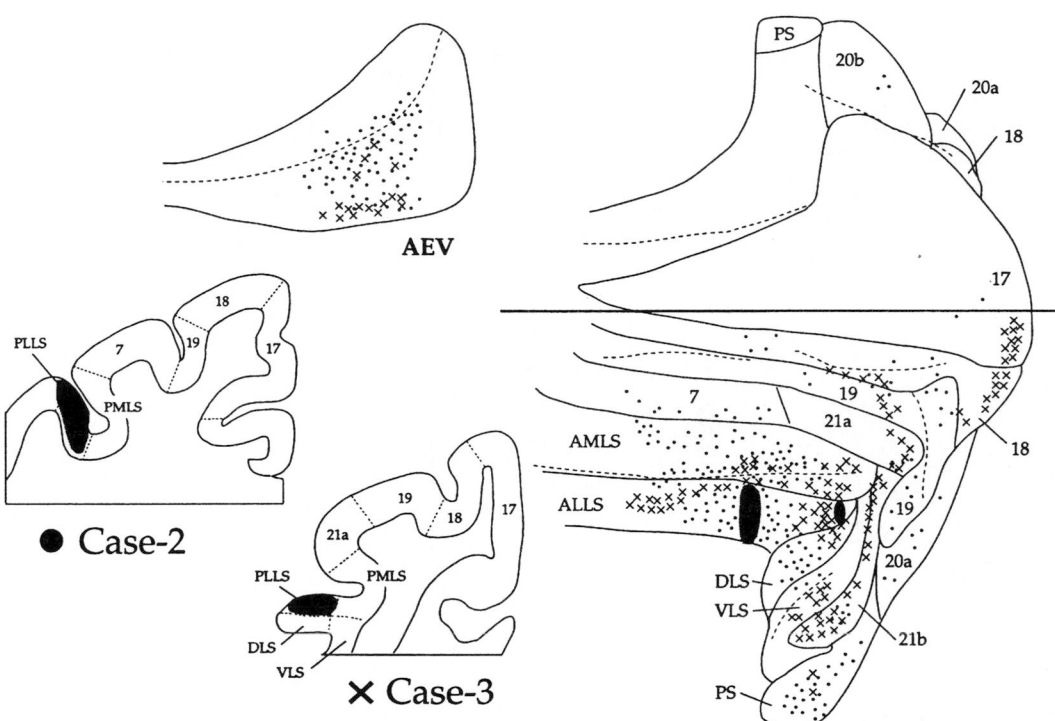

Fig. 3. The distribution of retrogradely-labeled neurons resulting from a WGA-HRP injection into different loci in PLLS (case 2 and case 3). The positions of labeled neurons from case 2 injections are depicted as filled circles whereas those from case 3 are shown as crosses.

tralaminar nuclei, and LGNd complex (Fig. 5). Dense retrograde and anterograde label was seen in the medial division of the LP–pulvinar complex (LPm) and LM-Sg, whereas only sparse label was found in the lateral division of the LP-pulvinar complex (LPl) and pulvinar. Moderate to weak label was observed in the posterior nuclear group, intralaminar nuclei, LGNd complex (C-laminae, MIN and GW). Although the pattern of thalamic label was similar following both ALLS and PLLS injections, there were several important distinctions. For example, label in the intralaminar nuclei (e.g., central lateral nucleus) and the posterior nuclear group was more pronounced following ALLS injections than with PLLS deposits. Labeling in LGNd complex was also noted only following PLLS injections.

Injection into AMLS or PMLS resulted in label in the intralaminar nuclei (particularly central lateral nucleus), LPl, LPm, pulvinar, LGNd com-

plex (C-laminae, MIN and GW). In these cases, moderate label was seen in the C-laminae compared to the very sparse label seen following injections into ALLS or PLLS. Moderate to weak label was seen in MIN and GW. In contrast to ALLS or PLLS injections, no label could be detected in the posterior nuclear group. The more rostral regions of LS (ALLS/AMLS) received heavier projections from the central lateral nucleus than did the most caudal regions (PLLS/PMLS).

Other subcortical projections

Although dense projections to the striatum arose from both ALLS/PLLS (Figs. 1h and 6) and AMLS/PMLS (Fig. 7), lateral bank injections resulted in much more striatal labeling than did medial bank injections. Within the caudate nucleus, a coarse topography of corticocaudate

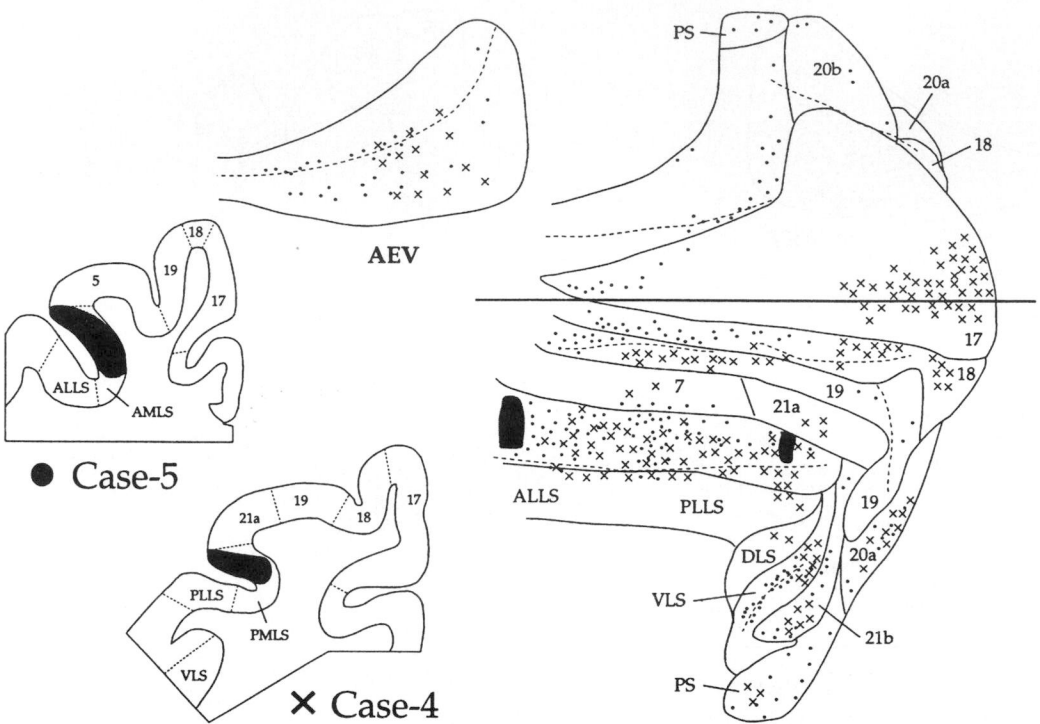

Fig. 4. The distribution of retrogradely-labeled neurons resulting from WGA-HRP injections into two different loci in the medial bank of LS (case 5 — ALLS and case 4 — PLLS). The positions of labeled neurons from case five injections are depicted as filled circles whereas those from case 4 are shown as crosses.

projections could be discerned such that neurons in more anterior regions of LS (ALLS/AMLS) projected to more rostral aspects of the caudate, whereas those from more posterior regions (PLLS/PMLS) terminated more caudally. Nevertheless, within this overall topography, no clear indication of retinotopy could be established. On the other hand, the pattern of corticofugal projections to the putamen suggested the possibility of a retinotopic organization. WGA-HRP deposits involving LS representations of the lower visual field resulted in labeling localized more medially than those found with injection sites centered on upper visual field representations.

A moderate to sparse amount of anterograde label was also seen in various regions of the pretectum, including the anterior and posterior pretectal nuclei, the nucleus of the optic tract (NOT), and the pretectal olivary nucleus following WGA-HRP or biocytin injections into

AMLS or PMLS. On the other hand, only sparse label could be detected following ALLS/PLLS injections (not illustrated). These results are consistent with previous findings (e.g., Kawamura et al., 1974).

Variable amounts of anterograde label could be found in all three nuclei of the accessory optic system (AON), the medial (MTN), lateral (LTN) and dorsal terminal nuclei (DTN). Of these, rather dense projections to LTN arose from ALLS/PLLS and AMLS/PMLS following either WGA-HRP or biocytin injections (Fig. 1g) while only sparse projections to MTN and DTN were observed following AMLS/PMLS injections (not illustrated). Thus, the present data is consistent with the previous reports on LS afferents to AON (e.g., Marcotte and Updyke, 1982).

Although axons from all subdivisions of LS terminated in both the superficial and deep laminae of SC, there were laminar differences in their

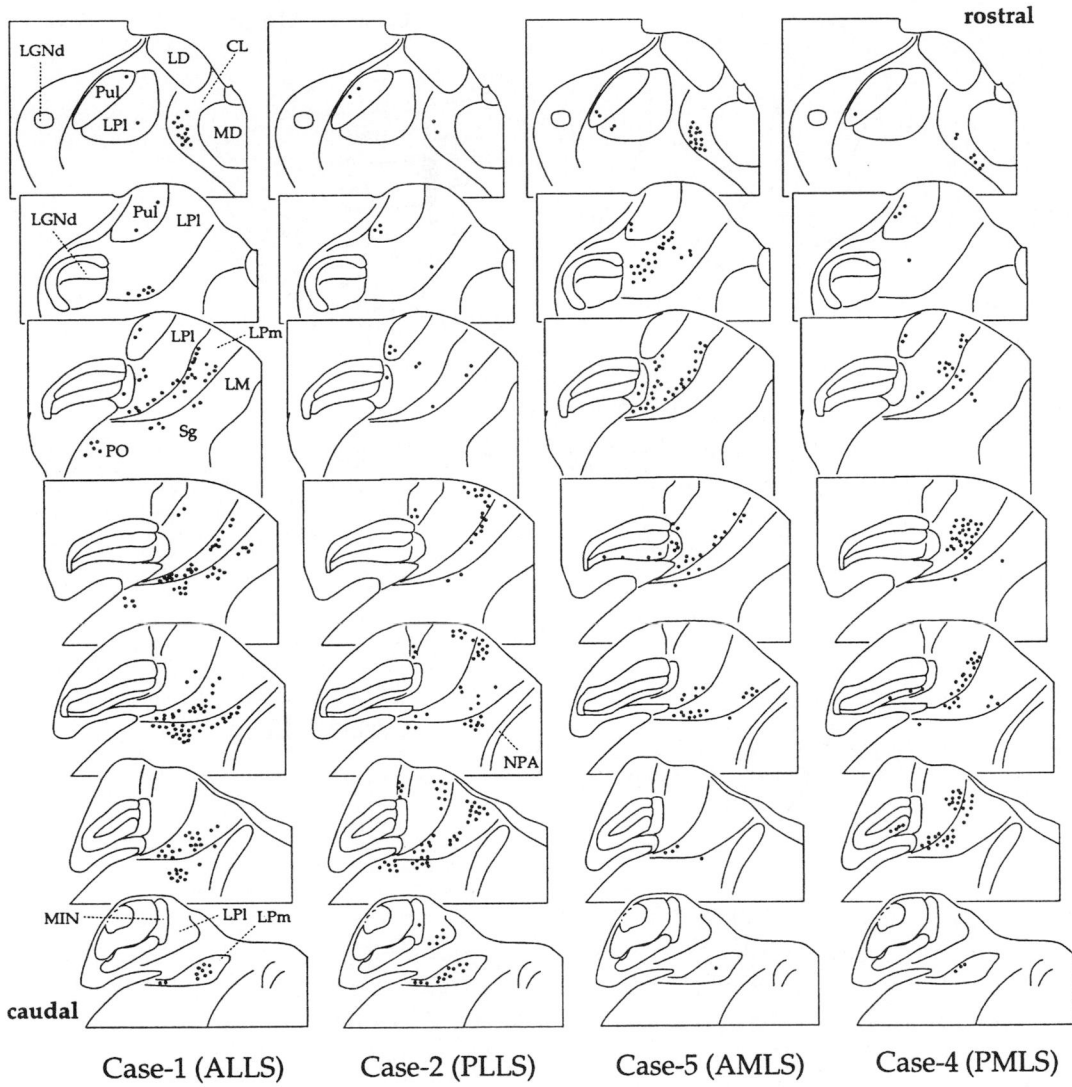

rostral

caudal

Case-1 (ALLS)   Case-2 (PLLS)   Case-5 (AMLS)   Case-4 (PMLS)

Fig. 5. The distribution of retrogradely-labeled thalamocortical neurons in the LP–pulvinar complex after four WGA-HRP injections into different subregions of LS.

distributions (not illustrated). For example, axons from AMLS/PMLS neurons terminated most densely in the superficial laminae but only modestly in the upper part of the deep laminae. In contrast, although axons from ALLS/PLLS neurons also terminated heavily in the superficial laminae (especially stratum opticum), many more labeled axons within the deep laminae were observed following lateral bank injections. In ad-

dition to this partial laminar organization, LS projections to both the superficial and deep laminae appeared to be topographically organized. Thus, the most rostral divisions (AMLS/ALLS) projected most densely to the lateral aspect of SC whereas the caudal divisions (VLS/DLS, not illustrated) terminated medially; regions in the middle were innervated from PMLS and PLLS (see Fig. 7, for example). Although the organiza-

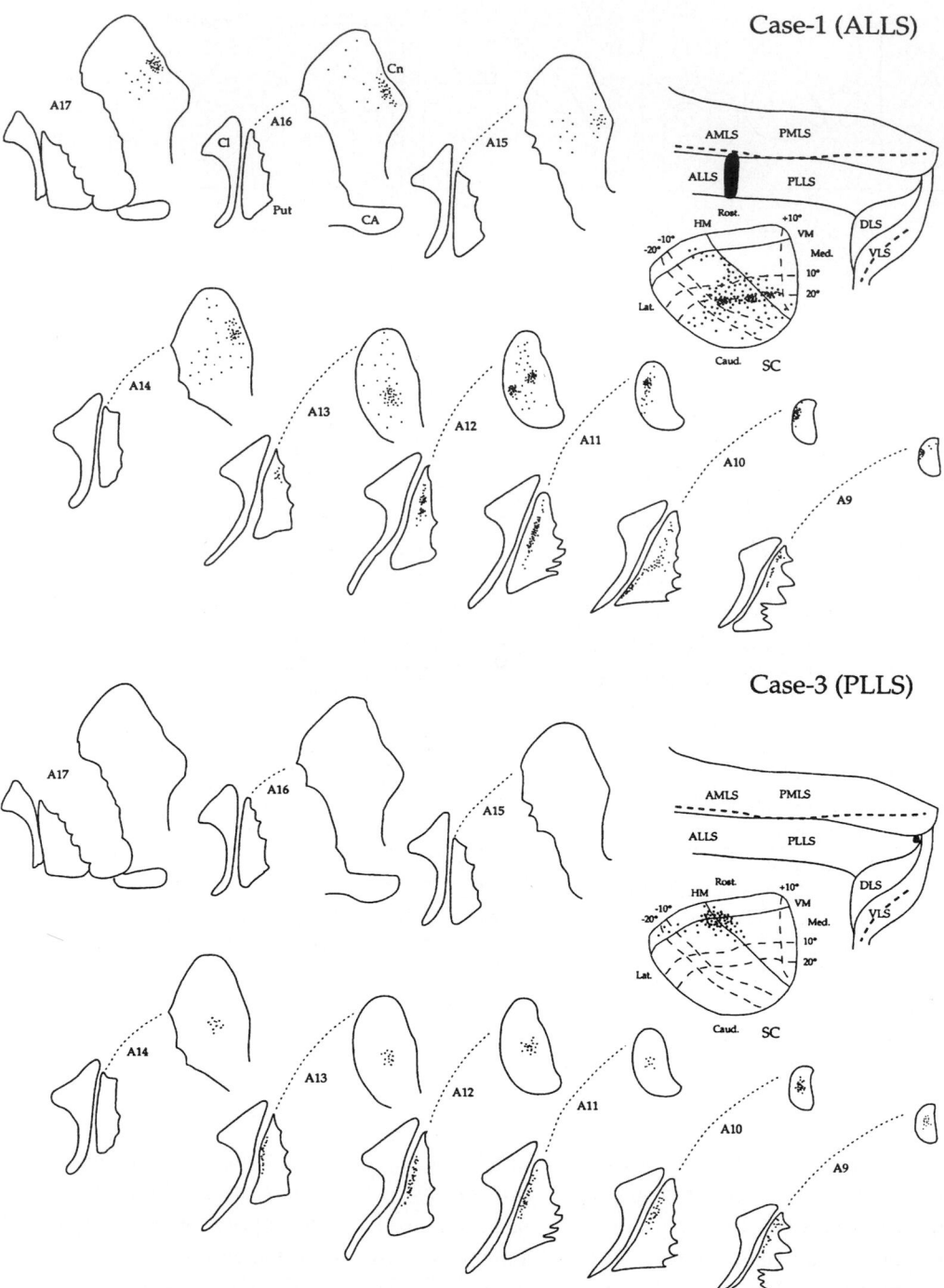

Fig. 6. The distribution of anterograde label in the striatum and superior colliculus after WGA-HRP injections into ALLS (upper) and PLLS (lower). The location of the injection sites are depicted on unfolded maps of LS (upper right). Coronal sections through the striatum at various AP levels (A17–A9) and a reconstruction of the dorsal view of the left superior colliculus are shown with the location of anterograde label depicted as dots. Similar conventions are used in Fig 7. In the caudate nucleus, LS afferents are topographically organized such that the more rostral portions of LS projects to more rostral aspects of the caudate. However, no

tion of corticotectal axons from individual subdivisions of LS did not display a precise visuotopy, fundamental relationships could be discerned between LS and SC. Thus, ALLS, which is biased toward the representation of the lower portion of the visual field, projected predominately to the lateral aspect of SC, where inferior visual receptive fields are located; the caudal-most regions of PMLS/PLLS, where central visual space is represented, project to the rostral SC, where representation of central visual space is located (Figs. 6 and 7).

Intrinsic connections of LS

In order to detail the intrinsic connections within regions of LS, the anterograde tracer biocytin, which can distinguish between axons and terminals (King et al., 1989), was injected stereotaxically into LS subregions at AP levels from +13 to −2. In all cases reported in the present study, the injection sites encompassed all laminae of the cortex. Fig. 9 depicts representative cases demonstrating the pattern of intrinsic connections from individual regions which are summarized in Fig. 10. In each case, labeled axons from a single, focal injection site could be seen at distances up to 8–10 mm distal to the center of the injection site. Medial bank injections (Fig. 9, cases 7, 8 and 11) produced dense label in the medial bank and fundal regions that extended rostrally and caudally from the injection site. Moderate label was also observed within the lateral bank. Thus, neurons within the medial bank not only have dense intra-areal connections with each other, but they also send projections into the lateral bank. Similarly, lateral bank injections (Fig. 9, cases 9, 10 and 13) resulted in label largely confined within the lateral bank that extended rostrally and cau-

dally from the injection sites. Although label was observed in the fundus, it was virtually restricted to the more lateral aspects of the fundus with little label in the medial bank. However, the lack of medial bank label that was evident following centrally placed lateral bank injections was not evident in the most rostrally or caudally placed lateral bank injections (Fig. 9, cases 9 and 10). In these cases, moderate label could be observed in the medial aspect of the fundus as well as in the medial bank. Following an injection into the fundus (Fig. 9, cases 12 and 14), anterograde label was distributed in both rostral and caudal directions and into both banks as well.

Biocytin labeled axons were not homogeneously distributed but, in some locations, were apportioned as clumps or patches. For example, following injections into the medial bank (Fig. 9, cases 7 and 11), several distinct regional densities of labeling were seen in the lateral bank and the fundus (indicated by arrows in Fig. 9).

With regards to the laminar distribution pattern of inter-areal terminals, all cases produced similar patterns which were independent of the rostrocaudal or mediolateral position of the injection site within LS subregions. For example, labeled axons and terminals from the rostral portion of ALLS (case 9) were distributed in both supra- and infragranular layers of LS (not illustrated). The terminal distribution of the caudal portion of PMLS deposit (case 8) was similar to those of case 9 described above.

Discussion

Extrinsic connections of LS

Cortical interconnections
   LS receives indirect retinal inputs from primary

clear indication of retinotopical organization was noted in this corticocaudate projection on the basis of these cases 1 and 3, and 5 and 6 (see Fig. 7), as well as others (not illustrated). On the other hand, the corticoputaminal projections show rather clear retinotopically organization: WGA-HRP deposits involving cortical representations of lower visual field result in labeling medially to those found in the higher visual field injection. These two cases show much more labeling in the striatum than those found in the following AMLS/PMLS cases.

242

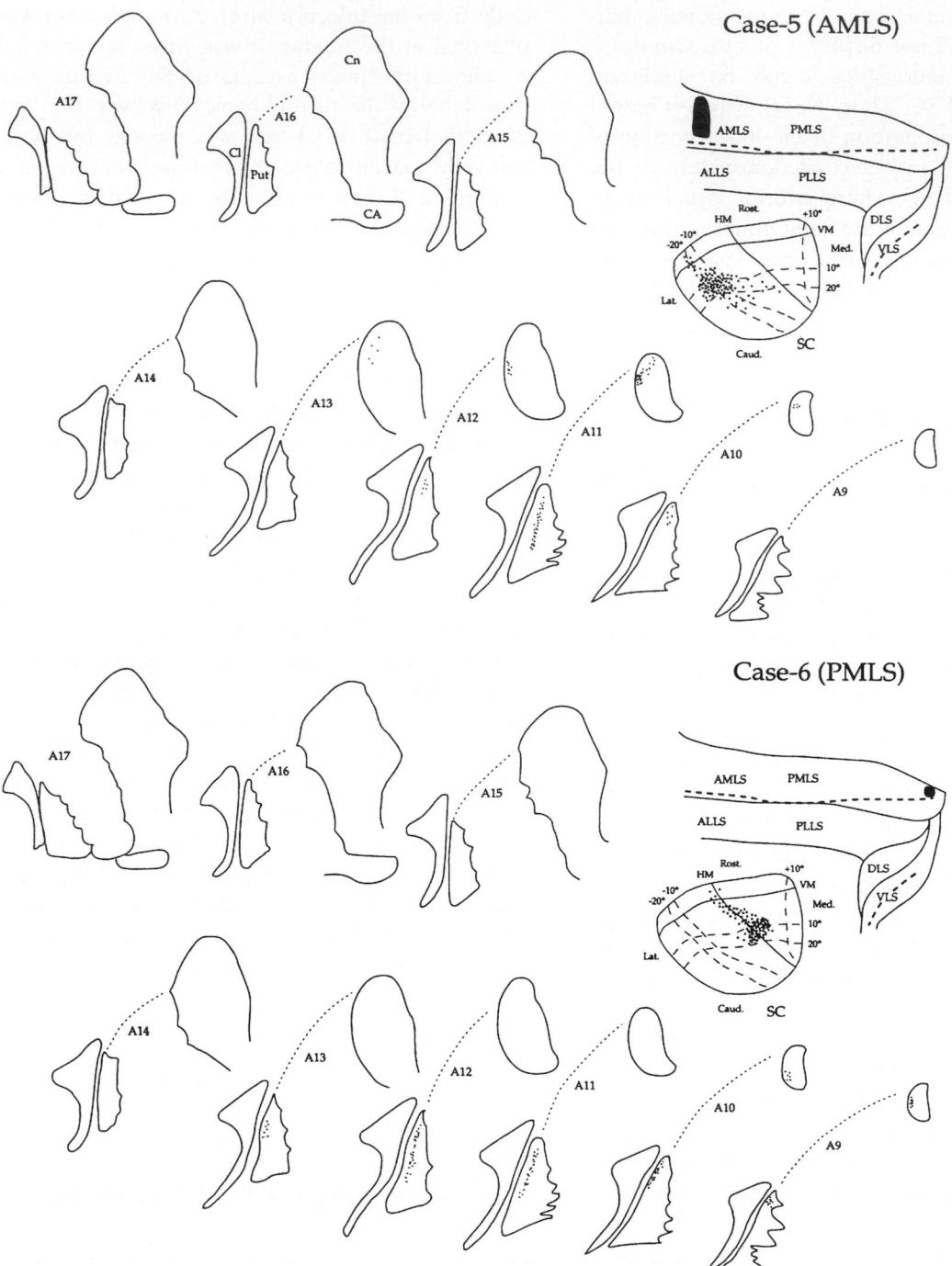

Fig. 7. The distribution of anterograde label in the striatum and superior colliculus after WGA-HRP injections into AMLS (upper) and PMLS (lower). Moderate to small number of label was present in the caudate nucleus and putamen. The projections to the caudate show the same rostral to caudal topographical relationship as illustrated in Fig. 6 with lateral bank injections. In putamen, anterogradely labeling, resulting from the tracer deposits involving cortical representations of lower visual field, trend to be distributed medially.

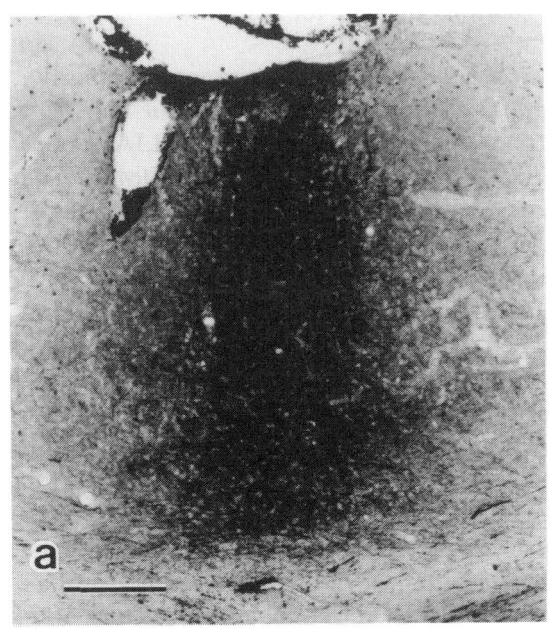

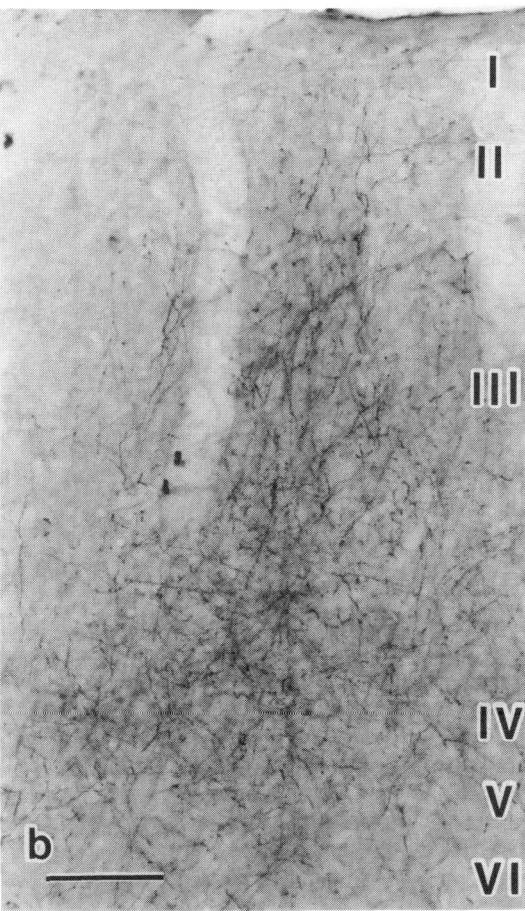

visual cortex (Rosenquist, 1985; Spear, 1991). The present data reveal, however, there are regional differences in the striate-LS projection. For example, dense projections to PMLS and the caudal-most aspects of PLLS arose from area 17 whereas only moderate to weak projections to AMLS/PLLS. No projections to ALLS could be observed. Compared with the more rostral regions of the lateral bank, the caudal portion of PLLS had a different pattern of connectivity (see Fig. 3), with dense interconnections with areas 17, 18 and 19. In contrast, caudal PLLS had only sparse connections with PS and AEV when compared with the more rostral subregions along the lateral bank. Thus, the caudal-most aspects of both PMLS and PLLS have similar patterns of connectivity that are distinct from the more rostral regions; this dichotomy is consistent with previous suggestions that the caudal-most regions of both PMLS and PLLS might be considered a single entity (cf. Sherk, 1986).

Moderate reciprocal connections with area 7 were noted between ALLS and the anterior aspects of PLLS. On the other hand, area 7 had very sparse connections with AMLS/PMLS as well as the caudal-most portion of PLLS. Olson and Lawler (1987) proposed that the posterior portion of area 7 (area 7p) is the feline homologue of the primate posterior parietal cortex and is involved in visual and oculomotor processes (e.g., Schlag-Rey and Schlag, 1977; Raczkowski and Rosenquist, 1981; Infante and Leiva, 1984). The dorsal division of the ventral anterior nucleus, which supplies afferents to area 7p, may act as a relay nucleus in an ascending pathway leading from the dentate nucleus to the parietal cortex (e.g., Sugimoto et al., 1981). Thus, LS may receive indirect cerebellar input from the dentate

Fig. 8. Photomicrographs showing a biocytin injection into the fundus at AP +7.5 in case 14 (a) and the resultant anterograde label in the fundus 2 mm rostral to the injection site (b). Scale a = 250 μm, b = 100 μm.

with the medial bank of LS (AMLS/PMLS), the LP-pulvinar complex is the predominate source of thalamic afferents to both medial and lateral banks of LS (see Rosenquist, 1985; Dreher, 1986 and references therein). As illustrated in Fig. 5, regions along the medial bank of the sulcus are connected primarily with LPl and some with pulvinar as well as LPm. Those areas on the lateral bank (ALLS/PLLS) are associated predominately with LPm and LM-Sg and, to a lesser extent, with pulvinar and LPl. The most lateral portion of LP-pulvinar complex, the pulvinar, is interconnected with the splenial visual area (e.g., Rosenquist, 1985) while the most medial portion of this complex, LM-Sg, is interconnected with AEV (e.g., Norita et al., 1986). Thus, although the interconnections between LS- and LP-pulvinar complex are more or less overlapping, it confirms the 'reverse medial-lateral organization between thalamus and cortex' previously recognized by Rosenquist (1985).

The posterior nuclear group is more closely associated with regions on the lateral bank than with those on the medial. Given the substantial auditory and somatosensory input that is known to terminate in the posterior group (see Imig and Morel, 1983; Stevens et al., 1993), it is likely that the lateral bank may be involved in non-visual and/or multimodal processing.

The more rostral regions of LS (ALLS/AMLS) receive heavier projections from the central lateral nucleus, which receives afferents from both the pretectum and the deep laminae of SC (Graham and Berman, 1981), than do the more caudal regions (PLLS/PMLS). While the central lateral nucleus presumably has a role in the generation of eye movements (Schlag-Ray and Schlag, 1977), it also serves as route a for ascending spinothalamic information to the motor cortex (e.g., Jones and Burton, 1974). Thus, the close association of ALLS/AMLS with eye movement- and other non-visual-related thalamic nuclei (see, the present results and Kaufman and Rosenquist, 1985) suggests that the most rostral regions of LS may serve as a multimodal region related to visuomotor behaviors, although there is no direct electrophysiological evidence concerning multimodal responses.

Other subcortical projections

Regions along the posterior aspects of the medial and lateral banks of LS, which have direct projections to the superficial and deep SC (Segal and Beckstead, 1984; Harting et al., 1992) are also the predominate source of extrastriate visual projections to the striatum (Burchinskaya et al., 1988; Norita et al., 1991; McHaffie et al., 1993; Updyke et al., 1993): afferents from both banks terminate in the caudal portion of the body and tail of the caudate as well as the posterolateral aspect of the putamen. These LS-recipient regions of the striatum, in turn, have distinct relationships with laminae of SC by way of striatofugal projections to different regions of the substantia nigra (SN). Thus, the caudate nucleus projects preferentially to the pars reticulata division of SN (SNr) (Royce and Laine, 1984), which gives rise, in turn, to nigrotectal fibers which innervate the intermediate and deep laminae of SC (Graybiel, 1978; Harting et al., 1988) where tectal output neurons are located (see Huerta and Harting, 1984 and references therein). The posterolateral aspect of the putamen, on the other hand, projects predominately to pars lateralis of SN (SNl), a region which gives rise to nigrotectal projections associated with the superficial laminae (Harting et al., 1988). Because striatal activity is known to modulate the activity of deep laminae output neurons in SC by way of double inhibitory GABA-ergic connections through SNr (e.g., Chevalier et al., 1985), cortically-processed visual information from LS can access SC via a 'presumed' indirect route. Because the deep laminae of SC are dependent upon corticofugal input from LS (Ogasawara et al., 1984: Hardy and Stein, 1988), LS may directly and 'indirectly' influence not only the deep laminae neurons (see McHaffie et al., 1993), which project to regions of the brainstem involved in moving the eyes, head and pinnae, but also influence superficial neurons which contribute to ascending tectothalamic pathways (e.g., Huerta and Harting, 1984).

Present results indicated that AON, which receives direct, and almost exclusively crossed, fibers from the retina (Simpson, 1984; Weber, 1985 and references therein), also receive substantial projections from LS. The results of this and previous studies (Berson and Graybiel, 1980; Marcotte and Updyke, 1982) have shown that LTN receives moderate projection from virtually all LS subregions. On the other hand, MTN and DTN receive only sparse to moderate projections mainly from AMLS/PMLS. Physiological studies indicate that LTN and MTN neurons respond preferentially to slow vertical motion of patterned stimuli (e.g., Lui et al., 1990) while DTN and NOT neurons respond selectively to medially directed horizontal motion of low velocity patterned stimuli (Collewijn, 1975; Simpson et al., 1979; Grasse and Cynader, 1984; Hoffmann and Distler, 1989). It follows that LTN and MTN are important in vertical OKN whereas DTN, together with NOT, are part of the pathway for horizontal OKN (see Simpson et al., 1988 and references therein). Thus, ALLS/PLLS may be a visual cortical region concerned with vertical OKN while AMLS/PMLS may be related to both.

Although the present results indicated that all LS regions project to various regions of the pretectal nuclei, we will mention only NOT projection, because NOT is an important visuo-motor relay between the retina and pre-oculomotor structures, mediating horizontal optokinetic nystagmus (e.g., Collewijn, 1975; Precht, 1981; Kato et al., 1988; Hoffmann et al., 1995), appears to receive fibers from both AMLS and PMLS; in contrast, few projections arise from ALLS and PLLS. A number of studies have been devoted to the efferent connections of NOT (cf. Kato et al., 1996). Thus, NOT sends axons to visuo-motor related structures such as the nucleus reticularis tegmenti pontis, dorsolateral pontine nucleus, nucleus prepositus hypoglossi, inferior olive, and vestibular nucleus.

Intrinsic connections of LS

Within LS itself, biocytin labeled axons were not homogeneously distributed but, in some locations, were apportioned as clumps or patches in subre-

gions of LS. The distribution of these inter-regional projections was independent of the rostrocaudal or mediolateral position of the injection site, with label distributed in both the supra- and infragranular layers. Furthermore, our WGA-HRP experiments indicated that ALLS, PLLS, AMLS, PMLS, and fundus all receive inter-areal afferent projections from both the supragranular and infragranular cortical layers. As reported previously in primates (Barbaresi et al., 1995, and references therein), feed-forward pathways (from lower to higher levels within a hierarchy) are believed to arise from the supragranular laminae and terminate mainly in infragranular laminae III and IV, whereas feed-back (from higher to lower) pathways arise mainly from infragranular laminae and terminate predominantly in supragranular laminae. In the cat, however, Symonds and Rosenquist (1984b) were not able to unequivocally detail such a scheme that was applicable to all visual areas of cortex. For example, area 17 projections to areas 18 and 19 arise mainly from supragranular laminae whereas the reciprocal projections from areas 18 and 19 included an infragranular component. Thus, they concluded that no unambiguous hierarchical relationship among the visual cortical areas in the cat appears to exist. From the present findings, together with a number of previous reports on a hierarchical point of view, as described above, the interconnections among LS subregions provide no clear evidence of simple hierarchical relationships between regions LS or may have feed-forward and feed-back pathways.

The distribution of labeled axons and terminals within LS itself, following focal injections of biocytin, was extensive, both within particular regions and particularly between regions. Often label could be observed at distances as great as 8–10 mm distal to an injection site. This rich plexus of connections provides an expansive anatomical network for modulating activity not only across LS regions but within individual regions as well. The inter-regional connectivity, however, was not equivalent but varied according to the bank injected and the specific AP level. For example, medial bank injections in the central aspects of

LS resulted in dense projections to the lateral bank, whereas injections in this same aspects of the lateral bank area produced only sparse medial bank label. In contrast, the most rostral aspects of ALLS and AMLS, as well as the most caudal portions of PLLS and PMLS, had moderate reciprocal interconnections. Overall, the main flow of inter-regional information within LS appears to be directed from visual regions on the medial bank to those on the lateral. Nevertheless, considerable lateral input may reach the medial bank from the lateral bank via fundus region. It is likely, therefore, that input derived from various visual cortical areas that project to the medial bank of LS have access to this inter-areal circuitry. This may provide a route by which visual cortical information can be relayed to other cortical and subcortical structures involved in visually-guided behaviors including AEV, striatum, AON, the pretectal nuclei, and the deep laminae of SC, despite the fact that these structures themselves do not receive substantial direct projections from the visual cortical areas that are associated with the medial bank.

Acknowledgements

The authors wish to thank Mr. Seiji Takahashi for his skillful technical assistance. This work was supported in part by a Grant-in-Aid (06680731) from the Japanese Ministry of Education, Science and Culture (MN), Nissan Science Foundation (MN) as well as NIH grant NS35008 (JGM).

References

Barbaresi, P., Guandalini, P. and Manzoni, T. (1995) Laminar pattern of termination of the ipsilateral cortical projection from SII to SI in cats. *J. Comp. Neurol.*, 360: 319–330.

Berson, D.M. and Graybiel, A.M. (1980) Some cortical and subcortical fiber projections to the accessory optic nuclei in the cat. *Neuroscience*, 5: 2203–2217.

Bishop, P.O., Kozak, W. and Vakkur, G.J. (1962) Some quantitative aspects of the cat's eye: axis and plane of reference, visual field co-ordinates and optics. *J. Physiol.*, 163: 466–502.

Burchinskaya, L.F., Zelenskaya, V.S., Cherkes, V.A. and Kolomiets, B.P. (1988) Pathways for transmission of visual and auditory information to the cat caudate nucleus. *Neurophysiology*, 19: 385–393.

Chevalier, G., Vacher, S., Deniau, J.M. and Desban, M. (1985) Disinhibition as a basic process in the expression of striatal functions. I. The striato-nigral influence on tecto-spinal/tecto-diencephalic neurons. *Brain Res.*, 334: 215–226.

Clare, M.H. and Bishop, G.H. (1954) Responses from an association area secondarily activated from optic cortex. *J. Neurophysiol.*, 17: 271–277.

Collewijn, H. (1975) Direction-selective units in the rabbit's nucleus of the optic tract. *Brain Res.*, 100: 489–508.

Dreher, B. (1986) Thalamocortical and corticocortical interconnections in the cat visual system: relation to the mechanisms of information processing. In: J.D. Pettigrew, K.J. Sanderson and W.R. Levick (Eds.), *Visual Neuroscience*, Cambridge University Press, Cambridge, pp. 290–314.

Graham, J. and Berman, N. (1981) Origins of the pretectal and tectal projections to the central lateral nucleus in the cat. *Neurosci. Lett.*, 26: 209–214.

Grant, S. and Shipp, S. (1991) Visuotopic organization of the lateral suprasylvian area and of an adjacent area of the ectosylvian gyrus of cat cortex: a physiological and connectional study. *Visual Neurosci.*, 6: 315–338.

Grasse, K.L. and Cynader, M.S. (1984) Electrophysiology of lateral and dorsal terminal nuclei of the cat accessory optic system. *J. Neurophysiol.*, 51: 276–293.

Graybiel, A.M. (1978) Organization of the nigrotectal connections: as experimental tracer study in the cat. *Brain Res.*, 143: 339–348.

Graybiel, A.M. and Berson, D.M. (1980) Histochemical identification and afferent connections of subdivisions in the lateralis posterior–pulvinar complex and related thalamic nuclei in the cat. *Neuroscience*, 5: 1175–1238.

Guillery, R.W., Geisert, E.E.Jr., Polley, E.H. and Mason, C.A. (1980) An analsis of the retinal afferents to the cat's medial interlaminar nucleus and to its rostral thalamic extension, the geniculate wing. *J. Comp. Neurol.*, 194: 117–142.

Hardy, S.C. and Stein, B.E. (1988) Small lateral suprasylvian cortex lesions produce visual neglect and decreased visual activity in the superior colliculus. *J. Comp. Neurol.*, 273: 527–542.

Harting, J.K., Huerta, M.F., Hashikawa, T., Weber, J.T. and van Lieshout, D.P. (1988) Neuroanatomical studies of the nigrotectal projection in the cat. *J. Comp. Neurol.*, 278: 615–631.

Harting, J.K., Updyke, B.V. and Van Lieshout, D.P. (1992) Corticotectal projections in the cat: anterograde transport studies of twenty-five cortical areas. *J. Comp. Neurol.*, 324: 379–414.

Heath, C.J. and Jones, E.G. (1971) The anatomical organization of the suprasylvian gyrus of the cat. *Ergebn. Anat. Entw.-Gesch.*, 45: 1–64.

Hoffmann, K.P. and Distler, C. (1989) Quantitative analysis of visual receptive fields of neurons in nucleus of the optic tract and dorsal terminal nucleus of the accessory optic system in macaque monkey. *J. Neurophysiol.*, 62: 416–428.

Hoffmann, K.P., Distler, C., Mark, R.F., Marotte, L.R., Henry, G.H. and Ibbotson, M.R. (1995) Neural and behavioral effects of early eye rotation on the optokinetic system in the Wallaby, Macropus eugenii. *J. Neurophysiol.*, 73: 727–735.

Huerta, M.F. and Harting, J.K. (1984) The mammalian superior colliculus: studies of its morphology and connections. In: H. Vanegas (Ed.), *Comparative Neurology of the Optic Tectum*, Plenum Publishing Corporation, pp. 687–773.

Hutchins, B. and Updyke, B.V. (1989) Retinotopic organization within the lateral posterior complex of the cat. *J. Comp. Neurol.*, 285: 350–398.

Imig, T.J. and Morel, A. (1983) Organization of the thalamocortical auditory system in the cat. *Annu. Rev. Neurosci.*, 6: 95–120.

Infante, C. And Leiva, J. (1984) Correlation between pulvinar-lateralis posterior unit activity and eye movements in the cat. *Exp. Neurol.*, 85: 453–460.

Jones, E.G. and Burton, H. (1974) Cytoarchitecture and somatic sensory connectivity of thalamic nuclei other than the ventrobasal complex in the cat. *J. Comp. Neurol.*, 154: 395–432.

Kato, I., Harada, K., Hasegawa, T. and Ikarashi, T. (1988) Role of the nucleus of the optic tract of monkeys in optokinetic nystagmus and optokinetic after-nystagmus. *Brain Res.*, 474: 16–26.

Kato, I., Watanabe, S., Sato, S. and Norita, M. (1996) Pretectofugal fibers from the nucleus of the optic tract in monkeys. *Brain Res.*, 705: 109–117.

Kaufman, E.S. and Rosenquist, A.C. (1985) Efferent projections of the thalamic intralaminar nuclei in the cat. *Brain Res.*, 335: 257–279.

Kawamura, S., Sprague, J.M. and Niimi, K. (1974) Corticofugal projections from the visual cortices to the thalamus, pretectum, and superior colliculus in the cat. *J. Comp. Neurol.*, 158: 339–362.

Kimura, A. and Tamai, Y. (1992) Sensory response of cortical neurons in the anterior ectosylvian sulcus, including the area evoking eye movement. *Brain Res.*, 575: 181–186.

King, M.A., Louis, P.M., Hunter, B.E. and Walker, D.W. (1989) Biocytin: a versatile anterograde neuroanatomical tract-tracing alternative. *Brain Res.*, 497: 361–367.

Lui, F., Biral, G.P., Benassi, C., Ferrari, R. and Corazza, R. (1990) Correlation between retinal afferent distribution, neuronal size, and functional activity in the guinea pig medial terminal accessory optic nucleus. *Exp. Brain Res.*, 81: 77–84.

Marcotte, R.R. and Updyke, B.V. (1982) Cortical visual areas of the cat project differentially onto the nuclei of the accessory optic system. *Brain Res.*, 242: 205–217.

Marshall, W.H., Talbot, S.A. and Ades, H.W. (1943) Cortical 'response' of the anesthetized cat to gross photic and electrical afferent stimulation. *J. Neurophysiol.*, 6: 1–14.

McHaffie, J.G., Norita, M., Dunning, D.D. and Stein, B.E. (1993) Corticotectal relationships: direct and "indirect" corticotectal pathways. *Prog. Brain Res.*, 95: 139–150.

Mesulam, M.-M., Hegarty, E., Barbas, H., Carson, K.A., Gower, E.C., Knapp, A.G., Moss, M.B. and Mufson, E.J. (1980) Additional factors influencing sensitivity in the tetramethyl benzidine method for HRP neurohistochemistry. *J. Histochem. Cytochem.*, 28: 1255–1259.

Mucke, L., Norita, M., Benedek, G. and Creutzfeldt, O. (1982) Physiologic and anatomic investigation of a visual cortical area situated in the ventral bank of the anterior ectosylvian sulcus of the cat. *Exp. Brain Res.*, 46: 1–11.

Norita, M., Mucke, L., Benedek, G., Albowitz, B., Katoh, Y. and Creutzfeldt, O.D. (1986) Connections of the anterior ectosylvian visual area (AEV). *Exp. Brain Res.*, 62: 225–240.

Norita, M., McHaffie, J.G., Shimizu, H. and Stein, B.E. (1991) The corticostriatal and corticotectal projections of the feline lateral suprasylvian cortex demonstrated with anterograde biocytin and retrograde fluorescent techniques. *Neurosci. Res.*, 10: 149–155.

Ogasawara, K., McHaffie, J.G. and Stein, B.E. (1984) Two visual corticotectal systems in cat. *J. Neurophysiol.*, 52: 1226–1245.

Olson, C.R. and Graybiel, A.M. (1983) An outlying visual area in the cerebral cortex of the cat. *Prog. Brain Res.*, 58: 239–245.

Olson, C.R. and Lawler, K. (1987) Cortical and subcortical afferent connections of a posterior division of feline area 7 (area 7p). *J. Comp. Neurol.*, 259: 13–30.

Palmer, L.A., Rosenquist, A.C. and Tusa, R.J. (1978) The retinotopic organization of lateral suprasylvian visual areas in the cat. *J. Comp. Neurol.*, 177: 237–256.

Precht, W. (1981) Visual-vestibular interaction in vestibular neurons: functional pathway organization. *Ann. N.Y. Acad. Sci.*, 374: 230–248.

Raczkowski, D. and Rosenquist, A.C. (1981) Retinotopic organization in the cat lateral posterior–pulvinar complex. *Brain Res.*, 221: 185–191.

Raczkowski, D. and Rosenquist, A.C. (1983) Connections of the multiple visual cortical areas with the lateral posterior–pulvinar complex and adjacent thalamic nuclei in the cat. *J. Neurosci.*, 3: 1912–1942.

Rosenquist, A.C. (1985) Connections of visual cortical areas in the cat. In: A. Peters and E.G. Jones (Eds.), *Cerebral Cortex*, Vol. 3, *Visual Cortex*, Plenum Press, New York, pp. 81–117.

Royce, G.J. and Laine, E.J. (1984) Efferent connections of the caudate nucleus, including cortical projections of the striatum and other basal ganglia: an autoradiographic and horseradish peroxidase investigation in the cat. *J. Comp. Neurol.*, 226: 28–49.

Schlag-Rey, M. and Schlag, J. (1977) Visual and presaccadic neuronal activity in thalamic internal medullary lamina of cat: a study of targeting. *J. Neurophysiol.*, 40: 156–173.

Segal, R.L. and Beckstead, R.M. (1984) The lateral suprasylvian corticotectal projection in cats. *J. Comp. Neurol.*, 225: 259–275.

250

Sherk, H. (1986) Location and connections of visual cortical areas in the cat's suprasylvian sulcus. *J. Comp. Neurol.*, 247: 1–31.

Simpson, J.I. (1984) The accessory optic system. *Ann. Rev. Neurosci.*, 7: 13–41.

Simpson, J.I., Soodak, R.E. and Hess, R. (1979) The accessory optic system and its relation to the vestibulocerebellum. *Prog. Brain Res.*, 50: 715–724.

Simpson, J.I., Giolli, R.A. and Blanks, R.H.I. (1988) The pretectal complex and the accessory optic system. In: J.A. Buttner-Ennever (Ed.), *Progress in Oculomotor Research, Neuroanatomy of the Oculomotor System*, Vol. 2, Elsevier Science Publishers, Ansterdam, pp. 335–364.

Spear, P.D. (1991) Functions of extrastriate visual cortex in non-primate species. From *Vision and Visual Dysfunction*, Vol. 4. In: A.G. Leventhal (Ed.), *The Neural Basis of Visual Function*, Macmillan Press, Basingstoke, England, pp. 339–370.

Spear, P.D., Miller, S. and Ohman, L. (1983) Effects of lateral suprasylvian visual cortex lesions on visual localization, discrimination, and attention in cats. *Behav. Brain Res.*, 10: 339–359.

Stevens, R.T., London, S.M. and Apkarian, A.V. (1993) Spinothalamocortical projections to the secondary somatosensory cortex (SII) in squirrel monkey. Brain Res., 631: 241–246.

Sugimoto, T., Mizuno, N. and Itoh, K. (1981) An autoradiographic study on the terminal distribution of cerebellothalamic fibers in the cat. *Brain Res.*, 215: 29–47.

Symonds, L.L. and Rosenquist, A.C. (1984a) Corticocortical connections among visual areas in the cat. *J. Comp. Neurol.*, 229: 1–38.

Symonds, L.L. and Rosenquist, A.C. (1984b) Laminar origins of visual corticocortical connections in the cat. *J. Comp. Neurol.*, 229: 39–47.

Updyke, B.V. (1993) Organization of visual corticostriatal projections in the cat, with observations on visual projections to claustrum and amygdala. *J. Comp. Neurol.*, 327: 159–193.

Weber, J.T. (1985) Pretectal complex and accessory optic system of primates. Brain Behav. Evol., 26: 117–140.

M. Norita, T. Bando and B. Stein (Eds.)
Progress in Brain Research, Vol 112
© 1996 Elsevier Science BV. All rights reserved.

CHAPTER 18

Areas PMLS and 21a of cat visual cortex are not only functionally but also hodologically distinct

B. Dreher*[1], R.L. Djavadian[2], K.J. Turlejski[2] and C. Wang[1]

[1]*Department of Anatomy and Histology, Institute for Biomedical Research, The University of Sydney, F13, N.S.W. 2006, Australia*
[2]*Department of Neurophysiology, Nencki Institute of Experimental Biology, 3 Pasteur Street, Warsaw 02-093, Poland*

In several cats, paired visuotopically matched injections of retrogradely transported fluorescent dyes, diamidino yellow (DY) and fast blue (FB), were made into two visuotopically organized, functionally distinct extrastriate cortical areas, the posteromedial lateral suprasylvian area (PMLS area) and area 21a respectively. After an appropriate survival time, the numbers of thalamic, claustral and cortical cells which were single-labelled with each dye as well as the numbers of cells in these structures labelled with both dyes (double-labelled cells) were assessed. The clear majorities of thalamic cells projecting to PMLS area (DY labelled cells) and to area 21a (FB labelled cells) were located in the ipsilateral lateral posterior-pulvinar complex with smaller proportions located in the laminae C and the medial intralaminar nucleus of the ipsilateral dorsal lateral geniculate nucleus and several nuclei of the rostral intralaminar thalamic group. Despite the fact that DY labelled (PMLS-projecting) and FB labelled (area 21a-projecting) cells in all thalamic nuclei were well intermingled, only 1–5% of retrogradely labelled thalamic cells projected to both areas (cells double-labelled with both dyes). Small proportions of retrogradely labelled cells were located in the ipsilateral and to a lesser extent the contralateral dorsocaudal claustra. The proportions of claustral neurons retrogradely labelled with both dyes varied from 4 to 9%.

Over half of the cortical neurons labelled retrogradely from area 21a or PMLS area were located in the supragranular layers of the ipsilateral area 17, with smaller proportions located in the supragranular layers of the ipsilateral areas 18 and 19 and even smaller proportions located in mainly but not exclusively, the infragranular layers of the ipsilateral areas 21b and 20a. Again despite strong spatial intermingling of neurons labelled with DY and these labelled with FB, the proportions of associational cortical neurons double-labelled with both dyes were small (2 to 5.5%). Finally, small proportions of neurons retrogradely labelled with DY or FB were located, mainly but not exclusively, in the supragranular layers of the contralateral areas 17, 18, 19 and 21a. Again, the proportions of the double-labelled neurons in the contralateral cortices were small (1–4.5%). Thus, the present study indicates that despite the fact that the diencephalic and telencephalic inputs to the visuotopically corresponding parts of area 21a and PMLS area originate from the same nuclei, areas and layers, the two areas receive their afferents from the largely separate populations of neurons.

Introduction

Apart from the primary visual cortices which receive their principal thalamic input from the main visual thalamic relay nucleus, the dorsal lateral geniculate nucleus (LGNd), the visual cortices of

virtually all mammals appear to consist of a number of more or less distinct 'extrastriate' areas, each of which containing at least partial representation(s) of the contralateral visual hemifield (Kaas and Krubitzer, 1991; Sereno and Allman, 1991; Spear, 1991; Coogan and Burkhalter, 1993). The establishment of homologies or even analogies, between particular cortical areas in different species of the same order or in species of different orders could provide some crucial insights

*Corresponding author. Fax: +61 2 3516556 or +61 2 3512813; e-mail: bogdand@anatomy.su.oz.au

concerning the steps involved in the evolution of the mammalian visual cortex. The task, however, is very difficult (Kaas and Krubitzer, 1991; Sereno and Allman, 1991; Krubitzer and Kaas, 1993; Payne, 1993) not least because the criteria for designating a particular region of the extrastriate visual cortex as a distinct cortical area, even in such widely studied orders with well-developed visual cortices as carnivores or primates, are frequently fairly limited. For example, in domestic cats, the part of the extrastriate visual cortex located in the middle suprasylvian gyrus (in the vicinity of the middle and posterior suprasylvian sulci) has been subdivided, on the basis of very limited and partially controversial criteria (see below), into two separate areas: the posteromedial lateral suprasylvian area (PMLS area; presumed homologue of the middle-temporal or MT area of primates, Zeki, 1974; Shipp and Grant, 1991; Payne, 1993) and area 21a. In particular, the original designation of area 21a and PMLS area as two distinct visual areas was based exclusively on: (1) their complex but somewhat controversial distinctness of their visuotopic organization and (2) rather limited information, available at the time, concerning their hodology (Heath and Jones, 1971; Palmer et al., 1978; Tusa and Palmer, 1980; Tusa et al., 1981). Not surprisingly the designation of areas 21a and PMLS as distinct visual areas has been seriously challenged. It has been pointed out, that: (1) the interpretation of the visuotopic organization of the region as indicating the existence of two distinct areas is somewhat arbitrary; (2) the two postulated areas are cytoarchitectonically very similar and (3) both areas receive their principal associational inputs from the supragranular laminae of area 17 and to a lesser extent the supragranular laminae of areas 18 and 19 (Montero, 1981; Sherk, 1986, 1988; Grant and Shipp, 1991). Further challenge to the designation of areas 21a and PMLS as two distinct areas comes from the fact that the patterns of the thalamo-cortical interconnections of both areas are also very similar (Rosenquist, 1985; Dreher, 1986; Sherk, 1986; Shipp and Grant, 1991). On the basis of the hodological similarities

and the overall pattern of the visuotopic organisation of the visual areas located around the lateral suprasylvian sulcus, a number of authors (Sherk, 1986, 1988; Shipp and Grant, 1991) consider PMLS area and at least part of the region designated as area 21a, as parts of a larger single visual cortical area originally described by Marshall and his colleagues (1943) and further delineated in 1954 by Clare and Bishop (also Turlejski and Michalski, 1975; Djavadian and Harutiunian-Kozak, 1983; Sherk, 1986; Afrikyan et al., 1991; Grant and Shipp, 1991; Mulligan and Sherk, 1993; Sherk and Mulligan, 1993).

Indeed, currently reported differences in the pattern of the thalamo-cortical and cortico-cortical connections between areas 21a and PMLS are very minor (see for reviews Rosenquist, 1985; Dreher, 1986) and at the moment at least, the argument in favour of the idea that areas 21a and PMLS constitute separate visual areas is based almost exclusively on reported major differences in the receptive field properties of cells in these two areas (Dreher, 1986; Mizobe et al., 1988; Wimborne and Henry, 1992; Dreher et al., 1993, 1996; Toyama et al., 1994; see also for review Spear, 1991). The question thus arises whether the functional differences between the two areas are based almost exclusively on the differences in their intrinsic neuronal cicuitries. It is however, not necessarily so. In particular, the above-mentioned great hodological similarities of areas 21a and PMLS do not necessarily imply that visuotopically corresponding parts of each area receive their thalamic and cortical inputs from the same subpopulations of neurons (for cat's areas 17, 18 and 19, Bullier et al., 1984a,b; Birnbascher and Albus, 1987; Ferrer et al., 1992). In order to test whether areas 21a and PMLS receive their thalamic and cortical inputs from the same or from largely distinct subpopulations of neurons, we have made in several animals paired visuotopically matched injections of retrogradely transported fluorescent dyes, diamidino yellow dihydrochloride (DY) and fast blue (FB), into PMLS area and area 21a respectively. After an appropriate survival time we have assessed the numbers of thalamic, claustral and cortical cells which were single-labelled with each dye and the numbers of cells in these structures labelled with both dyes (double-labelled cells). Preliminary communications, describing some of our findings, have been published elsewhere (Turlejski et al., 1992, 1994).

Methods

Animals and surgical procedures

Adult cats of either sex, weighing from 2.5 to 4.5 kg, were used. The animals, obtained from the Sydney University animal colony, were anaesthetized with intramuscular injections of xylazine (3 mg/kg) and ketamine hydrochloride (15 mg/kg). The surgical level of anaesthesia (indicated by lack of corneal reflex, lack of withdrawal reflex and lack of changes in heart rate when noxious stimuli were applied) was maintained by supplementary injections, at 30-min intervals, of half of the initial dose of ketamine and, at 60-min intervals, of half the initial dose of xylazine. Animals were mounted in a stereotaxic frame, their corneas protected with air-permeable contact lenses, the skin and muscles of the head dissected and openings in the skull as well as small openings in dura made over the areas 21a and PMLS of one hemisphere. Single pressure injections of 0.2–0.3 μl of the 4% aqueous solution of Fast Blue (Dr Illing, GMBH 7 Co.KG) and 0.2–0.3 μl of 3% aqueous solution of diamidino-yellow dihydrochloride were made stereotaxically into areas 21a and PMLS respectively, through glass micropipettes (external diameter of the tip 60–70 μm) attached to the needle of a microsyringe. The target regions for injections (representations of the central 5–10° of the contralateral visual hemifield) were determined on the basis of the available cortical visuotopic maps (Tusa et al. 1981; Grant and Shipp, 1991; Sherk and Mulligan, 1993). After completion of injections, the cortex was covered with dura and parafilm and the superficial muscles of the skull and skin were sutured. Antibiotics were applied topically and intramuscularly on a daily basis. To alleviate postoperative pain, the muscles and the inside of the skin over the skull opening were sprayed with oil-based solution of the long-lasting local anaesthetic lignocaine (Xylotox Extra).

Histological processing

Following the recovery from anaesthesia and survival of 6–8 days, the animals were deeply anaesthetized with intraperitoneal injections of 80 mg/kg of sodium pentobarbitone and perfused transcardially with about 1.2 l of warm (37°C) solution of 2.7% NaCl, followed by 1.2 l of 4% paraformaldehyde in 0.1 M phosphate buffer, pH 7.4, followed by 0.5 l of 10% and 1 l of 30% buffered sucrose. Before removal of the brain from the skull the head was mounted in the stereotaxic frame and brain cut coronally in situ at Horsley-Clarke plane A 8.5. After 3–4 days in the 30% sucrose solution, the two hemispheres were separated from the brainstem and sectioned on the freezing microtome into 40-μm thick coronal sections. Every third section was collected into the phosphate buffer, mounted on gelatinized slides and left uncovered to dry. Sections were kept in the freezer until examined, mapped and photographed under an appropriate fluorescent illumination. In addition, to allow better identification of areal and laminar boundaries in the cortex and thalamus, a number of alternate sections from each animal were counterstained, coverslipped and microscopically examined under normal light illumination (cf. Otsuka and Hassler, 1962 for identification of areas 17, 18 and 19; Sanides and Hoffmann, 1969; Tusa et al., 1981; Updyke, 1986 for identification of other cortical areas; Berson and Graybiel, 1978, 1983; Updyke, 1981; Berman and Jones, 1982; Hutchins and Updyke, 1989 for identification of the thalamic nuclei).

Data analysis

Sections were examined uncoverslipped using an Olympus BH-2 microscope equipped with a AH-RFL epifluorescence attachment and the set of

254

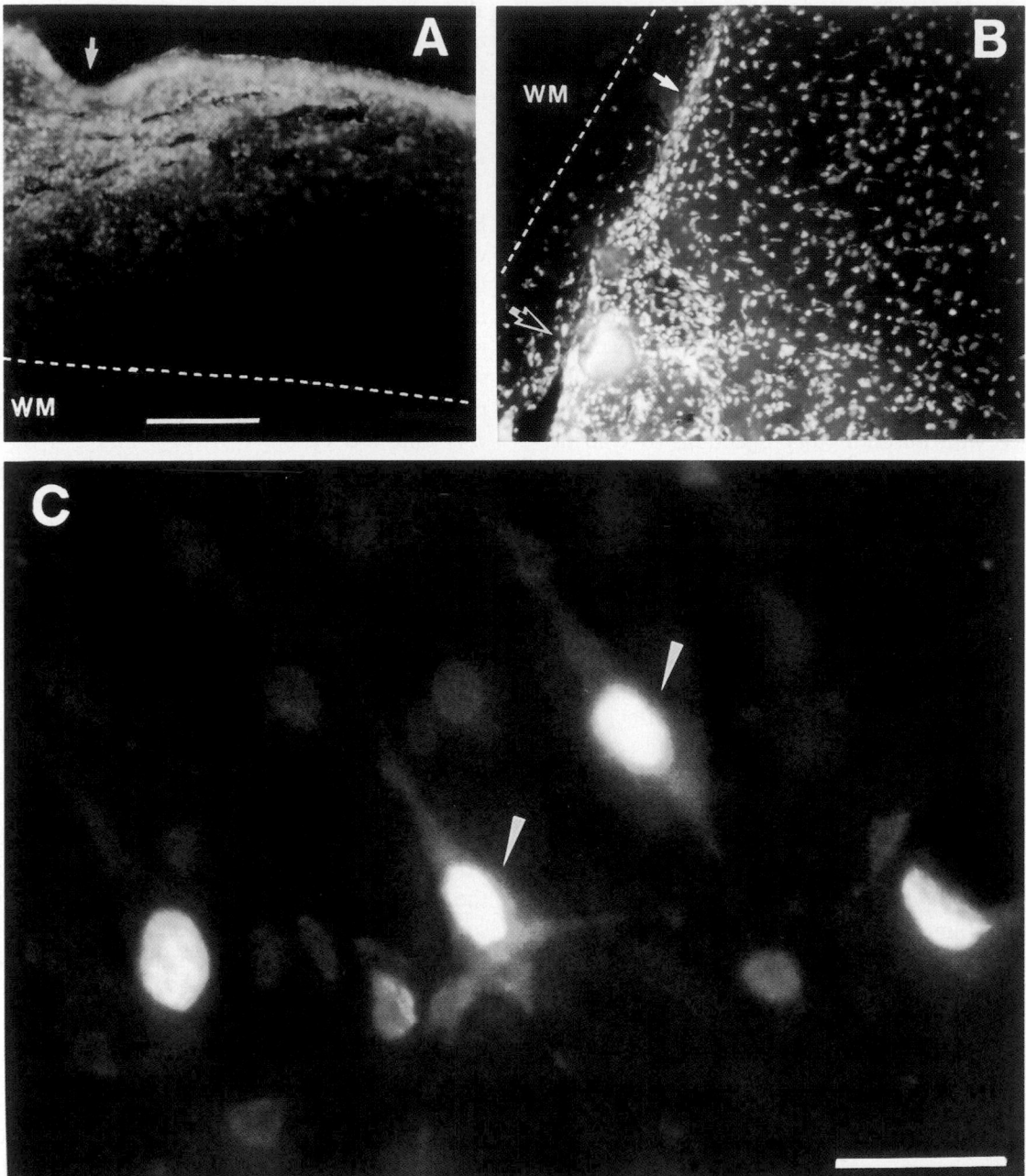

Fig. 1. (A) and (B) Fluorescence photomicrographs of coronal sections through the centers of injection sites of fast blue (FB) into area 21a and diamidino yellow (DY) into PMLS area of cat CR1. Area 21a was approached vertically and the verical arrow in A indicates the entrance point of the micropipette. PMLS area was approached at an angle of 45° laterally to the sagittal plane and about 35° rostrally to the coronal plane. The entrance point of the penetration into PMLS area was positioned about 0.7–0.8 mm medially from the suprasylvian sulcus. Arrows in B indicate micropipette tract. The middle suprasylvian sulcus is located laterally approximately parallel to the injection tract in B (Cat CR3, Figs. 2B,C). Note in both A and B that the dense core of the dye around the injection tracts does not extend into the white matter (WM). Note also retrogradely labelled cells below the dense core of the dyes. Scale bar in A represents 700 μm for A and 400 μm for B. (C) High magnification fluorescent photomicrograph of the retrogradely labelled cortical cells in layer 3 of area 17 ipsilateral to the injected areas 21a and PMLS. Note two double-labelled pyramidal cells (arrowed) with the cytoplasm labelled with FB and the nuclei labelled with DY. Note also labelling of basilar and apical dendrites with FB. Scale bar = 35 μm.

filters for violet light (excitation wavelength 300–450 nm, barrier filter 455 nm and below). Both the spread of the dyes and presence of the retrogradely labelled cells were carefully examined. The extent of the uptake region of the dye was estimated as the zone of high density of the dye surrounding injection tracks (Figs. 1A,B). The estimated region of uptake was then compared with the cyto-architectonic features of the area revealed in the neighbouring sections stained for cresyl violet and visuotopic maps of areas located in the vicinity of the middle and posterior suprasylvian sulci (Palmer et al., 1978; Tusa and Palmer, 1980; Tusa et al., 1981; Grant and Shipp, 1991). Neurons retrogradely labelled with FB exhibited bluish coloration of the cytoplasm while those labelled with DY have yellow nuclei. Finally, neurons 'double-labelled' with both DY and FB had yellow nuclei and bluish cytoplasm (Fig. 2C). Camera lucida drawings of selected sections from the regions of highest intermingling of FB and DY-labelled neurons were made and positions of all FB-labelled, all DY-labelled and all double-labelled cells were plotted with the aid of a computer-driven X-Y stage.

Results

Injection sites

In three animals the injections into area 21a (all of them FB injections) were restricted to the gray matter and did not spread outside the area (Tusa and Palmer 1980). The injection was smallest in cat CR1 where it spread rostrocaudally between the Horsley-Clarke coronal coordinates P1.8–2.5 and was limited to layers 1–4 (Fig. 1A). The injection was largest in cat CR3 where it spread rostrocaudally between Horsley-Clarke coordinates A0–P2 and involved most of the cortical thickness (Fig. 2). In these three animals, the injections into PMLS areas (all of them DY injections) were either completely restricted to area PMLS (cat CR1) or spread slightly into PLLS areas (cats CR3 and CR9). Again the injection was largest in cat CR3, where the dye spread rostrocaudally at the bottom of the middle suprasylvian sulcus at Horsley-Clarke coordinates A1–P1.5. The injection involved not only the representation of the central 5–10° of the contralateral visual hemifield in PMLS area but also a border zone of PLLS area where the same part of the contralateral visual hemifield is represented (Figs. 1B and 2; cf. Palmer et al., 1978).

Labelling of neurons in the vicinity of the injections

In addition to the retrograde labelling in the ipsilateral thalami and the ipsilateral and contralateral claustra and cortices (see below), labelled cells were found in layer 6 immediately below the injection site. Along the orientation of cortical layers the injection sites were surrounded by 'rings' of retrogradely labelled cells positioned mainly in the superficial layers (Figs. 1A,B). These rings however, did not spread beyond the injected areas and were usually wider in PMLS area (up to 1.5 mm from the dye permeated regions) than in area 21a.

Retrogradely labelled thalamic cells

Retrogradely labelled thalamic cells were restricted to the thalami ipsilateral to the injected cortices. Their total number varied substantially (from about 850 to about 1300 in the case of cells labelled with DY and from about 300 to about 850 in the case of cells labelled with FB). However, in all three animals studied, PMLS area-projecting thalamic cells and area 21a-projecting thalamic cells constituted about 16–19% of prosencephalic cells retrogradely labelled with DY or FB respectively (Dreher, 1986).

The LGNd complex

Consistent with our injections being targeted on those parts of areas 21a and PMLS where the central 5–10° are represented, retrogradely labelled cells were located in the vicinity of the border between the laminated LGNd and the medial interlaminar nucleus (MIN; Fig. 3A; cf. Sanderson, 1971). Labelled cells were found only in the parvocellular laminae C (C, C_1, C_2) and to

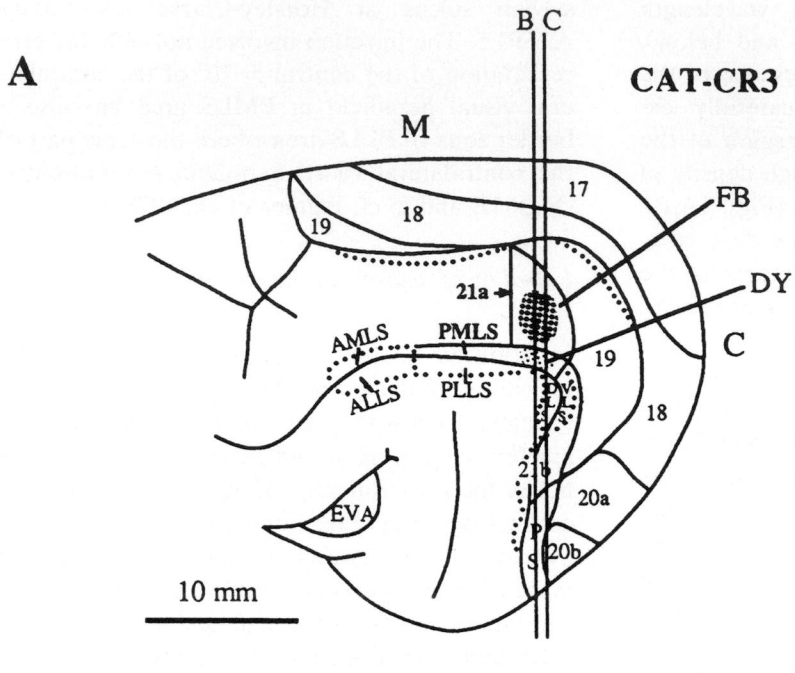

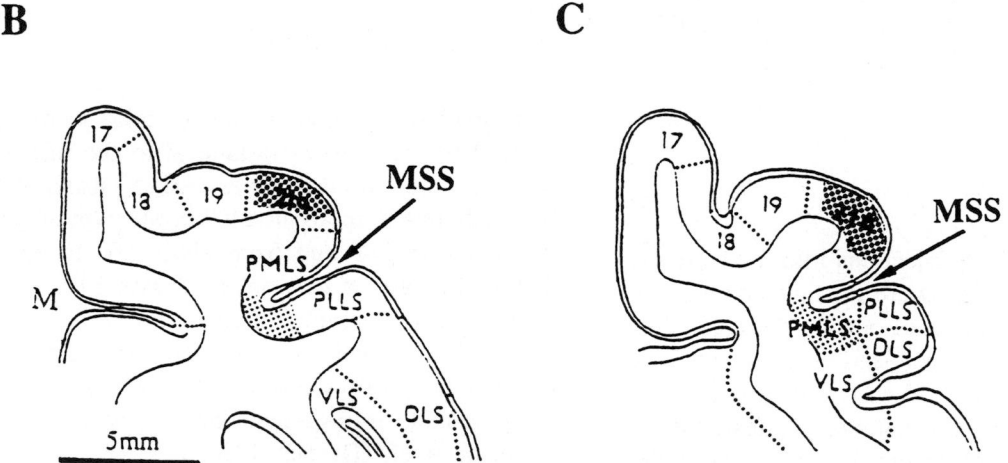

Fig. 2. (A) Latero-dorsal view of the left cerebral hemisphere of the cat with outlines of areas 17, 18, 19, 21a, 21b, 20a, 20b, as well as lateral suprasylvian areas (anteromedial, anterolateral, posteromedial and posterolateral, dorsal and ventral; AMLS, ALLS, PMLS, ALLS, PLLS, DLS and VLS, respectively), posterior suprasylvian (PS) and ectosylvian visual area (EVA). Delineation of cortical areas is based on publications of Tusa et al. (1981), Rosenquist (1985), Updyke (1986). M, medial; C, caudal. Hatchings within areas 21a and PMLS indicate respectively the spread of fluorescent FB and DY in cat CR3. (B) and (C) Outlines of two coronal sections through the left cerebral hemisphere of cat CR3 (at the levels indicated in A) with hatchings indicating the spread of FB through area 21a and DY through PMLS area. MSS — middle suprasylvian sulcus.

a lesser extent in the MIN (Figs. 3A,B) but not in the A laminae (Rosenquist, 1985; Dreher, 1986).

In all cases, however, the numbers of retrogradely labelled cells in the LGNd complex were small

(156–237 labelled cells in laminae C and only 26–38 labelled cells in the MIN). Overall, the LGNd cells projecting to areas 21a or PMLS constituted about 8–12% of retrogradely labelled thalamic cells (See also Dreher, 1986). Cells projecting to PMLS area (DY-labelled cells) were usually about twice as numerous as those projecting to area 21a and even in the regions of best intermingling of DY- and FB-labelled neurons, the double-labelled cells constituted only about 4% of retrogradely labelled cells (Fig. 3D).Thus, PMLS area and area 21a receive their direct

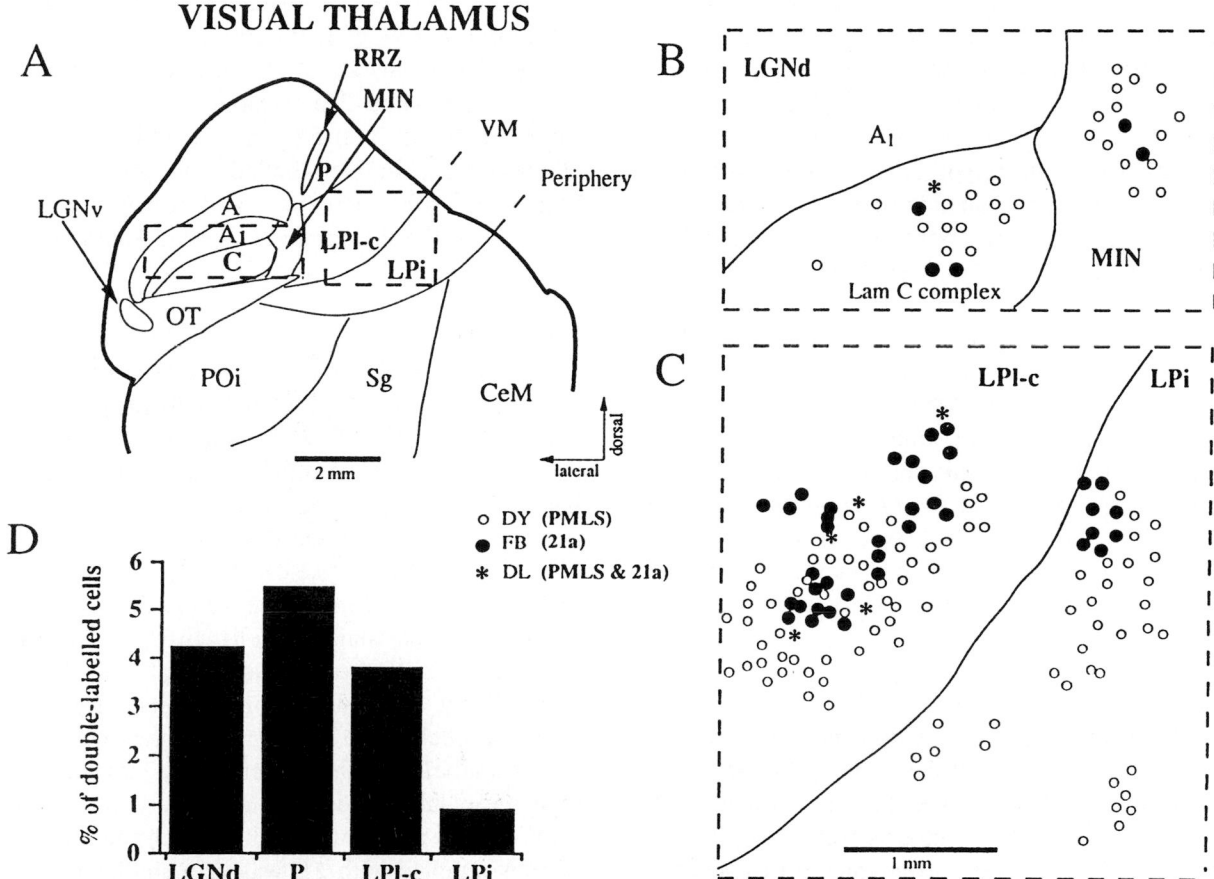

Fig. 3. Distribution of retrogradely labelled neurons in the ipsilateral visual thalamus following injections of fluorescent dyes DY and FB into areas PMLS and 21a respectively. (A) Camera lucida drawing of the dorsal part of the 40-μm coronal section through the thalamus of cat CR3 (approximately Horsley-Clarke level A 7.0; Berman and Jones, 1982). Divisions of visual thalamus are drawn after Raczkowski and Rosenquist (1981, 1983) and Hutchins and Updyke (1989). (B) and (C) are enlargements of the areas outlined by dashed lines in A. Note in B and C that labelled cells are located in the vicinity, but not at the border between C laminae complex and MIN or the border between LPl-c and LPi (where the vertical meridian is represented, Sanderson, 1971; Raczkowski and Rosenquist, 1981, 1983; Hutchins and Updyke, 1989). Note also that neurons labelled by DY (PMLS-projecting cells) are more numerous than these labelled by FB (area 21a-projecting cells) and that double-labelled cells (that is cells projecting to both PMLS area and area 21a) are very rare. (D) Histogram showing the overall percentage of double-labelled neurons in the visual thalamic nuclei of cat CR3. Abbreviations: CeM, centre median nucleus; LGNd, dorsal lateral geniculate nucleus (A, A1 and C laminae of LGNd); LGNv, ventral lateral geniculate nucleus; LPi, interjacent (or tecto-recipient) part of the lateral posterior nucleus; LPL-c, the lateral (or cortico-recipient) part of the lateral posterior nucleus (caudal layer); MIN, medial interlaminar nucleus of the LGNd complex; OT, optic tract; P, pulvinar; POi, intermediate division of posterior nucleus; RRZ, retino-recipient zone of pulvinar; Sg, suprageniculate nucleus; VM, representation of the vertical meridian.

LGNd inputs from largely distinct subpopulations of neurons in the C laminae and the MIN.

Pulvinar

Both the retino-recipient zone (Berman and Jones, 1977; Leventhal et al., 1980; called also the geniculate wing by Guillery et al., 1980) and the rest of the pulvinar contained fairly small numbers of DY- or FB-labelled cells. In the retino-recipient zone (RRZ), area 21a-projecting cells (FB-labelled) were about as numerous as PMLS-projecting cells (DY-labelled; cf. Raczkowski and Rosenquist, 1983; Dreher, 1986).They constituted about 7–12% of retrogradely labelled thalamic cells. By contrast, outside the RRZ, area 21a-projecting cells were about twice as numerous as those projecting to PMLS area (Berson and Graybiel, 1978, 1983; Raczkowski and Rosenquist, 1983; Sherk, 1986, see however Dreher, 1986). These neurons constituted respectively about 4–6% of retrogradely labelled thalamic cells projecting to area 21a but only about 2% of retrogradely labelled thalamic cells projecting to PMLS area. In the regions of the maximal intermingling of PMLS and area 21a-projecting cells, double-labelled cells constituted only about 5.5% of the population of DY- and FB-labelled cells (Fig. 3D).

Lateral posterior nucleus

Consistent with numerous previous studies (Rosenquist, 1985; Dreher, 1986; Sherk, 1986; Tong and Spear, 1986; Mizobe and Toyama, 1989; Miceli et al., 1991; Lomber et al., 1995) by far numerically the largest thalamic projections to areas 21a and PMLS originated from the lateral (or cortico-recipient) zone of the lateral posterior-pulvinar complex (LPl-c, Fig. 3). In the case of PMLS area-projecting cells retrogradely labelled LPl cells constituted 65–72% of the retrogradely labelled thalamic cells while in the case of area 21a they constituted 50–55% of the retrogradely labelled thalamic cells. However, even in the regions where DY-labelled and FB-labelled cells were maximally intermingled the double-labelled cells (i.e. cells projecting to both areas

21a and PMLS) were fairly rare and constituted only about 3.5–4.0% of all retrogradely labelled cells (Figs. 3C,D). Areas 21a and PMLS received also a direct input from the interjacent or tecto-recipient zone of the lateral posterior-pulvinar complex (LPi, Fig. 3). However, as indicated in Fig. 3C, the PMLS area-projecting cells (DY-labelled) were much more numerous than those projecting to area 21a (FB-labelled). They constituted about 15% of thalamic neurons projecting to PMLS but only about 3% of thalamic neurons projecting to area 21a. Despite the presence of clear zones of intermingling of DY- and FB-labelled cells, double-labelled cells projecting to both areas were very rare and constituted less than 1% of the retrogradely labelled LPi neurons (Fig. 3D).

Intralaminar nuclei

Consistent with earlier reports (for references see legend to Fig. 12), thalamic cells projecting to areas 21a or PMLS were also located in the rostral group of intralaminar nuclei. Cells projecting to PMLS area were usually, but not always, substantially more numerous than those projecting to area 21a (Figs. 4B–D). In the case of PMLS area they constituted about 5% of retrogradely labelled thalamic cells. In both the central medial and the central lateral nuclei, the retrogradely labelled neurons formed elongated bands parallel to the lateral wall of the nuclei and neurons labelled with FB were often intermingled with these labelled with DY (Figs. 4B,C). However, even in an animal CR3 in which the number of cells labelled with FB was very similar to that of cells labelled with DY, the double-labelled cells constituted not more than about 3% of retrogradely labelled neurons (Fig. 4D).

Ventral anterior nucleus and ventrolateral complex

Small proportions of the retrogradely labelled thalamic cells (about 1.5%) were located even more rostrally, mainly in the ventral anterior nucleus and the ventrolateral complex, although occasional labelled cells were also present in the lateral dorsal nucleus (Dreher, 1986). However,

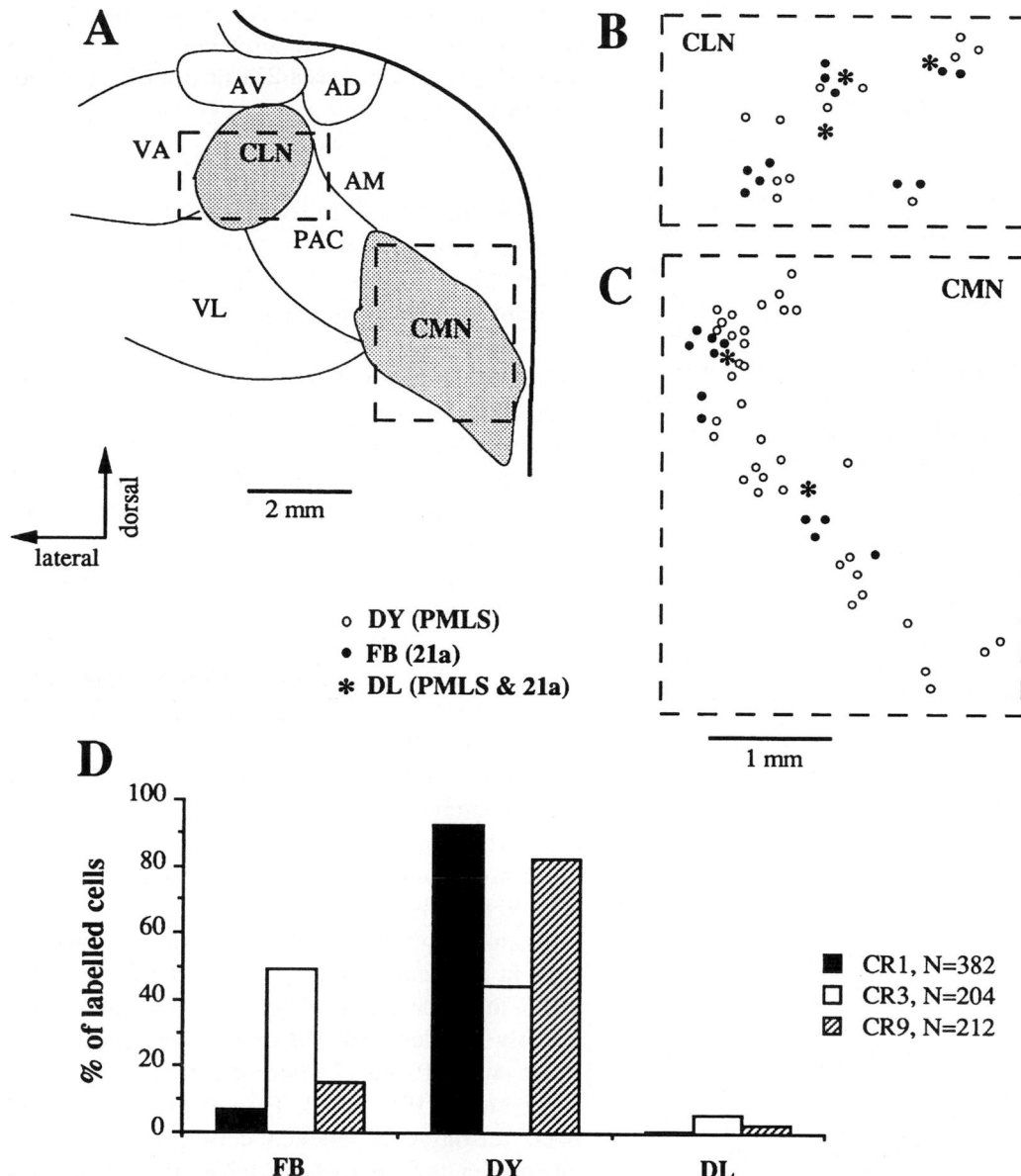

Fig. 4. Distribution of retrogradely labelled cells in the ipsilateral rostral intralaminar thalamic nuclei following injections of DY and FB into areas PMLS and 21a respectively. (A) Camera lucida drawing of the dorso-medial part of a coronal section through the ipsilateral thalamus of cat CR3 at Horsley-Clarke level of about A-11 (Berman and Jones, 1982). (B) and (C) are enlargements of the areas outlined by dashed lines in A. Note in C that neurons labelled by DY (PMLS-projecting) are somewhat more numerous than those labelled by FB (area 21a-projecting) and that in both B and C double-labelled cells (projecting to both PMLS area and area 21a) are rare. Each symbol represents one cell. (D) Histogram showing the proportions of DY, FB and double-labelled cells in the intralaminar nuclei of cats CR1, CR3 and CR9. Note that in cats CR1 and CR9 most cells are labelled with DY and that in all three cats the proportion of double-labelled cells is small. Abbreviations: AD, anterodorsal nucleus; AM, anteromedial nucleus; AV, anteroventral nucleus; CLN, central lateral nucleus; CMN, central medial nucleus; PAC, paracentral nucleus; VA, ventral anterior nucleus; VL, ventrolateral complex. Numbers in D indicate for each cat the numbers of retrogradely labelled cells in the region of intermingling of FB and DY-labelled cells. For other explanations see legend to Fig. 3.

no double-labelled cells were found in these nuclei.

Retrogradely labelled cells in the ipsilateral and contralateral claustra

Consistent with previous reports (LeVay and Sherk, 1981a; Norita, 1983), we observed both FB-labelled (area 21a-projecting) and DY-labelled (PMLS area-projecting) cells in the dorsocaudal claustrum, ipsilateral to the injected cortical areas (Figs. 5A–D). Retrogradely labelled claustral cells constituted 2–4% of all retrogradely labelled prosencephalic cells and were concentrated at the part of the visual claustrum where the central 10° of the contralateral visual field are represented (LeVay and Sherk, 1981b). Although PMLS projecting neurons were 2–3 times more numerous than those projecting to area 21a, the two sets of neurons were frequently well intermingled. However, even in the regions of maximal intermingling of the two populations, double-labelled cells constituted only 4–9% of the populations of labelled cells (Figs. 5A–D).

In addition, unlike in the case of the thalamic cells, there were FB-labelled and/or DY-labelled cells in the dorsocaudal claustrum contralateral to injected cortices (Norita, 1983; Minciacchi et al., 1985; Dreher, 1986). Retrogradely labelled cells in the contralateral claustrum: (1) were located in the topographically corresponding part of the dorsocaudal claustrum; (2) were much less numerous than those in the ipsilateral claustrum and (3) neurons projecting to PMLS area (DY-labelled) were about twice as numerous as those projecting to area 21a (FB-labelled). About 8% of neurons in the contralateral claustrum were double-labelled.

Retrogradely labelled cells in the ipsilateral cortical areas

In all three animals studied, PMLS area-projecting cortical cells and area 21a-projecting cortical cells constituted about 75–80% of prosencephalic cells retrogradely labelled with DY or FB respectively (Dreher, 1986). Furthermore, the distribution and relative numbers of retrogradely labelled cortical cells are in good agreement with those in numerous previously published studies (see legend to Fig. 12).

Area 17

About 60–75% of retrogradely labelled cortical neurons projecting to area 21a (FB-labelled) and about 45–60% of retrogradely labelled cortical neurons projecting to PMLS area (DY-labelled) were located in area 17 (Dreher, 1986; Sherk, 1986; Mizobe and Toyama, 1989). The labelled cells exhibited mostly the morphology of pyramidal cells (Fig. 2C) and over 95% of area 17 cells projecting to area 21a and over 90% of area 17 cells projecting to PMLS area were located in the supragranular layers 2–3 with the remainder scattered throughout layers 4–6 (Fig. 6; Gilbert and Kelly, 1975; Henry et al., 1978; Symonds and Rosenquist, 1984b; Dreher, 1986; Sherk, 1986; Mizobe and Toyama, 1989; Einstein and Fitzpatrick, 1991; Shipp and Grant, 1991). In two cats (CR1 and CR9), cells projecting to PMLS area and those projecting to area 21a were about equally numerous, while in the third cat (CR3) cells projecting to PMLS area were substantially more numerous than those projecting to area 21a (Fig. 9, top histogram).

Again, as in the case of the retrogradely labelled cells in the retinotopically or visuotopically organized thalamic nuclei (Figs. 3A,B,C), cells retrogradely labelled with FB or DY were located in those parts of area 17 where the central 5–10° are represented (Figs. 6A,B; Tusa et al., 1978). However, retrogradely labelled cells projecting to a given area tended to form fairly distinct patches (Symonds and Rosenquist, 1984a; Sherk, 1986; Shipp and Grant, 1991) and the patches of cells projecting to area 21a largely interdigitated with the clusters of cells projecting to PMLS area (Figs. 6D,E). Nevertheless, since the patches, especially those of FB-labelled cells, were not strictly discontinuous there was a substantial spatial overlap of the patches of cells projecting to each area (Fig. 6C). However, even in the regions of maximal overlap, the double-labelled cells constituted

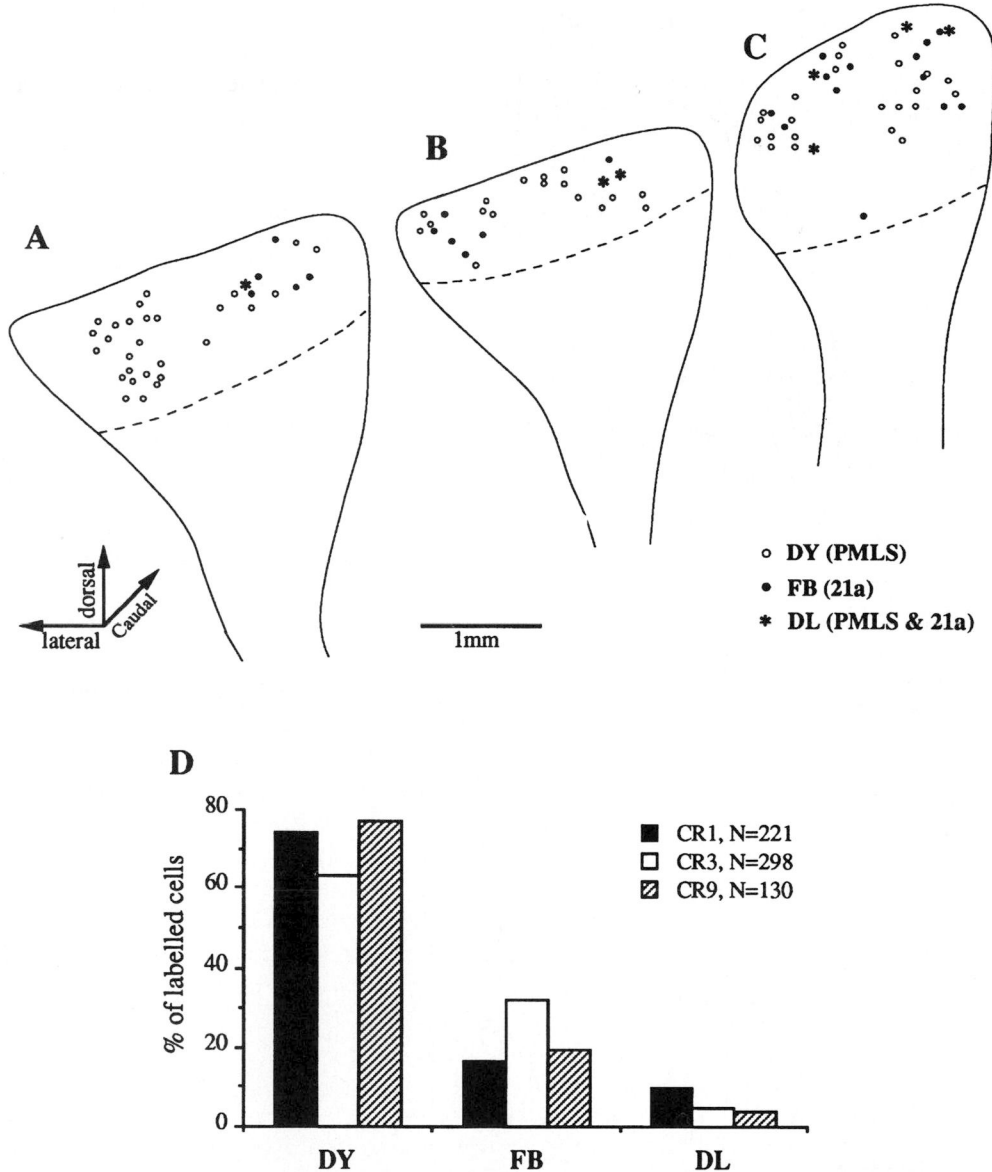

Fig. 5. Distribution of retrogradely labelled cells in the ipsilateral claustrum following injections of DY and FB into areas PMLS and 21a respectively. (A), (B) and (C) Camera lucida drawings of coronal sections through the claustrum of cat CR3 at Horsley-Clarke levels of about A-12.5 (A) to about A-10.5 (C). Each symbol represents one cell. Dashed lines indicate the lower borders of the visual claustrum, that is the part of the claustrum which contains the representation of the contralateral visual field (after LeVay and Sherk, 1981b). (D) Histogram showing, for all three investigated cats, the proportions of DY-labelled, FB-labelled and double-labelled cells (that is cells projecting to both PMLS area and area 21a) in the regions of the ipsilateral claustra where the DY-labelled and FB-labelled cells were intermingled. Note that in all three cats, DY-labelled cells were about three times as numerous as FB-labelled cells and that double-labelled cells were rare. For other explanations see legend to Fig. 3.

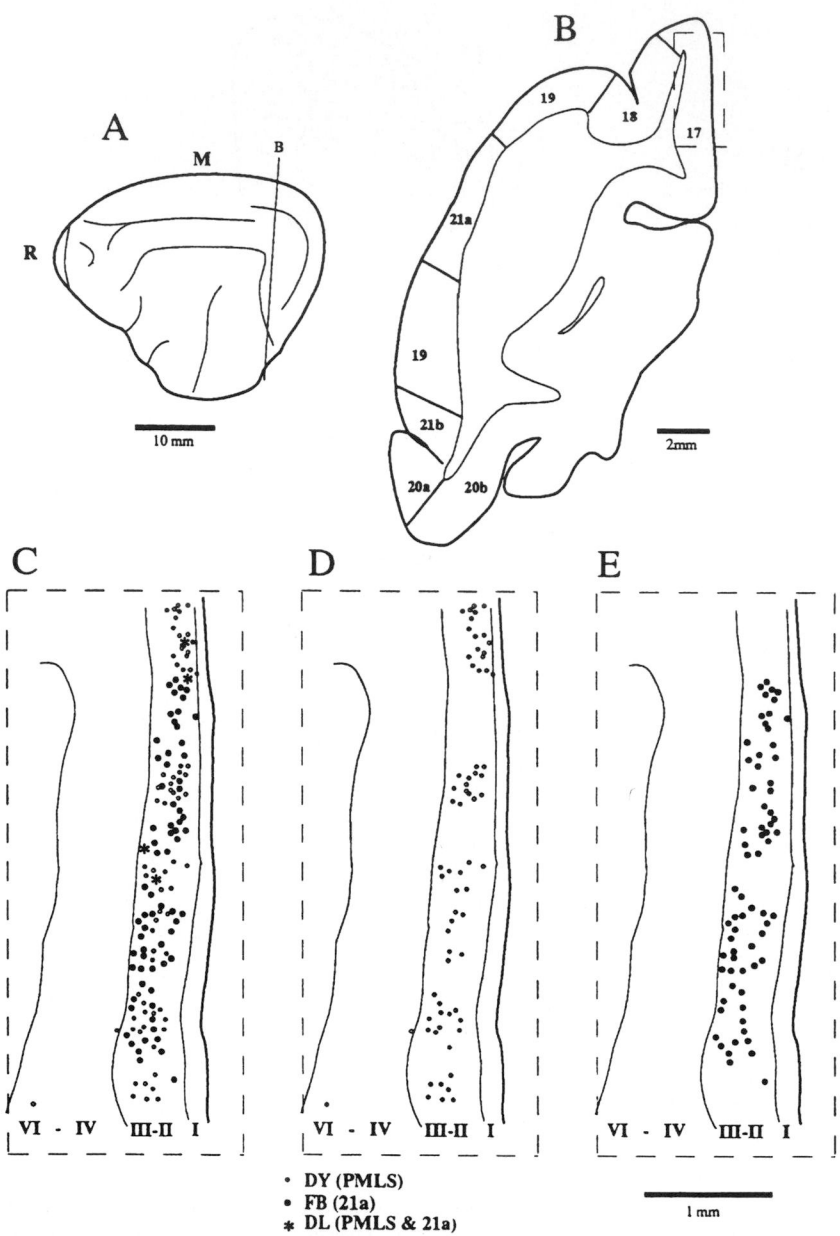

- • DY (PMLS)
- • FB (21a)
- * DL (PMLS & 21a)

1 mm

Fig. 6. Distribution of retrogradely labelled cells in the ipsilateral area 17 of cat CR3 following injections of DY and FB into areas PMLS and 21a respectively. (A) Outline of the latero-dorsal view of the left cerebral hemisphere. M, medial, R, rostral. (B) Camera lucida drawing of a coronal section through the ipsilateral cortex at the level B indicated in A. (C), (D) and (E) are the enlargements of the part of area 17 outlined by dashed lines in B. In C positions of all retrogradely labelled cells are plotted. In D and E positions of cells retrogradely labelled with DY (PMLS-projecting cells) and FB (21a-projecting cells) are plotted separately. Each symbol represents one cell. Retrogradely labelled cells are located in the part of area 17 where the region of approximately 3–10° of the contralateral visual hemifield along the horizontal meridian is represented (vertical meridian is represented at the border of areas 17 and 18; Tusa et al., 1978). Note that all, but two, retrogradely labelled cells are located in the supragranular layers 2 and 3. Note also in D and E the patchy distribution of retrogradely labelled cells (especially these labelled with DY) and the paucity of double-labelled cells (asterisks in C).

only about 2.5–5% of retrogradely labelled neurons (Figs. 6C and 9).

Area 18

About 15–25% of retrogradely labelled cortical neurons projecting to areas 21a or PMLS were located in area 18 (Dreher, 1986; Sherk, 1986; Mizobe and Toyama, 1989). About 90–95% of area 18 cells projecting to area 21a or PMLS area were located in the supragranular layers 2–3 with the remainder scattered mainly through layer 4, close to the border of layer 3 and to a lesser extent the infragranular layers 5–6 (Fig. 7; Gilbert and Kelly, 1975; Symonds and Rosenquist, 1984b; Dreher, 1986; Mizobe and Toyama, 1989). In two cats, cells projecting to PMLS area (DY-labelled) were substantially more numerous than those projecting to area 21a (FB-labelled; Fig. 9, the second histogram from the top).

As in the case of the retrogradely labelled cells in area 17, retrogradely labelled cells in area 18 were located in those parts of the area where the central 5–10° are represented (Figs. 7A,B; Tusa et al., 1979). Furthermore, as in area 17, there was patchy distribution of cells projecting to area 21a or PMLS (Figs. 7C–E). In the regions of maximal overlap the double-labelled cells (that is cells projecting to both areas) constituted only about 2–4% of retrogradely labelled neurons (Figs. 7C and 9).

Area 19

About 15–25% of retrogradely labelled cortical neurons projecting to areas 21a or PMLS area were located in area 19 (Dreher, 1986; Sherk, 1986; Mizobe and Toyama, 1989). As in the case of the retrogradely labelled cells in areas 17 and 18 (Figs. 6 and 7), retrogradely labelled cells in area 19 were located in those parts of the area where the central 5–10° are represented (Fig. 8; Tusa et al., 1979). Although the largest proportions (usually the majority) of the retrogradely labelled cells were located in the supragranular layers, substantial proportions of area 19 cells projecting to areas 21a/PMLS were located in the infragranular layers (especially layer 5) and

some were located in layer 4 (Figs. 8 and 10; Symonds and Rosenquist, 1984b; Dreher, 1986). In one of the cats (CR9) no retrogradely labelled cells were found in layer 6 (Fig. 10). In all three cats, cells projecting to PMLS area (DY-labelled) were about twice as numerous as those projecting to area 21a (FB-labelled; Fig. 9, the third histogram from the top). Patches of cells projecting to PMLS (DY-labelled) tended to interdigitate with patches of cells projecting to area 21a (FB-labelled); however, there was also a substantial spatial overlap of the two populations (Fig. 8). Cells projecting to both areas (double-labelled) were present in all layers except layer 1 (Fig. 8). In the regions of maximal spatial overlap of projections to areas 21a and PMLS, the double-labelled cells constituted 3–4.5% of retrogradely labelled neurons (Figs. 8 and 9).

Area 21b

About 5–10% of retrogradely labelled cortical neurons projecting to area 21a or PMLS area were located in area 21b (Symonds and Rosenquist, 1984b; Dreher, 1986; Sherk, 1986). However, in one cat (CR1) only DY-labelled (PMLS area-projecting) cells were found in area 21b. Most cells projecting to areas 21a or PMLS were located in the infragranular layer 5 with substantial proportions of cells located in the supragranular layers and only a few cells located in layers 4 and 6 (Figs. 8 and 10; cf. Symonds and Rosenquist, 1984b; Dreher, 1986; Mizobe and Toyama, 1989). In the regions of spatial overlap of the two projections, double-labelled cells constituted 4–5% of the retrogradely labelled cells (Figs. 8 and 9).

Area 20a

In two cats about 3–5% of the retrogradely labelled cortical neurons projecting to areas 21a or PMLS area were located in area 20a (cf. Dreher, 1986; Sherk, 1986; Mizobe and Toyama, 1989). In one cat no retrogradely labelled cells were found in this area. In the other two cats, the majorities (67–81%) of the retrogradely labelled cells were DY-labelled (PMLS area-projecting).

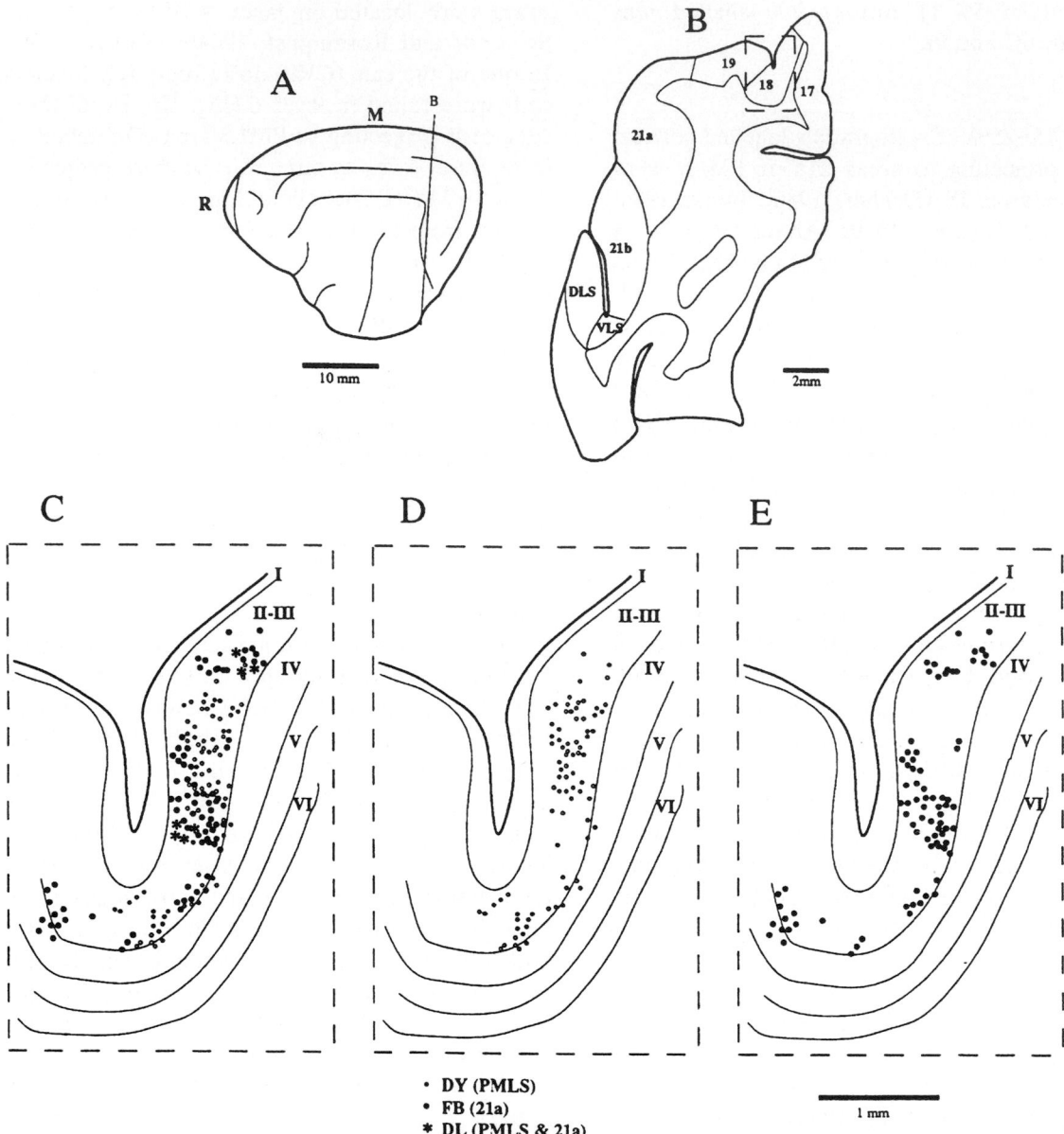

Fig. 7. Distribution of retrogradely labelled cells in the ipsilateral area 18 of cat CR3 following injections of DY and FB into areas PMLS and 21a respectively. (A) Outline of the latero-dorsal view of the left cerebral hemisphere. M, medial, R, rostral. (B) Camera lucida drawing of a coronal section through the ipsilateral cortex at the level B indicated in A. (C), (D) and (E) are the enlargements of the part of area 18 outlined by dashed lines in B. In C positions of all retrogradely labelled cells are plotted. In D and E positions of cells retrogradely labelled with DY (PMLS-projecting cells) and FB (21a-projecting cells) are plotted separately. Each symbol represents one cell. Retrogradely labelled cells are located in the part of area 18 where the region of approximately 3–10° of the contralateral visual hemifield along the horizontal meridian is represented (Tusa et al., 1979). Note that the great majority of the retrogradely labelled cells is located in the supragranular layers 2 and 3. Note also in D and E the patchy distribution of retrogradely labelled cells and the paucity of double-labelled cells (asterisks in C).

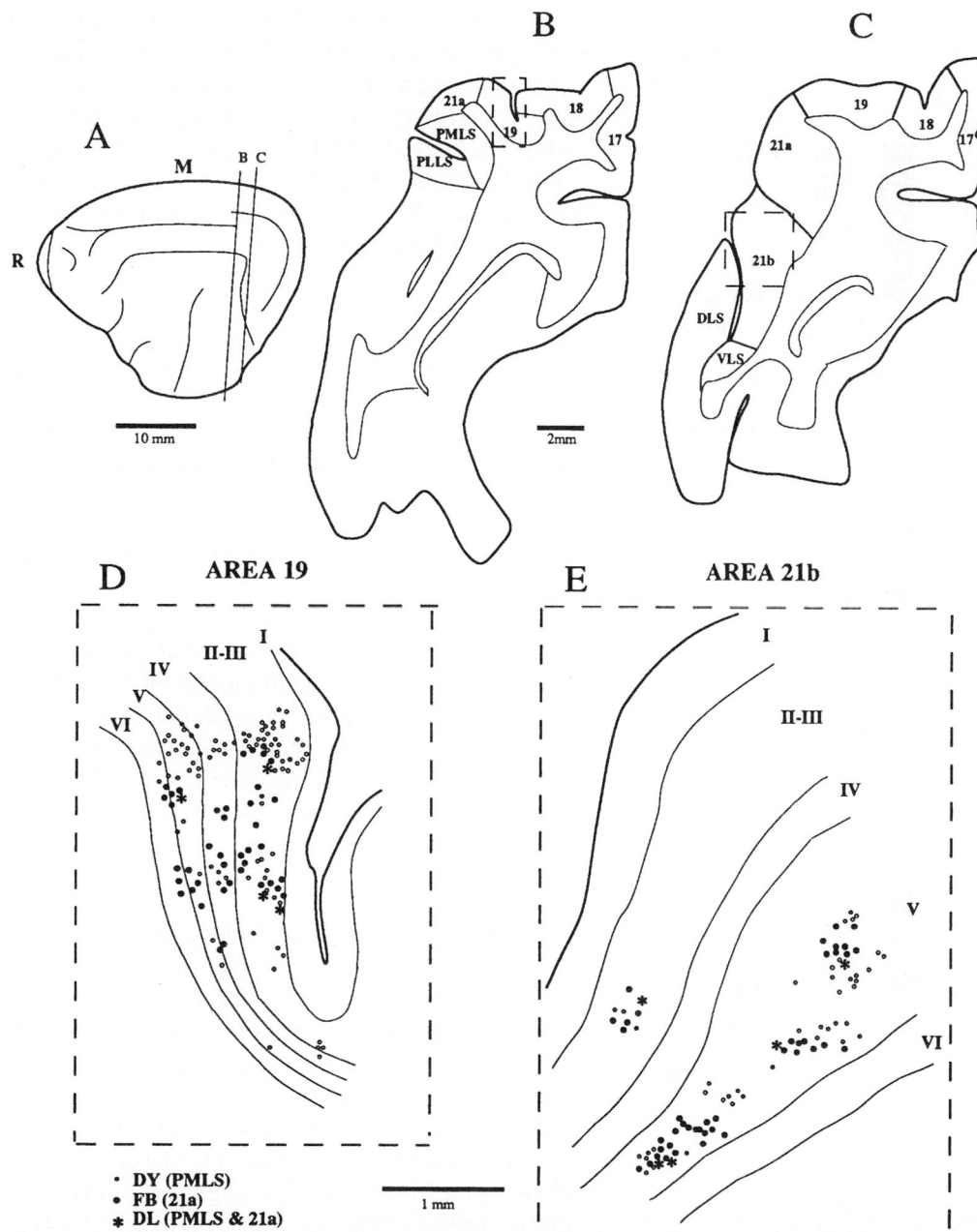

Fig. 8. Distributions of retrogradely labelled cells in the ipsilateral areas 19 and 21b of cat CR3 following injections of DY and FB into areas PMLS and 21a respectively. (A) Outline of the latero-dorsal view of the left cerebral hemisphere. M, medial, R, rostral. (B) and (C) Camera lucida drawings of a coronal section through the ipsilateral cortex at the levels B and C indicated in A. (D) and (E) are the enlargements of the parts of areas 19 and 21a respectively outlined by dashed lines in B and C. Note in D (area 19) that although the majority of the retrogradely labelled cells is located in the supragranular layers 2 and 3, substantial proportions of them are located in layers 4, 5 and 6 (cf. Fig. 9). Note in E (area 21b) that the great majority of retrogradely labelled cells are located in the infragranular layer 5 with only a few retrogradely labelled cells in the supragranular layers 2/3 (cf. Fig. 9). Note also in both D and E the patchy distribution of retrogradely labelled cells and the paucity of double-labelled cells (asterisks).

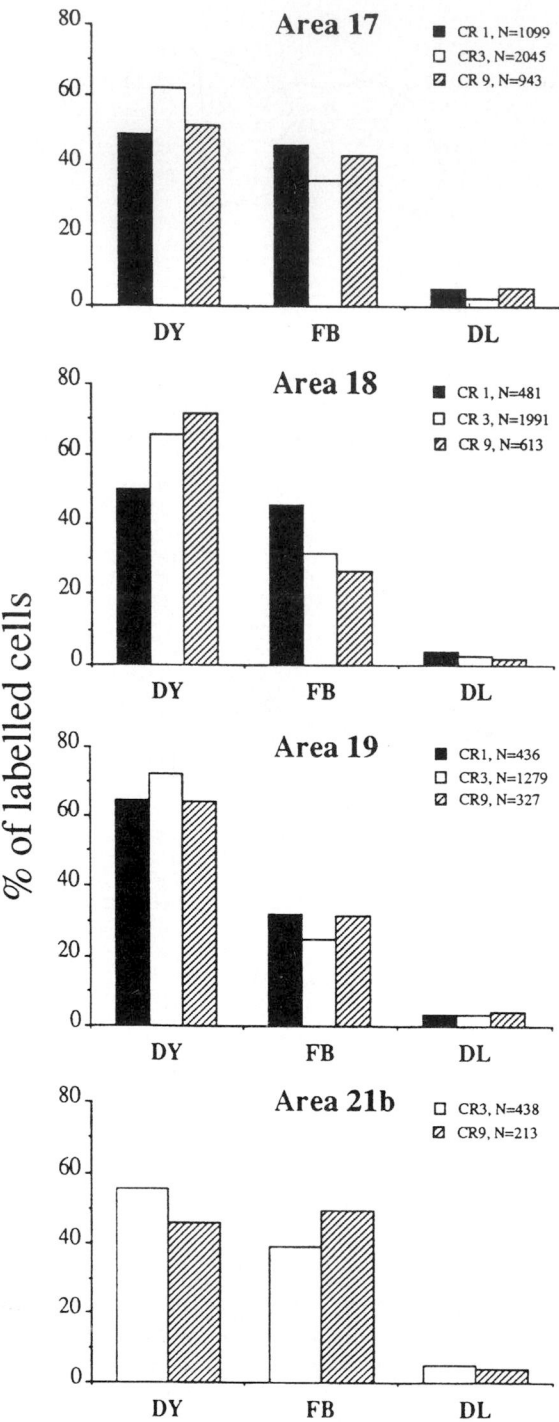

Fig. 9. Histograms showing the proportions of DY-labelled (PMLS-projecting), FB-labelled (area 21a-projecting) and double-labelled (projecting to both areas) cells, in the regions of the ipsilateral areas 17, 18, 19 and 21b where the DY-labelled and FB-labelled cells were intermingled. Note consistently low proportions of double-labelled cells in all the areas in all the animals. Note also, that in the histogram for area 21b data for cat CR1 are not shown, as in this animal no retrogradely labelled cells were found in this area.

Most retrogradely labelled cells were located in the infragranular layers 5 and 6 with smaller proportions located in the supragranular layers 2 and 3 and very few located in layer 4 (Symonds and Rosenquist, 1984b; Dreher, 1986; Mizobe and Toyama, 1989). In the regions of the spatial overlap of the two projections double-labelled cells constituted only 2–3.5% of the retrogradely labelled cells.

PS area

Only DY-labelled (PMLS area-projecting) cells were found in this area.

Retrogradely labelled cells in the contralateral cortical areas

Only in the cat CR3, substantial numbers of the retrogradely labelled cortical cells were found in the hemisphere contralateral to the injected areas 21a and PMLS. Cell retrogradely labelled with FB and/or DY were found not only in the homotopic area 21a but also in the heterotopic areas 17, 18 and 19. In addition, cells retrogradely labelled with DY (PMLS area-projecting cells) were found in the homotopic PMLS area and the posterolateral suprasylvian area (PLLS).

Areas 17 and 18

Area 17 contained a small number of cells single-labelled either with FB or DY and no double-labelled cells. Retrogradely labelled cells were found in the vicinity of the border with area 18 where the vertical meridian is represented. Over 95% of retrogradely labelled neurons were located in supragranular layer 3 and in layer 4, the remainder in the infragranular layers 5 and 6 (Keller and Innocenti, 1981; Segraves and Rosenquist, 1982a; Segraves and Innocenti, 1985; Dreher, 1986).

Area 18 contained about 300 retrogradely labelled cells located in the vicinity of the border with area 17. About 75% of them were located in layers 3 and 4 the remainder in layers 5 and 6. Cells labelled with DY were about three times as numerous as FB-labelled cells and in the region of the spatial intermingling of the two populations 1–2% of cells were double-labelled.

Area 19

About 600 cells located in the vicinity of the border with area 21a were retrogradely labelled with FB or DY. About 65% of FB-labelled cells were located in layers 3 and 4. This percentage was higher (about 80%) in the case of DY-labelled cells. The balance of retrogradely labelled cells

was located in the infragranular layers (Keller and Innocenti, 1981; Segraves and Rosenquist, 1982a; Dreher, 1986). Again, cells labelled with DY were about three times as numerous as FB-labelled cells and in the region of the spatial overlap of the two populations 1–2% of cells were double-labelled.

Area 21a

Contralateral area 21a contained about 2000 retrogradely labelled cells and FB-labelled (area 21a-projecting) cells were about twice as numerous as the DY-labelled (PMLS area-projecting) cells (Segraves and Rosenquist, 1982b; Dreher,

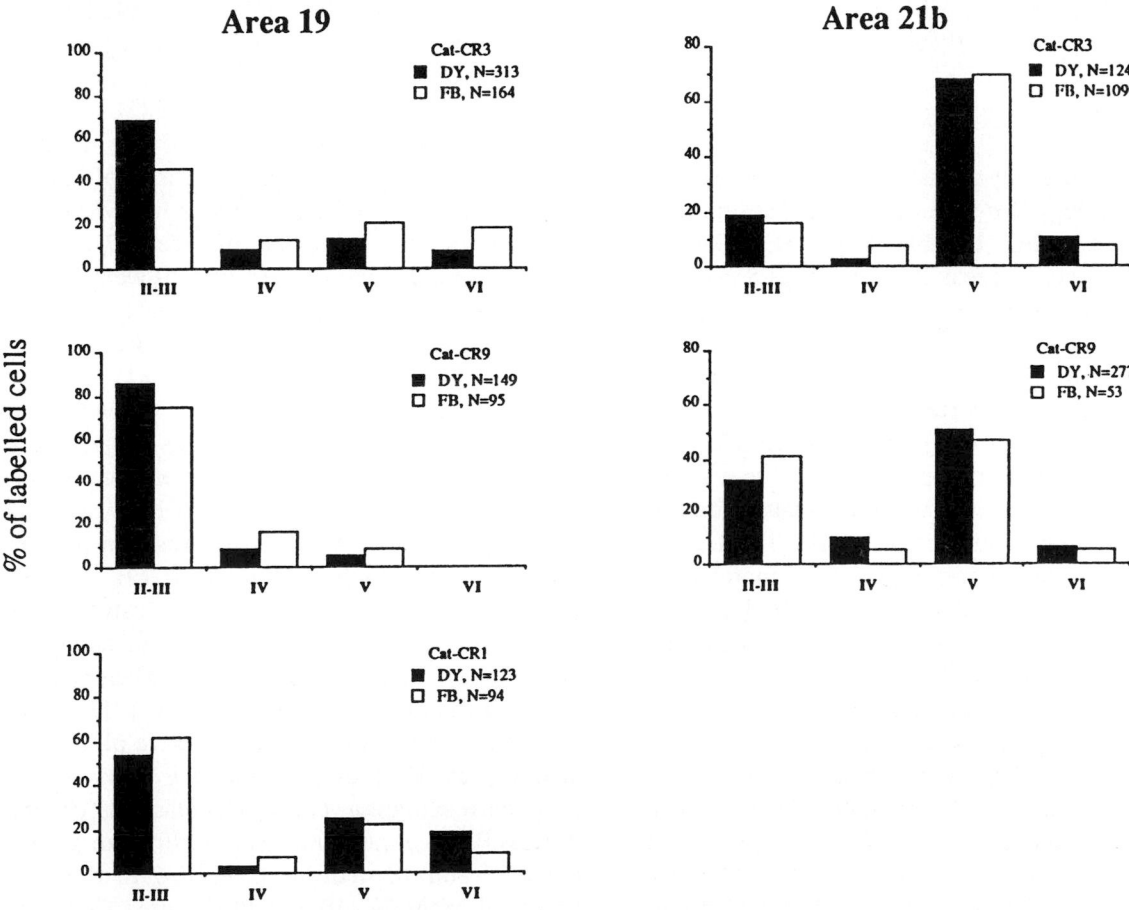

Fig. 10. Histograms showing laminar distributions of retrogradely labelled cells in the ipsilateral areas 19 and 21b following injections of DY and FB into respectively PMLS area and area 21a. Note that in all three cats, area 19 cells retrogradely labelled from the PMLS area or from area 21a were located predominantly in the supragranular layers 2 and 3 and that in cat CR9 there were no retrogradely labelled cells in the infragranular layer 6. Note also that in both cats in which there were retrogradely labelled cells in area 21b, DY- or FB-labelled cells were located predominantly in layer 5 and to a lesser extent in layers 2 and 3.

1986; Mizobe and Toyama, 1989). About 70% of FB-labelled cells were located in layers 3 and 4 (Dreher, 1986; Mizobe and Toyama, 1989). The proportion of DY-labelled cells in these layers was higher (about 85%). In the regions of spatial overlap of the two populations about 4.5% of cells were double-labelled.

Areas PMLS and PLLS

About 3500 cells labelled with DY were located in the caudal PMLS area in the region homotopic to the injection site. There were also some DY labelled cells in the most caudal parts of PLLS area. About 80% of the retrogradely labelled cells in PMLS area were located in layers 3 and 4 (Keller and Innocenti, 1981; Segraves and Rosenquist, 1982a; Dreher, 1986).

Discussion

In the present study we have confirmed numerous earlier findings (for references see Introduction and legend to Fig. 12) that visuotopically corresponding parts of areas 21a and PMLS receive: (1) their diencephalic and subcortical telencephalic afferents from the visuotopically corresponding parts of the same ipsilateral thalamic nuclei (mainly the cortico-recipient part of the lateral posterior nucleus) and visuotopically corresponding parts of the ipsilateral and to a lesser extent contralateral dorsocaudal claustrum; (2) their principal associational cortical afferents from the supragranular layers of visuotopically corresponding parts of areas 17, 18 and 19; and (3) heterotopic commissural afferents from layers 3 and 4 of areas 17, 18 and 19. On the other hand, even in the regions of a complete spatial overlap of the neurons projecting to area 21a or PMLS area the proportions of thalamic cells projecting to both areas (cells double-labelled with FB and DY) did not exceed 6%, the proportion of double-labelled claustral cells did not exceed 10% and the proportions of cortical cells projecting to both areas did not exceed 5%.

There is substantial evidence indicating that proportions of cells double-labelled with FB and DY (as estimated by us) do not dramatically underestimate the proportions of cells projecting to both areas. First, in an experiment in which a 2% solution of DY and a 2% solution of FB were sequentially injected into exactly the same point in area 18 of the cat, about 70% of retrogradely labelled cells in area 17 were double-labelled (Ferrer et al., 1988). In addition, the cells lying in patches in the middle of the labelled territory were almost all double-labelled (Ferrer et al., 1988). Both the survival time, processing of tissue and method of counting double-labelled cells were very similar to those in our experiments. Similarly, sequential injections of 2% solutions of FB and DY into exactly the same cortical regions of mice result in virtually 100% of double-labelled cells in the thalamic nuclei projecting to this cortical region (Segraves and Innocenti, 1985). Second, Birnbacher and Albus (1987) provided evidence that the proportions of the double-labelled cells in thalami and cortices of cats in which there were paired injections into visuotopically corresponding regions of areas 17 and 18, areas 17 and 19 or areas 18 and 19 were largely independent of the number and density of retrogradely labelled neurons. Third, despite the relatively large proportions (10–40%) of cells double-labelled with FB and DY among many thalamic, claustral and cortical cells projecting to visuotopically corresponding parts of pairs of cortical areas 17 and 18, areas 17 and 19 or areas 18 and 19, in many of those structures, the proportions of cells double-labelled with FB and DY were low, that is very similar to our estimates of proportions of cells projecting to both areas 21a and PMLS (Bullier et al., 1984a,b; Ferrer et al., 1992; Price and Ferrer, 1993). Very few cells double-labelled with FB and DY were also found among area 17 neurons projecting to visuotopically corresponding parts of areas 18/19 and PMLS/PLLS areas (Ferrer et al., 1992) or among areas 17 and 18 neurons projecting to the contralateral areas 17/18 and PMLS area (Segraves and Innocenti, 1985). It is also interesting to note in this context that although many LPl neurons projecting to PMLS area project also to the visuotopically corresponding parts of the ipsilateral area 17 (Tong and Spear, 1986; Miceli et al., 1991), a virtual absence of cells double-labelled with FB and DY was reported among LPl cells projecting to visuotopically corresponding parts of areas 17 and 21a (Wimborne et al., 1993).

Thus, it appears that two neighboring visual cortical areas, area 21a and PMLS area, receive their prosencephalic inputs from largely distinct subpopulations of cortical, thalamic and claustral neurons. Since the visuotopically corresponding regions of areas 21a and PMLS are separated by a several millimeters of PMLS area where more peripheral parts of the contralateral visual hemi-field are represented (Turlejski and Michalski, 1975; Palmer et al., 1978; Djavadian and Harutunian-Kozak, 1983; Grant and Shipp, 1991; Sherk and Mulligan, 1993) this result does not necessarily challenge 'the rule' that, the frequency of axonal bifurcation among the afferents to different pairs of cortical areas, is positively correlated with 'the relative proximity of the two target areas' (Bullier and Kennedy, 1987).

It is likely that the distinct populations of prosencephalic neurons which project to areas 21a and PMLS convey different types of information to each of the areas. In turn, the different type of information conveyed to the two areas is likely to underlie, at least partially, some of the functional differences between the two areas (for details see Figs. 11 and 12; for references see Introduction). For example, the differences in the composition of afferents to areas PMLS and 21a are likely to underlie the differences in the velocity preferences of cells in the two areas (Figs. 11 and 12; data from Dreher et al., 1996). In particular, the presence of strong Y-type input to PMLS neurons (Berson, 1985; Rauschecker et al., 1987a; Wang et al., 1995) and the paucity of such input to area 21a (Dreher et al., 1993) appear to underlie respectively the preference of over 20% of PMLS neurons for stimulus velocities exceeding 50°/s and the paucity of such neurons in area 21a (Fig. 11B; data from Dreher et al., 1996).

There are at least two lines of evidence suggesting that orientation selectivity of area 21a neurons is principally determined by the excitatory convergence of afferent orientation selective cortical neurons with similar optimal orientations and similar orientation tuning curves. First, although area 21a neurons have discharge fields which are at any given eccentricity substantially larger than those of their principal associational afferent cells in areas 17, 18 and 19 (Dreher, 1986), area 21a neurons exhibit strong orientation selectivity not only to long contours but also to contours which are significantly shorter than the discharge fields of area 21a neurons but similar in size to the discharge fields of area 17 neurons (Wimborne and Henry, 1992; Dreher et al., 1993). Second, reversible, bilateral or unilateral (ipsilateral) inactivation of areas 17 and 18, despite marked reduction in the magnitude of the responses of virtually all area 21a cells, does not result in a change of the shape of their orientation tuning curves (Michalski et al., 1993, 1994). Furthermore, not all orientations appear to be equally represented in area 21a. In particular, the majority of area 21a neurons responds optimally to the elongated stimuli oriented close to the verical meridian ($\pm 15°$) or close to 30° from the vertical meridian (Dreher et al., 1993). It is likely, therefore, that, as observed in the present study, distinct patches of area 17 cells (and to a lesser extent area 18), retrogradely labelled from visuotopically corresponding part of area 21a contain neurons with similar preferred optimal orientations. Indeed, it has been postulated in the past, that patchiness of intrinsic as well as extrinsic corticocortical projections in cats and primates reflects converging connections of cells sharing similar preferred orientations (LeVay, 1988; LeVay and Nelson, 1991; Katz and Calloway, 1992).

At any given eccentricity, the receptive fields of PMLS area neurons (like the receptive fields of area 21a neurons) are substantially larger than those of their principal cortical afferent neurons in areas 17, 18 and 19 (Dreher, 1986; Mulligan and Sherk, 1993; Dreher et al., 1996). Furthermore, when light bars of the length comparable to the length of cells' discharge field are applied,

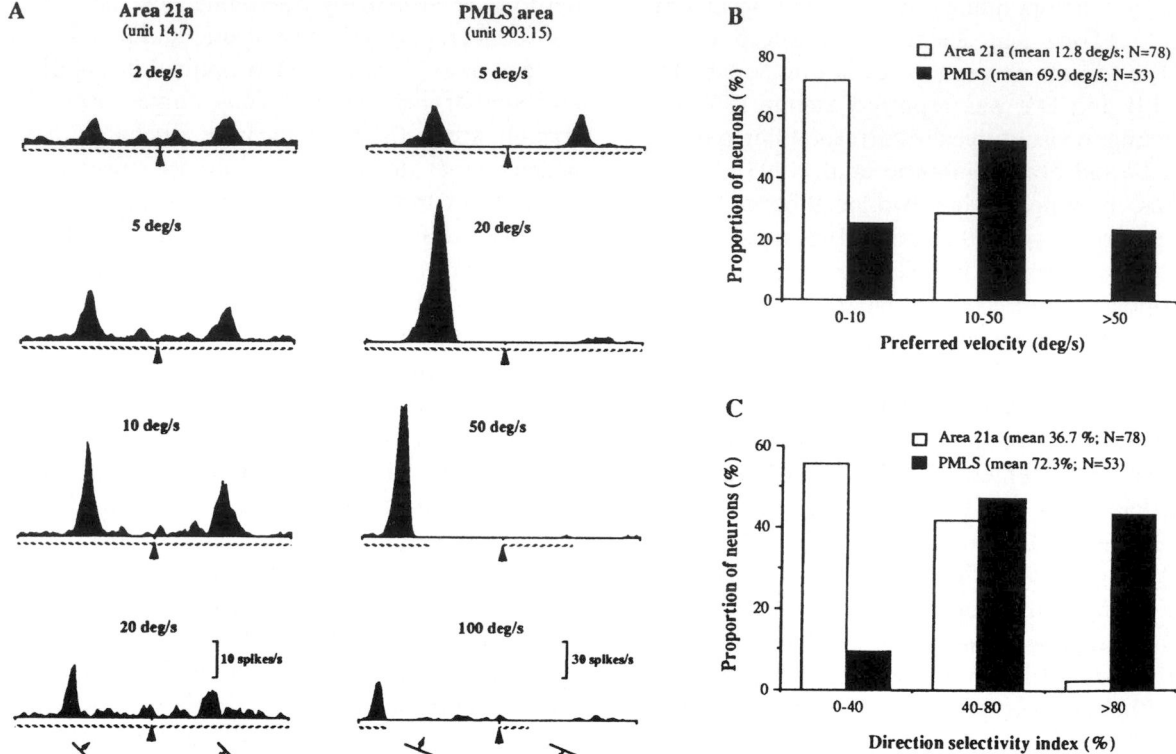

Fig. 11. (A) Peristimulus-time histograms for typical neurons in areas 21a and PMLS responding to an optimally oriented light bar moving at different velocities across the receptive field of a dominant eye (data from Dreher et al., 1996). The animals were anaesthetized with N_2O/O_2 (67/33%) gaseous mixture supplemented by the intravenous infusion of 1 mg/kg per h of sodium pentobarbitone, paralysed with gallamine triethiodide at the rate of 7.5 mg/kg per h and artificially ventilated. Action potentials were recorded extracellularly. Each response was compiled from the responses to 30 successive sweeps of optimally oriented elongated light bar (10° × 0.6°). In all area 21a records and in the top two records for PMLS area, in each sweep the stimulus moved forward (from the ordinate to the black arrowhead on the abscissa) and backward (from the arrowhead to the end of the abscissa) across the receptive field of the cell. Each sweep is compiled in 300 bins (150 bins in each direction). At velocities of 50°/s and 100°/s the stimulus moved only during the periods indicated by the hatched bars beneath the abscissae and then remained stationary for 400 ms (50°/s) or 800 ms (100°/s) outside the receptive field before moving again across the receptive field. Note that the PMLS neuron responds better at higher velocities (20 and 50°/s) at which it also exhibits a strong direction selectivity. (B) Histogram showing the distribution of preferred velocities for a sample of neurons recorded in areas 21a and PMLS (data from Dreher et al., 1996). Note that over 70% of area 21a neurons responded optimally at velocities not exceeding 10°/s and none responded optimally at velocities exceeding 50°/s. By contrast, most PMLS area neurons responded optimally at velocities exceeding 10°/s and over 20% responded optimally at velocities exceeding 50°/s. (C) Histogram showing direction selectivity indices (DI) for a sample of neurons recorded in areas 21a and PMLS (data from Dreher et al., 1996). The DI of a cell was calculated by the following formula: $DI = [(Rp - Rnp)/Rp] × 100\%$ where Rp and Rnp are the peak responses to the movement of optimally oriented elongated light slits at the preferred and non-preferred direction respectively. DI for a given cell was calculated for the responses at the optimal velocity. Note that DI of PMLS area neurons are significantly higher than those of area 21a neurons.

PMLS neurons exhibit a clear-cut orientation selectivity (Dreher et al., 1996). However, in view of the broadness of orientation tuning curves derived from the responses of PMLS neurons to moving short bars or spots (Spear, 1991; Dreher et al., 1996) it is unlikely that the orientation selectivity of PMLS neurons is principally determined by the excitatory convergence of orien-

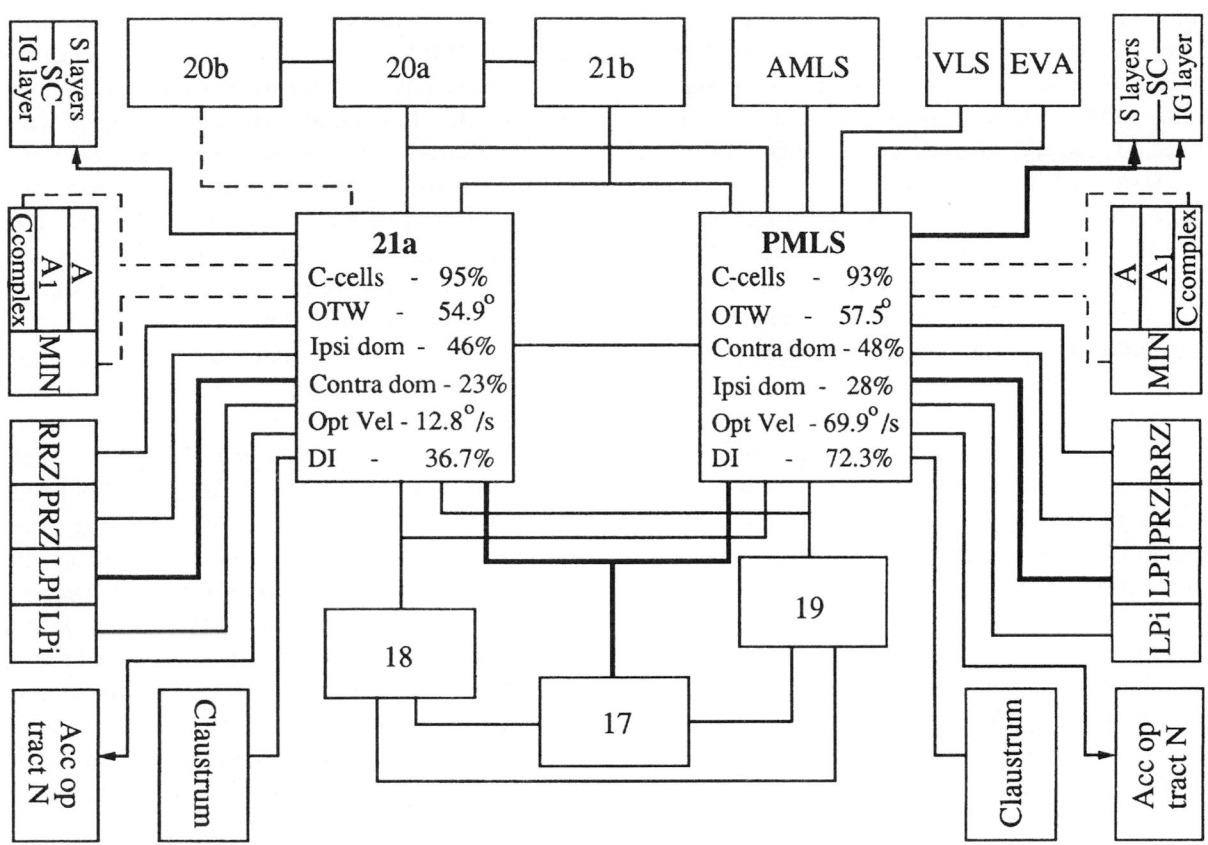

Fig. 12. Schematic diagram illustrating pattern of connections of areas 21a and PMLS. Strong connections are indicated by bold lines while weak or variable connections are indicated by the interrupted lines. The principal thalamic inputs to areas 21a and PMLS originate from the lateral division (or cortico-recipient) region of the lateral posterior-pulvinar complex (LPl). Both areas receive also direct thalamic input from the retino-recipient zone (RRZ) of the pulvinar, the LPi (interjacent or tecto-recipient zone) of the LP-pulvinar complex, the pretecto- recipient zone (PRZ) of the pulvinar and from the medial intralaminar nucleus (MIN) and laminae C of the dorsal lateral geniculate complex (for reviews see Rosenquist, 1985; Dreher, 1986; see also Sherk, 1986; Tong and Spear, 1986; Bruce and Stein, 1988; Mizobe and Toyama, 1989; Grant and Shipp, 1991, Miceli et al., 1991; the present study). Both areas are also reciprocally interconnected with rostral intralaminar thalamic nuclei (not illustrated; see Kaufman and Rosenquist, 1985 a,b; Dreher, 1986; cf. also the present study). Both areas 21a and PMLS receive their principal associational inputs from area 17 and weaker but substantial inputs from areas 18, 19 and 21b. Connections of areas 21a and PMLS with areas 17, 18, 19 and 21b as well as those with each other are reciprocal. In addition, area 21a is interconnected with area 20a and to a lesser extent area 20b while PMLS area is interconnected with the anteromedial lateral suprasylvian (AMLS) area, the ventral lateral suprasylvian (VLS) area, posterior suprasylvian area (PS, not illustrated in the diagram) and the ectosylvian visual (EVA) area (Miceli et al., 1985; Reinoso-Suarez and Roda, 1985; Rosenquist, 1985; Dreher, 1986; Norita, et al., 1986; Sherk, 1986; Updyke, 1986; Olson and Graybiel, 1987; Mizobe and Toyama, 1989; Grant and Shipp, 1991; the present findings). There are also reciprocal commissural connections with the contralateral, homotopic areas 21a and PMLS, heterotopic areas 17, 18, 19, and, albeit to a much lesser extent, with heterotopic areas 20a, 20b, 21b and PLLS (not illustrated; for references see Segraves and Rosenquist, 1982b; Rosenquist, 1985; Dreher, 1986; the present study). Both areas are also reciprocally interconnected with the ipsilateral and to a lesser extent the contralateral (not illustrated) dorsocaudal claustrum (Rosenquist, 1985; Dreher, 1986; cf. Updyke, 1993; the present study). Furthermore, both areas project directly to the striatum (longitudinal zone within the caudate nucleus and the posterolateral putamen; Updyke, 1993; not illustrated) and the accessory optic tract nuclei in the mesencephalon (Berson and Graybiel, 1980; Marcotte and Updyke, 1982). There are significant differences in the pattern of projections of areas 21a and PMLS to the ipsilateral superior colliculi (SC). Thus, while PMLS area projections to the SC terminate not only in the lower part of the superficial gray layer and the optic layer (S layers) but also in the dorsal part of the intermediate gray layer (IG layer), area 21a projects only to the lower part of the superficial gray layer and the optic layer (Berson, 1985; Dreher, 1986; Harting et al., 1992). Despite the strong similarity in the overall composition of the afferent inputs to the two areas, areas 21a and PMLS receive their afferents from largely distinct subpopulations

Continued overleaf

tation-selective cells with similar optimal orientations. On the other hand, the direction selectivity indices of PMLS neurons are not only significantly higher than those of area 21a neurons (Figs. 11 and 12; Dreher et al., 1993; 1996) but also significantly higher than those of their principal cortical afferent cells in areas 17 and 18 (Dreher, 1986; Sherk, 1989; Spear, 1991). Furthermore, not all directions of motion are equally represented in PMLS area. In particular, PMLS neurons either tend to prefer directions of motion away from the *area centralis*, that is, directions coinciding with direction of motion in the 'optic flow field', the pattern seen by locomoting observers when they fixate the point towards which they are heading (Rauschecker et al., 1987b; Spear, 1991) or the directions orthogonal to the radial-outward motion trajectories (Sherk et al., 1995). The question arises, whether PMLS neurons receive a convergent excitatory input from the subpopulations of direction-selective area 17 and 18 neurons which exhibit the same direction selectivity axis and the same preferred direction. Consistent with such notion, chronic or acute, bilateral or unilateral (ipsilateral), removal of areas 17 and 18 results in a dramatic reduction in the proportion of direction-selective PMLS neurons (Spear, 1988, see however Guedes et al., 1983). It is likely therefore, that the patchy distribution of neurons in areas 17 and 18 which project to visuotopically corresponding parts of PMLS area reflects convergence of neurons sharing similar direction selectivities. It remains to be established if there is a correlation between the distribution of area 17 neurons projecting to area 21a or PMLS area and the location of cytochrome-oxidase rich blobs in the supragranular layers of cat area 17, described recently by Murphy and her colleagues (1995).

The presumed functional distinctness of thalamic and telencephalic afferent inputs to areas 21a and PMLS and related to it distinct receptive field properties of neurons in each area, appear to underlie functional distinctness of, at least some, efferent projections from the two areas (Fig. 12). Thus, while the projection from area 21a to the ipsilateral superior colliculus (SC) terminates exclusively in the superficial layers, PMLS projections terminate in both the superficial SC layers and in the dorsal part of the intermediate gray layer (Dreher, 1986; Harting et al., 1992). Furthermore, like the deep collicular laminae, the corticotectal projections from the lateral suprasylvian cortex (including the PMLS area) are clearly implicated in visually guided behavior and play an important role in determining direction and velocity selectivities as well as the binocularity and spatial organization of receptive fields of the collicular neurons located in the deep laminae (Stein and Meredith, 1991). Second, the PMLS area, but not area 21a, is directly interconnected with the anteromedial and the ventral lateral suprasylvian areas as well as the ectosylvian visual area, that is, the areas which constitute some of the 'higher order' areas in the 'motion stream' of information processing (Dreher, 1986; Spear, 1991). By contrast, area 21a, unlike PMLS area, is interconnected with area 20b which has been implicated in the learning of discrimination of stationary patterns (Sprague et al., 1977).

Acknowledgements

We are grateful to Liam Burke and Maurice Ptito for their insightful comments on the manuscript. This study was supported by grants from the National Health and Medical Research Council of Australia and the Australian Research Council.

of neurons (the present study). These distinct inputs are probably, at least partially, responsible for the distinct receptive field properties of neurons in each area (listed in rectangles for areas 21a and PMLS- data from Dreher et al., 1996; cf. also Spear, 1991; Toyama et al., 1994). OTW, mean orientation tuning width; Opt vel, mean preferred velocity (cf. Fig. 11B); DI, mean direction selectivity index at the optimal velocities (cf. Fig. 11C). Note also that the majority of binocular cells in area 21a are dominated by the ipsilateral eye while the majority of binocular cells in PMLS area are dominated by the contralateral eye.

References

Afrikyan, M.B., Arutyunyan-Kozak, B.A., Dzhavadyan, R.L. and Kipriyan, T.K. (1991) Retinotopic organization of posterior suprasylvian area of the feline cerebral cortex. *Neurophysiology*, 23: 213–219.

Berman, N. and Jones, E.G. (1977) A retino-pulvinar projection in the cat. *Brain Res.*, 134: 237–248.

Berman, A.L. and Jones, E.G. (1982) The thalamus and basal telencephalon of the cat. *A Cytoarchitectonic Atlas with Stereotaxic Coordinates*. The University of Wisconsin Press, Madison, Wisconsin, pp. 164.

Berson, D.M. (1985) Cat lateral suprasylvian cortex: Y-cell inputs and corticotectal projection. *J. Neurophysiol.*, 53: 544–555.

Berson, D.M. and Graybiel, A.M. (1978) Parallel thalamic zones in the LP-pulvinar complex of the cat identified by their afferent and efferent connections. *Brain Res.*, 147: 139–148.

Berson, D.M. and Graybiel, A.M. (1980) Some cortical and subcortical fiber projections to the accessory optic nuclei in the cat. *Neuroscience*, 5: 2203–2217.

Berson, D.M. and Graybiel, A.M. (1983) Organization of the striate-recipient zone of the cat's lateralis posterior-pulvinar complex and its relations with the geniculostriate system. *Neuroscience*, 9: 337–372.

Birnbacher, D. and Albus, K. (1987) Divergence of single axons in afferent projections to the cat's visual cortical areas 17, 18, and 19: a parametric study. *J. Comp. Neurol.*, 261: 543–561.

Bruce, L.L. and Stein, B.E. (1988) Transient projections from the lateral geniculate to the posteromedial lateral suprasylvian visual cortex in kittens. *J. Comp. Neurol.*, 278: 287–302.

Bullier, J. and Kennedy, H. (1987) Axonal bifurcation in the visual system. *Trends Neurosci.*, 10: 205–210.

Bullier, J., Kennedy, H. and Salinger, W. (1984a) Bifurcation of subcortical afferents to visual areas 17, 18, and 19 in the cat cortex. *J. Comp. Neurol.*, 228: 309–328.

Bullier, J., Kennedy, H. and Salinger, W. (1984b) Branching and laminar origin of projections between visual cortical areas in the cat. *J. Comp. Neurol.*, 228: 329–341.

Clare, M.H. and Bishop, G.H. (1954) Responses from an association area secondarily activated from optic cortex. *J. Neurophysiol.*, 17: 271–277.

Coogan, T.A. and Burkhalter, A. (1993) Hierarchical organization of areas in rat visual cortex. *J. Neurosci.*, 13: 3749–3772.

Djavadian, R.L. and Harutiunian-Kozak, B.A. (1983) Retinotopic organization of the lateral suprasylvian area of the cat. *Acta Neurobiol. Exp.*, 43: 251–262.

Dreher, B. (1986) Thalamocortical and corticocortical interconnections in the cat visual system: relation to the mechanisms of information processing. In: J.D. Pettigrew, K.J. Sanderson and W.R. Levick, (Eds.), *Visual Neuroscience*, Cambridge University Press, Cambridge, UK, pp. 290–314.

Dreher, B., Michalski, A., Ho, R.H.T., Lee, C.W.F. and Burke, W. (1993) Processing of form and motion in area 21a of cat visual cortex. *Visual Neurosci.*, 10: 93–115.

Dreher, B., Wang, C., Turlejski, K.J., Djavadian, R.L. and Burke, W. (1996) Areas PMLS and 21a of cat visual cortex: two functionally distinct areas. *Cereb. Cortex*, 6 (in press).

Einstein, G. and Fitzpatrick, D. (1991) Distribution and morphology of area 17 neurons that project to the cat's extrastriate cortex. *J. Comp. Neurol.*, 303: 132–149.

Ferrer, J.M., Price, D.J. and Blakemore, C. (1988) The organization of corticocortical projections from area 17 to area 18 of the cat's visual cortex. *Proc. R. Soc. Lond. (Biol.)*, 233: 77–98.

Ferrer, J.M., Kato, N. and Price, D.J. (1992) Organization of association projections from area 17 to areas 18 and 19 and to suprasylvian areas in the cat's visual cortex. *J. Comp. Neurol.*, 316: 261–278.

Gilbert, C.D. and Kelly, J.P. (1975) The projections of cells in different layers of the cat's visual cortex. *J. Comp. Neurol.*, 163: 81–106.

Grant, S. and Shipp, S. (1991) Visuotopic organization of the lateral suprasylvian area and of an adjacent area of the ectosylvian gyrus of cat cortex: a physiological and connectional study. *Visual Neurosci.*, 6: 315–38.

Guedes, R., Watanabe, S. and Creutzfeldt, O.D. (1983) Functional role of association fibres for a visual association area: the posterior suprasylvian sulcus of the cat. *Exp. Brain Res.*, 49: 13–27.

Guillery, R.W., Geisert, Jr., E.E., Polley, E.H. and Mason, C.A. (1980) An analysis of the retinal afferents to the cat's medial intralaminar nucleus and to its rostral thalamic extension, the 'geniculate wing'. *J. Comp. Neurol.*, 194, 117–142.

Harting, J.K., Updyke, B.V. and Van Lieshout, D.P. (1992) Corticotectal projections in the cat: anterograde transport studies of twenty-five cortical areas. *J. Comp. Neurol.*, 324: 379–414.

Heath, C.J. and Jones, E.G.(1971) The anatomical organization of the suprasylvian gyrus of the cat. *Ergeb. Anat. Entwicklungsgesch.*, 45: 1–61.

Henry, G.H., Lund, J.S. and Harvey, A.R. (1978) Cells of the striate cortex projecting to the Clare-Bishop area of the cat. *Brain Res.*, 151: 154–158.

Hutchins, B. and Updyke, B.V. (1989) Retinotopic organization within the lateral posterior complex of the cat. *J. Comp. Neurol.*, 285: 350–398.

Kaas, J.H. and Krubitzer, L.A. (1991) The organization of extrastriate visual cortex. In: J. Cronly-Dilllon (Ed.), *Vision and Visual Dysfunction*, Vol.3. In: B.Dreher and S.R.Robinson (Eds.), *Neuroanatomy of the Visual Pathways and their Development*, Macmillan Press, Houndmills, Basingstoke, Hampshire, London, pp. 302–323.

Katz, L.C. and Callaway, E.M. (1992) Development of local cicuits in mammalian visual cortex. *Annu. Rev. Neurosci.*, 15: 31–56.

Kaufman, E.S. and Rosenquist, A.C. (1985a) Efferent projections of the thalamic intralaminar nuclei in the cat. *Brain Res.*, 335: 257–279.

Kaufman, E.S. and Rosenquist, A.C.(1985b) Afferent connections of the thalamic intralaminar nuclei in the cat. *Brain Res.*, 335: 281–296.

Keller, G. and Innocenti, G.M. (1981) Callosal connections of suprasylvian visual areas in the cat. *Neuroscience*, 6: 703–712.

Krubitzer, L.A. and Kaas, J.H. (1993) The dorsomedial visual area of owl monkeys: connections, myeloarchitecture, and homologies in other primates. *J. Comp. Neurol.*, 334: 497–528.

LeVay, S. (1988) The patchy intrinsic projections of visual cortex. *Prog. Brain Res.*, 75: 147–161.

LeVay, S. and Nelson, S.B. (1991) Columnar organization of the visual cortex. In: J. Cronly-Dilllon (Ed.), *Vision and Visual Dysfunction*, Vol. 4. In: A.G. Leventhal (Ed.), *The Neural Basis of Visual Function*, Macmillan Press, Houndmills, Basingstoke, Hampshire, London, pp. 266–315.

LeVay, S. and Sherk, H. (1981a) The visual claustrum of the cat. I. Structure and connections. *J. Neurosci.*, 1: 956–980.

LeVay, S. and Sherk, H. (1981b) The visual claustrum of the cat. II. The visual field map. *J. Neurosci.*, 1: 981–992.

Leventhal, A.G., Keens, J. and Törk, I. (1980) The afferent ganglion cells and cortical projections of the retinal recipient zone (RRZ) of the cat's 'pulvinar complex'. *J. Comp. Neurol.*, 194: 535–554.

Lomber. S., MacNeil, M.A. and Payne, B.R. (1995) Amplification of thalamic projections to the middle suprasylvian cortex following ablation of immature primary visual cortex in the cat. *Cereb. Cortex*, 5: 166–191.

Marcotte, R.R. and Updyke, B.V. (1982) Cortical visual areas of the cat project differentially onto the nuclei of the accessory optic system. *Brain Res.*, 242: 205–217.

Marshall, W.H., Talbot, S.A. and Ades, H.W. (1943) Cortical response of the anesthetized cat to gross photic and electrical afferent stimulation. *J. Neurophysiol.*, 6: 1–5.

Miceli, D., Repérant, J. and Ptito, M. (1985) Intracortical connections of the anterior ectosylvian and lateral suprasylvian visual areas in the cat. *Brain Res.*, 347: 291–298.

Miceli, D., Repérant, J., Marchand, L., Ward, R. and Vesselkin, N. (1991) Divergence of collateral axon branching in subsystems of visual cortical projections from the cat lateral posterior nucleus. *J. Hirnforsch.*, 32: 165–173.

Michalski, A., Wimborne, B.M. and Henry, G.H. (1993) The effect of reversible cooling of cat's primary visual cortex on the responses of area 21a neurons. *J. Physiol. (Lond.)*, 466: 133–156.

Michalski, A., Wimborne, B.M. and Henry, G.H. (1994) The role of ipsilateral and contralateral inputs from primary cortex in responses of area 21a neurons in cats. *Visual Neurosci.*, 11: 839–849.

Minchiacchi, D., Molinari, M., Bentivoglio, M. and Macchi, G.

(1985) The organization of the ipsi- and contralateral claustrocortical system in rat with notes on the bilateral projections in cat. *Neuroscience*, 16: 557–576.

Mizobe, K. and Toyama, K. (1989) Cortical and subcortical connectivity of area 21a of the cat. *Biomed. Res. Suppl.*, 3: 397–410.

Mizobe, K., Itoi, M., Kaihara, T. and Toyama, K. (1988) Neuronal responsiveness in area 21a of the cat. *Brain Res.*, 438: 307–310.

Montero, V.M. (1981) Topography of the cortico-cortical connections from the striate cortex of the cat. *Brain Behav. Evol.*, 18: 194–218.

Mulligan, K. and Sherk, H. (1993) A comparison of magnification functions in area 19 and the lateral suprasylvian visual area in the cat. *Exp. Brain Res.*, 97: 195–208.

Murphy, K.M., Jones, D.G. and Van Sluyters, R.C. (1995) Cytochrome-oxidase blobs in cat primary visual cortex. *J. Neurosci.*, 15: 4196–4208.

Norita, M. (1983) Claustral neurons projecting to the visual cortical areas in the cat: a retrograde double-labeling study. *Neurosci. Lett.*, 36: 33–36.

Norita, M., Mucke, L., Benedek, G., Albowitz, B., Katoh, Y. and Creutzfeldt, O.D. (1986) Connections of the anterior ectosylvian visual area (AEV). *Exp. Brain Res.*, 62: 225–240.

Olson, C.R. and Graybiel, A.M. (1987) Ectosylvian visual area of the cat: location, retinotopic organization, and connections. *J. Comp. Neurol.*, 261: 277–294.

Otsuka, R. and Hassler, R. (1962) Über Aufbau und Gliederung der Corticalen Sehesphäre bei der Katze. *Arch. Psychiatr. Nervenkr.*, 203: 212–234.

Palmer, L.A., Rosenquist, A.C. and Tusa, R.J. (1978) The retinotopic organization of lateral suprasylvian visual areas in the cat. *J. Comp. Neurol.*, 177: 237–256.

Payne, B.R. (1993) Evidence for visual cortical area homologs in cat and monkey. *Cereb. Cortex*, 3: 1–25.

Price, D.J. and Ferrer J.M.R. (1993) The incidence of bifurcation among corticocortical connections from area 17 in the developing visual cortex of the cat. *Eur. J. Neurosci.*, 5: 223–231.

Raczkowski, D. and Rosenquist, A.C. (1981) Retinotopic organization in the lateral posterior-pulvinar complex. *Brain Res.*, 221: 185–191.

Raczkowski, D. and Rosenquist, A.C. (1983) Connections of the multiple visual cortical areas with the lateral posterior-pulvinar complex and adjacent thalamic nuclei in the cat. *J. Neurosci.*, 3: 1912–1942.

Rauschecker, J.P., von Grünau, M.W. and Poulin, C. (1987a) Thalamo-cortical connections and their correlation with receptive field properties in the cat's lateral suprasylvian visual cortex. *Exp. Brain Res.*, 67: 100–112.

Rauschecker, J.P., von Grünau, M.W. and Poulin, C. (1987b) Centrifugal organization of direction preferences in the cat's lateral suprasylvian visual cortex and its relation to flow field processing. *J. Neurosci.*, 7: 943–958.

Reinoso-Suárez, F. and Roda, J.M. (1985) Topographical organization of the cortical afferent connections of the cortex of the anterior ectosylvian sulcus in the cat. *Exp. Brain Res.*, 59: 313–324.

Rosenquist, A.C. (1985) Connections of visual cortical areas in the cat. In: A. Peters and E.G. Jones (Eds.), *Cerebral Cortex*, Vol 3, Visual Cortex, Plenum Press, New York, NY, pp. 81–116.

Sanderson, K.J. (1971) The projection of the visual field to the lateral geniculate and medial intralaminar nuclei in the cat. *J. Comp. Neurol.*, 143: 101–118.

Sanides, F. and Hoffmann, J. (1969) Cyto- and myeloarchitecture of the visual cortex of the cat and of the surrounding integration cortices. *J. Hirnforsch.*, 11: 79–104.

Segraves, M.A. and Innocenti, G.I. (1985) Comparison of the distributions of ipsilaterally and contralaterally projecting corticocortical neurons in cat visual cortex using two fluorescent tracers. *J. Neurosci.*, 5: 2107–2118.

Segraves, M.A. and Rosenquist, A.C. (1982a) The distribution of the cells of origin of callosal projections in cat visual cortex. *J. Neurosci.*, 8: 1079–1089.

Segraves, M.A. and Rosenquist, A.C. (1982b) The afferent and efferent callosal connections of retinotopically defined areas in cat cortex. *J. Neurosci.*, 8: 1090–1107.

Sereno, M.I. and Allman, J.M. (1991) Cortical visual areas in mammals. In: J. Cronly-Dilllon (Ed.), *Vision and Visual Dysfunction,* Vol. 4. In: A.G. Leventhal (Ed.), *The Neural Basis of Visual Function*, Macmillan Press, Houndmills, Basingstoke, Hampshire, London, pp. 160–172.

Sherk, H. (1986) Location and connections of visual cortical areas in the cat's suprasylvian sulcus. *J. Comp. Neurol.*, 247: 1–31.

Sherk, H. (1988) Retinotopic order and functional organization in a region of suprasylvian visual cortex, the Clare-Bishop area. *Prog. Brain Res.*, 75: 237–244.

Sherk, H. (1989) Visual response properties of cortical inputs to an extrastriate area in the cat. *Vis. Neurosci.*, 3: 249–265.

Sherk, H. (1990) Functional organization of input from areas 17 and 18 to an extrastriate area in the cat. *J. Neurosci.*, 10: 2780–2790.

Sherk, H. and Mulligan, K.A. (1993) A reassessment of the lower visual field map in the striate-recipient lateral suprasylvian cortex. *Vis. Neurosci.*, 10: 131–158.

Sherk, H., Kim, J.-N. and Mulligan, K.A. (1995) Are the preferred directions of neurons in cat extrastriate cortex related to optic flow? *Vis. Neurosci.*, 12: 887–894.

Shipp, S. and Grant, S. (1991) Organization of reciprocal connections between area 17 and the lateral suprasylvian area of cat visual cortex. *Vis. Neurosci.*, 6: 339–355.

Spear, P.D. (1988) Influence of areas 17, 18, and 19 on receptive-field properties of neurons in the cat's posteromedial lateral suprasylvian visual cortex. *Prog. Brain Res.*, 75: 197–210.

Spear, P.D. (1991) Functions of extrastriate visual cortex in non-primate species. In: J. Cronly-Dillon (Ed.), *Vision and Visual Dysfunction,* Vol. 4. In: A.G. Leventhal (Ed.), *The Neural Basis of Visual Function*, Macmillan Press, Houndmills, Basingstoke, Hampshire, London, pp. 339–369.

Sprague, J.M., Levy, J., DiBerardino, A. and Berlucchi, G. (1977) Visual cortical areas mediating form discrimination in the cat. *J. Comp. Neurol.*, 172: 441–488.

Stein, B.E. and Meredith, M.A. (1991) Functional organization of the superior colliculus. In: J. Cronly-Dillon (Ed.), *Vision and Visual Dysfunction,* Vol. 4. In: A.G. Leventhal (Ed.), *The Neural Basis of Visual Function*, Macmillan Press, Houndmills, Basingstoke, Hampshire, London, pp. 85–110.

Symonds, L.L. and Rosenquist, A.C. (1984a) Corticocortical connections among visual areas in the cat. *J. Comp. Neurol.*, 229: 1–38.

Symonds, L.L. and Rosenquist, A.C. (1984b) Laminar origins of visual corticocortical connections in the cat. *J. Comp. Neurol.*, 229: 39–47.

Tong, L. and Spear, P.D. (1986) Single thalamic neurons project to both lateral suprasylvian visual cortex and area 17: a retrograde fluorescent double-labeling study. *J. Comp. Neurol.*, 246: 254–264.

Toyama, K., Mizobe, K., Akase, E. and Kaihara, T. (1994) Neuronal responsiveness in areas 19 and 21a, and the posteromedial lateral suprasylvian cortex of the cat. *Exp. Brain Res.*, 99: 289–301.

Turlejski, K. and Michalski, A. (1975) Clare-Bishop area in the cat: location and retinotopical projection. *Acta Neurobiol. Exp.*, 35: 179–188.

Turlejski, K, Djavadian, R.L. and Dreher, B. (1992) Diencephalic and telencephalic inputs to areas 21a and PMLS of cat visual cortex: a double label study. *Proc. Aust. Physiol. Pharmacol. Soc.*, 23: 133P.

Turlejski, K, Dreher, B. and Djavadian, R.L. (1994) Extent of collateralization among afferents to cat's areas PMLS, 21a and 18. *Eur. J. Neurosci. Suppl.*, 7: 189.

Tusa, R.J. and Palmer, L.A. (1980) Retinotopic organization of areas 20 and 21 in the cat. *J. Comp. Neurol.*, 193: 147–164.

Tusa, R.J., Palmer, L.A. and Rosenquist, A.C. (1978) The retinotopic organization of area 17 (Striate cortex) in the cat. *J. Comp. Neurol.*, 177: 213–236.

Tusa, R.J., Rosenquist, A.C. and Palmer, L.A. (1979) Retinotopic organization of areas 18 and 19 in the cat. *J. Comp. Neurol.*, 185: 657–678.

Tusa, R.J., Palmer, L.A. and Rosenquist, A.C. (1981) Multiple cortical visual areas: visual field topography in the cat. In: C.N.Woolsey (Ed.), *Cortical Sensory Organization,* Vol. 2. *Multiple Visual Areas*, Humana Press, Clifton, NJ, pp. 1–31.

Updyke, B.V. (1981) Multiple representations of the visual field. Corticothalamic and thalamic organization in the cat. In: C.N.Woolsey (Ed.), *Cortical Sensory Organization,* Vol. 2. *Multiple Visual Areas,* Humana Press, Clifton, NJ, pp. 83–101.

Updyke, B.V. (1986) Retinotopic organization within the cat's posterior suprasylvian sulcus and gyrus. *J. Comp. Neurol.*, 246: 265–280.

Updyke, B.V. (1993) Organization of visual corticostriatal projections in the cat, with observations on visual projections to claustrum and amygdala. *J. Comp. Neurol.*, 327: 159–193.

Wang, C., Dreher, B., Huxlin, K.R. and Burke, W. (1995) Convergence of Y and non-Y information channels in the PMLS area of cat visual cortex. *Proc. Aust. Physiol. Pharmacol. Soc.*, 26: 168P.

Wimborne, B.M. and Henry, G.H. (1992) Response characteristics of the cells of cortical area 21a of the cat with special reference to orientation specificity. *J. Physiol. (Lond.)*, 449: 457–478.

Wimborne, B.M., McCart, R.J. and Henry, G.H. (1993) Projections from the lateral division of the lateral posterior-pulvinar complex to area 21a and the striate cortex in the cat. *Brain Res.*, 603: 333–337.

Zeki, S.M. (1974) Functional organization of a visual area in the posterior bank of the superior temporal sulcus of the rhesus monkey. *J. Physiol. (Lond.)*, 236: 549–573.

M. Norita, T. Bando and B. Stein (Eds.)
Progress in Brain Research, Vol 112
© 1996 Elsevier Science BV. All rights reserved.

CHAPTER 19

Motion sensitivity and stimulus interactions in the striate-recipient zone of the cat's lateral posterior–pulvinar complex

C. Casanova* and T. Savard

Departments of Surgery-Ophthalmology and of Physiology and Biophysics, Faculty of Medicine, University of Sherbrooke, Sherbrooke, Quebec J1H 5N4, Canada

The cat's lateral posterior-pulvinar complex (LP-pulvinar) establishes reciprocal connections with the anterior ectosylvian visual (AEV) and lateral suprasylvian (LS) cortices; two regions which are believed to be involved in motion analysis. We have investigated the motion sensitivity of neurons in the LP-pulvinar complex by: (1) studying the respones properties of cells in the striate-recipient zone of the LP nucleus (LPl) to the drift of a two-dimensional texture pattern (visual noise); and (2) determining the extent to which the latter stimulus can modify the spatial frequency tuning function of LPl cells. Experiments were carried out on anesthetized normal adult cats. Almost all LPl cells (55 out of 63, 87%) responded to the motion of visual noise. For most units (39 out of 55, 71%), responses varied as a function of the direction of motion (bandwidth of 49°). One-third of the LPl units did not exhibit any preference for drift direction of noise. For practically all LPl cells, responses to noise varied as a function of drift velocity. Optimal velocities were distributed from 2 to 35°/s with a mean value of 27.5°/s (mean bandwidth of 2.5 octaves). The influence of visual noise on the spatial frequency tuning function of 22 LPl cells was also studied. For half of LPl cells, responses at all spatial frequencies were reduced when the grating and the texture pattern were moving in opposite directions (*anti phase* condition). This masking effect of noise was rarely observed when both stimuli were drifted in the same direction (*in phase* condition). These results suggest that the LP-pulvinar complex may be part of extrageniculate pathways involved in the analysis of motion of visual targets and/or the analysis of the relative movement between an object and its surrounding environment.

Introduction

Since the past decade or so, there has been increasing interest in the lateral posterior (LP) nucleus–pulvinar complex of higher mammals. This renewed interest comes in part from its putative role in visual attention (Petersen et al.,

---

*Corresponding author. École d'optométrie, Université de Montréal, C.P. 6128, Succ. Centre-ville, Montréal, Québec H3C 3J7, Canada; Tel.: +1 514 343 2407; fax: +1 514 343 2382; e-mail: casanovc@ere.umontreal.ca

1987; Desimone et al., 1990; Robinson and Petersen, 1992; Olshausen et al., 1993) even though the LP–pulvinar complex is likely to be involved in many other aspects of visual processing such as motion perception (Rauschecker, 1988; for reviews, see Casanova et al., 1991; Chalupa, 1991). This visual region of the posterior thalamus is unique since it does not receive its main visual input from the retina but rather from the striate and extrastriate cortical areas and from the mesencephalon (Updyke, 1977, 1981; Berson and Graybiel, 1978, 1983; Graybiel and Berson, 1980; Benedek et al., 1983; Raczkowski and Rosenquist,

1983; Abramson and Chalupa, 1985, 1988). Of interest is the fact that the connections between the LP-pulvinar and the visual cortex are reciprocal, thus placing this nuclei complex at the center of several cortico-thalamo-cortical loops. For example in cats, the LP–pulvinar complex is reciprocally connected with the posteromedial lateral suprasylvian cortex (PMLS) and the anterior ectosylvian visual cortex (AEV), two areas which are believed to be strongly involved in motion perception (Mucke et al., 1982; Rauschecker et al., 1987a; Benedek et al., 1988; Rauschecker, 1988; Spear, 1991). Any involvement of the LP–pulvinar complex in the analysis of moving objects requires that its constituting elements not only be sensitive to stimulus motion but also to specific parameters such as velocity and direction. Mason (1978) reported that cells in the cat's LP–pulvinar complex nucleus were more responsive and easier to drive with moving stimuli rather than with flashing stimuli. This author also showed (Mason 1978, 1981), along with other investigators (Chalupa et al., 1983; Chalupa and Abramson, 1988, 1989; Casanova et al., 1989) that a substantial number of cells in the lateral and medial part of the LP nucleus (LPl and LPm) were selective to the direction of motion and could respond to very high stimulus velocities.

To further ascertain the movement sensitivity of cells in the cat's LP–pulvinar complex, we studied the responses of neurons in the LPl (i.e. the striate-recipient zone) to the motion of a two-dimensional texture pattern (visual noise). This stimulus is characterized by its lack of specific orientation and its broad representation of spatial frequencies and has been successfully used to characterize motion sensitivity of neurons in various cortical areas (e.g. Hammond and MacKay, 1977; Hoffmann et al., 1980; Hamada, 1987; Orban et al., 1988; Casanova et al., 1995). Also, since visual responses of cells in cortical areas reciprocally connected with the LP-pulvinar (e.g. area 17, lateral suprasylvian (LS) and AEV; Gulyás et al., 1987; Von Grünau and Frost, 1983; Benedek et al., 1988) can be modulated by the presence of a textured background, we investigated the extent to which visual noise, presented in combination with a moving grating, can modify the spatial frequency tuning function of LPl cells. Parts of these results have been presented elsewhere (Casanova and Savard, 1996).

Methods

Animal preparation

Experiments were carried out on normal adult cats (2.5–3.5 kg) premedicated with acepromazine and atropine (1.0 and 0.2 mg/kg, respectively). General anesthesia was induced by intramuscular injection of Ketamine (25–35 mg/kg). Lidocaine hydrochloride (2%) was given at surgical wounds and pressure points. All animals were treated according to the guidelines of the Canadian Council on Animal Care. After cannulation of the cephalic vein, a tracheotomy was performed and the animal placed in a stereotaxic frame. During this period, ECG and the level of O_2 blood saturation were monitored. A heating pad was installed under the animal and the core temperature was kept constant at 37.5°C. Electrocardiogram and EEG were monitored throughout the experiment. The animal was paralyzed by injection of gallamine triethiodide (10 mg/kg per h) and artificially ventilated with a gas mixture of N_2O—O_2 (70:30) plus fluothane (0.25–0.5%). The end tidal CO_2 partial pressure was monitored by a capnometer and kept constant between 28 and 32 mmHg by adjusting the rate and stroke volume of the respiratory pump. The animal was continuously infused with a solution of 5% dextrose in lactated Ringer's and gallamine triethiodide. Pupils were dilated with atropine and nictitating membranes were retracted with local application of phenylephrine hydrochloride (10%). The eyes were protected with contact lenses of appropriate strength. A craniotomy was done over the LP–pulvinar complex and the dura retracted. The exposed cortex was covered with warm agar on which wax was melted to create a sealed chamber.

Single-unit recordings and visual stimulations

Varnished tungsten microelectrodes (A & M Systems) were used to record single-unit activity in the LP nucleus. In all experiments, a microelectrode was first lowered into the lateral geniculate nucleus (LGN). The geniculate visual field was mapped (Kaas et al., 1972) at two or three distinct coordinates to verify the accuracy of the stereotaxic adjustments. The signals were amplified, displayed on an oscilloscope, and played through an audio monitor. Neuronal activity was also passed through a window discriminator and fed to a computer for peristimulus time histogram (PSTH) acquisition (binwidth of 10 ms).

Manually controlled stimuli were projected on a translucent screen facing the animal onto which the position of the optic disks was plotted (Fernald and Chase, 1971). Almost all cells from which recordings were made had the center of their receptive fields located in the contralateral visual field within 30° from the area centralis ($\leq 30°$ and $\pm 20°$, respectively, for azimuth and elevation). Bars of variable width and length were used to characterize receptive field properties of the isolated unit such as ocular dominance and orientation selectivity. Each neuron was later classified on the basis of the unit's modulation index (MI) derived from its response to sine-wave drifting gratings. The MI was calculated by dividing the response at the first harmonic (AC component) by the mean firing rate (DC component) from which the spontaneous level was subtracted (Skottun and Freeman, 1984). For quantitative analysis, full screen sinusoidal drifting gratings of 50–60% contrast were displayed on a CRT (DataCheck 5117; mean luminance of 14 Cd/m^2) generally at 57 cm from the animal's eyes and subtending $28 \times 28°$ of visual angle. The stimuli were generated by a Picasso (Innisfree, Cambridge) interfaced with a Picasso controller (CED1708, Cambridge, UK) connected to a desktop computer. Before data acquisition, a search program was used to evaluate the optimal values of grating parameters for each cell. Then various properties were quantitatively studied. Orientation was varied over 360° in 12 or 24 steps of 30° and 15°, respectively. An orientation range of 360° included specification of the direction of motion. For example, 90° and 270° are both vertical and denote opposite directions of motion. The cell's direction selectivity was determined and quantified as follows: Direction Index (DI) = 1 − (response in the non-preferred direction/response in the preferred direction). Cells with a DI greater than 0.5 were regarded as selective to the direction of the stimulus motion. Cell responses at optimal orientation were also studied as a function of spatial and temporal frequencies and of grating contrast. In the following tests, cell responses to visual noise were studied. Under computer control, a full-screen moving random dot pattern was presented over 360°. This pattern was generated by a Dual Channel Velocity Field and Stereogram Generator (Innisfree) and had a maximal density of 256×256 elements (refresh rate of 200 Hz). At 57 cm, each element of the pattern subtended 0.11° of visual angle and represented the outcome of an independent Bernoulli trial, with equiprobable states. The 2D Fourier power spectrum of the texture frames showed that all spatial frequencies and all orientations had the same expected amount of power (see Fig. 1). Cells were regarded as tuned for noise direction when they responded to a specific range of directions only, and consequently when their selectivity could be assessed quantitatively by measuring the bandwidth of tuning (see Fig. 2B). Cells were not considered as band-pass tuned if all directions yielded the same response level. For a subset of LPl cells, the spatial frequency tuning function (at optimal orientation) was studied with and without noise background. The background was either stationary or moving at optimal speed and superimposed to the grating [transparent window]. In control conditions, the visual noise was stationary while the grating was drifted (presented at optimal orientation) at various spatial frequencies. In test conditions, the noise background moved either in the same direction as to the grating [in-phase motion] or in the opposite direction [anti-phase motion].

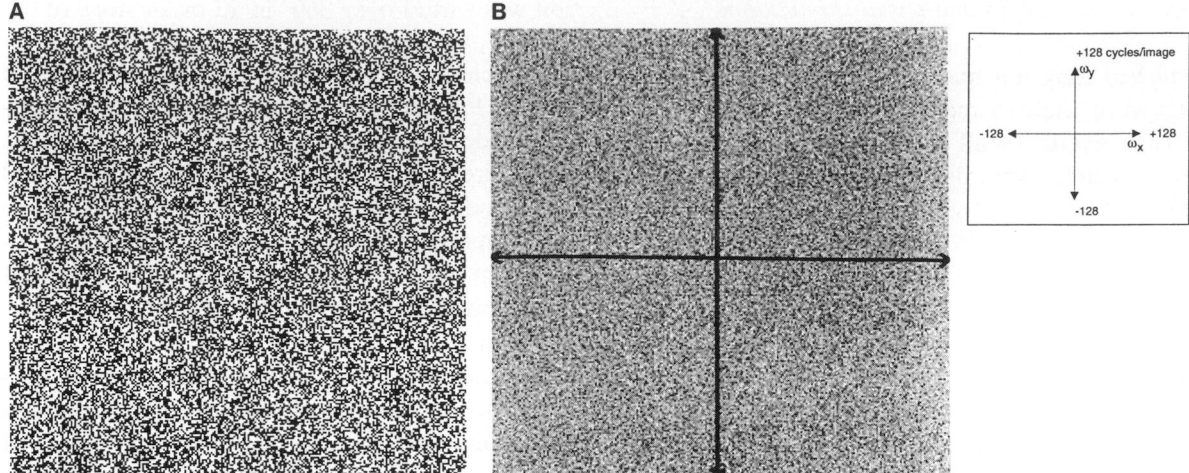

Fig. 1. (A) sample of the 256 × 256 pattern used in this study. (B) Fourier power spectrum of the pattern shown in A. Note that the pattern has a white power spectrum up to 128 cycles/image in both frequency axes.

During each test, the activity for a blank screen with the same luminance as that of the grating was recorded (spontaneous discharges). Unless specified otherwise, each stimulus presentation (including the blank screen) lasted 4 s and was repeated four times. Presentations were randomly interleaved and only the dominant eye was stimulated.

Histology

Electrolytic lesions were made along recording tracks. At the end of each experiment, the animal was killed by an intravenous overdose of Nembutal. The brain was removed from the skull and fixed in a solution of buffered formalin (10%). After 4–5 days, 100-μm serial sections of the brain were cut in the frontal plane with a Vibroslice. Every third section was stained to reveal acetylcholinesterase (AChE) activity in order to distinguish three major zones in the LP-pulvinar, i.e. the medial and lateral zones of the LP (LPm and LPl), and the pulvinar zone (Graybiel and Berson, 1980; Berson and Graybiel, 1983). The striate-recipient zone of LP appears pale compared to neighbouring subdivisions. The remaining sections were stained with cresyl violet and were matched with the AChE sections to confirm

the position of the recording tracks within the LP–pulvinar complex.

Results

Responses of LPl cells to visual noise

The visual properties of 63 LPl cells were quantitatively studied. The large majority of these cells (55 units; 87%) responded to the motion of visual noise. The profile of the evoked discharges varied from cell to cell. In most cases (38%), noise responses were sustained, i.e. their discharge rate was maintained at a constant level throughout stimulus presentation. For a second group of cells, responses were characterized by several discharge bursts (33%). The responses of the remaining units (29%) consisted of multiple peaks superimposed on a maintained discharge. A representative example of a texture-sensitive LPl cell is shown in Fig. 2. Part A illustrates the cell response to a moving texture field whose drift direction was varied over 360° in 12 steps of 30°. The unit responded with sustained discharges and was broadly tuned for direction (bandwidth of 55°). For most LPl units (39 out of 55, 71%), noise responses varied as a function of the direction of motion. The distribution of bandwidths, expressed as the half-width of the tuning curve at half-height, ranged between 19° and 101° with a mean $\pm$ SD of 49.3 $\pm$ 18.7°. Overall, LPl cells did not show any significant preference for horizontal, vertical or oblique direction of motion (uniform distribution; $\chi^2 = 2.6$, $P = 0.7$, df = 5). When stimulated with visual noise, 79% of LPl cells exhibited strong direction selectivity (DI > 0.5) along a given axis of movement. Fig. 3 shows the distribution of direction indices computed from responses to visual noise. Most values ranged between 0 and 1.9, with a mean of 0.92. How did noise responses compare with those evoked by drifting gratings? We found that the preferred direction of noise motion matched well with that of drifting gratings (Casanova and Savard, 1996). However, almost all LPl neurons were more broadly tuned for the direction of motion of visual noise than for the orientation of drifting gratings (see Fig. 2B). The mean bandwidths of tuning curves computed from gratings and noise direction tuning curves were significantly different (37.5 vs. 49.3°; t-test, $P = 0.005$). This result is expected if one considers the texture stimuli not in the space domain, as dot

patterns, but in the frequency domain, as patterns containing all spatial frequencies and orientations (see Casanova et al., 1995). Also, the selectivity to the stimulus direction was slightly more pronounced when LPl cells were stimulated by drifting noise rather than by gratings. The mean DIs were respectively 0.76 and 0.92 (medians of 0.85 and 0.95) for grating and noise (t-test, $P > 0.05$). Nevertheless, the number of cells classified as direction selective (DI > 0.5) was very similar whether considering the responses to noise or responses to gratings (79% vs 78%). This last observation suggests that the direction selectivity of LPl cells does not depend on stimulus type. Out of the 55 texture-sensitive units, a total of 16 were not tuned for drift direction, i.e. they responded similarly to all directions of noise motion. One may note that in this latter group of cells, more than 62% (10 out of 16) were not tuned for the drift of gratings. On the other hand, all cells tuned for noise direction also showed preferences for specific grating drift directions.

Finally, we studied the responses of 27 LPl cells as a function of noise velocity. Preferred veloci-

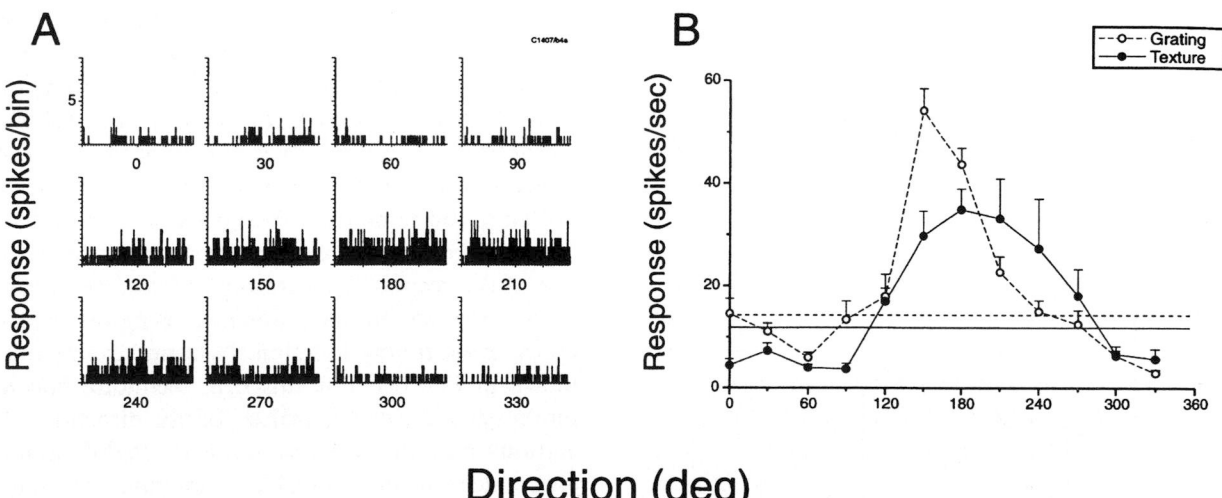

Fig. 2. Responses of a LPl cell to the drift of visual noise shown as: (A) PSTHs (duration of 4 s); and (B) corresponding tuning curves (DC component) as a function of direction. In (B), the direction tuning function for gratings is also shown (dashed line and empty symbols). The SEM is shown for each data point. The straight and dashed lines represent spontaneous activity levels during noise and grating test, respectively. Parameters of the gratings are: spatial frequency (SF) = 0.3 c/degree; temporal frequency (TF) = 3 Hz, contrast (C) = 0.6. The drift velocity of the noise is 7 degrees/s. Number of presentations is 4.

ties were mainly distributed from 2 to 35°/s with a mean value was 27.5°/s (median of 20.1). Example of tuning curves are shown in Fig. 4. As illustrated, most cells (78%) were band-pass, i.e. they showed reductions in their firing rates on both sides of the optimal drift velocity. The values of velocity bandwidth (full width at half-height) were distributed from 1.2 to 5 octaves with a mean of 2.5 octaves. Mean values $\pm$ SD of the lower and upper cut-offs were 4.3 ± 4 and $73 \pm 54°$/s. Only a small proportion of LPl cells (22%) were not band-pass tuned for drift velocity. Two of these units were of the low pass type, i.e. they exhibited no attenuation of their discharges at low velocities. One cell responded equally to all velocities tested, and for the three remaining cells, no attenuation of the discharges were observed at the high velocities used.

Stimulus interactions

We studied the influence of visual noise on the spatial frequency tuning function of 22 LPl cells. The grating was superimposed on either a station-

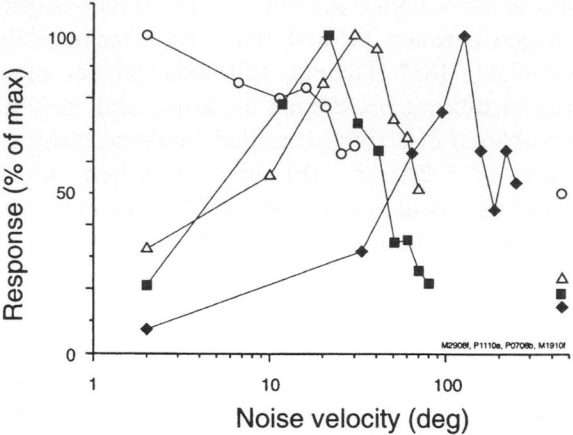

Fig. 4. Responses of four LPl cells as a function of noise velocity. The responses were normalized for illustration purposes. The symbols on the right side of the graph represent cell spontaneous activity levels.

ary or moving texture field. About half of the cells tested showed a significant change of their tuning function in presence of a moving texture pattern (there was no effect when the noise was stationary). A representative example is shown in Fig. 5. First, the spatial frequency tuning curves recorded in the two control conditions (drifting grating superimposed on a stationary noise), i.e. at the beginning and at the end of the test, were very comparable. When moving, the noise had a strong suppressive effect only when its direction of movement was opposite to that of the grating (anti-phase condition). No reduction in the responses was apparent when both stimuli were drifted in the same direction (in-phase condition). One may note, however, that there was a summation of the responses to grating and texture at low spatial frequencies only, thus increasing the bandwidth of the tuning function. A second example is shown in Fig. 6A. For this cell, the noise had a suppressive effect regardless of its direction of motion; the reduction was, however, slightly more pronounced in the anti-phase direction. As mentioned above, about half of LPl cells were affected by the presence of a moving background. More precisely, in the anti-phase condition, responses of nine cells were reduced, and two of them were enhanced while 11 units were not

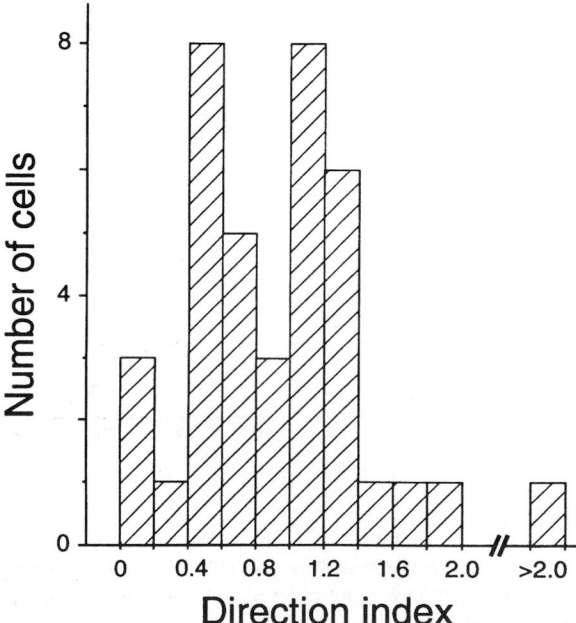

Fig. 3. Distribution of the noise direction index of LPl cells. Most cells exhibited a DI greater than 0.5.

affected. The suppressive effect observed when the two stimuli were moving in anti phase was not likely to be related to directional inhibitory mechanisms since it was also observed in cells which responded to the forward and reverse direction of the noise movement. Also, there were examples in which no inhibitory effect of the noise was apparent despite the fact that the cells were strongly direction-selective for the texture pattern (see Fig. 6B). No strong effects were observed when both stimuli moved in the same direction (in phase motion): we observed an increase of the responses in five cells and a suppressive effect was apparent for only three cells. For the remaining 14 units, the in-phase condition did not significantly modify cell discharges.

Discussion

Our results indicate that more than 85% of cells in the striate-recipient zone of the LP–pulvinar complex are sensitive to the drift of visual noise. These data are in agreement with the findings of Mason (1981) who reported that most cells in the striate-recipient zone of the LP nucleus were responsive to the motion of either a noise bar against a stationary noise background or a whole field of noise. Moreover, we found that for most units ($\sim 70\%$), responses vary as a function of the direction of motion. The resulting noise tuning function of LPl cells is broad with a mean bandwidth of 48°. Also, we found that the large majority of LPl cells exhibited strong directional preference along a given axis of movement and that almost all units were tuned to noise velocity and remained responsive to high velocities (mean high cut-off of 73°/s). Altogether, these results indicate not only that visual noise could be seen as a relevant tool for studying LPl physiology but also that LPl cells could be involved in the detection of moving objects.

Visual noise responses comparable to that of

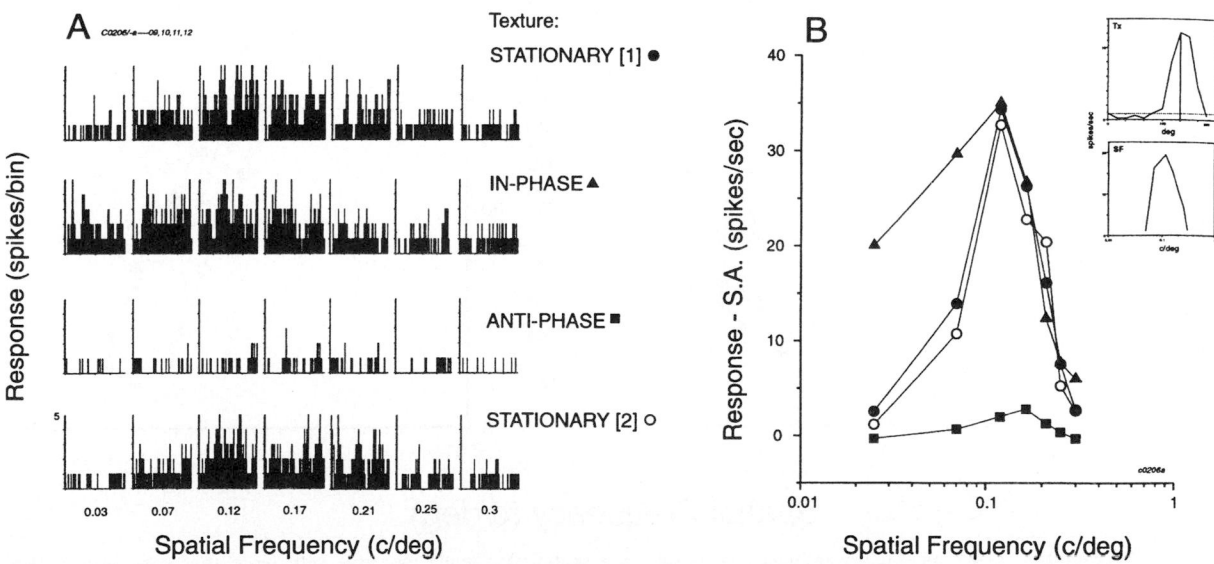

Fig. 5. Effect of visual noise on the spatial frequency tuning function of a LPl cell shown as PSTHs (A) and corresponding tuning curves (B). The traces 'stationary [1] and [2]' represent control conditions, i.e. the grating was drifted and superimposed on a stationary textured pattern. In-phase motion: when the two stimuli were moving in the same direction, there was an increase of the responses to low spatial frequencies only. Anti-phase motion: there was a strong masking effect of noise at all spatial frequencies when both stimuli were drifted in opposite directions. The upper inset represents noise direction tuning function. The vertical straight line indicates the direction at which the noise was drifted in the in-phase condition. The opposite direction (180° apart) was used during the anti-phase protocol. The lower inset shows the spatial tuning function in absence of noise.

LPl cells have been reported in cortical areas which are reciprocally connected with the LP–pulvinar complex. Two specific regions are of particular interest in the context of motion analysis, namely the AEV and PMLS cortices. Benedek et al. (1988) reported that two-thirds (66%) of neurons in the AEV cortex were also tuned to the direction of motion of large fields of visual noise. On the basis of cell properties (such as direction selectivity and optimal velocity) and afferent/efferent connectivity, these authors as well as Mucke et al. (1982) suggested that the AEV cortex is part of an extrageniculate pathway involved in motion detection. Despite the absence of a direct anatomical link between AEV and LPl (Norita et al., 1986), it is possible that the LPl is part of the same global system involved in the analysis of motion since information may flow between the two regions via indirect pathways (e.g. via intrinsic connections between the LPm and LPl, or via the superior colliculus or the LS cortex). However, although there are similarities

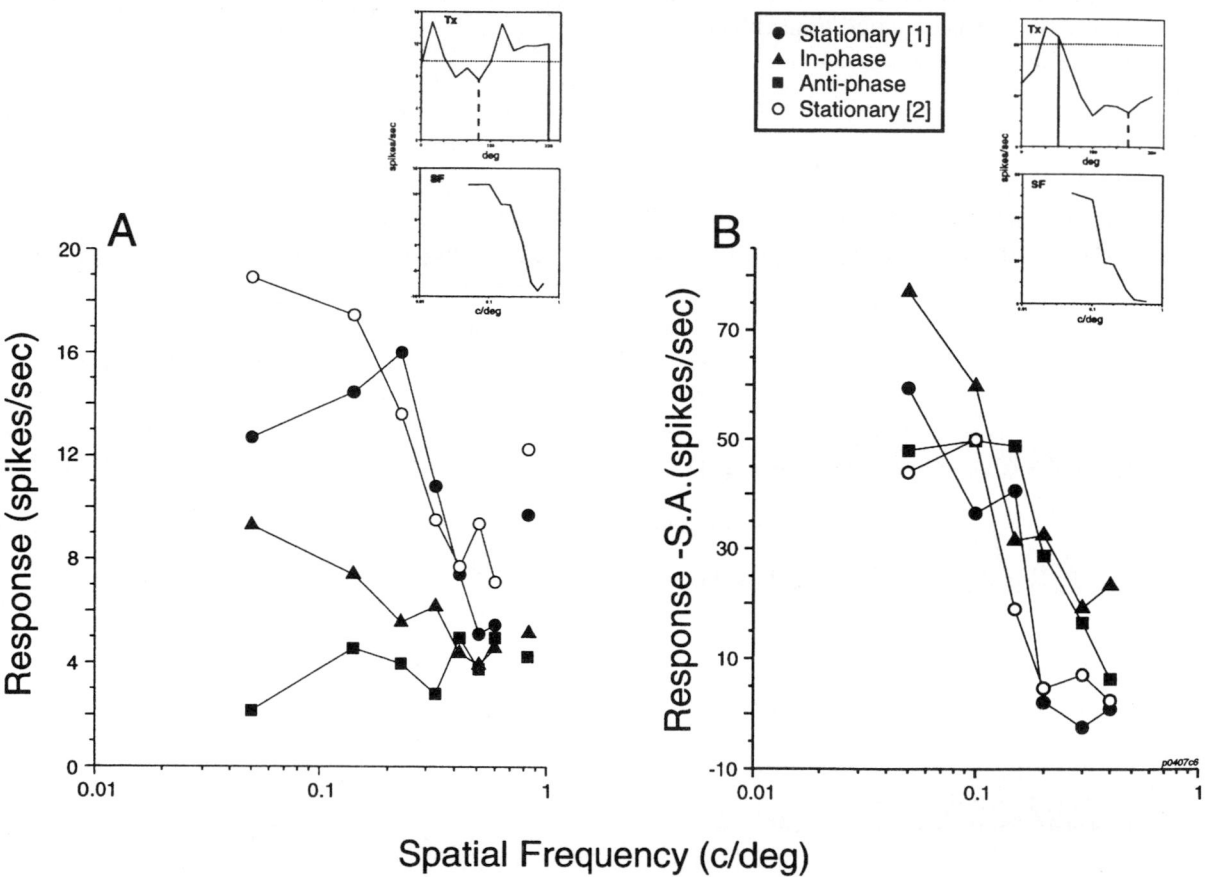

Fig. 6. (A) Example of the modulatory influence of noise on a spatial frequency low pass cell. Note that there was a clear attenuation of the discharges in both in- and anti-phase motions. Note that the reduction is more pronounced when the stimuli were moving in opposite directions. The symbols on the right side of the graph represent cell spontaneous activity levels during each test. (B) Example of a LPl cell not significantly affected by the addition of visual noise. Only a small increase of the response in the in-phase motion could be observed. The upper insets represent noise direction tuning function. The vertical straight and dashed lines indicate the direction at which the noise was drifted in the in-phase and anti-phase condition respectively. The lower insets show spatial tuning function in the absence of a textured pattern.

between properties of cells in LPl and the AEV (e.g., noise responsiveness, receptive field size), neurons in the AEV area are sensitive to higher velocities than that in the LPl, whether velocity was tested with visual noise as in the present study or with bars such as in Chalupa and Abramson (1989).

There are prominent reciprocal connections between the striate-recipient zone of LP and LS cortex (Updyke, 1981; Raczkowski and Rosenquist, 1983). Von Grunau and Frost (1983) initially reported that, overall, LS cells were not responsive to the drift of a visual noise. A large proportion of texture-sensitive cells (73%) were later described in this area by Hamada (1987). However, the textured pattern used in the latter study was rather coarse (1.6–4.8°) and most responses were likely to come from the stimulation of the cell's receptive field by clusters of pattern elements. It is of interest to note that data from our laboratory (unpublished observations) found that more than half (57%) of PMLS neurons were tuned to the direction of visual noise (pattern identical to the one used in the present study, i.e. element size of 0.11–0.22°). The PMLS cortex has also been associated with the analysis of motion (Spear, 1991) and it was proposed that the LS-LP loop may be involved in the detection and analysis of expanding flow-fields of motion (Rauschecker et al., 1987a,b; Rauschecker, 1988). The strong similarities between response properties of cells in PMLS and LPl support, to some extent, the above statement. However, our laboratory recently reported that the LPl do not strongly contribute to spatial and temporal sensitivity nor to direction selectivity of PMLS cells (Minville et al., 1995). This unanticipated result suggests that the physiological links between PMLS and LPl may not be as robust as was first believed on the basis of neuroanatomical identified connections. In this context, It would certainly be meaningful to determine the physiological nature of the projections between the LPl and the other regions of the LS cortex such as the PLLS. In fact, our understanding of the LP–pulvinar complex may largely depend on our

ability to clearly establish the functional significance of the bi-directional connections between this large extrageniculate nucleus and the visual cortex.

Our study also presents evidence that visual noise, when presented in conjunction with a drifting grating, can modulate the spatial contrast sensitivity of LPl cells. The main effect is a suppression of the discharges at all spatial frequencies when both stimuli are drifted in opposite directions (anti-phase). Von Grünau and Frost (1983) and Benedek et al. (1988) have previously shown that responses of neurons in PMLS and AEV to a discrete stimulus can be modulated by a textured background. In both studies the suppressive effect of noise was more apparent when the two stimuli were moving in phase. This last observation is somewhat at odds with our findings. This discrepancy may come from a difference in stimulus type (full-field grating and noise) or in the protocol used (in this study, both stimuli were superimposed as in Nordmann et al., 1992; Blackwell, 1994). It may also reflect the fact that LPl computes the relative movement of an object differently in a complex visual environment, allowing a more accurate and efficient global motion analysis along with the remaining regions of the extrageniculate system which could lead to appropriate behaviour.

In conclusion, these results suggest that the LPl may be part of an extrageniculate system involved in the analysis of motion and perhaps, in more complex behaviour such as orientation and movement of the animal's body (Benedek et al., 1988; Rauschecker, 1988). The mechanisms by which neuronal information is relayed, integrated, and analysed by the cortico-LP-cortical loops still need to be uncovered. This represents a major task when one considers the multiple reciprocal connections which exist between the LP–pulvinar complex and all visual areas of the brain. On the basis of the results whereby LPl has little influence on the responsiveness of PMLS cells (Minville et al., 1995), it seems unlikely, at first view, that a given subregion of the LP-pulvinar alone would be essential to drive extrastriate cortical areas. It may well be, however, that the LP-pulvinar provides a unique platform in which immediate and parallel computations made in various areas could be evaluated and compared in order to prepare better strategies to be used by the visual system as a whole [for discussion, see Creutzfeldt (1988); Mumford (1991)].

Acknowledgements

We are grateful to J.G. Daugman for providing the 2D Fourier power spectrum of texture frames presented in Fig. 1. This work was supported by grant No. MT-10962 of MRC of Canada to (CC), and by installation grants from FRSQ and FCAR. (TS) was supported in part by a fellowship from FCAR and NSERC.

References

Abramson, B.P. and Chalupa, L.M. (1985) The laminar distribution of the cortical connections with the tecto- and cortico-recipient zones in the cat's lateral posterior nucleus. *Neuroscience*, 15: 81–95.

Abramson, B.P. and Chalupa, L.M. (1988) Multiple pathways from the superior colliculus to the extrageniculate visual thalamus of the cat. *J. Comp. Neurol.*, 271: 397–418.

Benedek, G., Norita, M. and Creutzfeldt, O.D. (1983) Electrophysiological and anatomical demonstration of an overlapping striate and tectal projection to the lateral posterior-pulvinar complex of the cat. *Exp. Brain Res.*, 52: 157–169.

Benedek, G., Mucke, L., Norita, M., Albowitz, B. and Creutzfeldt, O.D. (1988) Anterior ectosylvian visual area (AEV) of the cat: physiological properties. *Prog. Brain Res.*, 75: 245–255.

Berson, D.M. and Graybiel, A.L. (1978) Parallel thalamic zones in the LP–pulvinar complex of the cat identified by their afferent and efferent connections. *Brain Res.*, 147: 139–148.

Berson, D.M. and Graybiel, A.M. (1983) Organization of the striate-recipient zone of the cat's lateralis posterior-pulvinar complex and its relations with the geniculostriate system. *Neuroscience*, 9: 337–372.

Blackwell, K.T. (1994) The effect of noise on contrast sensitivity. *Invest. Ophthalmol. Vis. Sci. (suppl.)*, 35: 1368.

Casanova, C. and Savard, T. (1996) Responses to moving texture patterns of cells in the striate-recipient zone of the cat's lateral posterior-pulvinar complex. *Neuroscience*, 70: 439–447.

Casanova, C., Freeman, R.D. and Nordmann, J.P. (1989) Monocular and binocular response properties of cells in the striate-recipient zone of the cat's lateral posterior-pulvinar complex. *J. Neurophysiol.*, 62: 544–557.

Casanova, C., Nordmann, J.P. and Molotchnikoff, S. (1991) Le complexe noyau latéral postérieur-pulvinar des mammifères et la fonction visuelle. *J. Physiol. (Paris)*, 85: 44–57.

Casanova, C., Savard, T., Nordmann, J.P., Molotchnikoff, S. and Minville, K. (1995) Comparison of the responses to moving texture patterns of simple and complex cells in the cat's area 17. *J. Neurophysiol.*, 74: 1271–1286.

Chalupa, L.M. (1991) Visual function of the pulvinar. In: A.G. Leventhal (Ed.), *The Neural Basis of Visual Function*, Vol. 4, CRC Press, Boca Raton, pp. 141–159.

Chalupa, L.M. and Abramson, B.P. (1988) Receptive-field properties in the tecto- and striate-recipient zones of the cat's lateral posterior nucleus. *Prog. Brain Res.*, 75: 85–94.

Chalupa, L.M. and Abramson, B.P. (1989) Visual receptive fields in the striate-recipient zone of the lateral posterior-pulvinar complex. *J. Neurosci.*, 9: 347–357.

Chalupa, L.M., Williams, R.W. and Hughes, M.J. (1983) Visual response properties in the tecto-recipient zone of the cat's lateral posterior-pulvinar complex: A comparison with the superior colliculus. *J. Neurosci.*, 3: 2587–2596.

Creutzfeldt, O.D. (1988). Extrageniculo-striate visual mechanisms: compartmentalization of visual functions. *Prog. Brain Res.*, 75: 307–320.

Desimone, R., Wessinger, M., Thomas, L. and Schneider, W. (1990) Attentional control of visual perception: cortical and subcortical mechanisms. *Cold Spring Harbor Symp. Quant. Biol.*, LV: 963–971.

Fernald, R. and Chase, R. (1971) An improved method for plotting retinal landmarks and focusing the eyes. *Vision Res.*, 11: 95–96.

Graybiel, A.M. and Berson, D.M. (1980) Histochemical identification and afferent connections of subdivisions in the lateralis–pulvinar complex and related thalamic nuclei in the cat. *Neuroscience*, 5: 1175–1238.

Gulyás, B., Orban, G.A., Duysens, J. and Maes, H. (1987) The suppressive effect of moving textured backgrounds on responses of cat striate neurons to moving bars. *J. Neurophysiol.*, 57: 1767–1791.

Hamada, T. (1987) Neural response to the motion of textures in the lateral suprasylvian area of cats. *Behav. Brain Res.*, 25: 175–185.

Hammond, P. and MacKay, D.M. (1977) Differential responsiveness of simple and complex cells in cat striate cortex to visual texture. *Exp. Brain Res.*, 30: 275–296.

Hoffmann, K.P., Morrone, C.M. and Reuter, J.H. (1980) A comparison of the responses of single cells in the LGN and visual cortex to bar and noise stimuli in the cat. *Vision Res.*, 20: 771–777.

Kaas, J.H., Guillery, R.W. and Allman, J.M. (1972) Some principles of organization in the dorsal lateral geniculate nucleus. *Brain Behav. Evol.*, 6: 253–299.

Mason, R. (1978) Responses of cells in the dorsal lateral geniculate complex of the cat to textured visual stimuli. *Exp. Brain Res.*, 25: 323–326.

Mason, R. (1981) Differential responsiveness of cells in the visual zones of the cat's LP-pulvinar complex to visual stimuli. *Exp. Brain Res.*, 43: 25–33.

Minville, K., Savard, T. and Casanova, C. (1995) Inactivation of the striate-recipient zone of LP influences the responses of only a few cells in PMLS cortex. *Soc. Neurosci. Abstr.*, 21: 906.

Mucke, L., Norita, M., Benedek, G. and Creutzfeldt, O.D. (1982) Physiologic and anatomic investigation of a visual cortical area situated in the ventral bank of the anterior ectosylvian sulcus of the cat. *Exp. Brain Res.*, 46: 1–11.

Mumford, D. (1991) On the computational architecture of the neocortex. I. The role of the thalamo-cortical loop. *Biol. Cybern.*, 65: 135–145.

Nordmann, J.P., Freeman, R.D. and Casanova, C. (1992) Contrast sensitivity in amblyopia: masking effects of noise. *Invest. Ophthalmol. Vis. Sci.*, 33: 2975–2985.

Norita, M., Mucke, L., Benedek, G., Albowitz, G., Katoh, Y. and Creutzfeldt, O.D. (1986) Connections of the anterior ectosylvian visual area (AEV). *Exp. Brain Res.*, 62: 225–240.

Olshausen, B.A., Anderson, C.H. and Van Essen, D.C. (1993) A neurobiology model of visual attention and invariant pattern recognition based on dynamic routing of information. *J. Neurosci.*, 13: 4700–4719.

Orban, G.A., Gulyás, B. and Spileers, W. (1988) Influence of moving textured backgrounds on responses of cat area 18 cells to moving bars. *Prog. Brain Res.*, 75: 137–145.

Petersen, S.E., Robinson, D.L. and Morris, J.D. (1987) Contributions of the pulvinar to visual spatial attention. *Neuropsychologia.*, 25: 97–105.

Raczkowski, D. and Rosenquist, A.C. (1983) Connections of the multiple visual cortical areas with the lateral posterior–pulvinar complex and adjacent thalamic nuclei in the cat. *J. Neurosci.*, 3: 1912–1942.

Rauschecker, J.P. (1988) Visual function of the cat's LP/LS subsystem in global motion analysis. *Prog. Brain Res.*, 75: 95–107.

Rauschecker, J.P., Von Grüneau, M.W. and Poulin, C. (1987a) Centrifugal organization of direction preferences in the cat's lateral suprasylvian visual cortex and its relation to flow field processing. *J. Neurosci.*, 7: 943–9580.

Rauschecker, J.P., Von Grüneau, M.W. and Poulin, C. (1987b) Thalamo-cortical connections and their correlation with receptive field properties in the cat's lateral suprasylvian visual cortex. *Exp. Brain Res.*, 67: 100–112.

Robinson, D.L. and Petersen, S.E. (1992) The pulvinar and visual salience. *TINS*, 15: 127–132.

Skottun, B.C. and Freeman, R.D. (1984) Stimulus specificity of binocular cells in the cat's visual cortex: ocular dominance and the matching of left and right eyes. *Exp. Brain Res.*, 56: 206–216.

Spear, P.D. (1991) Functions of extrastriate visual cortex in non-primate species. In: A.G. Leventhal (Ed.), *The Neural Basis of Neural Function*, Vol. 4, CRC Press, Boca Raton, pp. 339–370.

Updyke, B.V. (1977) Topographic organization of the projections from cortical areas 17, 18 and 19 onto the thalamus, pretectum and superior colliculus in the cat. *J. Comp. Neurol.*, 173: 81–122.

Updyke, B.V. (1981) Projection from visual areas of the middle suprasylvian sulcus onto the lateral posterior complex and adjacent thalamic nuclei in cat. *J. Comp. Neurol.*, 201: 477–506.

Von Grünau, M. and Frost, B.J. (1983) Double-opponent-process mechanisms underlying RF structure of directionally specific cells of cat lateral suprasylvian visual area. *Exp. Brain Res.*, 49: 84–92.

M. Norita, T. Bando and B. Stein (Eds.)
Progress in Brain Research, Vol 112
© 1996 Elsevier Science BV. All rights reserved.

CHAPTER 20

Comparisons of cross-modality integration in midbrain and cortex

Barry E. Stein* and Mark T. Wallace

Department of Neurobiology and Anatomy, Bowman Gray School of Medicine / Wake Forest University, Winston-Salem, NC 27157-1010, USA

Multisensory neurons are abundant in the superior colliculus and anterior ectosylvian cortex of the cat. Despite the fact that these areas receive inputs from different regions, and are likely to be involved in different functional roles, their multisensory neurons have many fundamental similarities. They all have multiple receptive fields, one for each sensory input, and these receptive fields overlap one another. It is this spatial correspondence among receptive fields that determines the manner in which both populations of neurons integrate the inputs they receive from different sensory channels. Several principles of integration characterize both cortical and midbrain multisensory neurons, and these constancies in the fundamentals of cross-modality integration are likely to provide a basis for coherence at different levels of the neuraxis. Yet there are also obvious differences in these populations of multisensory neurons. Cortical receptive fields are significantly larger than those in the midbrain, have a lower incidence of suppressive surrounds, and exhibit less cross-modality inhibitory interactions than in the midbrain. Presumably, these differences reflect a greater emphasis on non-spatial aspects of cross-modality integration in cortex than is required by the orientation and localization functions mediated by the superior colliculus.

Introduction

A fundamental task of the central nervous system is to integrate information from multiple sources in order to best prepare the organism to evaluate and respond to an external event. Of particular concern here are the neural mechanisms by which the nervous system integrates information from different sensory channels. This is made possible, in large part, by the convergence of sensory inputs from different modalities onto common sets of neurons. Such 'multisensory' convergence takes place at many sites in the central nervous system (see Stein and Meredith, 1993 for a recent review), and the intersensory effects that arise from the integration of these inputs are most dramati-cally evident when combinations of related stimuli alter the probability of detecting, localizing, and/or reacting to stimuli (e.g., Sumby and Pollack, 1954; Hershenson, 1962; Morell, 1968a,b, 1972; Bernstein et al., 1969; Fidell, 1970; Simon and Craft, 1970; Nickerson, 1973; Andreassi and Greco, 1975; Posner et al., 1976; Welch and Warren, 1980, 1986; Bertelson and Radeau, 1981; Gielen et al., 1983; Stein et al., 1989; Perrott et al., 1990; Hughes et al., 1994; Frens et al., 1995; Munoz and Carneil, 1995; Peck et al., 1995). Also noteworthy are the many perceptual and behavioral anomalies that result from the combination of discordant intersensory signals (e.g. McGurk and MacDonald, 1976; Reisberg et al., 1981; Stein et al., 1989).

Only recently have the changes in the individual neurons that underlie the synthesis of different sensory inputs begun to be explored systemat-

---

* Corresponding author.

ically. The best-studied of these neurons are those in the deeper aspects of the superior colliculus (SC) of the cat (see Stein and Meredith, 1993), and more recently in the deep SC of guinea pig (King and Palmer, 1985) and macaque monkey (Frens and van Opstal, 1995; Wallace et al., 1996), which receive combinations of inputs from the visual, auditory, and somatosensory modalities. However, there are also many multisensory neurons in cortex.

In the cat, the region of cortex surrounding the anterior ectosylvian sulcus (AES) has been described as a 'polysensory' area, where inputs from several sensory modalities converge (Graybiel, 1972; Roda and Reinoso-Suarez, 1983; Reinoso-Suarez and Roda, 1985; Benedek et al., 1996). This region is composed of three modality-specific regions: a visual area, referred to as the anterior ectosylvian visual area (AEV: Mucke et al., 1982; Olson and Graybiel, 1987); a somatosensory area, referred to as the fourth somatosensory cortex (SIV: Clemo and Stein, 1982); and an auditory area, referred to as Field AES (Clarey and Irvine, 1986). Distributed near the borders of these unimodal regions are many multisensory neurons (Clemo et al., 1991) which are capable of integrating their multiple sensory inputs (Wallace et al., 1992) and which do not directly influence SC neurons (Wallace et al., 1993).

The purpose of the present study was to compare the manner in which visual, auditory and somatosensory inputs are integrated in individual neurons of the SC and AES to determine whether common principles of multisensory integration are operative at very different levels of the neuraxis. Portions of the data used in these comparisons are taken from prior investigations (e.g. Wallace et al., 1992, 1993), but many of the observations have not been reported elsewhere.

Methods

Identical anesthetic conditions were maintained and identical stimuli were presented to superior colliculus (SC) and anterior ectosylvian sulcus (AES) neurons, so that their properties could be compared directly. The procedures used were fundamentally the same as those described in detail in several previous publications (e.g. see Meredith and Stein, 1986a,b; Wallace et al., 1993; Wallace and Stein, 1994) and are only briefly described here.

A hollow cylinder was implanted over the SC ($n = 4$) or AES ($n = 3$) in cats deeply anesthetized with sodium pentobarbital (40 mg/kg, i.p.). The animals recovered after 7–10 days before the first recording session. During recording sessions the animal was anesthetized with halothane (0.75–2.0%) and paralyzed with pancuronium bromide (0.5–1.0 mg/kg per h). 'Search' stimuli were regularly presented during an electrode traverse and when a neuron was isolated, its responses to visual, auditory and somatosensory stimuli were examined. After the effective modalities were determined, its various receptive fields were mapped using hand-held or electronically-generated stimuli. Visual receptive fields were determined using a pantoscope to present bars or spots of light onto a translucent hemisphere. Manually-controlled brushes and von Frey hairs were used to map somatosensory receptive fields, and broad-band noise bursts from hoop-mounted speakers were used to determine auditory receptive fields. Computer-controlled stimuli were then used for all quantitative analyses.

When testing for cross-modality interactions, modality-specific stimuli ('single-modality tests') and stimuli from two modalities ('combined-modality tests') were presented in an interleaved manner at long (12–15 s) interstimulus intervals. A statistical criterion ($P < 0.05$, paired t-test) was used to determine whether a multisensory interaction occurred. Specifically, the number of impulses elicited by the combined-modality test had to differ significantly from that evoked by the most effective single-modality stimulus. The magnitude of this interaction was determined by the formula:

$$CM - SM_{max}/SM_{max} \times 100 = \%Interaction$$

where CM = response to the combined-modality

stimuli and SM_{max} = response to the most effective single-modality stimulus.

At the end of a recording session the animal was allowed to recover from paralysis, at which point the anesthetic was discontinued. Once the animal could breath and locomote independently, it was returned to its home cage. Following the final recording session, the animal was sacrificed with an overdose of barbiturate and perfused intracardially with 10% formaldehyde. Standardized histological techniques were used to reconstruct and plot the location of recorded neurons.

Results

A total of 233 neurons in the SC and 181 neurons in the AES were evaluated here. A substantial proportion of SC (54%) and AES (26%) neurons responded to stimuli from two or more modalities (Fig. 1). In the SC these multisensory neurons were found primarily in stratum griseum intermediale; in the AES they were found primarily in or near the caudal aspect of the fundus and in the caudal aspect of the AES. Multisensory neurons in the AES were most prevalent at the borders between unimodal regions, and because the recording sites were most frequently near the borders of auditory and visual areas, the incidence of visual-auditory neurons in this sample may be spuriously high (see Wallace et al., 1992). Nevertheless, the overall incidence of multisensory neurons in the AES was still substantially lower than that found in the SC (Fig. 1).

It was interesting to note that the response properties of multisensory SC and AES neurons were similar to one another and also indistinguishable from those of their unimodal neighbors. In both regions unimodal and multisensory somatosensory neurons responded best to rapid, low-amplitude displacements of the cutaneous surface, and neurons responsive to stimulation of 'deep' receptors were far less common. Unimodal and multisensory auditory neurons responded best to brief (100–150 ms) broadband noise bursts and had similar binaural response characteristics. Unimodal and multisensory visual neurons responded best to small (< 4°) bars or spots of light moving at intermediate velocities (20–150°/s), and had similar incidences of directional selectivity.

Previously, several 'principles' of multisensory integration were identified in SC neurons (see Stein and Meredith, 1993 for a review). These principles were based on how cross-modality interactions were altered as the spatial and temporal relationships between two different sensory

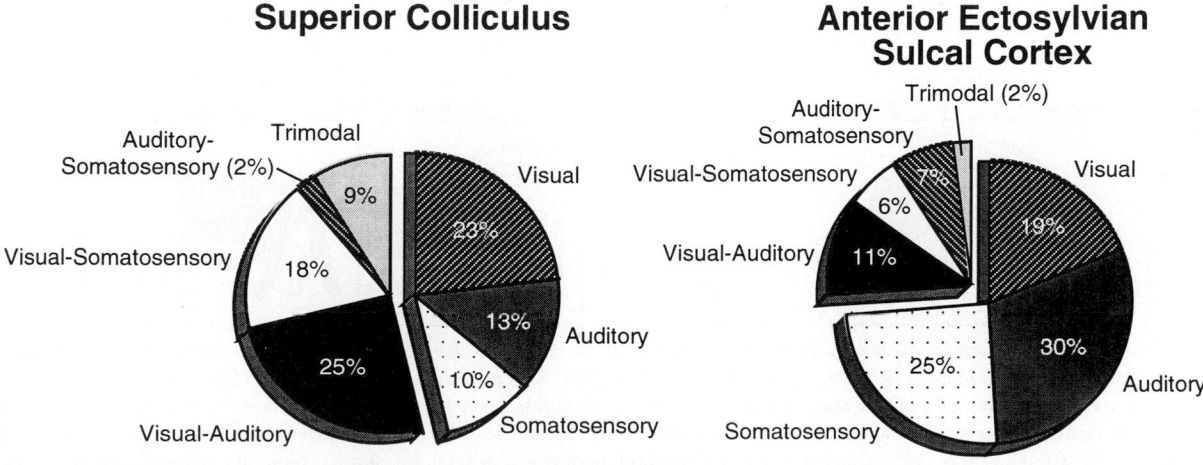

Fig. 1. Distribution of sensory convergence patterns in superior colliculus and anterior ectosylvian cortex.

stimuli were altered. They were used here as a convenient guide to the specific comparisons to be made between the cross-modality interactions in the SC and AES.

Spatial properties of multisensory integration

As previously noted (see Discussion), the different unimodal receptive fields of multisensory neurons in the SC exhibit a high degree of spatial correspondence (Fig. 2). A similar correspon-

dence was apparent in AES neurons (Fig. 2), but was unexpected given the lack of obvious spatiotopic organization in its unimodal visual and auditory regions (see Mucke et al., 1982; Clarey and Irvine, 1986; Olson and Graybiel, 1987; Rauschecker and Korte, 1993). Despite the similarities in cross-modality receptive field correspondence, the receptive fields of AES multisensory neurons were generally far larger than their counterparts in the SC (also see Wallace et al., 1992).

SUPERIOR COLLICULUS

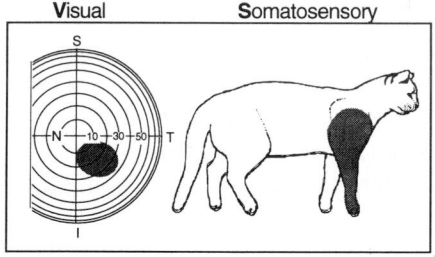

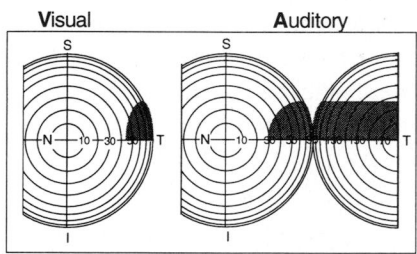

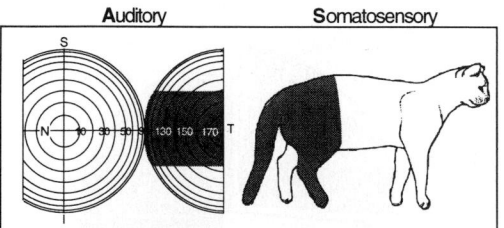

ANTERIOR ECTOSYLVIAN SULCUS

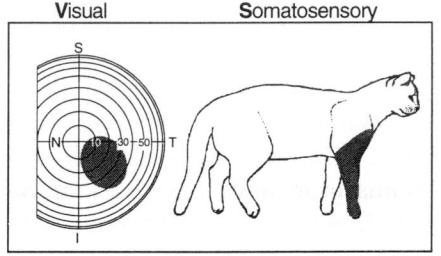

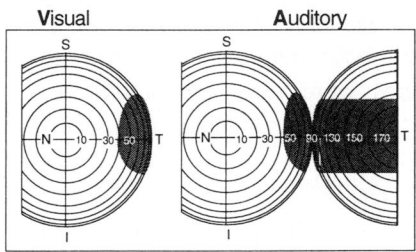

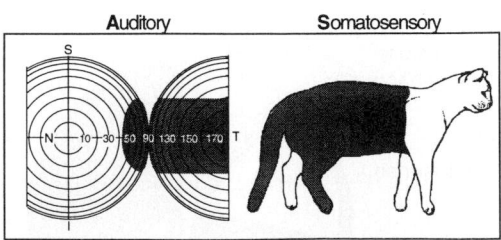

Fig. 2. Receptive field correspondence in multisensory neurons in SC and AES. Representative visual-somatosensory, visual-auditory and auditory-somatosensory neurons are shown for each region. Receptive fields are shaded. Auditory receptive fields are plotted on a representation of auditory space in which the sphere represents space in front of the interaural plane, and the attached hemisphere represents space behind this plane on the contralateral side. The joining of the sphere and hemisphere is along the interaural line (0° elevation, 90° azimuth). Caudal ipsilateral auditory space is not shown. Note the cross-modality spatial correspondence between the receptive fields in each of the multisensory neurons.

The functional role of receptive field overlap was examined in SC and AES neurons by presenting pairs of stimuli from different modalities (e.g., one visual, one auditory) simultaneously, or at predetermined intervals. In each example, one stimulus was designated the 'test' stimulus and remained at the same position within its receptive field. The second, or 'modulating', stimulus, was presented at a number of different positions, both within and outside its receptive field or at various preset intervals before or after the test stimulus.

Responses to the test stimulus were enhanced significantly when the modulating stimulus was presented simultaneously within its receptive field (Fig. 3). However, when the modulating stimulus was located outside the excitatory borders of its receptive field, the enhancement it produced was lost, and in many instances in the SC, and in some instances in the AES, enhancement was replaced by significant inhibition. This 'spatial' principle of multisensory integration was observed in each of the multisensory categories encountered in the SC and AES: visual-auditory, visual-somatosensory, and auditory-somatosensory, and trimodal. Despite the presence of inhibition in

some AES neurons (see Fig. 3), its incidence and magnitude were typically far less than are seen in the SC.

Temporal properties of multisensory integration

Sensory responses in both SC and AES were characterized by significant differences in modality-specific response latencies. Auditory latencies were the shortest (SC: mean = 14 ms, range 8–40 ms; AES: mean = 19 ms, range 12–50 ms), visual latencies were the longest (SC: mean = 91 ms, range 58–140 ms; AES: mean = 104 ms, range 72–165 ms), and somatosensory latencies were intermediate (SC: mean = 23 ms, range 10–40 ms; AES: mean = 28 ms, range 16–62 ms). Nevertheless, the matching of response latencies was not essential to produce a cross-modality interaction. In fact, any given unimodal stimulus altered activity in SC and AES neurons for a long enough period of time to produce a reasonably long window during which the presentation of a stimulus from another modality would produce a significant interaction. In general, maximal interactions occurred when the periods of maximal unimodal

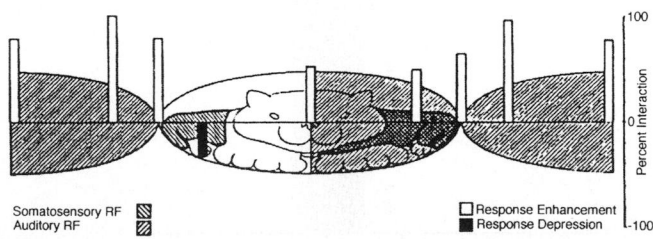

Fig. 3. Multisensory integration in SC and AES neurons depends on the spatial relationships among stimuli and their receptive fields. An example of this spatial principle is shown for a visual-auditory SC neuron (left) and for an auditory-somatosensory AES neuron (right). Receptive fields are shown in hatching. For the SC neuron, the visual stimulus was presented at the same location within its receptive field on all trials. The auditory stimulus was systematically varied, and its location on the interaural plane (i.e., 0° elevation) is shown by the base of the bars. The height of the bars represent the magnitude of the interaction. Open bars depict response enhancement and closed bars depict response depression. Similarly, for the AES neuron the somatosensory stimulus was held at one location within the receptive field while the location of the auditory stimulus was systematically varied along the interaural axis. Note for both examples that response enhancement is obtained when the stimuli are coincident in space and within their respective receptive fields.

activity overlapped. As the temporal interval between the stimuli increased by presenting either stimulus progressively earlier or later than the other, the magnitude of the interaction decreased (Fig. 4).

In some cases, when the temporal interval between the two stimuli was particularly long, enhancement was replaced by depression. This was observed in both SC ($n = 9$) and AES ($n = 4$) neurons and suggests that many sensory stimuli have an inhibitory phase trailing their initial excitation. When the inhibitory phase of one stimulus overlaps the excitatory phase of another, an inhibitory interaction results.

Inverse effectiveness / multiplicative interactions

In both the SC and the AES the enhancements that resulted from combinations of two different sensory stimuli were generally far greater than would be predicted by a simple additive model. Rather, they seemed multiplicative, so that the combination of a stimulus in one modality (e.g. visual) that evoked three impulses with a stimulus from another modality (e.g. auditory) that evoked two impulses might produce a discharge train of 10 impulses. Furthermore, the magnitude of the interaction depended on the effectiveness of the unimodal stimuli. In all cases examined, combina-

SUPERIOR COLLICULUS

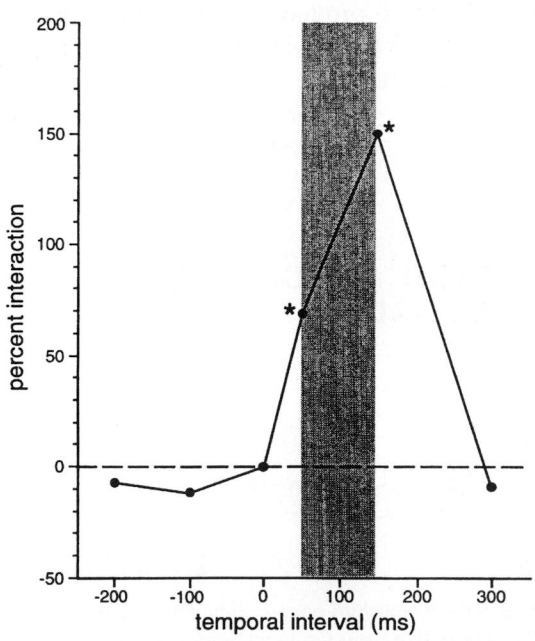

ANTERIOR ECTOSYLVIAN SULCAL CORTEX

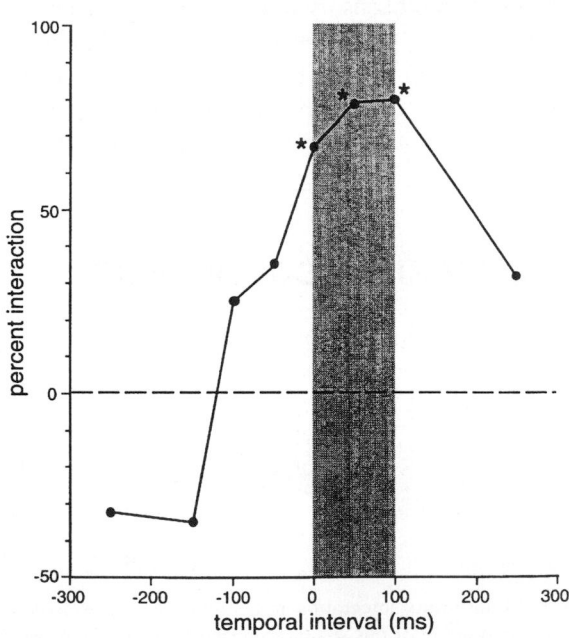

Fig. 4. Multisensory integration in both SC (left) and AES (right) depends on the timing of sensory inputs. In both of these visual-auditory neurons the percent interaction (enhancement or depression) is plotted as a function of the temporal interval between the stimuli. A temporal interval of 0 represents the simultaneous onset of the stimuli. Negative numbers represent the time in milliseconds that the auditory stimulus preceded the visual stimulus and positive numbers represent the interval at which the visual stimulus preceded the auditory. Asterisks denote statistically significant ($P < 0.05$) interactions, and the shading shows the temporal 'window' during which significant interactions occurred. Note the similarity in the curves and temporal windows for neurons in SC and AES.

tions of weakly effective unimodal stimuli produced the greatest response enhancements in both SC and AES neurons. This 'inverse effectivenes' principle is apparent from the plots of unimodal effectiveness vs. interactive product shown in Fig. 5.

Preservation of receptive field properties

Although the effects of combining stimuli from different sensory modalities could significantly alter the responses of SC and AES neurons, they did not substantially alter their characteristic receptive field properties. Thus, receptive field borders did not change, nor did such characteristics as direction selectivity, velocity selectivity, binaural/binocular category, adaptation rate, etc. Rather, the enhancement and depression effects were primarily evident as a change in the absolute level of activity evoked.

Discussion

The results illustrate substantial similarities in the properties of multisensory neurons in the SC and AES. Although cortical receptive fields are generally larger than those in the SC, in both regions of the brain the individual receptive fields of multisensory neurons are in spatial correspondence, thereby forming a common foundation for integrating multisensory information. Furthermore, in both areas response enhancements are produced by spatially coincident stimuli from different sensory modalities and show the same general dependencies on the timing of these different inputs. Yet, despite the impressive magnitude of these interactions, they do not substantially alter the selectivity of constituent unimodal receptive fields. This is likely to provide the basis for coherence in the processing of multisensory information throughout the nervous system.

Although intersensory receptive field correspondence in the SC has been known for some time (e.g. Wickelgren and Sterling, 1969; Stein et al., 1975, 1976; Meredith and Stein, 1986a,b; Wallace et al., 1993), and is entirely consistent with the general correspondence of the different sensory maps in this structure (see Stein and Meredith, 1993 for further details), this is not true for the AES. Although the AES does contain a soma-

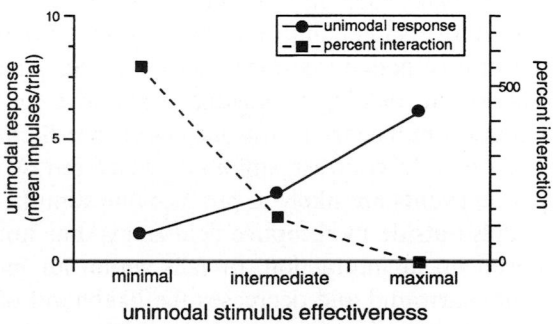

Fig. 5. Multisensory integration is inversely related to unimodal stimulus effectiveness. This 'inverse effectiveness' principle is demonstrated in representative visual-auditory neurons from the SC (left) and AES (right). Unimodal responses and multisensory interactions are plotted as a function of three levels of unimodal stimulus effectiveness. Note the strong similarities between the midbrain and cortical neurons.

totopic map, referred to as SIV (Clemo and Stein, 1982, 1983), there is no apparent spatiotopic auditory map, and either no visuotopic representation or a very crude one (see Mucke et al., 1982; Olson and Graybiel, 1987; Rauschecker and Korte, 1993). Consequently, the presence of cross-modality receptive field register in AES multisensory neurons was somewhat surprising.

Yet, a spatial correspondence among the different receptive fields of the same multisensory neuron appears to be a general feature of the vertebrate brain. It is apparent in the midbrain of both mammalian and non-mammalian species (Drager and Hubel, 1976; Knudsen, 1982; Hartline, 1984; Stein, 1984; King, 1993), as well as in various polysensory regions of rat (Ramachandran et al., 1993; Barth et al., 1995) and primate (Bruce et al., 1981; Duhamel et al., 1989; Watanabe and Iwai, 1991; Stein et al., 1992) cortex. It is likely that the widespread nature of this overlap among the receptive fields of multisensory neurons reflects the most efficient way to integrate and enhance the salience of information derived from the same event. Such enhancement is especially useful when the individual stimuli are weak and difficult to detect or ambiguous, and is reflected in the inverse relationship between the effectiveness of the unimodal stimuli and the magnitude of the multisensory enhancement produced when they are combined.

Because stimuli produced by the same environmental event generally originate from the same point in space, they also fall within the excitatory fields of the same multisensory neurons. When multisensory activity is enhanced in the SC, the likelihood of eliciting SC-mediated attentive and orientation behaviors is also increased (see Stein et al., 1989). In contrast, stimuli derived from two separate events are likely to produce one stimulus that falls outside its receptive field and is thus not involved in an interaction, or falls within an inhibitory surround and decreases the likelihood of an SC-mediated response. Apparently, active mechanisms are at work to ensure that compensatory changes occur in non-visual modalities (especially the auditory system) to maintained

intersensory receptive field alignment even when small movements of the eyes occur (Jay and Sparks, 1984, 1987a,b; Peck et al., 1995, Hartline et al., 1995). Presumably, this ensures that the spatial principles by which multisensory neurons enhance or depress sensory inputs remain largely intact.

In the AES, however, the larger excitatory receptive fields and the lower incidence and magnitude of surround inhibition indicate that multisensory spatial processing is less rigorous than in the SC. One might expect, then, that if this is reflective of multisensory cortex in general, some higher order functions are less dependent on the spatial alignment of stimuli than are the orientation functions of the SC.

Indeed, a number of multisensory interactions appear to be less influenced by the relative spatial positions of the two stimuli than would be expected based on the properties of SC neurons and of SC-mediated behavior. For example, in human subjects an auditory stimulus is capable of increasing the perceived intensity of a visual stimulus even when it is not at the same spatial location (Stein et al., 1996). Similarly, when a subject is required simply to detect and react to a visual stimulus, a non-coincident auditory stimulus can readily enhance performance (Watkins and Feeher, 1965), and will do so even when the two stimuli are in opposite hemifields (Hughes et al., 1994). A similar phenomenon is evident in speech perception, where localizing the source of speech is influenced by the relative spatial positions of auditory and visual cues, but the ability of the cues to influence the perceived sound is independent of their position (Fisher and Pylyshyn, 1994).

That different multisensory neuronal populations may carry on their tasks independently, and need not interact with one another directly, is especially evident by the differences between the multisensory neurons in the AES and SC. Despite the fact that they are activated concurrently by the same stimuli, AES multisensory neurons neither receive inputs from, nor send projections to, the SC (Wallace et al., 1993). The segregation of these two populations of multisensory neurons was surprising at first, given the observation that robust corticotectal projections originate from the AES (Stein et al., 1983). It is now evident, however, that this projection arises nearly exclusively from unimodal neurons (Wallace et al., 1993).

Our current knowledge of the physiological properties of the many different populations of multisensory neurons is rudimentary. Nevertheless, the present observations indicate that there are many similarities in the principles of multisensory integration that subserve the generation of immediate orientation behaviors, such as those mediated by the SC, and the modulation of 'higher order' cortical processes, such as those leading to perception and cognition. Undoubtedly, there will be specialized multisensory properties in some areas of the brain to underlie specialized functions. Nevertheless, we suggest that a fundamental core of multisensory properties, such as those found here to be shared by the SC and AES, will prove to be widespread in the central nervous system. This will ensure that a given multisensory event is either enhanced or degraded in many areas simultaneously, thereby maintaining perceptual and behavioral coherence across structures.

Acknowledgements

We thank Nancy London for her technical assistance. This work was supported by NIH grant NS 22543.

References

Andreassi, J.L. and J.R. Greco (1975) Effects of bisensory stimulation on reaction time and the evoked cortical potential. *Physiol. Psychol.* 3: 189–194.

Barth, D.S., Goldberg, N., Brett, B. and Di, S. (1995) The spatiotemporal organization of auditory, visual and auditory-visual evoked potentials in rat cortex. *Brain Res.* 678: 177–190.

Benedek, G., Fischer-Szatmari, L., Kovaks, G., Perenyi, J. and Katoh, Y.Y. (1996) Visual, somatosensory and auditory modality properties along the feline suprageniculate-AES-insular pathway. *Prog. Brain Res.*, in press.

Bernstein, I.H., Clark, M.H. and Edelstein, B.A. (1969) Effects of an auditory signal on visual reaction time. *J. Exp. Psychol.*, 80: 567–569.

Bertelson, P. and Radeau, M. (1981) Cross-modal bias and perceptual fusion with auditory-visual spatial discordance. *Percept. Psychophys.*, 29: 578–584.

Bruce, C.R., Desimone, R. and Gross, C.G. (1981) Visual properties of neurons in a polysensory area in superior temporal sulcus of monkey. *J. Neurophysiol.* 46: 369–384.

Clarey, J.C. and Irvine, D.R.F. (1986) Auditory response properties of neurons in the anterior ectosylvian sulcus of the cat. *Brain Res.*, 386: 12–19.

Clemo, H.R. and Stein, B.E. (1982) Somatosensory cortex: A 'new' somatotopic representation. *Brain Res.*, 235: 162–168.

Clemo, H.R. and Stein, B.E. (1983) Organization of a fourth somatosensory area of cortex in cat. *J. Neurophysiol.*, 50: 910–925.

Clemo, H.R., Meredith, M.A., Wallace, M.T. and Stein, B.E. (1991) Is the cortex of cat anterior ectosylvian sulcus a polysensory area? *Soc. Neurosci. Abstr.*, 17: 1585.

Drager, U.C. and Hubel, D.H. (1976) Topography of visual and somatosensory projections to mouse superior colliculus. *J. Neurophysiol.*, 39: 91–101.

Duhamel, J.-R., Colby, C.L. and Goldberg, M.E. (1989) Congruent visual and somatosensory response properties of neurons in the ventral intraparietal area (VIP) in the alert monkey. *Soc. Neurosci. Abstr.*, 15: 162.

Fidell, S. (1970) Sensory function in multimodal signal detection. *J. Acoust. Soc. Am.*, 47: 1009–1015.

Fisher, B.D. and Pylyshyn, Z.W. (1994) The cognitive architecture of bimodal event perception: a commentary and addendum to Radeau (1994). *Cahiers Psychol.*, 13: 92–96.

Frens, M.A. and van Opstal, A.J. (1995) Audio-visual integration in the superior colliculus of the awake monkey. *Soc. Neurosci. Abstr.*, 21: 1194.

Frens, M.A., van, A.J. Opstal and van der Willigen, R.F. (1995) Spatial and temporal factors determine audio-visual interactions in human saccadic eye movements. *Percept. Psychophys.*, 57: 802–816.

Gielen, S.C., Schmidt, R.A. and van der Heuvel, P.J.M. (1983) On the nature of intersensory facilitation of reaction time. *Percept. Psychophys.*, 34: 161–168.

Graybiel, A.M. (1972) Some ascending connections of the pulvinar and nucleus lateralis posterior of the thalamus in the cat. *Brain Res.*, 44: 90–125.

Hartline, P.H. (1984) The optic tectum of reptiles: neurophysiological studies. In: H. Vanegas (Ed.), *Comparative Neurology of the Optic Tectum*, Plenum, New York, pp. 601–618.

Hartline, P.H., Pandey Vimal, R.L., King, A.J. Kurylo, D.D. and Northmore, D.P.M. (1995) Effects of eye position on auditory localization and neural representation of space in superior colliculus of cats. *Exp. Brain Res.*, 104: 402–408.

Hershenson, M. (1962) Reaction time as a measure of intersensory facilitation. *J. Exp. Psychol.*, 63: 289–293.

Jay, M.F. and Sparks, D.L. (1984) Auditory receptive fields in primate superior colliculus shift with changes in eye position. *Nature*, 309: 345–347.

Jay, M.F. and Sparks, D.L. (1987a) Sensorimotor integration in the primate superior colliculus. I. Motor convergence. *J. Neurophysiol.*, 57: 22–34.

Jay, M.F. and Sparks, D.L. (1987b) Sensorimotor integration in the primate superior colliculus. II. Coordinates of auditory signals. *J. Neurophysiol.*, 57: 35–55.

King, A.J. (1993) A map of auditory space in the mammalian brain; neural computation and development. *Exp. Physiol.*, 78: 559–590.

King, A.J. and Palmer, A.R. (1985) Integration of visual and auditory information in bimodal neurones in the guinea-pig superior colliculus. *Exp. Brain Res.*, 60: 492–500.

Knudsen, E.I. (1982) Auditory and visual maps of space in the optic tectum of the owl. *J. Neurosci.*, 2: 1177–1194.

McGurk, H. and MacDonald, J. (1976) Hearing lips and seeing voices. *Nature*, 264: 746–748.

Meredith, M.A. and Stein, B.E. (1986a) Visual, auditory, and somatosensory convergence on cells in superior colliculus results in multisensory integration. *J. Neurophysiol.*, 56: 640–662.

Meredith, M.A. and Stein, B.E. (1986b) Spatial factors determine the activity of multisensory neurons in cat superior colliculus. *Brain Res.*, 365: 350–354.

Miller, J.O. (1982) Divided attention: Evidence for coactivation with redundant signals. *Cogn. Psychol.*, 14: 247–279.

Miller, J.O. (1986) Time course of coactivation in bimodal divided attention. *Percept. Psychophys.*, 40: 331–343.

Morrell, F. (1972) Visual system's view of acoustic space. *Nature*, 238: 44–46.

Mucke, L., Norita, M., Benedek, G. and Creutzfeldt, O. (1982) Physiologic and anatomic investigation of a visual cortical area situated in the ventral bank of the anterior ectosylvian sulcus of the cat. *Exp. Brain Res.*, 179: 1–11.

Munoz, D.P. and Corneil, B.D. (1995) Evidence for interactions between target selection and visual fixation for saccade generation in humans. *Exp. Brain Res.*, 103: 168–173.

Nickerson, R.S. (1973) Intersensory facilitation of reaction time: energy summation or preparation enhancement? *Psychol. Rev.*, 80: 489–509.

Olson C.R. and Graybiel, A.M. (1987) Ectosylvian visual area of the cat: location, retinotopic organization, and connections. *J. Comp. Neurol.*, 261: 277–294.

Peck, C.R., Baro, J.A. and Warder, S.M. (1995) Effects of eye position on saccadic eye movements and on the neural responses to auditory and visual stimuli in cat superior colliculus. *Exp. Brain Res.*, 103: 227–242.

Perrot, D.R., Saberi, K., Brown, K. and Strybel, T.Z. (1990) Auditory psychomotor coordination and visual search performance. *Percept. Psychophys*, 48: 214–226.

Posner, M.I., Nissen, M.J. and Klein, R.M. (1976) Visual dominance: an information-processing account of its origins and significance. *Psychol. Rev.*, 83: 157–171.

Raab, D. (1962) Statistical facilitation of simple reaction times. *Trans. N.Y. Acad. Sci.*, 24: 574–590.

Ramachandran, R., Wallace, M.T., Clemo, H.R. and Stein, B.E. (1993) Multisensory convergence and integration in rat cortex. *Soc. Neurosci. Abstr.*, 19: 1447.

Rauschecker, J.P. and Korte, M. (1993) Auditory compensation for early blindness in cat cerebral cortex. *J. Neurosci.*, 13: 4538–4548.

Reinoso-Suarez, F. and Roda, J.M. (1985) Topographic organization of the cortical afferent connections to the cortex of the anterior ectosylvian sulcus in the cat. *Exp. Brain Res.*, 59: 313–324.

Reisberg, D., Scheiber, R. and L. Potemken (1981) Eye position and the control of auditory attention. *J. Exp. Psychol.: Human Percept. Perf.*, 7: 318–323.

Roda J.M. and Reinoso-Suarez, F. (1983) Topographical organization of the thalamic projections to the cortex of the anterior ectosylvian sulcus in the cat. *Exp. Brain Res.*, 49: 131–139.

Stein, B.E. (1984) Multimodal representation in the superior colliculus and optic tectum. In: H. Vanegas (Ed.), *Comparative Neurology of the Optic Tectum*, Plenum, New York, pp. 819–841.

Stein, B.E. and Meredith, M.A. (1993) *The Merging of the Senses*, MIT Press, Cambridge, MA.

Stein, B.E., Magalhaes-Castro, B. and Kruger, L. (1975) Superior colliculus: visuotopic-somatotopic overlap. *Science*, 189: 224–226.

Stein B.E., Magalhaes-Castro, B. and Kruger, L. (1976) Relationship between visual and tactile representations in cat superior colliculus. *J. Neurophysiol.*, 39: 401–419.

Stein, B.E., Spencer, R.F. and Edwards, S.B. (1983) Corticotectal and corticothalamic efferent projections of SIV somatosensory cortex in cat. *J. Neurophysiol.*, 50: 896–909.

Stein, B.E., Meredith, M.A., Huneycutt, W.S. and McDade, L. (1989) Behavioral indices of multisensory integration: orientation to visual cues is affected by auditory stimuli. *J. Cogn. Neurosci.*, 1: 12–24.

Stein, B.E., Meredith, M.A. and Wallace, M.T. (1993) The visually responsive neuron and beyond: Multisensory integration in cat and monkey. *Prog. Brain Res.*, 95: 79–90.

Stein, B.E., London, N., Wilkinson, L.K. and Price, D.D. (1996) Enhancement of perceived visual intensity by auditory stiumuli: a psychophysical analysis. *J. Cogn. Neurosci.*, in press.

Watanabe, J. and Iwai, E. (1991) Neuronal activity in visual, auditory and polysensory areas in the monkey temporal cortex during visual fixation task. *Brain Res. Bull.*, 26: 583–592

Wallace, M.T. and Stein, B.E. (1994) Cross-modal synthesis in the midbrain depends on input from association cortex. *J. Neurophysiol.*, 71: 429–432.

Wallace, M.T., Meredith, M.A. and Stein, B.E. (1992) Integration of multiple sensory modalities in cat cortex. *Exp. Brain Res.*, 91: 484–488.

Wallace, M.T., Meredith, M.A. and Stein, B.E. (1993) Converging influences from visual, auditory, and somatosensory cortices onto output neurons of the superior colliculus. *J. Neurophysiol.*, 69: 1797–1809.

Wallace, M.T., Wilkinson, L.K. and Stein, B.E. (1996) Representation and integration of multiple sensory inputs in primate superior colliuclus. *J. Neurophysiol.*, in press.

Watkins, W.H. and Feehrer, C.E. (1965) Acoustic facilitation of visual detection. *J. Exp. Psychol.*, 70: 332–333.

Welch, R.B. and Warren, D.H. (1986) Intersensory interactions. In: K.R. Boff, Kaufman, L. and Thomas, J.P. (Eds.), *Handbook of Perception and Human Performance*, Vol. I, *Sensory Processes in Perception*, John Wiley, New York, pp. 1–36.

Wickelgren, B.G. and Sterling, P. (1969) Influence of visual cortex on receptive fields in the superior colliculus of the cat. *J. Neurophysiol.*, 32: 16–23.

Zahn, J.R., Abel, L.A. and L.F. Dell'Osso (1978) Audio-ocular response characteristics. *Sens. Proc.*, 2: 32–37.

M. Norita, T. Bando and B. Stein (Eds.)
Progress in Brain Research, Vol 112
© 1996 Elsevier Science BV. All rights reserved.

CHAPTER 21

Sensory organization of the superior colliculus in cat and monkey

Mark T. Wallace* and Barry E. Stein

Department of Neurobiology and Anatomy, Bowman Gray School of Medicine / Wake Forest University, Winston-Salem, NC 27157-1010, USA

The sensory representations and response properties of neurons in the multisensory layers of the superior colliculus (SC) of the cat and monkey are strikingly similar. In both species, significant numbers of unimodal visual, auditory, and somatosensory neurons are intermixed, and share modality-specific stimulus selectivities. In addition, these neurons have receptive fields that are organized in similar map-like representations, with the different maps sharing the same axes and, thus, exhibiting a characteristic cross-modality correspondence. Both species also have a large contingent of multisensory neurons that synthesize information from different sensory modalities using a common set of spatial, temporal, and physical principles. These findings suggest that, despite differences in the ontogeny and behavioral repertoires of cat and monkey, there is a substantial conservation in the organization and functional properties of SC neurons. Apparently, these functional features of the SC are adaptive in a wide variety of ecological situations.

Introduction

The superior colliculus (SC) plays a major role in directing attentive and orientation movements (Sprague and Meikle, 1965; Schneider, 1969; Casagrande et al., 1972; Stein, 1984; Sparks, 1986; Stein and Meredith, 1993). It receives sensory inputs from visual, auditory and somatosensory modalities (Stein, 1984; Wallace et al., 1993), and has direct connections to brainstem and spinal cord regions which control movements of the eyes, ears, and head (Grantyn and Grantyn, 1982; Huerta and Harting, 1984; Meredith and Stein, 1985; Moschovakis and Karabelas, 1985; Vidal et al., 1988; Meredith et al., 1992; Wallace et al., 1993). The behavioral role of the SC is markedly facilitated by its access to multiple sensory channels because of the increase in the probability that at least one of these channels will be activated by, and carry sufficient information to detect and localize, an important environmental event. However, the SC also contains a population of very specialized neurons. These 'multisensory' neurons, which have been most extensively studied in the cat, are not only capable of responding to sensory inputs from more than one modality, but are also able to synthesize these inputs and thereby enhance the salience of the signals they carry. The properties of these neurons and the contributions they are believed to make to overt behavior have been reviewed by Stein and Meredith (1993).

Despite a wealth of information concerning the sensory/multisensory and integrative capabilities of cat SC neurons, surprisingly little is known about these features in the SC of other animals. In primates, numerous studies have examined the visual and premotor properties of SC neurons (Schiller and Koerner, 1971; Cynader and Berman,

* Corresponding author.

1972; Goldberg and Wurtz, 1972; Robinson, 1972; Schiller and Stryker, 1972; Wurtz and Goldberg, 1972; Updyke, 1974; Mohler and Wurtz, 1976; Marrocco and Li, 1977; Albano et al., 1978; Sparks, 1978; Moors and Vendrik, 1979; Jay and Sparks, 1984, 1987a,b), but little characterization of the non-visual and multisensory properties of its constituent neurons has been done. Perhaps it is because the monkey SC has been used so extensively as a model to understand the visuomotor properties of its neurons that our ignorance of SC sensory and multisensory properties seems so notable.

Recently, however, studies of the sensory and multisensory properties of the rhesus monkey have been conducted using the same methods and procedures as in the cat (Wallace et al., 1996). This strategy allows systematic comparisons among species, which are vital in assessing the value of any single neural model. The present report makes such comparisons directly by drawing on a number of observations in which the same visual, auditory, somatosensory or multisensory properties were examined in cat and monkey using equivalent methods and procedures. Some of the data necessary for these comparisons are drawn from prior investigations (e.g. see Wallace and Stein, 1994; Wallace et al., 1993, 1996), but many of the observations have not been reported elsewhere.

Materials and methods

Data were acquired from a total of four adult rhesus monkeys (*Macaca mulatta*) and four adult cats (*Felis domesticus*). Unless otherwise stated, stimulus, recording and analysis procedures were identical for the cat and monkey. However, for practical reasons, the monkey data were acquired in a single acute recording session from each animal, while the cat data were acquired during a series of chronic recordings. All studies were carried out in accordance with NIH guidelines (publication 86-23) and were also in compliance with the guidelines of the Bowman Gray School of Medicine/Wake Forest University Institutional

Animal Care and Use Committee. This institution is accredited by the American Association for the Accreditation of Laboratory Animal Care. The methods used were similar to those described by Wallace et al. (1993, 1996) and are more briefly summarized here.

Surgical preparation

Anesthesia was induced with ketamine hydrochloride (10 mg/kg). During surgery, anesthesia was maintained with halothane (2–4%) and body temperature was maintained with a heating pad. A craniotomy exposed the cortex overlying the SC, and a hollow cylinder/head mount was affixed to the skull (McHaffie and Stein, 1983) to support it and thereby avoid obstructing the eyes, pinnae, face, or body surface.

Recording

Anesthesia during recording was maintained with either halothane (0.5–1.0%) or ketamine hydrochloride (5–10 mg/kg per h; i.v.). No differences in response properties were noted between these anesthetics. Fluids (lactated Ringers; 4–6 c^3/h), paralytics (pancuronium bromide; 0.5–1.0 mg/kg per h) and anesthetic were delivered intravenously. The animal was respired and end tidal CO_2 was maintained between 3.6 and 4.2%. EEG, heart rate, salivation, and lacrimation were monitored. The optic discs were mapped onto a translucent hemisphere on which receptive fields were also mapped, and contact lenses were applied to correct refractive errors. Neural responses were recorded with tungsten microelectrodes (1–3 $M\Omega$ at 1 kHz).

Sensory classification

Visually-responsive neurons were sought using a variety of moving and stationary flashed stimuli. Bars and spots of light (53 cd/m^2 against a background of 2.7 cd/m^2) could be moved in all directions across the receptive field. Somatosensory-responsive neurons were sought using mechanical taps, strokes with a camel's hair brush, manual compression of the skin, rotation of joints, and thermal stimuli. Cutaneous receptive fields were studied quantitatively using computer-controlled mechanical stimuli delivered from a probe tip mounted to a modified moving-coil vibrator (Ling 102A shaker). Auditory-responsive neurons were identified by their responses to a variety of complex auditory stimuli that included hisses, clicks, claps, whistles, and broad-band (200–20 000 Hz) noise bursts. They were categorized with computer-controlled broad-band sound bursts delivered from a set of hoop-mounted speakers positioned 15 cm from each ear.

Multisensory tests

A multisensory neuron was defined as one that responded to stimuli from more than one modality, or whose responses to one modality were altered by the presence of a stimulus from a second modality. Responses to each 'single-modality' stimulus alone and to the multisensory combination (e.g. visual-auditory) were quantitatively determined by presenting the stimulus or stimulus combination in the relevant fields 8–16 times at 10–30-s intertrial intervals, with the different conditions interleaved. Generally, during multisensory trials the two sensory stimuli were presented simultaneously or within 50–100 ms of one another.

Data acquisition and analysis

Neuronal responses were assessed by determining the number of impulses evoked and by calculating the mean numbers of impulses, standard deviations and standard errors of the mean. Responses were analyzed statistically to determine whether a significant (two-tailed t-test, $P < 0.05$) change (increase, enhancement; decrease, depression) in the number of impulses occurred with combined stimuli when compared to the most effective single-modality stimulus.

Results

Sensory convergence patterns

The present report focuses on the deep layers of the SC (stratum griseum intermediale and below), which, like the overlying superficial layers, contain unimodal visual neurons. However, deep layers also contain many unimodal auditory and somatosensory neurons, as well as 'multisensory' neurons capable of responding to all of these sensory modalities. All possible combinations of multisensory neurons were encountered, and their distributions in cat and monkey are shown in Fig. 1. Despite the mixture of modality representations in the SC, it is apparent from these distributions that visually-responsive neurons are particularly well-represented. In cat 75% and in monkey 63% of all sensory-responsive neurons responded to visual stimuli. Although the visual receptive fields of multisensory neurons were larger than their unimodal counterparts (see below), their visual properties were virtually indistinguishable from one another, so visually-responsive neurons will be treated here as a single population.

Visual response properties

The visual properties of cat and monkey SC neurons were quite similar in their high degree of binocularity, tendency to habituate to repetitions of the identical stimulus, incidence of direction selectivity, and spatial summation/inhibition (Table 1). Similarly, the majority of these neurons in both species (cat, 75%; monkey, 81%) responded exclusively to moving stimuli, with a substantial proportion (cat, 49%; monkey, 44%) preferring movement at specific speeds. Such velocity selectivity, and the population distributions, are shown in Fig. 2. As is evident from this figure, neurons with broadly tuned velocity profiles comprised the majority (cat, 51%; monkey, 56%) of these populations. Even the response latencies were similar in the two species. In cat, mean response latency

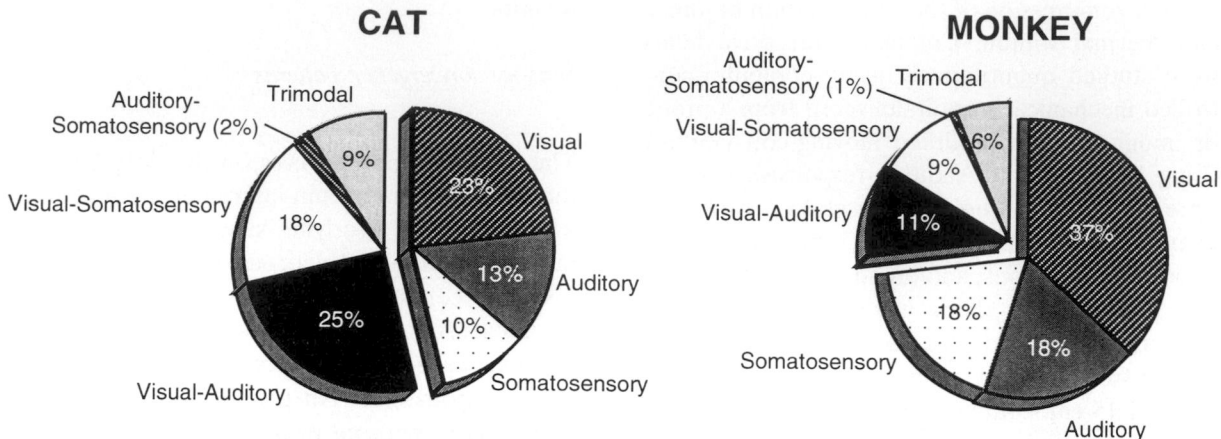

Fig. 1. Distribution of sensory convergence patterns in the deep SC of cat and monkey. Portions of the figure adapted from Wallace et al. (1996). Adapted by permission.

was 91 ms, with a range of 58–140 ms; in monkey, the mean value was slightly greater (97 ms), but the range (68–132 ms) was similar.

One characteristic of the transition from superficial to deep layers is a substantial increase in visual receptive field size. This was observed in both species, but was most dramatically apparent in monkey, where superficial receptive fields representing foveal visual space could be as small as 0.3° in diameter while corresponding receptive fields in deep layers were typically 10–20° in diameter. In cat, receptive fields representing the area centralis were approximately 5° in diameter in the superficial layers versus 15–25° in the deep layers. In both species, receptive fields became progressively larger at sites representing more temporal portions of the visual field and multisensory receptive fields were generally larger than unimodal receptive fields at the same visual eccentricities. Nevertheless, the superficial/deep laminar contrasts in receptive field size were noted throughout the visual field.

Auditory response properties

In cat, slightly less than half (49%), and in monkey somewhat more than one-third (36%), of the neu-

rons responded to auditory stimuli. Like their visual counterparts, multisensory auditory receptive fields proved to be larger than unimodal receptive fields, but their response characteristics were fundamentally the same and are, therefore, treated here as a single population.

Auditory-responsive SC neurons exhibited little frequency specificity and responded best to broad-band sound bursts in both species. The latency to a contralateral stimulus was short, with a mean latency in cat of 14 ms (range: 8–40 ms), and in monkey of 18 ms (range: 8–44 ms). Binaural neurons were substantially more common than monaural neurons in both species (cat, 77%; monkey, 85%).

Auditory receptive fields, like the visual receptive fields described above, varied widely in size, with unimodal neurons having significantly smaller receptive field diameters than multisensory neurons (cat: interaural extent of receptive field in unimodal neurons, 43°; multisensory, 77°; monkey: unimodal, 47°; multisensory, 65°). Receptive fields increased substantially in size as progressively more caudal SC sites were sampled. Whereas receptive fields in the rostral SC of cat generally represented a circumscribed region of frontal auditory space 30–45 in azimuthal extent, receptive fields in the caudal SC were substantially larger,

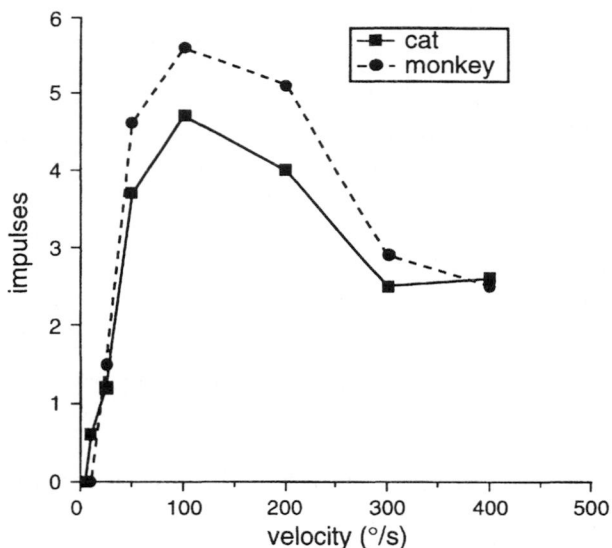

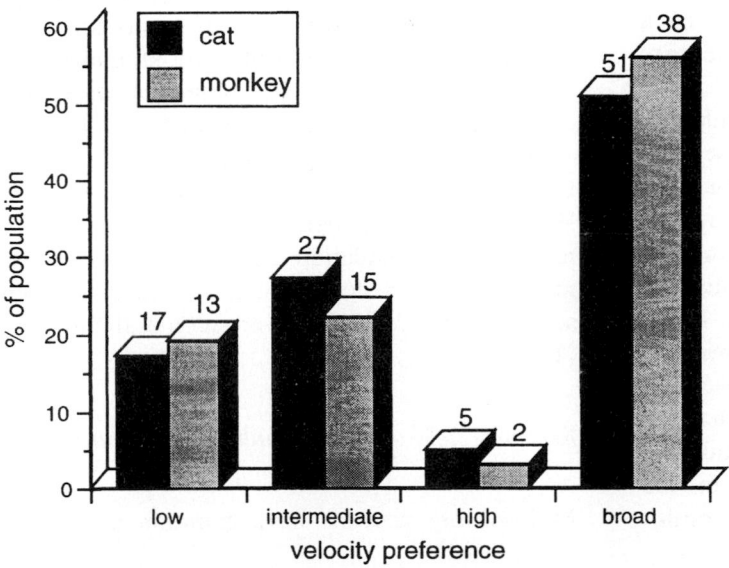

Fig. 2. Velocity selectivity of visually-responsive SC neurons in cat and monkey. Top panel shows profiles for one representative neuron from cat and one from monkey. Both neurons preferred stimuli moving at intermediate velocities (optimal velocity, 100°/s). Bottom panel shows the population distribution for both species. Neurons are classified as having a preference for low ($< 20°/s$), intermediate ($20-150°/s$) or high ($> 150°/s$) stimulus velocities. However, the largest population were broadly-tuned neurons, showing little specificity for stimulus velocity. Note the similarities in the distributions for the two species. Values above bars indicate number of neurons.

TABLE 1

Visual response properties in cat and monkey SC

	Response to flash	Habituation	Direction selectivity	Spatial summation	Spatial inhibition	Binocularity
Cat	25% (31/122)	70% (80/115)	29% (31/107)	20% (21/103)	10% (11/105)	89% (63/71)
Monkey	19% (13/68)	76% (38/50)	15% (10/68)	25% (14/56)	16% (9/56)	93% (42/45)

generally subtending greater than 90° of azimuth. A similar tendency was seen in monkey.

Somatosensory response properties

In both species, somatosensory-responsive neurons comprised more than one-third of the deep-layer sensory population (cat, 39%; monkey, 34%). As in the case of visually-responsive and auditory-responsive neurons, the absence of a difference between unimodal and multisensory neurons allowed somatosensory-responsive neurons to be dealt with here as a single population.

Once again species differences in sensory properties were comparatively small, and somatosensory-responsive SC neurons in cat and monkey had many of the same fundamental features (Table 2). Thus, their response latencies were very similar (cat: mean, 23 ms, range 10–40 ms; monkey: mean, 26 ms, range 12–65 ms), receptive fields were generally restricted to the contralateral body surface, were smallest on the face and forelimb, and were largest on the trunk and hindlimb. Consistent with observations in the other modalities, somatosensory-responsive multi-sensory neurons had significantly larger receptive fields than their unimodal counterparts in both cat (84 cm^2 vs. 37 cm^2) and monkey (361 cm^2 vs. 131 cm^2). Similarly, in both species, high threshold mechanical and thermal stimuli were ineffective, with the majority (cat, 73%; monkey, 84%) of neurons best driven by high velocity (> 100 mm/s) deflections of the hair and/or skin (see Table 2 and Fig. 4). In general, responses to such a stimulus were transient, and a sustained response could be elicited only by a stimulus moving continuously across the cutaneous surface.

A difference between cat and monkey somatosensory neurons was found in the distribution of receptor types represented in the SC. Whereas in cat, nearly a quarter of the somatosensory population responded selectively to manipulation of deep tissue, comparatively few such neurons (2/37; 5%) were seen in the monkey (Table 2).

Receptive field overlap in multisensory neurons

Multisensory neurons were common in cat and

TABLE 2

Somatosensory response properties in cat and monkey SC

	Velocity preference		Adaptation		Receptor type		
	Low	High	Slow	Rapid	Hair	Skin	Deep
Cat	27% (15/56)	73% (41/56)	7% (4/58)	93% (54/58)	45% (27/60)	32% (19/60)	23% (14/60)
Monkey	16% (6/37)	84% (31/37)	0% (0/37)	100% (37/37)	65% (24/37)	30% (11/37)	5% (2/37)

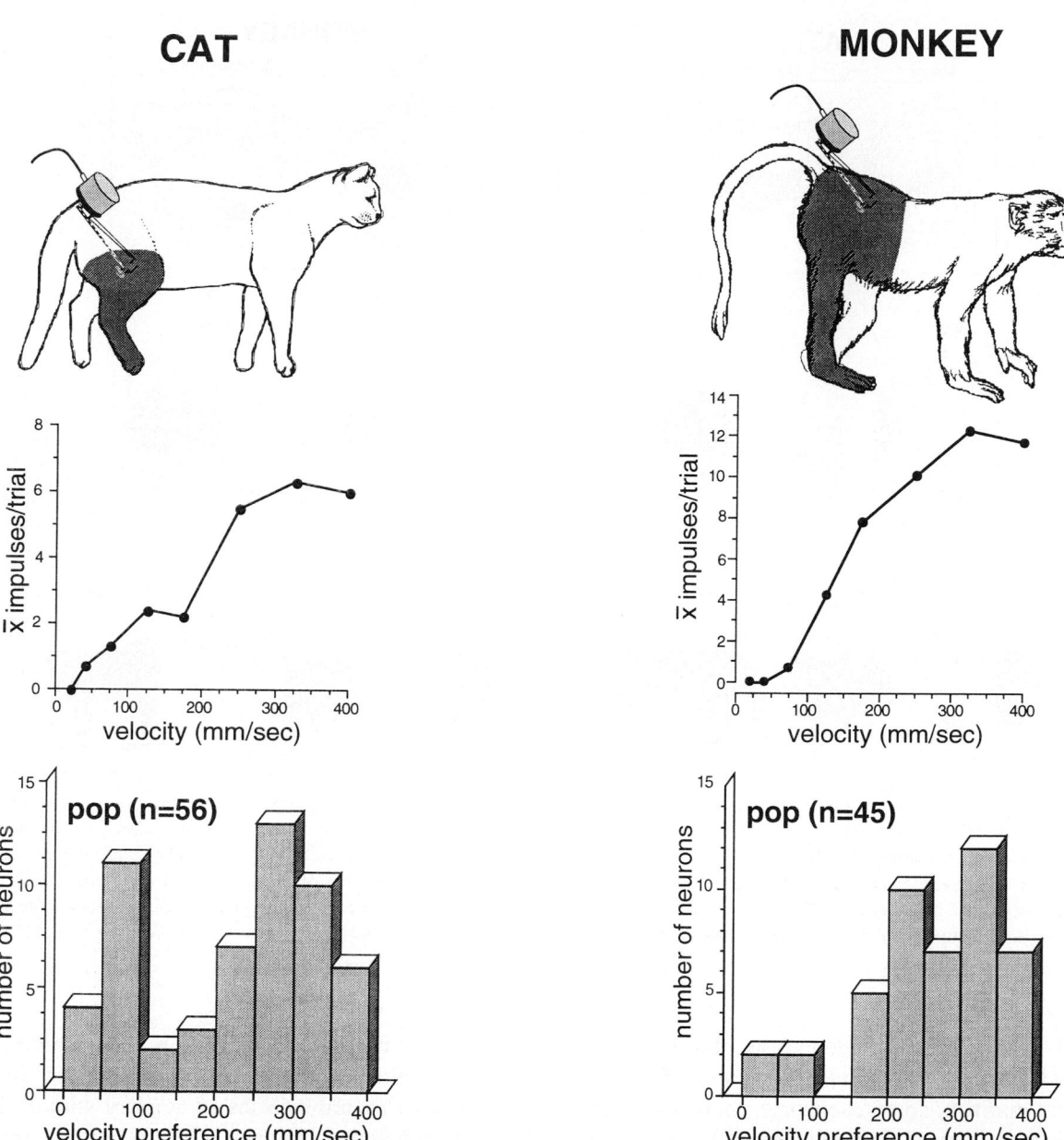

Fig. 3. Velocity selectivity in somatosensory-responsive SC neurons in cat and monkey. Top panels show receptive fields (shading) and the probe used in studying velocity selectivity for a representative neuron in each species. The response profiles of these neurons are shown in the middle panels, where mean response is plotted as a function of stimulus velocity. Bottom panels show the similar velocity preferences for somatosensory-responsive populations in both species. Portions of the figure adapted from Wallace et al. (1996). Adapted by permission.

monkey (Fig. 1). In cat, such neurons made up over half (54%) of the deep-layer sensory population, and in monkey such neurons comprised

greater than one-quarter (27%) of this population. In nearly all cases the different receptive fields of a given multisensory neuron exhibited

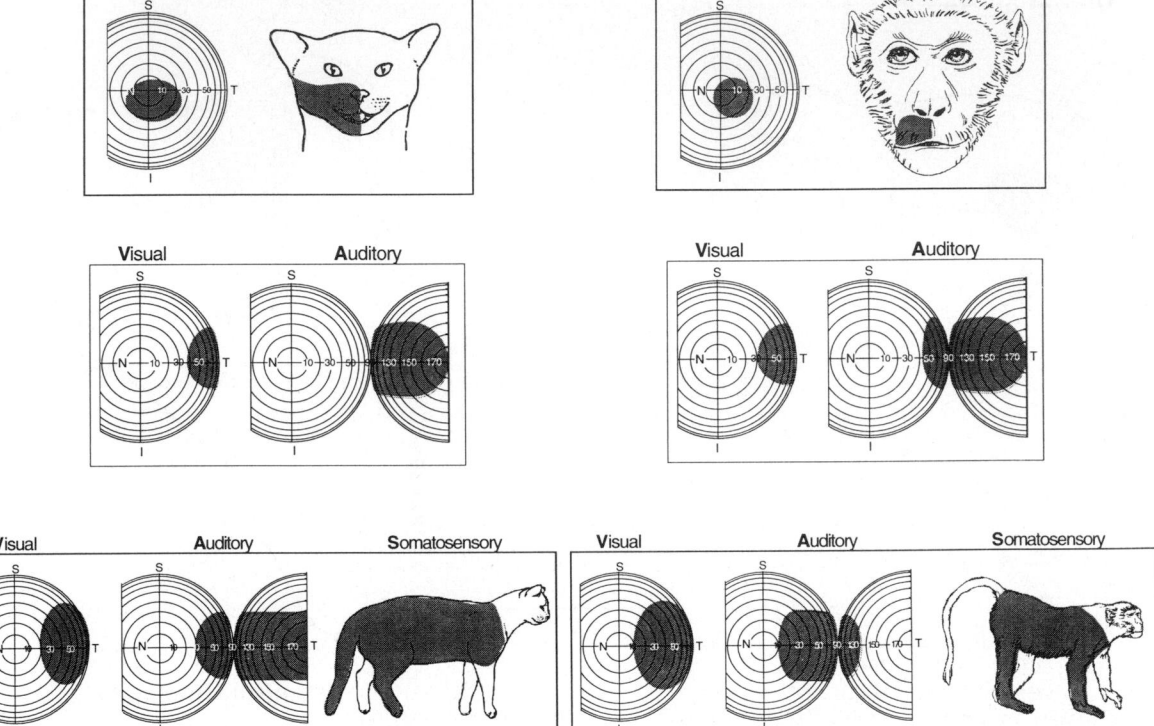

Fig. 4. Receptive field correspondence in multisensory neurons in cat and monkey SC. A representative visual-somatosensory, visual-auditory and trimodal neuron is shown for each species. Receptive fields are shown in shading. Auditory receptive fields are plotted on a representation of auditory space in which the sphere represents space in front of the interaural plane, and the attached hemisphere represents space behind this plane on the contralateral side. The joining point of the sphere and hemisphere is along the interaural line (0° elevation, 90° azimuth). Ipsilateral caudal space is not shown. In each of the multisensory neurons, note the good cross-modality spatial correspondence between the receptive fields. Portions of the figure adapted from Wallace et al. (1996). Adapted by permission.

good spatial correspondence, as shown by the bimodal examples illustrated in Fig. 5. Trimodal (visual-auditory-somatosensory) neurons were similarly organized. However, given the large size of their receptive fields, the spatial fidelity of their intersensory correspondence was less compelling.

Multisensory integration

Perhaps most apparent from the various analyses conducted here was the fundamental similarity in the nature of multisensory integration in SC neu-

rons. Sixty-two neurons (cat: $n = 44$; monkey: $n = 18$) were examined quantitatively for their responses to combinations of sensory stimuli. In the overwhelming majority of these neurons (cat: 37/44 or 84%; monkey: 15/18 or 83%) substantial differences between responses to unimodal and multimodal stimuli were noted. In the examples shown in Fig. 6, a modest response to each of the unimodal stimuli is seen. However, when the two stimuli were combined, the neuron's response was significantly enhanced, and far exceeded a simple sum of the two unimodal responses. In contrast, the combination of two stimuli from the

CAT

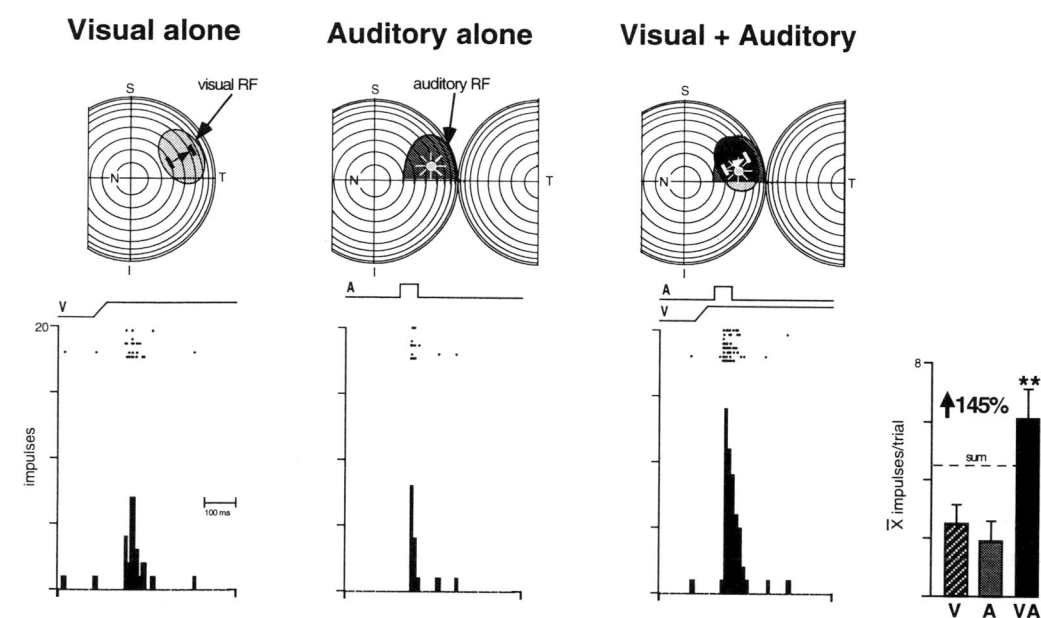

Visual alone   **Auditory alone**   **Visual + Auditory**

MONKEY

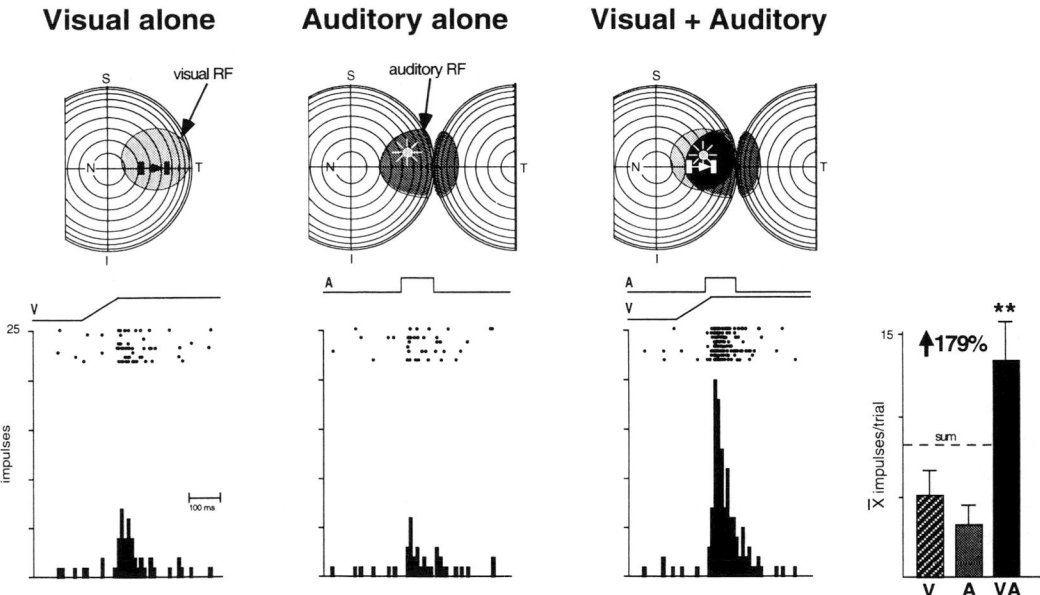

Visual alone   **Auditory alone**   **Visual + Auditory**

Fig. 5. Multisensory enhancement in cat and monkey. The receptive fields and responses of a representative visual-auditory neuron are shown for each species. Plotted on the top are the neurons' receptive fields and the location of the stimuli within these fields. Visual stimuli were moving bars of light and auditory stimuli were broad-band noise bursts delivered from a stationary speaker. The region of receptive field overlap is shown in black. Bottom panels are made up of rasters, histograms and bar graphs showing each neuron's response to unimodal and multisensory stimuli. In both examples, modest responses are elicited when the unimodal

Continued overleaf

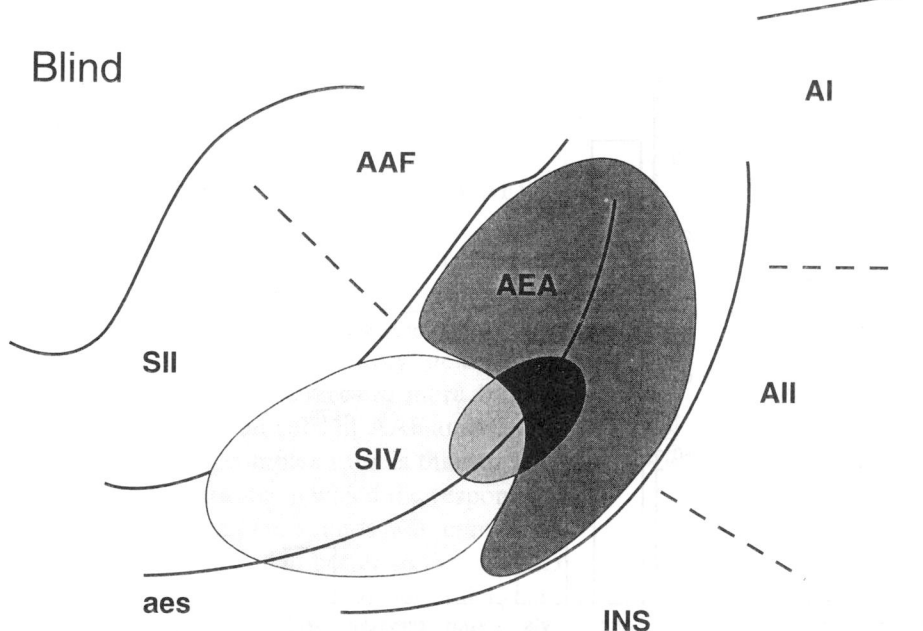

Blind

AI

AAF

AEA

SII

AII

SIV

aes

INS

Fig. 5. Expansion of auditory and somatosensory representations at the expense of visual area AEV in the AES region of visually deprived cats (compare to Fig. 1B in normal cats). The expansion was derived from careful track reconstructions in four visually deprived cats whose AES region was completely mapped by semichronic recording (Rauschecker, 1995).

mal controls, were again excluded from the analysis. Electrode track reconstructions from BD cats differed from those of normal cats in several respects. First, visual responses were significantly reduced. Few, if any, cells with a purely visual response were present in the ventral bank and fundus of the AES or elsewhere. Any visual neurons that were still encountered in this area virtually always had an additional auditory input.

While the lack of visual responsiveness was not totally unexpected in view of the severely reduced responsiveness in the striate cortex of BD cats (Wiesel and Hubel, 1965; Singer and Tretter, 1976), the second major finding did come as a surprise to us: unlike in striate cortex, only very few unresponsive neurons were encountered. Rather, in the same region of the AE cortex where visual cells had been found in control cats, auditory and, to a lesser extent, somatosensory responses were now abundant. In addition, the amount of multi-modal convergence was significantly higher. The results from individual BD cats

are displayed in Fig. 4B. Altogether, 64 out of 303 neurons (21.1%) were unresponsive (NCR), 77.8% (186/239) gave unimodal and 22.2% (53/239) gave bimodal responses.

Compared to normal controls, BD cats have (i) a smaller proportion of visually responsive neurons ($\chi^2 = 183.8$, $P < 0.0001$, χ^2-test), (ii) an increased proportion of auditory neurons ($\chi^2 = 21.8$, $P < 0.0001$), (iii) a significant proportion of neurons with somatosensory responses ($\chi^2 = 20$, $P < 0.0001$), (iv) many more bimodal units, especially such with an auditory input ($\chi^2 = 44.7$, $P < 0.0001$), and (v) an unchanged proportion of unresponsive neurons ($\chi^2 = 1.9$, $P > 0.1$).

Multimodality in the AES region

The principal finding of our studies was that in a region of cat cortex in which visual responses normally predominate, area AEV in the anterior ectosylvian (AE) sulcus (Mucke et al., 1982; Olson and Graybiel, 1983), most neurons become briskly responsive to non-visual stimuli after long-term binocular deprivation (BD) from birth. Furthermore. auditory neurons in the AE region of BD cats were more sharply tuned to the spatial location of a sound source than in AE of normal controls (Korte and Rauschecker, 1993). Finally, a higher proportion of bimodal neurons is found in AES of BD animals. This raises questions about the extent of overlap that normally exists in this part of cortex between inputs from different sensory modalities and the degree of convergence onto single neurons.

The issue of multisensory convergence in the AES region has been a matter of some controversy in the past. While it is undisputed that neurons responsive to different sensory modalities can be found in neighboring regions within the AES, it is not clear if these different modalities exist (i) in separate areas (Mucke et al., 1982; Clemo and Stein, 1983; Meredith and Clemo, 1989; Olson and Graybiel, 1983, 1987), (ii) in largely unimodal clusters within the same areas (Clarey and Irvine, 1986, 1990a; Wallace et al., 1992), or (iii) as single neurons with multimodal input (Berman, 1961; Minciacchi et al., 1987; Jiang et al., 1990). Our data from normal animals seem to support the first two possibilities (predominantly unimodal areas or neurons), but do not exclude input from other modalities that remains sub-threshold under our recording conditions. Indeed, some of the studies that found high proportions of multimodal or intermingled unit responses in the AES region used chloralose anesthesia, which is known to reveal subliminal input. In addition, different electrodes or recording techniques may contribute to the discrepancy in results. It is important to note, however, that the difference between normal and BD cats in our study cannot be explained by any of these factors, because identical techniques were used in all animals.

Mechanisms of sensory substitution

The existence of sub-threshold input could provide a convenient mechanism for an increase of non-visual and multimodal responsiveness in the AES region of visually deprived cats, because it would require only a modification of existing synaptic input. Such changes in the efficacy of Hebb-type synaptic connections have been postulated previously as the basis for developmental plasticity in primary visual cortex (see Rauschecker, 1991, for review). It appears conceivable that similar rules could allow plastic changes throughout neocortex during early postnatal development, or even beyond (Merzenich et al., 1990; Kaas, 1991; Recanzone et al., 1993). Depending on sensory-driven activity, certain inputs would be maintained or strengthened, while others would recede or fail to develop. It is entirely possible that the amount of multimodal convergence changes even during normal development: while a kitten may be born with a high amount of multimodal convergence, predominantly unimodal territories may get carved out by crossmodal competition later. In such a model, visual deprivation would disturb the input balance to the AES region during development and provide a better chance for the survival of existing non-visual inputs.

As an alternative to this notion, the cross-modal intrusion of formerly visual territory by functional non-visual input, i.e. sprouting, has to be considered (Darian-Smith and Gilbert, 1994). At first sight, the following would seem to speak against this possibility: we found a significant number of brisk somatosensory responses in this rather caudal part of the AES region (between A10 and A13) in BD cats. Generally, somatosensory neurons have been reported for more rostral positions, between A15 and A18 (Carreras and Andersson, 1963; Clemo and Stein, 1983). Any growth of somatosensory axons, therefore, would have to occur over a distance of several millimeters. However, comparable within-modality expansions have been reported recently in the somatosensory cortex of adult monkeys after deafferentation (Pons et al., 1991). The possibility of a cross-modal invasion of AEV by sprouting from neighboring regions as a result of visual deprivation, therefore, cannot be discounted.

No increase in the proportion of unresponsive neurons was found in BD cats, nor did we encounter any silent areas along the electrode tracks. This suggests that, regardless of which of the above two mechanisms is responsible, existing visual inputs were replaced or supplemented by non-visual inputs as a result of cross-modal competition. In striate cortex, the proportion of visually unresponsive neurons is vastly increased after long-term binocular eyelid suture (Wiesel and Hubel, 1965; Singer and Tretter, 1976). There is no evidence from animal studies that any of these neurons become responsive to non-visual stimuli. The changes in AE cortex described in the present study are, therefore, less likely to originate at these more peripheral levels. On the other hand, activation of primary visual cortex by somatosensory stimuli during Braille reading has recently been reported in early-blind humans (Pons, 1996; Sadato et al., 1996). Such effects could be subserved by the stabilization of transitory projections to visual cortex from other areas, as they have been reported in neonatal cats (Innocenti and Clarke, 1984).

Other cases of cross-modal plasticity have been reported only after drastic interventions. If their normal target structures are surgically removed, optic tract fibers can be rerouted to auditory or somatosensory thalamus, rendering neurons there and in the corresponding cortices responsive to visual stimuli (Métin and Frost, 1989; Sur et al., 1990).

Sensory substitution in blind humans

Recent results from blind humans correspond extremely well with the findings in visually deprived cats and confirm the validity of that animal model for studies of early blindness and compensatory plasticity (see Neville, 1990, and Rauschecker, 1995, for recent reviews). Not only do visual cortical areas in blind humans get activated by nonvisual stimuli (Veraart et al., 1990; Kujala et al., 1992; Pascual-Leone et al., 1993; Rösler et al., 1993; Uhl et al., 1993; Sadato et al., 1996), which mirrors the expansion of non-visual into visual territories observed in visually deprived cats. It has also been shown unequivocally now in a large number of subjects that blind humans do show superior capacities in non-visual modalities, such as sound localization (Muchnik et al., 1991), which had long been conjectured (Griesbach, 1899; Rice, 1970; Juurmaa and Suonio, 1975; Warren, 1978; Bagdonas et al., 1980; Niemeyer and Starlinger, 1981), but had also been disputed (Axelrod, 1959; Fisher, 1964; Spigelman, 1972). Again, this result matches that in visually deprived cats of an improved sound localization capacity (Rauschecker and Kniepert, 1994). The finding, both in cats and humans, as well as previously in monkeys (Hyvarinen et al., 1981; Carlson et al., 1987), that brain regions designed for the processing of vision can become activated by auditory or somatosensory stimuli shows the extreme plasticity of the brain in adapting to changes in its environment (see Rauschecker, 1995, for a summary of these results). However, these findings also pose an interesting philosophical question: what is the kind of percept that a blind individual experiences when a 'visual' area becomes activated by an auditory or tactile stimulus? Do blind individuals 'see' their environment with their tactile senses, as has previously been suggested by the term 'facial vision' (Supa et al., 1944)? Do they 'see' sounds in ways similar to a sonar system (Kellogg, 1962)? Or does the visual area simply get transformed into an auditory or somatosensory representation by the new type of input? To phrase it differently, is the percept determined by the type of sensory input or by the (functionally preordained) brain region that receives it? Undoubtedly, an auditory stimulus will still be perceived as a sound by blind individuals, because primary auditory regions are also activated. However, does the coactivation of 'visual' regions add anything to the quality of this sound that is not normally perceived, or does the expansion of auditory territory simply enhance the acuteness of perception for auditory stimuli?

A common code for sensorimotor integration

If we consider the projection targets rather than the inputs of a brain region, the answer to the above questions may perhaps be found more easily. For a behavioral reaction to a particular stimulus to be adequate, the same code has to be used at the interface, regardless of sensory modality (Sparks, 1986). Thus, a cortical module at any processing level applies the same type of operation to different sets of input and transforms them into a specific response. In the case of AES, input from different sensory modalities arrives in the same cortical region before being relayed on to the SC, where the information converges onto single neurons (Wallace et al., 1992; Wallace and Stein, 1994).

Neighboring cortex areas presumably share particular functional aspects, partly defined by their common projection targets. In cat AES, the function shared by all sensory modalities may be motion in space, or other more general aspects of spatial processing having to do with locomotion (Rauschecker and Korte, 1993). Therefore, a common code for spatial information has to be used that can be interpreted by subcortical structures, such as the SC. A compensatory expansion of AEA at the expense of AEV thus results in finer resolution of auditory behaviors (Rauschecker and Kniepert, 1994) rather than in a reinterpretation of auditory signals as visual. On the other hand, it appears likely that the sensorimotor integration occurring in structures such as the SC (Stein and Meredith, 1993) and the motor feedback provided to them (Held and Hein, 1963) is instrumental in calibrating spatial perception without vision (Jones, 1975; Hartlage, 1976).

Summary and conclusions

Long-term visual deprivation from birth leads to reorganization of the anterior ectosylvian region of the cat's cortex. A visual area in the fundus of AES, area AEV, in which neurons normally respond briskly to visual stimuli, is taken over almost completely by auditory and somatosensory inputs. Auditory neurons are sharply tuned to the location of a sound source in azimuth. A higher than normal proportion of neurons in the AES of visually deprived cats responds to multimodal stimuli.

Takeover of visual regions by non-visual inputs can be explained by the same mechanisms that are invoked for developmental plasticity within the visual system: neural activity and competition between different inputs leads to changes in synaptic efficacy. Additional sprouting cannot be excluded.

The radical changes of sensory modality within the AES, caused by visual deprivation, suggest that the same code is used by visual, auditory and somatosensory maps in this region to represent the sensory environment and lead to sensorimotor transformation.

Acknowledgements

The present chapter is largely based on data previously reported by Rauschecker and Korte (1993) and summarized in a wider context by Rauschecker (1995). The help of Martin Korte in collecting these data is gratefully acknowledged.

References

Axelrod, S. (1959) Effects of Early Blindness. New York: Am. Found. Blind.

Bagdonas, A.P., Kodryunas, R.B. and Linyauskaite, A.I. (1980) Psychoacoustic functions in visually normal and impaired and blind subjects. *Hum. Physiol.*, 6: 108–113.

Benedek, G., Mucke, L., Norita, M., Albowitz, B. and Creutzfeldt, O.D. (1988) Anterior ectosylvian visual area (AEV) of the cat: physiological properties. *Prog. Brain Res.*, 75: 245–255.

Berman, A.L. (1961) Interaction of cortical responses to somatic and auditory stimuli in anterior ectosylvian gyrus of cat. *J. Neurophysiol.*, 24: 608–620.

Brenner, E. and Rauschecker, J.P. (1990) Centrifugal motion bias in the cat's lateral suprasylvian visual cortex is independent of early flow field exposure. *J. Physiol.*, 423: 641–660.

Carlson, S., Pertovaara, A. and Tanila, H. (1987) Late effects of early binocular visual deprivation on the function of Brodmann's area 7 of monkeys (*Macaca arctoides*). *Dev. Brain Res.*, 33: 101–111.

Carreras, M. and Andersson, S.A. (1963) Functional properties of neurons of the anterior ectosylvian gyrus of the cat. *J. Neurophysiol.*, 26: 100–126.

Clarey, J.C. and Irvine, D.R.F. (1986) Auditory response properties of neurons in the anterior ectosylvian sulcus of the cat. *Brain Res.*, 386: 12–19.

Clarey, J.C. and Irvine, D.R.F. (1990a) The anterior ectosylvian sulcal auditory field in the cat: I. An electrophysiological study of its relationship to surrounding auditory cortical fields. *J. Comp. Neurol.*, 301: 289–303.

Clarey, J.C. and Irvine, D.R.F. (1990b) The anterior ectosylvian sulcal auditory field in the cat: II. A horseradish peroxidase study of its thalamic and cortical connections. *J. Comp. Neurol.*, 301: 304–324.

Clemo, H.R. and Stein, B.E. (1983) Organization of a fourth somatosensory area of cortex in the cat. *J. Neurophysiol.*, 50: 910–923.

Darian-Smith, C. and Gilbert, C.D. (1994) Axonal sprouting accompanies functional reorganization in adult cat striate cortex. *Nature*, 368: 737–740.

Duffy, C. and Wurtz, R.W. (1990) Response of MST neurons to optic flow stimuli with shifted centers of motion. *J. Neurosci.*, 15: 5192–5208.

Fisher, G.H. (1964) Spatial localization by the blind. *Am. J. Psychol.*, 77: 2–13.

Graziano, M.S., Andersen, R.A. and Snowden, R.J. (1994) Tuning of MST neurons to spiral motions. *J. Neurosci.*, 14: 54–67.

Griesbach, H. (1899) Vergleichende Untersuchungen über die Sinnesschärfe Blinder und Sehender. *Pflügers Arch.*, 74: 577–638.

Hartlage, L.C. (1976) Development of spatial concepts in visually deprived children. *Percept. Motor Skills.*, 42: 255–258.

Held, R. and Hein, A. (1963) Movement-produced stimulation in the development of visually-guided behavior. *J. Comp. Physiol. Psychol.*, 56: 872–876.

Hyvärinen, J., Carlson, S. and Hyvärinen, L. (1981) Early visual deprivation alters modality of neuronal responses in area 19 of monkey cortex. *Neurosci. Lett.*, 4: 239–243.

Innocenti, G.M. and Clarke, S. (1984) Bilateral transitory projection from auditory cortex in kittens. *Dev. Brain Res.*, 14: 143–148.

Jiang, H., Leporé, F., Ptito, M. and Guillemot, J.-P. (1990) Modality specificity of neuronal responses in the anterior ectosylvian cortex (AEC) of cats. *Soc. Neurosci. Abstr.*, 16: 1221.

Jones, B. (1975) Spatial perception in the blind. *Br. J. Psychol.*, 66: 461–472.

Juurmaa, J. and Suonio, K. (1975) The role of audition and motion in the spatial orientation of the blind and the sighted. *Scand. J. Psychol.*, 16: 209–216.

Kaas, J.H. (1991) Plasticity of sensory and motor maps in adult mammals. *Annu. Rev. Neurosci.*, 14: 137–167.

Kellogg, W.N. (1962) Sonar system of the blind. *Science*, 137: 399–404.

Knight, P.L. (1977) Representation of the cochlea within the anterior auditory field (AAF) of the cat. *Brain Res.*, 130: 447–467.

Korte, M. and Rauschecker, J.P. (1993) Auditory spatial tuning of cortical neurons is sharpened in cats with early blindness. *J. Neurophysiol.*, 70: 1717–1721.

Kujala, T., Alho, K., Paavilainen, P., Summala, H. and Näätänen, R. (1992) Neural plasticity in processing sound location by the early blind: an event-related potential study. *Electroencephalogr. Clin. Neurophysiol.*, 84: 469–472.

Lappe, M. and Rauschecker, J.P. (1993) A neural network for the processing of optic flow from ego-motion in man and higher mammals. *Neural Comput.*

Lappe, M. and Rauschecker, J.P. (1994) On heading detection from optic flow. *Nature*, 369: 712–713.

Meredith, M.A. and Clemo, H.R. (1989) Auditory cortical projection from the anterior ectosylvian sulcus (field AES) to the superior colliculus in the cat: an anatomical and electrophysiological study. *J. Comp. Neurol. T*, 289: 687–707.

Merzenich, M.M., Recanzone, G.H., Jenkins, W.M. and Grajski, K.A. (1990) Adaptive mechanisms in cortical networks underlying cortical contributions to learning and non-declarative memory. *Cold Spring Harbour Symp. Quant. Biol.*, 55: 873–887.

Métin, C. and Frost, D.O. (1989) Visual responses of neurons in somatosensory cortex of hamsters with experimentally induced retinal projections to somatosensory thalamus. *Proc. Natl. Acad. Sci.*, 86: 357–361.

Minciacchi, D, Tassinari, G. and Antonini, A (1987) Visual and somatosensory integration in the anterior ectosylvian cortex of the cat. *Brain Res.*, 410: 21–31.

Muchnik, C., Efrati, M., Nemeth, E., Malin, M. and Hildesheimer, M. (1991) Central auditory skills in blind and sighted subjects. *Scand. Audiol.*, 20,: 19–23.

Mucke, L., Norita, M., Benedek, G. and Creutzfeldt, O.D. (1982) Physiologic and anatomic investigation of a visual cortical area situated in the ventral bank of the anterior ectosylvian sulcus of the cat. *Exp. Brain Res.*, 46: 1–11.

Neville, H.J. (1990) Intermodal competition and compensation in development. In: Diamond, A. (Ed.), *The Development and Neural Bases of Higher Cognitive Functions*, N.Y. Acad. Sci. Press, pp. 71–91.

Niemeyer, W. and Starlinger, I. (1981) Do the blind hear better? Investigations on auditory processing in congenital or early acquired blindness. II. Central functions. *Audiology*, 20: 510–515.

Norita, M., Mucke, L., Benedek, G., Albowitz, B., Katoh, Y. and Creutzfeldt, O.D. (1986) Connection of the anterior ectosylvian visual area (AEV). *Exp. Brain Res.*, 62: 225–40.

Olson, C.R. and Graybiel, A.M. (1983) An outlying visual area in the cerebral cortex of the cat. *Progr. Brain Res.*, 58: 239–245.

Olson, C.R. and Graybiel, A.M. (1987) Ectosylvian visual area of the cat: location, retinotopic organization, and connections. *J. Comp. Neurol.*, 261: 277–294.

Palmer, L.A., Rosenquist, A.C. and Tusa, R.J. (1978) The retinotopic organization of lateral suprasylvian visual areas in the cat. *J. Comp. Neurol.*, 177: 237–256.

Pascual-Leone, A., Cammarota, A., Wassermann, E.M., Brasil-Neto, J.P., Cohen, L.G. and Hallett, M. (1993) Modulation of motor cortical outputs to the reading hand of braille readers. *Ann. Neurol.*, 34: 33–37.

Pons, T.P. (1996) Novel sensations in the congenitally blind. *Nature*, 380: 479–480.

Pons, T.P., Garraghty, P.E., Ommaya, A.K., Kaas, J.H., Taub, E., and Mishkin, M. (1991) Massive cortical reorganization after sensory deafferentation in adult macaques. *Science*, 252: 1857–1860.

Rauschecker, J.P. (1988) Visual function of the cat's LP/LS subsystem in global motion processing. *Progr. Brain Res.*, 75: 95–108.

Rauschecker, J.P. (1991) Mechanisms of visual plasticity: Hebb synapses, NMDA receptors, and beyond. *Physiol. Rev.*, 71: 587–615.

Rauschecker, J.P. (1995) Compensatory plasticity and sensory substitution in the cerebral cortex. *Trends Neurosci.*, 18: 36–43

Rauschecker, J.P. and Korte, M. (1993) Auditory compensation for early blindness in cat cerebral cortex. *J. Neurosci.*, 13: 4538–4548.

Rauschecker, J.P. and Kniepert, U. (1994) Enhanced precision of auditory localization behavior in visually deprived cats. *Eur. J. Neurosci.*, 6: 149–160.

Rauschecker, J.P. and Kreiter, A. (1988) Banded connections between areas 17 and PMLS in cat visual cortex: possible functional role for global motion processing. *Soc. Neurosci. Abstr.*, 14: 602.

Rauschecker J.P., von Grünau M.W. and Poulin, C. (1987a) Centrifugal organization of direction preferences in the cat's lateral suprasylvian visual cortex and its relation to flow field processing. *J. Neurosci.*, 7: 943–958.

Rauschecker J.P., von Grünau M.W. and Poulin, C. (1987b) Thalamo-cortical connections and their correlation with receptive field properties in the cat's lateral suprasylvian visual cortex. *Exp. Brain Res.*, 67: 100–112.

Sadato, N., Pascual-Leone, A., Grafman, J., Ibanez, V., Deiber, M.-P., Dold, G., Hallett, M. (1996) Activation of the primary visual cortex by Braille reading in blind subjects. *Nature*, 380: 526–528.

Spigelman, M.N. (1972) A comparative study of the effects of early blindness on the development of auditory-spatial learning. In: Jastrzembska Z.S. (Ed.), *The Effects of Blindness and Other Impairments on Early Development*, New York: Am. Found. Blind, pp. 29–45.

Stein, B.E. and Meredith, M.A. (1993) *The Merging of the Senses*, MIT Press.

Supa, M., Cotzin, M. and Dallenbach, K.M. (1944) 'Facial vision': the perception of obstacles by the blind. *Am. J. Psychol.*, 57: 133–183.

Sur, M., Pallas, S.L. and Roe, A.W. (1990) Cross-modal plasticity in cortical development: differentiation and specification of sensory neocortex. *Trends Neurosci.*, 13: 227–233.

Tian, B. and Rauschecker, J.P. (1994) Processing of frequency-modulated sounds in the cat's anterior auditory field. *J. Neurophysiol.*, 71: 1959–1975.

Uhl, F., Franzen, P., Podreka, I., Steiner, M. and Deecke, L. (1993) Increased regional cerebral blood flow in inferior occipital cortex and cerebellum of early blind humans. *Neurosci. Lett.*, 150: 162–164.

Veraart, C., De Volder, A., Wanet-Defalque, M.C., Bol, A., Michel, C. and Goffinet, A.M. (1990) Glucose utilization in human visual cortex is abnormally elevated in blindness of early onset, but decreased in blindness of late onset. *Brain Res.*, 510: 115–121.

Vidyasagar, T.R. (1978) Possible plasticity in the rat superior colliculus. *Nature*, 275: 140–141.

Wallace, M.T., Meredith, M.A. and Stein, B.E. (1992) Integration of multiple sensory modalities in cat cortex. *Exp. Brain Res.*, 91: 484–488.

Wallace, M.T. and Stein, B.E. (1994) Cross-modal synthesis in the midbrain depends on input from cortex. *J. Neurophysiol.*, 71: 429–432.

Warren, D.H. (1978) Perception by the blind. In: Carterette, E.C. and Friedman M.P. (Eds.), *Handbook of Perception, Vol X: Perceptual Ecology*, Academic Press, pp. 65–90.

Wiesel, T.N. and Hubel, D.H. (1965) Comparison of the effects of unilateral and bilateral eye closure on cortical unit responses in kittens. *J. Neurophysiol.*, 28: 1026–1040.

M. Norita, T. Bando and B. Stein (Eds.)
Progress in Brain Research, Vol 112
© 1996 Elsevier Science BV. All rights reserved.

CHAPTER 23

Visual, somatosensory and auditory modality properties along the feline suprageniculate-anterior ectosylvian sulcus/insular pathway

György Benedek[1]*, László Fischer-Szatmári[1], Gyula Kovács[1], János Perényi[1] and Yoshimitsu Y. Katoh[2]

[1]*Department of Physiology, Albert Szent-Györgyi Medical University, P.O. Box 1192, H-6720 Szeged, Hungary*
[2]*Department of Anatomy, Fujita-Gakuen Medical University, Toyoake, Japan*

Physiological properties of single units were investigated in the suprageniculate nucleus (SG) and in the cerebral cortex along the anterior ectosylvian sulcus (AES), including the insular cortex. The recording was performed with the aid of carbon-filled glass micropipetts in barbiturate-anesthetized cats. The main findings of the study can be summarized as follows.

1. The physiological properties of the cells in the suprageniculate nucleus and in the AES/insular cortex exhibited striking similarities in a series of aspects: (a) The frequencies of occurrence of uni-, bi- and trimodal cells were similar. (b) The majority of the unimodal cells (75% in the AES/insular region and 65% in the SG) has visual sensitivity in both structures. The bimodal and trimodal cells were also dominated by visual sensitivity. (c) The somatosensory and auditory modalities were similarly present in both structures, although less frequently than the visual one. (d) No systematic topologi-

cal organization was found in either structure. (e) The visual, somatosensory and auditory receptive fields were uniform and covered a fairly large proportion of the personal space.

2. Statistical comparison of some physiological properties of cells situated deep in the AES with those of cells in the insular cortex revealed differences as follows: (a) The insular cortex contained significantly more bi- and trimodal cells than the sulcal areas. (b) Cells in the insular cortex preferred significantly lower stimulus velocities and larger stimuli than cells in the depths of the AES.

These results seem to support the notion of a suprageniculate-AES/insular thalamo-cortical multisensory entity. Additionally, the physiological differences between the sulcal AES cortex and gyral insula are in agreement with the morphological differences found earlier in the afferentation of these areas (Norita et al., 1986, 1991).

Introduction

The concept of the parallel organization of sensory pathways includes the notion that the processing of different aspects of sensory information is based on the existence of several anatomically separate pathways and cortical sensory areas (Schneider, 1969; Lennie, 1980; Ungerleider and Mishkin, 1982; Goodale and Milner, 1992; Merigan and Maunsell, 1993). The cat brain seems to be a useful model for studying parallel sensory processing since an almost complete separation of geniculo-striate and extrageniculate pathways was found here (see Rosenquist, 1985; Garey et al., 1991 for reviews) The descriptions of the anterior ectosylvian visual area (AEV) (Mucke et al., 1982; Olson and Graybiel, 1983, 1987; Benedek et al., 1986, 1988; Benedek and Hicks, 1988; Hicks et al., 1988; Tamai and Miyashita 1989a,b; Kimura

---

*Corresponding author.

and Tamai, 1992), revealed it to be a particularly interesting extrageniculate area since it seems the only known visual cortical area without afferents from the lateral geniculate complex (Rosenquist, 1985). After the first reports on the existence of the AEV as a unimodal visual area (Mucke et al., 1982; Olson and Graybiel, 1983, 1987), various findings appeared on the modality properties of neurons along the AES; unimodal somatic and auditory and also multimodal sensory areas have been described (Clemo and Stein, 1982; Minciacchi et al., 1987, Clarey and Irvine, 1986; Hicks et al., 1988; Benedek and Hicks, 1988; Wallace et al., 1992; Jiang et al., 1994). Anatomical studies have described two main sources of afferentation towards sensory areas along the AES (Mucke et al., 1982; Norita et al., 1986, 1991; Olson and Graybiel, 1987; Tamai and Miyashita, 1989b). Cortico-cortical pathways, originating mostly from the medial bank of the lateral suprasylvian sulcus, provide afferentation to the AES region. Additionally, thalamic efferents, originating mostly from the suprageniculate nucleus (SG) and the medial part of the lateral posterior nucleus (LPm), play a role in the afferentation of AES cortical areas.

The major aim of the present study was to estimate the importance of afferentation coming from the SG in determining the physiological properties of AES neurons. First we compared the physiological characteristics of SG cells with those described earlier in cells along the AES. In this study we used stimuli which we found to be effective in the AES cortex in our earlier studies. Then we compared the physiological properties of neurons in the insular cortex, which has been described as a major cortical target of suprageniculate efferents with those of neurons lying deep in the sulcal regions of the AES. These deep, sulcal neurons receive heavy afferentation from more dorsal thalamic nuclei, such as the LPm.

Materials and methods

Experiments were conducted on a total of 21 adult cats ranging in weight from 2.5 to 3.5 kg. The experimental protocol had been accepted by the Ethical Committee for Animal Research of Albert Szent-Györgyi Medical University. The anesthesia was initiated with ketamine hydrochloride (i.m. 30 mg/kg) and was maintained with barbiturate (pentobarbital sodium 1.8–2.5 mg/kg per h). The trachea and the femoral vein were cannulated, wound edges and pressure points were treated generously with procaine hydrochloride (1%) and the animals' heads were placed within a Kopf stereotaxic headholder frame. The cortex was exposed over the AES or over the thalamus and it was covered with a 4% solution of 38°C agar dissolved in saline. The animals were then paralyzed by intravenous injection of gallamine triethiodide (20 mg/kg). A solution containing gallamine (8 mg/kg per h), dextrose (10 mg/kg per h) and dextran (50 mg/kg per h) in Ringer's solution was infused continuously at a rate of 3 ml/h. Artificial respiration was maintained. The end-tidal CO_2 level was kept constant between 3.5 and 4.5%. Body temperature was maintained between 37 and 38°C. The contralateral eye was treated with neosynephrine (1%) and atropine (1%) and covered with a contact lens. The ipsilateral eye was covered during visual stimulation. Electrophysiological recordings were made extracellularly by means of glass microelectrodes filled with carbon fibers (Armstrong-James and Millar, 1979). The exact position of the electrode tip in the thalamus was determined from a comparison of the actual receptive field position with the LGN atlas of Sanderson (1971). This correction permitted an accurate aiming at the SG through use of coordinates A: 4–5.5, L: 5.5, V: 18–20.

We searched for single neurons and then we presented visual, auditory, somatosensory and noxious stimuli in a semirandom order. For visual stimulation, bars of light of 1.5° width and various lengths (2°, 10° or 50°) were swept across a tangent screen in the preferred direction. Neuronal activity was recorded and correlated with the movement of the light stimulus. The optimal speed was selected as the one evoking the highest activity of the neuron. At least ten trials were run in every test. A neuron was considered to be responsive if the firing rate around the response was at least twice as high as those during the resting period. Auditory stimuli included natural stimuli, such as hand clapping, noises and computer-controlled clicks (80 dB) delivered from a movable speaker placed 20 cm from the animal's head. Somatosensory stimuli were vertical deflections of the hair or rapid indentation of the skin. Noxious stimulation was performed by applying electric shocks (1.0 ms, 5–30 V) through a pair of electrodes placed into the cavity of the 2nd dental incisor, contralateral to the recording. The evoked spike activity of the neuron was recorded during five repetitions.

Successful electrode penetrations were marked by electrolytic lesions. Histological sections were stained with neutral red and in the SG with both neutral red and acetylcholinesterase staining (Hardy et al., 1976). SG cells were localized by taking into account the electrolytic lesions and the microdriver readings made during recordings.

Results

Physiological properties of SG neurons

Altogether 256 single units were investigated in the SG. Cell locations were verified histologically. Modality properties of SG neurons were studied in 142 neurons. We could find neurons responsive to visual, auditory, tactile somatosensory and nociceptive somatosensory stimulation. The majority of the cells were sensitive to visual stimulation. One hundred and eleven of the 141 neurons responded to moving visual stimuli. Unimodal visual response properties were found in 74 single units, while visual responsiveness was accompanied by auditory sensitivity in nine neurons and by somatosensory sensitivity in 13 neurons. Trimodal (visual, tactile somatosensory and auditory) sensitivity was found in four neurons. Unimodal auditory responsiveness was observed in 16 neurons, and unimodal tactile somatosensory responsiveness in 20 neurons. We detected three neurons that responded only to noxious stimulation of the tooth pulp, and one neuron with bimodal visual plus nociceptive properties. One neuron

was bimodal in the sense that it responded to both auditory and tactile somatosensory stimulation. The neurons with different types of modality properties were evenly distributed in the SG; no clustering of cells with any receptive field properties could be seen (Fig. 1).

Visual sensitivity was characterized by a sensitivity to moving stimuli in a huge receptive field that consistently included the area centralis (Fig. 2). The receptive field covered most parts of the lower contralateral quadrant and extended 10–30° into the ipsilateral hemifield too. This receptive field structure was quite homogenous among the SG cells, and accordingly no retinotopic organization could be established in the SG.

Forty-three SG cells of the 111 cells tested (40%) displayed a directional sensitivity. Velocity sensitivity was characterized by a preference for stimuli moving at a considerable velocity. All the neurons evidently preferred stimuli moving at a velocity higher than 10°s. When we compared the effectiveness of 20°/s and 120°/s in 113 cells, 61 cells preferred 120°/s to the lower velocity, 32 cells preferred 20°/s to the higher velocity and 20 cells showed no preference difference between 20°/s and 120°/s. A length-sensitivity study indicated cells with both end-inhibiting and end-summating properties, although no dominance of any type of cells was found.

The somatosensory cells were activated by stimulation of the whole contralateral body surface and some had receptive fields extending to the ipsilateral body surface, too. No receptive field organization could be detected. We found only four cells among 141 single units tested that exhibited sensitivity to noxious stimulation of the tooth pulp. No correspondence was observed between tactile somatosensation and pain sensation in any unit.

A total of 30 units was identified that possessed auditory sensitivity. The auditory cells responded to a wide scale of acoustic stimulations. The units produced a train of discharges upon stimulation after a latency of 11–18 ms. The auditory receptive field covered a rather wide angle:, sometimes the whole personal space. No preferred type or frequency of stimulation was observed.

Sensory properties of neurons lying along the AES and in the insular cortex

Altogether 164 neurons with sensory properties were recorded extracellularly and identified histologically in the cortex lying along the AES and in the insular cortex (Fig. 3). Since the physiological properties of these neurons have been extensively described, we concentrated on an analysis of the differences between subregions in the AES cortex. In order to facilitate a comparison between insular cells in the gyral cortex and those lying deep in the AES (sulcal AES, sAES), we drew an arbitrary border that was determined by a line connecting the most lateral extensions of the white matter in the actual sections (see insert in Table I for an illustration).

According to this division, 123 cells were recorded in the gyral surface of the insula and 41 cells in the depths of the sAES. This arbitrary, anatomical division, however, seems to have some physiological relevance, too (Table 1). There was a much higher percentage of multimodal neurons in the insula than in the deep part of the sulcus. Of the 123 insular cells, 45 were bimodal and three were trimodal, while among the 41 cells found in the deep part of the sulcus, only four bimodal cells and one trimodal cell were found.

Both regions showed a predominance of visual modality. Almost all (35 of the 36) unimodal sAES neurons were visual; only one auditory cell was found. The four bimodal cells demonstrated visual-auditory sensitivity in two cases, and visual-somatosensory as well as auditory-somatosensory responsivity in one case each.

Of the 123 insular cells, 110 were sensitive to visual stimulation, 62 unimodally, 45 bimodally

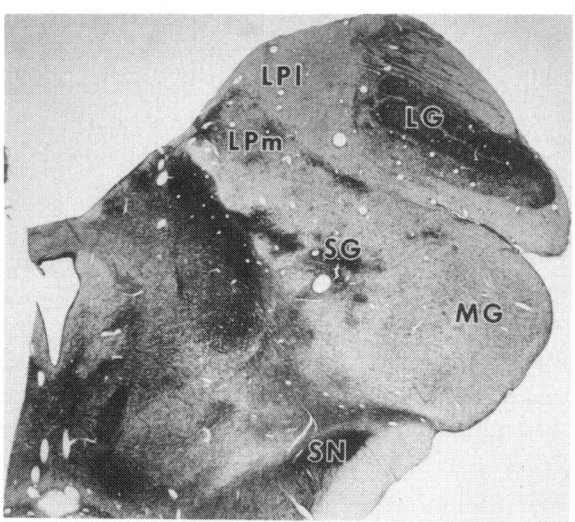
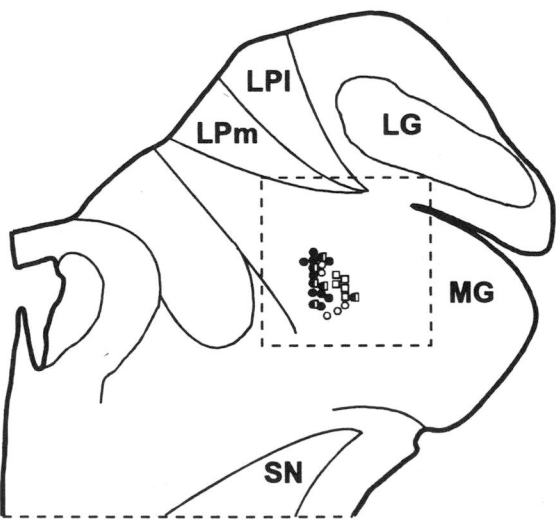

Fig. 1. The location and modality sensitivity of single units recorded in the suprageniculate nucleus of the cat. Left side: Coronal section of the thalamus with acetylcholinesterase staining after Hardy et al. (1976). The SG is represented by cholinesterase negative and cholinesterase positive patches medial to the medial geniculate body and ventrolaterally to the lateral posterior complex. Right side: schematic representation of the histological section on the left side. Borders of the major nuclei are marked by continous lines, while the largest extension of the recordings is marked by the stippled box. Neurons, recorded at this anterio–posterior level (A5.5) are marked as follows: visual neurons ●, auditory □, somatosensory ○ and nociceptive ■ unimodal neurons. (Bimodal neurons are indicated by the corresponding combination of half-symbols.) Abbreviations: LPl, lateral part of the lateral-posterior nucleus; LPm, medial part of the lateral-posterior nucleus; LG, lateral geniculate body; MG, medial geniculate body; SN, substantia nigra.

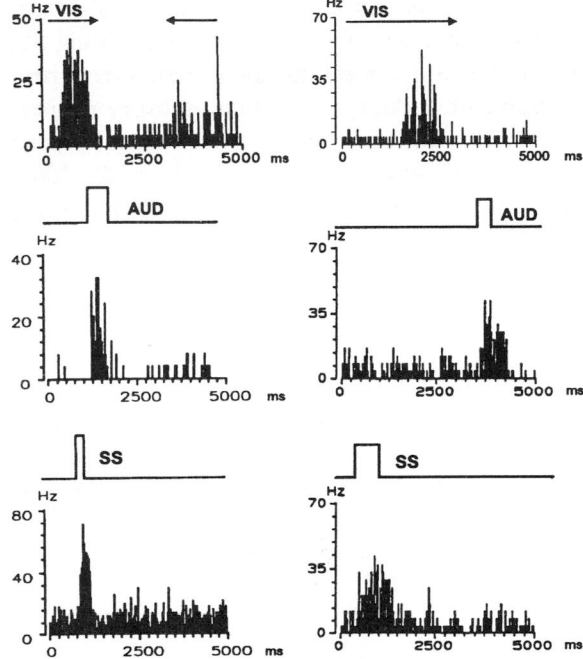

Fig. 2. Peristimulus histograms recorded in the suprageniculate nucleus (left side) and in the AES/insular cortex (right side) upon visual (top), auditory (middle) or somatosensory (bottom) stimulation. Abbreviations: Vis, visual stimulation with a $1.5 \times 10°$ bar of light, moving along the neurons' preferred trajectory of motion; AUD, auditory stimulation; SS, tactile somatosensory stimulation of the skin; and rectangles above histograms indicate stimulation.

(42 visual-auditory and three visual-somatosensory) and three trimodally (visual-auditory-somatosensory). The chi-square test revealed a significant difference between the sAES and the insular cortex with regard to the occurrence of multimodal cells ($\chi^2 = 10.12$, $P < 0.015$, df = 1).

As described earlier, the visual sensitivity in the AES/insular cortex was characterized by a sensitivity to moving stimuli. The velocity preference seemed to be distinctive between the sAES and insular cells. Of the 36 unimodal visual cells in the sAES, 30 clearly preferred high ($> 100°/s$) stimulus velocities to a moderately low ($20°/s$) velocity. Two neurons showed no velocity preference. No neuron was found that preferred clearly low ($< 5°/s$) velocities. In contrast, 36 of the 75 visual unimodal cells in the insula had a prefer-

ence for the highest ($> 100°/s$) velocities tested, and 26 neurons preferred the $20°/s$ stimulation to the higher velocities. The chi-square test indicated that this difference between the velocity preferences of the cells was significant ($\chi^2 = 19.37$, $P < 0.00001$, df = 1).

Discussion

The present study has demonstrated the unique sensory characteristics of the neurons in both the SG and the AES/insular cortex. Single unit recordings revealed particular response characteristics, of which we emphasize the following.

First, the thalamic and the cortical region studied displayed common multisensory characteristics. This multisensory character of the areas manifested itself in the presence of unimodal neurons that responded to either visual or somatosensory or auditory stimuli and in the presence of single units responding to two or even three modalities. This multisensory character was earlier described in the AES/insular cortex (Minciacchi et al., 1987; Hicks et al., 1988; Jiang et al., 1994), although the area investigated was mostly a limited part of the whole anatomical entity. A similar multisensory character of the SG, however, has not been described previously. Earlier studies dealt mainly with the modality properties of the whole posterior nuclear complex and the pulvinar–lateral posterior complex ((Rose and Woolsey, 1958; Poggio and Mountcastle, 1960; Wepsic, 1966; Aitkin, 1973; Berkley, 1973; Fish and Chalupa, 1979). Hence, the number of SG cells studied was low and the lack of definite borders around the SG made interpretation of the findings rather difficult. To our knowledge our study is the first physiological report in which the area of the SG is explored with multimodal challenges between well-defined borders. Interestingly, the visual character of this nucleus was not revealed by the early studies. Later, several authors (Hicks et al., 1984; Krupa et al., 1984; Hutchins and Updyke, 1989) described the visual properties of the SG, but the multisensory character of the SG was not detected.

It is interesting that the occurrence of uni-modal and polymodal cells is rather similar in the SG and in the AES cortex. Altogether 75% of the SG cells were found to be unimodal and about 25% multimodal, data quite close to those were found by Minciacchi et al., (1987) in the AES cortex, and by others (Hicks et al., 1988; Benedek and Hicks, 1988) in the insular cortex. Further, the predominance of visual sensitivity among SG and AES/insular neurons and the sparser representation of auditory and somatosensory properties in both the thalamic and the cortical region (Clarey and Irvine, 1986; Minciacchi et al., 1987; Hicks et al., 1988; Benedek and Hicks, 1988;

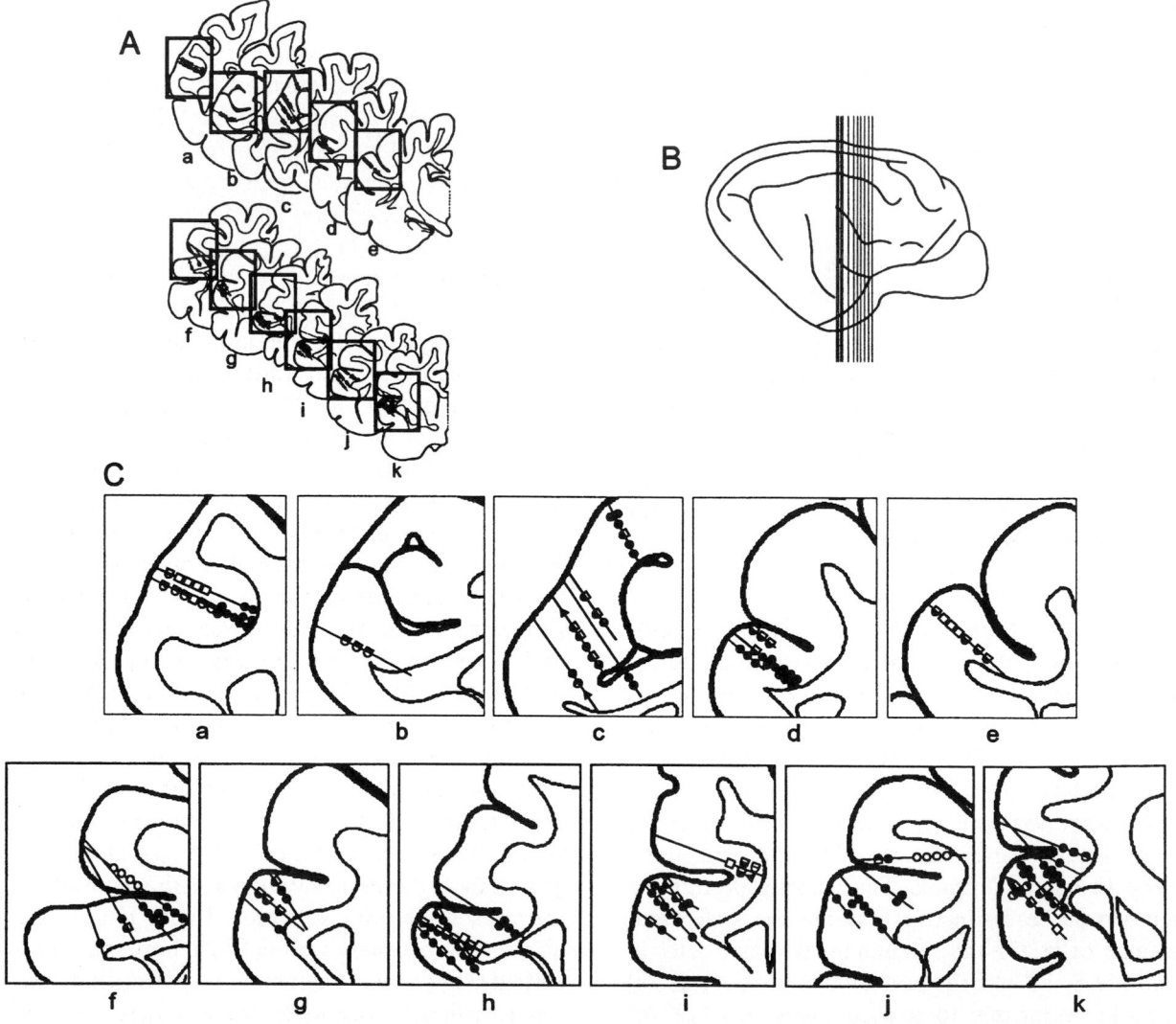

Fig. 3. The location and modality sensitivity of neurons with sensory properties in the feline AES/insular cortex. Insert on the top right (B) shows the positions of the coronal sections depicted on the top left (A). Bottom panels (C) are the magnifications of the AES/insular cortical regions in the corresponding coronal sections. Symbols: visual ●; auditory □; somatosensory ○; and nociceptive ■ unimodal neurons. Bimodal neurons are indicated by the corresponding combination of half-symbols. Trimodal ▲ neurons are indicated by triangles.

TABLE 1

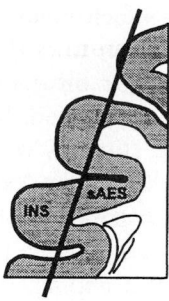

		sAES n=41	INS n=123
UNIMODAL	VIS	35	62
	AUD	1	13
	SS	0	0
BIMODAL	VIS - AUD	2	42
	VIS - SS	1	3
	AUD - SS	1	0
TRIMODAL	VIS - AUD - SS	1	3

Proportion of uni-, bi- and trimodal neurons and their modality properties in the gyral insular (INS) and in the sulcal AES (sAES) regions. The division between the two regions was made according to the arbitrary line indicated on the schematic drawing above. Abbreviations: Vis, visual sensitivity; SS, somatosensory sensitivity; Aud, auditory sensitivity

Wallace et al., 1992; Jiang et al., 1994) is a common property.

The second common characteristics of single units recorded in the SG and along the AES/insular cortex is their particular receptive field property. Both structures include neurons with extremely large receptive fields, which makes any organization rather difficult or almost impossible. This receptive field property in the AES/insular cortex has been described for visual (Mucke et al., 1982; Benedek et al., 1986, 1988; Hicks et al., 1988a), auditory (Clarey and Irvine, 1986; Middlebrooks et al., 1994) or somatosensory (Minciacchi et al., 1987) neurons. Poggio and Mountcastle (1960) similarly found huge somatosensory, noci-

ceptive receptive fields in the SG. A loose receptive field organization has been described in the SG by Hicks et al. (1986), while the auditory receptive fields in SG have not been yet extensively investigated. These particular physiological properties of the neurons seem to be gaining importance in an anatomical context. The SG and the insular cortex are reciprocally interconnected (Guldin and Markowitsch, 1984; Hicks et al., 1986; Norita et al., 1991). These thalamo-cortical reciprocally interconnected structures are closely connected with the deeper layers of the superior colliculus (SC) which provides a heavy afferentation to the SG (Heath and Jones, 1971; Huerta and Harting, 1984; Hicks et al., 1986; Abramson and Chalupa, 1988; Katoh and Benedek; 1995 Katoh et al., 1995), while the AES/insular cortex sends a dense cortico-tectal efferentation towards the deeper layers of the SC (Stein, 1978; Tortelly et al., 1980; Stein et al., 1983; Norita et al., 1986, 1991; Miyashita and Tamai, 1989). It is worth mentioning that the posteromedial lateral suprasylvian area (PMLS), the other primary source of afferentation to the AES/insular cortex, is also heavily connected to this part of the SC (Tortelly et al., 1980; Berson and McIlwain, 1983; Grant and Shipp, 1991). The physiological properties of the deeper layer of the SC have been extensively described (see Stein and Meredith, 1993, for a review). This region includes an overlapping multimodal representation, with large receptive fields, motion and directional sensitivity, and seems to play an important role in the multisensory integration of signals important in orientation as well as body and eye-movements. The close connection between these structures may explain the similarities in the physiological response properties of AES/insular cells, SG neurons and neurons in deeper layers of the SC. SC neurons are well known for their predominance of visual sensitivity and for the rather large receptive fields that nevertheless exhibit some crude, but well-lineated topical organization (Stein and Arigbede, 1972; Gordon, 1973; Meredith et al., 1991). Additionally, both the SC and the AES/insular cortex have been described as sites for the integration of multisensory signals (Meredith and Stein, 1986; Stein and Meredith, 1990; Wallace et al., 1992; Wallace and Stein, 1994).

A comparison between the physiological properties of neurons lying in the deep, sulcal part of the AES is also important as a matter of anatomical connections. The deep sulcal part of the AES, especially in its caudal aspect in the AEV proper receives a heavy afferentation from the medial part of the lateral posterior nucleus (LPm), while neurons in the gyral aspect of the AES/insular cortex have no input from the LPm, but receive their thalamic afferentation predominantly from the SG (Norita et al., 1986, 1991). A difference in the cortical afferentation of these two regions has also been noted by Squatrito et al. (1981). We attempted to distinguish between the sulcal and gyral parts of the AES/insular cortex drawing an arbitrary dividing line. This comparison revealed distinct differences between the neurons in the gyral part of the sulcus, that includes most of the insular cortex and the neurons in the deep, sAES cortex, which includes the AEV proper. The occurrence of multimodal cells was significantly higher in the gyral part of the insular cortex than in the deeper, sAES parts. This distinction corresponds to the earlier ones that found a unimodal, visual area along the deep ventral bank of the AES and a multimodal area along the gyral, insular cortex (Mucke et al., 1982; Olson and Graybiel, 1983, 1987; Minciacchi et al., 1887; Hicks et al., 1988ab). Similarly, the neurons in the deep, sAES showed a preference for stimuli moving at rather high speed, while gyral, insular neurons preferred moderately high stimulus velocities.

Thus, the differing physiological properties of cortical structures lying along the AES are in concert with their differing thalamic and cortical afferentations. The AEV, which obtains a predominant thalamic afferentation from the LPm, has almost entirely unimodal, mostly visual properties. This is in line with the predominant visual properties of the LPm (Mason, 1978, 1981). Those parts of the cortex that receive predomi-

nantly suprageniculate afferentation from the thalamus display a much higher extent of multimodality. The sensitivity towards the stimulus velocity reveals a similar divergence between the two cortical structures studied. Insular, gyral cells prefer less extreme stimulus velocities than cells in the AEV and in the other deep, sulcal parts of the AES. This corresponds to the velocity sensitivity properties found by us in the SG.

These physiological findings, together with the well-known anatomical connections, lend support to the notion of a colliculo-suprageniculate-insular pathway that runs parallel to the well-known primary sensory thalamo-cortical pathways. The behavioral function of this tecto-thalamo-cortical pathway has not yet been delineated, but it evidently provides multisensory information about changes in the environment without information about the discriminating character of the stimuli.

Acknowledgement

This work was supported by OTKA (Hungary) Grant No: T016959 and by a US-Hungarian Joint Fund (Grant No: 223).

References

Abramson, B.P. and Chalupa L.M. (1988) Multiple pathways from the superior colliculus to the extrageniculate visual thalamus of the cat. *J. Comp. Neurol.*, 271: 397–418.

Aitkin, L.M. (1973) Medial geniculate body of the cat: responses to tonal stimuli of neurons in medial division. *J. Neurophysiol.*, 36: 275–283.

Armstrong-James, M. and Millar, J. (1979) Carbon fibre microelectrodes. *J. Neurosci. Methods*, 1: 279–287.

Benedek, G. and Hicks, T.P. (1988) The visual insular cortex of the cat: organization, properties and modality specificity. *Prog. Brain Res.*, 75: 271–277.

Benedek, G., Jang E.K. and Hicks T.P. (1986) Physiological properties of visually responsive neurones in the insular cortex of the cat. *Neurosci. Lett.*, 64: 269–274.

Benedek, G., Mucke, L., Norita, M., Albowitz, B. and Creutzfeldt, O.D. (1988) Anterior ectosylvian visual area (AEV) of the cat: physiological properties. *Prog. Brain Res.*, 75: 245–255.

Berkley, K.J. (1973) Response properties of cells in ventrobasal and posterior group nuclei of the cat. *J. Neurophysiol.*, 36: 940–952.

Berson, D.M. and McIlwain, J.T. (1983) Visual cortical inputs to deeper layers of cat's superior colliculus. *J. Neurophysiol.*, 50: 1143–1155.

Clarey, J.C. and Irvine, R.F. (1986) Auditory response properties of neurons in the anterior ectosylvian sulcus of the cat. *Brain Res.*, 386: 12–19.

Clemo, H.R. and Stein, B.E. (1982) Somatosensory cortex: a 'new' somatotopic representation. *Brain Res.*, 235: 162–168.

Fish, S.E. and Chalupa, L.M. (1979) Functional properties of pulvinar-lateral posterior neurons which receive input from the superior colliculus. *Exp. Brain Res.*, 36: 245–287.

Garey, L.J., Dreher, B. and Robinson, S.R. (1991) The organization of the visual thalamus. In: B. Dreher and S.R. Robinson (Eds.), *Neuroanatomy of the Visual Pathways and their Development, Vision and Visual Dysfunction*, Vol. 3, Macmillan Press, London, pp. 176–234.

Goodale, M.A. and Milner, A.D. (1992) Separate visual pathways for perception and action. *Trends Neurosci.*, 15: 20–25.

Gordon, B.G. (1973) Receptive fields in the deep layers of the cat superior colliculus. *J. Neurophysiol.*, 36: 157–178.

Grant, S. and Shipp, S. (1991) Visuotopic organization of the lateral suprasylvian area and of an adjacent area of the ectosylvian gyrus of cat cortex: a physiological and connectional study. *Vis. Neurosci.*, 6: 315–338.

Guldin, W.O. and Markowitsch, H.J. (1984) Cortical and thalamic afferent connections of the insular and adjacent cortex of the cat. *J. Comp. Neurol.*, 229: 393–418.

Hardy, H., Heimer, L., Switzer, R. and Watkins, D. (1976) Simultaneous demonstration of horseradish peroxydase and acetylcholinesterase. *Neurosci. Lett.*, 3: 1–5.

Heath, C. and Jones, E.G. (1971) An experimental study of ascending connections from the posterior group of thalamic nuclei in the rat. *J. Comp. Neurol.*, 141: 397–426.

Hicks, T.P., Watanabe, S., Miyake, A. and Shoumura, K. (1984) Organization and properties of visually responsive neurons in the suprageniculate nucleus of the cat. *Exp. Brain Res.*, 55: 359–367.

Hicks, T.P., Stark, C.A. and Fletcher W. (1986) Origins of afferents to visual suprageniculate nucleus in the cat. *J. Comp. Neurol.*, 246: 544–554.

Hicks, T.P., Benedek, G. and Thurlow, G.A. (1988a) Modality specificity of neuronal responses within the cat's insula. *J. Neurophysiol.*, 60: 422–437.

Hicks, T.P., Benedek, G. and Thurlow, G.A. (1988b) Organization and properties of neurons in a visual area within the insular cortex of the cat. *J. Neurophysiol.*, 60: 397–421.

Jiang, H., Lepore, F., Ptito, M. and Guillemot, J.P. (1994) Sensory modality distribution in the anterior ectosylvian cortex (AEC) of cats. *Exp. Brain Res.*, 97: 404–414.

Huerta, M.F. and Harting J.K. (1984) The mammalian superior colliculus: studies of its morphology and connections. In: H. Vanegas (Ed.), *Comparative Neurology of the Optic Tectum*, Plenum, New York, pp. 687–773.

Hutchins, B. and Updyke, B.V. (1989) Retinotopic organization within the lateral posterior complex of the cat. *J. Comp. Neurol.*, 285: 350–398.

Katoh, Y.Y. and Benedek, G. (1995) Organization of the colliculo-suprageniculate pathway in the cat: a wheat-germ agglutinin-horseradish peroxydase study. *J. Comp. Neurol.*, 352: 381–397.

Katoh, Y.Y., Benedek, G. and Deura, S. (1995) Bilateral projections from the superior colliculus to the suprageniculate nucleus in the cat: a WGA-HRP/double fluorescence tracing study. *Brain Res.*, 669: 298–302.

Kimura, A. and Tamai, Y. (1992) Sensory response of cortical neurons in the anterior ectosylvian sulcus, including the area evoking eye movement. *Brain Res.*, 575: 181–186.

Krupa, M., Maire-Lepoivre, E. and Imbert, M. (1984) Visual properties of neurons in the suprageniculate nucleus of the cat. *Neurosci. Lett.*, 51: 13–18.

Lennie, P. (1980) Parallel visual pathways: a review. *Vision Res.*, 20: 561–594.

Mason, R. (1978) Functional organization in the cat's pulvinar-complex. *Exp. Brain Res.*, 31: 51–66.

Mason, R. (1981) Differential responsiveness of cells in the visual zones of the cat's LP pulvinar complex to visual stimuli. *Exp. Brain Res.*, 43: 25–33.

Meredith, M.A. and Stein, B.E. (1986) Visual, auditory, and somatosensory convergence on cells in superior colliculus results in multisensory integration. *J. Neurophysiol.*, 56: 640–662.

Meredith, A.M., Clemo, H.R. and Stein, B.E. (1991) Somatotopic component of the multisensory map in the deep laminae of the cat superior colliculus. *J. Comp. Neurol.*, 312: 353–370.

Merigan, W.H. and Maunsell, J.H. (1993) How parallel are the primate visual pathways. *Annu. Rev. Neurosci.*, 16: 369–402.

Middlebrooks, J.C., Clock, A.E., Xu, L. and Green, D.M. (1994) A panoramic code for sound location by cortical neurons. *Science*, 264: 842–844.

Minciacchi, D., Tassinari, G. and Antonini, A. (1987) Visual and somatosensory integration in the anterior ectosylvian cortex of the cat. *Brain Res.*, 410: 21–31.

Miyashita, E. and Tamai, Y. (1989) Subcortical connections of frontal 'oculomotor' areas in the cat. *Brain Res.*, 502: 75–87.

Mucke, L., Norita, M., Benedek, G. and Creutzfeldt, O. (1982) Physiologic and anatomic investigation of a visual cortical area situated in the ventral bank of the anterior ectosylvian sulcus of the cat. *Exp. Brain Res.*, 46: 1–11.

Norita, M., Mucke, L., Benedek, G., Albowitz, B., Katoh, Y. and Creutzfeldt, O.D. (1986) Some efferent and afferent connections of the anterior ectosylvian visual area (AEV). *Exp. Brain Res.*, 62: 225–240.

Norita, M., Hicks, T.P. Benedek, G. and Katoh Y. (1991) Afferent and efferent connections of the feline insular visual cortex. *J. Hirnforsch.*, 32: 119–134.

334

Olson, C.R. and Graybiel, A.M. (1983) An outlying visual area in the cerebral cortex of the cat. *Prog. Brain Res.*, 58: 239–245.

Olson, C.R. and Graybiel, A.M. (1987) Ectosylvian visual area of the cat: location, retinotopic organization and connections. *J. Comp. Neurol.*, 261: 277–294.

Poggio, G.F. and Mountcastle, V.B. (1960) A study of the functional contributions of the lemniscal and spinothalamic systems to somatic sensibility. Central nervous mechanisms in pain. *Bull. Johns Hopkins Hosp.*, 106: 266–316.

Rose, J.E. and Woolsey, C.N. (1958) Cortical connections and functional organization of the thalamic auditory system of the cat. In: H.F. Harlow and C.N. Woolsey (Eds.), *Biological and Biochemical Bases of Behavior*, University of Wisconsin Press, Madison, pp. 127–150.

Rosenquist, A.C. (1985) Connections of visual cortical areas in the cat. In: A. Peters and E.G. Jones (Eds.), *Cerebral Cortex. Visual Cortex*, Vol. 3, Plenum Press, New York, pp. 81–117.

Sanderson, K.J. (1971) The projection of the visual field to the lateral geniculate and medial intralaminar nuclei in the cat. *J. Comp. Neurol.*, 143: 101–118.

Schneider, G.E. (1969) Two visual systems. *Science*, 163: 895–902.

Squatrito, S., Galleti, C., Maioli, M.G. and Battaglini, P.P. (1981) Cortical visual input to the orbito-insular cortex in the cat. *Brain Res.*, 221: 71–79.

Stein, B.E. (1978) Development and organization of multimodal representation in cat superior colliculus. *Fed. Proc.*, 37: 2240–2245.

Stein, B.E. and Arigbede, M.O. (1972) Unimodal and multimodal response properties of neurons in the cat's SC. *Exp. Neurol.*, 36: 179–196.

Stein, B.E. and Meredith, M.A. (1990) Multisensory integration. Neural and behavioral solutions for dealing with stimuli from different sensory modalities. *Ann. N.Y. Acad. Sci.*, 608: 51–70.

Stein, B.E. and Meredith, M.A. (1993) *The Merging of Senses*, MIT Press, Cambridge, MA.

Stein, B.E., Spencer, R.F. and Edwards, S.B. (1983) Corticotectal and corticothalamic efferent projections of SIV somatosensory cortex in cat. *J. Neurophysiol.*, 50: 896–909.

Tamai, Y. and Miyashita, E. (1989a) Eye movements following cortical stimulation in the ventral bank of the anterior ectosylvian sulcus of the cat. *Neurosci. Res.*, 7: 159–163.

Tamai, Y. and Miyashita, E. (1989b) Subcortical connections of an 'oculomotor' region in the ventral bank of the anterior ectosylvian sulcus in the cat. *Neurosci. Res.*, 7: 249–256.

Tortelly, A., Reinoso-Suarez, F. and Llamas, A. (1980) Projections from non-visual cortical areas to the superior colliculus demonstrated by retrograde transport of HRP in the cat. *Brain Res.*, 1988: 543–549.

Ungerleider, L.G. and Mishkin, M. (1982) Two cortical visual systems. In: D.J. Ingle, M.A. Goodale and R.J.W. Mansfield (Eds.), *Analysis of Visual Behavior*, MIT Press, Cambridge, MA, pp. 549–586.

Wallace, M. and Stein, B.E. (1994) Cross-modal synthesis in the midbrain depends on input from cortex. *J. Neurophysiol.*, 71: 429–432.

Wallace, M.T., Meredith, M.A. and Stein, B.A. (1992) Integration of multiple sensory modalities in the cat cortex. *Exp. Brain Res.*, 91: 484–488.

Wepsic, J.G. (1966) Multimodal sensory activation of cells in the magnocellular medial geniculate nucleus. *Exp. Neurol.*, 15: 299–318.

M. Norita, T. Bando and B. Stein (Eds.)
Progress in Brain Research, Vol 112
© 1996 Elsevier Science BV. All rights reserved.

CHAPTER 24

The development of topographically-aligned maps of visual and auditory space in the superior colliculus

Andrew J. King*, Jan W.H. Schnupp, Simon Carlile, Adam L. Smith and Ian D. Thompson

University Laboratory of Physiology, Parks Road, Oxford, OX1 3PT, UK

The role of the superior colliculus in attending and orienting to sensory stimuli is facilitated by the presence within this midbrain nucleus of superimposed maps of different sensory modalities. We have studied the steps involved in the development of topographically-aligned maps of visual and auditory space in the ferret superior colliculus. Injections of fluorescent beads into the superficial layers showed that the projection from the contralateral retina displays topographic order on the day of birth (P0). Recordings made from these layers at the time of eye opening, approximately 1 month later, revealed the presence of an adult-like map of visual space. In contrast, the auditory space map in the deeper layers emerged gradually over a much longer period of postnatal life. In adult ferrets in which one eye had been deviated laterally just before eye opening, the auditory spatial tuning of single units recorded in the contralateral superior colliculus was shifted by a corresponding amount, so that the registration of the visual and auditory maps was maintained. Chronic application of the NMDA-receptor antagonist MK801 disrupted the normal development of the auditory space map, but had no effect on the visual map in either juvenile or adult animals, or on the auditory map once it had matured. These findings indicate that visual cues may play an instructive role, possibly via a Hebbian mechanism of synaptic plasticity, in the development of appropriately tuned auditory responses, thereby ensuring that the neural representations of both modalities share the same coordinates. Changes observed in the auditory representation following partial lesions of the superficial layers at P0 suggest that these layers may provide the source of the visual signals responsible for experience-induced plasticity in auditory spatial tuning.

Introduction

The ability to re-direct attention and orientate to novel sensory stimuli requires that those stimuli are first localized. To achieve its sensorimotor function, the superior colliculus (SC) contains multiple, topographically-aligned maps of sensory space, as well as motor maps that mediate movements of the peripheral sense organs (for review see Stein and Meredith, 1993). This arrangement would appear to provide an efficient means by which any unimodal or multimodal stimulus can activate a common set of premotor neurons, leading to an appropriate orientation response that is directed toward the location of the sensory target.

The sensory receptive fields of neurons recorded in the intermediate and deep layers of the SC can occupy a substantial fraction of a hemifield, and the size and position of the different modality receptive fields of multisensory neurons are rarely precisely the same. However, within these large receptive fields, sensory neurons in the SC tend to respond most strongly to stimuli presented at much more restricted regions of space. When these so-called best positions or best areas are used to define the neurons' spatial preferences, the close correspondence between the different sensory representations, and particu-

---
*Corresponding author. Tel.: +44 1865 272523; Fax: +44 1865 272469; email:ajk@physiol.ox.ac.uk

larly between the maps of visual and auditory space (Knudsen, 1982; King and Palmer, 1983; Middlebrooks and Knudsen, 1984; King and Hutchings, 1987), becomes apparent.

Although they appear to share a common set of coordinates within the SC, the various sensory maps found there are constructed in different ways. The visual and somatosensory maps are derived from point-to-point projections from the retina and body surface respectively, whereas the map of auditory space is first generated within the brain by neurons that exhibit systematic variations in sensitivity to combinations of sound localization cues that result from the acoustical properties of the ears and head (King and Carlile, 1995). Because the sensory systems of different modalities initially represent spatial information in different coordinate frames, establishing and maintaining the alignment of the different sensory maps in the SC is not straightforward. For example, the registration of the maps should be degraded if an animal moves its eyes or pinnae relative to the head. However, recordings in awake primates (Jay and Sparks, 1987) and cats (Hartline et al., 1995; Peck et al., 1995) have shown that the responses of many auditory neurons in the SC are altered as the direction of gaze changes, in a manner that may at least partially transform the auditory space representation from ear and head-centred coordinates into retinocentric coordinates.

Setting up and maintaining the alignment of the maps during development is also problematic because of growth-related changes in the relative geometry of different sense organs. Numerous studies have shown that achieving a common topographic organization for the representation of visual and auditory space in the SC results from the developmental plasticity of auditory spatial tuning (reviewed by King and Carlile, 1995; Knudsen and Brainard, 1995). In this paper, we investigate the role of visual experience in this process. In particular, we examine the possibility that the visual map in the superficial layers provides the instructive signals for the development of deeper layer auditory responses that are tuned to corresponding regions of space.

Methods

All procedures were carried out on ferrets that were born and reared in the laboratory animal house.

Anatomical experiments

On the day of birth (P0), ferrets were anaesthetized with Saffan (Alphaxalone/Alphadolone acetate i.m.). A craniotomy was made to expose one SC, which, at birth, is not yet overlaid by the cortex, and the dura removed. Calibrated borosilicate tubing pipettes were used to make pressure injections of green or rhodamine (red) latex microspheres (Lumafluor, NY; diluted 1:10 in sterile saline) into the SC. One tracer was injected into the rostral part of the nucleus and the other into the caudal region. The space above the midbrain was filled with absorbable gelatine sponge (Sterispon) and the wound margins were sutured together. A local anaesthetic was applied to the scalp and, following recovery from anaesthesia, the animals were returned to their mother. Following a survival period of 24 h, the animals were re-anaesthetized and perfused transcardially with phosphate buffered saline (PBS) followed by 4% paraformaldehyde in 0.1 M phosphate buffer (pH 7.4). A small incision was made into the dorsal pole of the cornea to indicate the orientation of the eye. The eyes were removed, opened up, and the retina carefully dissected away from surrounding structures. The brain was also removed, sunk in 30% sucrose in 0.1 M phosphate buffer, and sectioned parasagitally at 50 μm on a freezing microtome.

Strabismus surgery

A simple outward deviation (exotropic strabismus) of the left eye was surgically induced by removal of the medial rectus muscle under halothane and nitrous oxide anaesthesia on P27–28. The eyelids were temporarily re-closed by suturing with absorbable chromic collagen, as they normally remain closed until about P32. The right eye was enucleated at the same time. The animals were returned to their home cages until they were fully grown.

Elvax implants

Full details are given in Schnupp et al. (1995). Briefly, Elvax 40P pellets were impregnated with tritiated(+)-10,11-dihydro-5-methyl-5H-dibenzo(a,d)-cyclohepten-5,10-imine maleate (^{3}H-MK801) to a final concentration of either 1 mM or 10 mM. P25–27 ferrets were anaesthetized with Saffan (2 ml/kg). After making a craniotomy and aspirating the cortex above the midbrain, 400-μm thick Elvax slices were placed on the dorsal surface of the SC. The cavity was filled with Sterispon, the bone flap replaced, and the wound margins sutured together. The animals were allowed to survive until P61–70 when they were used for electrophysiological recording. Elvax slices containing MK801 were also implanted in three adult ferrets at P108–115. These animals were allowed to recover and returned to the animal house for 5–6 weeks until the terminal recording experiment was carried out on P136–158.

Superior colliculus lesions

A partial lesion of the superficial layers of the SC was made in Saffan-anaesthetized ferrets on the day of birth. Following exposure of the midbrain, the superficial layers were sectioned medio-laterally and the caudal region removed by aspiration. The extent of the lesion varied between animals from about 40–90% of the total area of the superficial layers. At the conclusion of the surgery the animals were recovered and returned to their mothers. When they reached adulthood, the animals were prepared for recording.

Electrophysiological recording

Full details are given in King and Hutchings (1987) and Schnupp et al. (1995). Juvenile ferrets were anaesthetized with either Saffan (2 ml/kg) or a mixture of Domitor (medetomidine hydrochloride, 250 μm/kg) and ketamine (60 mg/kg). Anaesthesia was then maintained by continuous i.v. infusion of these agents. All the adult ferrets used in this study were first anaesthetized with Saffan and then paralysed with Flaxedil (gallamine triethiodide). Anaesthesia and paralysis were maintained with sodium pentobarbital (1 mg/kg per h) and Flaxedil (20 mg/kg per h) respectively. In all ferrets, the trachea was cannulated, body temperature was maintained at 39°C, and the electrocardiogram was monitored. The electroencephalogram and end-tidal CO_2 were also monitored continuously where paralysis was used. A craniotomy was made above the cortex overlying the SC and a minimal metal headholder, which supported the animal from behind, was attached to the skull. The juvenile ferrets were not paralysed, so eye position was stabilized with fine sutures that connected the conjunctiva to the surrounding skin.

All recordings were carried out in an anechoic chamber. The cortex above the midbrain was left intact, except where Elvax sheets had previously been implanted. A flashing LED or filament bulb was used as the visual stimulus. The auditory stimulus consisted of 100-ms broadband noise bursts delivered from a KEF T27 loudspeaker. Both visual and auditory stimuli could be presented from any direction with respect to the animal's head by means of a robotic hoop that was controlled from outside the chamber. Neural activity was recorded with a tungsten microelectrode that was lowered into the brain using a remotely-controlled, motorized microdrive. The neural signals were filtered, amplified and digitized. In most cases, single units were discriminated on the basis of spike amplitude alone or spike shape (see Schnupp et al., 1995). In one set of experiments, stimulus-evoked responses were quantified by applying a fast Fourier transform to the amplified neural signal, in order to estimate the power spectral density in multi-unit clusters (see King and Carlile, 1994 for details). The magnitude of the stimulus-evoked response was estimated for both single-unit and power spectral density measurements by subtracting the values in a spontaneous, control window from those in the response window.

Electrolytic lesions were made in each electrode track to allow subsequent reconstruction of the recording sites. At the end of the recording session, the animal was deeply anaesthetized and perfused transcardially with PBS followed by 10% formaldehyde in PBS. The brainstem was removed, cryoprotected and sectioned on a freezing microtome.

Results

Development of the visual map in the superficial layers

We have examined the evolution of the visual map in the superficial layers of the SC by using fluorescent tracers to define the topographic order in the retinocollicular projection and by mapping the visual receptive fields of the neurons found there. Fig. 1 shows the pattern of retrograde labelling in a flattened whole-mount of the retina following injections of rhodamine and green latex beads into the contralateral SC at P0. The rhodamine beads, which were injected into rostro-lateral SC, were concentrated in cell bodies in the inferior temporal region of the retina. Multiple injections of green beads into caudo-medial regions of the SC resulted in labelling of ganglion cells in the superior retina, particularly on the nasal side. Although labelled neurons were found in large regions of the retina, there was very little overlap in the distribution of ganglion cells containing each tracer. This shows that the retinocollicular projection already possesses topographic order on the day of birth.

Eye opening in ferrets occurs naturally at around P32. Although the optics were cloudy at this time, we had no difficulty in mapping the visual receptive fields of neurons in the superficial layers. The variation in the azimuthal centre of the receptive fields with recording site is shown in

Fig. 2 for data pooled from three ferrets at P33–37. The visual azimuths varied systematically from the anterior midline near the rostral end of the SC to 130° into the contralateral hemifield at the caudal end. The map of visual space remained unchanged in animals recorded in the second and third postnatal months and closely resembled that found in normal, adult ferrets (see Fig. 4).

Development of the auditory map in the deeper layers

The auditory responses recorded in ferrets at just over one month of age were mostly rather poorly tuned for sound location and there was little indication of any topographic order in the representation (King, 1993; King and Carlile, 1995). Consequently, the close alignment of the visual and auditory maps, which is characteristic of adult animals (see Fig. 5), was not apparent in the juvenile ferrets. This is illustrated in Fig. 3 in which the preferred sound azimuths of multi-unit responses recorded in the deeper layers are plotted against the visual azimuths of superficial layer units recorded in the same vertical electrode tracks. During the second postnatal month, the auditory responses became more sharply tuned for sound location and the registration with the overlying visual map gradually emerged.

Effect of early eye deviation on the development of the visual and auditory maps

A comparison of the map of visual azimuth in the superficial layers of the SC in normal adult ferrets and in animals reared with a lateral deviation of the contralateral eye is shown in Fig. 4. The relative magnification of this dimension of the visual field is not uniform as a slightly greater area of the SC is used for representing anterior visual space than for more peripheral receptive fields. Consequently, more of the variance in the relationship between visual receptive field position and SC recording site was accounted for by a second-order polynomial function. In the strabismic ferrets, measurements with a reversible ophthalmoscope revealed that the left eye was deviated laterally by 15–20°. At each recording site in the SC, the head-centred visual receptive field positions in these animals tended to be more peripheral than those mapped in the normals. The polynomial function that provided the best fit to these data was virtually identical to that fitted to the normal data, except that it was shifted laterally at its midpoint by 15°, which is very close to the mean difference in eye position between the two groups.

We have previously reported that, compared to the responses in normal ferrets, the preferred sound directions of auditory units recorded in these animals were shifted laterally by a corresponding amount (King et al., 1988). We have now extended these observations by concentrating on the difference between the azimuthal coor-

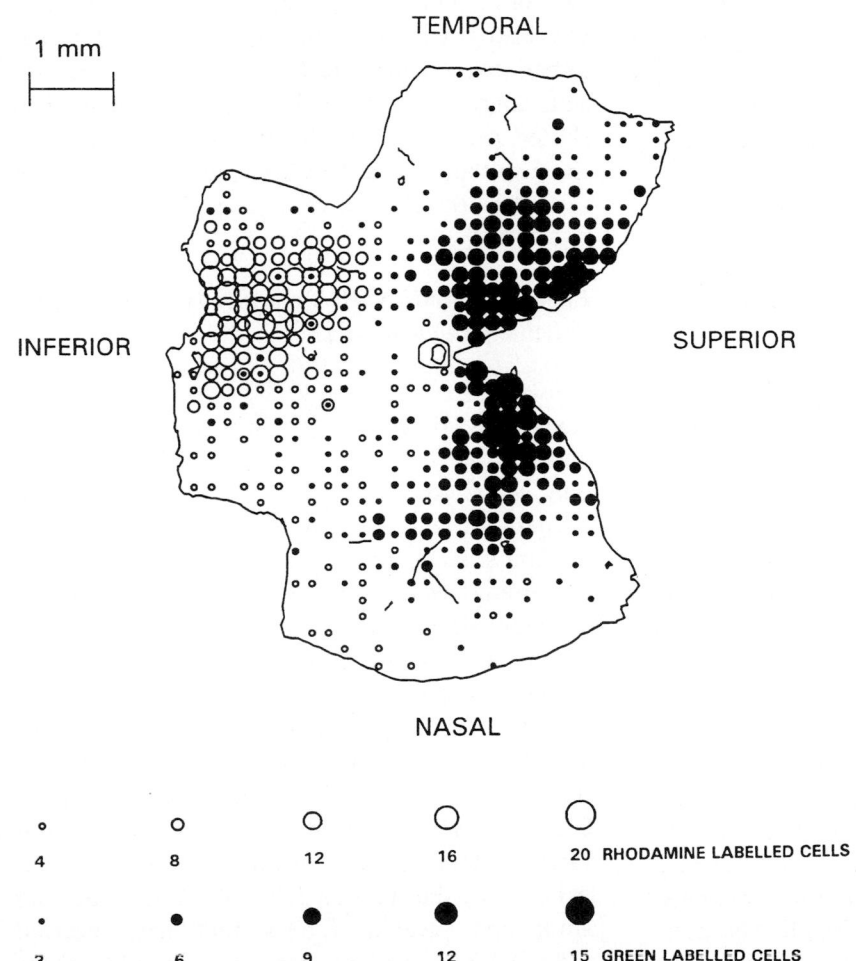

TEMPORAL

1 mm

INFERIOR

SUPERIOR

NASAL

°	O	O	O	O	
4	8	12	16	20 **RHODAMINE LABELLED CELLS**	

·	•	●	●	●	
3	6	9	12	15 **GREEN LABELLED CELLS**	

Fig. 1. Flattened whole-mount of the retina showing the distribution of retrogradely-labelled ganglion cells following a single 50-nl injection of rhodamine beads into the rostro-lateral region and multiple injections of green beads (total of 75 nl) into the caudo-medial region of the contralateral SC in a P0 ferret. The retina was sampled at 200-μm intervals with 75 μm × 75 μm grid boxes. The open circles represent cells containing rhodamine beads, while the filled circles represent cells labelled by green beads. The number of labelled cells in the grid boxes sampled in each part of the retina is shown by the size of the circles, as indicated below the figure.

340

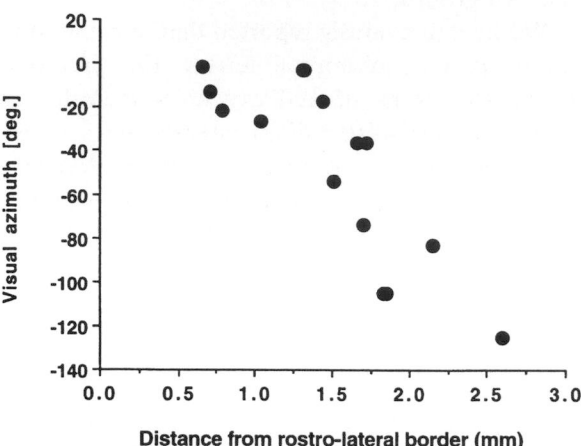

VISUAL MAP AT EYE OPENING

Fig. 2. Representation of visual azimuth in the superficial layers of the SC in ferrets recorded at P33–37. The best azimuths of multi-unit responses are plotted against the distance of each recording site from the rostro-lateral border of the nucleus. These data are expressed in this and subsequent figures in head-centred co-ordinates. 0° lies directly in front of the animal and negative numbers denote the hemifield contralateral to the recording site.

dinates of visual and auditory units recorded in the same electrode penetrations. We have considered only those auditory single units that were tuned to single positions in space at sound levels of around 25 dB above threshold. In doing so, we have excluded a small proportion of the units recorded in both the normal and strabismic animals that were classified as either being too broadly tuned to be attributed a single best position or as having bilobed azimuth response profiles (see Schnupp et al., 1995 for details). A comparison of the visual and auditory best azimuths is illustrated in Fig. 5. The spatial representations of the two modalities were closely aligned in both normal and strabismic animals. The superficial layer visual azimuth was subtracted from the auditory azimuth of each unit recorded in the deeper layers to yield auditory-visual misalignment values. These are plotted as histograms in Fig. 5. For both groups, the histograms appeared to be normally distributed and clustered around 0. Following the chronic change

in eye position in the strabismic ferrets, the visual receptive fields were centred at more contralateral positions than in normal animals. Had the auditory map remained uncompensated, we would have expected to see a systematic shift in auditory-visual misalignment of about 15° toward positive values in the strabismic ferrets. Instead, we observed a mean misalignment of nearly −6°, although the mean values were not significantly different in the two groups of animals (t_{95} = 1.51; P = 0.14). By comparing the variance of the misalignments, we found that the widths of the histograms were not significantly different either (two-tailed F test; P = 0.08). These results suggest that the map of auditory space had shifted rostrally within the SC to compensate for the displacement in the visual field representation caused by the chronic change in eye position.

Effect of chronic NMDA-receptor blockade on the visual and auditory maps

Electrophysiological recordings were made from animals that had received MK801-Elvax implants only if it was established during the surgical preparation on the day of recording that the implant still covered the whole of at least one SC. Recordings were commenced unilaterally within an hour of removing the Elvax. Compared to control animals, we noted no discernible change in the visual multi-unit activity recorded in the superficial layers of the SC. The topographic organization of the representation of visual azimuth is shown for adult and juvenile ferrets in Fig. 6. The best azimuths of the visual receptive fields varied systematically between the rostral and caudal ends of the nucleus. The visual maps in both adult and juvenile ferrets that had received MK801-Elvax implants closely resembled those found in unoperated, normal control animals recorded at corresponding ages.

In contrast to the normal visual map in the superficial layers, we found that chronic release of the NMDA receptor antagonist MK801 did affect the auditory responses in the deeper layers. In all animals treated with Elvax containing MK801, the proportion of auditory units tuned to single regions of space was significantly reduced compared to unoperated, age-matched animals or to animals that had received Elvax implants containing dimethyl sulfoxide (DMSO) as drug-free controls (Schnupp et al., 1995). We observed an age-dependent effect on the topography of the auditory representation. In the adult MK801 group, plotting the azimuthal best positions of these tuned units against the visual best azimuths of the superficial layer responses revealed a close alignment between the two maps, as further illustrated by a narrow auditory-visual misalignment histogram (Fig. 7). The spread (variance) in misalignment values was not significantly different from that found in either normal, unoperated adults (two-tailed F test, $P - 0.48$) or drug-free, adult controls ($P = 0.84$). In contrast, the best azimuths of the tuned units recorded in the SC of juvenile ferrets reared with MK801-Elvax implants were poorly correlated with the histolog-

ical coordinates of their recording sites. Because the visual map in these animals was essentially normal (Fig. 6), a comparison of visual and auditory best azimuths showed that there was considerable scatter in the relationship between the two representations (Fig. 7). The variance in the auditory-visual misalignment values for these animals was significantly greater than that found in both juvenile, unoperated ($P < 0.02$) and drug-free ($P < 0.01$) control groups.

Effect of partial lesions of the superficial layers on the development of the visual and auditory maps

Visual responses recorded from the remaining rostral region of the superficial layers in the SC of adult ferrets in which the caudal pole had been aspirated on P0 appeared to have normal receptive fields. These units were tuned to azimuthal locations that covered a restricted region of the anterior hemifield and fell either within or very close to the normal range of visual azimuths represented in this part of the nucleus in adult

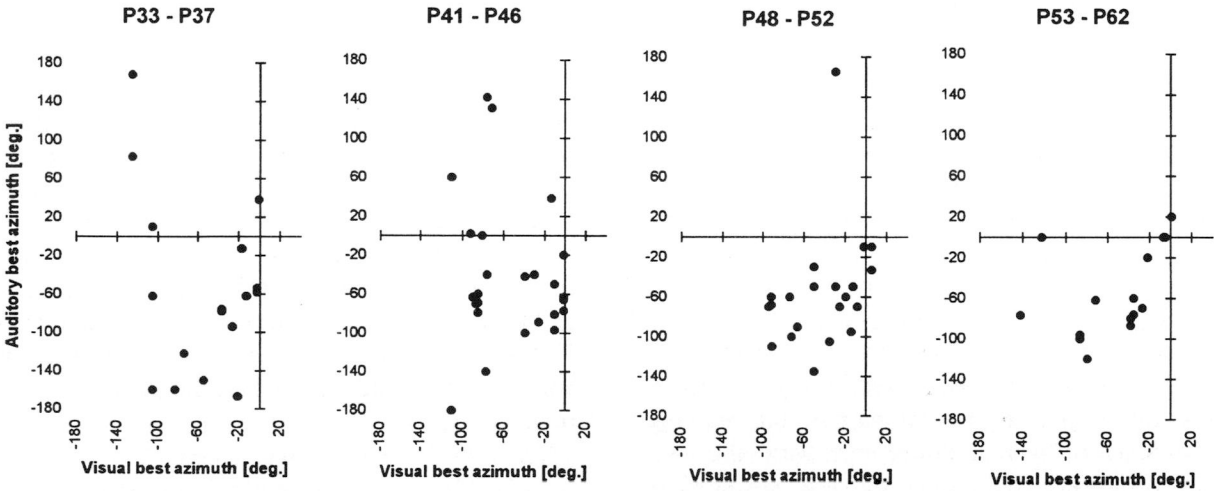

Fig. 3. Development of topographically-aligned visual and auditory representations in the SC. These data were obtained from juvenile ferrets at the postnatal ages indicated at the top of the figure. For each vertical electrode penetration, the visual best azimuth of multi-unit activity recorded in the superficial layers is plotted against the auditory best azimuth of multi-unit responses recorded in the deeper layers. These estimates of spatial selectivity are based on power spectral density measurements, whereas the number of sound-evoked spikes from single units is used in the remainder of the figures. We have previously shown that the two recording techniques produce equivalent results (King and Carlile, 1994).

342

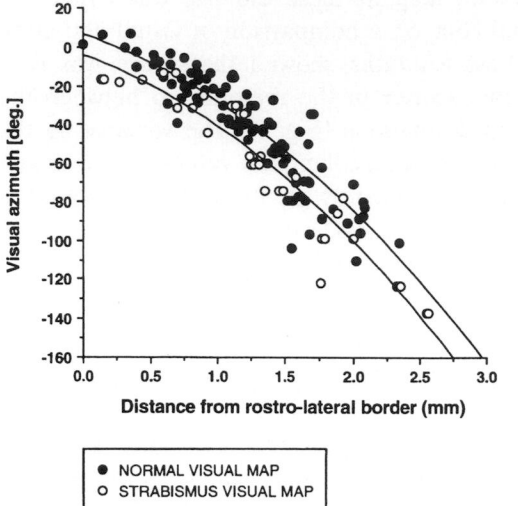

Fig. 4. Effect of lateral deviation of the eye on the representation of visual azimuth in the superficial layers of the SC. The filled circles represent data from normal, adult ferrets, while the open circles indicate the responses recorded in adult ferrets in which an exotropic strabismus had been surgically induced in the contralateral eye on P27–28. The lines correspond to the second-order polynomial functions fitted to each set of data. The visual receptive fields of the SC neurons remained aligned with the retina; when expressed in head-centred co-ordinates they became displaced relative to the normal group by an amount equivalent to the lateral shift in eye position.

animals (Fig. 8A). The extent of the visual field representation was therefore determined by the size of the lesion. The electrode penetrations from which these units were recorded were then extended into the deeper layers. The auditory units we encountered there were tuned to very similar anterior locations. Again their best azimuths fell within the normal range of values associated with the rostral part of the SC (Fig. 8B). Some of these auditory units were also visually responsive. The auditory best positions of both unimodal and bimodal neurons were in close correspondence with the receptive fields of the visual units recorded in the overlying superficial layers, and the variance of the auditory-visual misalignments did not differ significantly from

that measured for the equivalent rostral region of the SC in normal animals ($P = 0.13$).

We determined the extent of the lesioned area both histologically and by the absence of characteristically strong visual drive in the caudal region of the SC. Most of the response properties of the auditory units recorded beneath this lesioned area appeared to be normal and the majority were tuned to single spatial regions. Despite the absence of visual activity in the superficial layers, approximately one third of these deep layer auditory units were also visually responsive (presumably as a result of descending cortical inputs). The azimuthal best positions of the units are plotted against the rostro-caudal coordinates of their recording sites in Fig. 9. The great majority were tuned to sound locations within the contralateral hemifield. But whereas many had best azimuths that fell within the normal range, there were also numerous units with best azimuths not normally seen in the corresponding regions of the SC. Because of the lack of a visual map in the superficial layers against which to compare these preferred sound directions, we calculated the angular difference between the observed auditory best azimuths and the value predicted for each recording site from the polynomial function that provided the best fit to the representation of sound azimuth in normal, adult ferrets. The variance of the differences between observed and predicted best azimuths for the auditory units for which the superficial layers were missing was significantly greater than the variance associated with either the rostral region with intact superficial layers in the same animals ($P \ll 0.01$) or the normal control group ($P \ll 0.01$). Interestingly, the auditory best azimuths appeared to be equally scattered in both unimodal and bimodal (visual-auditory) neurons.

Discussion

Maps of visual and auditory space in the SC share common co-ordinates. As well as providing a basis for delivering modality-independent spatial

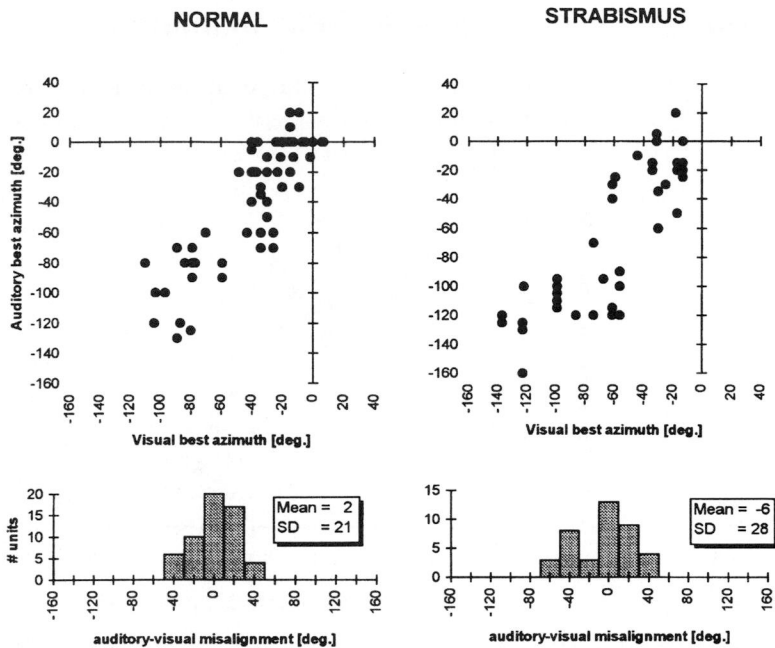

NORMAL

STRABISMUS

Fig. 5. Effect of lateral deviation of the eye on the registration of the visual and auditory representations in the SC. The data on the left are from normal, adult ferrets and those on the right are from adult ferrets in which an exotropic strabismus had been surgically induced in the contralateral eye on P27–28. The histograms show the angular difference between the best azimuths of tuned auditory units recorded in the deeper layers and those of visual multi-unit responses recorded in the superficial layers of the same vertical electrode penetrations. Neither the mean nor the variance of the auditory-visual mis-alignments differ significantly between the two groups. Adapted from King et al. (1988).

signals to the neural pathways that control orientation behaviour, the alignment of the visual and auditory representations allows multisensory cues to be integrated in ways that may improve the accuracy of orienting behaviour (Stein and Meredith, 1993). In this report, we have examined the developmental steps that lead to the formation of spatially matched visual and auditory maps in the ferret SC. We have shown that the visual map in the superficial layers matures some weeks before the higher-order map of auditory space in the deeper layers, and that the activity-dependent process of auditory map development is regulated by visual signals that may arise from the superficial layers.

Development of the visual map

We found that the retinocollicular projection has topographic order at P0. Changes in the number

of retinal ganglion cells (Henderson et al., 1988; Cucchiaro, 1991; Thompson and Morgan, 1993) and in the specificity of this projection (Snider and Chalupa, 1993) continue to occur until the end of the first postnatal week. Nevertheless, by the time of eye opening at around P32, the retinocollicular pathway should be topographically mature. This is supported by the results of our recordings from juvenile ferrets within a few days of the eyes opening, which revealed the presence of an adult-like representation of visual space in the superficial layers. In adult animals in which an exotropic strabismus of the contralateral eye had been induced just before normal eye opening, we found that the visual receptive fields were centred on locations that were, on average, displaced laterally by 15° compared to the normal map. We can therefore conclude that the retinocollicular projection was unaltered by this

Chronic MK-801 in adulthood

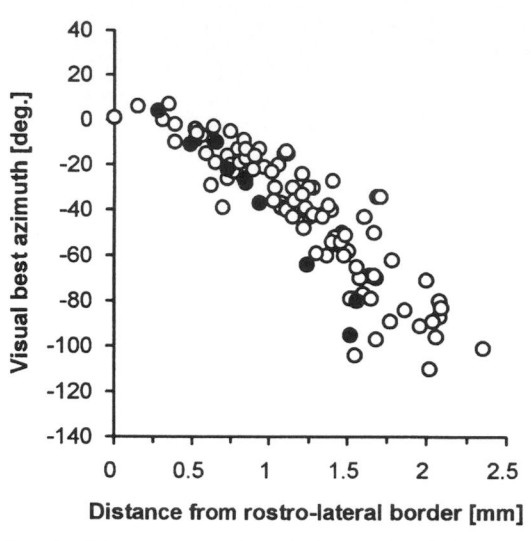

Chronic MK-801 in infancy

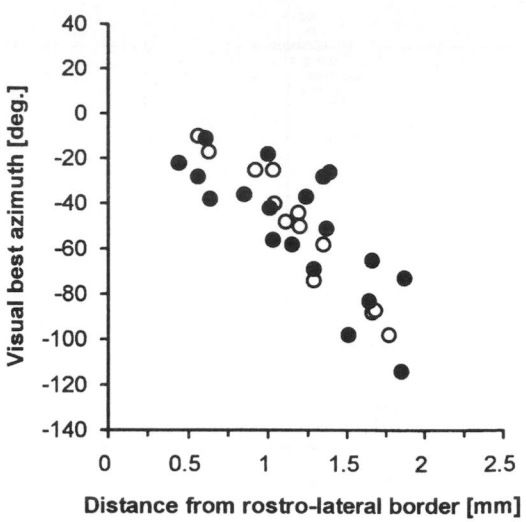

Fig. 6. The representation of visual azimuth in the superficial layers of adult (left) and juvenile (P61–70, right) ferrets. The azimuthal best positions of multi-unit responses are plotted against the distance of each recording site from the rostro-lateral border of the SC. The filled circles represent data from animals in which MK801-Elvax implants had been placed on the dorsal surface of the SC 5–6 weeks before recording. The open circles represent the data obtained from normal, unoperated animals. Adapted from Schnupp et al. (1995).

procedure, as the lateral shift in the visual field representation matched the change in eye position.

We were also unable to detect any developmental change in the visual map in the superficial layers following chronic treatment with the NMDA receptor antagonist MK801. Other reports have indicated a role for NMDA receptors in the refinement of retinocollicular topography (Cline and Constantine-Paton, 1989; Simon et al., 1992). However, in contrast to these studies, we implanted the Elvax on the SC after the early postnatal changes in the projection had been completed. The early maturation of the retinocollicular pathway in the ferret may also explain why the visual map in the SC did not appear to become reorganized following a partial ablation of the superficial layers on the day of birth. In neonatal hamsters, partial SC lesions can result in compression of the entire contralateral visual field onto the remaining portion of the superficial lay-

ers (see Finlay, this volume). However, we found that, in ferrets, visual units recorded in this region were tuned to the same azimuthal locations as in normal animals. The visual map was therefore abruptly truncated at the caudal end of the intact superficial layers. Whereas the ferret retinocollicular projection is topographically organized at P0, the hamster pathway is still disordered at this age (Thompson and Cordery, 1994). The capacity for reorganization of the pattern of connections from the retina to the SC following manipulation of neural activity or reduction in target volume therefore appears to be limited by the developmental maturity of the projection at the time of the surgery.

Visual experience induced plasticity in the developing map of auditory space

In contrast to visual responses in the superficial layers, the auditory responses recorded in the

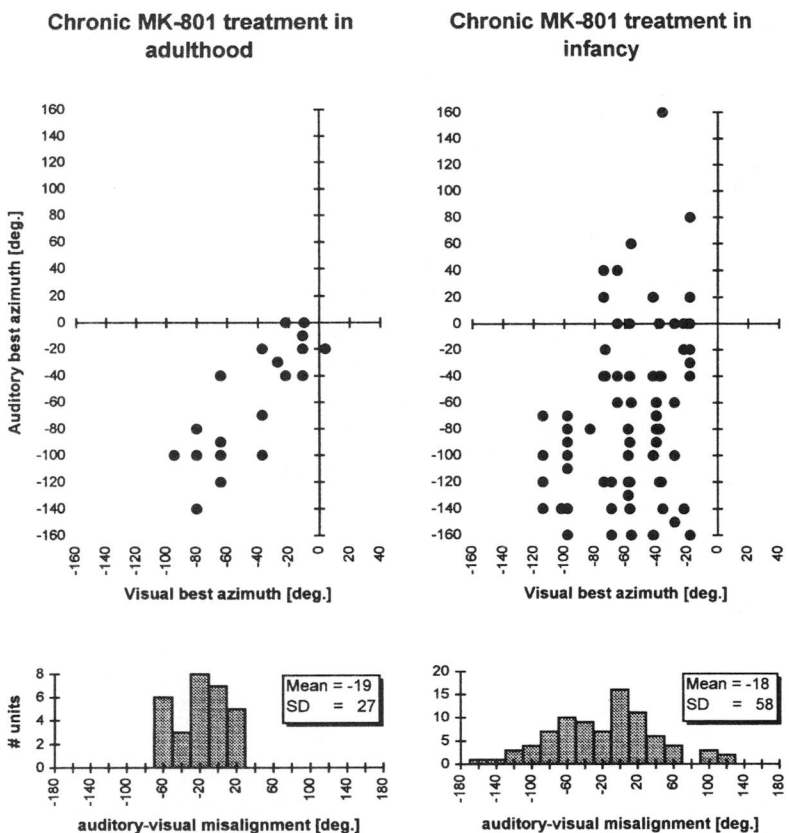

Fig. 7. Relationship between the representations of visual azimuth in the superficial layers and auditory azimuth in the deeper layers of the SC of adult (left) and juvenile (P61–70, right) ferrets that had received MK801-Elvax implants 5–6 weeks before recording. The histograms plot the angular difference between the best azimuths of auditory and visual units recorded in each electrode track. The histogram is significantly wider in the juvenile animals. Adapted from Schnupp et al. (1995).

intermediate and deep layers were initially poorly tuned for sound location and exhibited very little order in the distribution of azimuthal locations to which they responded most strongly. The topographic order in the auditory representation, and the alignment with the overlying visual map, gradually emerged over a course of several weeks following the onset of hearing. A similar result, albeit with a different time course, has been reported for the guinea pig (Withington-Wray et al., 1990). This delayed appearance of the auditory space map can be understood in terms of the way in which it is constructed. Whereas the visual map results from a topographic projection from the retina, an equivalent projection from the

cochlea produces a neural map of sound frequency, not sound source position. The topographic representation of auditory space must therefore be computed within the brain by tuning the neurons to different localization cue values. In mammals, SC neurons are sensitive to monaural spectral cues and interaural level differences (Hirsch et al., 1985; Wise and Irvine, 1985; Middlebrooks, 1987; Carlile and King, 1994; King et al., 1994). We have shown that, in the ferret, the spatial pattern of these cues continues to change for a period of several weeks after the onset of hearing (King and Carlile, 1995). Thus, the acoustical basis for the map of auditory space is continually changing during much of the period over

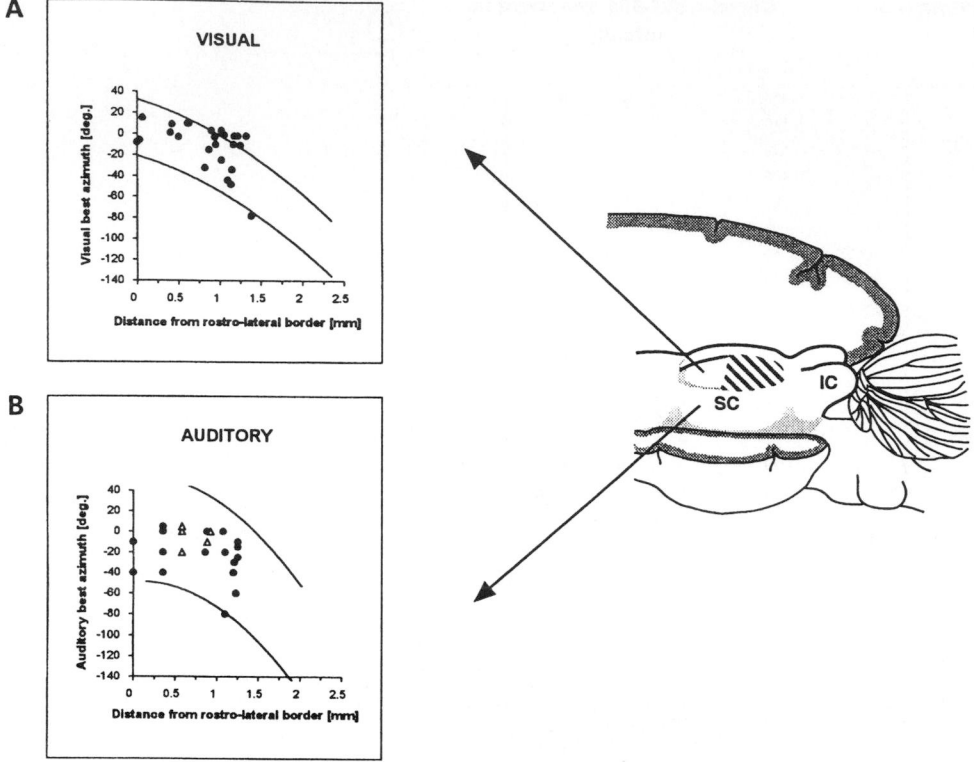

Fig. 8. (A) Visual representation in the remaining superficial layers of the SC in adult ferrets following aspiration of the caudal portion of these layers at P0. (B) Auditory representation in the deeper layers of the same vertical electrode tracks. The azimuthal best positions are plotted against the distance of their recording sites from the rostro-lateral border of the SC. The open triangles in (B) represent the auditory best azimuths of units that were also visually responsive, whereas the filled circles indicate the values obtained from unimodal auditory neurons. The two lines in each panel represent two standard deviations on either side of the polynomial function that provides the best fit to the visual and auditory maps in normal ferrets. The hatched area of the SC in the parasagittal view on the right indicates the aspirated region of the superficial layers.

which the topographic order in the representation emerges.

The correlation between the developmental time course for the localization cues and the map of auditory space suggests that the establishment of the topographic alignment of the visual and auditory maps may, at least in part, be limited by the maturation of these cues, which, in turn, depends on growth of the head and outer ears. However, sensory experience also plays a crucial role in refining the auditory map, so that a close correspondence with the visual map is achieved in spite of individual variations in the auditory localization cue values and in the relative position of the eyes and ears. For example, distorting the binaural localization cues by plugging one ear in infancy leads to a compensatory adjustment in the auditory map so that it remains precisely aligned with the visual map (Knudsen, 1985; King et al., 1988). That visual cues may play a dominant role in aligning the two maps is suggested by our finding that displacement of the visual receptive fields relative to the head, caused by lateral deviation of the contralateral eye in infancy, is followed by an equivalent change in the auditory representation. Rotation of the eye disrupts the development of the auditory map, but to a much lesser extent if the animals are visually deprived

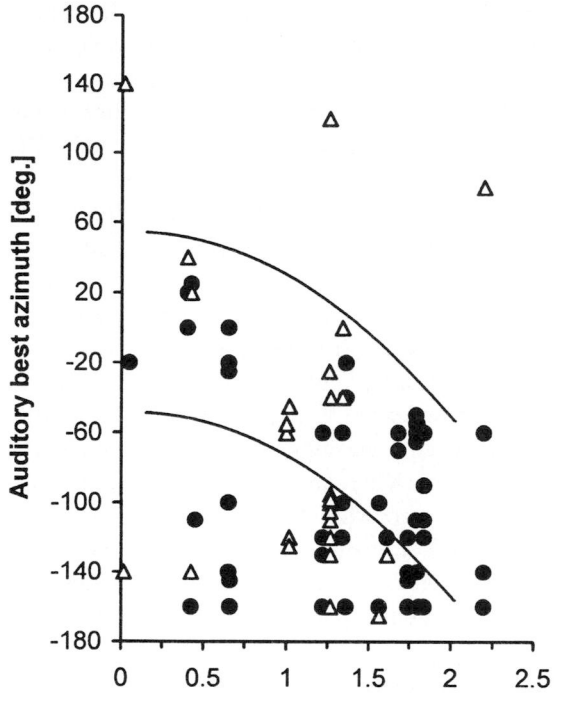

Fig. 9. Auditory representation in the adult ferret SC following aspiration of the caudal region of the superficial layers at P0. The auditory best positions of single units recorded in the intermediate and deep layers are plotted against the distance of their recording sites from the rostro-lateral border of the SC for those electrode tracks where the superficial layers were missing. The open triangles represent the auditory best azimuths of units that were also visually responsive, whereas the filled circles indicate the values obtained from unimodal auditory neurons. The two lines each represent two standard deviations on either side of the polynomial function that provides the best fit to the map of sound azimuth in normal ferrets.

at the same time, suggesting that the shift in auditory spatial tuning results from the displacement of the visual receptive fields rather than the abnormal eye position (King and Carlile, 1995). Optical displacement of the visual field in barn owls reared with prisms mounted in front of their eyes also leads to a corresponding change in the auditory space map (Knudsen and Brainard, 1991). On the other hand, if the visual image is blurred (by eyelid suture), the auditory representation is

degraded to some extent (Knudsen et al., 1991; Withington, 1992; King and Carlile, 1993). Taken together, these observations suggest that a crude map of auditory space can develop in the absence of visual cues. However, when available, visual information can influence the development of the auditory map, and adjust its topographic order to match that of the visual map in the SC.

Basis for visual instruction of the developing auditory space map

A Hebbian mechanism of synaptic plasticity could provide the basis for establishing the alignment of the visual and auditory maps in the SC. Support for this possibility is provided by our finding that chronic application of MK801 during the period when the topographic order in the auditory representation normally emerges, results in a degraded map and, consequently, a much poorer alignment with the unchanged visual map. On the other hand, the registration of the maps was unaltered by treatment of adult animals with MK801-Elvax implants. The activation of NMDA-type glutamate receptors may provide a means of detecting temporal correlations among simultaneously active synaptic inputs (Fox and Daw, 1993), and thereby act as a trigger for the activity-dependent steps that lead to the emergence of a precise and unambiguous representation of auditory space in the SC. Feldman et al. (1995) have also reported that NMDA-receptor mediated currents may be particularly involved in expressing the visually-guided changes in the auditory receptive fields of neurons in the barn owl optic tectum. The involvement of these receptors in the SC is consistent with a Hebbian mechanism in which more accurate visual signals could lead to the selective strengthening of auditory inputs that convey information from the same region of space.

Source of instructive visual signals

Most auditory neurons in the barn owl optic tectum are also visually responsive (Knudsen,

1982). Optical displacement of the visual field has been reported to shift the auditory receptive fields of not only these units but also those of apparently unimodal units recorded in the external nucleus of the inferior colliculus, which projects topographically to the tectum (Brainard and Knudsen, 1993). The only equivalent study in mammals has shown that dark rearing disrupts the development of auditory spatial selectivity in the guinea pig SC, but has no effect on the ICX (Binns et al., 1995). By using [^{3}H]MK801, we were able to determine the area over which this NMDA-receptor antagonist is released (Schnupp et al., 1995). Our results suggest that MK801 released from the Elvax penetrated at least as far as 800 μm below the SC surface and therefore reached both the superficial and intermediate layers of the nucleus. Although further experiments will be required to verify this, we currently believe that the changes that we have observed in the developing auditory map are most likely to be localized to the SC itself, rather than the sources of afferent input.

Many auditory neurons in the SC also receive visual inputs, which most likely originate from extrastriate visual areas of the cortex (reviewed in Stein and Meredith, 1993). However, their visual receptive fields tend to be much larger than those of neurons recorded in the superficial layers. Moreover, the proportion of deep SC neurons that receive multisensory inputs is much smaller in early postnatal life (Wallace et al., 1993). If the site of developmental visual-auditory interaction does reside within the SC, an alternative source of guiding visual input would appear to be needed. Although the superficial and deeper layers of the SC have been regarded as anatomically and physiologically distinct regions (Casagrande et al., 1972; Edwards et al., 1979), connections are now known to exist between them (e.g. Mooney et al. 1984, 1992; Behan and Appell, 1992). Our finding that the auditory representation was disrupted only in the region beneath the ablated superficial layers is consistent with the possibility that those layers may provide local signals that calibrate the development of the underlying audi-

tory neurons. That bimodal neurons in the region of the SC where the superficial layers were missing also tended to exhibit aberrant auditory best positions would seem to indicate that other visual afferents to the SC were unable to rescue the spatial tuning of the auditory responses.

At first sight, the lack of a measurable visual input to all acoustically-responsive neurons, particularly during early life, would appear to question the significance of synchronized activity in the superficial layers for the development of auditory spatial tuning. However, visual stimuli can alter the responses of these neurons to sound, even if, by themselves, they do not evoke a change in firing rate (King and Palmer, 1985). It may therefore be necessary to employ more sensitive cross-correlation techniques to reveal the influence of activity in the superficial layers on the responses of neurons in the deeper layers.

Acknowledgements

We are grateful to Pat Cordery for excellent technical assistance and to the Wellcome Trust for financial support. Andrew King is a Wellcome Senior Research Fellow (grant number 031316 · 90/Z) and Jan Schnupp is a Wellcome Prize Student (039456 · 93/Z). Simon Carlile was supported by a Beit Memorial Fellowship and Adam Smith by a Wellcome Vision Research Training fellowship (034242 · 91/Z).

References

Behan, M. and Appell, P.P. (1992) Intrinsic circuitry in the cat superior colliculus: projections from the superficial layers. *J. Comp. Neurol.*, 315: 230–243.

Binns, K.E., Withington, D.J. and Keating, M.J. (1995) The developmental emergence of the representation of auditory azimuth in the external nucleus of the inferior colliculus: the effects of visual and auditory deprivation. *Dev. Brain Res.*, 85: 14–24.

Brainard, M.S. and Knudsen, E.I. (1993) Experience-dependent plasticity in the inferior colliculus: a site for visual calibration of the neural representation of auditory space in the barn owl. *J. Neurosci.*, 13: 4589–4608.

Carlile, S. and King, A.J. (1994) Monaural and binaural spectrum level cues in the ferret: acoustics and the neural representation of auditory space. *J. Neurophysiol.*, 71: 785–801.

Casagrande, V.A., Harting, J.K., Hall, W.C., Diamond, I.T. and Martin, G.F. (1972) Superior colliculus of the tree shrew: a structural and functional subdivision into superficial and deep layers. *Science*, 177: 444–447.

Cline, H.T. and Constantine-Paton, M. (1989) NMDA receptor antagonists disrupt the retinotectal topographic map. *Neuron*, 3: 413–426.

Cucchiaro, J.B. (1991) Early development of the retinal line of decussation in normal and albino ferrets. *J. Comp. Neurol.*, 312: 193–206.

Edwards, S.B., Ginsburgh, C.L., Henkel, C.K. and Stein, B.E. (1979) Sources of subcortical projections to the superior colliculus in the cat. *J. Comp. Neurol.*, 184: 309–30.

Feldman, D.E., Brainard, M.S. and Knudsen, E.I. (1996) Newly learned auditory responses mediated by NMDA receptors in the owl inferior colliculus. *Science*, 270: 525–528.

Fox, K. and Daw, N.W. (1993) Do NMDA receptors have a critical function in visual cortical plasticity? *Trends Neurosci.*, 16: 116–122.

Hartline, P.H., Pandey Vimal, R.L., King, A.J., Kurylo, D.D. and Northmore, D.P.M. (1995) Effects of eye position on auditory localization and neural representation of space in superior colliculus of cats. *Exp. Brain Res.*, 104: 402–408.

Henderson, Z., Finlay, B.L. and Wikler, K.C. (1988) Development of ganglion cell topography in ferret retina. *J. Neurosci.*, 8: 1194–1205.

Hirsch, J.A., Chan, J.C.K. and Yin, T.C.T. (1985) Responses of neurons in the cat's superior colliculus to acoustic stimuli. I. Monaural and binaural response properties. *J. Neurophysiol.*, 53: 726–745.

Jay, M.F. and Sparks, D.L. (1987) Sensorimotor integration in the primate superior colliculus. II. Coordinates of auditory signals. *J. Neurophysiol.*, 57: 35–55.

King, A.J. (1993) The Wellcome Prize Lecture. A map of auditory space in the mammalian brain: neural computation and development. *Exp. Physiol.*, 78: 559–590.

King, A.J. and Carlile, S. (1993) Changes induced in the representation of auditory space in the superior colliculus by rearing ferrets with binocular eyelid suture. *Exp. Brain Res.*, 94: 444–455.

King, A.J. and Carlile, S. (1994) Responses of neurons in the ferret superior colliculus to the spatial location of tonal stimuli. *Hear. Res.*, 81: 137–149.

King, A.J. and Carlile, S. (1995) Neural coding for auditory space. In: M. S. Gazzaniga (Ed.), *The Cognitive Neurosciences*, MIT Press, pp. 279–293.

King, A.J. and Hutchings, M.E. (1987) Spatial response properties of acoustically responsive neurons in the superior colliculus of the ferret: a map of auditory space. *J. Neurophysiol.*, 57: 596–624.

King, A.J. and Palmer, A.R. (1983) Cells responsive to free-field auditory stimuli in guinea-pig superior colliculus: dis-

tribution and response properties. *J. Physiol. Lond.*, 342: 361–381.

King, A.J. and Palmer, A.R. (1985) Integration of visual and auditory information in bimodal neurones in the guinea-pig superior colliculus. *Exp. Brain Res.*, 60: 492–500.

King, A.J., Hutchings, M.E., Moore, D.R. and Blakemore, C. (1988) Developmental plasticity in the visual and auditory representations in the mammalian superior colliculus. *Nature*, 332:73–76.

King, A.J., Moore, D.R. and Hutchings, M.E. (1994) Topographic representation of auditory space in the superior colliculus of adult ferrets after monaural deafening in infancy. *J. Neurophysiol.*, 71: 182–194.

Knudsen, E.I. (1982) Auditory and visual maps of space in the optic tectum of the owl. *J. Neurosci.*, 2: 1177–1194.

Knudsen, E.I. (1985) Experience alters the spatial tuning of auditory units in the optic tectum during a sensitive period in the barn owl. *J. Neurosci.*, 5: 3094–3109.

Knudsen, E.I. and Brainard, M.S. (1991) Visual instruction of the neural map of auditory space in the developing optic tectum. *Science*, 253: 85–87.

Knudsen, E.I. and Brainard, M.S. (1995) Creating a unified representation of visual and auditory space in the brain. *Annu. Rev. Neurosci.*, 18: 19–43.

Knudsen, E.I., Esterly, S.D. and du Lac, S. (1991) Stretched and upside-down maps of auditory space in the optic tectum of blind-reared owls; acoustic basis and behavioral correlates. *J. Neurosci.*, 11: 1727–1747.

Middlebrooks, J.C. (1987) Binaural mechanisms of spatial tuning in the cat's superior colliculus distinguished using monaural occlusion. *J. Neurophysiol.*, 57: 688–701.

Middlebrooks, J.C. and Knudsen, E.I. (1984) A neural code for auditory space in the cat's superior colliculus. *J. Neurosci.*, 4: 2621–2634.

Mooney, R.D., Bradley, G.K., Jacquin, M.F. and Rhoades, R.W. (1984) Dendrites of the deep layer, somatosensory superior colliculus neurons extend into the superficial laminae. *Brain Res.*, 324: 361–365.

Mooney, R.D., Huang, X. and Rhoades, R.W. (1992) Functional influence of interlaminar connections in the hamster's superior colliculus. *J. Neurosci.*, 12: 2417–2432.

Peck, C.K., Baro, J.A. and Warder, S.M. (1995) Effects of eye position on saccadic eye movements and on the neuronal response to auditory and visual stimuli in cat superior colliculus. *Exp. Brain Res.*, 103: 227–242.

Schnupp, J.W.H., King, A.J., Smith, A.L. and Thompson, I.D. (1995) NMDA-receptor antagonists disrupt the formation of the auditory space map in the mammalian superior colliculus. *J. Neurosci.*, 15: 1516–1531.

Simon, D.K., Prusky, G.T., O'Leary, D.D.M. and Constantine-Paton, M. (1992) *N*-methyl-D-aspartate receptor antagonists disrupt the formation of a mammalian neural map. *Proc. Natl. Acad. Sci. USA*, 89: 10593–10597.

Snider, C.J. and Chalupa, L.M. (1993) Specificity of the retinocollicular pathway in developing cat and ferret. *Soc. Neurosci. Abstr.*, 19: 454.

Stein, B.E. and Meredith, M.A. (1993) *The Merging of the Senses*, MIT Press, Cambridge, MA.

Thompson, I.D. and Cordery, P.M. (1994) The development of the retino-collicular map in the Syrian hamster. *Soc. Neurosci. Abstr.*, 20: 1704.

Thompson, I.D. and Morgan, J.E. (1993) The development of retinal ganglion cell decussation patterns in postnatal pigmented and albino ferrets. *Eur. J. Neurosci.*, 5: 341–356.

Wallace, M.T., Meredith, M.A. and Stein, B.E. (1993) Development of multisensory integration in cat superior colliculus. *Soc. Neurosci. Abstr.*, 19: 240.

Wise, L.Z. and Irvine, D.R.F. (1985) Topographic organization of interaural intensity difference sensitivity in deep layers of cat superior colliculus: implications for auditory spatial representation. *J. Neurophysiol.*, 54: 185–211.

Withington, D.J. (1992) The effect of binocular lid suture on auditory responses in the guinea-pig superior colliculus. *Neurosci. Lett.*, 136: 153–6.

Withington-Wray, D.J., Binns, K.E. and Keating, M.J. (1990) The developmental emergence of a map of auditory space in the superior colliculus of the guinea pig. *Dev. Brain Res.*, 51: 225–36.

M. Norita, T. Bando and B. Stein (Eds.)
Progress in Brain Research, Vol 112
© 1996 Elsevier Science BV. All rights reserved.

CHAPTER 25

What do developmental mapping rules optimize?

Meijuan Xiong and Barbara L. Finlay*

Department of Psychology, Cornell University, Ithaca, NY 14853, USA

Convergence ratios between pre- and postsynaptic cells in the visual system vary widely between cell classes, areas of the visual field, between individuals and between species. Proper stabilization of the convergence and divergence of single visual neurons is critical for visual integration generally, and for specific functions such as those of rod and cone pathways, or the center and peripheral regions of the visual field. In early development, retinal ganglion cells, target cells and all their processes are produced in excess and stabilize at certain mature values. The intent of the investigations described here is to determine what features of cell connectivity are stabilized over normal variability by these developmental processes and how such stabilization is accomplished, using the developing mammalian retinotectal system as an example.

Orderly compression of the retinotopic map into a half

tectum was induced by a partial tectal ablation at birth in hamsters, increasing the ratio of retinal ganglion cells to superior colliculus target cells. The convergence problem is solved in this case by undersampling the spatial array with respect to normal, preserving local spatial resolution, but potentially reducing sensitivity or introducing aliasing artifacts. Receptive field sizes of single neurons are indistinguishable from normal, and reduction of branching of presynaptic axon arbors is the mechanism of the remapping. Behaviorally, though the entire visual field is still represented in the remaining colliculus, the solution has a cost in decreased probability and increased latency to orient to visual stimuli, particularly in the peripheral visual field. The generality of this solution for retinal and other central convergence regulation problems is evaluated.

Introduction

During development, the visual system must solve functional problems that are defined at the level of visual information processing, using mechanisms that are defined at the level of interactions, growth, and trophic requirements of cells. These problems must often be solved in the absence of any immediate evaluation of visual function, leaving the tuning of developmental processes only under the control of natural selection. In most cases of early nervous system development, there is substantial variability in the initial deployment of neurons and their processes, but a reasonably stable adult solution is produced. What features of visual organization are optimized in early de-

velopment, and how is that control expressed through the interactions of developing neurons?

The particular case we will discuss is the control of convergence between populations of neurons in the visual system. The total number of neurons projecting to a number of neurons in a target structure sets the boundaries of convergence, but within these limits, any solution is possible: an afferent neuron could establish a synaptic connection with every neuron in the target population, or with only one. Close to this range of variability is seen in the visual system, from the one-to-one convergence of retinal neurons in the primate fovea to the extended distribution of wide-field amacrine cells. Differences in convergence ratios are directly related to the different functions of the retina, within individuals and across species. The low convergence ratio from cones to ganglion cells in the central retina preserves high acuity in photopic conditions but

*Corresponding author. Tel.: +1 607 2556394; fax: +1 607 2558433; e-mail: blf2@cornell.edu

sacrifices sensitivity, whereas the high convergence ratio from rods to ganglion cells in the peripheral retina allows high sensitivity in scotopic conditions but loses spatial resolution. Different functional cell classes, such as the X/Y, color-opponent or motion sensitive classes require different organization of input convergence. Across species, highly developed central vision versus panoramic vision require variable convergence across the retina in the first case, and more consistent convergence in the second case.

What might happen if an array of target neurons is reduced by some fraction, with the requirement that the entire input array must be represented in the target structure? Two classes of solutions are possible. The number of neurons in the target population projecting to any one target neuron could increase: a cell usually receiving four inputs might accept eight, spread over a larger area of visual field. This solution lowers the spatial frequency represented by the targets, blurring spatial resolution, but preserves some measure of sensitivity by keeping the number of target neurons potentially reporting activity in some visual field location as high as possible. Alternatively, the number of neurons projecting to a target neuron could be held constant, but the number of neurons in the target structure representing any particular location in the visual field could be reduced. This preserves high frequency selectivity, but spatial localization may be compromised. To adequately sample a simple sine wave, a sampling array must at least give one sampling point for every peak and trough of the sine wave, and for a given sampling array with a regular sampling interval (d), the highest frequency sine wave (1/2 d) which can be unambiguously represented, is referred to as the Nyquist limit (Rowe, 1991). Spatial undersampling may occur when the frequency of the stimulus is above the Nyquist limit. For example, when spatial frequency of the sine wave is higher than the Nyquist limit of the sampling array, the output of the sampling array is ambiguous. A high frequency sine wave would appear to the system to come from a lower frequency, a phenomenon

called aliasing. Additionally, since fewer target neurons can report the stimulus, sensitivity of the target neuron population to a visual stimulus of constant size is reduced.

Variation in population convergence ratios almost certainly occurs commonly in normal development, and certainly occurs over evolution. Which solution does the visual system favor, and how is it executed? Note that the 'best' solution might differ by cell class, e.g. photopic cells might choose blur, and scotopic cells spatial undersampling.

Some examples of convergence variation and their relation to measured acuity

Sterling et al. (1988) demonstrated that the convergence ratio between cones and ganglion cells set visual resolution close to the Nyquist limit. They determined the convergence of cones onto ganglion cells by tracing dendrites through serial ultrathin sections from electron micrographs in the cat. In the cat, 16 cones converge onto four cone bipolar cells which then converge onto one ON-beta ganglion cell. The Nyquist limit for the cone array is 18.7 cycles/degree, however, the 4:1 convergence of cones onto cone bipolars reduces the resolution of the cone signal by a factor of four to about 9.3 cycles/degree, which is very close to the Nyquist limit for the cone bipolar (CBb1) array alone (the Nyquist limit for the cone bipolar array alone is 9.2 cycles/degree). In contrast, the rod system appears to have a much higher convergence ratio: 1500 rods converge onto 100 rod bipolars which can innervate five AII amacrines, eventually innervating one on-beta ganglion cell. The Nyquist limit for rods in the cat is 64 cycles/degree, which would be reduced to about 14 cycles/degree by a 20:1 convergence onto rod bipolars. This higher convergence ratio compared to cone-bipolar convergence sacrifices resolution but gains higher sensitivity. Behavioral measurements of visual resolution are consistent with the anatomical data. The cat beta ganglion cells can resolve gratings as high as 9 cycles/degree (Cleland et al., 1979).

Convergence ratios have considerable variability from one region to another region in the retina within individuals and across species. In the rabbit, there is a potential convergence of seven rod bipolars onto one AII amacrine cell at the peak visual streak, 11–16 rod bipolars in the inferior retina, and as many as 40–75 rod bipolars may converge on each AII amacrine cell in the superior retina (Vaney et al., 1991). In the primate fovea, one cone connects to one ganglion cell (Wässle et al., 1990), and the ratio of cones to ganglion cells increases continuously toward the peripheral retina where at 10 mm eccentricity there are 16 cones for each ganglion cell (Wässle and Boycott, 1991).

In the visual pathway beyond the retina such as the projection from the retina to the visual cortex through the LGN or the SC, there are also various convergence solutions. An example of variation in convergence ratios has been reported in the macaque (Schein and de Monasterio, 1987; reviewed in Finlay and Pallas, 1989). In the geniculocortical projections, the convergence ratio between parvocellular cells and layer 4 visual cortical cells remains relatively constant from central to peripheral visual field. However, the convergence ratio of magnocellular cells upon layer 4 cells increases from center to periphery by about a factor of seven.

How are these different convergence ratios set during development and what is the structural basis for variations in the convergence ratio? A large, but fully definable set of developmental mechanisms could be responsible (Fig. 1).

Developmental processes underlying array matching

Fundamental developmental mechanisms that establish convergence ratios are cell generation, cell death, variation in axonal arbor, synaptic density and variation in dendritic arbor. Each of these could serve as an independent or dependent variable: for example, excess neurogenesis might upregulate synaptic density in a target population, or depression of the growth of synaptic arbor

might increase afferent cell death. Cellular recognition and response to changing convergence conditions could take many forms, including trophic interactions, competitive interactions between afferent neurons or any number of types of activity-dependent stabilization.

Cell generation

At a population level, numerical matching may be set by neurogenesis alone. This has most often been described in invertebrates, in which individual cells are identifiable and in which cell death may play a limited role (see review by Williams and Herrup, 1988). In vertebrates, even though substantial cell death may reduce both afferent and target populations, in some cases, the initial afferent/target neuron ratios persist unchanged (Sperry, 1990).

Cell death

Overproduction of cells and subsequent cell death is a common feature of developing nervous systems. Cell death does appear to stabilize particular ratios of afferent-to-target neurons in some interconnecting populations. Perhaps the best example of this is the numerical relationship between Purkinje cells and their granule cell inputs. Herrup and Sunter (1987) have studied granule cell death in chimeric mutants of mouse and have shown that numerical imbalance of Purkinje cells and granule cells is corrected linearly by cell death in granule cells. It is unclear whether this is a general rule. Doubling the size of target tissue in a case where half of the afferent cells normally die does not prevent all cell death (reviewed in Oppenheim, 1991); similarly, large removals of target can result in no change in afferent neuron survival, due to axon remodeling (reviewed in Finlay, 1992).

Axonal remodeling and collateral elimination

Often in early vertebrate development, axonal terminal arbors are widespread, and later excess branches are retracted and axonal arbors are refined. To cite a well-known example, the retino-geniculate axonal arborizations are initially diffuse and then later restrict their territory and segregate into eye-specific laminae (Shatz and Sretavan, 1986). Callosal projections from sensory cortex are very diffuse in neonates and later are pruned to the adult form (Innocenti et al., 1977). This developmental process does not function to regulate the number of afferents and target cells, but rather the types and spatial extent of input a postsynaptic neuron receives.

Dendritic and synaptic remodeling

Dendritic arbor and synaptic density have been shown to be quite malleable in development, principally in cases where changes in total activity or learning have occurred (Greenough and Bailey, 1988). Developmental sculpting of functional systems by hormones often targets the dendritic arbor of target cells as a means of regulating the number of cells and connectivity of a multiple-neuron circuit (Sengelaub, 1986)

Experimental manipulation of afferent / target cell ratios: compression and expansion of the retinotectal projection

'Compression' of projections into experimentally reduced targets has been demonstrated in both regenerating and developing nervous systems. After removal of part of the tectum, the retina is able to compress its projection into a smaller than normal tectal volume while preserving topographic order in the goldfish (Gaze and Sharma, 1970) and frog (Udin, 1977). In the hamster, early partial tectum ablation results in compression of the retinotectal projections (Jhaveri and Schneider, 1974; Finlay et al., 1979a) (Fig. 2). Similarly, a hemi-retina can expand in an orderly way into a complete tectum in goldfish (Schmidt et al., 1978). In the hamster, neonatal enucleation of one eye results in the loss of the normal massive contralateral projection to the superior colliculus, and an expanded ipsilateral projection from the remaining eye (Finlay et al., 1979b) (Fig. 2).

Compression of the retinotectal projection of the hamster can be demonstrated both electro-physiologically and anatomically. Multi-unit recording in the tectal fragment reveals that the visual field is represented in an orderly way in the residual tectum and the size of multi-unit receptive fields is increased (Finlay et al., 1979) (Fig. 3, left). The area of retina projecting to a zone of defined size in the tectum is larger in 'compressed' animals (Pallas and Finlay, 1991) (Fig. 4). In contrast, the size of single unit receptive fields

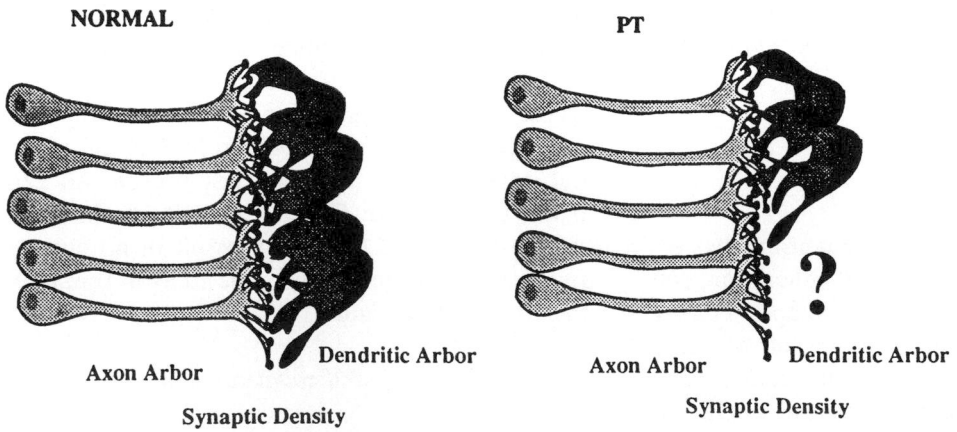

NORMAL   **PT**

Axon Arbor   Dendritic Arbor   Axon Arbor   Dendritic Arbor

Synaptic Density   Synaptic Density

Fig. 1. Schematic of a size disparity experiment.

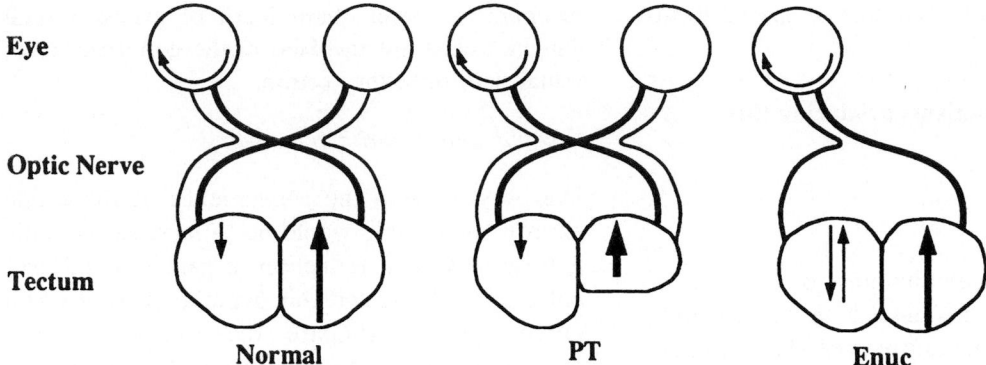

Fig. 2. Schematic of (left) the normal mapping of the two retinas on the normal colliculus of a hamster; (center) a hamster with a partial lesion of the superior colliculus on the day of birth and (right) a hamster with one eye enucleated at birth.

remains constant (Pallas and Finlay, 1989) (Fig. 3, right). Moreover, a number of properties of single neuron responses, such as selectivity for stimulus velocity and preferred stimulus size also remains normal. Therefore, the convergence from retinal ganglion cells to the tectal fragment is increased

at the population level, but it remains the same at the single neuron level. This solution to compression of a projection thus retains the spatial resolution of single tectal neurons, but reduces the number of neurons representing any particular location in the visual field. Spatial resolution is

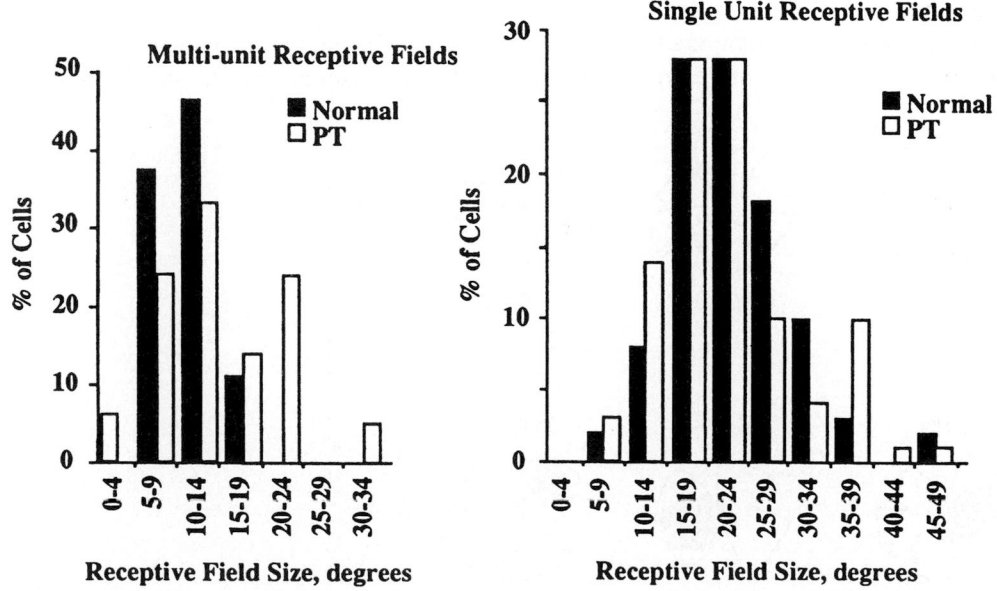

Fig. 3. Comparison of frequency histogram of the receptive field diameters of multi-unit evoked response in normal and partially-ablated superior colliculi (left) and the same for single-unit evoked responses (right).

maintained at the potential cost of spatial localization and sensitivity.

Developmental mechanisms producing this solution

Cell death

After partial unilateral ablations of 20–75% of the caudal tectum in neonatal hamsters resulting in a compression of the entire contralateral visual field on to the remaining tectal fragment, there was only a minor increase in cell death (8%) in the contralateral retina and no change in cell death rates within the tectum (Wikler et al., 1986). This means that a half-sized tectum would potentially face double the amount of retinal inputs (Fig. 4).

The 'population match' hypothesis for cell death (reviewed in Hamburger and Oppenheim, 1982; Oppenheim, 1991), that neuronal death serves to quantitatively match the numbers of the projecting neuronal populations to the numbers of their targets is thus not supported in this case. Violations of this hypothesis, however, are commonplace (reviewed in Finlay, 1992). The residual tectum accommodates an apparently larger than normal ratio of retinal inputs, indicating that

mechanisms other than death of excess retinal ganglion cells are involved in the compression of visual field onto the tectum.

Synaptic and dendritic remodeling

One way to keep the convergence at the single neuron level stable would be to increase synaptic density without a reduction in ganglion cell axonal arbors. We tested this hypothesis by electron microscopy to compare the synaptic density between normal and partially ablated tectum. The synaptic density did not increase in the tectal fragment (Xiong and Finlay, 1993). Since the volume of neuropil to neurons is unaltered, we can also infer that the number of synaptic sites per postsynaptic neuron is constant. A similar result is found in the regenerating compressed retinotectal projection of goldfish (Murray et al., 1982; Hayes and Meyer, 1988). Thus, synaptic density is not up-regulated by excess input and is not responsible for the constant spatial convergence at the single neuron level.

Retinal ganglion cell axonal arbor size and complexity

Using in vitro HRP injection to reconstruct retinal ganglion cell axonal arbors in the partial

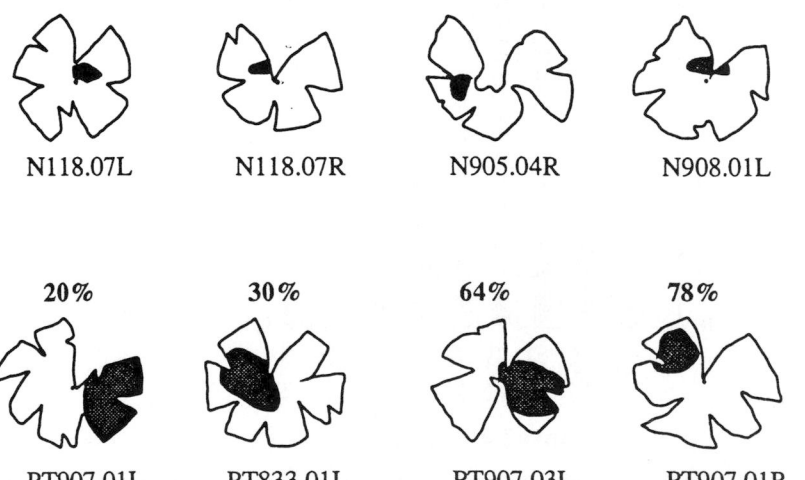

Fig. 4. Examples of the area labeled in the retina by injection of a constant amount of HRP in four normal superior colliculi (upper row) and four partially ablated colliculi (bottom row).

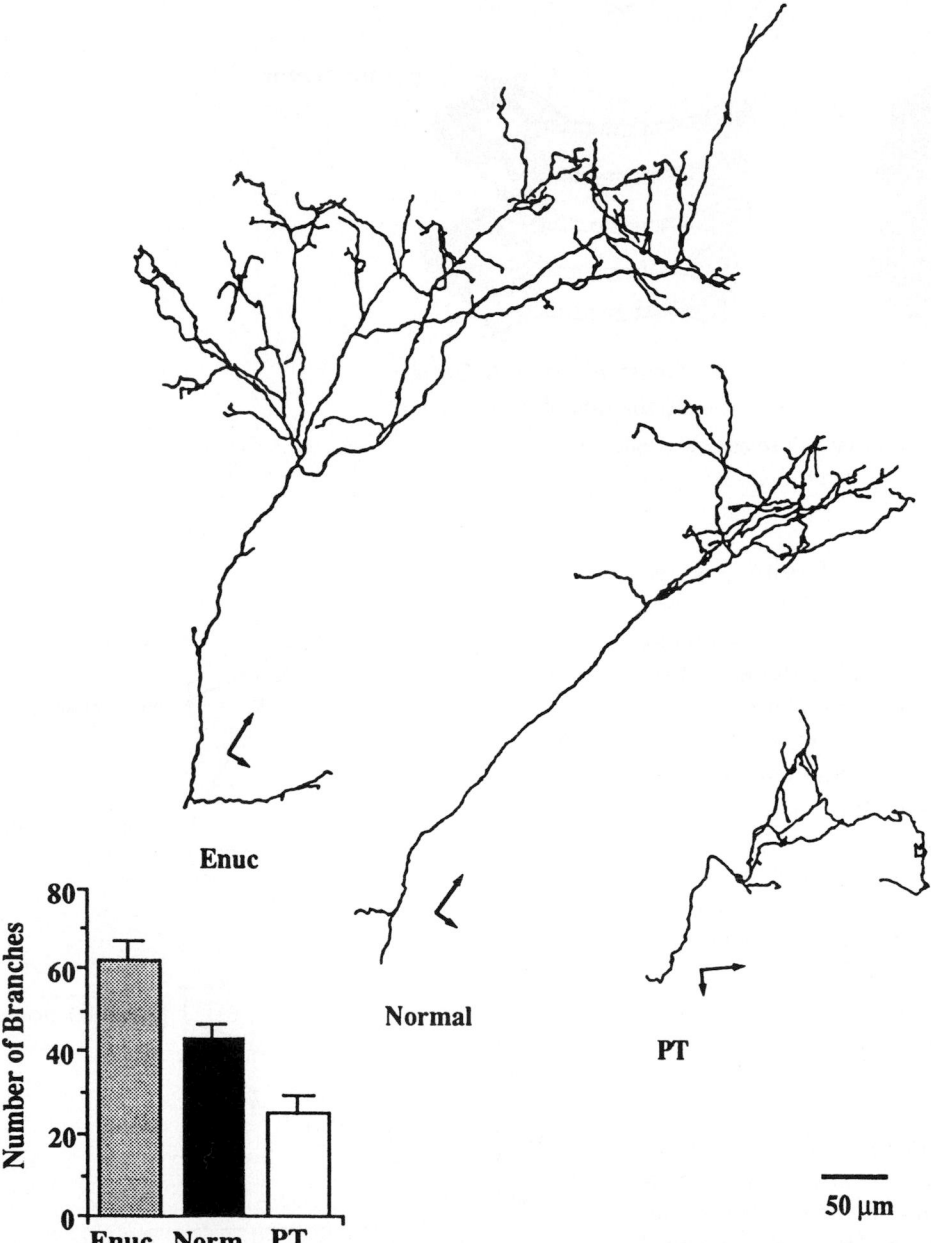

Fig. 5. Examples of axon arborization after early expansion of the retinotectal projection (left), the normal case (middle), or compression (right).

tectum, we saw that the retinal ganglion cell axonal arbor area was markedly reduced from normal (Fig. 5). The basic geometry of the arbor is unchanged: branch density within the area covered by the arbor is normal, but the number of branches and the area over which the arbor spreads are reduced (Xiong et al., 1994) (Fig. 6).

Many studies have shown extensive axonal arbor remodeling during initial growth and regeneration of axons in the central nervous system

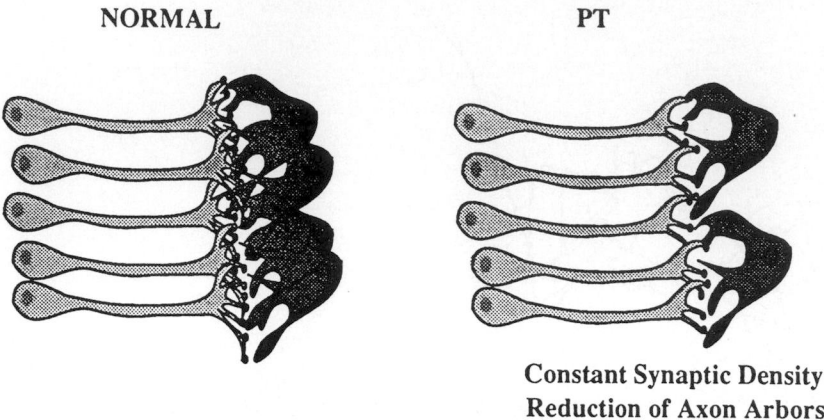

NORMAL   PT

**Constant Synaptic Density
Reduction of Axon Arbors**

Fig. 6. Schematic drawing of the hamster's solution to a reduction in postsynaptic terminal sites in the retinotectal position.

(Shatz, 1990). The most studied system is the retinotectal projections in fish and frogs, in which the structure of retinal axonal arbors is extremely dynamic (Reh and Constantine-Paton, 1984; O'Rourke et al., 1990, 1994). Remodeling of retinal axon terminal arbors contributes both to the refinement of the topographical map and also helps maintain retinotopography as the entire retinal projection shifts in the tectum to compensate for differences in growth patterns in the retina and tectum (Gaze et al., 1979; Reh and Constantine-Paton, 1984; Easter and Stuermer, 1984; O'Rourke et al., 1990, 1994).

Behavioral consequences of compression of the retinotectal projection

In general, the most dramatic deficits after colliculus ablation are related to attending, localizing, and orienting to sensory stimuli (Stein and Meredith, 1993). Evidence from physiological studies also suggests that neurons in the superior colliculus have sacrificed the capacity for accurate spatial resolution and detailed analysis of objects, but serve optimally for the detection of direction and velocity of a moving stimulus. Common response properties for all visual neurons in the superficial layers of the mammalian superior colliculus have been described (Vanegas, 1984).

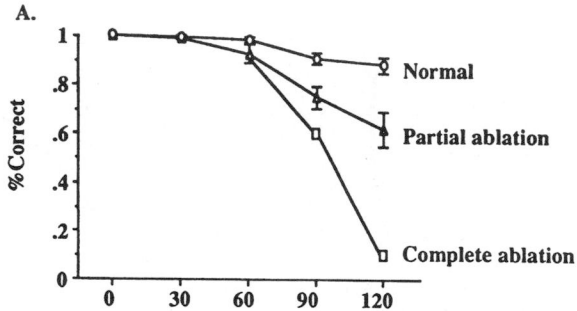

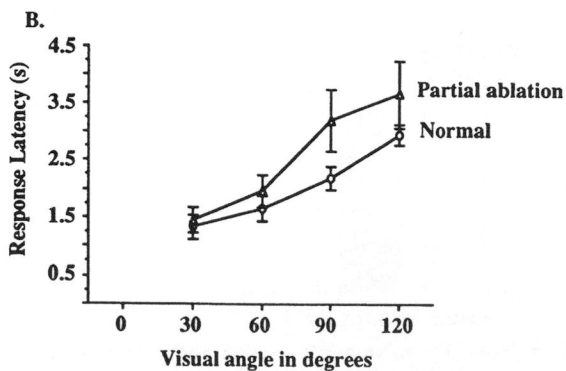

Fig. 7. (A) Comparisons of percentage of correct responses in a sunflower seed orientation task among normal, partial tectum-ablated and complete tectum-ablated hamsters. X-axis: visual angle in degrees. Y-axis: percentage of correct responses. (B) Comparison of response latency of sunflower seed orientation between partial tectum-ablated and normal hamsters. Y-axis, response latency; open circles, normal; open triangles, partial ablations; open square, complete ablation.

Neurons have receptive fields larger than those found in the geniculo-cortical system without separate on and off regions. Most neurons respond to both binocular and monocular stimuli, moving stimuli at relatively low velocities are preferred and some cells exhibit orientation specificity. Neurons in the tectum also respond optimally to stimuli far smaller than their receptive fields.

In the hamster, one behavior dependent on the superior colliculus is orientation to the head and body toward small objects, particularly in the peripheral visual field. In the central visual field, hamsters without a superior colliculus can still notice and approach food items (Schneider, 1969; Mort et al., 1980; Finlay et al., 1980). We used orientation to sunflower seeds to assess orientation behavior of the hamster. The hamster was first trained to face straight ahead on a small platform, and the sunflower seed was presented 2 cm away from several directions, both in the horizontal planes, and 45° above the hamster's head.

All 'compressed' animals perform as well as normal hamsters in their central visual field, the part of the field not dependent on the colliculus, but the response degrades towards the visual periphery (Fig 7a). Their performance level, however, is higher than that in animals with a complete tectal ablation (data replotted from Mort et al., 1980; Figs. 7a,b). In the peripheral visual field, both normal and ablated animals show greater latency to turn their heads in order to pursue sunflower seeds. Colliculus-ablated animals have even greater latency in organizing an orientation response to sunflower seeds than normal animals (Fig. 7b). Therefore, while hamsters with compressed projections show spared orienting behavior throughout their visual field, consistent with the representation of the full visual field in their colliculus, they fail to detect the stimulus a greater proportion of the time and take longer to respond when they do respond.

Activity dependence and convergence ratios

Summarizing these results, when retinal ganglion cells confront a smaller-than-normal tectum in development, preservation of spatial resolution at the level of the single neuron is the result, reducing the number of cells representing any particular receptive field location. This remapping is achieved primarily by reduction of retinal ganglion cell axonal arbors (Fig. 6). Behaviorally, while orientation is preserved over the entire visual periphery, it is slower and less accurate. We hypothesize that the compression of the retinotectal pathway has impaired the recruitment of the motor response of orienting. When a sensorimotor transformation is performed, the collicular output signal which is coded spatially has to be transformed to a drive for motoneurons which are coded temporally. Recruitment of this driving force may be impaired due to fewer cells in the superior colliculus representing each retinal location. Alternatively, spatial sampling may be distorted. Such neuronal scrambling could come from irregularity of distances in the sampling array, size of sampling elements, or spatial dislocation of connections. Hess and Field (1993) have argued that the loss of resolution in the human peripheral visual field comes from uncalibrated disarray rather than from spatial undersampling. Topographical order is somewhat subnormal in the representation of the visual periphery, which could also lead to slow and irregular responses.

Activity-dependent mechanisms, based on the highly correlated patterns of activity in neighboring retinal ganglion cells, are responsible for much of axonal remodeling and refinement of the topographical maps in the developing visual system (Udin, 1988). It seems likely that activity dependent stabilization can account for these results directly. A decreased number of target neurons does not change the pattern of correlation in its input array, and if a single postsynaptic cell is designed to select a certain level of correlation, the size of the input array will be irrelevant to the number of cells detected. A different setting for spatially-based correlation can produce receptive fields of different sizes for different functional classes of cells. The goal of sensory coding is to represent the external world, not the features of the nervous system sensing it, and a mechanism that maintains spatial selectivity independent of numbers of cells in the nervous system has this useful feature.

Acknowledgements

Supported by NIH RO1 NS19245. We thank John Niederer for technical assistance.

References

Cleland, B.G., Harding, T.H. and Tulunay-Keesey, U. (1979) Visual resolution and receptive field size: examination of two kinds of retinal ganglion cells. *Science*, 205: 1015–1017.

Easter, S.S. and Steurmer, C.A.O. (1984) An evaluation of the hypothesis of shifting terminals in the goldfish optic tectum. *J. Neurosci.*, 4: 1052–1063.

Finlay, B.L. (1992) Cell death and the creation of regional differences in neuronal numbers. *J. Neurobiol.*, 23: 1159–1171.

Finlay, B.L. and Pallas, S.L. (1989) Control of cell number in the developing visual system. *Prog. Neurobiol.*, 32: 207–234.

Finlay, B.L., Schneps, S.E. and Schneider, G.E. (1979) Orderly compression of the retinotectal projection following partial tectal ablation in the newborn hamster. *Nature*, 280: 153–154.

Finlay, B.L., Wilson, K.G. and Schneider, G.E. (1979) Anomalous ipsilateral retinal projections in Syrian hamsters with neonatal lesions: topography and functional capacity. *J. Comp. Neurol.*, 183: 721–740.

Finlay, B.L., Sengelaub, D.R., Berg, A.T. and Cairns, S.J. (1980) A neuroethological approach to hamster vision. *Behav. Brain Res.*, 1: 479–496.

Gaze, R.M. and Sharma, S.C. (1970) Axial differences in the reinnervation of the goldfish tectum by regenerating optic fibers. *Exp. Brain Res.*, 10: 71–180.

Gaze, R.M., Keating, M.J., Ostberg, A. and Chung, S.H. (1979) The relationship between retinal and tectal growth in larval *Xenopus*: implications for the development of the retinotectal projection. *J. Embryol. Exp. Morphol.*, 53: 103–143.

Greenough, W.T. and Bailey, C.H. (1988) The anatomy of a memory: convergence of results across a diversity of tests. *Trends Neurosci.*, 11: 142–147.

Hamburger, V. and Oppenheim, R.W. (1982) Naturally occurring neuronal death in vertebrates. *Neurosci. Comment.*, 1: 39–55.

Hayes, W.P. and Meyer, R.L. (1988) Optic synapse number but not density is constrained during regeneration onto a surgically halved tectum in goldfish: HRP-EM evidence that optic fibers compete for fixed numbers of postsynaptic sites on the tectum. *J. Comp. Neurol.*, 274: 539–559.

Herrup, K. and Sunter, K. (1987) Numerical matching during cerebellar development. *J. Neurosci.*, 7: 829–837.

Hess, R.F. and Field, D.J. (1994) Is the spatial deficit in strabismic amblyopia due to loss of cells or an uncalibrated disarray of cells? *Vision Res.*, 34: 3397–3406.

Innocenti, G.M. Fiore, L. and Caminiti, R. (1977) Exuberant projection into the corpus callosum from the visual cortex of newborn cats. *Neurosci. Lett.*, 4: 237–242.

Jhaveri, S.R. and Schneider, G.E. (1974) Neuroanatomical correlates of spared or altered function after brain lesions in the newborn hamster. In: *Plasticity and Recovery of Function in the Central Nervous System*, Academic Press, New York, pp. 65–109.

Mort, E., Finlay, B.L. and Cairns, S. (1980)The role of the superior colliculus in visually-guided locomotion and visual orienting in the hamster. *Physiol. Psychol.*, 8: 20–28.

Murray, M., Sharma, S. and Edwards, M.A. (1982) Target regulation of synaptic number in the compressed retinotectal projection of goldfish. *J. Comp. Neurol.*, 209: 374–385.

Oppenheim, R.W. (1991) Cell death during development of the nervous system. *Annu. Rev. Neurosci.*, 14: 453–502.

O'Rourke, N.A. and Fraser, S.E. (1990) Dynamic changes in optic fiber terminal arbors lead to retinotopic map formation: an in vivo confocal microscope study. *Neuron*, 51: 159–171.

O'Rourke, N.A., Cline, H.T. and Fraser, S.E. (1994) Rapid remodeling of retinal arbors in the tectum with and without blockade of synaptic transmission. *Neuron*, 12: 921–934.

Pallas, S.L. and Finlay, B.L. (1989) Conservation of receptive field properties of superior colliculus cells after developmental rearrangements of retinal input. *Vis. Neurosci.*, 2: 121–135.

Pallas, S.L. and Finlay, B.L. (1991) Compensation for population size mismatches in the hamster retinotectal system: alterations in the organization of retinal projections. *Vis. Neurosci.*, 6: 271–281.

Reh, T.A. and Constantine-Paton, M. (1984) Retinal ganglion cell terminals change their projection sites during larval development of *Rana pipiens*. *J. Neurosci.*, 4: 422–457.

Rowe, M. H. (1991) Functional organization of the retina. In: B. Dreher and S.R. Robinson (Eds.), *Neuroanatomy of Visual Pathways and Their Development*, Vol. 3, CRC Press, Boca Raton, FL.

Schein, S.J. and de Monasterio, F.M. (1987) Mapping of retinal and geniculate neurons onto striate cortex of macaque. *J. Neurosci.*, 7: 996–1009.

Schmidt, J.T., Cicerone, C.M. and Easter, S.S. (1978) Expansion of the half retinal projection to the tectum in goldfish: an electrophysiological and anatomical study. *J. Comp Neurol.*, 177: 257–278.

Schneider, G.E. (1969) Two visual systems. *Science*, 163: 895–902.

Sengelaub, D.R., Dolan, R.P. and Finlay, B.L. (1986) Cell generation, death and retinal growth in the development of the hamster retinal ganglion cell layer. *J. Comp Neurol.*, 246, 527–543.

Shatz, C.J. and Sretavan, D.W. (1986) Interactions between ganglion cells during the development of the mammalian visual system. *Annu. Rev. Neurosci.*, 9: 171–207.

Shatz, C. (1990) Impulse activity and the patterning of connections during CNS development. *Neuron*, 5: 745–756.

Sperry, D.G. (1990) Variation and symmetry in the lumbar and thoracic dorsal root ganglion cell populations of newly metamorphosed *Xenopus laevis*. *J. Comp. Neurol.*, 292: 54–64.

Stein, B.E. and Meredith, M.A. (1993) *The Merging of the Senses*, MIT Press, Cambridge.

Sterling, P., Freed, M.A. and Smith, R.G. (1988) Architecture of rod and cone circuits to the on-beta ganglion cell. *J. Neurosci.*, 8: 623–642.

Udin, S.B. (1977) Rearrangements of the retinotectal projection in *Rana pipiens* after unilateral caudal half-tectum ablation. *J. Comp. Neurol.*, 173: 561–582.

Udin, S.B. and Fawcett, J.W. (1988) The formation of topographic maps. *Annu. Rev. Neurosci.*, 11, 289–328.

Vanegas, H. (1984) *Comparative Neurology of the Optic Tectum*, Plenum, New York.

Vaney, D.I., Gynther, I.C. and Young, H.M. (1991) Rod-signal interneurons in the rabbit retina 2. AII amacrine cells. *J. Comp. Neurol.*, 310: 154–169.

Wässle, H. and Boycott, B.B. (1991) Functional architecture of the mammalian retina. *Physiol. Rev.*, 71: 447–480.

Wässle H., Grunert, U., Rohrenbeck, J. and Boycott, B.B. (1990) Retinal ganglion cell density and cortical magnification factor in the primate. *Vis. Res.*, 30: 1897–1912.

Wikler, K.C., Kirn, J., Windrem, M.S. and Finlay, B.L. (1986) Control of cell number in the developing visual system: III. Partial tectal ablation. *Dev. Brain Res.*, 28,: 23–32.

Williams, R.W. and Herrup, K. (1988) The control of neuron number. *Annu. Rev. Neurosci.*, 11: 423–454.

Xiong, M.J. and Finlay, B.L. (1993) Changes in synaptic density after developmental compression or expansion of retinal input to the superior colliculus. *J. Comp Neurol.*, 330: 455–463.

Xiong, M.J., Pallas, S.L., Lim, S. and Finlay, B.L. (1994) Regulation of retinal ganglion cell axon arbor size by target availability: mechanisms of compression and expansion of the retinotectal projection. *J. Comp. Neurol.*, 344: 581–597.

M. Norita, T. Bando and B. Stein (Eds.)
Progress in Brain Research, Vol 112
© 1996 Elsevier Science BV. All rights reserved.

CHAPTER 26

The effect of damage of the brachium of the superior colliculus in neonatal and adult hamsters and the use of peripheral nerve to restore retinocollicular projections

K.-F. So[1]*, H. Sawai[2], S. Ireland[1], D. Tay[1] and Y. Fukuda[3]

[1]*Neuroregeneration Laboratory, Department of Anatomy, The University of Hong Kong, 5 Sassoon Road, Hong Kong*
[2]*Department of Welfare System and Health Science, Okayama Prefectural University, Kuboki, Souja 719-11, Japan*
[3]*Department of Physiology, Osaka University Medical School, Suita 565, Japan*

Using horseradish peroxidase (HRP) tracing technique, we were able to confirm the critical age in hamsters as reported previously (So et al., 1981). Thus, following transection of the retinal fibers at the brachium of the superior colliculus (BSC) on postnatal-day 4 (P4) or later, no retinocollicular projections were observed in the adult stage. However, the retinal fibers were observed to reinnervate the superior colliculus (SC) if the BSC was cut on P3 or earlier. Physiological recording showed a close to normal retinocollicular map following a BSC damage on P0. Although retinal fibers did not reinnervate the SC following a BSC cut on or after P4, they could be observed to grow along a membrane over the damaged site. Bridging the site of BSC damage in adult hamsters using a segment of peripheral nerve (PN), retinal fibers labelled with WGA-HRP were observed to reinnervate the SC along the PN graft and visual evoked responses could be recorded in the SC showing the PN graft is effective in restoring damaged central visual pathways in adult mammals.

Introduction

It is well known that the loss of neurons or damage of their connective pathways in the visual system of adult mammals, unlike that of developing mammals, result in irrevocable loss of vision. Thus, studies on the regeneration and plasticity of the visual system during development may offer useful approaches to repair the visual function following a similar lesion in adult stage. The

hamster is a particularly suitable species for the study of the plasticity of visual pathways following damage during development because the hamster is relatively immature at birth (gestation is 15.5–16 days) compared to other animals and many of the retinal fibers are still developing postnatally (Woo et al., 1985; So et al., 1990) permitting easier experimental manipulation.

We have previously demonstrated the presence of critical age in restoration of hamster's retino-tectal pathways using behavioral and anatomical techniques (So et al., 1981). The right brachium of the superior colliculus (BSC) was transected in hamsters on postnatal day 0, 3, 4, 5, or 6. When the animals were 12 weeks old, their turning movements elicited by stimuli in various parts of

────────
*Corresponding author. Neuroregeneration Laboratory, Department of Anatomy, The University of Hong Kong, 5 Sassoon Road, Hong Kong, fax: +852 28170857; e-mail: hrmaskf@hkucc.hku.hk

the visual field were tested. The retinal projections of both eyes were later traced with degeneration and autoradiographic techniques. In animals with transection of the right BSC on postnatal day 4 (P4) or later, no optic fibers crossed the cut. These animals showed no turning toward stimuli in the left visual field but turn readily to the left with vibrissa stimulation. In these cases, evidence of sprouting of retinofugal fibers was found in the diencephalon, namely, in the outer lamina of the ventral nucleus of the lateral geniculate body, and in the nucleus lateralis posterior. In animals with a lesion of the right BSC on the day of birth or P3, the optic fibers crossed the cut and terminate in the superior colliculus (SC). Most of these animals showed nearly normal turning toward stimuli in all parts of the visual field. These results indicate the importance of the retinocollicular projections rather than retinothalamic connections on visually elicited turning. Moreover, our study also indicated a critical age for recovery of the turning behavior from BSC lesion. This behavior is spared if the lesion is made sufficiently early before a critical age. When the cut is made at later ages, the turning behavior is lost.

The lack of regrowth across the site of transection after the critical age does not imply an absolute loss of growth potential with age. We (So and Aguayo, 1985; So et al., 1986) and others (Berry et al., 1986; Vidal-Sanz et al., 1987) have shown that damaged axons of the retinal ganglion cells (RGCs) in adult mammals can be induced to grow into a grafted segment of peripheral nerve (PN). In addition, we have recently shown that an additional transplantation of a short segment of PN into the vitreous of the eye of hamsters can further enhance the regenerative ability of the axotomized RGCs (Lau et al., 1994; Ng et al., 1995). We, therefore, are interested to study if damaged retinal axons after a critical age can regrow into the SC through a PN bridged across the BSC cut. Thus, in this paper, we will review the results from three sets of experiments. First, we have conducted a series of studies aiming to confirm the critical age following a BSC cut using modern neuroanatomical tracing technique (Ireland, 1991). Second, we have, using physiological technique, investigated the topography of the animals following a BSC cut on P0 (Ireland, 1991). Third, we have studied whether a segment of PN bridged across the BSC transection site can enhance the regeneration of damaged retinocollicular projections (Sawai et al., 1995).

Materials and Methods

Golden Syrian hamsters (*Mesocricetus auratus*) were used in the present study.

Experiment 1: Tracing of retinocollicular projections after transection of the BSC in developing and adult hamsters

The right BSC was transected in P0 ($n = 6$), P3 ($n = 8$), P5 ($n = 4$), P10 ($n = 5$) and 1 month ($n = 8$) old hamsters. Hypothermia was used as the anesthesia for the developing animals. Intraperitoneal injection of sodium pentobarbitone (6 mg/100 g body wt) was used for all operations in adult animals. The method of BSC transection was similar to that described previously (So et al., 1981). At 2 months of age, 0.2–1 μl of 25% HRP (Type VI, Sigma) solution was injected into the posterior chamber of the left eye of all operated animals to trace the retinofugal projections. The animals were perfused with Karnosky's fixative and a 1:2 series of sagittal sections at 60 μm was obtained using a freezing microtome. One series of the sections was reacted for HRP histochemistry using tetramethyl-benzidine (TMB) as the chromogen (Mesulam, 1978). The remaining series was used to stain for myelin using the Loyez method (Sidman et al., 1971).

Experiment 2: Retinotopic organization of the SC in normal hamsters and in hamsters following a BSC cut on P0

Ten adult hamsters at 10 weeks of age were used. Five of these were normal animals and the other five had a right BSC cut on day 0. The animals were anaesthetized with an intraperitoneal injection of a mixture of chloral hydrate (12.6 mg/100 g body wt) and sodium pentobarbitone (3 mg/100 g body wt). The details of the recording method have been published (Finlay and So, 1979). A glass-coated tungsten microelectrode 1–3 μm at the tip was used for recording single or multiple unit responses from the SC. The visual receptive field of the units were used to assess the retinocollicular topography. Small lesions to allow subsequent histological reconstruction of electrode penetrations were made at the end of each recording session.

Experiment 3: Physiological and anatomical studies of the retinocollicular projections following a BSC cut and PN grafting in adult hamsters

Physiological studies

The left BSC was transected in 21 adult hamsters and 16 of them received a PN transplantation at the site of the transection; the remaining five did not receive the PN graft and were used as controls. Autologous PN segments, 2–3 mm in length, were dissected out from the right peroneal nerve and used for grafting. Electrophysiological data were obtained from these animals 3–75 weeks later. The animals were anaesthetized with an intraperitoneal injection of urethane (1.2 g/kg body wt) and silver ball electrodes were used to record field potentials in the left SC in response to diffuse flash stimuli applied to the right eye. The details of the transplantation and recording techniques have already been described (Sawai et al., 1995).

Anatomical studies

The left BSC was cut in 19 adult hamsters and 12 of them received a PN graft at the site of the BSC cut and the remaining seven did not receive the graft and served as controls. The right eye was injected with 8 μl of 10% WGA-HRP (Toyobo) 8–12 weeks after the surgery to trace the retinocollicular projections into the left SC. The animals were perfused 36–48 h later. Frozen sections of the brains were obtained and reacted

with HRP histochemistry using TMB as the chromogen (Mesulam, 1978).

Results

Experiment 1: Tracing of retinocollicular projections after transection of the BSC in developing and adult hamsters

Following transection of the BSC in P0 or P3 hamsters, HRP labelled retinofugal fibers were observed to reinnervate the right SC when the animals were examined at 2 months of age. The retinocollicular fibers were seen in the entire rostrocaudal extent of the right SC in the P0 cases (Fig. 1C). The right SC of P3 cases were also heavily innervated by retinal fibers except in two cases with only sparse retinocollicular projections to the right SC. No innervation of the SC by retinal fibers was observed in any of the cases with transection of the BSC on P5 or later.

No scar tissue could be observed in the animals with a BSC cut on P0. The only evidence of the cut was the disruption of the retinal fibers at the rostral SC as indicated by the HRP labels (Fig. 1C, arrow) and myelinated fibers (Fig. 1D, arrow). Abnormal routing of the fibers could be observed (Fig. 1D, double arrows). Some scar tissues were seen in the animals with a BSC cut on P3. In animals with BSC cut on P5 or later, scar tissue or a gap (Fig. 2A) was evident at the site of damage.

Following damage of the BSC at all ages, a membrane was observed to form on top of the right SC. Retinal fibers were seen to grow along these membranes and they were myelinated as indicated by the Loyez stain (Figs. 1C–F and 2). Thus, retinal fibers grew into these membranes even in animals with a BSC cut on P5 or at 1 month of age (Fig. 2). However, these retinal fibers were not seen to reinnervate the SC.

Experiment 2: Retinotopic organization of the SC in normal hamsters and in hamsters following a BSC cut on P0

In all the control animals, the topography of the retinocollicular map was similar to those previ-

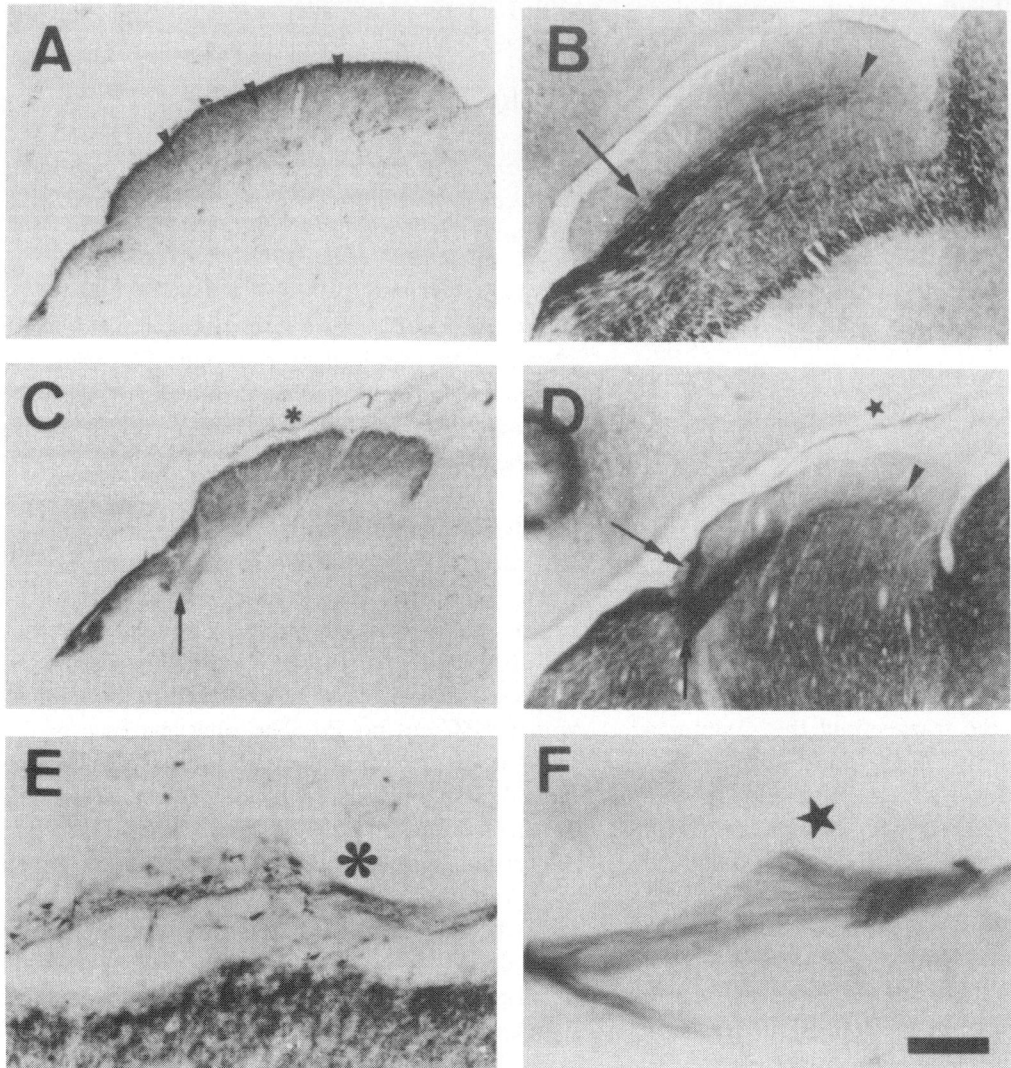

Fig. 1. (A) Photograph of a sagittal section showing the retinocollicular projections from a normal adult hamster. Arrowheads denote the location of HRP labelled retinal fibers and terminals in the superficial grey layer of the SC. (B) Photograph of a sagittal section from a normal adult hamster demonstrating the myelinated fibers in the SC stained by the Loyez method. The arrow points to the rostral part of the optic fiber layer where the bundles of fibers appear quite thick but they become much thinner caudally (arrowhead). (C) Photograph of a sagittal section showing the retinocollicular projections in an animal with a BSC cut on P0. The SC is smaller compared to the normal animal (A). There is a slight disruption of the distribution of the HRP reaction products at the rostral end of the SC (arrow) which might correspond to the lesion site. (D) Photograph of an adjacent section of C showing the disruption of the fibers (arrow) and the abnormal routing of the fibers at the rostral SC (double arrows). The optic fiber layer at the caudal SC (arrowhead) appears similar to normal. (E) Photograph showing retinal fibers located in a membrane on top of the SC (*). The same membrane is also shown in C (*). (F) Photograph showing the fibers in the membrane are myelinated (☆). The same membrane is shown in D (☆). Scale bar in F for A–D is 500 μm and E and F is 50 μm.

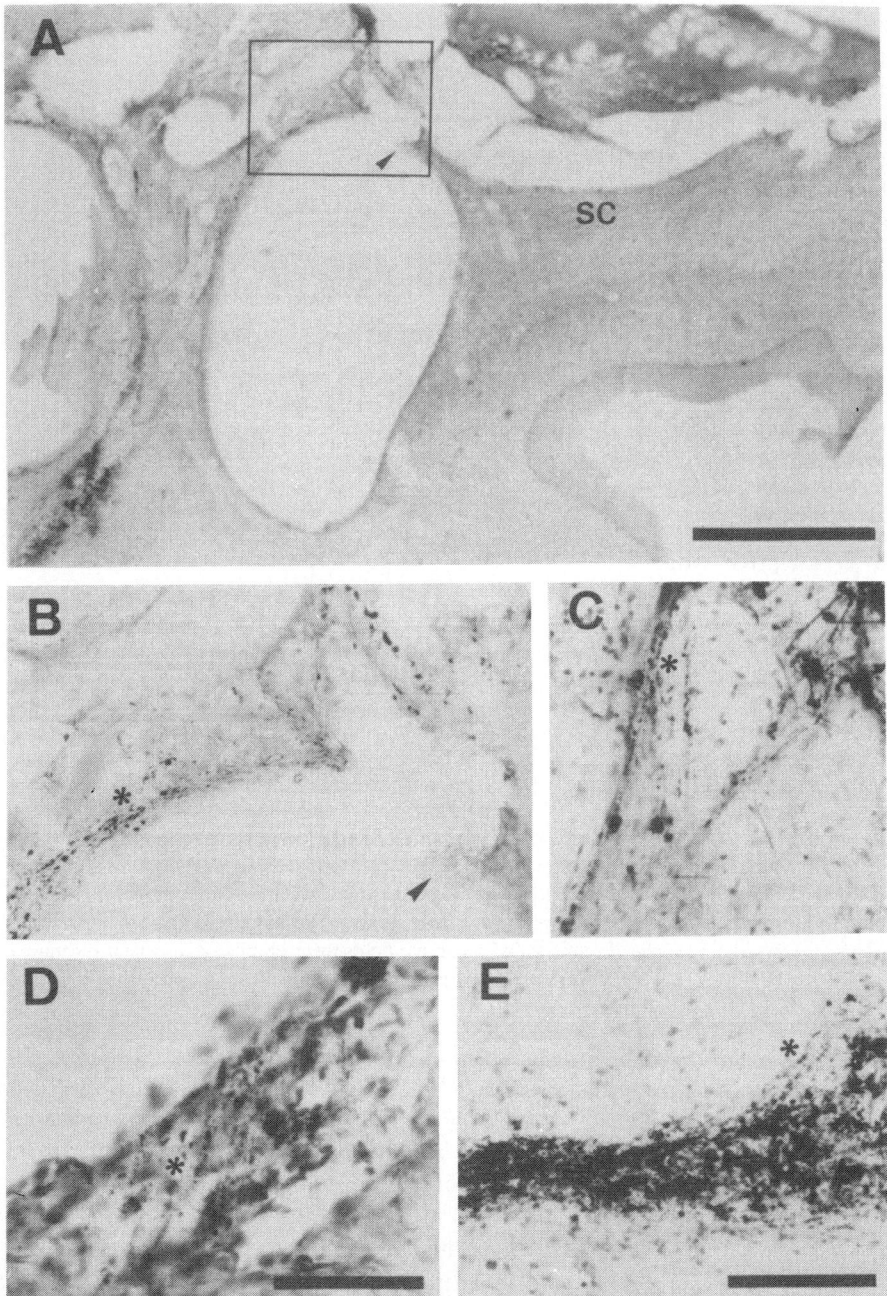

Fig. 2. (A) Photograph showing the lesion site of the SC from an animal with a BSC cut at 1 month old. A membrane is observed forming over the lesion site. (B) Photograph showing the membrane outlined by the rectangle in A. Note that HRP labelled retinal fibers can be seen in the membrane. However, no fibers were seen in the membrane attached to the SC (arrowhead in A and B) C–D. Photographs of similar membranes containing HRP labelled retinal fibers from three different animals with BSC cut at 1 month old. Scale bar in A is 500 μm , D is 25 μm and (E) is 50 μm (magnification of B and C is same as E).

ously reported (Tiao and Blakemore, 1976; Finlay et al., 1978). The nasal and temporal fields were represented in rostral and caudal SC, respectively whereas the upper and lower visual fields in the medial and lateral SC, respectively (Fig. 3). The topographic map was similar in the animals with a BSC cut on P0 (Fig. 4) although abnormalities were observed. Thus, reversal of the receptive fields was seen and no receptive field was recorded in some parts of the SC (Fig. 4).

Experiment 3: Physiological and anatomical studies of the retinocollicular projections following a BSC cut and PN grafting in adult hamsters

Anatomical studies

Retinal fibers labelled with WGA-HRP were

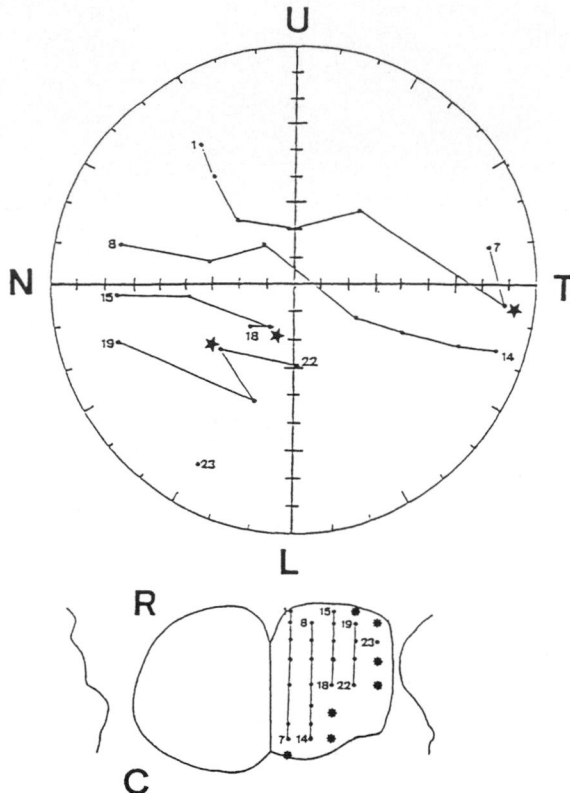

Fig. 4. Position of electrode penetration in the SC of a hamster with a BSC cut on P0 (lower figure) and corresponding visual receptive fields (upper figure). Only the centers of the receptive field are shown. The stars denote the sites where reversal of receptive fields were observed. No visual responses were observed in some part of the SC (*). Scale bar is 1 mm.

observed in the left SC in seven of the 12 grafted animals (Fig. 5A) whereas no reinnervation was observed in the rest of the grafted animals or in the seven control animals with no graft (Fig. 5B). In animals with no retinocollicular projections, labelled retinal fibers could be seen in the optic tract indicating the success of the anterograde transport of the WGA-HRP.

Physiological studies

Visual responses could be recorded from the grafted and control animals prior to the BSC cut but not immediately afterwards. At the time of recording, the grafted animals were classified into successfully or unsuccessfully grafted animals

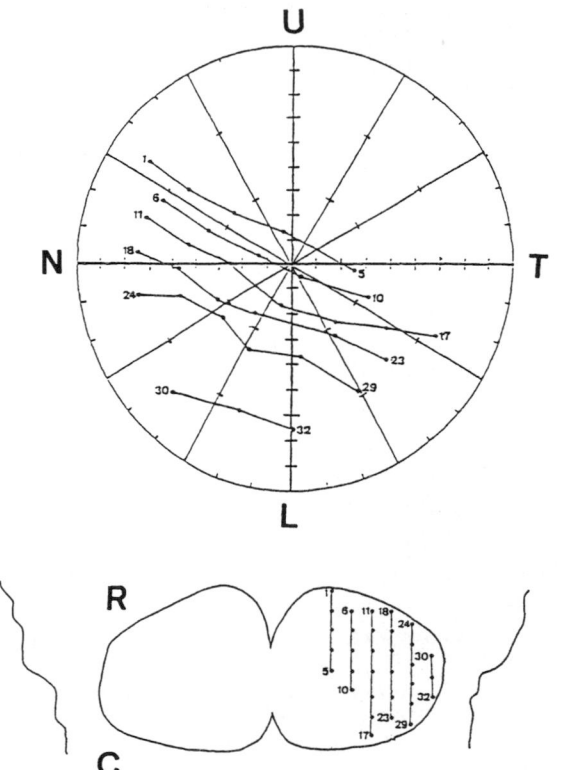

Fig. 3. Position of electrode penetration in the SC of a normal adult hamster (lower figure) and corresponding contralateral visual receptive fields (upper figure). Only the centers of the receptive field are shown. Scale bar is 1 mm. U, upper; L, lower; N, nasal; T, temporal; R, rostral and C, caudal.

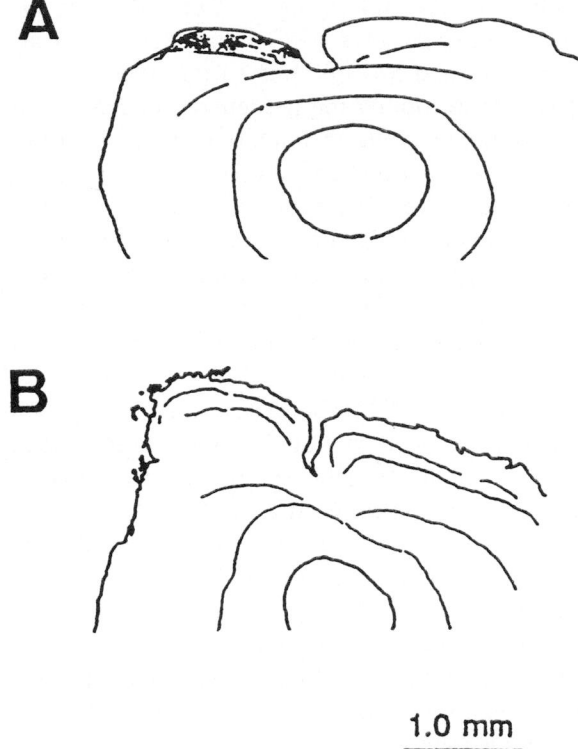

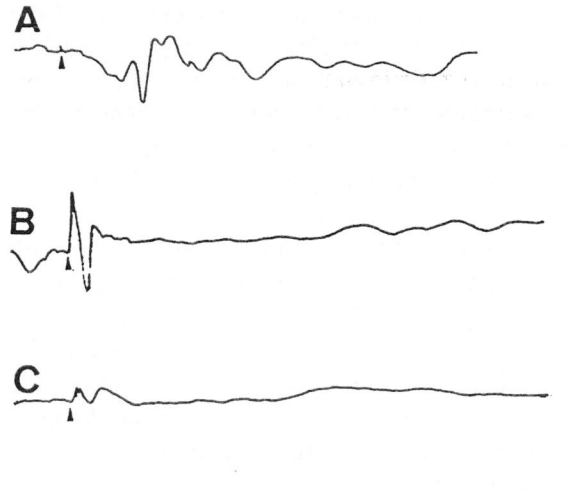

Fig. 6. (A) Averaged evoked potentials in response to diffuse flash of light (arrowhead) 11 weeks after transection of BSC and PN grafting. The responses were recorded from the central surface of the SC. (B) The lack of visual evoked potentials from the SC of an unsuccessfully grafted animal. (C) No visual evoked potentials were observed from an animal with BSC cut. In each record 100 responses were averaged. Upward deflection: positive; Calibration bar is 50 ms and 50 mV in A, and 50 ms and 30 mV in B and C.

1.0 mm

Fig. 5. (A) Drawing of the SC from a successfully grafted animal illustrating the distribution of the WGA-HRP labelled retinal fibers in the SC. (B) Drawing of the SC from a BSC cut animals showing the lack of retinal fibers in the SC.

based on whether the PN was detached from the SC and/or the BSC. Using this criteria, eight animals were grouped as successfully PN grafted and eight as unsuccessfully PN grafted animals. In response to diffuse flash stimuli applied to the contralateral eye, multiple negative evoked field potentials with latencies of 55–200 ms were recorded from all successfully grafted (Fig. 6A), but not in unsuccessfully grafted (Fig. 6B) (with the exception of one animal) or control animals (Fig. 6C).

Discussion

Our results support and extend the previous findings (So et al., 1981) that there is a critical age after which transection of the BSC does not result in reinnervation of the SC. However, these

transected retinal fibers are shown to regrow along a membrane on top of the SC even after the critical age. A PN graft bridging the transected optic fibers and the SC at the damaged site leads to physiologically functional reinnervation of the SC suggesting that PN can be used to restore damaged connections in the CNS even when the site of damage is far from the cell bodies.

There are intrinsic and extrinsic factors which may affect the ability of the damaged retinal axons to cross a site of transection after the critical age of P3. It has been proposed that events occurring in the retina before the critical age might be important in determining the growth potential of most of the retinal axons in hamsters (Chen et al., 1995). A critical period for the regeneration of fibers from RGCs across a retinal lesion is also found in an opossum (MacLaren and Taylor, 1995). It is not clear what are the important events occurring in the retina at the critical age. A decline in the production of trophic factors by glia cells or a down-regulation of the receptors to these factors in the RGCs might be important. Extrinsic factors which might affect regeneration after the critical age include the formation of scar at the site of transection and occurrence of inhibitory molecules (Schwab et al., 1993).

The retinal fibers reinnervating the SC following a BSC cut on P3 are most likely regenerating fibers because the number of retinal fibers in the optic nerve in hamsters starts declining on P2 (Tay et al., 1986). However, the fibers innervating the SC following a BSC cut on P0 could be a combination of regenerating and newly growing axons. This is supported by the results from our physiological experiments on animals with a BSC cut on day 0. Assuming the pattern of generation of RGCs in hamsters is similar to that of the mouse (Dräger, 1985) so RGCs in the hamsters which are born first should be generated in the central retina and RGCs which are generated later should be located in the peripheral retina. Axons from the central retina should have reached the SC by the time the BSC was cut on P0 since RGCs from the central retina are known to be born as early as E10.5 (Sengelaub et al., 1986) Since receptive fields can be recorded from the central and peripheral retina from these animals, it suggests that some of the retinal fibers reinnervating the SC are regenerating fibers.

Although there are mistakes observed in the retinotopic map of the animals with a BSC cut on day 0, the topography of the map is quite similar to that of the normal. This explains why most of the animals with a BSC cut on P0 can exhibit close to normal turning (So et al., 1981) in response to visual stimuli.

Although retinal fibers did not reinnervate the SC after the critical age but they were observe to grow in a membrane on top of the SC. A similar finding was previously reported in the rat following a BSC cut (Harvey et al., 1986) supporting the finding that some of the retinal axons can still regrow even late in development. This point is clearly illustrated in the study in which PN was used to induce regeneration of retinal axons in adult rats (So and Aguayo, 1985; Berry et al., 1986; Vidal-Sanz et al., 1987), hamsters (So et al., 1986; Cho and So, 1992) and cats (Watanabe et al., 1993). Although the percentage of regenerating RGCs in adult mammals is small, it shows that a PNS environment, unlike that of the CNS, can support regeneration of CNS axons. The Schwann cells in the PN are probably the important cell type for supporting regeneration because they have been shown to secrete many different type of trophic factors including BDNF (Meyer et al., 1992), NGF (Heumann et al., 1987; Meyer et al., 1992) and CNTF (Sendtner et al., 1992). This might also explain why the PN bridging the site of BSC cut in the present experiment can enhance retinal axons to reinnervate the SC.

There have not been any previous studies to attempt to replace the retinocollicular projections with a segment of PN graft following damage of the BSC because it is generally believed that retinal axons transected far from the somata could not regenerate. Richardson and his colleagues (Richardson et al., 1982) in their pioneering study in the rat demonstrated that very few, if any, optic axons sectioned just rostral to the chiasm, a site far away from the RGCs, regenerated into a PN graft. This result is confirmed in a recent study in the hamster (Lau et al., 1994). The results of the above studies taken together demonstrate an inverse relationship between the distance of axotomy and the regenerative ability of the RGCs into the PN graft. Such relationship has also been observed in other parts of the CNS in which significantly more respiratory axons regenerated into a PN graft implanted to the medulla oblongata when the implantation was closer to the central respiratory cells (Lammari-Barreault et al., 1991). The reason for this phenomenon is not clear at the moment but it may be related to the inhibitory effects of some of the nonneuronal environment in the CNS, including astrocytes (Bähr et al., 1995)), oligodendrocytes and CNS myelin (Schwab et al., 1993). Thus, the longer the surviving CNS axons, the more the non-neuronal environment and therefore the bigger the inhibitory effect on axonal regeneration.

It is apparent that this lack of axonal regeneration has been overcome to a certain extent by the present experimental design and it is important to discuss the possible differences between the present and previous studies involving the retinal pathways. One of the major differences is collateral projections. It is known that most of the retinocollicular projections send collaterals to the lateral geniculate nucleus (Chalupa and Thompson, 1980). Transecting the brachium of the SC should still preserve these collaterals. However, this would not be the case if the retinal axons were damaged in the eye or in the optic nerve. This difference is important because it is known that the loss of RGCs is much lower after a lesion of the BSC (Perry et al., 1982) than a section of the optic nerve or tract, presumably due to the remaining collateral projections.

It is possible that the PN (see above) and the collaterals can contribute trophic factors to the RGCs. For example, the collaterals can transport retrogradely any trophic factors produced by the cells in the lateral geniculate nucleus back to the RGCs which might be important in overcoming some of the inhibitory effect following a distal axotomy. This idea is supported in a recent study in which a segment of PN (a trophic source) transplanted intravitreously close to the RGCs cell bodies can enhance the regenerative response of RGCs following transection of the optic nerve either at 2 mm (Ng et al., 1995) or 6–8 mm (Lau et al., 94) from the eye. Further studies will be needed to investigate these various possibilities.

Acknowledgements

The research conducted in Hong Kong was supported by The Hong Kong Research Grant Council and research grants from The University of Hong Kong to KFS and DT. The research work carried out in Japan was supported by Grants-in-Aid for Scientific Research from the Ministry of Education, Culture and Science of Japan, and a Grant from Uehara Foundation for Science Promotion.

References

Bähr, M., Przyrembel, C. and Bastmeyer, M. (1995) Astrocytes from adult rat optic nerves are nonpermissive for regenerating retinal ganglion cell axons. *Exp. Neurol.*, 131: 211–220.

Berry, M., Rees, L. and Sievers, J. (1986) Unequivocal regeneration of rat optic nerve axons into sciatic nerve isografts. In: G. D. Das and R. B. Wallace (Eds.), *Neural Transplantation and Regeneration*, Springer-Verlag, New York, pp. 63–79.

Chalupa, L.M. and Thompson, I. (1980) Retinal ganglion cell projections to the superior colliculus of the hamster demonstrated by the horseradish peroxidase technique. *Neurosci. Lett.*, 19: 13–19.

Chen, D.F., Jhaveri, S. and Schneider, G.E. (1995) Intrinsic changes in developing retinal neurons result in regenerative failure of their axons. *Proc. Natl. Acad. Sci. USA,* 92: 7287–7291.

Cho, E.Y.P. and So, K.-F. (1992). Characterization of the sprouting of axon-like processes from retinal ganglion cells after axotomy in adult hamsters: a model using intravitreal implantation of a peripheral nerve. *J. Neurocytol.*, 21: 589–603.

Dräger, U.C. (1985) Birth dates of retinal ganglion cells giving rise to the crossed and uncrossed optic projections in the mouse. *Proc. R. Soc. London*, 224: 57–77.

Finlay, B. and So, K.-F. (1979). Altered retinotectal topography in hamsters with neonatal tectal slits. *Neurosci.*, 4: 1119–1128.

Finaly, B.L., Schneps, S.E., Wilson, K.G. and Schneider, G. E. (1978) Topography of visual and somatosensory projections to the superior colliculus of the golden hamster. *Brain Res.*, 142: 223–235.

Heuman, R., Korsching, S., Bandtlow, C. and Thoenen, H. (1987) Changes of nerve growth factor synthesis in nonneuronal cells in responses to sciatic nerve transection. *J. Cell Biol.*, 104: 1623–1631.

Harvey, A.R., Gan, S.K. and Dyson, S.E. (1968) Regrowth of retinal axons after lesions of the brachium and pretectal region in the rat. *Brain Res.*, 368: 141–147.

Ireland, S.M.L. (1991) The plasticity of the visual system following damage of the brachium of the superior colliculus in neonatal and adult hamsters: an anatomical and physiological study. Ph.D. thesis, The University of Hong Kong, p. 152

Lammari-Barreault, N., Rega, P. and Gauthier, P. (1991) Axonal regeneration from central respiratory neurons of the adult rat into peripheral nerve autografts: effect of graft location within the medulla. *Neurosci. Lett.*, 125: 121–124.

Lau, K.C., So, K.-F. and Tay, D. (1994). Intravitreal transplantation of a segment of peripheral nerve enhances axonal regeneration of retinal ganglion cells following distal axotomy. *Exp. Neurol.*, 128: 211–215.

MacLaren, R.E. and Taylor, J.S.H. (1995) A critical period for axon regrowth through a lesion in the developing mammalian retina. *Eur. J. Neurosci.*, 7: 2111–2118.

Mesulam, M.M. (1978) Tetramethyl benzidine for horseradish peroxidase neurohistochemistry: a noncarcinogenic blue reaction product with superior sensitivity for visualizing neural afferents and efferents. *J. Histochem. Cytochem.* 26: 106–117.

Meyer, M., Matsuoka, I., Wetmore, C., Olson, L. and Thoenen, H. (1992) Enhanced synthesis of brain-derived neurotrophic factor in the lesioned peripheral nerve: different mechanisms are responsible for the regulation of BDNF and NGF mRNA. *J. Cell Biol.*, 119: 45–54.

Ng, T.F., So, K.-F. and Chung, S.K. (1995). Influence of peripheral nerve grafts on the expression of GAP-43 in regenerating retinal ganglion cells in adult hamsters. *J. Neurocytol.*, 24: 487–496.

Perry, V.H. and Cowey, A.A. (1982) A sensitive period for ganglion cell degeneration and the formation of aberrant retino-fugal connections following tectal lesions in rats. *Neurosci.*, 7: 583–594.

Richardson, P.M., Issa, V.M.K. and Sheime, S. (1982) Regeneration and retrograde degeneration of axons in the rat optic nerve. *J. Neurocytol.*, 11: 949–966.

Sawai, H., Fukuda, Y., Sugioka, M., Morigiwa, K., Sasaki, H. and So, K.-F. (1996). Functional and morphological restoration of the lesioned retinocollicular pathways by peripheral nerve autografts in adult hamsters. *Exp. Neurol.*, 137: 94–104.

Schwab, M.E., Kapfhammer, J.P. and Bandtlow, C. E. (1993) Inhibitors of neurite growth. *Ann. Rev. Neurosci.*, 16: 565–595.

Sendtner, M., Stockli, K.A. and Thoenen, H. (1992) Synthesis and localization of ciliary neurotrophic factor in the sciatic nerve of adult rat after lesion and during regeneration. *J. Cell Biol.*, 118: 139–148.

Sengelaub, D.R., Dolan, R.P. and Finlay, B.L. (1986) Cell generation, death, and retinal growth in the development of the hamster retinal ganglion cell layer. *J. Comp. Neurol.*, 246: 527–543.

Sidman , R.L., Angevine, Jr., J.B. and Tater-Pierce, E. (1971) *Atlas of the Mouse Brain and Spinal Cord.*, Harvard University Press, Cambridge.

So, K.-F. and Aguayo, A.J. (1985). Lengthy regrowth of cut axons from ganglion cells after peripheral nerve transplantation into the retina of adult rats. *Brain Res.*, 328: 349–354.

So, K.-F., Schneider, G.E. and Ayres, S. (1981). Lesions of the brachium of the superior colliculus in neonate hamsters: correlation of anatomy with behavior. *Exp. Neurol.*, 72: 379–400.

So, K.-F., Xiao, Y.-M. and Diao, Y.-C. (1986). Effects on the growth of damaged ganglion cell axons after peripheral nerve transplantation in adult hamsters. *Brain Res.*, 377: 168–172.

So, K.-F., Campbell, G. and Lieberman, A.R. (1990) Development of the mammalian retinogeniculate pathway:target finding, transient synapses and binocular segregation. *J. Exp. Biol.*, 153: 109–124.

Tay, D., So, K.-F., Jen, L.S. and Lau, K.C. (1986). The postnatal development of the optic nerve in hamsters: an electron microscopic study. *Dev. Brain Res.*, 30: 268–273.

Tiao, Y.C. and Blakemore, C. (1976) Functional organization in the superior colliculus of the golden hamster. *J. Comp. Neurol.*, 168: 483–504.

Vidal-Sanz, M., Bray, G.M., Villegas-Perez, M.P., Thanos, S. and Aguayo, A. J. (1987) Axonal regeneration and synapse formation in the superior colliculus by retinal ganglion cells in the adult rat. *J. Neurosci.*, 7: 2894–2909.

Watanabe, M., Sawai, H. and Fukuda, Y. (1993) Number, distribution, and morphology of retinal ganglion cells with axons regenerated into peripheral nerve graft in adult cats. *J. Neurosci.*, 13: 2105–2117.

Woo, H.H., Jen, L.S. and So, K.-F. (1985). The postnatal development of retinocollicular projections in normal hamsters and in hamsters with one eye removed at birth: a horseradish peroxidase tracing study. *Dev. Brain Res.*, 20: 1–13.

M. Norita, T. Bando and B. Stein (Eds.)
Progress in Brain Research, Vol 112
© 1996 Elsevier Science BV. All rights reserved.

CHAPTER 27

A proposed reorganization of the cortical input-output system

Yasuhiko Tamai

Department of Physiology, Wakayama Medical College, Wakayama 640, Japan

This paper proposes a new concept for interpretation of the cerebral cortex. In this schema, the cortex is an 'imaging system' which plays an integral part in the 'functional circuit' of the living body. This functional circuit is made up of five components: the 'imaging system' of the cortex, the 'image forming system' of the basal ganglia, thalamus and cerebellum, the 'image-acting system' of the brainstem and spinal cord, the 'sensory system' of the sense organs and the 'external words.' An image constructed in the cortex yields a purposeful movement by way of its outputs through the image-acting system. The image to drive the purposeful movement is formed and accumulated in the cortex by trial and error under the control of the image-forming system.

In the proposed organization outlined here, the cortex can be divided into two components, both of which have their own input and output for forming and realizing the image. One is comprised of specific areas for analyzing proper sensations from the sense organs through a specific area of the thalamus (e.g., retina to lateral geniculate to primary visual cortex). The other system is composed of non-specific areas analyzing more diffuse input from various sensory organs through non-specific areas of the thalamus. These two systems work in concert to form fundamental and detailed images for purposeful movements through tight intercortical connections. Whereas the outputs from non-specific areas yield common behaviors regardless of sensory modality, outputs from the specific areas controls movements related to a specific sensation. Thus, it is proposed that these two systems work cooperatively: the former constructs the image and causes the initial movements, and the latter controls the movements to realize the image from the changing information of transient conditions.

This model differs from other models of cortical organization which segregate cortex on the basis of sensory-related input regions and motor-related output regions. Existing data is consistent with the proposed organizational scheme. Although it may be difficult to prove specific aspects of this new model, I believe that the hypotheses contained within it will serve as an important foundation for driving future research.

Introduction

When considering the function of the human body, it is virtually impossible to separate the sensory and motor systems from one another. Despite this, there has been surprisingly little research which has focused on the interactions of these systems — generally scientists have chosen to study sensory issues or motor issues. As a result of this, cortex has generally been subdivided on the basis of sensory or motor designations (e.g. visual cortex, auditory cortex, primary motor cortex, etc.). Most of the existing models of brain function are based on this form of functional localization, and they fail to provide satisfactory explanations of our behavior, especially how our movements occur and why our movements become purposeful to the external world. The solution to these problems requires a new concept of the brain linking the sensory and motor systems to the external world. The present study proposes a new model of the brain to understand the input and output systems of the cortex.

Three principles of the behavior

Let us consider a relatively simple behavior —

374

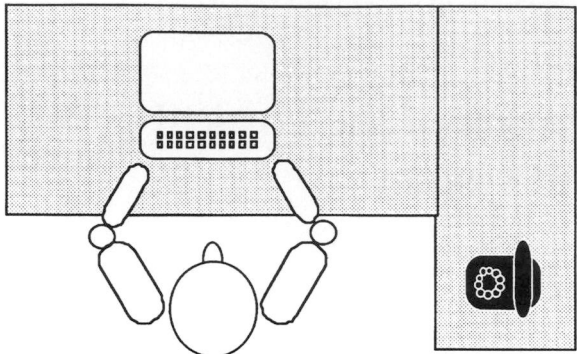

Fig. 1. The sequence of behavior in answering the telephone. There is a common pattern to the series of movements — turning toward the phone, picking up the receiver and placing the receiver to the ear (see text).

answering the telephone. In this example, the phone is located near by to make the behavior simple (Fig. 1). When the bell rings, I turn toward the phone, pick up the receiver and place it to my ear. There is a common element in these series of movements. I first turned because I constructed an image of the telephone from the sound of the bell. My movement continued until the telephone came into sight. The turning movements were smaller if the phone was beside me, and were larger if the phone was located behind me. Performing these movements does not require serious thought, but does need some information about the direction and speed of the movement required for the task. The movements automatically stop when the telephone comes into sight.

The subsequent movements, reaching out my hand to pick up the receiver and placing the receiver to my ear, are also performed by constructing an image of receiver and ear, respectively. Each movement lasts until I touch the receiver and I feel the receiver at my ear, respectively. Thus, I present the first principle of behavior (Fig. 2).

Principle 1

Purposeful behavior is generated by the construction of an image and stops when the image is satisfied

That is, our actions represent a response to an image constructed in the brain, and movements continue until the image is realized, and the action will stop when the image is completed. From the input perspective, there are two types of information: the initial information by which the image is constructed and subsequent information with which the image is compared. From the output perspective, there are two movements: the initial movements triggered by the image and the subsequent movements to complete the image. Thus, I present the second principle of the behavior.

Principle 2

There are two systems in the brain: one system performs an image and another system completes the image

The sequential movement of answering the phone was not programmed at the moment when

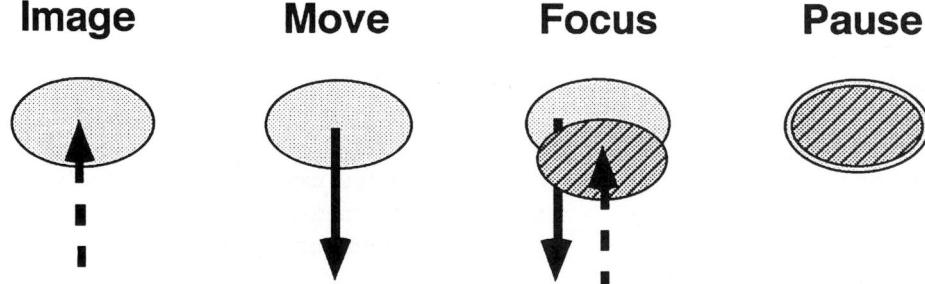

Fig. 1. Schematic drawing of the theory that a purposeful behavior is generated by an image and stops when the image is satisfied. The information flow is from left to right. Initial information is constructed as an image and the image drives the movement. The movement pauses when the image is satisfied by subsequent information (See text).

the phone rang. Instead, the pattern of these purposeful movements has already been accumulated as images in the brain gathered through experience and training. However, the information and movements after the initial action of answering the phone are situation-dependent for completing the action needed to realize the image and will not have been accumulated.

There is one more important point in this example. The sound of the bell itself is objective. However, the behavior of answering the phone is subjective because I am the one who generated the image of answering the phone in response to the sound of the bell. That is, the sound of the bell itself does not contain the intention to answer the phone, but I generated the image of answering the phone or I selected the behavior of answering the phone in response to the sound of the bell.

Let us discuss this point in further detail in the following situation (Fig. 3). There is a desk in the center of a room, and it is necessary to detour around this desk in order to reach an identical desk in the corner of the room on which sits the telephone. In this situation, the center desk is standing in my way, therefore, visual information from the desk defines it not as a desk but as an obstruction that I must detour around. Conversely, the second desk, even though it has the same shape and color as the first, provides me with information to stop here. Thus, the meaning of sensory information depends on immanent requirements of the individual who is behaving purposefully. In other words, sensory information during purposeful behavior is filtered and selected based on our inner condition. All information is not recognized blindly but intended or expected information is recognized according to the purpose. Thus, I present the third principle of behavior.

Principle 3

Information is actively selected during purposeful behavior

Thus, we select information to construct an image and to complete the image. The result is

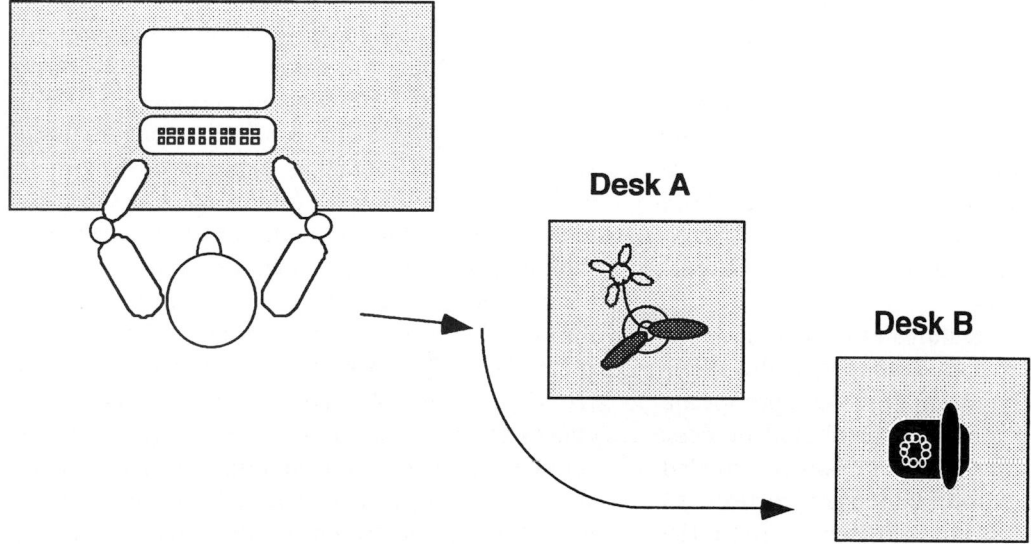

Fig. 3. Active selection of information. There are two identical desks in the room. For the person who wants to answer the phone, the desk located in the center of the room is an obstructing object requiring a detour while the other desk in the corner of the room is the objective (see text).

that our behavior becomes purposeful to various situations in the external world.

A new model of the nervous system

In the first section, I described a few simple principles of our behavior using the example of answering the telephone. In current accepted dogma, the series of movements required to answer the phone is programmed into the brain. The sound of the bell generates distinct patterns of activity in auditory areas of the brainstem, which ascend to the thalamus and cortex. The signal is next passed on to association areas, which develop the intention of answering the phone. Finally, the series of movements is performed by means of signals generated in motor cortex and passed down to the brainstem and spinal circuitry which ultimately drive the muscles responsible for the movement (Allen and Tsukahara, 1974; Popper and Eccles, 1981; Paillard, 1982). In this type of sensorimotor scenario, different regions of the brain are assigned a particular functional role in this process.

In considering the cerebral cortex, it seems difficult to envisage how such functional specialization could be conferred by it's anatomical organization. Most of the neocortex is ordered in a highly laminated fashion, with input fibers and output cells being organized in a highly stereotyped manner (Chow and Leiman, 1970). On the basis of this, it's difficult to conceptualize how certain areas would be best suited to deal with sensory inputs, others in the formulation of intention, and finally others in the control of outputs which drive movement.

The organizational features of the cerebral cortex seem better suited to dealing with the transformation of sensory information into appropriate motor signals by means of a distributed process. In such a scenario, input information is funneled to more than the appropriate sensory cortices, an image and intention are constructed from the patterning of activity throughout the cortical mantle, and the outputs of a number of cortical zones provide the signals necessary to drive the

purposeful movement required at that moment. This idea will be discussed further in the following session using the schematic drawing in Fig. 4.

The cortex can be divided into two principal areas. One is comprised of specific areas devoted to the analysis of sensation from specific sense organs such as the eye, ear and skin. The inputs to these regions come via specific areas of the thalamus (routes III–IV in Fig. 4A). The other cortical area is a non-specific area which analyzes more diffuse inputs from the various sensory organs. Inputs to these regions come via non-specific areas of the thalamus (routes VII–V). Input information is dealt with by these two areas of cortex, which results in the formation of a meaningful image. Thus, I call these cortical zones the 'imaging system'. It seems likely that the former signals (i.e. specific cortex) relate to objective information and that the latter signals (i.e. non-specific cortex) may relate to fundamental information among various sensations. The pattern of activity in these regions of cortex becomes a meaningful image when the pattern coincides with a pattern accumulated previously. Some images then result in the generation of a purposeful movement.

These two areas of cortex have their own outputs (route I). Output from the non-specific area is necessary to maintain purposeful movement until the image is realized, while output from the specific area is important to support the timely execution of the purposeful movements. Once we move, information about the external world and the internal condition must be continually updated. These updates are sent to the cortex through the sensory organs (routes III to IV and routes III, VII to V) and the cerebellum (routes III, VIII, VI and V). Whereas the sensory organs convey information about the external world, the cerebellum conveys information as to the internal state of the body. This external and internal information is integrated in the thalamus along with information from the basal ganglia. Thus, the basal ganglia and thalamus integrate three types of the information: external information from the sensory organs (routes IV and VII), internal information from the cerebellum (route VI), and information from the cortex (route V). By means of this pattern of connectivity, the basal ganglia and thalamus can reference the image in the cortex and the transient changes in information which accompany the movement. In such a feedback circuit, new information is actively biased by the preceding internal image.

In the early stages of development, the basal ganglia and thalamus play a major role in the patterning and execution of movement. As devel-opment continues, the cortex accumulates information regarding output patterns of movement and takes a more prominent role in the genera-tion of the movement. Finally, the cortex assumes the principal role in this process as accumulated images and movements result in purposeful be-havior. Thus, I call the basal ganglia, the thala-mus (including parts of the hypothalamus) and the cerebellum an 'image-forming system'. The image-forming system serves to collect and ex-tract useful information. If the behavior is pur-poseful, the patterns of activity associated with it

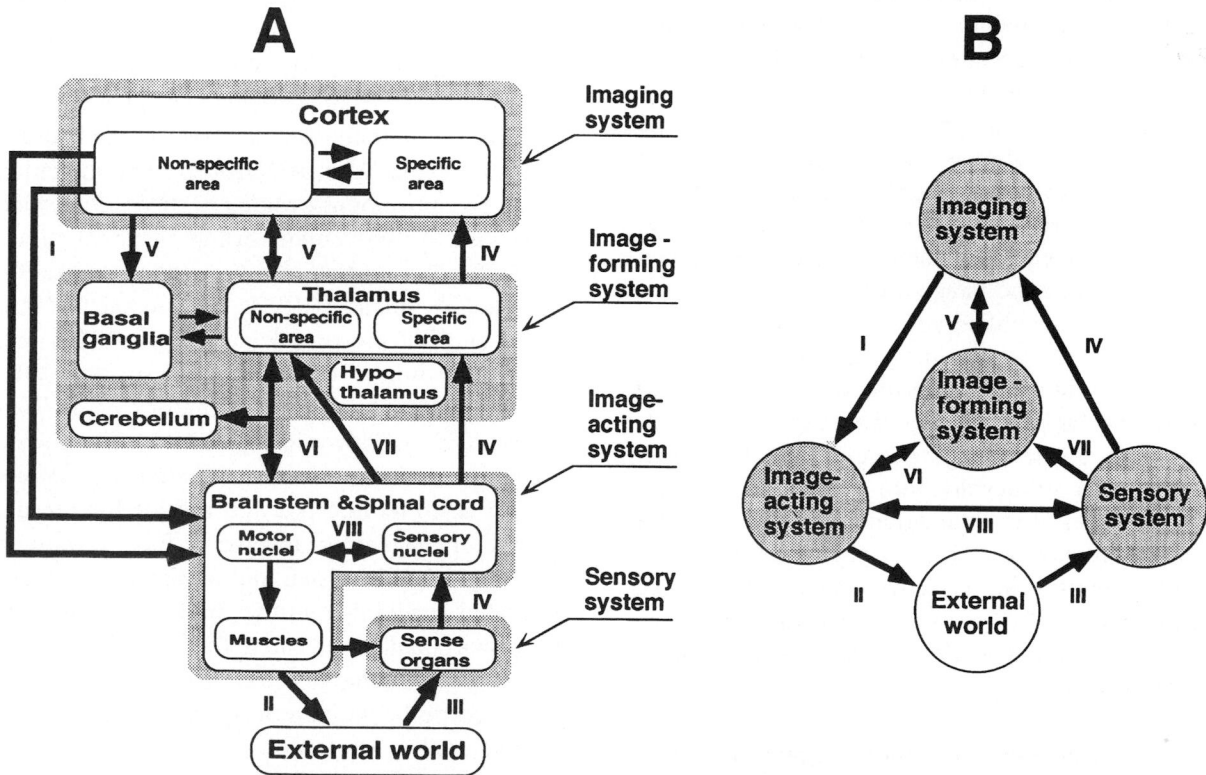

Fig. 4. Organization of the nervous system. There are four systems in the brain: the 'imaging system' in the cortex, the 'image-forming system' in the basal ganglia, thalamus and cerebellum, the 'image-acting system' in the brainstem and spinal cord, and the 'sensory system' of the sense organs (A). These systems form a 'functional circuit' with the external world (B). An image constructed in the imaging system causes a purposeful movement through the image-acting system. Images of purposeful movements are accumulated in the imaging system by trial and error under control of the image-forming system. The image-forming system has information from the other three systems in the functional circuit and plays an important role in forming the image for purposeful movements to the external world. The imaging system (i.e. cortex) has two sources of the information, direct information from the sensory system and indirect information through the image-forming system. These two inputs work in concert to form fundamental and detailed images to realize purposeful movements.

are allowed to accumulate in cortex. The hypothalamus and the limbic system may play an important role in the selection of information. As alluded to, the image-forming system may be the most active during the early stages of development and may not be as active in the adult. Cortical outputs ultimately result in the execution of a movement by way of influencing the motor nuclei in the brainstem and the spinal cord (route II). This system is called here an 'image-acting system'. Fundamental movements such as standing, walking and running, as well as postural reflexes may be built in to this system during development.

The circuits in the cerebrum, the brainstem nuclei, the cerebellum and the thalamus (Sasaki, 1979) may play an important role in the accumulation of accurate movements. Once movements become smooth and accurate, signals from the cortex become simple and fundamental. In the early stages of development, cortical outputs may be much more complex. However, after a movement or series of movements become fluid and is so analyzed by the image-forming system, cortical outputs are simplified and the movement is controlled mainly by the image-acting system. The imaging system of the cortex, the image-forming system in the basal ganglia, thalamus and cerebellum, the image-acting system in the brainstem and spinal cord develop into a 'functional circuit' which drives our interactions with the external world. Damage to or lack of any of these component systems leads to movement disorders. The relationships of these systems and their interface with the external world are shown schematically in the Fig. 4B.

Extension of input and output signals in the cortex

As previously mentioned, most experiments concerning the input and output signals of the cortex have focused on either sensory or motor cortical areas, respectively. Such an experimental emphasis results in the impression that input and output signals are integrated in different cortical areas. The experiments described here are based on the hypothesis that input information is spread across a broad area of the cortex, and that each cortex which receives such information has it's own integrated pattern of activity which results in image formation and output activity which contributes to the execution of a movement.

Experiments were carried out using awake cats in which microelectrodes were chronically implanted in the cerebral cortex. Implantation of the microelectrodes for recording extracellular single unit activity and silver wire electrodes for recording the electrooculogram (EOG) were performed under chloralose anesthesia (60–100 mg/kg i.v.). A special microelectrode with miniature manipulator was designed for this experiment (Tamai et al., 1991). With this device, single unit discharges from several cortical areas was recorded simultaneously in awake and unrestrained animals. Microelectrodes were implanted in five eye movement-evoking cortices, confirmed by evoking eye movements using electrical stimulation. The five cortices are the medial wall under the cruciate sulcus (CRU), the medial and the lateral bank of the presylvian sulcus (PRE_M and PRE_L), the fundus of the coronal sulcus (COR) and the ventral bank of the anterior ectosylvian sulcus (AES) as shown in Fig. 5B. Fig. 5A shows a schematic drawing of the experimental set-up. The cat is sitting in a box and waiting for a visual or auditory signal to receive a food reward, a small dry fish. The visual and auditory stimuli are generated randomly and the food reward appears through a small hole when the stimulus is presented. Unit discharges, EOG and guide signals were recorded simultaneously in a data recorder and analyzed by computer.

Figs. 6 and 7 show the activity of typical neurons in these areas during visually-guided and auditory-guided saccades, respectively. Neurons were simultaneously recorded through five electrodes implanted in the five eye movement-evoking cortices shown in Fig. 5B. Neuronal activity was normalized to the onset of the sensory stimulus and the initiation of the saccades to form the

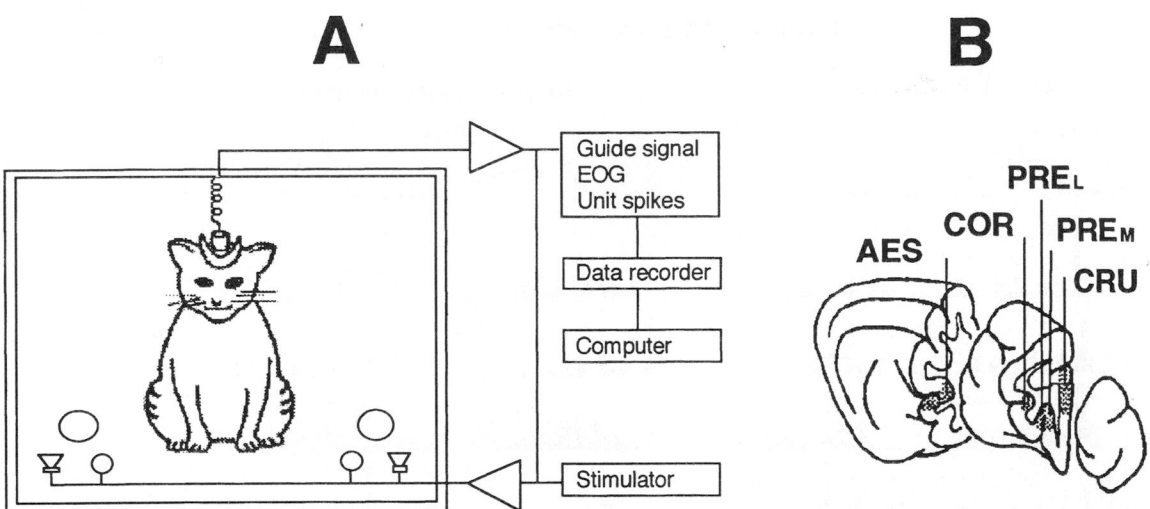

Fig. 5. Schematic drawing of the experimental set-up (A) and location of the five implanted electrodes (B). Electrodes are located in the medial wall under the cruciate sulcus (CRU), the medial and the lateral bank of the presylvian sulcus (PRE$_M$ and PRE$_L$), the fundus of the coronal sulcus (COR) and the ventral bank of the anterior ectosylvian sulcus (AES). Saccades were evoked by electrical stimulation. The stimulus signal, EOG and unit discharges were simultaneously recorded using a data recorder and analyzed by computer.

spike histograms. Many neurons exhibited peaks in the spike histogram by both methods of normalization. However, there appear to be two distinct types of neurons in these areas, the 'sensory-locked neuron' and the 'movement-locked neuron.' The sensory-locked neuron exhibited a peak in activity time-locked to the onset of the sensory stimulus (arrow; left column) and did not exhibit a peak normalized to the initiation of the saccade (right column). In contrast, the movement-locked neuron showed peak activity before the saccade and temporally-linked to the initiation of the saccade, and showed no peak time-locked to the onset of the sensory stimulus. Thus, it appears that the former neurons are more related to the sensory stimulus and the later neurons more related to the motor response (i.e. saccade). In each of the cortices examined, we see examples of sensory-locked and movement-locked neurons which respond to either visual or auditory stimulation. These results indicate that each of these cortices has information from at least two different sensory modalities as information related to the generation of a saccade.

Another point of interest was whether movement-locked neurons fire prior to all saccades. The answer is no. Rather, it appears that neurons in these cortices fire cooperatively or alternatively when different saccades are made (Fig. 8). In a pair of neurons observed in COR (COR N1) and AES (AES N1), there is a peak in the spike histogram before 7 of 14 saccades (large arrows in Fig. 8A). In contrast, there is no such time-locked peak in the remaining half of the trials (small arrows in Fig. 8A). The activity of this pair of neurons in two different areas was well correlated for many of the same saccades (exceptions: saccades 11 and 12). In another pair of neurons, COR (COR N2) and PRE$_L$ (PRE$_L$ N1), activity was also time-locked to the saccade in 6 of 14 trials (large arrows in Fig. 8B). However, in this neuronal pair no correlation is observed between their firing for the same saccades; the time-locked discharges are seen to different saccades in each case. In such a case, these two neurons appear to alternate their firing for different saccades. These results suggest that individual output may not be sufficient to cause saccades. Rather, the output

Light-guided saccade

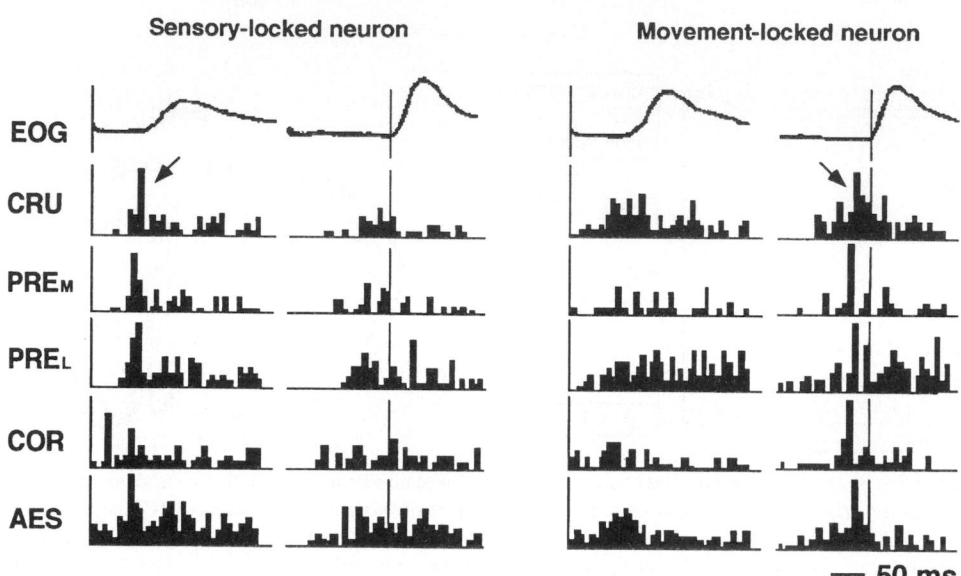

Fig. 6. Spike histograms of typical neurons during visually-guided saccades. All neurons were recorded simultaneously through five electrodes implanted in each cortex as shown in Fig. 5B. The uppermost traces (EOG) are an average of the electrooculogram obtained during these saccades. Spike histograms were normalized to the stimulus (left column) or to the initiation of the saccade (right column). In the sensory-locked neuron, the peak in activity appeared following the visual stimulus and was well correlated with stimulus onset (arrow in left column). Such a peak was not seen in relation to the initiation of the saccade. However, the movement-locked neuron exhibited a peak in activity before the saccade and well correlated with the initiation of the saccade (arrow in right column), but did not show a peak related to the visual stimulus. Note that both sensory-locked and movement-locked neurons were observed in all cortices.

from a number of cortices may play an integral role in the generation of saccades.

Let us summarize the input and output integration involved in saccades: (1) various cortices induce saccades following electrical stimulation (Schlag and Schlag-Rey, 1970; Guitton and Mandl, 1978; Tamai et al., 1984; Tamai et al., 1989); (2) these cortices have their own descending pathways to subcortical structures related to eye movements, such as the superior colliculus, pontine nucleus and brain stem reticular formation (Miyasita and Tamai, 1989; Tamai and Miyasita, 1989); (3) input information from various sensory modalities reaches these cortices; (4) there are output-related neurons that discharge before saccades in each of these cortices; (5) these output-

related neurons do not fire to all saccades; and (6) the output-related neurons discharge alternatively or cooperatively among the cortices. These results suggest that the input and output information needed in order to evoke a saccade is not restricted to a particular area of cortex, but rather is distributed over a broad region of cortex which encompasses a number of distinct regions. Such an idea may explain the discrepancy in the data describing neural activity in the so-called frontal eye field (FEF) (Bizzi, 1968; Goldberg and Bushnell, 1981). Here, activity does not always occur prior to the generation of a saccade, calling into question the presumptive role of the FEF as a cortical 'oculomotor center.' In our hypothesis, neurons in the FEF are not always required to fire before the saccade, since it is the pattern of activity in a number of cortices which ultimately drives eye movements.

Relation to previous studies of the nervous system

Classically, the visual system can be divided into two principal components: the geniculo-striate visual (GSV) and extrageniculate visual (EGV) systems (Trevarten, 1968; Diamond and Hall, 1969; Schneider, 1969; Pasik and Pasik, 1980). The ascending pathways of the GSV and EGV were summarized and applied simply to the specific and non-specific areas in the thalamus and cortex of the proposed new model in the visual system (Fig. 9). However, special attention should be given to the output pathways of this model. In this scheme, there are two major outputs, one exiting the striate cortex and the other exiting the extrastriate cortex (Fig. 9, I). In the new model presented in this paper, we say that the specific (i.e. modality-related such as striate cortex) cor-

tices and non-specific (i.e. associational) cortices have a separate system of outputs. In this view, whereas striate (i.e. specific) outputs may be more related to the detailed analysis of a visual object, extrastriate (i.e. non-specific) outputs may be more related to attentional phenomena.

In the auditory system, the traditional pathway takes information from the cochlea to the auditory cortex. This ascending pathway includes many nuclei and has a complex course in the brainstem, and terminates in primary auditory cortex by way of projections from the medial geniculate body (Davis, 1951; Nedzelnitesky, 1973). Most attention has been focused on this medial geniculate-cortex pathway, and relatively little work has examined auditory processing in thalamic regions other than the medial geniculate (Huang and Lindsley, 1973; Phillips and Irvine, 1979). Such an ascending pathway may project to the non-specific area of the cortex and form a dual system with the specific area (i.e. primary auditory cortex). In addition, such systems may interact with other

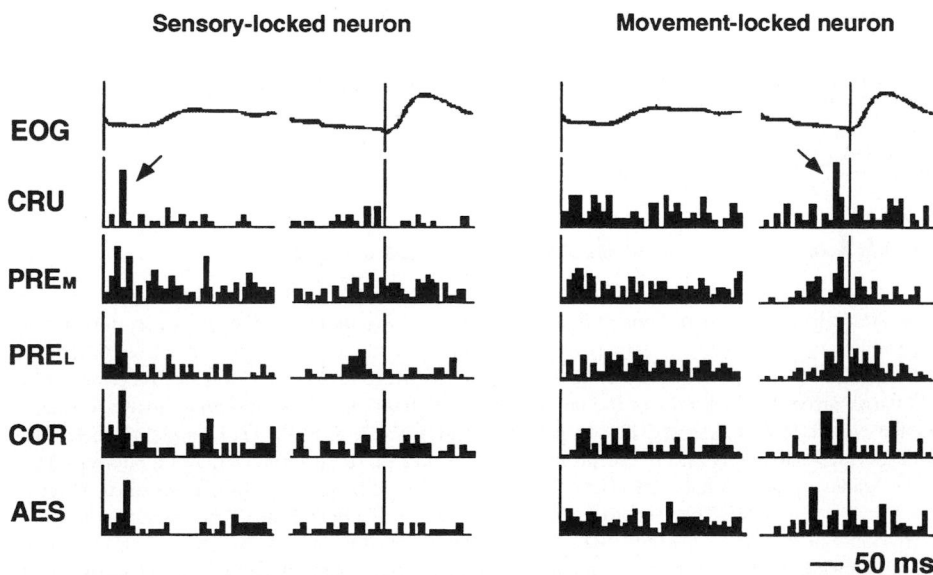

Fig. 7. Spike histograms of typical neurons during auditory-guided saccades under the same conditions as described in Fig. 6.

modalities (i.e. vision) at these levels. Although the output circuitry of these auditory systems is not well established, one can propose an organization parallel to that seen in the visual system, where outputs of the specific areas are involved in the detailed analysis of and response to the auditory world, and outputs of the non-specific areas are more involved in auditory attention.

In this context, the most noteworthy system is the somatosensory system. The major output of this system has been classically described for a long time as the so-called motor cortex. Motor cortical activity forms the basis for skeletomotor movements. For example, in our example cited at the beginning of this paper, to turn toward the telephone at the sound of the bell, outputs from the somatosensory system must control the neck muscles which serve to turn the head toward the direction of the phone. Furthermore, other somatosensory information is necessary for standing and walking toward the phone. Somatic information projects to the cortex through two ascending pathways: one via the ventral posterior nucleus of the thalamus and another pathway via a non-primary thalamic nucleus (Curry and Gordon, 1972; Huang and Lindsley, 1973). In the first pathway, there is a direct projection to the motor cortex (Tamai et al., 1984; Water et al., 1985). Information from the periphery and carried through this direct pathway is very important in

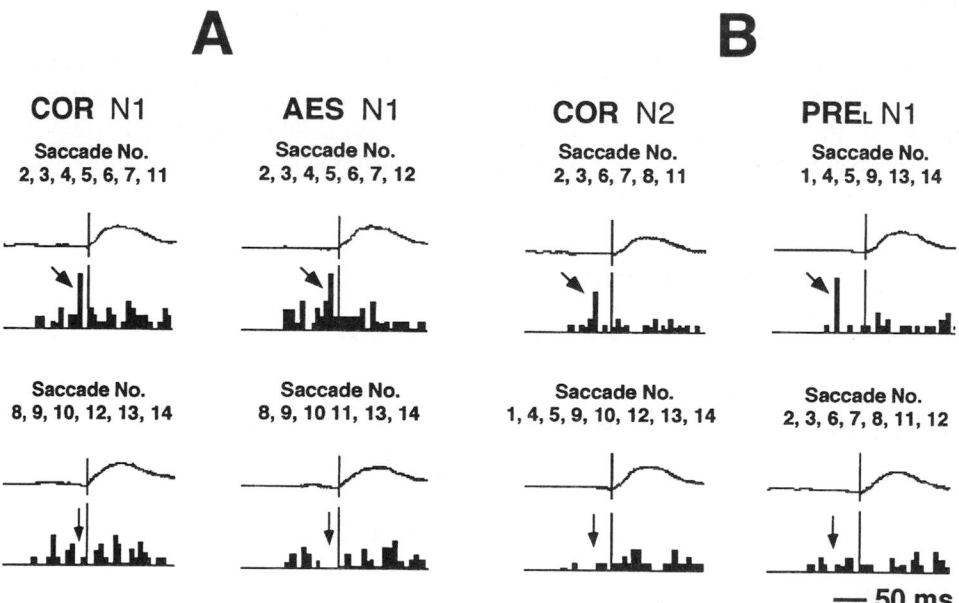

A   **B**

COR N1   **AES** N1   **COR** N2   **PRE**L N1

Saccade No.
2, 3, 4, 5, 6, 7, 11

Saccade No.
2, 3, 4, 5, 6, 7, 12

Saccade No.
2, 3, 6, 7, 8, 11

Saccade No.
1, 4, 5, 9, 13, 14

Saccade No.
8, 9, 10, 12, 13, 14

Saccade No.
8, 9, 10 11, 13, 14

Saccade No.
1, 4, 5, 9, 10, 12, 13, 14

Saccade No.
2, 3, 6, 7, 8, 11, 12

— **50 ms**

Fig. 8. Example of cooperative or alternative firing in movement-locked neurons among different cortical areas during the performance of saccades. (A) Cooperatively firing neurons: the activity of the neuron in the fundus of the coronal sulcus (COR N1) was time-locked to saccade Nos. 2, 3, 4, 5, 6, 7 and 11 evoked by an auditory stimulus (large arrows), but was not time-locked to saccade Nos. 8, 9, 10, 12, 13 and 14 (small arrows). The activity of the neuron in the ventral bank of the anterior ectosylvian sulcus (AES N1) was time-locked to saccade Nos. 2, 3, 4, 5, 6, 7 and 12, but not to saccades No. 1, 8, 9, 10, 11, 13 and 14. These two neurons discharged cooperatively to the same saccades with the exception of Nos. 11 and 12. (B) Alternatively firing neurons: The activity of this pair of neurons in the fundus of the coronal sulcus (COR N2) and the lateral bank of the presylvian sulcus (PRE_L N1) were related to different visually-evoked saccades. The response of the COR neuron (COR N2) was time-locked to saccade Nos. 2, 3, 6, 7, 8 and 11, while the response of the PRE_L neuron (PRE_L N1) was time-locked to saccade Nos. 1, 4, 5, 9, 13 and 14 (large arrows). Neither neuron showed a time-locked response to the remaining saccades (small arrows). All neurons were recorded simultaneously for the same saccades.

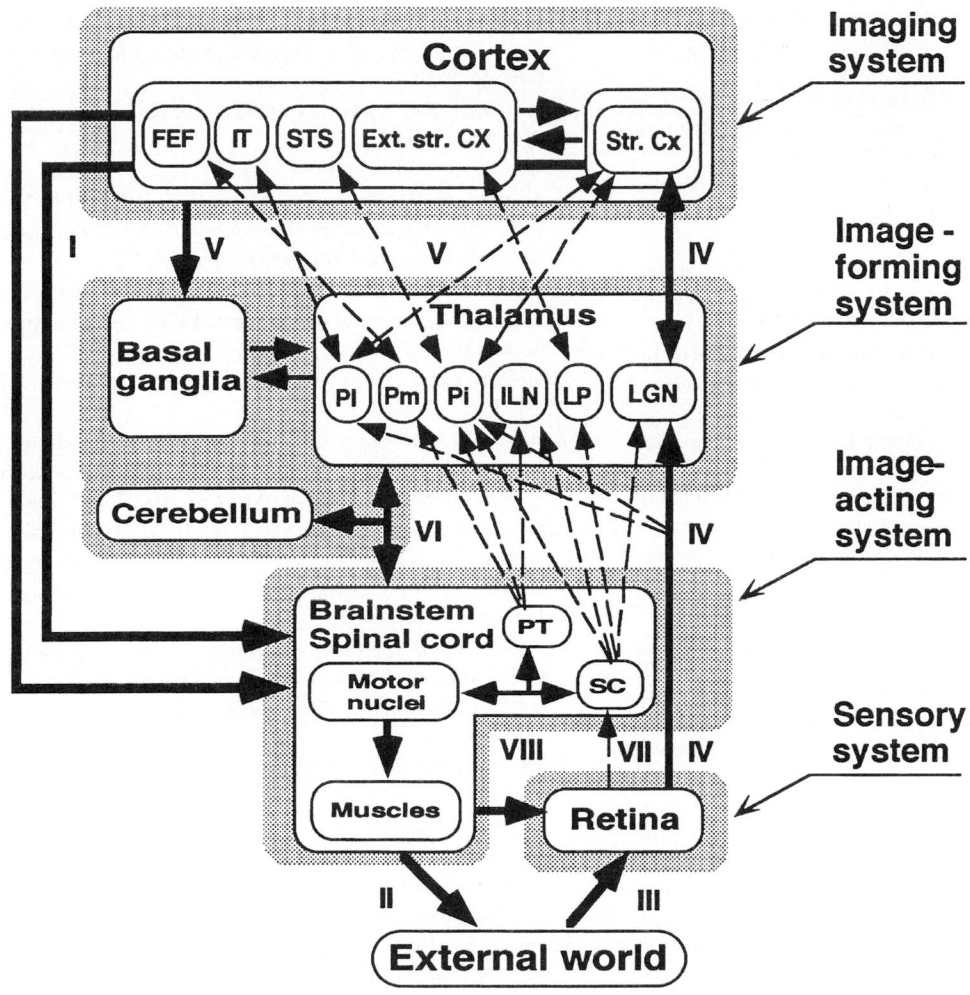

Fig. 9. A proposed model of the visual system (see text). Str. Cx, striate cortex; Ext. str. CX, extra striate cortex; STS, superior temporal sulcus; IT, inferior temporal cortex; FEF, frontal eye field; LGN, lateral geniculate nucleus; LP, nucl. lateralis posterior; ILN, nucl. intralaminaris; Pi, inferior pulvinar; Pm, medial pulvinar; Pl, lateral pulvinar; SC, superior colliculus; PT, pretectum.

the control of movement of the extremities (Asanuma and Arrisian, 1984). In contrast, the latter pathway may correspond to the extragenic-ulate pathways of the visual and auditory systems, and may be more involved in the attentional aspects of somatosensation. Both pathways likely contribute to the generation of a harmonious and purposeful movement.

Acknowledgements

I thank my friend Dr. K. Kanda for his support of these experiments, Miss T. Tujinaka for training and taking care of the animals over a long period.

References

Allen, G.I. and Tsukahara, N. (1974) Cerebrocerebellar communication system. *Physiol. Rev.*, 54: 957–1006.

Asanuma, H. and Arrisian, K. (1984) Experiments on functional role of peripheral input to motor cortex during voluntary movements in the monkey. *J. Neurophysiol.*, 52: 212–227.

Bizzi, E. (1968) Discharge of frontal eye field neurons during saccadic and following eye movement in unanesthetized monkeys. *Exp. Brain Res.*, 6: 69–80.

Chow, K.L. and Leiman, A.L. (1970) Aspects of the structural and functional organization of the neocortex. *Neurosci. Res. Progr. Bull.*, 8: 157–220.

Curry, M.J. and Gordon, G. (1972) The spinal input to the posterior group in the cat. An electrophysiological investigation. *Brain Res.*, 44: 417–437

Davis, H. (1951) Psychophysiology of hearing and deafness. In: S.S. Stevens (Ed.), *Handbook of Experimental Psychology*, Wiley, pp. 1116–1142.

Diamond, I.T. and Hall, W.C. (1969) Evolution of neocortex. *Science*, 164: 251–262.

Goldberg, M.E. and Bushnell, M.C. (1981) Behavioral enhancement visual responses in monkey cerebral cortex. II. Modulation in frontal eye fields specially related to saccades. *J. Neurophysiol.*, 46: 773–787.

Guitton, D. and Mandl, G. (1978) Frontal 'oculomotor' area in alert cat. I. Eye movements and neck activity evoked by stimulation. *Brain Res.*, 149: 295–312.

Huang, C.C. and Lindsley, D.B. (1973) Polysensory responses and sensory interaction in pulvinar and related posterolateral thalamic nuclei in cat. *Electroenceph. Clin. Neurophysiol.*, 34: 265–280.

Miyashita, E. and Tamai, Y. (1989) Subcortical connection of frontal 'oculomotor' areas in the cat. *Brain Res.*, 502: 75–87.

Nedzelnitesky, V. (1973) Neuroanatomy of the auditory system. *Arch. Otolaryngol.*, 98: 397–413.

Phillips, D.P. and Irvine, D.R.F. (1979) Acoustic input to single neurons in pulvinar-posterior complex of cat thalamus. *J. Neurophysiol.*, 42:123–136.

Paillard, J. (1982) Apraxia and the neurophysiology of motor control. *Phil. Trans. R. Soc. Lond.*, B298: 111–134.

Pasik, T. and Pasik, P. (1980) Extrageniculostriate vision in primates. In: S. Lessell and J.T.W. van Dalem (Eds.), *Neuroophthalmology*, Vol. 1, Excerpta Medica, Amsterdam, Oxford, pp. 95–119.

Popper, K.R. and Eccles, J.C. (1981) *The Self and it's Brain*, Springer, Berlin.

Sasaki, K. (1979) Cerebro-cerebellar circuit in cats and monkeys. In: M. Ito, N. Tsukahara, K. Kubota and K. Yagi (Eds.), *Integrated Control Functions of the Brain*, Kodansha-Sci., Tokyo; Elsevier, Amsterdam, pp.128–138.

Schlag, J. and Schlag-Rey, M. (1970) Induction of oculomotor responses by electrical stimulation of the prefrontal cortex in the cat. *Brain Res.*, 22: 1–13.

Schneider, G.E. (1969) Two visual systems. *Science*, 163: 895–902.

Tamai, Y. and Miyashita, E. (1989) Subcortical connections of an 'oculomotor' region in the ventral bank of the anterior ectosylvian sulcus in the cat. *Neurosci. Res.*, 7: 249–256.

Tamai, Y., Water, R. and Asanuma, H. (1984) Caudal cuneate nucleus projection to the direct thalamic relay to motor cortex in cat: an electrophysiological and anatomical study. *Brain Res.*, 323: 360–364.

Tamai, Y., Fujii, T., Nakai, M., Komai, N and Tsujimoto, T. (1984) Monocular movement of contralateral eye following cortical stimulation in the coronal sulcus of cat. *Brain Res.*, 324: 138–141.

Tamai, Y., Miyasita, E. and Nakai, M. (1989) Eye movements following cortical stimulation in the ventral bank of the anterior ectosylvian sulcus of the cat. *Neuroscience Res.*, 7: 159–163.

Trevarthen, C.B. (1968) Two mechanisms of vision in primate. *Psychol. Forsh.*, 31: 299–337.

Water, R.S., Tamai, Y. and Asanuma, H. (1985) Caudal cuneate nucleus projection to the direct thalamic relay to motor cortex: an electrophysiological study. *Brain Res.*, 360: 361–365.

M. Norita, T. Bando and B. Stein (Eds.)
Progress in Brain Research, Vol 112

CHAPTER 28

Neural bases of residual vision in hemicorticectomized monkeys

Maurice Ptito*[1,2], Marc Herbin[1,2], Denis Boire[1] and Alain Ptito[2]

[1]*Département de Psychologie and centre de recherche en sciences neurologiques, Université de Montréal,
CP 6128, Montréal H3C 3J7*
[2]*Department of Neurology and Neurosurgery, Montreal Neurological Institute, McGill University, 3801 University St., Montreal, Canada*

In this series of studies, we have attempted to characterize anatomically the organization of the retinofugal pathways in monkeys that underwent the surgical removal in infancy of the entire left cerebral hemisphere. Hemidecordication in baby monkeys produced a transneuronal retrograde degeneration of the retinal ganglion cells (RGCs) that affected mainly the foveal rim. Although the density of RGCs in this region was drastically diminished, the soma sizes of the surviving cells remained normal. The lateral geniculate nucleus (dLGN) ipsilateral to the removed cortex was dramatically reduced in size although it still showed normal layering. There was a marked reduction in the number of neurons in both the parvocellular and magnocellular layers and a heavy gliosis. By contrast, the superior colliculus ipsilateral to the lesion was remarkably well preserved: although slightly reduced in volume, it showed little gliosis and a metabolic activity, as revealed by cytochrome oxidase histochemistry, similar to the superior colliculus contralateral to the lesion. Behavioral perimetry indicated a partial sparing of vision up to 45° in the 'blind' hemifield. We argue that the preservation of the retino-tectal pathway mediates most of the residual visual functions found in the 'blind field' of hemispherectomized human subjects.

Introduction

Massive damage to the primary visual cortex (striate cortex, cytoarchitectonic area 17, area V1) produces a persistent contralateral hemianopia in humans as well as monkeys (cf. Cowey and Stoerig, 1991; Ptito et al., 1991b). In humans, despite their constant denial of having seen anything (the blindsight phenomenon, Weiskrantz et al., 1974; Weiskrantz, 1986), using a forced choice procedure, a number of residual visual capacities have been revealed in the visual hemifield contralateral to the damaged striate cortex. For example, these patients could orient their gaze towards a stimulus moving in their blind field (Poppel et al., 1973; Weiskrantz et al., 1974), localize stationary targets by pointing with their finger (Weiskrantz et al., 1974; Perenin and Jeannerod, 1978), detect and discriminate flicker, movement (Barbur et al., 1980; Blythe et al., 1987; Magnussen and Mathiesen, 1989) as well as line orientations and shapes (Perenin, 1978; Weiskrantz, 1987). They could also perform discriminations, in their blind hemifield, based on the spectral composition of the stimuli (Stoerig and Cowey, 1989, 1992). These residual abilities have been attributed to spared remnants of striate cortex (Fendrich et al., 1992) or to the various extra-geniculo-striate cortical pathways originating in the transneuronally partially degenerated hemiretinae which send visual information via the subcortical retino-recipient nuclei to extrastriate visual cortical areas (Cowey and Stoerig, 1991). Recent imaging (Celesia et al.,

*Corresponding author. Tel.: +1 514 3432330; fax: +1 514 3435787 e-mail: ptito@ere.umontreal.ca

1991) and psychophysical (Fendrich et al., 1992; Kasten et al., 1995) studies of human patients support the idea that spared remnants of striate cortex can account for the residual vision in the scotomatous hemifield.

In the past few years, evidence for the pivotal role of extra-striate cortices in blindsight has been accumulating in monkeys with striate cortex lesions. For example, when the retrograde tracer enzyme horseradish peroxidase (HRP) was placed in the extra-striate cortices (mainly area V4) of a unilaterally destriated monkey, retrogradely labelled cells were found in the dLGN ipsilateral to the lesioned hemisphere (Kisvàrday et al., 1991). The projections from the dLGN to extra-striate cortices have already been documented in normals monkeys (cf. Yukie and Iwai, 1981; Bullier and Kennedy, 1983) and seem to survive lesions of area V1. Moreover, intraocular injection of HRP in the same animals indicated the presence of retinogeniculate terminals in both the parvocellular and magnocellular layers of the degenerated dLGN (Kisvàrday et al., 1991). In the macaque monkey, Rodman and her colleagues (1989) had previously shown that cells in area MT remained visually responsive and directionally selective following lesions or inactivation of visuotopically corresponding parts of the visual cortex (cf. also Girard et al., 1992). Likewise, Gross and his colleagues (Bruce et al., 1986; Gross, 1991) found that the early combined inactivation of area V1 and superior colliculus in monkeys silences the responses of Superior Temporal Polysensory neurons to visual stimuli. Similarly, functional imaging studies in human subjects have also indicated that the presumed equivalent in humans of the middle temporal area (MT or V5) of the macaque was active even in the absence of V1 when moving targets were presented in the blind hemifield (Barbur et al., 1993).

It seems then that extra-striate pathways survive the destruction of striate cortex and could be involved in mediating some visual functions in the blind hemifield such as motion detection and wavelength discrimination. Others have suggested (Cowey and Stoerig, 1991; Ptito et al., 1991b) that

the retinal projection via the colliculus and pulvinar to the associative visual cortices might be per se sufficient to mediate such residual vision in the blindfield. Indeed, the residual detection abilities observed in striatectomized monkeys disappear following the destruction of the ipsilateral superior colliculus (Mohler and Wurtz, 1977; Rodman et al., 1990). The superficial layers of the superior colliculi of the monkey like those of all mammals have been shown to receive direct input from the retina as well as the striate cortex and other numerous extra-striate areas and contain a complete retinotopic and visuotopic representation of the contralateral visual hemifieldfield (see for review, Stein and Meredith, 1991). Furthermore, cells in the superficial layers of the superior colliculi respond to flashing, moving or 'jerking' displacement of visual stimuli (cf. Moors and Vendrick, 1979) while the electrical stimulation of the intermediate layers triggers a saccadic eye movement towards the specific region in the visual field corresponding to its retinotopic representation (Robinson and Wurtz, 1976). Several studies suggest that the superior colliculi contribute to qualitatively different aspects of vision than those in which the cortical visual areas are involved. Patients with midbrain pathologies show reduced latencies of orienting responses (Heywood and Ratcliff, 1975; Posner et al., 1982) and collicular lesions in several mammalian species impair orienting responses towards visual stimuli, reduce responsiveness to novel stimuli (cf. Dreher et al., 1965; Schneider, 1976; Milner et al., 1978; Goodale and Murison, 1979) and produce deficits in movement discrimination (Anderson and Symmes, 1969; Ptito et al., 1976; Collin and Cowey, 1980). One way to specify the role of this other visual system is the use of a model whereby the entire cerebral hemisphere is removed. In this case massive retrograde degeneration should affect the entire dorsal lateral geniculate nucleus (dLGN) and the pulvinar leaving retinofugal projections to the midbrain structures. If this was the case, hemispherectomized subjects should be able to perform tasks in their blind field which are 'collicular' in nature. Indeed, such patients could point accurately to moving, stationary (Perenin and Jeannerod, 1978; Ptito et al., 1991b; Braddick et al., 1992) as well as flashing targets (Ptito et al., 1991b) in their blind field and discriminate rapid movement and differences in velocities (Ptito et al., 1991b). Moreover, spatial summation across the vertical meridian was also possible suggesting an interaction between the superior colliculi (Tomaiuolo et al., 1994). The patients could not however respond to motion-in-depth (King et al., in press) nor discriminate the direction of motion (Perenin, 1991; Ptito et al., 1991b).

The goal of the present series of experiments was to examine the neural basis of residual vision using the monkey as a model and to extend previous findings in hemispherectomized cats and humans. We first evaluated behaviorally some visual functions (e.g. orienting responses and stimulus localization) in both the intact and affected hemifields following complete hemispherectomy. Second, we investigated the anatomy of the retinofugal projections in these same animals. Particular attention was given to the organization of the retina, the dLGN and the superior colliculus (SC). We were able to demonstrate a certain amount of visual recovery in the blind field and show that the SC ipsilateral to the lesion undergoes a much lesser degeneration than the ipsilateral dLGN. The striking preservation of the SC supports the notion that, following early hemicorticectomy, these midbrain structures are likely to underlie some residual visual functions, such as detection and localization of visual targets, in the blind hemifield.

Materials and methods

Subjects

Seven infant monkeys (*Cercopithecus aethiops sabeus*) were used. The monkeys were born and reared in the facilities of the Biomedical Primate Research Center of St-Kitts (West Indies). They underwent the surgery at a median age of 16 weeks post-natal.

Surgery

Following an initial dose of Atropine (Atro-Sol, 0.2 mg/kg) the monkeys were deeply anaesthetized with Nembutal (30 mg/kg). A craniotomy over the left hemisphere was performed and the dura was retracted. Under direct viewing through a dissecting microscope, the hemisphere was gently retracted from the midline with cotton-tipped applicators. Using a suction pipette, the hemisphere was separated from the diencephalon and the caudate nucleus (extra-thalamo-caudate surgery). Most of the hemisphere was removed in one block and great care was taken to remove all of the ipsilateral residual cortex, especially on the ventral surface. The remaining structures were covered with surgical pads and the dura was sown into place. The temporal muscle was then attached to the midline and the skin was sutured. The animals were given an immediate injection of antibiotics (Penbritin) and upon recovery they were returned to their mothers. All monkeys received post-operative injections of antibiotics for a period of 10 days. The monkeys were separated from their mothers after 6 months and placed in commune in a large screened (3 m × 2 m × 3 m) area. Neurobehavioral assessment of visual functions was carried out 736 days following surgery using tests similar to those used by Ptito et al. (1976) in the rhesus monkey and have been reported elsewhere (Ptito et al., 1991b). The demarcation of the extent of the visual field is reported here. Following the behavioral evaluation, the monkeys were prepared for the neuroanatomical studies.

Short-term post-surgical behavior

The monkeys became alert and active a few days after surgery. They clung to their mother and had their eyes wide open. Once the mother was put to sleep in order to retrieve the baby monkey to carry on testing, it showed a strong grip on the mother's fur. When put on the floor, there was a striking tendency to circle towards the side of the lesion. The head and body turning bias were very strong. In addition, when the experimenter put his index finger in the palm of the hand ipsilateral to the ablated hemisphere and lifted the animal, the grip was strong enough to hold the monkey above the floor for more that 30 s. Interestingly enough, the reflex was still preserved for the contralateral hand. When the animal was held by the scruff of the neck, the contalateral limbs showed abnormal muscle tone and were just hanging compared to the strong resistance of the extended ipsilateral limbs. The asymmetries between the two sides of the body were notable also at the sensory level. The monkeys never responded to tactile stimulation on the contralateral side of the body. No withdrawal responses could be observed to sensory stimulations applied to the right limbs or the right side of the face as compared to the immediate response obtained for the left side. Recovery was fairly rapid and some of the aforementioned neurological deficits, mostly motor ones, subsided within a few weeks to a point that it was difficult to distinguish the lesioned from the normal monkeys in their home cage.

Perimetry

The extent of the visual field for both eyes was measured according to the technique used by Sherman (1974) in the cat and adapted to the monkey. The animal was placed in a restraining chair positioned at the center of a perimeter. The monkey was trained to fixate a target (3° of visual angle) at the center of the perimeter positioned 27 cm from the eyes. A second stimulus (a morsel of fruit on a stick about 1 cm^2 in size) was then randomly introduced in the visual field at various eccentricities (14 at 15° steps : 0°, 15°, 30°, 45°, 60°, 75° and 90°, left and right) and the monkey had to orient its gaze in response to this new stimulus. It was then rewarded with the piece of fruit that served as the stimulus. Each session consisted of ten trials at each stimulus position for a total of 140 trials. The monkey was tested for 6 days in order to obtain 60 trials for stimulus

position and a total number of 840 trials. The perimetry evaluation was performed every 6 months for a period of 2 years. Fixation of the target and the gaze shift (e.g. orientation of the eyes in the direction of the target) elicited by the new stimulus were continuously recorded with a video camera. Results were calculated from the analysis of the films for each animal by two indepependant observers.

Results

Fig. 1 illustrates the mean percentage of correct responses (clear orientation of the gaze towards the target) by the young-lesioned monkeys for each visual angle tested in both hemifields. Orienting responses were apparent in the hemifield contralateral to the lesion 6 months after surgery and did not change over time. These animals responded 53%, 35% and 16% of the time to stimuli presented at 15°, 30° and 45°, respectively, in the blind hemifield. No responses were elicited beyond 45°. In the normal hemifield occasional errors (absence of orienting reponses to the target) were seen only in the far periphery, a result usually found in normal animals. The extent of the binocular visual field is represented by the shaded area and contrasts with what has been obtained in adult-lesioned monkeys and humans where there is a clear and persistent contralateral homonomous hemianopia (Ptito et al., 1991b).

Neuroanatomical studies

Following the completion of the behavioral evaluation, the animals were subjected to a series of neuroanatomical techniques in order to study the retinofugal pathways. We have looked at the organization of the retina using whole-mount retinae stained with cresyl violet and we have described quantitatively the cytoarchitecture of the superior colliculi and the lateral geniculate nucleus (both ipsi- and contralateral to the lesion) following intra-ocular injection of horseradish peroxidase.

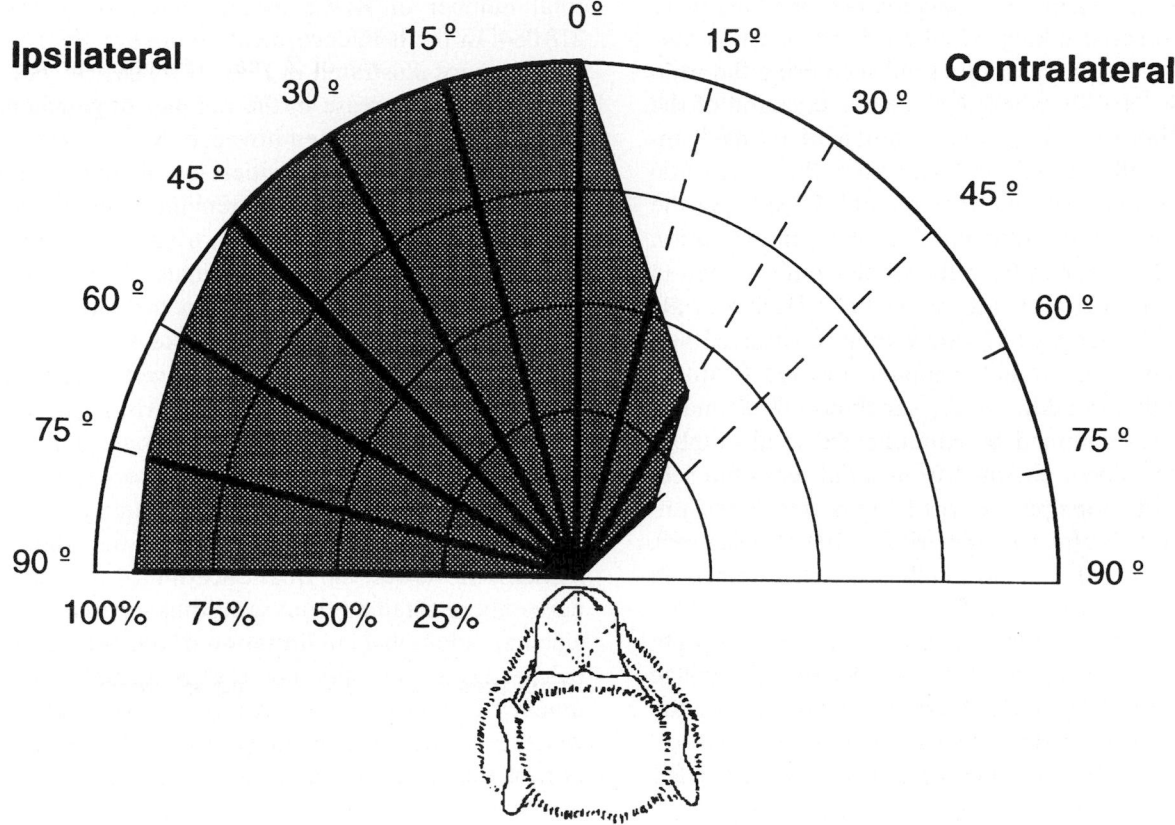

Fig. 1. Visual perimetry for young-hemispherectomized monkeys. The responses to a novel stimulus are plotted with respect to the position in which the target was presented. Each line represents the percentage of appropriate orienting responses. The extent of the binocular visual field is represented by the shaded area.

Retina

Subjects

Retinae were obtained from two monkeys operated 4-month post-natally and two normal adult monkeys.

Retinal processing

The monkeys were first kept for 2 h in a dark room; they were then deeply anaesthetized by an intramuscular injection of a mixture of Rompun (2% thiazine clorohydrate) and Ketalar (5% Ketamine clorohydrate) and transcardially perfused with a 10% buffered saline solution followed by 4% buffered paraformaldehyde. The eyes were oriented, removed and stored in 4% paraformaldehyde at 4°C. The retinae were dissected free from the eye cup in a saline bath and floated onto a gelatinized slide, optic fiber layer uppermost (Stone, 1981). A series of small radial cuts were made to aid in flattening the retina. The mounted retinae were air dried overnight in humid air with formaline vapors and stained with cresyl violet. The quantification of ganglion cell density was done under a light-microscope using a 40 × objective in samples of 0.0121 mm² areas 2-mm apart. In order to avoid under-estimation of cell numbers in the foveal region which could be due to the multilayering of ganglion cells, this region was cut out, embedded in Epon and sectioned at 4 μm. The ganglion cell numbers were re-estimated taking samples 0.5-mm apart. A correction factor was then established using the ratio NHSi/NWMi (where NHSi is the equation of the function describing the variation of retinal ganglion cells (RGC) defined from the horizontal sections in a specific region and NWMi being the equation of the function describing the variation of RGCs defined from the wholemount retinae in the same region (Ptito et al., 1995; Herbin et al., 1995). Isodensity maps were then constructed and plotted using retinal mapping software (Ptito et al., 1995). In addition, the ganglion cells estimates were re-examined by counting the total number of optic fibers obtained from serial semi-thin sections of four optic nerves (counts were done under a light microscope with a 100 × objective). The sizes of ganglion cells were determined in central (4° from the fovea) and peripheral (16°, 32° and 48°) retinal regions. Cells were brought into focus under a 40 × objective and the image was seized by a CCD camera connected to the microscope. Morphometric analysis (circular diameter) was subsequently performed on a CRT monitor using an MCID/M1 image analysis system (Imaging Research Inc.).

Results

Retinal ganglion cells distribution

Cell counts in the central retina and on equally spaced polar coordinates radiating from the central area were used to construct isodensity maps for normal and hemidecorticate monkeys (Fig. 2). Fig. 2A illustrates the ganglion cell distribution in the right and Fig. 2B in the left retinae of a normal monkey. The ganglion cells density was highest in the central retina and dropped off monotonically to a relatively low density in the extreme periphery.

Cell concentrations reached a peak of about 45 000 ganglion cells/mm^2 at the foveal rim and dropped off rapidly within 3-mm eccentricity (10 000 ganglion cells/mm^2). At 5 mm and 12 mm from the center, cell density was 3860 cells/mm^2 and 483 cells/mm^2, respectively. The

total number of RGCs in each hemifovea was 218 084. In the hemidecordicate monkeys, the isodensity maps illustrated in Figs. 2C,D clearly indicated a sharp decrease in the number of ganglion cells in the affected hemifovae only (121 280 or 72% reduction). Cell densities in the perifoveal region and in the periphery remained similar to those in normal animals. The horizontal sections of the retinae of operated animals showed the clear presence of the foveal pit. In the foveal rim however, unlike the normal monkeys, the ganglion cells in the lesioned animals were reduced to a single row of cells in comparison to the multilayering found in the adjacent tissue (Fig. 3). The reduction of the ganglion cells layering in the central fovea of hemidecorticate monkeys confirmed the cells counts acquired from the wholemounts. Moreover, the quantification of the optic nerve fibers obtained from semi-thin sections provided an additional confirmation of the two previous methods. The optic nerve of a normal vervet monkey contains on the average 1 130 000 fibers which compares reasonably well with the estimation of the total number of retinal ganglion cells of about 1 230 000. In the hemidecorticate animal, the estimated total number of RGCs was 860 000 while the average number of optic nerve axons was about 744 000. The reduction in the total number of RGCs in the lesioned monkeys was essentially due to the massive loss of cells in the foveal region.

Size distribution

Frequency histograms (Fig. 4) show that the size distribution of RGCs in the normal monkey is trimodal although there is a certain amount of overlap between the modes. In the foveal rim (4° from the foveola), RGCs appeared to form a uniform population on the basis of soma size, the average circular diameter of their perikarion being 7.3 μm (SD = 0.85). In the perifovea (16° from the foveola) and periphery (32° and 48°), average soma size increased with eccentricity being 11 μm (SD = 2.08), 13.6 μm (SD = 2.38) and 14.3 μm (SD = 2.79), respectively. Presumed

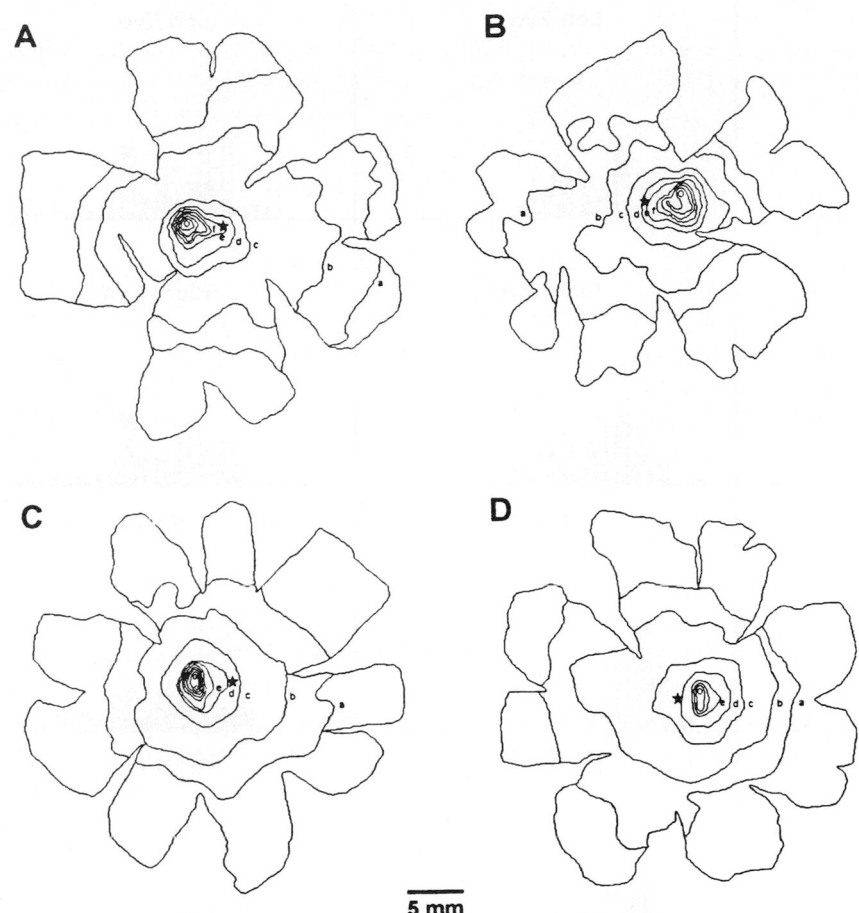

Fig. 2. Retinal ganglion cell isodensity maps in normal (A, right eye; B, left eye) and hemidecorticate (C, right eye; D, left eye) monkeys. Isodensity curves are labeled from the periphery towards the fovea: a = 500 RGC/mm²; b = 1000 RGC/mm²; c = 5000 RGC/mm²; d = 10000 RGC/mm²; e = 20000 RGC/mm²; f = 25000 RGC/mm²; the compact bands in the center are the areas of highest density and have the following values: 30000 RGC/mm²; 35000 RGC/mm²; 40000 RGC/mm²; 45000 RGC/mm².

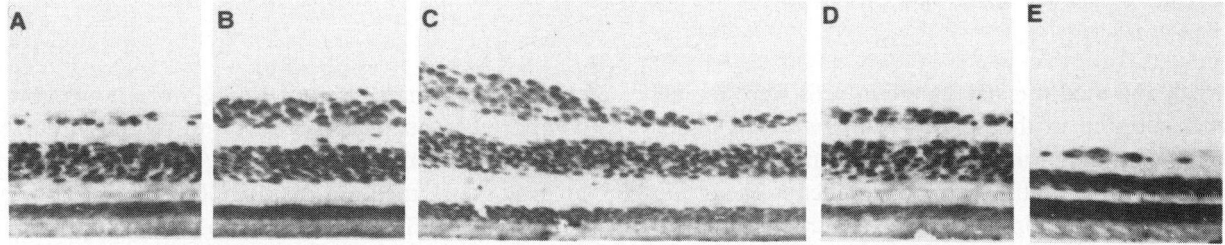

Fig. 3. Cresyl violet-stained cross-section through the right retina of an hemidecorticate monkey: A = 48° eccentricity temporally; B = 16° eccentricity temporally; C = foveal region (from 4° eccentricity temporally to 4° eccentricity nasally); D = 16° eccentricity nasally; E = 48° eccentricity nasally.

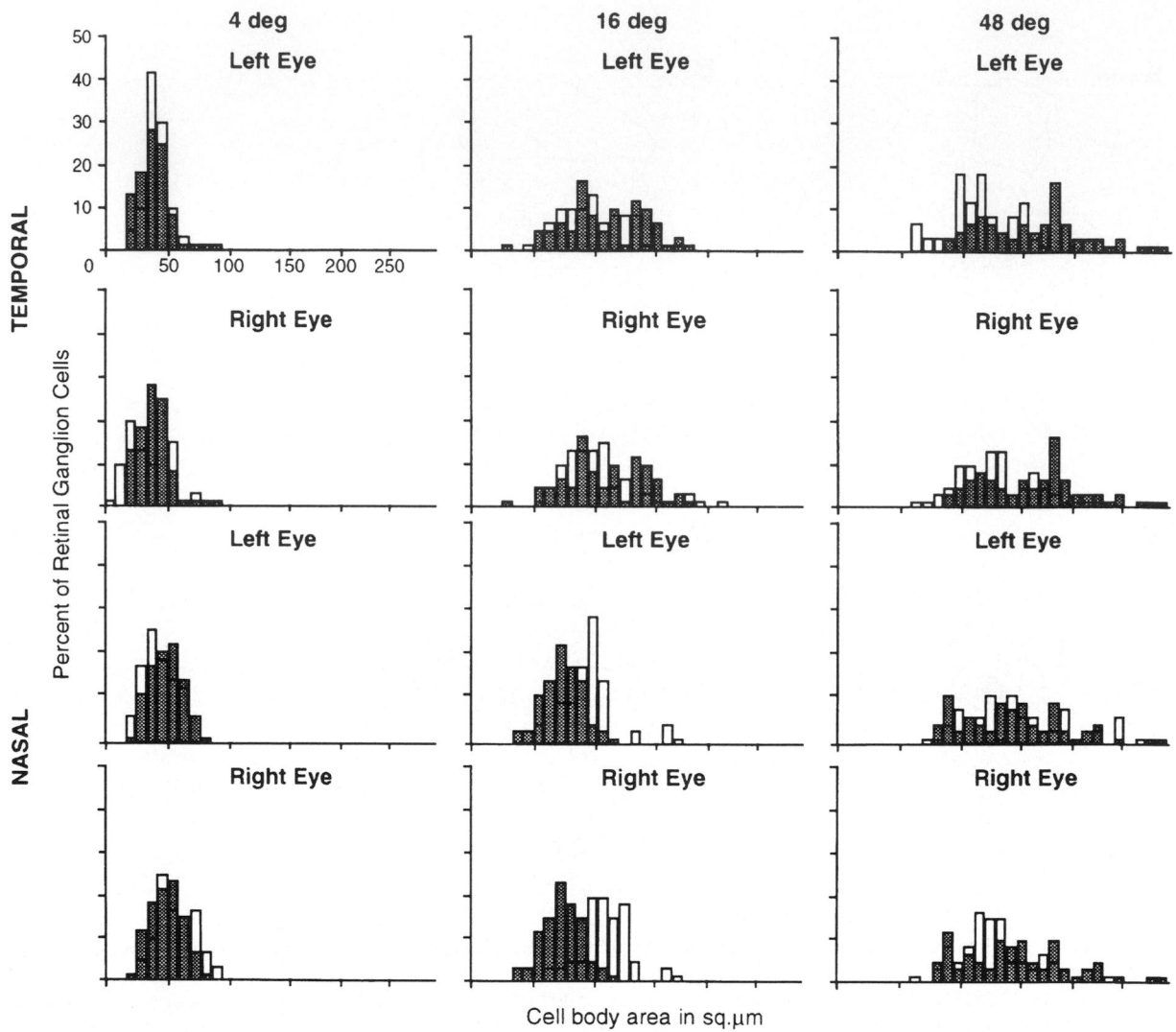

Fig. 4. Cell body size histograms of the ganglion cell layer for the temporal and nasal hemiretinae of the left and right eyes of a normal and an hemidecorticate monkeys measured at various eccentricities. The filled histograms represent the size distribution in a normal retina whereas the empty ones indicate the size distribution in the retina of a lesioned monkey.

small and medium size ganglion cells were mainly concentrated in the central retina whereas larger cells were found in the periphery. In hemidecorticate monkeys, there was no significant change in the mean ganglion cells size. In particular, at the foveal rim somata sizes of RGCs seemed identical to that in normal monkeys although cell density and packing effect were dramatically reduced. Similarly, in the perifovea and periphery, no

changes in somata sizes of RGCs were noticeable compared to normals. However, in the nasal perifoveal region of the right retina, there seemed to be a loss of medium size RGCs. Comparisons between the affected hemiretina of each eye with the corresponding normal hemiretina also seemed to indicate a loss of medium size ganglion cells.

In conclusion, hemidecordication in baby monkeys produces a transneuronal retrograde degeneration of the retina that affects mainly the foveal rim. There was a drastic reduction of the density of RGCs whereas the soma size of the surviving cells remained normal. Although we could not classify the surviving ganglion cells in terms of Pα (or A) cells, Pβ (or B) cells or Pγ (or C and ϵ) cells on the basis of their morphology (cf. Leventhal et al., 1981; Perry and Cowey, 1981), it is reasonable to assume, in light of previous studies in destriated monkeys, that the Pβ RGCs (Weller et al., 1979; Dineen and Hendrickson, 1981; Weller and Kaas, 1984; Cowey et al., 1989) presumed equivalent of the medium size ganglion cells were, in our sample, selectively affected.

Lateral geniculate nucleus and superior colliculus

Subjects

Four male monkeys taken from the behavioral sample and operated as infants were used to study the retino-geniculate projections. At the age of injection of the retrograde tracer they were 4 years old.

Tissue preparation

Under deep general anaesthesia (mixture of Rompun, 2% thiazine chlorohydrate and Ketalar, 5% Ketamine chlorohydrate, i.m.), a 50 μl solution of HRP (Sigma, type VI; 50%) was injected over a 10-min period into the vitreous body of the eye ipsi- or contralateral to the lesion using a sterile Hamilton syringe and under ophtalmological viewing of the fundus. Five days later, the animals were given a lethal dose of sodium pentobarbital and perfused through the heart with a 0.9% PBS solution followed by 4% paraformaldehyde, 10% and 20% sucrose. The brain was stereotaxically cut in the frontal plane into three blocks by two coronal cuts, rostrally at the level of the optic chiasm and caudally at the level of the superior colliculus (SC). The blocks were placed for 2 days in 30% sucrose; the blocks containing the dLGN and the SC were cut in coronal sec-

tions 40 μm thick on a freezing microtome. Alternate series of sections were processed for HRP histochemistry using tetramethylbenzidine (TMB) and ammonium molybdate as chromogens. Alternate sections were stained with cresyl violet and for cytochrome oxidase histochemistry.

Results

Lateral geniculate nucleus (pars dorsalis)

Quantitative analysis

Neuronal and glial cell bodies were drawn through a camera lucida at a magnification of 100 $\times$ over samples of 4270 μm^2. Samples were distributed evenly throughout the dLGN surface. Only neurons with clearly visible nucleoli were taken into account. Cells situated on the boundaries of the counting fields were also included. Cell body circle diameter was measured using a Zeiss Videoplan planimeter. Samples were taken on Nissl and on cytochrome oxidase stained sections. Due to the very low density of the neuronal cell population in the dLGN ipsilateral to the hemidecortication, only occasional neuronal cell bodies were included in the samples. In order to sample a larger area and to give a better estimate of neuronal cell densities in the dLGN ispsilateral to the lesion, sampling of neuronal cell bodies was carried out on the cytochrome oxidase stained sections. In addition to some patchy diffuse granular precipitates seen in the cytochrome oxidase stained sections, several moderately to intensely stained cell bodies were observed. These cells are slightly elongated and appear to bear several primary dendrites. Their general appearance strongly suggest that these cells are in fact neurons. Moderately to intensely stained cell bodies were also observed in many sites in our material. These cells were clearly observed in several nuclei on the same sections on which the sampling of dLGN was carried out (very clear cytochrome oxidase stained neurons were seen in the nearby substantia nigra and nucleus geniculatus lateralis pars ventralis). All cytochrome oxidase stained cells were counted and measured and cells density was

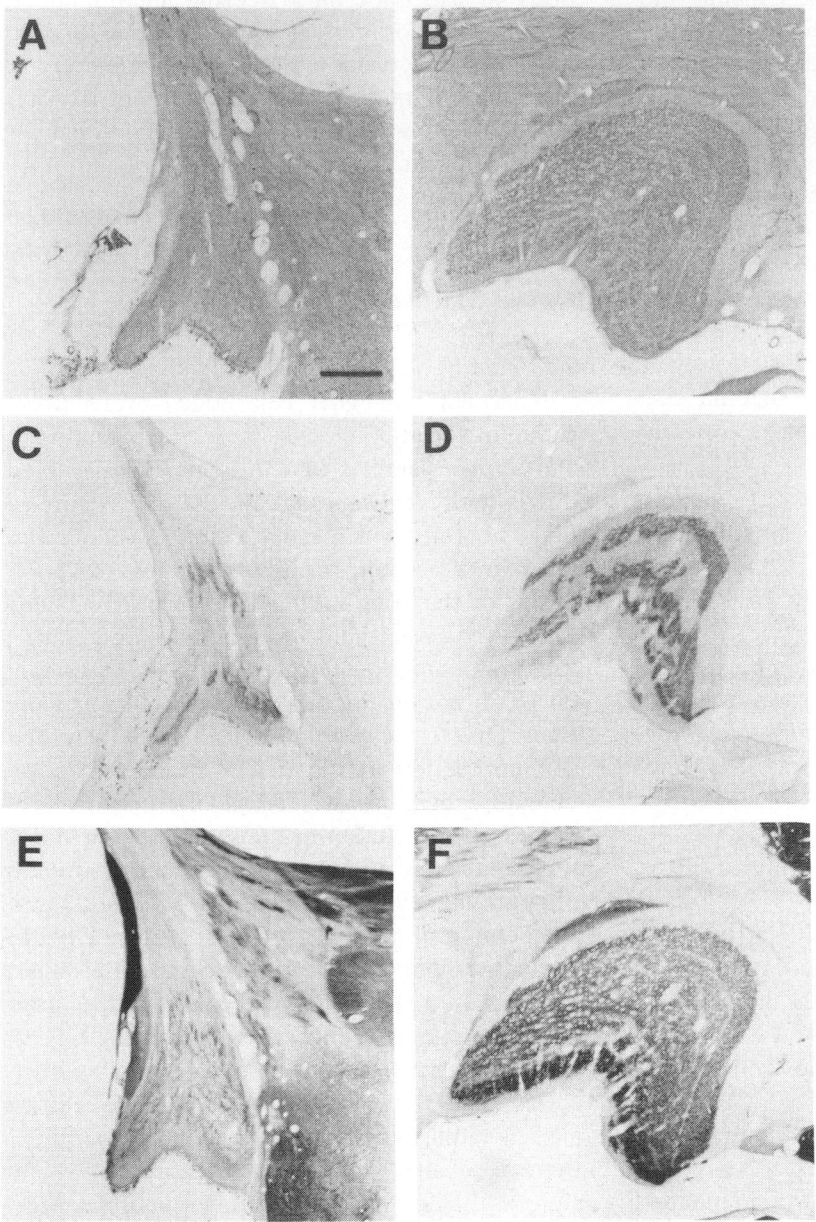

Fig. 5. Coronal section of the dLGN of an hemicorticectomized monkey. Adjacent sections through the left (A, C and E; ipsilateral to the lesion) and right (B, D and F) dLGN are shown following cresyl violet staining (A and B), HRP histochemistry after a right intraocular injection of HRP (C and D) and cytochrome oxydase histochemistry (E and F). Note the massive reduction in the size of the dLGN ipsilateral to the lesion (A vs. B). Clear anterograde labeling is observed in both dLGNs. In C labeling is seen in magnocellular layer 1 and parvocellular layers 4 and 6 of the left dLGN. In D labeling is seen in magnocellular layer 2 and parvocellular layers 3 and 5 of the right dLGN.Cytochrome oxidase histochemistry reveals intense oxydative metabolism in the magnocellular layers of the dLGN contralateral to the lesion (F) comparable to that observed in a normal monkey. Note the depressed activity throughout the dLGN ipsilateral to the lesion (E) and also the very faint staining of the magnocellular layers.

estimated by dividing the number of cells by the area of the dLGN of each section.

Density and cell size

Although the dLGN ipsilateral to the lesion was still present and visible, it underwent a significant size reduction estimated by volumetry to be approximately 70% (Fig. 5A). Both the parvocellular and magnocellular layers showed a significant decrease in neuronal cell density (Figs. 6 and 7, see details in legends). Figs. 8A,B illustrate the dramatic loss of geniculate neurons in all layers of the dLGN ipsilateral to the ablated cortices. The dLGN contralateral to the lesion showed in both subdivisions a neuronal density similar to that in a normal animal. By contrast, gliosis was heavy in the ipsilateral dLGN, the glial cell density being much higher in both the parvocellular and magnocellular layers than the contralateral dLGN and the normal control (Figs. 8C,D).

Morphometric analysis of the cresyl violet

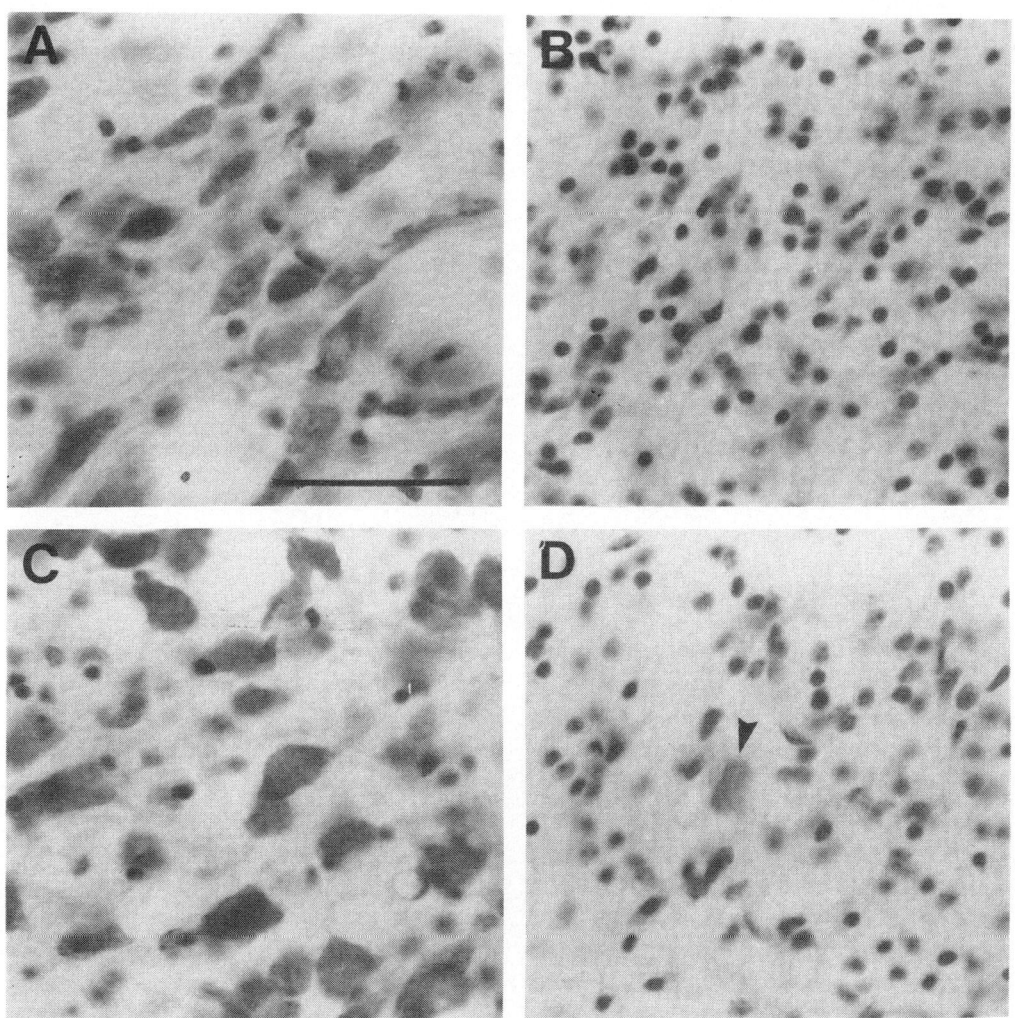

Fig. 6. Cresyl violet-stained sections through the right (A, parvocellular and C, magnocellular) and left (B, parvocellular and D, magnocellular) dLGN of an hemicorticectomized monkey. Scale bar: 50 μm for all panels. Note the heavy gliosis and the massive neuronal loss in the dLGN ipsilateral to the lesion (B and D). A remaining neuronal profile is indicated by the arrow in D.

396

stained cells (Fig. 9) in the parvocellular layers of the normal (Fig. 9A) and the hemidecorticate monkeys (Figs. 9B,C) showed similar size distribution. Cell bodies with circle diameter between 3 and 6 μm were glial cells whereas those with circle diameter greater than 6 μm were presumed to be neurons. Only a few neurons responding to the latter criterion were captured in sampling the cresyl violet stained sections. Because neuronal cell density was very low, the sampled area utilized in the quantitative analysis was too small to give a reliable estimate of the size distribution of

neurons in this highly gliotic structure of hemidecordicate monkeys. In order to circumvent this problem, neuronal cell densities and morphometry were carried out on cytochrome oxidase stained sections (Fig. 10). Morphometric analysis of the surviving neurons in the parvocellular layers of the dLGN ipsilateral to the lesion (Fig. 10B) indicated that their size distribution was similar to that found in the contralateral dLGN (Figs. 9C and 10A) and in normal monkeys (Fig. 9A).

Anterograde transport of HRP injected into

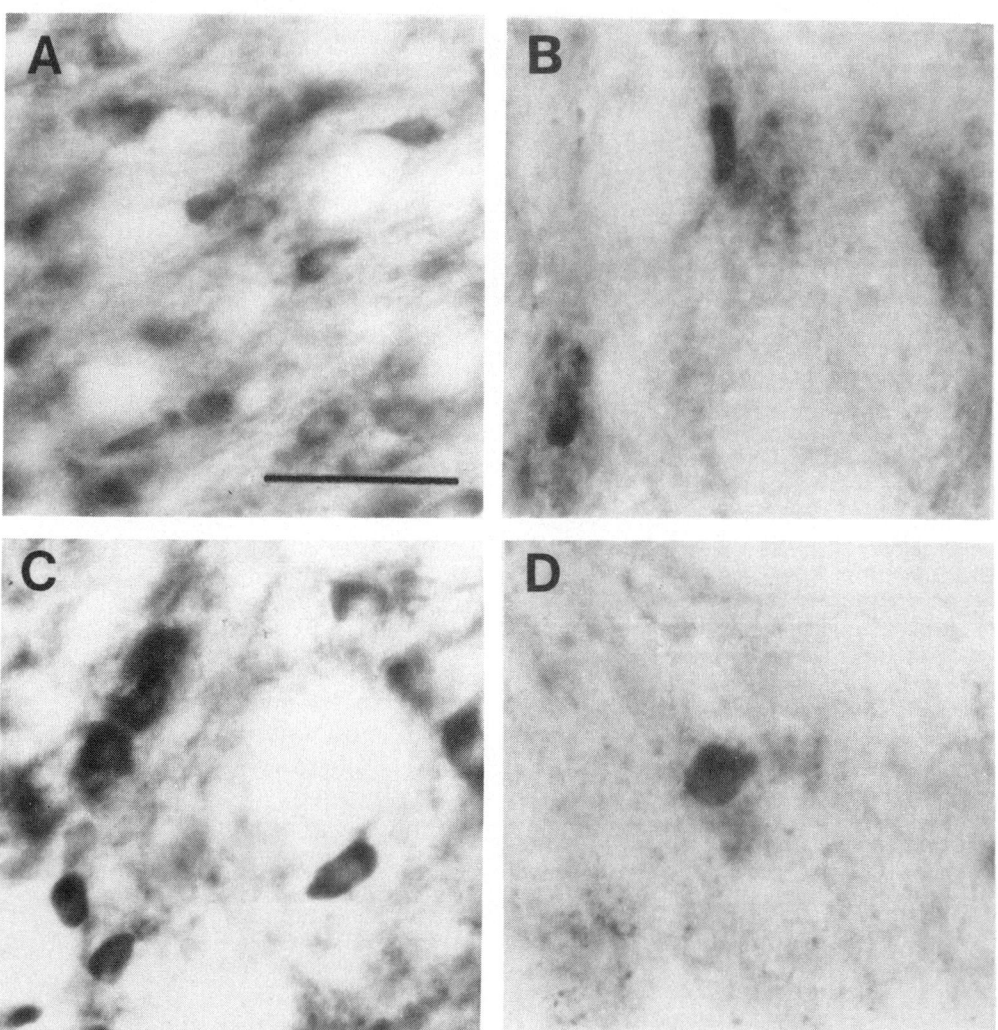

Fig. 7. Cytochrome oxydase stained sections through the right (A, parvocellular and C, magnocellular) and left (B, parvocellular and D, magnocellular) dLGN of a hemidecorticate monkey. Scale bar: 50 μm for all panels. Note the intense staining of the neuronal profiles in both dLGNs contralateral (A and C) and ipsilateral (B and D) to the lesion.

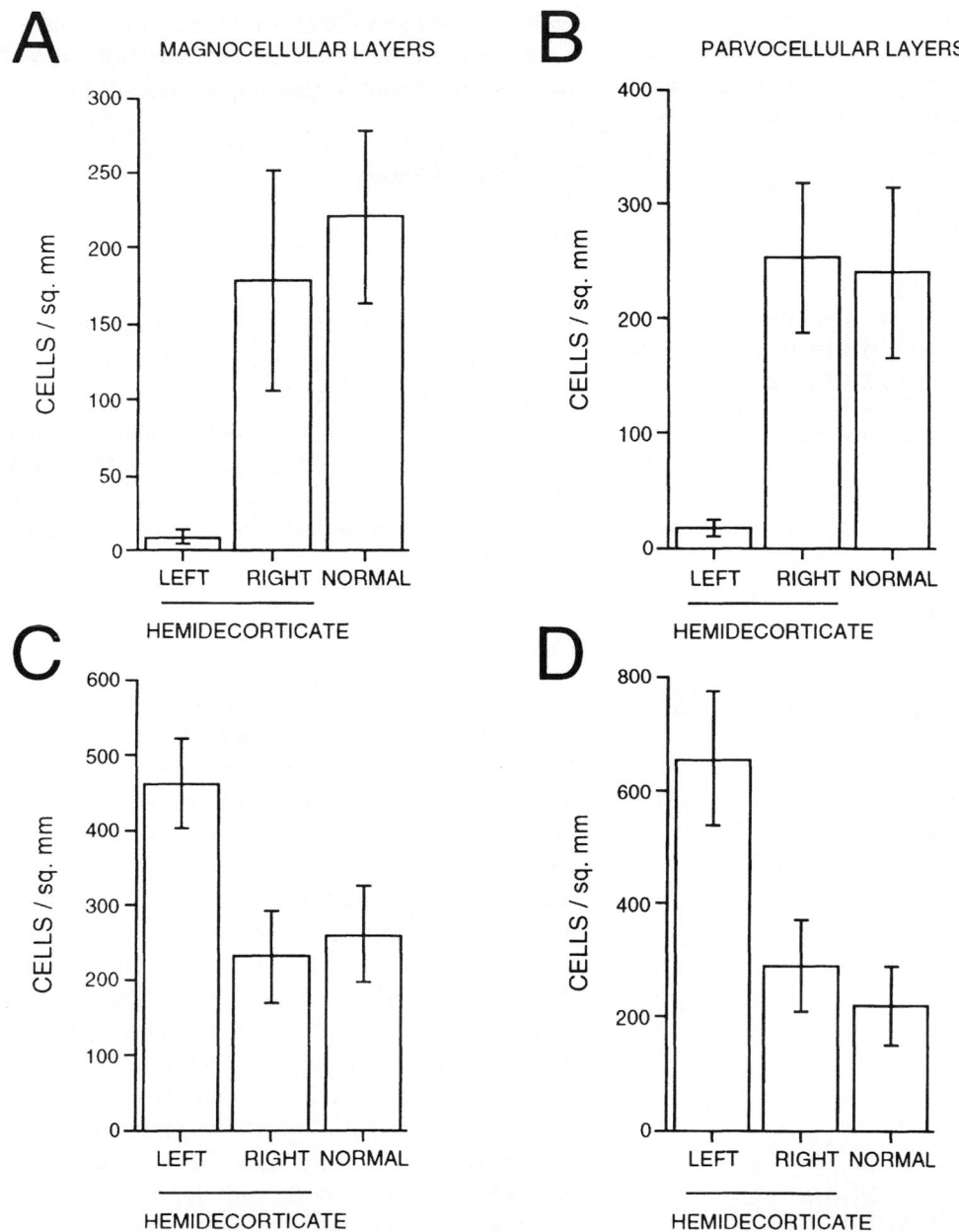

Fig. 8. Neuronal (A, magnocellular and B, parvocellular) and glial (C, magnocellular and D, parvocellular) cell densities in the ipsi- and contralateral dLGN of an hemidecorticate and a normal monkey.

the eye ipsi- or contralateral to the lesion produced terminal labeling in the appropriate layers of both dLGN (Figs. 5C,D). The paucity of neurons in the parvocellular and magnocellular layers of the dLGN contrasted with the abundance of the labelled retinal terminals. It is not possible at this stage to determine whether these terminals have post-synaptic contacts on dLGN neurons projecting to other cerebral structures or on interneurons. It seems from the work of Kisvàrday

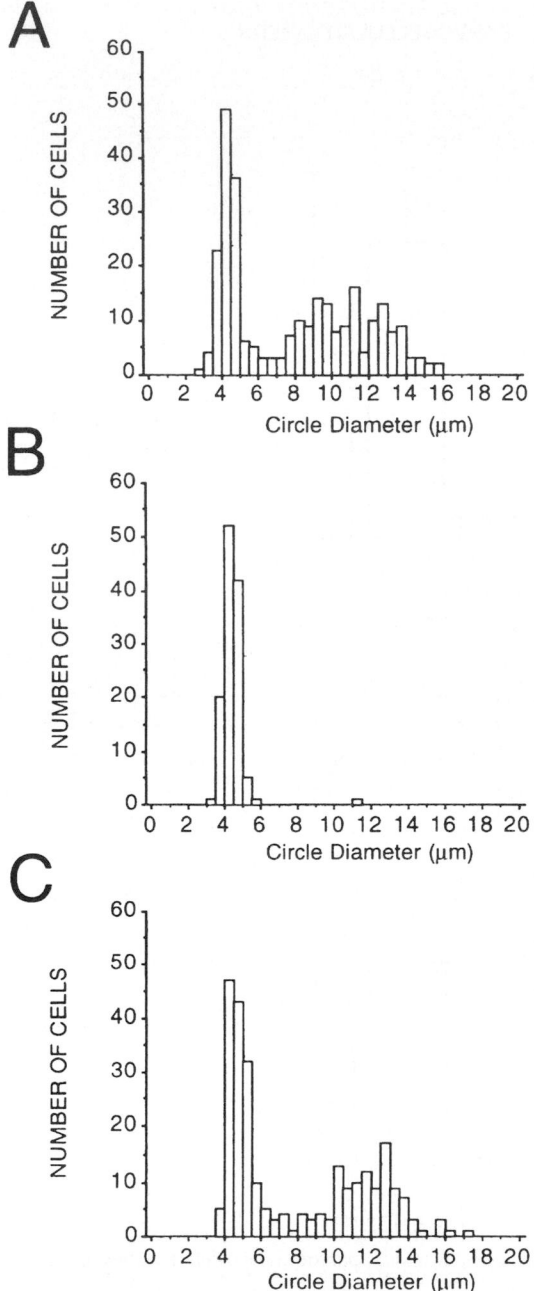

Fig. 9. Frequency distribution of circle diameters of cresyl violet-stained cells in the parvocellular layers of the dLGN of A, normal animal; B, ipsilateral and C, contralateral to the lesion. The large population of small cells (3–6 μm in circle diameter) represents glial cells. There is no difference between the cell populations in the parvocellular layers of both the dLGN of a normal animal and the contralateral one of a lesioned animal. Due to the small sampling area and the low density of cells in the dLGN ipsilateral to the lesion (B), a few neurons only were included in the samples (see Fig. 10).

and his colleagues (1991) on monkeys with unilateral lesions of the striate cortex that both types of neurons are found in the degenerated dLGN.

Superior colliculus

Method

Neuronal and glial cell bodies were drawn through a camera lucida at a magnification of 40 × over samples of 25 165 μm². Adjacent samples were taken along a transect from the collicular surface to the central grey. Only neurons with clearly visible nucleoli were taken into account. Cells situated on the boundaries of the counting fields were also included.

Quantitative analysis

The decrease in the volume of the superior colliculus ipsilateral to the lesion was about 20% which contrasts with the much greater reduction observed in the dLGN on the same side. The analysis of Nissl stained sections indicated that there were no significant differences in neuronal densities between the superior colliculi ipsilateral and contralateral to the hemispherectomy and the normal monkeys (Figs. 11A,C). Slight gliosis was however observed in the superior colliculus ipsilateral to the hemispherectomy. Figs. 11B,D illustrate the increase in the number of glial cells in the superficial gray layers. The metabolic activity as revealed by cytochrome oxidase histochemistry was similar in both superior colliculi. Anterograde transport of HRP injected into the eye ipsi- or contralateral to the lesion produced terminal labelling in the strata optici of both superior colliculi. The superior colliculus seems then to be much less affected by the hemispherectomy than the ipsilateral dLGN.

Discussion

When hemispherectomy is performed in adult

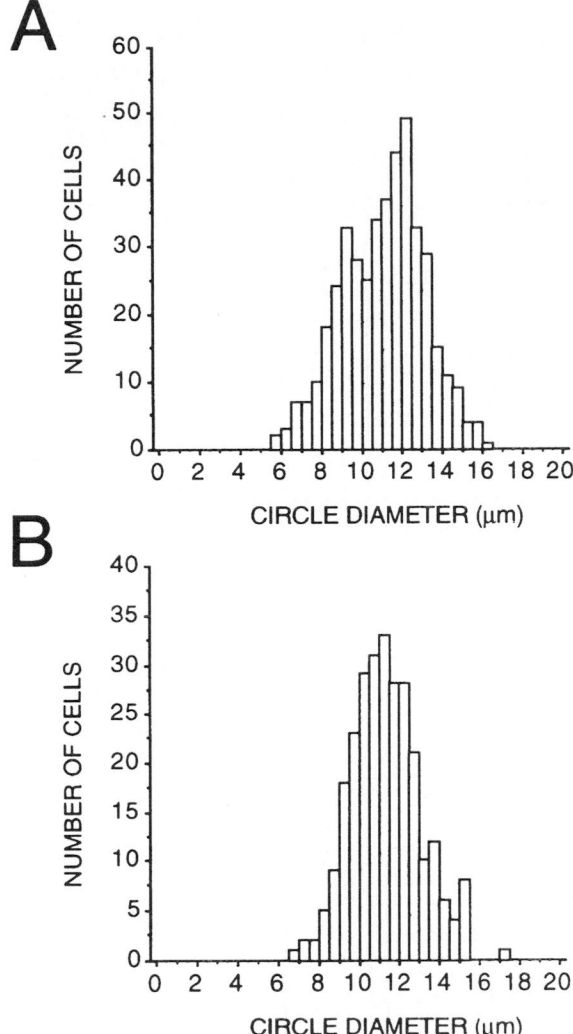

Fig. 10. Frequency distribution of cytochrome oxidase (CO) stained cells in the parvocellular layers of the dLGN of a lesioned monkey. A, contralateral and B, ipsilateral to the lesion. Note that glial cells (circle diameter < 6 μm) are not stained with CO. Cell populations revealed with CO are equivalent to those derived from cresyl-violet stained sections. Size distribution of neurons in the dLGN contralateral to the lesion (A) revealed with CO is the same as that obtained with cresyl violet-stained material (Fig. 9C) once the glial cells are excluded from the distribution. Cell populations in the dLGN ipsilateral to the lesion is comparable to what has been obtained with CO staining (A) and cresyl violet (Fig. 9C) in the contralateral dLGN and in normal monkeys (Fig. 9A).

cats, Old and New World monkeys or humans, there is a persistent contralateral homonymous

hemianopia (Hovda and Villablanca, 1990; Ptito et al., 1991a; Ptito et al., 1991b). By contrast, this same lesion made in our newborn monkeys resulted in an incomplete contralateral visual field impairment. In fact, our monkeys could orient their gaze to targets presented within 45° in their blind hemifield. This is consistent with findings by Hovda and Villablanca (1990) in neonatally hemispherectomized cats using a lick-suppression task. Their animals, tested in a refined behavioral apparatus, also showed a sparing of visual function up to 45° eccentricity. In addition, Wessinger and his colleagues (1994) using a retinal image stabilizer found islands of vision up to 6° from the vertical meridian, in the blind hemifield of hemispherectomized human patients. When the visual stimuli were presented within these areas, the patients could detect them significantly above chance level, while their performance dropped to chance level when the stimuli fell outside these patches. Because of the early presence of a brain pathology in these subjects, the sparing of visual functions in the blind hemifield has been attributed to an expansion of the normal zone of nasotemporal overlap (Wessinger et al., 1994; Wessinger et al., in press). The expansion of the normal zone of naso-temporal overlap has also been observed in cats and monkeys with an early section of the optic tract (Leventhal et al., 1988a,b; 1993).

It is also possible that these residual visual abilities are subserved by the preservation of a direct projection from the retina to the superior colliculus and possibly to other subcortical structures (see review by Cowey and Stoerig, 1991). Since retinal projections to these other midbrain structures are numerically smaller and in some cases still controversial (see review by Cowey and Stoerig, 1991), we have concentrated our efforts on the superior colliculus. We have found in the present study that the superior colliculus ipsilateral to the lesion undergoes little reduction in volume, shows a small amount of gliosis and has a metabolic activity similar to the contralateral one. These results are similar to those obtained on hemispherectomized neonate and adult cats

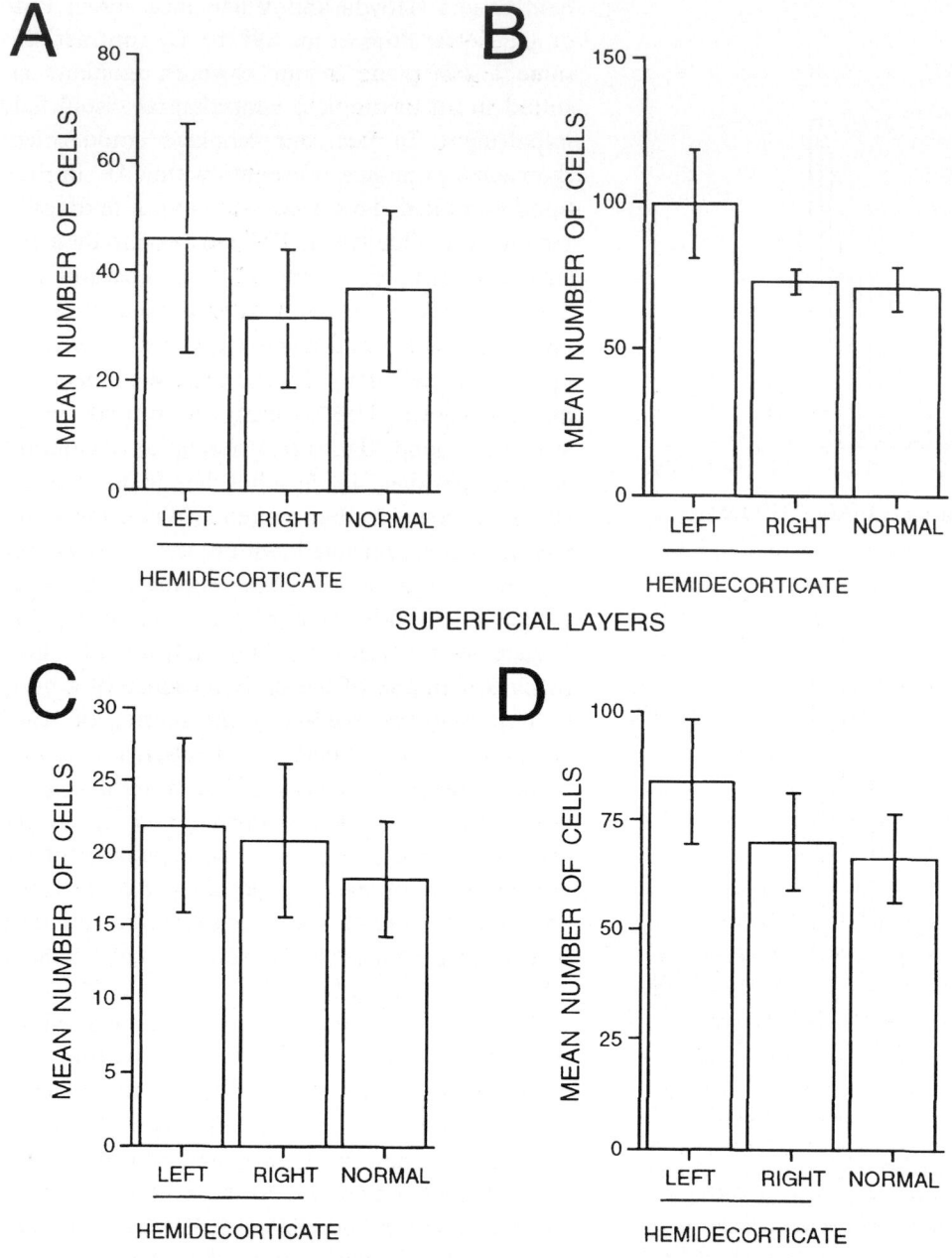

Fig. 11. Neuronal (A, superficial gray layers and C, intermediate gray layers) and glial (B, superficial gray layers and D, intermediate gray layers) cell densities in the left and right superior colliculi of an hemidecordicate monkey and a normal animal. Note the slight gliosis of the superficial gray layers of the left superior colliculus (ipsilateral to the lesion).

(Hovda and Villablanca, 1990). These authors have reported that, following neonatal hemispherectomy, the superior colliculus ipsilateral to the lesion undergoes very little neuronal degeneration, contains more large cells and shows less gliosis than following adult hemispherectomy; furthermore, an asymetry in the oxydative metabolism of both superior colliculi, the ipsilateral one being more markedly reduced, has been observed. The reduced degeneration and the sustained oxidative metabolism of the ipsilateral superior colliculus could be due to the preservation of the retino-tectal projections as shown in our monkeys and/or the formation of a new crossed corticofugal projection from the remaining hemisphere onto the superior colliculus ipsilateral to the lesion as demonstrated in the neonatally hemispherectomized cat (Hovda and Villablanca, 1989). This midbrain structure could thus act as a functional subcortical endstation for the analysis of visual information since we have no evidence at this moment that the information is relayed to the intact hemisphere. Indeed, intra-ocular injections of tritiated proline have produced labeling of retinal terminals in midbrain structures only (Boire et al., 1995).

Our results on the lateral geniculate nucleus lend more support to the latter conclusion namely that the superior colliculus could subserve by itself residual vision in the blind field. Indeed, the dLGN ipsilateral to the lesion undergoes a massive retrograde degeneration accompanied by a marked reduction in its volume, a drastic loss of neurons in both the parvocellular and magnocellular layers and a severe gliosis. Retinal projections are, however, still maintained since we have found labelled retinal terminals in the appropriate layers of the ipsilateral dLGN following intraocular injections of HRP. We have no explanation at this moment for the paucity of neurons and the large number of retinal terminals in the parvocellular and magnocellular retinal recipient layers of the dLGN ipsilateral to the lesion. It would seem like these retinal projections onto the degenerated dLGN have no or very few post-synaptic targets. It might also be that the surviving cells in the degenerated dLGN on which retinal ganglion cells project are mainly interneurons although some projection neurons are still present (see Fig. 6D). These surviving cells might be responsible for the low but nonetheless presence of oxidative metabolic activity. These results ob-

tained on hemispherectomized monkeys are at variance with those obtained by Shook and Villablanca (1991) in the hemispherectomized cat. These authors have found that the dLGN ipsilateral to the lesion in neonatally hemispherectomized kittens showed a greater neuronal sparing than that in adult-lesioned cats. This greater neuronal sparing was mainly seen in the medial region of the dLGN which corresponds to the central part of the retina. This sparing was moreover most pronounced in laminae A following early hemispherectomy, the average size of surviving neurons was larger after neonatal than after adult lesions and there was a minimal ipsilateral gliosis (Shook and Villablanca, 1991). This remarkable preservation of the dLGN might explain the functional recovery of visual functions previously reported by Villablanca and his colleagues (1986) as well as by others (Norsell, 1983) in hemispherectomized kittens but cannot be used to explain the residual visual functions reported in hemispherectomized humans (see Ptito et al., 1991a) and monkeys (Ptito et al., 1991b) since this structure undergoes a massive retrograde degeneration in both species (Ueki, 1966; Boire et al., 1995).

Conclusion

The results obtained in our experiments namely that following early removal of the entire cerebral hemisphere in monkeys, some residual visual functions are still present in the blind hemifield (e.g. detection and orientation of visual stimuli). Moreover, anatomical investigation has revealed that the lateral geniculate nucleus ipsilateral to the lesion is dramatically reduced in size, shows a marked reduction in its number of neurons in both the parvocellular and magnocellular layers and exhibits heavy gliosis. By contrast, the superior colliculus ipsilateral to the lesion is remarkably well preserved: it is slightlty reduced in volume and shows little gliosis. Our results are commensurate with the two visual systems hypothesis put forward by Schneider (1969, 1975). The collicular system is thought to be associated with the detection and the localization of stimuli in the visual field (where?) whereas the geniculo-striate system is responsible for the recognition of these stimuli (what?). The disappearance of the latter through retrograde degeneration after massive cortical damage and the remarkable survival of the former could provide an explanation for the various data reported on hemispherectomized patients. Indeed, the visual functions remaining in the blind field of such patients concern mainly the phenomenal detection and localization of visual targets whose attributes are collicular in nature (e.g. flashing, moving or 'jerking' displacement of visual stimuli, movement discrimination, contrast sensitivity and orienting responses) (see introduction). It has also been shown that other 'higher level' visual functions such as discriminations based on wavelength, form, direction of motion are no longer possible (Ptito et al., 1991b; Faubert et al., 1994; King et al., in press) in the blind field of hemispherectomized patients.

Acknowledgements

This work was supported by the National Science and Engineering Research Council of Canada (MP), The Ministère de l'Education du Québec (Fonds FCAR) (MP and AP), and the Cordeau post-doctoral fellowship (MH). The authors are indebted to professor Bogdan Dreher for helpful criticisms on the manuscript.

References

Anderson, K.V. and Symmes, D. (1969) The superior colliculus and higher visual functions in the monkey. *Brain Res.*, 13: 37–52.

Barbur, J.L., Ruddock, K.H. and Waterfield, V.A. (1980) Human visual responses in the absence of the geniculo-calcarine projection. *Brain*, 103: 905–928.

Barbur, J.L., Watson, J.D.G., Frackowiak, R.S.J. and Zeki, S. (1993) Conscious visual perception without V1. *Brain*, 116: 1293–1302.

Blythe, I.M., Kennard, C. and Ruddock, K.H. (1987) Residual vision in patients with retrogeniculate lesions of the visual pathways. *Brain*, 110: 887–905.

Boire, D., Herbin, M., Ptito, A. and Ptito, M. (1995) The dorsal lateral geniculate nucleus of the vervet after neonatal hemispherectomy. *Soc. Neurosci. Abs.*, 21: 1306.

Braddick, O., Atkinson, J., Hood, B., Harkness, W.Jackson, G. and Vargha-Khadem, F. (1992) Possible blindsight in infants lacking one cerebral hemisphere. *Nature*, 360: 461–463.

Bruce, C.J., Desimone, R. and Gross, C. (1986) Both striate cortex and superior colliculus contribute to visual properties of neurons in superior temporal polysensory area of macaque monkeys. *J. Neurophysiol.*, 55: 1057–1075.

Bullier, J. and Kennedy, H. (1983) Projection of the lateral geniculate nucleus onto cortical area V2 in the macaque monkey. *Exp. Brain Res.*, 53: 168–172.

Collin, N.G. and Cowey, A. (1980) The effects of ablation of frontal eye-fields and superior coliculi on visual stability and movement discrimination in the rhesus monkey. *Exp. Brain Res.*, 40: 251–260.

Celesia, G.G., Bushnell, D., Cone Toleikis, S. and Brigell, M.G. (1991) Cortical blindness and residual vision: is the 'second' visual system in humans capable of more than rudimentary visual perception? *Neurology*, 41: 862–869.

Cowey, A., and Stoerig, P. (1991) The neurobiology of blindsight. *TINS*, 14(4): 140–145.

Cowey, A., Stoerig, P. and Perry, V.H. (1989) transneuronal retrograde degeneration of retinal ganglion cells after damage to striate cortex in macaque monkeys: selective loss of Pβ cells. *Neuroscience*, 29: 65–80.

Dreher, B., Marchiavafa, P.L. and Zemicki, B. (1965) Studies on the visual fixation reflex. II. The neural mechanisms of the fixation reflex in normal and and pre-trigeminal cats. *Acta Biol. Exp.*, 25: 207–217.

Dineen, J.I. and Hendrickson, A.E. (1981) Age-correlated differences in the amount of degeneration after striate cortex lesions in monkeys. *Invest. Ophthalmol. Vis. Sci.*, 21: 749–752.

Faubert, J., Diaconu, V., Ptito, M. and Ptito, A. (1995) Modeling visual scatter in the human eye: implications for residual vision. *Invest. Ophthalmol. Vis. Sci.*, 36: s633.

Fendrich, R., Wessinger, C.M. and Gazzaniga, M. (1992) Residual vision in a scotoma: Implications for blindsight. *Science*, 258: 1489–1491.

Girard, P., Salin, A. and Bullier, J. (1992) Response selectivity of neurons in area MT of the macaque monkey during reversible inactivation of area V1. *J. Neurophysiol.*, 67: 1–10.

Goodale, M.A., and Murison, R.C.C. (1975) The effects of lesions of the superior colliculus on locomotor orientation and the orienting reflex in the rat. *Brain Res.*, 88: 243–261.

Gross, C.G. (1991) Contribution of striate cortex and superior colliculus to visual function in area MT, the superior temporal polysensory area and inferior temporal cortex. *Neuropsychologia*, 29: 497–515.

Herbin, M., Boire, D., Ptito, M. and Ptito, A. (1995) Organization of the retina following cerebral hemispherectomy in the vervet monkey. *Soc. Neurosci. Abstr.*, 21: 452.

Heywood, S. and Ratliff, G. (1975) Long-term oculo-motor consequences of unilateral colliculectomy in Man. In: G. Lennerstrand and P. Bach-y-Rita (Eds.), *Basic mechanisms of ocular motility and their clinical implications.* Pergamon Press, Oxford, pp. 561–564.

Hovda, D.A. and Villablanca, J.R. (1989) Depth perception in cats after cerebral hemispherectomy: comparisons between neonatal- and adult-lesioned animals. *Behav. Brain Res.*, 32: 231–240.

Hovda, D.A. and Villablanca, J.R. (1990) Sparing of visual field perception in neonatal but not adult cerebral hemispherectomized cats. Relationship with oxidative metabolism of the superior colliculus. *Behav. Brain Res.*, 37: 119–132.

Kasten, E., Wüst, S. and Sabel, B.A. (1995) Stability and variability of visual field defects in brain damaged patients. *Soc. Neurosci. Abstr.*, 21: 1652.

King, S.M., Frey, S., Villemure, J.G., Ptito, A. and Azzopardi, P. (1996) Perception of motion-in-depth in patients with partial or complete cerebral hemispherectomy. *Behav. Brain Res.*, in press.

Kisvàrday, Z.F., Cowey, A., Stoerig, P. and Somogyi, P. (1991) Direct and indirect retinal input into degenerated dorsal lateral geniculate nucleus after striate cortex removal in monkey: implications for residual vision. *Exp. Brain Res.*, 86: 271–292.

Leventhal, A.G., Rodieck, R.W. and Dreher, B. (1981) Retinal ganglion cell classes in the Old World monkey: morphology and central projections. *Science*, 213: 1139–1142.

Leventhal, A.G., Ault, S.G. and Vitek, D.J. (1988a) The naso-temporal division in primate retina: the neural bases of macular sparing and splitting. *Science*, 240: 66–67.

Leventhal, A.G., Schall, J.D. and Ault, S.G. (1988b) Extrinsic determinants of retinal ganglion cell structure in the cat. *J. Neurosci.*, 8: 2028–2038.

Leventhal, A.G., Thompson, K.G. and Liu, D. (1993) Retinal ganglion cells within the foveola of New World (*Saimiri sciureus*) and Old World (*Macaca fascicularis*) monkeys. *J. Comp. Neurol.*, 338: 242–254.

Milner A.D., Foreman, N.P. and Goodale, M.A. (1978). Go-left go-right discrimination performance and distractability following lesions of prefrontal cortex or superior colliculus in stumptail macaques. *Neuropsychologia*, 16: 381–390.

Mohler, C.V. and Wurtz, R.H. (1977) Role of striate cortex and superior colliculus in visual guidance of saccadic eye movements in monkeys. *J. Neurophysiol.*, 40: 74–94.

Moors, J. and Vendricks, A.J.H. (1979) Responses of single units in the monkey superior colliculus to moving stimuli. *Exp. Brain Res.*, 35: 349–369.

Norsell, K. (1983) Visually guided behaviour of cats in the

absence of retino-geniculo-cortical pathways. Acta Ophthalmol. (Suppl. Copenhagen), 160: 1–99.

Perenin, M.T. (1991) Discrimination of motion direction in perimetrically blind fields. *NeuroReport*, 2: 397–400.

Perenin, M.T. and Jeannerod, M. (1978) Visual function within the hemianopic field following early cerebral hemidecortication in man. I. Spatial localization. *Neuropsychologia*, 16: 1–13.

Perry, V.H. and Cowey, A. (1981) The morphological correlates of X- and Y-like retinal ganglion cells in the retina of monkeys. *Exp. Brain Res.*, 43: 226–228.

Pöppel, E., Held, R. and Frost, D. (1973) Residual visual function after brain wounds involving the central visual pathways in man. *Nature*, 243: 295–296.

Posner, M.I., Cohen, Y. and Rafal, R.D. (1982) Neural systems control of spatial orientation. *Phil. Trans. R. Soc. London, B*, 298: 187–198.

Ptito, M., Dumont, M., Cardu, B. and Leporé, F. (1976) Etude neurocomportementale sur le singe colliculectomisi. *Cortex*, 12: 88–99.

Ptito, A., Lassonde, M., Leporé, F. and Ptito, M. (1987) Visual discrimination in hemispherectomized patients. Neuropsychologia, 25: 869–879.

Ptito, A., Lepore, F., Ptito, M. and Lassonde, M. (1991a) Target detection and movement discrimination in the blind field of hemispherectomized patients. *Brain*, 114: 869–879.

Ptito, M., Leporé, F., Michel, E, and Guillemot, J.-P. (1991b) Partial recovery of functions following telencephalic hemispherectomy in infant monkeys (*C. aethiops sabeus*). *IBRO Abstr.*, 27: 179.

Ptito, M., Herbin, H., Boire, D. and Montfort, S. (1995) The vervet (Cercopithecus aethiops sabeus) retina. *Soc. Neurosci. Abstr.*, 21: 453.

Robinson, D.L. and Wurtz, R.H. (1976) Use of an extraretinal signal by monkey superior colliculus neurons to distinguish real from self-induced stimulus movement. *J. Neurophysiol.*, 32: 852–870.

Rodman, H.R., Gross. C.G. and Albright, T.D. (1989). Afferent basis of visual response properties in area MT of the macaque. I. Effects of striate cortex removal. *J. Neurosci.*, 10: 2033–2050.

Rodman, H.R., Gross. C.G. and Albright, T.D. (1990). Afferent basis of visual response properties in area MT of the macaque. II. Effects of superior colliculus removal. *J. Neurosci.*, 10: 1054–1064.

Schneider, G.E. (1969) Two visual systems. *Science*, 163: 895–902.

Schneider, G.E. (1975) Two visuomotor systems in the syrian hamster. *Neurosci. Res. Prog. Bull.*, 13: 255–258.

Sherman, S.M. (1974) Visual fields of cats with cortical and tectal lesions. *Science*, 185: 355–357.

Shook, B.L. and Villablanca, J.R. (1991) Quantitative cytoarchitectural analysis of cellular degeneration in the dorsal lateral geniculate nuclei of cats and kittens with cerebral hemispherectomy. *Exp. Neurol.*, 111: 80–94.

Stein., B.E. and Meredith, M.A. (1991) Functional organization of the superior colliculus. In: J. Cronly-Dillon (Ed.), *Vision and Visual Dysfunction. Vol. 4*, A.G. Leventhal (Ed.), The Neural Basis of Visual Function, McMillan Press, Houndmills, basingstoke, Hampshire, London, pp. 85–110.

Stoerig, P. (1987). Chromaticity and achromaticity — Evidence for a functional differentiation in visual field defects. *Brain*, 110: 869–886.

Stoerig, P. and Cowey, A. (1989) Residual target detection as a function of stimulus size. *Brain*, 112: 1123–1139.

Stoerig, P. and Cowey, A. (1992) Wavelength discrimination in blindsight. *Brain*, 115: 425–444.

Stone, J. (1981) The wholemount handbook. A guide to the preparation and analysis of retinal wholemounts. Maitland, Sydney.

Timney, B. and Lansdown, G. (1989) Binocular depth perception, visual acuity and visual fields in cats following neonatal section of the optic chiasm. *Exp. Brain Res.*, 74: 272–278.

Tomaiuolo, F., Ptito, A., Paus, T. and Ptito, M. (1994) Spatial summation across the vertical meridian in hemispherectomized patients. *Soc. Neurosci. Abstr.*, 20: 1580.

Ueki, K. (1966) Hemispherectomy in the human with special reference to the preservation of function. *Prog. Brain Res.*, 21: 285–338.

Villablanca, J.R., Burgess, J.W. and Olmstead, C.E. (1986) Recovery of function after neonatal or adult hemispherectomy in cats: I. Time course, movement, posture and sensorimotor tests. *Behav. Brain Res.*, 19: 205–226.

Weller, R.E. and Kaas, J.H. (1981) Developmental changes and suceptibility to retinal ganglion cell loss after lesions of visual cortex in primates and other mammals. In: J. Stone, B. Dreher and D.H.Rapaport (Eds.), *Development of visual pathways in mammals*. New York, Alan Liss, pp. 289–302.

Weller, R.E., Kaas, J.H. and Wetzel, A.B. (1979) Evidence for the loss of X cells of the retina after long-term ablations of the visual cortex in monkeys. *Brain Res.*, 160, 134–138.

Wessinger, C.M., Fendrich, R., Ptito, A., Villemure, J.G. and Gazzaniga, M.S. (1996) Residual vision with awareness in the field contralateral to a partial or complete functional hemispherectomy. Neuropsychologia (in press).

Weiskrantz, L. (1986) Blindsight: a case study and implications, Oxford, Clarendon Press. Weiskrantz, L., Warrington, E.K., Sanders, M.D. and Marshall, J. (1974) Visual capacity in the hemianopic field following a restricted occipital ablation. *Brain*, 97: 709–728.

Weiskrantz, L. (1987) Residual vision in a scotoma — A follow-up study of 'form' discrimination. *Brain*, 110: 77–92.

Yukie, M. and Iwai, E. (1981) Direct projection from the dorsal lateral geniculate nucleus to the prestriate cortex in macaque monkeys. *J. Comp. Neurol.*, 201: 81–97.

M. Norita, T. Bando and B. Stein (Eds.)
Progress in Brain Research, Vol 112

Extrageniculostriate vision in humans: investigations with hemispherectomy patients

C. Mark Wessinger[1], Robert Fendrich*[1], Michael S. Gazzaniga[1], Alain Ptito[2] and J-G Villemure[2]

[1]*Center for Neuroscience, University of California, Davis, CA 95616, USA*
[2]*Montreal Neurological Institute and Hospital, Montreal, Quebec H3A 2B4, Canada.*

Residual vision was assessed in the blind hemifield of one hemispherectomized and one partially hemispherectomized patient, using an interval two alternative forced choice detection task. Fixation instabilities were controlled by retinal stabilization. In both patients, residual vision was found in the hemianopic field close to the vertical meridian. This residual vision is largely confined to a band not wider than 3°, but there is a local region in each patient where it extends more than 3° from the meridian. However, more than 6° from the vertical meridian we found no indication of residual function in either patient. Within the band of spared vision, subjects are aware of stimuli and can perform simple shape discriminations. Visual acuity profiles argue against an explanation based on eccentric fixation. Explanations based on the retino-tectal pathway or on retinal naso-temporal overlap are possible.

Introduction

In humans, damage to striate visual cortex normally produces a profound state of phenomenal blindness in those visual field regions which map to the lesioned cortex (Holmes, 1918; Horton and Hoyt, 1991). However, patients sometimes demonstrate residual visual function within these apparently blind regions. In 1917, Riddoch reported cases where the ability to detect moving objects was preserved within scotomas (Riddoch, 1917). In 1951, Bender and Kreiger reported that some patients could detect and localize stimuli within clinically blind regions if they were tested in darkness (Bender and Kreiger, 1951). The surviving visual function in these early studies was attributed to inefficient processing by remnants of spared cortex.

In 1973, Pöppel et al. reported the first quantitative study of preserved oculomotor function within perimetrically blind fields (Pöppel et al., 1973). When patients were asked to saccade to stimuli flashed within their field defects, the magnitude of their saccades was found to be correlated with the eccentricity of the stimuli. This was the case despite the fact that these patients claimed they could not actually see these stimuli. While acknowledging that remnants of spared cortex could be mediating this result, Pöppel et al. speculated that the residual vision in their patients might be mediated by the retinotectal visual pathway. This premise has elicited a great deal of interest.

In 1974, Weiskrantz coined the term 'blindsight' to refer to visual abilities that were preserved in the absence of phenomenal awareness (Weiskrantz et al., 1974). Subsequently, a variety of residual visual abilities have been reported (see Weiskrantz, 1986 for review). In these investigations, the absence of phenomenal vision has gen-

*Corresponding author. Tel.: +1 916 754 5213; fax: +1 916 7578827; e-mail: rmfendrich@ucdavis.edu

erally been taken to demonstrate a failure of the primary visual pathway, and the preserved functions are attributed to secondary visual pathways (for alternate explanations; Campion et al., 1983; Celesia et al., 1991; Gazzaniga et al., 1994). The retino-tectal pathway is the one most frequently proposed, although arguments for a geniculo-extrastriate pathway have also been made (Cowey and Stoerig, 1991).

The conjecture that the retinotectal pathway can mediate residual vision following geniculostriate lesions gains substantial support from primate research. Following the total ablation of striate cortex, monkeys can (with reduced sensitivity) localize stimuli (Humphrey, 1974; Keating, 1975; Pasik and Pasik, 1982) and be trained to make discriminations based on brightness, shape and wavelength (Schilder, Pasik and Pasik, 1971). Subsequent removal of the superior colliculus eliminates these abilities (Mohler and Wurtz, 1977).

Nevertheless, there are grounds for questioning the hypothesis that secondary pathways mediate residual vision in humans. If blindsight is the result of the operation of such pathways, one would expect the phenomenon to be a common occurrence in cases of striate lesions. In fact, blindsight is a relatively rare phenomenon. Moreover, when residual vision is found, it is generally difficult to rule out the possibility that the surviving visual functions are a consequence of remnants of spared cortex as the early investigators proposed. It is noteworthy that in monkeys sparing of more than 2% of the striate cortex can virtually eliminate indications of any visual deficit (Keating, 1975), at least when eye motions are not constrained.

In addition, there are recent demonstrations that blindsight can be restricted to small and isolated islands in the blind field of hemianopic patients (Fendrich et al., 1992, 1993; Wessinger, 1995). Figs. 1A,C show conventional clinical perimetric fields for stroke patients C.L.T. and F.N., respectively. Figs. 1B,D show fields for these patients obtained using a criterion free interval 2-alternative forced choice (i2afc) method. This testing consisted of having the subjects report

during which of two 600-ms intervals, delimited by tones, a stimulus was flashed.

Confidence ratings taken on a 5-point scale during the i2afc testing reveal why the islands of preserved detection found with this method were not found by conventional perimetry. These ratings indicate that for stimulus presentations within as well as outside these islands, the subjects were not aware of seeing a stimulus appear in either test interval and believed they were guessing. Since one would not expect a secondary visual system to produce a patchy distribution of residual abilities, it is likely that the islands of preserved function found in these patients reflect corresponding islands of functioning cortical tissue. In cases of blindsight, the absence of phenomenal vision therefore cannot be regarded as an exclusive marker for vision based on non- striate pathways.

In hemispherectomy patients who have undergone complete removal or deafferentation of a cerebral hemisphere, remnants of surviving cortex cannot be the basis of any residual blind-field abilities. Work reported in this volume by Maurice Ptito et al. demonstrates that in monkeys neonatal hemispherectomy produces a massive degeneration of the ipsilesional lateral geniculate nucleus but only a modest reduction in the size of the ipsilesional superior colliculus (SC). Hemi-

(a)

30°

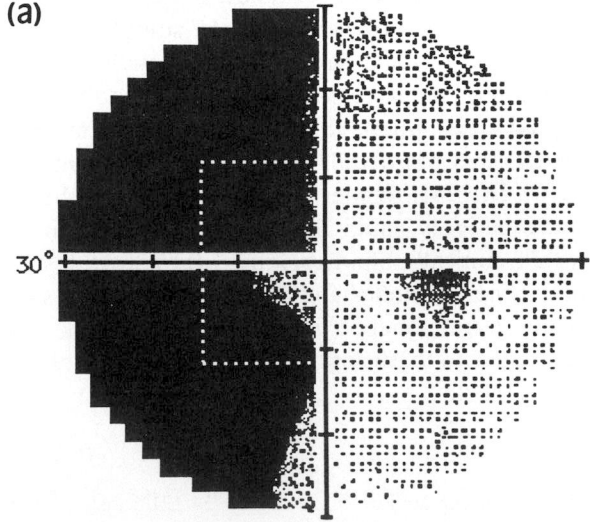

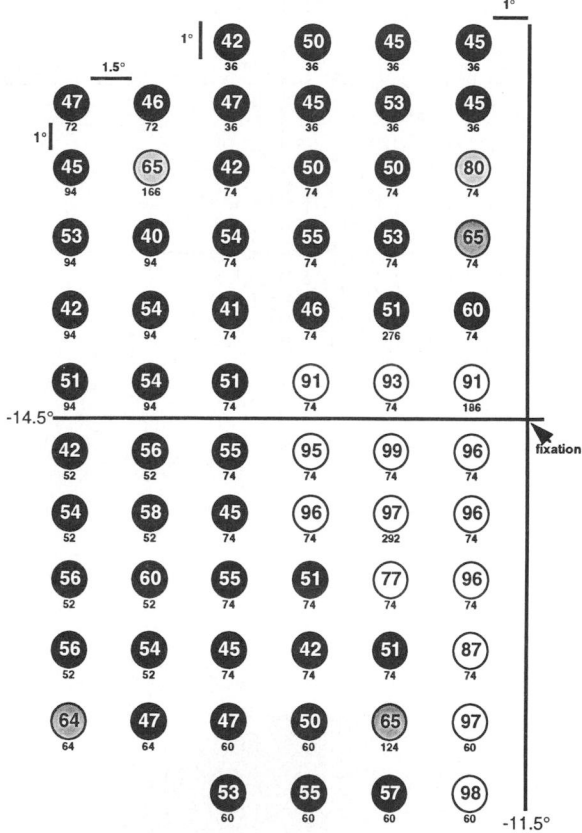

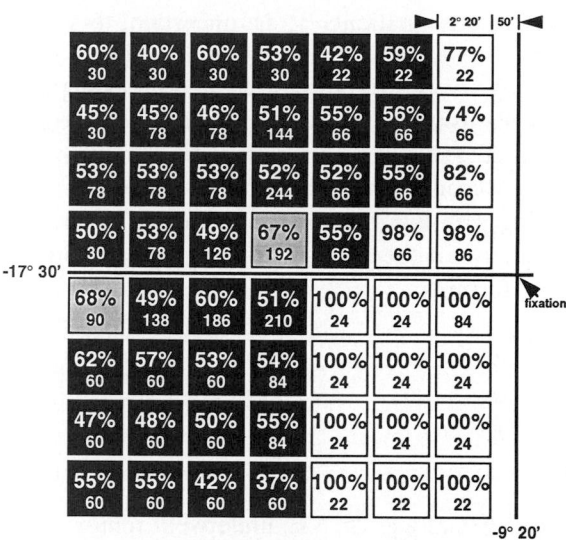

Fig. 1. (a) C.L.T.'s visual fields obtained with a Humphrey automated perimeter using the 30-2 program. Shaded portions are the regions that have been diagnosed clinically blind. The dashed border in the left visual field surrounds the approximate area tested with stabilized field mapping. (b) Results of stabilized field mapping in C.L.T.'s left visual field. Each of 68 different test locations are represented by a circle. The number in the circle represents the percentage of correct trials for that location. The number below each circle indicates the total number of trials for that location. White circles show test locations where performance was largely unimpaired. Stippled circles are locations where detection was impaired but greater than chance with a Bonferroni correction. Gray circles show locations where detection was greater than chance but not with a Bonferroni correction. (c) Visual fields for F.N. obtained with a Humphrey automated perimeter. The dashed outline represents the area tested with stabilized field mapping. (d) Schematic diagram detailing results of stabilized field mapping with F.N. Each square represents one of 56 test locations. Presented at each location is the percentage of correct detection performance (%) and the number of trials (*n*). Areas of chance performance are represented by black squares, areas of above chance performance with awareness are shown in white, and above chance performance in the absence of awareness are indicated by stippled squares.

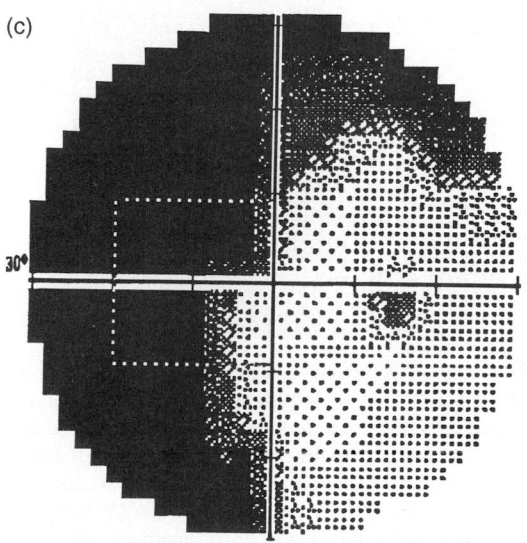

spherectomy patients are therefore especially well suited for demonstrating residual visual abilities that can be mediated by retinal afference to the superior colliculus and pulvinar. Previous investigations have, in fact, reported that these patients are capable of detecting stimuli and making simple discriminations within their blind field (Perenin and Jeannerod, 1978; Perenin, 1978; Ptito et al., 1987; 1991).

However, in previous investigations of hemispherectomy patients it is uncertain if stimuli were always fully confined to a subject's blind field. We were therefore interested in testing such patients using methods that would guarantee that the accurate and stable retinal placement of stimuli. Recently, we were able to carry out such testing with two patients operated on by Dr. J-G Villemure at the Montreal Neurological Institute and Hospital.

Methods

Subjects

S.E.

In 1990, at age 25, S.E. underwent removal of a congenital porencephalic cyst to alleviate intractable seizures. Surgery resulted in a right temporo-parieto-occipital lobectomy. Subsequent to surgery, S.E. is seizure free and exhibits a dense left homonymous hemianopia without macular sparing. A Humphrey right eye visual field is shown in Fig. 2A. S.E. functions within the average range of intelligence (FSIQ 93).

J.B.

In 1983, at age 18, a subtotal hemispherectomy was performed on J.B. to control seizures. The posterior portion of a congenital porencephalic cyst occupying the left hemisphere was removed. Seizures, initially halted, began recurring 3 months later. In 1985, the anterior portion of the cyst was removed eliminating the seizures. This two-stage procedure resulted in a complete left functional hemispherectomy. The frontal and occipital poles were left in place to prevent superficial hemosiderosis and reduce hydrocephalus, but they are totally disconnected from the rest of the brain (Villemure, 1991). Subsequent to this surgery, J.B. has been seizure free and demonstrates a dense right homonymous hemianopia without macular sparing. A Humphrey right eye visual field is shown in Fig. 2B. J.B. functions just below the average range of intelligence (FSIQ 88).

Stimuli were generated with a Macintosh IIcx computer and displayed on a Macintosh color

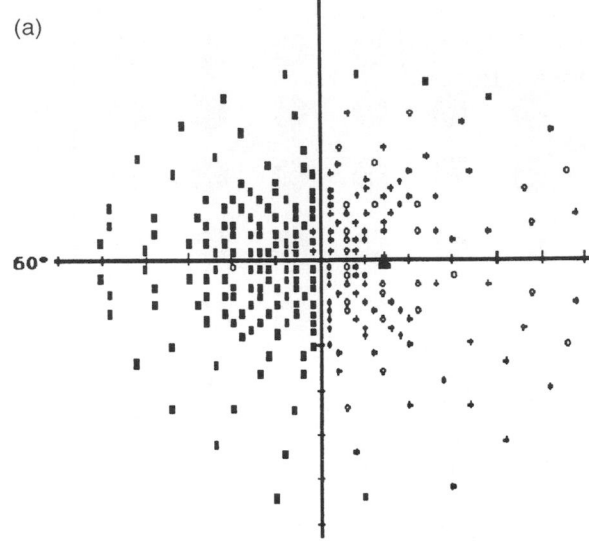

(a)

60°

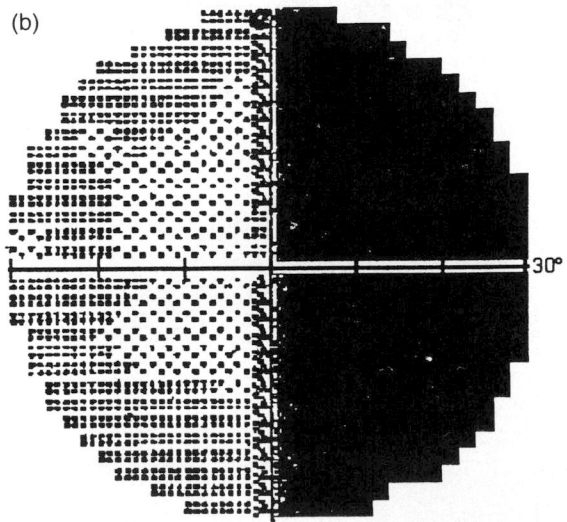

(b)

30°

Fig. 2. Automated perimetry demonstrating field defects in each subject subsequent to surgery. Only right eye fields are shown since our testing was performed with the right eye. (a) Humphrey full field screening perimetry of ±60° of S.E.'s central visual field. Open circles are locations of detected stimuli indicating areas of vision, filled squares are locations of undetected stimuli indicating areas of blindness. (b) Humphrey central 30-2 threshold perimetry of ±30° of J.B.'s central visual field. Stippling indicates areas of seeing; dark area indicates area of blindness.

monitor. A forehead rest and bite bar were used to position the subject's head. Subject's viewed displays monocularly with their right eye, which was monitored with a Purkinje image eyetracker. Displays were retinally stabilized with a mirror image deflector system attached to the eyetracker. The combined image deflector eyetracker system has a resolution of 1' of arc, and a response time of less than 2 ms. The ability to retinally stabilize stimuli allowed us to use extended and repetitive stimulus presentations to a constant retinal position irrespective of a subject's eye motions. The effective distance of the perimetric screen when viewed through the stabilizer optics was 57 cm.

Detection

We tested an area in the central portion of each subject's hemianopic field with a grid of $36 \times 2.5°$ squares. This grid extended laterally from 1° to 13.5° from the vertical meridian, and vertically 10° above and below the horizontal meridian. To minimize light scatter, the squares were dark (< 1 cd/m^2) against a light (10 cd/m^2) background. As in our previous testing, we used an i2afc method to minimize criterion effects. On each trial, tones defined two consecutive 600-ms intervals. During one interval, the target was flashed three times for 100 ms. Subjects reported the interval containing the target presentation with a keyboard press. Four locations within each subject's seeing field were also tested to confirm the task was being properly performed. On a subset of the trials, verbal confidence ratings were collected using a 5-point scale with 1 designating a guess.

Shape discrimination

Additional experiments assessed the ability of our subjects to discriminate shapes within and outside the zone with spared detection. Pairs of stimuli were presented simultaneously at mirror symmetric locations in each subject's blind and seeing field. Stimuli were displayed continuously during a tone delimited 300-ms interval. The subjects reported if the stimuli were the same or different.

In separate trial blocks, simple and complex stimuli were presented. The simple set consisted of a black 2° square or diamond (the square rotated 90°); the complex set consisted of 78 line drawings from the Snodgrass picture set (Snodgrass and Vanderwart, 1980). Complex figures did not subtend more than 5° of visual angle. With both subjects, the simple stimuli were centered 'close' (2.5°) or 'far' (4.75°) from the vertical meridian, and 2.75° above or below the horizontal meridian. With S.E., complex stimuli were presented 3.6° or 7° from the vertical meridian, and 3.6° above or below the horizontal meridian. With J.B., the complex stimuli were presented only at the 'close' locations. For both the simple and complex stimuli, the medial edge of the stimulus was always separated from the vertical meridian by at least 1°. All of the 'close' stimuli fell within S.E.'s zone of sparing, while the 'upper close' stimuli fell within J.B.'s zone of sparing.

Shape identification

We also assessed the ability of our subjects to name simple and complex stimuli presented for 300 ms within and outside the zone with spared detection. The stimuli and positions were the same as those used in the discrimination experiment, but stimuli were presented only to one visual field.

Results

Figs. 3a,b show the obtained stabilized field maps for S.E. and J.B., respectively. In both subjects, a zone of residual vision was found along the vertical meridian where the stimulus presentations were detected well above chance. This band is generally not wider than 3.5°, but in both subjects it extends farther outward at one field location, although not beyond 6°. In J.B. this band occurs both above and below the horizontal meridian; in J.B. it occurs only in the upper visual field. Outside this band of sparing, we found no indication of any residual ability to detect the flashed stimuli.

Both subjects were aware of their residual vision. The mean confidence values within the band of sparing were 3.5 and 4.7 for S.E. and J.B., respectively, in contrast to values of 1.3 and 1.2 for areas with no sparing. Thus, in the area we tested neither subject exhibited detection without awareness or true 'blindsight'.

Results of the shape discrimination experiments for both subjects are presented in Table 1. When stimuli were placed within the region with spared detection both subjects readily performed this task, but they were poorer with complex stimuli. Performance did not differ from chance when stimuli were placed outside the spared region.

Results of the identification experiment are presented in Table 2. Both subjects could identify the square and diamond when they were presented within the zone of sparing, but could not identify these figures outside this zone. However,

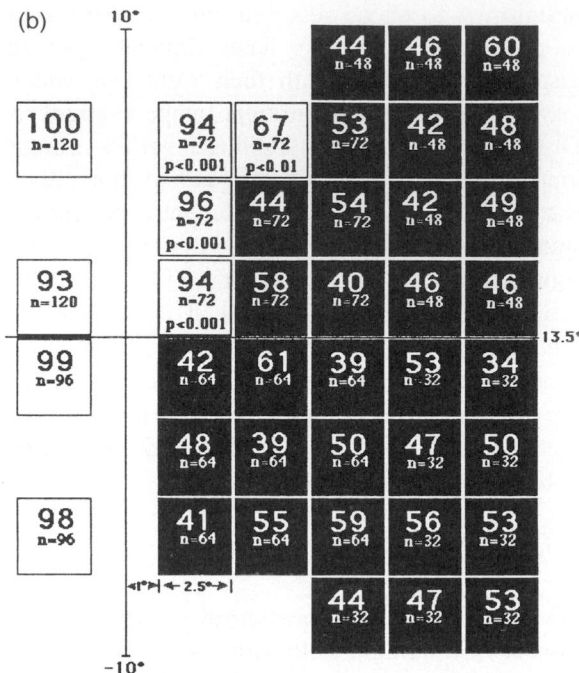

Fig. 3. Schematic representations of stabilized visual field detection results for S.E. (3a) and J.B. (3b). Open squares represent locations of above chance performance, filled squares indicate locations of chance performance. Values presented at each location are percent correct detections, number of trials run (n), and significant P values.

neither subject was able to identify the complex shapes at any position within their blind field. Identification performance was at ceiling in the seeing field of both subjects.

Discussion

The fact that residual vision was limited to a region proximal to the vertical meridian suggests that our results could be attributed to eccentric fixation. Although stimuli were stabilized, if our subjects fixated eccentrically during the calibration of the eyetracker, stimuli presented to their blind field could have fallen within their seeing field. However, the fact that J.B.'s residual vision was limited to his superior visual field argues against this artifact, as does the irregular distribu-

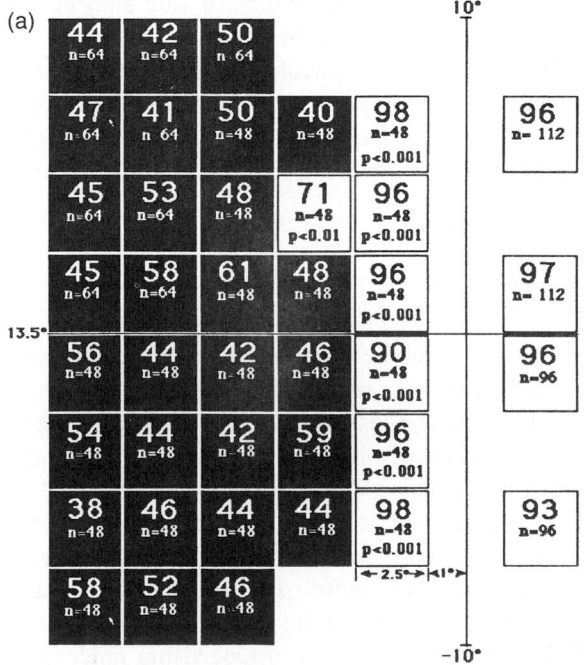

TABLE 1

Results of shape discrimination experiment

Subject	Upfar	Upclose	Downfar	Downclose
Simple stimuli				
J.B.	31/60	58/60[a]	31/60	36/60
SE	18/40	37/40[a]	20/40	35/40[a]
Complex stimuli				
J.B.	11/26	20/26[a]	12/26	9/26
S.E.	13/26	12/26	13/24	65/104[c]

[a]$P < 0.001$; [b]$P < 0.01$; [c]$P < 0.02$.
Values shown are number correct over total trials.

TABLE 2

Results of stimulus identification experiment

Subject	Upfar	Upclose	Downfar	Downclose	Control
Simple stimuli					
J.B.	12/20	20/20[a]	9/20	11/20	80/80
SE	24/48	43/48[a]	29/48	44/48[a]	95/96
Complex stimuli					
J.B.	–	*0/12*	–	*0/12*	*12/12*
SE	*0/8*	*0/8*	*0/8*	*0/8*	*16/16*

Note: [a]$P < 0.001$.

tion of residual vision along the lateral edge of the zone of sparing with both subjects. In addition, we obtained acuity profiles across the subjects point of fixation using a Tübinger perimeter (Johnson et al., 1979). Both showed an acuity peak at the fixation point, with acuity becoming non-measurable 1° into their hemianopic field, a pattern indicative of foveal fixation.

Since it does not appear that eccentric fixation can account for the observed residual vision, why was that residual vision restricted to regions proximal to the vertical meridian? One possibility is that these abilities are subserved by a zone of naso-temporal overlap. Psychophysical studies in humans (Fendrich and Gazzaniga, 1989; Sugishita et al., 1993; McFadzean et al., 1994) and anatomi-

cal studies in monkeys (Bunt et al., 1977; Leventhal et al., 1988a; Fukuda et al., 1989) suggest the zone of naso-temporal overlap is normally narrower than the band of sparing we found in our subjects [although in a recent psychophysical investigation with a callosotomy patient, Fendrich et al. (1996) found evidence of this band more than 1° from the vertical meridian]. However, in the absence of a normal pathway during development, the zone containing uncrossed retinal projections from the nasal hemiretina and crossed projections from the temporal hemiretina may have expanded. This would provide the intact hemisphere with increased access to the ipsilateral visual field. Recent developmental studies in the cat and monkey are in accord with this possibility. In the cat, abnormal development following a unilateral optic tract section results in a significant increase in the number and size of contralaterally projecting alpha and beta (but not gamma) cells within the temporal retina (Leventhal et al., 1988b,c; Schall et al., 1988). Such developments tend to favor the central visual field (Payne et al., 1993). In the neonatal monkey the normally sharp naso-temporal division (Lia et al., 1988; Leventhal et al., 1989) begins to break down during abnormal development caused by a retinal lesion (Leventhal et al., 1993).

While these considerations favor an interpretation based on naso-temporal overlap, they do not rule out the possibility that retino-collicular projections could be the source S.E. and J.B.'s residual vision, conceivably via intertectal connections to their remaining hemisphere. Recent quantitative cytoarchitecture and cytochrome oxydase activity analyses of the effects of early hemispherectomy in the vervet monkey further support this notion, showing that the ipsilesional SC is affected by the loss of cortical input but remains a functional subcortical endstation for analysis of visual information. Additionally, cellular integrity and metabolism of the SC are much less affected than those of the dorsal LGB (Boire et al., 1995; Ptito et al., 1995). Adaptive rearrangements involving tectal projections might be largely confined to regions close the vertical meridian due to the functional importance of the central visual field. A more detailed appraisal of the character of the residual vision in these patients may help to better define the pathway which subserves their midline sparing. If, for instance, they can discriminate color or motion direction within their zone of sparing, this will argue in favor of an interpretation in terms of naso-temporal overlap. If they cannot make these discriminations, the argument for a retinotectal pathway will gain support. It should be noted, however, that if it can be shown that our subject's residual abilities are mediated by the retinotectal system, the fact that they were aware of their residual vision will further erode the case that conscious perception is tied to geniculostriate function.

A final caveat concerns the possibility that specific tests for orienting behaviors might reveal residual abilities not related to detection or discrimination. Given the established role of the colliculus in primate oculomotor control (Mohler and Wurtz, 1976), a residual influence on saccadic orienting responses seems a plausible possibility. As previously noted, it was Pöppel et al.'s report of the effect of blind field stimuli on the magnitude of saccadic responses which initiated the formal study of human blindsight. It remains to be established, however, whether in humans even oculomotor behaviors can be influenced by retino-collicular afferents in the absence of cortical support. Hemispherectomy patients should provide a valuable resource for addressing this question.

References

Bender, M.B. and Kreiger, H.P. (1951) Visual function in perimetrically blind fields. *Arch. Neurol. Psychiatry*, 65: 72–79.

Boire, D., Herbin, M., Ptito, A. and Ptito, M. (1995) The dorsal lateral geniculate nucleus of the vervet monkey after neonatal hemispherectomy. *Soc. Neurosci. Abstr.*, 21: 1306.

Bunt A.H., Minckler, D.S. and Johanson, G.W. (1977) Demonstration of bilateral projections to the central retina of the monkey with horseradish peroxidase neuronography. *J. Comp. Neurol.*, 171: 619–630.

Campion, J., Latto, R. and Smith, Y.M. (1983) Is blindsight an effect of scattered light, spared cortex and near-threshold vision? *Behav. Brain Sci.*, 6: 423–486.

Celesia, G.G., Bushnell, D., Cone Toleikis, S. and Brigell, M.G. (1991) Cortical blindness and residual vision: is the 'second' visual system in humans capable of more than rudimentary visual perception? *Neurology*, 41: 862–869.

Cowey, A. and Stoerig, P. (1991) The neurobiology of blindsight. TINS, 14: 140–145. Fendrich, R. and Gazzaniga, M.S. (1989) Evidence of foveal splitting in a callosotomy patient. *Neuropsychologia*, 27: 273–281.

Fendrich, R., Wessinger, C.M. and Gazzaniga, M.S. (1996) Naso-temporal overlap at the retinal vertical meridian: investigations with a callosotomy patient. *Neuropsychologia*, 34: 637–646.

Fendrich, R. Wessinger, C.M. and Gazzaniga, M.S. (1993) Sources of blindsight — reply to Stoerig and Weiskrantz. *Science*, 261: 493–495.

Fendrich, R., Wessinger, C.M. and Gazzaniga, M.S. (1993) Residual vision in a scotoma: implications for blindsight. *Science*, 258: 1489–1491.

Fukuda, Y., Sawai, H., Watanabe, M., Wakakuwa, K. and Morigiwa, K. (1989) Naso-temporal overlap of crossed and uncrossed retinal ganglion cell projections in the Japanese monkey (*Macaca fuscata*). *J. Neurosci.*, 9: 2353–2373.

Gazzaniga, M.S. Fendrich, R. and Wessinger, C.M. (1994). Blindsight reconsidered. *Curr. Directions Psychol. Sci.*, 3: 93–96.

Holmes, G. (1918) Disturbances of vision by cerebral lesions. *Br. J. Ophthalmol.*, 2: 353–384.

Horton, J.C. and Hoyt, W.F. (1991) The representation of the visual field in human striate cortex. *Arch. Ophthalmol.*, 109: 816–824.

Humphrey, N.K. (1974) Vision in monkey after removal of striate cortex: a case study. *Perception*, 3: 241–255.

Johnson, C.A., Keltner, J.L., and Balestrery, F. (1979) Acuity profile perimetry. *Arch. Ophthalmol.*, 97: 684–689.

Keating, E.G. (1975) Effects of prestriate and striate lesions on the monkeys ability to locate and discriminate visual forms. *Exp. Neurol.*, 47: 16–25.

Leventhal, A.G., Ault, S.J. and Vitek, D.J. (1988a) The Naso-temporal division in primate retina: The neural bases of macular sparing and splitting. *Science*, 240: 66–67.

Leventhal, A.G., Schall, J.D. and Ault, S.J. (1988b) Extrinsic determinants of retinal ganglion cell structure in the cat. *J. Neurosci.*, 8: 2028–2038.

Leventhal, A.G., Schall, J.D., Ault, S.J., Provis, J.M. and Vitek, D.J. (1988c) Class-specific cell death shapes the distribution and pattern of central projection of cat retinal ganglion cells. *J. Neurosci.*, 8: 2011–2027.

Leventhal, A.G., Ault, S.J., Vitek, D.J. and Shou, T. (1989) Extrinsic determinants of retinal ganglion cell development in primates. *J. Comp. Neurol.*, 286: 170–189.

Leventhal, A.G., Thompson, K.G. and Liu, D. (1993) Retinal ganglion cells within the foveola of new world (*Saimiri sciureus*) and old world (*Macaca fascicularis*) Monkeys. *J. Comp. Neurol.*, 338: 242–254.

Lia, B., Snider, C.J. and Chalupa, L.M. (1988) The Naso-temporal division of the retinal ganglion cell decussation pattern in the fetal rhesus monkey. *Soc. Neurosci. Abstr.*, 14: 308.

McFadzean, R., Bronsnahan, D., Hadley, D., Mutlukan, E. (1994) Representation of the visual field in the occipital striate cortex. *Br. J. Ophthalmol.*, 78: 185–190.

Mohler, C.W. and Wurtz, R.H. (1977). Role of striate cortex and superior colliculus in visual guidance of saccadic eye movements in monkeys. *J. Neurophysiol.*, 40: 74–94.

Pasik, P. and Pasik, T. (1982) Visual functions in monkeys after total removal of visual cortex. In: W.D. Neff, (Ed.), *Contributions to Sensory Physiology*, Academic Press, New York Vol. 7, pp. 147–200.

Payne, B.R., Foley, H.A. and Lomber, S.G. (1993) Visual cortex damage-induced growth of retinal axons into the lateral posterior nucleus of the cat. *Vis. Neurol.*, 10: 747–752.

Perenin, M.T. (1978) Visual function within the hemianopic field following early cerebral hemidecortication in man. II. Pattern Discrimination. *Neuropsychologia*, 16: 697–708.

Perenin, M.T. and Jeannerod, M. (1978). Visual function within the hemianopic field following early cerebral hemidecortication in man. I. Spatial localization. *Neuropsychologia*, 16: 1–13.

Pöppel, E., Held, R. and Frost, D. (1973) Residual visual functions after brain wounds involving the central visual pathways in man. *Nature*, 243: 295–296.

Ptito, A., Lassonde, M. Lepore, F. and Ptito, M. (1987) Visual discrimination in hemispherectomized patients. *Neuropsychologia*, 25: 869–879.

Ptito, A., Lepore, F., Ptito, M., and Lassonde, M. (1991) Target detection and movement discrimination in the blind field of hemispherectomized patients. *Brain*, 114: 497–512.

Ptito, A., Boire, D., Herbin, M. and Ptito, M. (1995) Quantitative cytoarchitecture and cytochrome oxydase activity in the superior colliculus of the vervet monkey after neonatal hemispherectomy. *Soc. Neurosci. Abstr.*, 21: 816.

Riddoch, G. (1917) Dissociation of visual perceptions due to occipital injuries, with especial reference to appreciation of movement. *Brain*, 40: 15–57.

Schall, J.D., Ault, S.J., Vitek, D.J. and Leventhal, A.G. (1988) Experimental induction of an abnormal ipsilateral visual field representation in the geniculocortical pathway of normally pigmented cats. *J. Neurosci.*, 8: 2039–2048.

Schilder, P., Pasik, T. and Pasik, P. (1971). Extrageniculate vision in the monkey. II. Demonstrations of brightness discrimination. *Br. Res.*, 32: 383–398.

Snodgrass, J.G. and Vanderwart, M. (1980) A standardized set of 260 pictures: Norms for name agreement, image agreement, familiarity, and visual complexity. *J. Exp. Psychol.: Human Learn. Mem.*, 6: 174–215.

Sugishita, M., Hemmi, I., Sakuma, I., Shiokawa, Y. (1993) The Problem of macular sparing after unilateral occipital lesions. *J. Neurol.*, 241: 1–9.

Villemure, J.G. (1991) In: F. Andermann (Ed.), *Chronic Encephalitis and Epilepsy*, Butterworth-Heinemann, Boston, pp. 235–241.

Weiskrantz, L. (1986) *Blindsight: A Case Study and Implications*, Oxford University Press, Oxford.

Weiskrantz, L., Warrington, E.K., Sanders, M.D. and Marshall, J. (1974) Visual capacity in the hemianopic field following a restricted occipital ablation. *Brain*, 97: 709–728.

Wessinger, C.M. (1995) *Residual visual abilities following damage to primary visual cortex*. Dissertation, University of California, Davis.

M. Norita, T. Bando and B. Stein (Eds.)
Progress in Brain Research, Vol 112
© 1996 Elsevier Science BV. All rights reserved.

CHAPTER 30

Visual inputs to cerebellar ventral paraflocculus during ocular following responses

K. Kawano[*][1], A. Takemura[1], Y. Inoue[1], T. Kitama[1], Y. Kobayashi[2] and M.J. Mustari[3]

[1]*Neuroscience Section, Electrotechnical Laboratory, 1-1-4, Umezono, Tsukubashi, Ibaraki 305, Japan*
[2]*ATR Human Information Processing Research Laboratory, Kyoto 619-02, Japan*
[3]*University of Texas Medical Branch, Galveston, TX, USA*

The activity of cerebellar Purkinje cells was recorded in the ventral paraflocculus (VPFL) of alert monkeys during ocular following responses induced by brief movements of the visual scene. The mossy fiber input evoked 'simple-spikes' which increased their activity during either ipsiversive or downward motion of the visual scene. On the other hand, the 'complex-spikes' evoked by the climbing fiber input increased their activity during either contraversive or upward movement.

To further define the sources of the visual input to the VPFL, we recorded single units in the dorsolateral pontine nucleus (DLPN), which is known to project to the VPFL as mossy fibers, and in the pretectal nucleus of the optic tract (NOT), which is known to project to the inferior olive whose neurons project to the cerebellum as climbing fibers. In both areas, most neurons were directionally selective, and responded to the moving visual scene with very short latencies (~ 40–60 ms). These results suggest that both DLPN and NOT neurons deliver information concerning movements of the visual scene to the VPFL through different pathways.

Introduction

Ocular following responses are short-latency tracking movements of the eyes elicited by sudden movements of a large-field visual stimulus (Miles et al., 1986). The floccular/ventral parafloccular lobes of the cerebellum have been suggested to play an important role in the generation of these, as well as other, slow tracking eye movements (Miles et al., 1986). Purkinje cells (P-cells) in the cerebellum are differentially responsive to two distinct afferent systems: mossy fiber inputs produce 'simple-spikes' whereas climbing fiber inputs evoke 'complex-spikes' (Ec-

cles et al., 1966; Thach, 1968). We have previously reported that the simple-spikes of P-cells in the ventral paraflocculus (VPFL) of the cerebellum are activated during ocular following (Shidara and Kawano, 1993). In addition, our preliminary studies have shown that the complex-spike activity of P-cells is also modulated in association with movements of the visual scene and ocular following (Kobayashi et al., 1995). We discuss here the possible sources of visual inputs to the VPFL and summarize experiments in which we have investigated the responses of neurons in several regions of the brain to the visual stimuli that elicit ocular following (Kawano et al., 1992, 1994; Shidara and Kawano, 1993; Takemura et al., 1995). We have recorded single unit activity in the dorsolateral pontine nucleus (DLPN), which is known to project to the VPFL as mossy fibers, and in the

---

*Corresponding author. Tel.: +81 298 585845; fax: +81 298 585849.

pretectal nucleus of the optic tract (NOT), which is known to project to the inferior olive. These inferior olive neurons project, in turn, to the cerebellum as climbing fibers (Blanks, 1988; Stein and Glickstein, 1992; Mustari et al., 1994). Most of the neurons in these areas are directionally selective and respond to a moving visual scene with very short latencies (~ 40–60 ms).

Methods

All data were collected from Japanese monkeys (*Macaca fuscata*) using equipment and procedures already reported in previous studies (Kawano et al., 1992, 1994; Shidara and Kawano, 1993). Briefly, monkeys were trained prior to surgery to fixate on a small spot of light. Under sodium pentobarbital anesthesia and aseptic conditions, each monkey was implanted with a head holder, which allowed the head to be fixed in the standard stereotaxic position during the experiments, with a cylinder for microelectrode recording and scleral search coils for measuring eye movements (Judge et al., 1980). Animals faced a translucent tangent screen (85° along the vertical and horizontal meridians), on which a random dot pattern was back-projected. The positions of both eyes were monitored with the electro-magnetic induction technique (Fuchs and Robinson, 1966), and single units were recorded using conventional methods. After isolating a single unit, we first observed neuronal responses to a visual scene moving at 80°/s in each of eight directions to determine the preferred direction. We then studied the unit's responses to moving the visual scene in the unit's preferred direction at different speeds (10, 20, 40, 80, 160°/s). After sorting these data according to the stimulus conditions (speeds, direction of movement), peristimulus histograms were constructed. The firing rate was averaged over the time interval between 40 ms and 140 ms, measured from the onset of the stimulus motion; this was used as a measure of neuronal response amplitude. The procedures reported here were

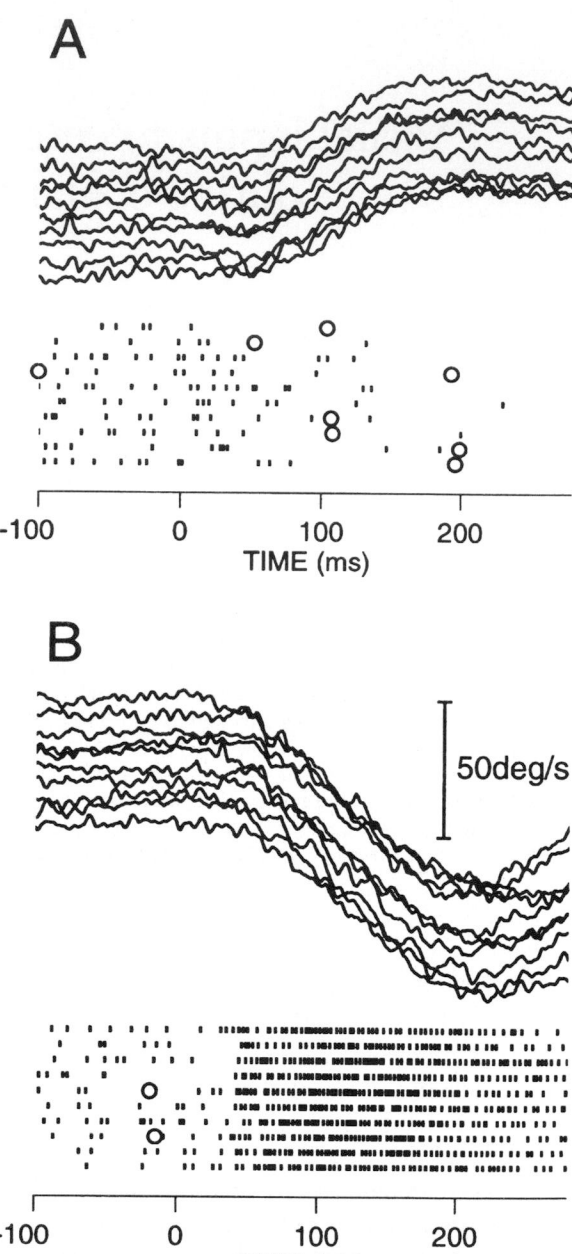

Fig. 1. Sample of ocular responses and activity of a single P-cell to 80°/s upward (A) and downward (B) test ramps. In the upper column, vertical eye velocity was shown. In the lower column, P-cell activity was shown in raster formats. The circles in the raster display indicate complex spike firing and short lines simple spike firing. All data were aligned on the onset of stimulus motion.

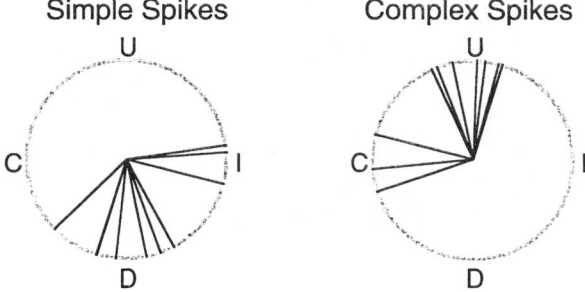

Fig. 2. Comparison of the directional preferences of the simple- and complex-spike activity of ten P-cells.

approved by the Electrotechnical Laboratory Animal Care and Use Committee.

Results

Fig. 1 shows typical ocular following responses to several stimulus presentations of 80°/s upward (A) and downward (B) test ramps. Also shown, in raster form, is the activity of a P-cell in the ventral paraflocculus (VPFL). Each line in the raster represents a single trial and shows the time of occurrence of the P-cell discharges. Two kinds of activities could be observed: low-frequency complex spikes (circles) and high-frequency simple spikes (short vertical tick marks). When the visual scene was moved upward, the simple-spike activity of this P-cell decreased, and it's complex-spike activity increased. On the other hand, when

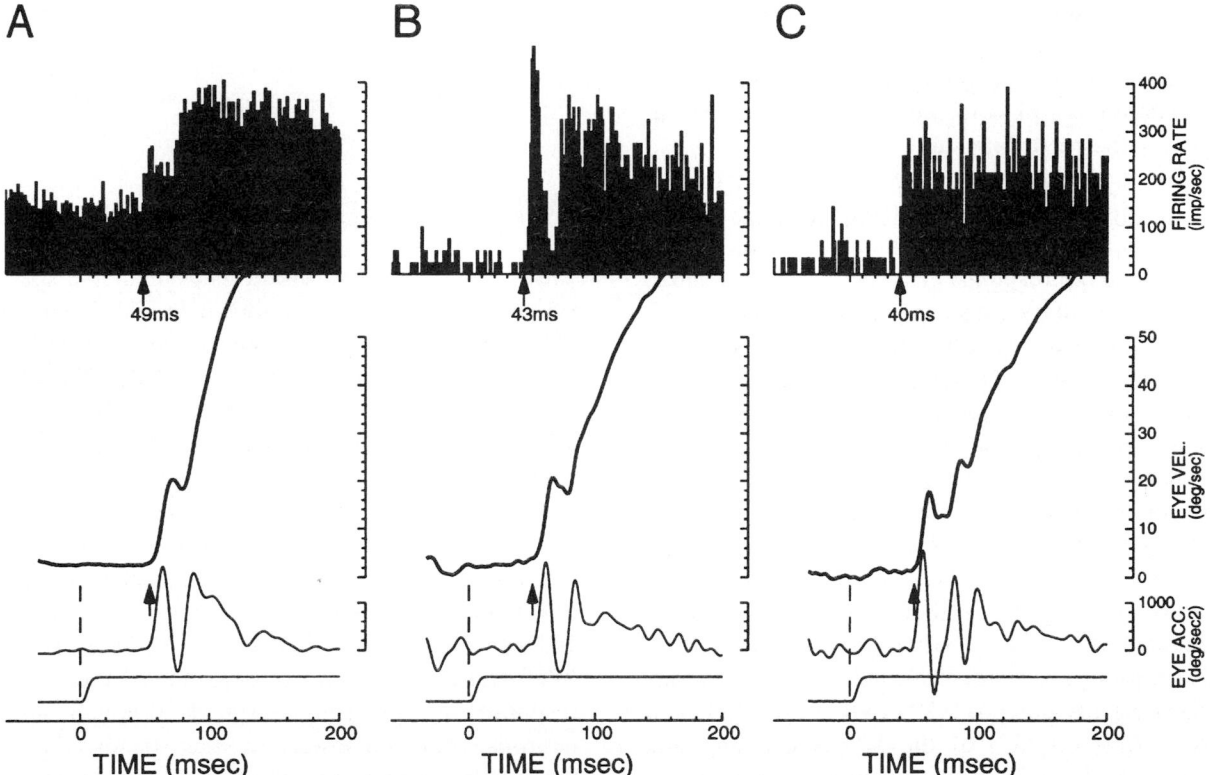

Fig. 3. Comparison of the response properties of simple-spike activity of a P-cell in the VPFL (A), a DLPN neuron (B), and an MST neuron (C) during ocular following (Kawano et al., 1992, 1994; Shidara and Kawano, 1993). The visual scene was moved downward at 160°/s in each case. These particular neurons were selected because all preferred downward motion at high speeds. Traces from top to bottom: peristimulus time histogram, the averaged vertical eye velocity, eye acceleration, and stimulus speed. All traces are aligned on stimulus onset (vertical dashed lines). Arrows indicate the latencies of the neuronal activity and the ocular following responses.

418

the visual scene moved downward, the simple-spike activity increased, and it's complex-spike activity decreased. To determine the effect of directional preference on simple- and complex-spike activity, we studied neuronal responses to a visual scene moving at 80°/s in each of eight directions. For the P-cell shown in Fig. 1, the directional preference of simple-spikes was downward while that of complex-spikes was upward. The directional preferences of ten P-cells, whose complex-spike activity could be clearly discriminated from their simple-spike activity during ocular following, are shown in Fig. 2. The directional preferences of simple- and complex-spike activity for each P-cell were always roughly opposite. In seven P-cells, simple-spikes responded most vigorously to downward motion and complex-spikes to upward motion. In the remaining three P-cells, simple-spikes responded most vigorously to movements directed ipsilateral to the recording site and complex-spikes to motion directed contralateral to the recording site.

To correlate the contribution of visual input from different afferent sources with P-cell simple-spike activity during ocular following, we compared the response properties of simple-spikes with those of neurons in the DLPN and MST. The pattern of visually-evoked activity for three neurons, which responded vigorously to downward movement of the visual scene at 160°/s, are shown in Fig. 3. All of these neurons started to increase their firing frequency before the onset of the ocular responses. The latency of ocular following ranged from 50 to 60 ms. The latency distributions of DLPN and MST neurons are shown in Fig. 4, together with those from the simple-spike activity of P-cells in the VPFL (Kawano et al., 1992, 1994; Shidara and Kawano, 1993). Eighty percent (120/150) of the MST neurons, 73% (56/77) of the DLPN neurons, and 46% (11/24) of the P-cells were activated less than 50 ms after the onset of the stimulus motion. The differences in the response latencies of neurons in these three regions are consistent with previous anatomical reports, suggesting that the MST neurons provide visual information to the

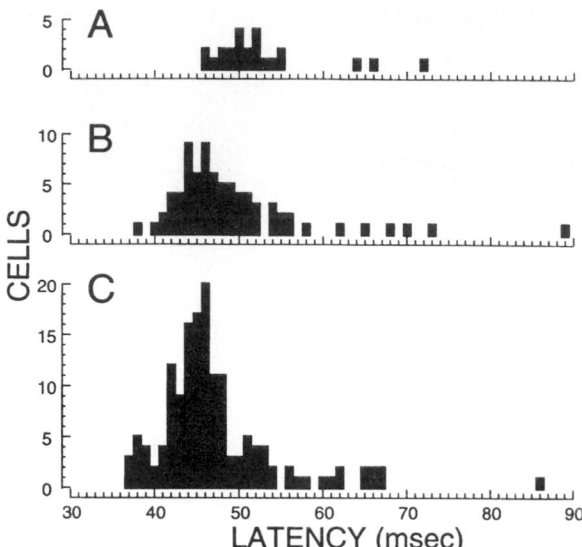

Fig. 4. Comparison of the latencies of activation of simple-spikes of P-cells in the VPFL (A), neurons in DLPN (B), and neurons in MST (C) (Kawano et al., 1992, 1994, 1996; Shidara and Kawano, 1993). Histograms show the latency distributions.

VPFL via the DLPN neurons (Glickstein et al., 1985; Langer et al., 1985; Tusa and Ungerleider, 1988). The directional preferences of simple-spike activity of 37 P-cells are shown in Fig. 5 (upper column) together with those of the neurons in MST and DLPN (Kawano et al., 1992, 1994; Shidara and Kawano, 1993). Unlike P-cells in the VPFL, neurons in MST and DLPN showed a wide range of directional preferences, and anisotropies were minor. The simple-spike activity of 24 P-cells, whose dependence on speed was examined (Fig. 5, lower column), showed their best responses at high stimulus speeds (especially 160°/s) (Shidara and Kawano, 1993). Most of the MST and DLPN neurons also showed their best responses at high stimulus speeds although some neurons preferred lower stimulus speeds (Fig. 5, lower column) (Kawano et al., 1992, 1994). These results indicate that the response properties of neurons in MST and DLPN were quite similar and extended over a broad range of directions and speeds than those of the simple-spikes of P-cells in the VPFL.

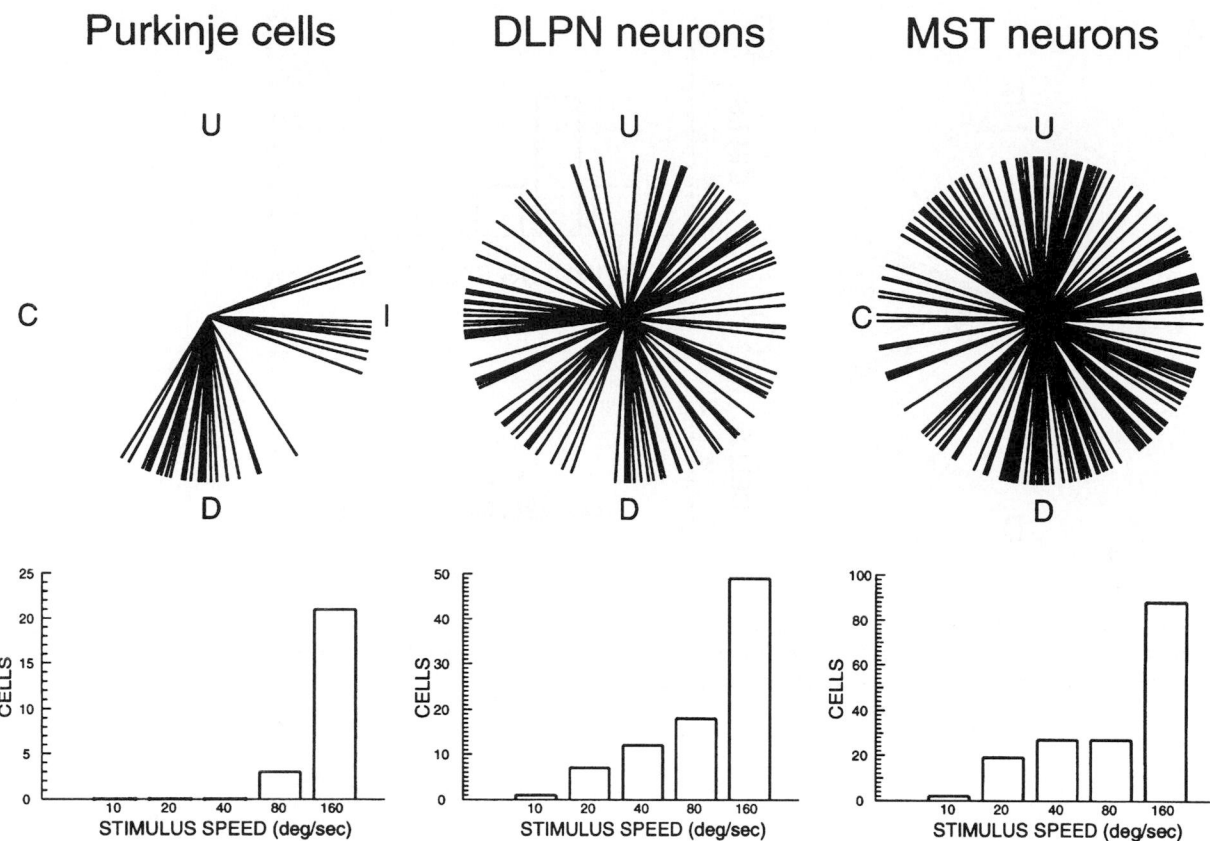

Fig. 5. Comparison of the response properties of neurons in three regions (from left to right: simple-spike activity of VPFL P-cells, DLPN neurons, and MST neurons) (Kawano et al., 1992, 1994, 1996; Shidara and Kawano, 1993). Upper panels, polar plots of the estimated preferred directions of individual neurons. Lower panels, distributions of the preferred speeds of the neurons (stimulus speeds used were 10, 20, 40, 80, and 160°/s).

It is known that complex spikes are generated in P-cells by climbing fiber afferents which originate from the neurons in the inferior olive (Thach, 1968). To understand the origin of the complex-spike activity in the VPFL, we recorded single unit activity in the pretectal nucleus of the optic tract (NOT) because it is known that these neurons project to the inferior olive (Simpson et al., 1988).

Fig. 6 summarizes the response properties of NOT neurons during ocular following (Takemura et al., 1995). All of the NOT neurons preferred ipsiversive movements (Fig. 6A, cf. Mustari and Fuchs, 1990), and, unlike MST or DLPN neurons, NOT neurons tended to prefer relatively slow

stimulus speeds (Fig. 6B). As shown in Fig. 6C, 74% (20/27) of NOT neurons were activated less than 50 ms after the onset of the stimulus motion. On average, response latencies of NOT neurons were longer than those of MST neurons. This result is consistent with the idea that NOT neurons receive visual inputs from MST neurons during ocular following.

Discussion

The results suggest that while both DLPN and NOT neurons convey information concerning movement of the visual scene to the VPFL, they do so via different pathways. Fig. 7 shows the

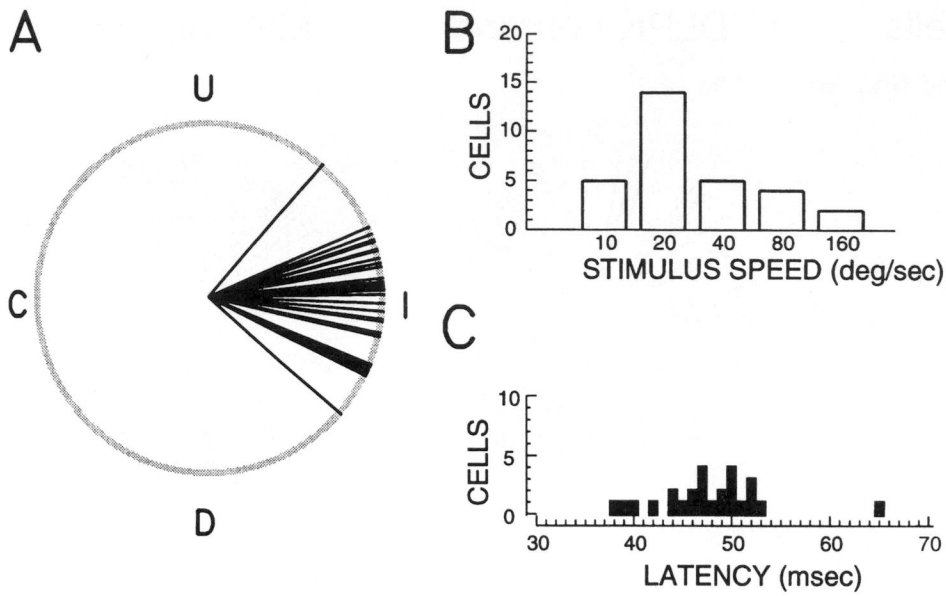

Fig. 6. Response properties of NOT neurons to large-field visual motion that elicited ocular following. (A) distribution of the preferred directions of the individual neurons. (B) Distribution of the preferred speeds of the neurons. (C) Latency distribution of the neurons (the stimulus speeds used for this measure was 80°/s).

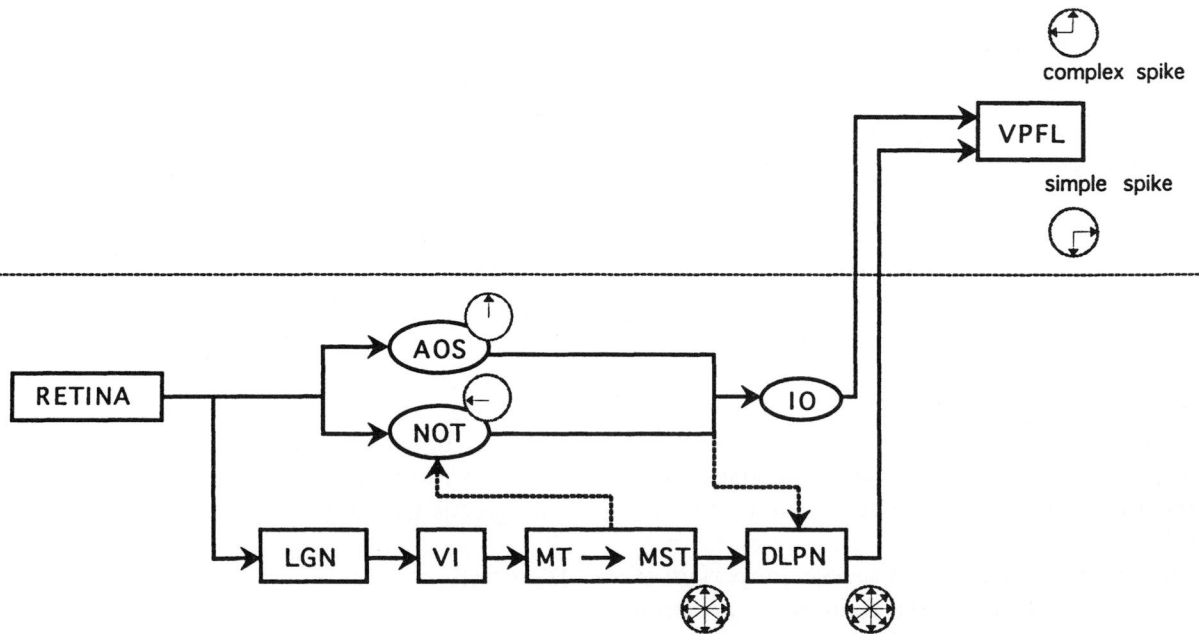

Fig. 7. Putative neural pathways for generating ocular following responses. AOS, accessory optic system; DLPN, dorsolateral pontine nucleus; IO, inferior olive; LGN, lateral genuclate nucleus; NOT, the pretectal nucleus of optic tract; MST, medial superior temporal area; MT, middle temporal area; VPFL, ventral paraflocculus. The broken line indicates the mid-line, and the VPFL is on the right side and the other areas are on the left side. The directional preferences of neurons are indicated inside circles.

putative pathways mediating visual inputs to VPFL. Visual information reaches MT and MST via the retina, LGN, and visual cortex (Maunsell and Newsome, 1987). Neurons in MT and MST analyze the movements of the visual scene and then relay this information to the DLPN and the NOT (Glickstein et al., 1985). The DLPN projects directly to VPFL as mossy fibers (Langer et al., 1985) whereas NOT projects first to the inferior olive and thence to VPFL as climbing fibers (Blank, 1988; Simpson et al., 1988; Mustari, et al., 1994). In P-cells of the VPFL, visual information from two different sources (i.e. the mossy fibers and climbing fibers), is integrated into a motor command to drive the eyes. NOT neurons prefer ipsiversive movements and project via the ipsilateral inferior olive to the contralateral VPFL (Blank, 1988); this is consistent with the finding that complex-spikes with a preference for horizontal motion prefer contraversive movement. For the remaining P-cells, whose complex-spikes show a preference for upward movement, the visual afferents to their climbing fibers may originate from the accessory optic system. This is consistant with the observation that neurons in the lateral terminate nucleus of the accessory optic system also prefer upward movement (Mustari and Fuchs, 1989) and appear to provide input from the accessory optic system to the inferior olive (Simpson et al., 1988).

There are still remaining problems to be resolved. One is the discrepancy between the speed preference of NOT neurons and complex spikes. Our preliminary results reveal that the complex spikes of most P-cells prefer high stimulus speeds (Kobayashi et al., 1995). A systematic study of speed selectivity for both complex spikes and NOT neurons is needed to address this question. Another important issue is to determine why the direction-preference of simple spikes is opposite to that of complex spikes. Because mossy fibers do not show the directional biases observed in MST and DLPN neurons (Kawano and Shidara, 1993), the direction-preference of simple spikes may originate in Purkinje cells themselves

as a result of long-term influences from complex spikes.

Acknowledgements

We thank Dr. F.A. Miles for his comments on this manuscript.

References

Blanks, R.H.I. (1988) Cerebellum. In: J.A. Büttner-Ennever (Ed.), *Neuroanatomy of the Oculomotor System, Reviews of Oculomotor Research*, Vol. 2, Elsevier, Amsterdam, pp. 225–272.

Eccles, J.C., Llinas, R. and Sasaki, K. (1966) Parallel fiber stimulation and the responses induced thereby in the Purkinje cells of the cerebellum. *Exp. Brain Res.*, 1: 17–39.

Fuchs, A.F. and Robinson, D.A. (1966) A method for measuring horizontal and vertical eye movement chronically in the monkey. *J. Appl. Physiol.*, 21: 1068–1070.

Judge, S.J., Richmond, B.J. and Chu, F.C. (1980) Implantation of magnetic search coils for measurement of eye position: an improved method. *Vision Res.*, 20: 535–538.

Glickstein, M., May, J. and Mercer, B.E. (1985) Corticopontine projection in the macaque: the distribution of labeled cortical cells after large injections of horse-radish peroxidase in the pontine nuclei. *J. Comp. Neurol.*, 235: 343–359.

Kawano, K. and Shidara, M. (1993) The role of the ventral paraflocculus in ocular following in the monkey. In: N. Mano, I. Hamada and M. DeLong (Eds.), *Role of the Cerebellum and Basal Ganglia in Voluntary Movement*, Elsevier, pp 189–196.

Kawano, K., Shidara, M. and Yamane, S. (1992) Neural activity in dorsolateral pontine nucleus of alert monkey during ocular following responses. *J. Neurophysiol.*, 67: 680–703.

Kawano, K., Shidara, M., Watanabe, Y. and Yamane, S. (1994) Neural activity in cortical area MST of alert monkey during ocular following responses. *J. Neurophysiol.*, 71: 2305–2324.

Kawano, K., Shidara, M., Takemura, A., Inoue, Y., Gomi, H. and Kawato, M. (1996) Inverse-dynamics representation of eye movement by cerebellar during short-latency ocular following responses. *Ann. NY Acad. Sci.*, 781: 314–321.

Kobayashi, Y., Kawano, K., Takemura, A., Inoue, Y., Kitama, T., Gomi, H. and Kawato, M. (1995) Inverse-dynamics representation of complex spike discharges of Purkinje cells in monkey cerebellar ventral paraflocculus during ocular following responses. *Soc. Neurosci. Abstr.*, 21: 140.

Langer, T., Fuchs, A.F., Scudder, C.A. and Chubb, M.C. (1985) Afferents to the flocculus of the cerebellum in the rhesus macaque as revealed by retrograde transport of horseradish peroxidase. *J. Comp. Neurol.*, 235: 1–25.

Maunsell, J.H.R. and Newsome, W.T. (1987) Visual processing in monkey extrastriate cortex. *Annu. Rev. Neurosci.*, 10: 363–402.

Miles, F.A., Kawano, K. and Optican, L.M. (1986) Short-latency ocular following responses of monkey. I. Dependence on temporospatial properties of visual input. *J. Neurophysiol.*, 56: 1321–1354.

Mustari, M.J. and Fuchs, A.F. (1989) Response properties of single units in the lateral terminal nucleus of the accessory optic system in the behaving primate. *J. Neurophysiol.*, 61: 1208–1220.

Mustari, M.J. and Fuchs, A.F. (1990) Discharge patterns of neurons in the pretectal nucleus of the optic tract (NOT) in the behaving primate. *J. Neurophysiol.*, 64: 77–90.

Mustari, M.J., Fuchs, A.F., Kaneko, C.R.S. and Robinson, F. (1994) Anatomical connections of the primate pretectal nucleus of the optic tract. *J. Comp. Neurol.*, 349: 111–128.

Shidara, M. and Kawano, K. (1993) Role of Purkinje cells in the ventral paraflocculus in short-latency ocular following responses. *Exp. Brain Res.*, 93: 185–195.

Simpson, J.I., Giolli, R.A. and Blanks, R.H.I. (1988) The pretectal nuclear complex and accessory optic system. In: J.A. Büttner-Ennever (Ed.), *Neuroanatomy of the Oculomotor System, Reviews of Oculomotor Research*, Vol. 2, Elsevier, Amsterdam, pp. 335–364.

Stein, J.F. and Glickstein, A.F. (1992) Role of the cerebellum in visual guidance of movement. *Physiol. Rev.*, 72: 967–1017.

Takemura, A., Inoue, Y., Kawano, K. and Mustari, M.J. (1995) Ocular following response related neuronal activity in the nucleus of the optic tract of monkey. *Jpn. J. Physiol.*, in press.

Thach, W.T. (1968) Discharge of Purkinje and cerebellar nuclear neurons during rapidly alternating arm movements in the monkey. *J. Neurophysiol.*, 31: 785–797.

Tusa, R.J. and Ungerleider, L.G. (1988) Fiber pathways of cortical areas mediating smooth pursuit eye movements in monkeys. *Ann. Neurol.*, 23: 174–183.

M. Norita, T. Bando and B. Stein (Eds.)
Progress in Brain Research, Vol 112
© 1996 Elsevier Science BV. All rights reserved.

CHAPTER 31

Context dependent discharge characteristics of saccade-related Purkinje cells in the cerebellar hemispheres of the monkey

N. Mano*, Y. Ito and H. Shibutani

Department of Neurophysiology, Tokyo Metropolitan Institute for Neuroscience, Musashidai 2-6, Fuchu-shi, Tokyo 183, Japan

In the previous paper (Mano et al, 1991), we reported the discharge patterns of saccade-related Purkinje cells during visually guided saccade task, which were recorded from posterior cerebellar hemisphere, the Crus IIa. In the present study, we analysed these P-cell's simple spike activity during the spontaneous saccade in inter-trial intervals (ITI) of visually guided saccade task, comparing with the activity during the visually guided saccade. We found that the modulation of simple spike discharges during spontaneous saccade was weaker than the modulation during the visually triggered saccade.

We recorded single unit discharges of Purkinje cells from cerebellar posterior hemisphere (Crus IIa) in awake Japanese monkeys (*Macaca fuscata*), trained to perform simple reaction time saccade task gazing at a small light rear-projected on to a tangent screen 54 cm in front of the monkey. Horizontal and vertical eye positions were measured by a corneal search coil method. Comparison of simple spike activity associated with spontaneous saccade during ITI to the activity during visually triggered saccade clarified that the discharge patterns of simple spikes are basically the same during both types of the saccades, but the amount of the phasic modulation (increase or decrease of discharge rate) were larger for all directions (up, down, left and right) during visually guided saccade than that during spontaneous saccade in all saccade-related Purkinje cells so far examined in two monkeys. The modulation, however, cannot be assumed to have been induced by the visual stimulus per se. Because, the maximum increase of simple spike discharge rate aligned at saccade onset is larger than that aligned at target jump. And, the half width of the change was wider when aligned at target light jump than when aligned at the onset of saccades, in all the four directions, indicating the changes of the firing rate were more time-locked to the onset of saccadic eye movements than to the triggering visual stimulus.

The present findings suggest that the cerebellar hemisphere plays a more important role in the control of externally triggered voluntary eye movements than in the control of self-initiated, self-paced eye movements. We discussed these findings combining with previous findings on limb movement-related P-cells (Mano et al, 1980, 1986, 1989), from view point of the general role of the cerebellar hemisphere in the control of voluntary movements.

Introduction

It is well established that the cerebellar hemisphere plays an important role in the control of voluntary limb movements, based on the clinico-pathological studies in humans (Holmes, 1917; Gilman et al., 1981), and on the lesion or recording studies in experimental animals (Dow and Moruzzi, 1958; Ito, 1984).

On the other hand, the cerebellar hemisphere has been forgotten for nearly two decades since Ron and Robinson (1973) reported the effects of electrical stimulation of posterior lobes in evoking saccadic and smooth eye movements in the monkey. During the studies on the role of cerebellar hemispheres in controlling voluntary wrist

---
*Corresponding author. Tel.: +81 3 0423 253881, ext. 4207; fax: +81 3 0423 218678; e-mail: manonori@truin.ac.jp

movements (Mano and Yamamoto, 1980; Mano et al., 1986, 1989), we found many Purkinje cells (P-cells) in the posterior lobe of the cerebellar hemisphere, whose unitary activities changed, closely time-locked to spontaneous saccadic eye movements. We reported the localization and discharge patterns of these 'saccade-related P-cells' in the posterior cerebellar hemisphere during visually guided saccades (Mano et al., 1991).

In this paper, we will report that the saccade-related P-cells change their simple-spike activity during spontaneous saccade with similar pattern as that during visually guided saccade, but, quantitatively, the change is weaker than that during visually triggered saccade. We will also report the results of the quantitative analysis of the discharge characteristics about time-locking and directionality.

Finally, from the point of view of the general role of the cerebellar hemisphere in controlling the voluntary movements, we will discuss the common and different discharge characteristics of eye movement- and limb movement-related P-cells, combining the results of the present and previous studies (Mano and Yamamoto, 1980; Mano et al., 1986, 1989; Mano et al., 1991).

Methods

Details were described in the previous paper (Mano et al., 1991). In short, two Japanese adult monkeys (*Macaca fuscata*) were trained to perform visually guided eye movements by looking at a small light projected on the tangent screen 57 cm in front of the monkeys (Wurtz, 1969; Wurtz and Goldberg, 1972). To begin the task, when a small red light came on at the center of the screen, the monkey had to press a key by hand and to gaze at the light. After his gaze fixation on the light for 1 s, it jumped 20°, or began to move with constant velocity of 10°/s, to the peripheral direction of up, down, left or right in a random fashion. When the monkey kept his gaze on the jumped or moving light for 1 s, the light changed its color from red to green. Then the monkey had

to release the pressed key to receive a drop of fruit juice as a reward.

The animal was seated in a primate chair. The head was firmly fixed to the chair by a fixation pedestal attached on the skull by small screws and dental resin, and was located at the center of magnetic fields (Robinson, 1963; Fuchs and Robinson, 1966). Discharges of single P-cells were recorded by an elgiloy microelectrode (Suzuki and Azuma, 1976), which was advanced by hydraulic X-Y coordinate microdrive into the cerebellar cortex through occipital lobe of the cerebrum 20–30° obliquely from sagittal plane. P-cells were identified by the presence of two kinds of spikes; simple (SS) and complex spikes (CS) (Mano et al., 1989). Horizontal and vertical eye positions were recorded through a scleral search coil chronically implanted on the eye (Judge et al., 1980).

During recording sessions, discharges of P-cells were monitored on oscilloscope and by audio-monitor. The original spike discharges and other event signals were recorded simultaneously on magnetic tape by 14 channel data recorder for later offline analysis. After recording sessions, these data were analyzed by a microcomputer. We used the off-line analysis program DSYS developed by H. Suzuki for the analysis of spontaneous saccades during inter-trial intervals and visually guided saccades.

After the recording sessions, the animals were sacrificed under deep anesthesia (Pentobarbital 50 mg/kg i.p.), perfused transcardially with saline and then 10% formalin containing 2% ferrocyanide. Marking lesions were stained blue by this additive (Suzuki and Azuma, 1976; Mano et al., 1989).

Results

Somatotopy of the cerebellar hemisphere

Eye movement-related P-cells were found within a restricted area of posterior cerebellar cortex (Fig. 1) where we could hear by audio-monitor

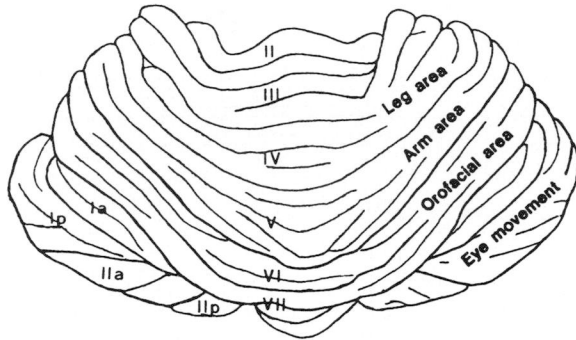

Fig. 1. Somatotopy of the cerebellar hemisphere (dorsal view) observed by single unit activity in alert monkeys. Top, rostral; bottom, caudal.

the phasic modulation of background multiple unit discharges well time-locked to spontaneous saccades (so-called saccade swishes). The P-cells isolated in this area changed their simple spike discharge rate well related to both spontaneous and visually triggered saccades. But, they did not change their discharge rate during smooth pursuit eye movements, nor did they respond to the on or off of target light per se, and their discharge rate change was not related to the activities of neck EMGs. We termed these P-cells as 'saccade-related P-cells' (Mano et al., 1991).

The majority of saccade-related P-cells were located in crus IIa, and some P-cells in the deep layer of crus I. Combining the results of the present studies on the eye movements with the observations during the studies on the wrist movement (Mano and Yamamoto, 1980; Mano et al., 1986, 1989), Fig. 1 summarizes the somatotopy of the cerebellar hemisphere based on single P-cell activity in alert monkeys related to leg, arm, oro-facial and eye movements.

Time locking to the saccadic eye movement

Many of the saccade-related P-cells changed their discharge rate 20–80 ms prior to the onset of saccades, as we reported before (Mano et al., 1991). To know whether the discharge rate change preceding the saccade onset was related to the saccadic eye movement per se or the change was evoked by the visual stimulus of target light jump, we compared the maximum change of simple-spike discharges and the half-width of the phasic modulation on the histogram of the discharges aligned at the onset of the target jump from center to the periphery (Fig. 2A) and at the onset of saccadic eye movements (Fig. 2B). The maximum frequency change was calculated by the difference between the peak frequency during the saccade and the control frequency 500 ms prior to the visual target jump on the histogram. As shown by the sample P-cell in Fig. 2, the maximum increase (134.3 simple spikes/s) was larger in the saccade-onset aligned histogram (Fig. 2B) than that (105.7 simple spikes/s) in the target jump aligned-histogram (Fig. 2A). And the half width of the increase (the width of the histogram at the half increase of the discharge rate) was narrower in the former (Fig. 2B: 80 ms) than in the latter (Fig. 2A: 100 ms).

The pattern of the phasic discharge rate modulation was qualitatively the same in either directions of the saccade task: left, right, up and down, as we reported in the previous paper (see Fig. 3 in Mano et al., 1991). Quantitative analysis of the histograms as illustrated in Fig. 2 clarified that the larger change and narrower half-width in the saccade onset-aligned histogram (Fig. 2B) than those in the target-aligned histogram (Fig. 2A) are also true in all four directions of the saccade (Fig. 3). The average value of the maximum increases for four directions was 134.2 simple spikes/s for the saccade onset aligned and 100.6 for the target jump aligned histograms. Further, this result was also true for all saccade-related P-cells so far analysed: ten P-cells which increased simple spike discharges for all directions; and three P-cells which decreased the discharges.

Thus, we may conclude that the phasic discharge rate modulation during the visually guided saccade is more related to the saccadic eye movement than to the visual stimulus of the target light jump. We recently found a few P-cells which showed short latency response to the visual stimu-

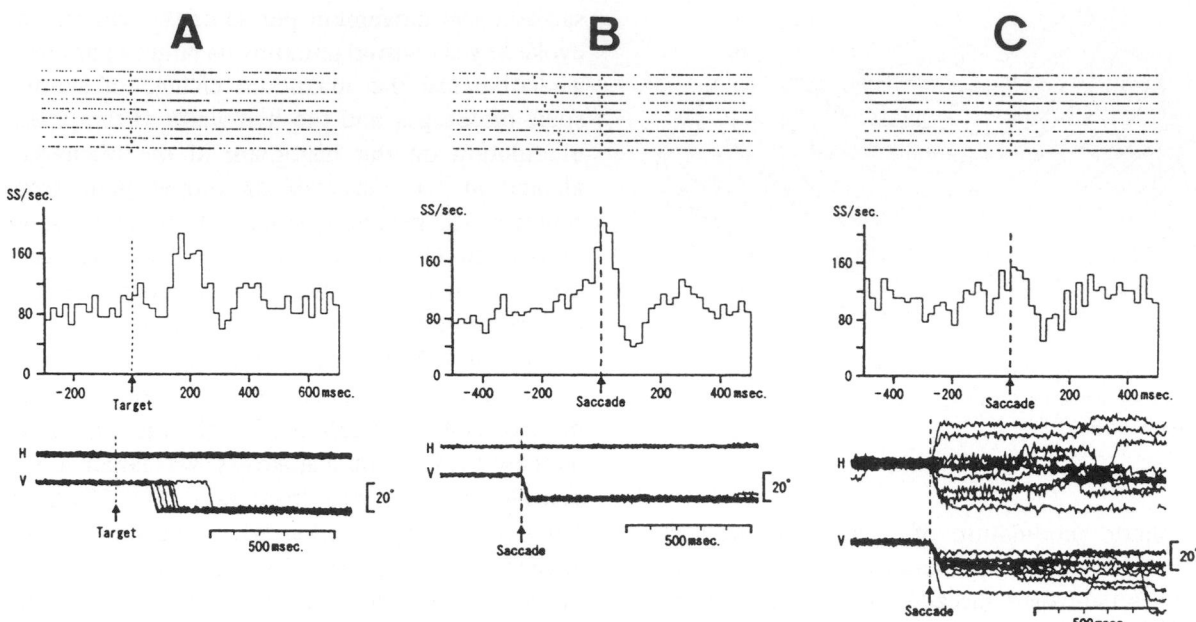

Fig. 2. Discharges of a saccade-related P-cell during visually guided and spontaneous saccades. The raster and histogram of simple spike discharges and eye positions during six trials are illustrated aligned at the jump of target light (A), at the onset of visually guided saccade (B) and at the onset of spontaneous saccade (C). In the raster display, each dot represents a simple spike, and the bars in (B) about 150–220 ms before the saccade represent the moments of target light jumps from center to 20° downward. H, horizontal eye positions (up, leftward eye movement); V, vertical eye positions (up, upward eye movement). Time scale is same for the rasters and histograms. The minus sign of the time scale represents the time prior to the event used for the alignment.

lus in addition to the saccade-related discharge rate change. These cells, however, were located in the more posterior area different from the Crus IIa (unpublished observation). Therefore, the above described conclusion may safely be confined to the saccade-related P-cells in the Crus IIa and Crus I rather than to be generalized to the cerebellar hemisphere.

Non-reciprocal discharge in the opposite directions

In a given P-cell recorded from the hemisphere, the discharge patterns during saccades were qualitatively the same in all directions (see Fig. 3 in Mano et al., 1991). Quantitatively, however, there were some differences in the amount of modulation depending on the direction (Fig. 3A), namely, there was a preferred direction, but no P-cell showed reciprocal discharge pattern, i.e. all P-cells changed their discharge rate in the same direction (increase or decrease) during saccades to opposite directions of eye movements, up vs.

down, left vs. right.

Visually guided vs. spontaneous saccade

To compare the P-cell discharges during the spontaneous saccade to those during visually triggered saccade, we collected the spontaneous saccades during the inter-trial intervals aligned at the saccade onset (Fig. 2C). In the strict sense, the P-cell activity during the spontaneous saccade should be compared to that during the visually triggered saccade, by collecting the spontaneous saccade of the same amplitude and of the same directions as the trained saccade. But, the amplitude was on average about 20°, and because the directionality of the discharge pattern during the triggered saccades to four directions were similar (Fig. 3) and the background discharges did not significantly differ depending on the eye positions (see Fig. 3 in Mano et al., 1991), we might be justified to compare them (Fig. 2C vs. Fig. 2B) as a first approximation.

The maximum change (72.6 simple spikes/s) obtained from the histogram aligned at the onset of the spontaneous saccade (Fig. 2C) was smaller than that (134.3 simple spikes/s) aligned at the onset of visually triggered saccades (Fig. 2B). This was true for all four directions of the saccade Fig. 4A: the average value for the four directions of the spontaneous saccade being 77.5 simple spikes/s versus 134.2 simple spikes/s. This was also true for other P-cells so far analyzed (Fig. 4B).

The present results as shown in Fig. 4 presumably suggest that the saccade-related P-cells in the hemisphere are more concerned with the visually guided eye movement than with the self-paced spontaneous saccade.

Discussion

Common and different discharge characteristics during eye and wrist movements

Present series of our studies (Mano et al., 1991;

Mano et al., 1993) clarified that the P-cells related to voluntary eye movements in the cerebellar hemisphere share several discharge characteristics with the P-cells related to voluntary wrist movements (Mano and Yamamoto, 1980; Mano et al., 1986, 1989). First, P-cells located in the lateral hemisphere begin to change their discharge frequencies prior to the onset of EMG activity, in both limb and eye movements. The change was more time-locked to the onset of movements than to the visual trigger stimulus. This suggests that the lateral cerebellar hemisphere of the posterior lobe play a role in the initiation or preprogramming of voluntary eye movements, similarly like the lateral hemisphere of the anterior lobe in voluntary limb movements. Second, P-cells change their activity phasically in relation to the dynamic phases of both limb and eye movements, and the tonic activity of P-cells is poorly or not at all correlated to maintained position of either limb or eye. Third, the phasic activity is non-reciprocal or bi-directional for the opposite directions of both limb and eye move-

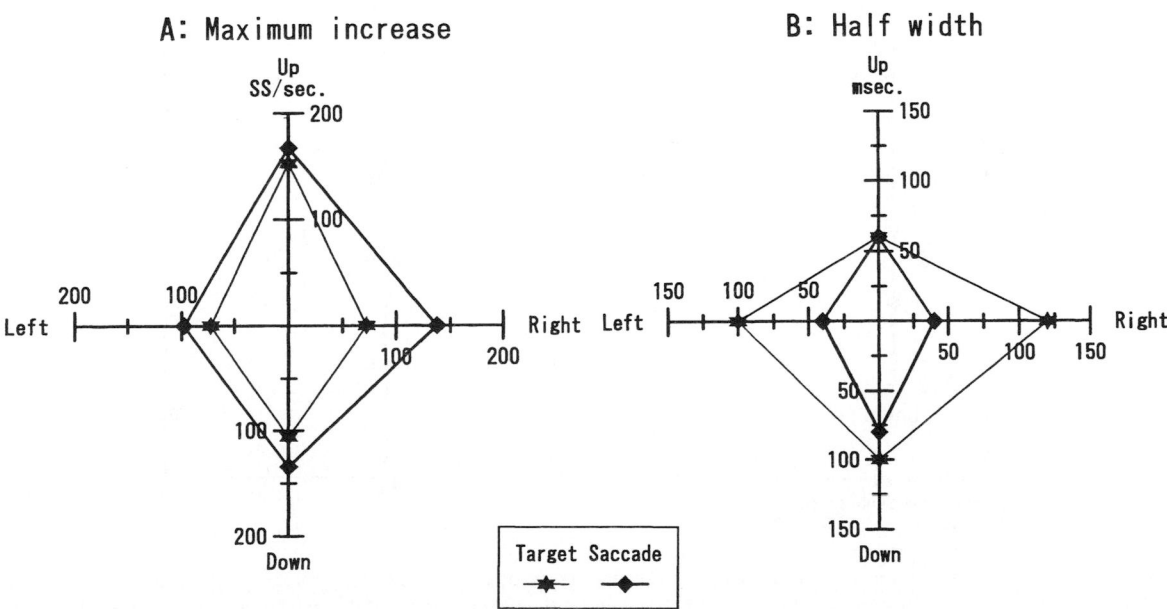

Fig. 3. Comparison of the maximum increase (A) and half width (B) of simple spike discharges between target aligned and saccade onset aligned histograms during visually guided saccade task toward the four directions (up, down, left and right). SS, simple spike. The same P-cell as illustrated in Fig. 2.

ments. This is a quite different feature from the reciprocal or unidirectional discharges of saccade-related neurons in the oculomotor nucleus (Fuchs and Luschei, 1970; Robinson, 1970), pontine reticular formation (Sparks and Travis, 1971; Luschei and Fuchs, 1972), nucleus tegmenti pontis (Crandall and Keller, 1985), superior colliculus (Wurtz and Goldberg, 1972; Sparks, 1978), and fastigial oculomotor regions of the cerebellum (Ohtsuka and Noda, 1990, 1991). This suggests that the cerebellar hemisphere processes the information of higher motor function than simple motor execution. Finally, this suggestion is further supported by another common characteristics that the cerebellar hemisphere is more active

with externally guided voluntary movements than with self-paced movements in both eye and limb movements (Mano et al., 1986, 1989).

A clear difference in the discharge characteristics of wrist and eye movement-related P-cells was that wrist-related P-cells changed their activity with both rapid and slow movements (Mano and Yamamoto, 1980), whereas eye-movement-related P-cells located in the cerebellar hemisphere change activity only with rapid (saccadic) eye movements, but not with slow pursuit eye movements (Mano et al., 1991).

Thus, there are both common and different discharge characteristics for the control of voluntary limb and eye movements. One has to take

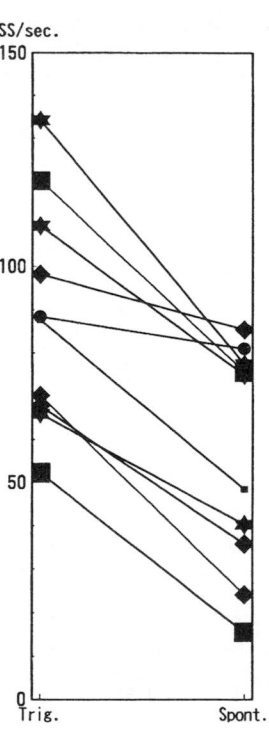

Fig. 4. Comparison of the maximum increase of simple spike discharges of a P-cell during visually triggered and spontaneous saccade (A). The same P-cell as illustrated in Figs. 2 and 3. B compares the average value of the maximum increases for ten P-cells during visually triggered and spontaneous saccades. Each point represents an average value in each P-cell of the increases during the saccade to four directions. Other conventions are the same as in Fig. 3.

into consideration these two aspects in interpreting the function of a motor center controlling the voluntary movements.

Visually triggered versus self-initiated voluntary movement

As for the control mechanism of eye movements, there have been many reports on the different neuronal activity depending on the context of movements. In the frontal eye field (FEF), the neuronal activity did not precede the onset of spontaneous saccade, and FEF was interpreted not to be involved in the initiation of a voluntary eye movement (Bizzi, 1967, 1968). Later, it was found that FEF discharges prior to the onset of visually triggered saccade, but the discharge prior to the spontaneous saccade was very weak or absent (Bruce and Goldberg, 1985). On the other hand, the supplementary eye field (SEF) was found where neuronal discharge clearly preceded the onset of spontaneous saccade (Schlag and Schlag-Rey, 1985).

In previous studies on the wrist movements (Mano et al., 1986), we reported that the complex spike discharged during visually guided but not during self-paced wrist movements, and the simple spike discharges changed more vigorously during the visually triggered movements. Although we have not yet analyzed the complex spike discharges during the saccade, the simple spike modulation during self-initiated saccade was weaker than the modulation during visually triggered saccade (Figs. 2B,C; Fig. 4). This result is similar to the report (Ohtsuka and Noda, 1992) that the fastigial oculomotor neurons discharged vigorously during vision- and memory-guided saccades but discharged very weakly during spontaneous saccades.

These findings suggest that the cerebellum plays a context-dependent role in the control of voluntary eye and limb movements, a major role in controlling externally guided movements and a minor role in the self-paced movements.

References

Bizzi, E. (1967) Discharge of frontal eye field neurons during eye movements in unanesthetized monkeys. *Science*, 157: 1588–1590.

Bizzi, E. (1968) Discharge of frontal eye field neurons during saccadic and following eye movements in unanesthetized monkeys. *Exp. Brain Res.*, 6: 69–80.

Bruce, C.J. and Goldberg, M.E. (1985) Primate frontal eye fields. I. Single neurons discharging before saccades. *J. Neurophysiol.*, 53: 603–635.

Crandall, W.F. and Keller, E.L. (1985) Visual and oculomotor signals in nucleus reticularis tegmenti pontis in alert monkey. *J. Neurophysiol.*, 54: 1326–1345.

Dow, R.S. and Moruzzi, G. (1958) *The Physiology and Pathology of the Cerebellum*, The University of Minnesota Press, Minneapolis.

Fuchs, A.F. and Luschei, E.S. (1970) Firing patterns of abducens neurons of alert monkeys in relationship to horizontal eye movement. *J. Neurophysiol.*, 33: 382–392.

Fuchs, A.F. and Robinson, D.A. (1966) A method for measuring horizontal and vertical eye movement chronically in the monkey. *J. Appl. Physiol.*, 21: 1068–1070.

Gilman, S., Bloedel, J.R. and Lechtenberg, R. (1981) Disorders of the Cerebellum. *Contemp. Neurol. Ser.*, 21: 1–415.

Holmes, G. (1917) The symptoms of acute cerebellar injuries due to gunshot injuries. *Brain*, 40: 461–535.

Ito, M. (1984) *The Cerebellum and Neural Control*, Raven Press, New York.

Judge, S.J., Richmond, B.J. and Chu, C. (1980) Implantation of magnetic search coils for measurement of eye position: an improved method. *Vision Res.*, 20: 535–538.

Luschei, E.S. and Fuchs, A.F. (1972) Activity of brain stem neurons during eye movements of alert monkeys. *J. Neurophysiol.*, 35: 455–461.

Mano, N. and Yamamoto, K. (1980) Simple-spike activity of cerebellar Purkinje cells related to visually guided wrist tracking movements in the monkey. *J. Neurophysiol.*, 43: 713–728.

Mano, N., Kanazawa, I. and Yamamoto, K. (1986) Complex-spike activity of cerebellar Purkinje cells related to wrist tracking movement in the monkey. *J. Neurophysiol.*, 56: 137–158.

Mano, N., Kanazawa, I. and Yamamoto, K. (1989) Voluntary movements and complex-spike discharges of cerebellar Purkinje cells. *Exp. Brain Res. Ser.*, 17: 265–280.

Mano, N., Ito, Y. and Shibutani, H. (1991) Saccade-related Purkinje cells in the cerebellar hemispheres of the monkey. *Exp. Brain Res.*, 84: 465–470.

Mano, N., Kanazawa, Y., Ito, Y. and Shibutani, H. (1993) Discharge characteristics of saccade- versus wrist-related Purkinje cells in the cerebellar hemispheres of the monkey. *Excerpta Med. Int. Congr. Ser.*, 1024: 183–193.

430

Ohtsuka, K. and Noda, H. (1990) Direction-selective saccade-burst neurons in the fastigial oculomotor region of the macaque. *Exp. Brain Res.*, 81: 659–662.

Ohtsuka, K. and Noda, H. (1991) Saccadic burst neurons in the oculomotor region of the fastigial nucleus of macaque monkeys. *J. Neurophysiol.*, 65: 1422–1434.

Ohtsuka, K., and Noda, H. (1992) Burst discharges of fastigial neurons in macaque monkeys are driven by vision- and memory-guided saccades but not by spontaneous saccades. *Neurosci. Res.*, 15: 224–228.

Robinson, D.A. (1963) A method of measuring eye movement using scleral search coil in a magnetic field. *IEEE Trans. Biomed. Electron.*, 40: 137–145.

Robinson, D.A. (1970) Oculomotor unit behavior in the monkey. *J. Neurophysiol.*, 33: 393–404.

Ron, S. and Robinson, D.A. (1973) Eye movements evoked by cerebellar stimulation in the alert monkey. *J. Neurophysiol.*, 36: 1004–1022.

Schlag, J. and Schlag-Rey, M. (1985) Unit activity related to spontaneous saccades in frontal dorsomedial cortex of monkey. *Exp. Brain Res.*, 58: 208–211.

Sparks, D.L. (1978) Functional properties of neurons in the monkey superior colliculus: coupling of neuronal activity and saccade onset. *Brain Res.*, 156: 1–16.

Sparks, D.L. and Travis, R.P. (1971) Firing patterns of reticular formation neurons during horizontal eye movements. *Brain Res.*, 33: 477–481.

Suzuki, H. and Azuma, M. (1976) A glass-insulated 'elgiloy' microelectrode for recording unit activity in chronic monkey experiments. *Electroenceph. Clin. Neurophysiol.*, 41: 93–95.

Wurtz, R.H. and Goldberg, M.E. (1972) Activity of superior colliculus in behaving monkey. III. Cells discharging before eye movements. *J. Neurophysiol.* 35: 575–586.

M. Norita, T. Bando and B. Stein (Eds.)
Progress in Brain Research, Vol 112
© 1996 Elsevier Science BV. All rights reserved.

CHAPTER 32

Further evidence for the specific involvement of the flocculus in the vertical vestibulo-ocular reflex (VOR)

Kikuro Fukushima*, Shinki Chin, Junko Fukushima, Masaki Tanaka and Sergei Kurkin

Department of Physiology, Hokkaido University School of Medicine, West 7, North 15, Sapporo 060, Japan

We examined the simple-spike activity of floccular Purkinje (P) cells during sinusoidal pitch rotation and vertical optokinetic stimuli in alert, head-fixed cats. The great majority of pitch-responding P cells also responded to optokinetic stimuli with increased activity when the directions of the resultant eye movements were the same. During rapid modification of the VOR induced by visual pattern movement, modulation amplitudes of the cells tested increased together with the eye velocity increase. Maximal activation directions of these cells studied during vertical rotation in many planes were near the vertical canal planes. These results suggest that the activity of the majority of pitch-responding P cells contains a vertical eye velocity component during vestibular or optokinetic stimuli in addition to canal inputs during pitch rotation.

Introduction

In order to maintain stable visual perception, the brain uses several eye movement subsystems (Robinson, 1981). Among them, the vestibulo-ocular reflex (VOR) detects semi-circular canal- and otolith-signals and converts them into appropriate eye muscle responses for stabilization of retinal images (see Robinson, 1981 for review). The vestibular end-organs code head movement as a signal close to head velocity so that the vestibular command must be converted neurally into an eye position signal. The major portion of the temporal conversion is performed by Robinson's common integrator which integrates (mathematically) velocity to position not only for vestibular signals, but also for other signals for conjugate eye movements (Robinson, 1981). Temporal conversion of vertical semi-circular canal signals in the VOR requires vestibular commissural connections related to the vertical canals (Shimazu and Precht, 1966; Kasahara and Uchino, 1974; Uchino et al., 1986) and the interstitial nucleus of Cajal (Anderson et al., 1979; Anderson, 1981; Fukushima, 1987, 1991 for review). However, these circuits alone are still insufficient to explain the conversion completely (Fukushima et al., 1992 for review).

The involvement of the cerebellum, particularly the flocculus in this signal conversion process has been suggested in cats, rabbits and chinchilla (Carpenter, 1972; Robinson, 1974; Hassul et al., 1976; Precht and Anderson, 1979; Ito et al., 1982; Flandrin et al., 1983; Godaux and Vanderkelen,

*Corresponding author. Tel.: +81 11 7065038, ext 5039; fax: +81 11 7065041; e-mail:kikuro@med.hokudai.ac.jp

1984; see Ito, 1984 for review). In contrast, results obtained in alert monkeys suggest that the flocculus does not play a major role in the VOR itself (Lisberger and Fuchs, 1978; Miles et al., 1980; Zee et al., 1981; Lisberger et al., 1994), although it does in it's plasticity and modification by visual inputs (see Ito, 1984 for review; also Partsalis et al., 1995; Zhang et al., 1995).

In alert cats, we have shown that the flocculus contains many Purkinje (P) cells that respond to pitch rotation during the vertical VOR and that the majority of them receive excitation from the contralateral posterior canal. Since chemical deactivation of these areas specifically impairs downward VOR during up-pitch rotation, we suggested that these P cells are involved in the temporal conversion of posterior canal signals in the VOR (Fukushima et al., 1993, 1994; Fukushima and Kaneko, 1995). In order to understand how the flocculus could be involved in the signal conversion in the vertical VOR, we examined properties of simple-spike activities of floccular P cells to sinusoidal pitch rotation and vertical optokinetic stimuli.

Methods

Alert cats were used. The animal's head was attached painlessly to the stereotaxic apparatus and pitched 21° nose down from the stereotaxic horizontal plane to bring the horizontal canals in the earth horizontal plane and vertical canals close to the earth vertical plane (Blanks et al., 1972). The inter-aural midpoint of the animal's head was brought close to the axis of vertical as well as horizontal rotation. The cat's eyes were positioned near the center of the cubic induction coil frame (60-cm side length), and a tangent screen was attached to the apparatus in the frontal plane that subtended 90° by 90° of visual angle. An optokinetic pattern (5° wide, horizontally oriented black stripes separated by 10°) together with a red laser spot (0.7° in diameter) were back-projected on the screen using a video-projector, and both were moved sinusoidally vertically while the turn-table was stationary. Occasio-

nally, these stimuli were presented singly, by turning one off, or the optokinetic pattern was moved while the laser spot remained stationary. Recordings were performed in a dimly lit room. Single-unit activity was recorded extracellularly in the flocculus and later confirmed histologically. To examine cell properties further, responses to pitch rotation (0.5 Hz $\pm 5°$) were also examined for some cells while vertical optokinetic stimuli were presented at the same frequency and amplitude either in phase (VOR suppression condition) or 180° out of phase with the turntable (VOR enhancement condition), similar to the method used for visual-vestibular interaction in the horizontal VOR (see Ito, 1984 for review). Activity of some cells was also studied during sinusoidal pitch rotation (0.5 Hz $\pm 5°$) without visual stimuli at several yaw positions in order to examine maximal activation directions (MAD, Baker et al., 1984). MAD was defined as the orientation angle at which maximum gain was obtained. Gain was calculated from the best-fit, sinusoidal gain curve as in previous studies (Fukushima et al., 1993, 1995).

Eye-coil signals were calibrated by sinusoidal whole-body rotation (0.2 Hz, 10° peak-to-peak amplitude) of the cat first in the pitch plane and then in the horizontal plane. All data were analyzed as previously described (Fukushima et al., 1995).

Results

Since the purpose of our study was to understand how the cerebellar flocculus is involved in the processing of semicircular canal signals in the vertical VOR, we searched for P cells whose simple-spike activity was modulated during the vertical VOR induced by pitch rotation (0.5 Hz $\pm$ 5°).

Comparison of pitch and vertical optokinetic stimuli

Of the 52 P cells whose activity was modulated by pitch rotation in a dimly lit room, 36 (69%) increased activity during up-pitch, while the remaining 16 (31%) increased activity during down-pitch.

Of the 52, activity of 34 P cells was also examined during vertical optokinetic stimuli. None, except one, of 34 cells showed eye position related activity during inter-saccadic intervals.

Thirty of the 34 P cells (88%) also responded to vertical optokinetic stimuli. The great majority of these cells (24/30 = 80%; 13 up- and 11 down-pitch P cells) discharged to pitch and optokinetic stimuli as shown in Fig. 1 for two representative cells. The cell shown in Figs. 1A,B was excited during up-pitch (Fig. 1A), and increased its activity during downward optokinetic stimulation (Fig. 1B); whereas the cell shown in Figs. 1C,D exhibited a down-pitch response (Fig. 1C), and its activity increased during upward optokinetic stimuli (Fig. 1D). The response magnitudes of these cells to pitch and optokinetic stimuli were similar, although eye movement responses were smaller during optokinetic stimuli (note the smaller calibration bars for the eye traces, $\dot{V}$. All of the 13 up-pitch P cells increased activity during downward optokinetic stimuli and all of the 11 down-pitch P cells increased activity during up-optokinetic stimuli with their response phases near the stimulus velocity. The distributions of response gains of up- and down-pitch P cells were similar with overall mean gain of 1.8 spikes/s per degree/s.

Figs. 2A,B summarize gain and phase responses of a representative P cell during optokinetic stimulation using constant-amplitude stimulation ($\pm5°$). Simultaneously recorded eye movement

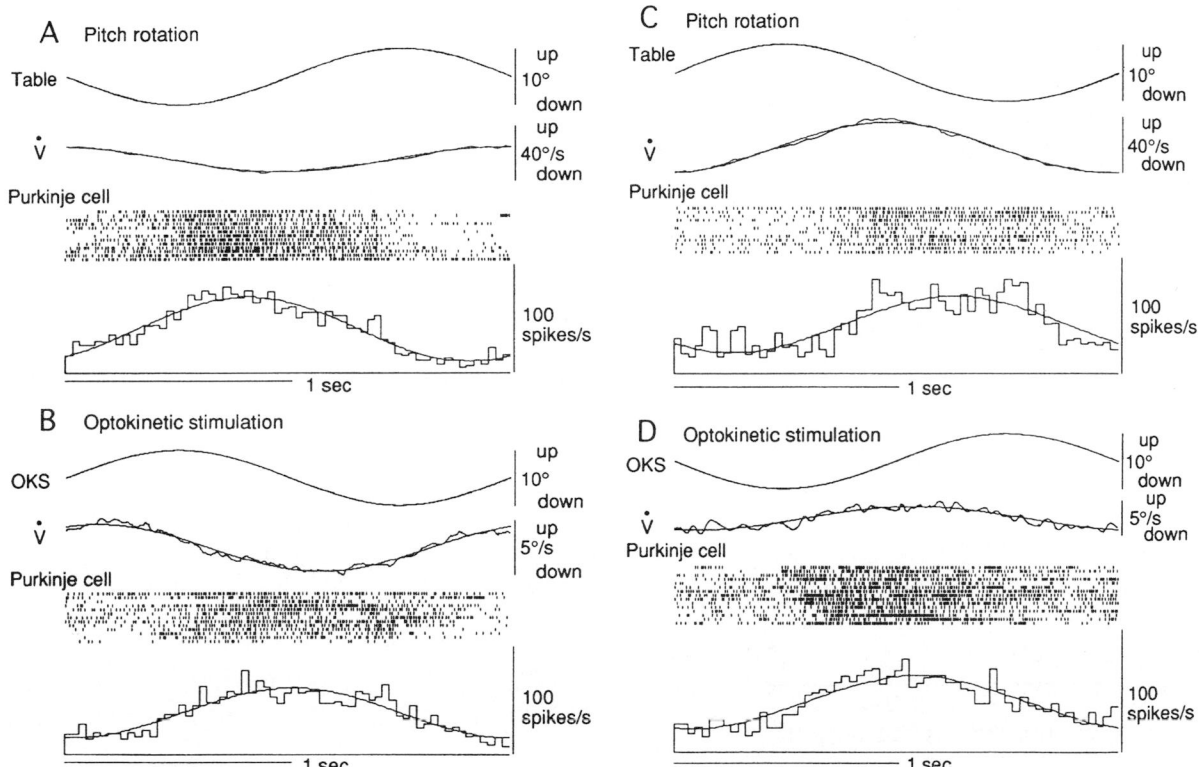

Fig. 1. Responses of two representative P cells (A,B,C,D) to sinusoidal pitch rotation (A,C) and vertical optokinetic stimuli (B,D). A and C: traces are averaged turntable position (Table), 'de-saccaded' vertical eye velocity ($\dot{V}$) and overlaid sine curve fit, spike rasters and histogram with sine curve fit. B and D: similar arrangement for averaged optokinetic stimulus-position (OKS), 'de-saccaded' vertical eye velocity ($\dot{V}$) with sine curve fit, spike rasters and histogram with sine curve fit.

responses are also plotted for a comparison (open squares). P cell gain decreased (A) and phase lagged at higher stimulus frequencies (B) in parallel with eye movement.

To examine whether the apparent eye velocity response of P cell during sinusoidal optokinetic stimuli may have been due to visual inputs, we presented a stationary spot while moving the visual pattern vertically as illustrated in Figs. 2C,D. The consistent P cell responses seen while the cats followed the optokinetic stimuli (Fig. 2C) disappeared when the cats fixated the stationary laser spot (Fig. 2D), ė and P cell discharge returned to its irregular spontaneous levels and was unrelated to the stimulus. For three P cells, sinu-

soidal optokinetic stimuli were also presented horizontally; none of them showed a consistent response. These results show that the responses evoked by sinusoidal optokinetic stimuli were not produced by visual inputs, but were related mainly to the resultant vertical eye movement, although we do not exclude the possibility that visual inputs may have contributed to their responses in some minor fashion.

To obtain further information on the discharge characteristics of these P cells during pitch and optokinetic stimuli, we examined their behavior during rapid modification of the pitch VOR induced by visual inputs. An example is presented in Fig. 3 for a down-pitch/up-optokinetic P cell.

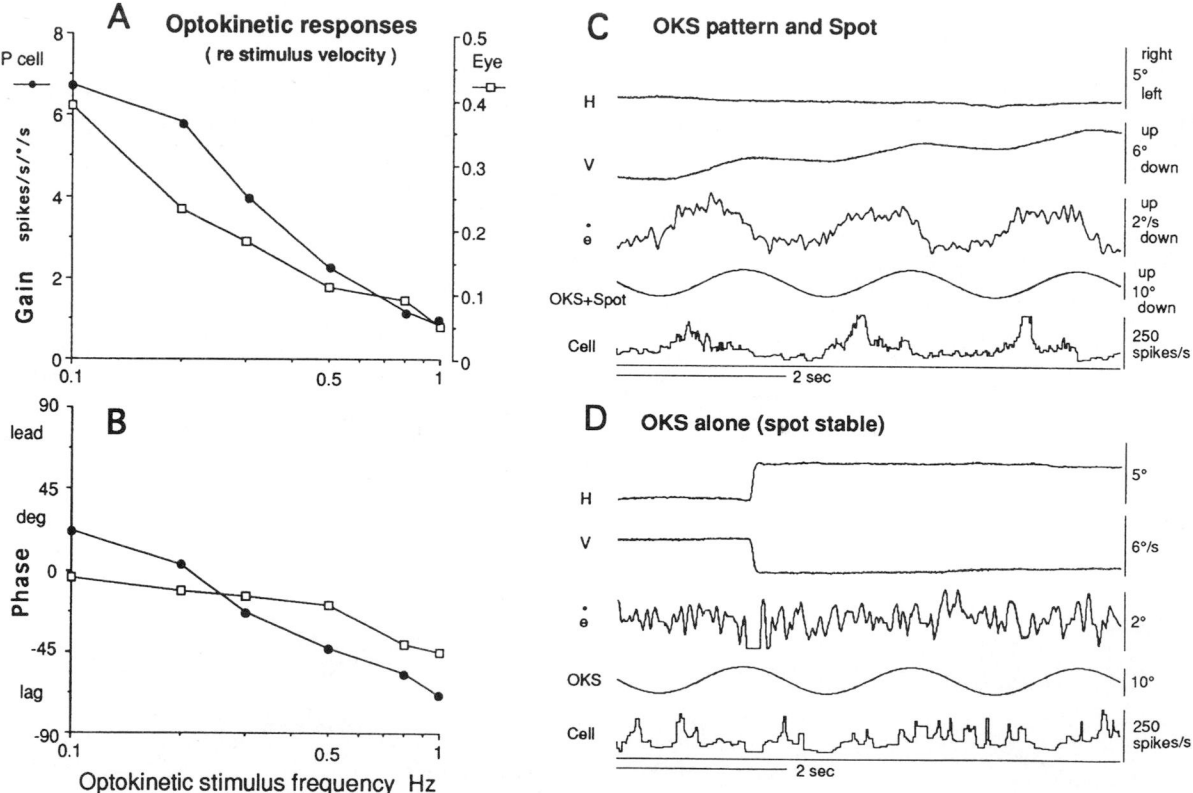

Fig. 2. Responses of a P cell to sinusoidal vertical optokinetic stimuli. A and B: P cell and eye responses (re: stimulus velocity) at different stimulus frequencies. C and D: Response of the same P cell to sinusoidal vertical optokinetic stimuli while the cat followed the pattern (C) or fixated a stationary spot (D). The optokinetic pattern and spot were moved together in C; the spot was stationary against the moving pattern in D. Traces in C and D are single horizontal eye-position (H), vertical eye-position (V), vertical eye velocity (ė), stimulus position (OKS and/or Spot) and cell discharges. Saccade in D was removed in eye velocity trace.

Compared to its response during pitch rotation without visual stimuli (Fig. 3A, control), the cell response consistently increased ($\sim 110\%$) when the optokinetic stimuli were presented 180° out of phase with the turn-table (Fig. 3B, VOR enhancement condition), whereas the P cell response decreased when the optokinetic stimuli were presented in phase with it (Fig. 3C, VOR suppression condition). Fig. 3D plots peak-to-peak P cell discharge as a function of maximum eye velocity of the simultaneously recorded VOR in the three stimulus conditions. Although there is a clear positive linear correlation between the discharge and eye velocity, since our cats could not suppress the VOR efficiently, we do not know whether this relationship is linear at all eye velocities (cf., Lisberger and Fuchs, 1978).

Maximal activation directions of pitch-responding Purkinje cells

We reported earlier, by examining MADs (Baker et al., 1984), that up-pitch P cells receive excitation primarily from the contralateral posterior canal and that down-pitch P cells receive excitation primarily from the anterior canal of either ipsilateral or contralateral side (Fukushima et al., 1993). To examine whether P cells that responded to optokinetic stimuli in the present study behave similarly, we examined MADs for some P cells to vertical rotation with a constant frequency and amplitude (0.5 Hz, 5°) while changing the horizontal position of the turntable (Fig. 4B). As in previous studies, P cells responded to vertical but not horizontal rotation. Phase and gain values (re: table position) of a representative up-pitch/down-optokinetic P cell are plotted in Fig. 4A as a function of horizontal orientation angles (illustrated in Fig. 4B). Positive gains were correlated with phase lags, and negative gains with phase leads (cf., Baker et al., 1984). The gain curves are fit well with a least-squares sinusoid. Phase values were mostly constant except for an

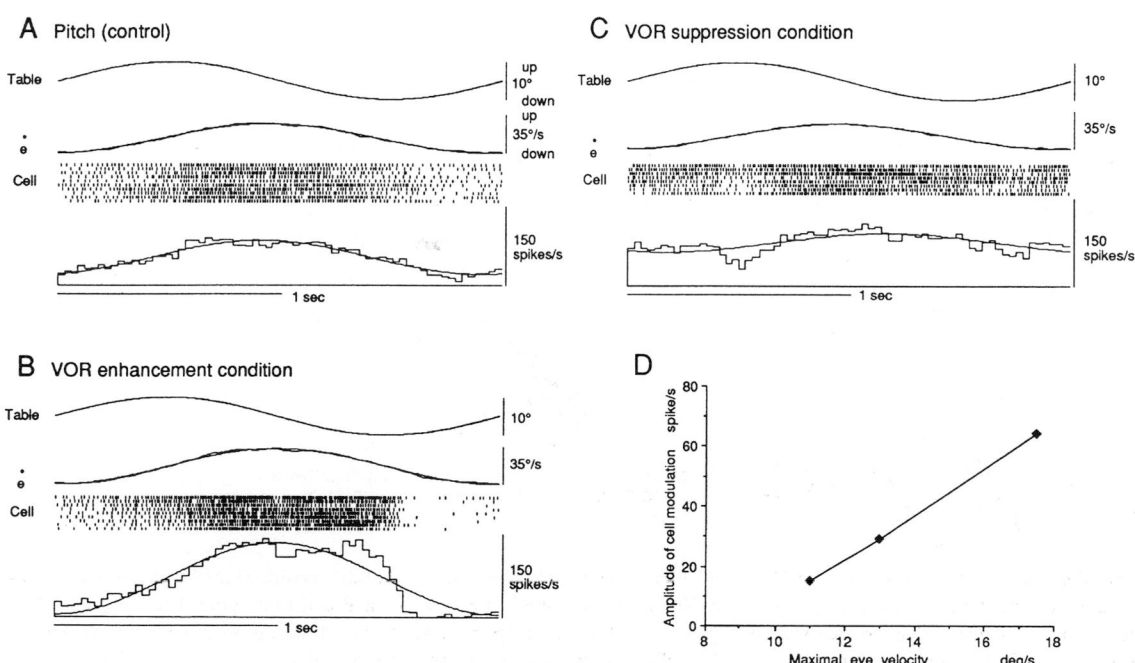

Fig. 3. Responses of a P cell during rapid modification of pitch VOR induced by visual inputs. Pitch rotation (at 0.5 Hz ± 5°) without visual stimuli (A), with visual pattern stimuli that were presented 180° out of phase (B) or in phase (C) with the turn-table. D: quantitative comparison of response modulation in the three conditions.

abrupt shift near the null regions. These response properties are typical of responses evoked by inputs to semicircular canals with their MADs near the contralateral posterior canal (Fig. 4A, MAD 135°, null −45°). The distributions of MADs of the P cells examined are plotted in Fig. 4C (dots) together with MADs reported earlier (Fig. 3A of Fukushima et al., 1993). MADs of all P cells that showed an up-pitch/down-optokinetic pattern were in the quadrant near the plane of the contralateral posterior canal, whereas MADs of down-pitch/up-optokinetic P cells were in the quadrants where major excitation comes from anterior canal inputs.

Effects of muscimol infusion into up-pitch responding floccular areas in alert cats

In order to obtain further information on how the flocculus could be involved in the signal conversion in the vertical VOR, we injected a GABA agonist muscimol (1 μg) into the region where we recorded most of our up-pitch P cells. An example is shown in Fig. 5. After muscimol-infusion, the downward VOR induced by nose-up pitch was greatly depressed at low frequencies (arrows in Fig. 5B, note the reduced range for the scale at the right), and virtually abolished at 0.05 Hz (not shown), although it was still clearly evoked at higher frequencies as previously reported (Fukushima et al., 1993).

Vertical optokinetic responses were examined at the same frequency (Figs. 5C,D), despite the fact that up and down eye velocity components were induced almost symmetrically before infusion (Fig. 5C), down eye velocity components were virtually abolished after infusion (Fig. 5D). Post-saccadic exponential centripetal drift also appeared after muscimol infusion as can be seen following saccades without visual stimuli while the turn-table was stable (Fig. 5E). The mean time constant of such postsaccadic drift was 1.2 s.

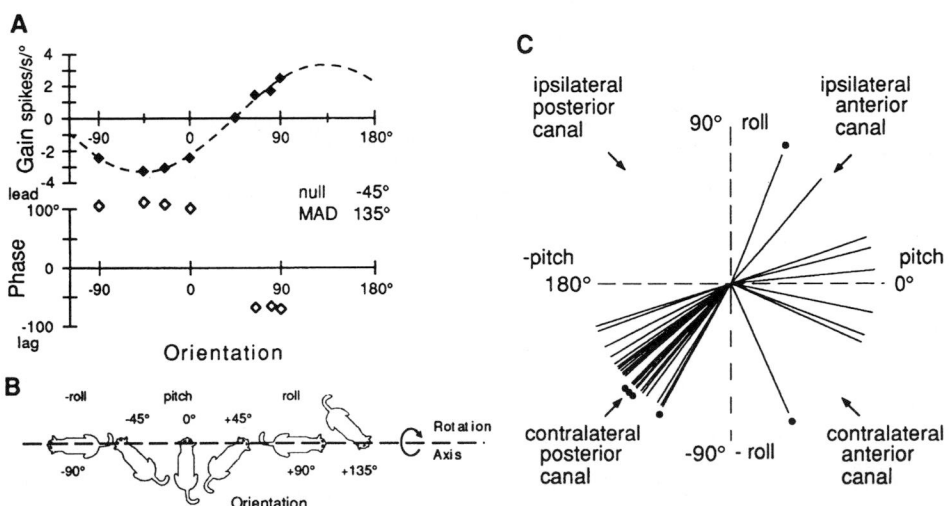

Fig. 4. Sinusoidal pitch responses of floccular P cells. A: Plots of gain and phase values (re: table position) during sinusoidal vertical rotation against different horizontal orientation angles (indicated schematically in B) of a P cell that showed an up-pitch/down optokinetic response pattern. Phase values were not plotted for low gain responses. C: Maximal activation direction of floccular P cells responding to pitch rotation during vertical VOR. Cells marked with dots were those tested with vertical optokinetic stimuli. Other cells were taken from previous results (Fukushima et al., 1993). Arrows indicate functional planes of individual vertical canals (Robinson, 1982).

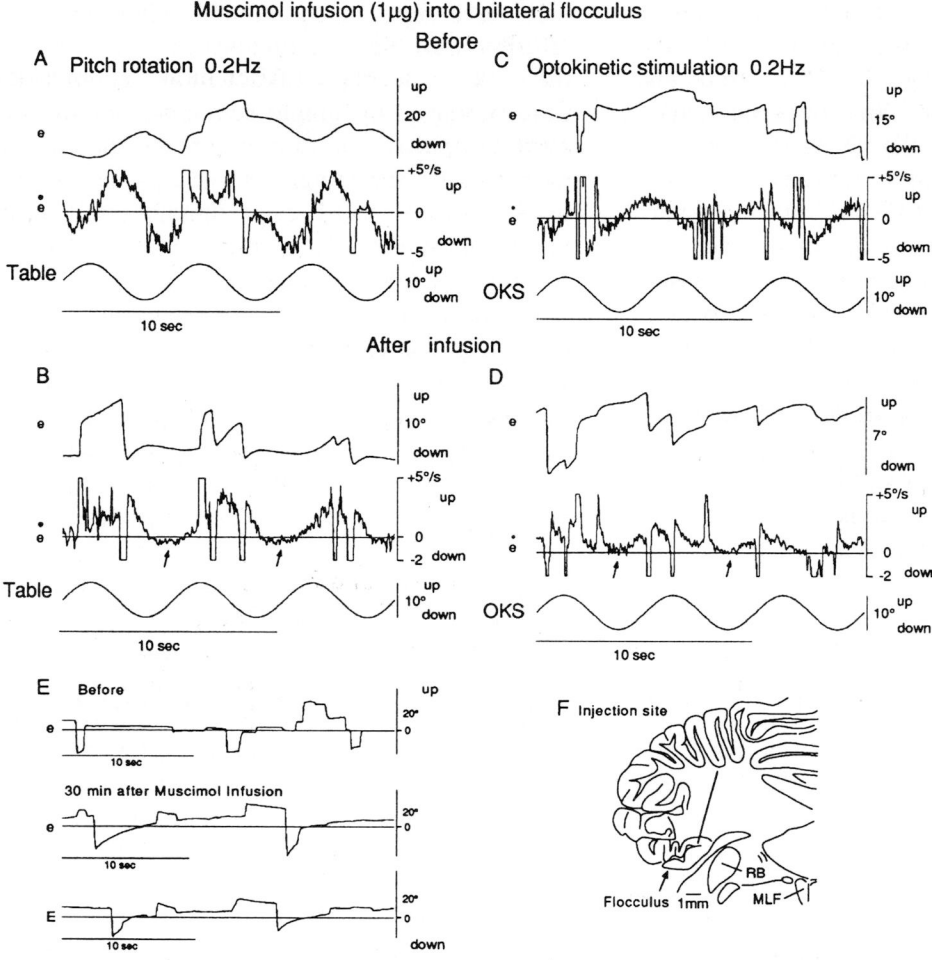

Fig. 5. Effects of muscimol infusion into unilateral flocculus where up- pitch responding P cells were recorded. For further explanation, see text. RB-restiform body; MLF-medial longitudinal fasciculus. Traces in A and C are single vertical eye-position (e), vertical eye velocity (ė), and stimulus position (OKS and Table). Vertical eye-position traces are shown in E.

Discussion

The great majority of pitch-responding P cells in this study responded to sinusoidal vertical optokinetic stimuli and most of those (80%) showed a characteristic discharge pattern to the two stimuli (Fig. 1); their activity increased when the resultant eye movements during the two stimuli were in the same direction (either up or down). This response pattern is similar to the pattern reported in rabbits and rats in the horizontal system during vestibular and optokinetic stimuli, and vestibular and retinal slip inputs have been reported to be responsible for that response pattern (Precht, 1983; Ito, 1984 for review; Blanks and Precht, 1983; Nagao, 1990). In this study, however, optokinetic responses of our cells cannot be explained by visual inputs alone, since clear modulation of P cell discharge was observed only when the cats followed optokinetic stimuli, but not during fixation of a spot against the moving optokinetic pattern (Figs. 2C,D). P cell optokinetic responses were not induced by eye position sensitivity, since only one of 34 cells tested showed eye position sensitivity (e.g. Fig. 2D; cf., Noda and Suzuki, 1979). Moreover, since response magnitudes of some of our cells changed in parallel with eye velocity changes induced by visual-vestibular interaction (Fig. 3), eye velocity components must have contributed to their responses. We do not exclude the possibility that visual inputs may have contributed minimally to their responses during optokinetic responses (cf., Noda and Warabi, 1986; Nagao, 1990).

The major response type in alert monkey flocculus is the gaze velocity P cells (GVP). (Miles and Fuller, 1975; Lisberger and Fuchs, 1978; Miles et al., 1980; Shidara and Kawano, 1993; Lisberger et al., 1994). Although their activity has not been studied during pitch rotation, it has been assumed that downward GVP cells receive excitation from the anterior canal (King and Leigh, 1982). Our MAD analysis suggests that our P cells that showed an up pitch/down optokinetic response pattern received excitation primarily from the contralateral posterior canal (Fig. 4). This seems contradictory to the interpretation in the monkey data. However, since our cats could not suppress the VOR sufficiently during pitch rotation (Fig. 4C), our results are insufficient to conclude whether or not our cells are different from monkey GVP cells (Lisberger and Fuchs, 1978; Shidara and Kawano, 1993). In order to answer this question, it will be necessary to examine P cell behavior in well-trained animals.

Muscimol infusion into the floccular up-pitch responding areas impaired vertical components of the VOR and optokinetic response (Fig. 5). Although postsaccadic exponential centripetal drift seen after downward saccades (Fig. 5E) must have contributed in this impairment, the drift alone cannot explain the lack or significant decrease of downward components for the following reasons: (1) similar impairment was observed even when the experimenter presented a stationary laser spot 10–20° above the straight-ahead position to shift the animals' gaze upward (Fukushima, unpublished observation; and (2) there was no clear increase in upward components after muscimol infusion (also Fig. 5 of Fukushima et al., 1993); if the lack or decrease of downward components were produced by simple summation of the upward drift, the upward components should also increase correspondingly. Rather, phase advance in the vertical VOR even though small, after muscimol infusion (Fukushima et al., 1993) is consistent with our interpretation.

We have suggested that the temporal transformation of the semicircular canal signals in the vertical VOR requires the flocculus in addition to the already proposed integrator circuits for vertical eye movement that include the interstitial nucleus of Cajal and vestibular commissural connections related to the vertical semicircular canals (Shimazu and Precht, 1966; Galiana and Outerbridge, 1984; Fukushima, 1987; Fukushima et al., 1992, 1993, 1994; Chimoto et al., 1992; Fukushima and Kaneko, 1995). More specifically, we have suggested that many floccular P cells that receive excitation from the contralateral posterior canal (Fukushima et al., 1993) would inhibit VOR interneurons related to the anterior canal (Ito et al., 1977; Hirai and Uchino, 1984; Sato and Kawasaki, 1990), thus compensating for the functionally insufficient vestibular commissural inhibition from posterior canal, type I neurons-to-VOR interneurons related to the anterior canal (Uchino et al., 1986). Our results showing that the vertical VOR as well as the vertical optokinetic response was impaired following muscimol infusion into up-pitch responding floccular areas (Fig. 5) is consistent with this interpretation.

Acknowledgements

We gratefully acknowledge Prof. M. Kato for his support of these experiments, Ms. Y. Kobayashi for technical assistance, and Dr. C.R.S. Kaneko of the Regional Primate Research Center, University of Washington, Seattle, USA, for his valuable comments on the manuscript. This work was supported, in part, by grants from the Japanese Ministry Science and Culture (06260201, 06680782, 07252201, 07680886), Toyota Riken and Marna

Cosmetics. Present addresses: Shinki Chin, Department of Ophthalmology, School of Medicine, Hokkaido University; Junko Fukushima, College of Medical Technology, Hokkaido University. Sergei Kurkin is on leave from the A.B. Kogan Research Institute for Neurocybernetics, Rostov State University, Rostov-on-Don, Russia.

References

Anderson, J.H. (1981) Behavior of the vertical canal VOR in normal and INC-lesioned cats. In: A.F. Fuchs and W. Becker (Eds.), *Progress in Oculomotor Research. Developments in Neuroscience*, Vol, 12, Elsevier, North-Holland, pp. 395–401.

Anderson, J.H., Precht, W. and Pappas, C., Changes in the vertical vestibuloocular reflex due to kainic acid lesions of the interstitial nucleus of Cajal. *Neurosci. Lett.*, 14: 259–264, 1979.

Baker, J., Goldberg, J., Hermann, G., and Peterson, B. W. (1984) Optimal response planes and canal convergence in secondary neurons in vestibular nuclei of alert cats. *Brain Res.*, 294: 133–137.

Blanks, R.H.I. and Precht, W. (1983) Responses of units in the rat cerebellar flocculus during optokinetic and vestibular stimulation. *Exp. Brain Res.*, 53: 1–15.

Blanks, R.H., Curthoys, I.S. and Markham, C.H. (1972) Planar relationships of semicircular canals in the cat. *Am. J. Physiol.*, 223: 55–62.

Carpenter, R.H.S. (1972) Cerebellectomy and the transfer function of the vestibulo-ocular reflex in the decerebrate cat. *Proc. Roy. Soc. B*, 181: 353–374.

Chimoto, S., Iwamoto, Y. and Yoshida, K. (1992) Projection of vertical eye movement related neurons in the interstitial nucleus of Cajal to the vestibular nucleus in the cat. *Neurosci. Res.*, 15: 293–298.

Crawford, D., Cadera, W. and Vilis, T. (1991) Generation of torsional and vertical eye position signals by the interstitial nucleus of Cajal. *Science*, 252: 1551–1553.

Flandrin, J.M., Courjon, J.H., Jeannerod, M. and Schmid, R. (1983) Effects of unilateral flocculus lesions on vestibulo-ocular responses in the cat. *Neuroscience*, 8: 809–817.

Fukushima, K. (1987) The interstitial nucleus of Cajal and its role in the control of movements of head and eyes. *Progr. Neurobiol.*, 29: 107–192.

Fukushima, K. (1991) The interstitial nucleus of Cajal in the midbrain reticular formation and vertical eye movement. *Neurosci. Res.*, 10: 159–187.

Fukushima, K and Fukushima, J. (1990) Activity of eye-movement-related neurons in the region of the interstitial nucleus of Cajal during sleep. *Neurosci. Res.*, 9: 126–139.

Fukushima, K. and Kaneko, C.R.S. (1995) Vestibular integra-

tors in the oculomotor system. *Neurosci. Res.*, 22: 249–258.

Fukushima, K., Fukushima, J. Harada, C., Ohashi, T. and Kase, M. (1990) Neuronal activity related to vertical eye movement in the region of the interstitial nucleus of Cajal in alert cats. *Exp. Brain Res.*, 79: 43–64.

Fukushima, K., Kaneko, C.R.S. and Fuchs, A.F. (1992) The neuronal substrate of integration in the oculomotor system. *Progr. Neurobiol.*, 39: 609–639.

Fukushima, K., Buharin, E.V. and Fukushima, J. (1993) Responses of floccular Purkinje cells to sinusoidal vertical rotation and effects of muscimol infusion into the flocculus in alert cats. *Neurosci. Res.*, 17: 297–305.

Fukushima, K., Buharin, E.V. and Fukushima, J. (1994) Specific involvement of the cerebellar flocculus in the integration process in the vertical vestibulo-ocular reflex (VOR): Vertical canal-response of floccular Purkinje cells and effects of muscimol infusion in alert cats. In: U. Büttner, A.F. Fuchs, T. Brandt and D.S. Zee (Eds.), *Contemporary Ocular Motor and Vestibular Research*, Georg Thieme Verlag, Stuttgart, New York, pp. 250–252.

Fukushima, K., Ohashi, T., Fukushima, J. and Kaneko, C.R.S. (1995) Discharge characteristics of vertical vestibular plus saccade neurons in the rostral midbrain of alert cats. *J. Neurophysiol.*, 73: 2129–2143.

Galiana, H.L. and Outerbridge, J.S. (1984) A bilateral model for central neural pathways in vestibuloocular reflex. *J. Neurophysiol.*, 51: 210–241.

Godaux, E. and Vanderkelen, B. (1984) Vestibulo-ocular reflex, optokinetic response and their interactions in the cerebellectomized cat. *J. Physiol.* (*Lond.*), 346: 155–170.

Hassul, M. Daniels, P.D., and Kimm, J. (1976) Effects of bilateral flocculectomy on the vestibulo-ocular reflex in the chinchilla. *Brain Res.*, 118: 339–343.

Hirai, N. and Uchino, Y. (1984) Floccular influence on excitatory relay neurones of vestibular reflexes of anterior semicircular canal origin in the cat. *Neurosci. Res.*, 1: 327–340.

Ito, M. (1984) The cerebellum and neural control. Raven Press, New York, pp. 133–374.

Ito, M., Nishimaru, N. and Yamamoto, M. (1977) Specific patterns of neuronal connections involved in the control of the rabbit's vestibulo-ocular reflexes by the cerebellar flocculus. *J. Physiol.* (London), 265: 833–854.

Ito, M., Jastreboff, P.J. and Miyashita, Y. (1982) Specific effects of unilateral lesions in the flocculus upon eye movements in albino rabbits. *Exp. Brain Res.*, 45: 233–242.

Kasahara, M. and Uchino, Y. (1974) Bilateral semicircular canal inputs to neurons in cat vestibular nuclei. *Exp. Brain Res.*, 20: 285–296.

King, W.M. and Leigh, R.J. (1982) Physiology of vertical gaze. In: G. Lennerstrand, D.S. Zee and E.L. Keller (Eds.), *Functional Basis of Ocular Motility Disorders*, Pergamon press, Oxford, pp. 267–276.

Lisberger, S.G. and Fuchs, A.F. (1978) Role of primate flocculus during rapid behavioral modification of vestibuloocular reflex. I. Purkinje cell activity during visually guided horizontal smooth-pursuit eye movements and passive head rotation. *J. Neurophysiol.*, 41: 733–763.

Lisberger, S.G., Pavelko, T.A., Bronte-Stuart, H.M. and Stone, L.S. (1994) Neural basis for motor learning in the vestibuloocular reflex of primates. II. Changes in the responses of horizontal gaze velocity Purkinje cells in the cerebellar flocculus and ventral paraflocculus. *J. Neurophysiol.*, 72: 954–973.

Miles, F.A. and Fuller, J.H. (1975) Visual tracking and the primate flocculus. Science, 189: 1000–1003.

Miles, F.A., Fuller, J.H., Braitman, D.J. and Dow, B.M. (1980) Long-term adaptive changes in primate vestibuloocular reflex. III. Electrophysiological observations in flocculus of normal monkeys. *J. Neurophysiol.*, 43: 1437–1476.

Nagao, S. (1990) Eye velocity is not the major factor that determines mossy fiber responses of rabbit floccular Purkinje cells to head and screen oscillation. *Exp. Brain Res.*, 80: 221–224.

Noda, H. and Suzuki, D.A. (1979) The role of the flocculus of the monkey in saccadic eye movements. *J. Physiol.* (*Lond.*), 294: 317–334.

Noda, H. and Warabi, T. (1986) Discharges of Purkinje cells in monkey's flocculus during smooth-pursuit eye movements and visual stimulus movements. *Exp. Neurol.*, 93: 390–403.

Partsalis, A.M., Zhang, Y. and Highstein, S.M. (1995) Dorsal Y group in the squirrel monkey. II. Contribution of the cerebellar flocculus to neuronal responses in normal and adapted animals. *J. Neurophysiol.*, 73: 632–650.

Precht, W. and Anderson, J.H. (1979) Effects of cerebellec

tomy on the cat's vertical vestibuloocular reflex. *Pflügers Arch.*, 382: 51–55.

Robinson, D.A. (1974) The effect of cerebellectomy on the cat's vestibulo-ocular integrator. *Brain Res.*, 71: 195–207.

Robinson, D.A. (1981) Control of eye movements. In: J.M. Brookhart and J.M. Mountcastle (Eds.), *Handbook of Physiology*, American Physiological Society, Bethesda, pp. 1275–1320.

Robinson, D.A. (1982) The use of matrices in analyzing the three- dimensional behavior of the vestibulo-ocular reflex. *Biol. Cybern.*, 46: 53–66.

Sato, U. and Kawasaki, T. (1990) Operational unit responsible for plane- specific control of eye movement by cerebellar flocculus in cat. *J. Neurophysiol.*, 64: 551–564.

Shimazu, H. and Precht, W. (1966) Inhibition of central vestibular neurons from the contralateral labyrinth and its mediating pathway. *J. Neurophysiol.*, 29: 467–492.

Shidara, M. and Kawano, K. (1993) Role of Purkinje cells in the ventral paraflocculus in short-latency ocular following responses. *Exp. Brain Res.*, 93: 185–195.

Uchino, Y., Ichikawa, T., Isu, N., Nakashima, H. and Watanabe, S. (1986) The commissural inhibition on secondary vestibulo-ocular neurons in the vertical semicircular canal system in the cat. *Neurosci. Lett.*, 70: 210–216.

Zhang, Y., Partsalis, A.M. and Highstein, S.M. (1995) Properties of superior vestibular nucleus flocculus target neurons in the squirrel monkey. II. Signal components revealed by reversible flocculus inactivation. *J. Neurophysiol.*, 73: 2279–2293.

Zee, D.S., Yamazaki, A., Butler, P.H. and Gücer, G. (1981) Effects of ablation of flocculus and paraflocculus on eye movements in primate. *J. Neurophysiol.*, 46: 878–899.

Subject Index